25 *Years*
Previous Solved Papers

ESE 2025

UPSC ENGINEERING SERVICES EXAMINATION

Preliminary Examination

MECHANICAL ENGINEERING

Objective Solved Papers

Volume-I

- Topicwise presentation
- Thoroughly revised & updated

Also useful for
State Engineering Services Examinations, Public Sector Examinations & Other Competitive Examinations

MADE EASY PUBLICATIONS

www.madeeasypublications.org

MADE EASY Publications Pvt. Ltd.

Corporate Office: 44-A/4, Kalu Sarai (Near Hauz Khas Metro Station), New Delhi-110016

Contact: 9021300500

E-mail: infomep@madeeasy.in

Visit us at: www.madeeasypublications.org

ESE-2025 : Preliminary Examination
Mechanical Engineering : Volume-1
Topicwise Objective Solved Questions : (2000-2024)

1st Edition : 2007
2nd Edition : 2008
3rd Edition : 2009
4th Edition : 2010
5th Edition : 2011
6th Edition : 2012
7th Edition : 2013
8th Edition : 2014
9th Edition : 2015
10th Edition : 2016
11th Edition : 2017
12th Edition: 2018
13th Edition: 2019
14th Edition: 2020
15th Edition: 2021
16th Edition: 2022
17th Edition: 2023

18th Edition: 2024

Director's Message

B. Singh (Ex. IES)

Engineering is one of the most chosen graduating field. Taking engineering is usually a matter of interest but this eventually develops into "purpose of being an engineer" when you choose engineering services as a career option.

Train goes in tunnel we don't panic but sit still and trust the engineer, even we don't doubt on signalling system, we don't think twice crossing over a bridge reducing our travel time; every engineer has a purpose in his department which when coupled with his unique talent provides service to mankind.

I believe *"the educator must realize in the potential power of his pupil and he must employ all his art, in seeking to bring his pupil to experience this power"*. To support dreams of every engineer and to make efficient use of capabilities of aspirant, MADE EASY team has put sincere efforts in compiling all the previous years' ESE-Pre questions with accurate and detailed explanation. The objective of this book is to facilitate every aspirant in ESE preparation and so, questions are segregated chapterwise and topicwise to enable the student to do topicwise preparation and strengthen the concept as and when they are read.

I would like to acknowledge efforts of entire MADE EASY team who worked hard to solve previous years' papers with accuracy and I hope this book will stand up to the expectations of aspirants and my desire to serve student fraternity by providing best study material and quality guidance will get accomplished.

B. Singh (Ex. IES)
CMD, MADE EASY Group

Volume-I

MECHANICAL ENGINEERING

Objective Solved Questions

of UPSC Engineering Services Examination

Contents

Sl.	Topic	Pages
1.	Fluid Mechanics	1-130
2.	Thermodynamics	131-206
3.	Heat Transfer	207-273
4.	Internal Combustion Engines	274-319
5.	Refrigeration and Air-conditioning	320-387
6.	Turbo Machinery	388-500
7.	Power Plant Engineering	501-544
8.	Renewable Sources of Energy	545-557

UNIT

I

Fluid Mechanics

Syllabus

Fluid Mechanics: Basic concepts and Properties of fluids, Manometry, Fluid statics, Buoyancy, Equations of motion, Bernoulli's equation and applications, Viscous flow of incompressible fluids, Laminar and Turbulent flows, Flow through pipes and head losses in pipes.

Contents

Sl.	Topic	Page No.
1.	Fluid Properties	2
2.	Fluid Pressure and Measurement	18
3.	Hydrostatic Forces on Surface	27
4.	Buoyancy, Floatation and Liquids in Relative Equilibrium	37
5.	Fluid Kinematics	49
6.	Fluid Dynamics and Flow Measurement	65
7.	Flow Through Pipes	79
8.	Vortex Flow	92
9.	Laminar Flow	96
10.	Turbulent Flow	104
11.	Boundary Layer Theory	106
12.	Open Channel Flow	114
13.	Drag and Lift	117
14.	Dimensional Analysis	122
15.	Flow over Notches and Weirs	129

OOOO

1 Fluid Properties

1.1 **Assertion (A):** If a cube is placed in a liquid with two of its surfaces parallel to the free surface of the liquid, then the pressures on the two surfaces which are parallel to the free surface, are the same.
Reason (R): Pascal's law states that when a fluid is at rest, the pressure at any plane is the same in all directions.
(a) Both **A** and **R** are individually true and **R** is the correct explanation of **A**
(b) Both **A** and **R** are individually true but **R** is not the correct explanation of **A**
(c) **A** is true but **R** is false
(d) **A** is false but **R** is true [ESE : 2000]

1.2 The shear stress developed in a lubricating oil, of viscosity 9.81 poise, filled between two parallel plates 1 cm apart and moving with relative velocity of 2 m/s is
(a) 20 N/m^2 (b) 196.2 N/m^2
(c) 29.62 N/m^2 (d) 40 N/m^2
[ESE : 2001]

1.3 A capillary tube is inserted in mercury kept in an open container.
Assertion (A): The mercury level inside the tube shall rise above the level of mercury outside.
Reason (R): The cohesive force between the molecules of mercury is greater than the adhesive force between mercury and glass.
(a) Both **A** and **R** are individually true and **R** is the correct explanation of **A**
(b) Both **A** and **R** are individually true but **R** is not the correct explanation of **A**
(c) **A** is true but **R** is false
(d) **A** is false but **R** is true [ESE : 2001]

1.4 In the phenomenon of cavitation, the characteristic fluid property involved is
(a) surface tension
(b) viscosity
(c) bulk modulus of elasticity
(d) vapour pressure [ESE : 2002]

1.5 The capillary rise at 20°C in clean glass tube of 1 mm diameter containing water is approximately
(a) 15 mm (b) 50 mm
(c) 20 mm (d) 30 mm
[ESE : 2002]

1.6 Match **List-I** (Type of fluid) with **List-II** (Variation of shear stress) and select the correct answer using the codes given below the lists:

List-I
A. Ideal fluid
B. Newtonian fluid
C. Non-Newtonian fluid
D. Bingham plastic

List-II
1. Shear stress varies linearly with the rate of strain
2. Shear stress does not vary linearly with the rate of strain
3. Fluid behaves like a solid until a minimum yield stress beyond which it exhibits a linear relationship between shear stress and the rate of strain
4. Shear stress is zero

Codes:

	A	B	C	D
(a)	3	1	2	4
(b)	4	2	1	3
(c)	3	2	1	4
(d)	4	1	2	3

[ESE : 2002]

1.7 The equation of a velocity distribution over a plate is given by $u = 2y - y^2$, where u is the velocity in m/s at a point y meter from the plate measured perpendicularly. Assuming μ = 8.60 poise, the shear stress at a point 15 cm from the boundary is
(a) 1.72 N/m^2 (b) 1.46 N/m^2
(c) 14.62 N/m^2 (d) 17.20 N/m^2
[ESE : 2002]

1.8 **Assertion (A):** In general, viscosity in liquids increases and in gases it decreases with rise in temperature.

Reason (R): Viscosity is caused by intermolecular forces of cohesion and due to transfer of molecular momentum between fluid layers; of which in liquids the former and in gases the later contribute the major part towards viscosity.

(a) Both **A** and **R** are individually true and **R** is the correct explanation of **A**
(b) Both **A** and **R** are individually true but **R** is not the correct explanation of **A**
(c) **A** is true but **R** is false
(d) **A** is false but **R** is true **[ESE : 2002]**

1.9 Match **List-I** (Rheological equation) with **List-II** (Types of fluids) and select the correct answer using the codes given below the lists:

List-I

A. $\tau = \mu\left(\frac{du}{dy}\right)^n, n = 1$

B. $\tau = \mu\left(\frac{du}{dy}\right)^n, n < 1$

C. $\tau = \mu\left(\frac{du}{dy}\right)^n, n > 1$

D. $\tau = \tau_0 + \mu\left(\frac{du}{dy}\right)^n, n = 1$

List-II

1. Bingham plastic
2. Dilatant fluid
3. Newtonian fluid
4. Pseudo-plastic fluid

Codes:

	A	B	C	D
(a)	3	2	4	1
(b)	4	1	2	3
(c)	3	4	2	1
(d)	4	2	1	3

[ESE : 2003]

1.10 Match **List-I** (Flows over or inside the systems) with **List-II** (Type of flow) and select the correct answer using the codes given below the lists:

List-I

A. Flow over a sphere
B. Flow over a long circular cylinder
C. Flow in a pipe bend
D. Fully developed flow in a pipe at constant flow rate

List-II

1. Two dimensional flow
2. One dimensional flow
3. Axisymmetric flow
4. Three dimensional flow

Codes:

	A	B	C	D
(a)	3	1	2	4
(b)	1	4	3	2
(c)	3	1	4	2
(d)	1	4	2	3

[ESE : 2003]

1.11 An oil of specific gravity 0.9 has viscosity of 0.28 stokes at 38°C. What will be its viscosity in Ns/m^2?

(a) 0.2520 (b) 0.0311
(c) 0.0252 (d) 0.0206

[ESE : 2004]

1.12 Consider the following statements:

1. Viscosity 2. Surface tension
3. Capillarity 4. Vapour pressure

Which of the above properties can be attributed to the flow of jet of oil in an unbroken stream?

(a) 1 only (b) 2 only
(c) 1 and 3 (d) 2 and 4

[ESE : 2005]

1.13 A vertical clean glass tube of uniform bore is used as a piezometer to measure the pressure of liquid at a point. The liquid has a specific weight of 15 kN/m^3 and a surface tension of 0.06 N/m in contact with air. If for the liquid, the angle of contact with glass is zero and the capillary rise in the tube is not to exceed 2 mm, what is the required minimum diameter of the tube?

(a) 6 mm (b) 8 mm
(c) 10 mm (d) 12 mm **[ESE : 2006]**

1.14 When the pressure on a given mass of liquid is increased from 3.0 MPa to 3.5 MPa, the density of the liquid increases from 500 kg/m^3 to 501 kg/m^3. What is the average value of bulk modulus of liquid over the given pressure range?

(a) 700 MPa (b) 600 MPa
(c) 500 MPa (d) 250 MPa

[ESE : 2006]

1.15 If the relationship between the shear stress τ and the rate of shear strain $\left(\frac{du}{dy}\right)$ is expressed as

$$\tau = \mu\left(\frac{du}{dy}\right)^n$$

then the fluid with exponent $n > 1$ is known as which one of the following?

(a) Bingham plastic
(b) Dilatant fluid
(c) Newtonian fluid
(d) Pseudoplastic fluid

[ESE : 2007]

1.16 What are the dimensions of kinematic viscosity of a fluid?

(a) LT^{-2} (b) L^2T^{-1}
(c) $ML^{-1}T^{-1}$ (d) $ML^{-2}T^{-2}$

[ESE : 2007]

1.17 **Assertion (A):** Blood is a Newtonian fluid.
Reason (R): The rate of strain varies non-linearly with shear stress for blood.

(a) Both **A** and **R** are individually true and **R** is the correct explanation of **A**
(b) Both **A** and **R** are individually true but **R** is not the correct explanation of **A**
(c) **A** is true but **R** is false
(d) **A** is false but **R** is true

[ESE : 2007]

1.18 In an experiment, the following shear stress-time rate of shear strain values are obtained for a fluid:

Time rate of shear strain (1/s)	:	0	2	3	4
Shear stress (kPa)	:	0	1.4	2.6	4

How can the fluid be classified?

(a) Newtonian fluid (b) Bingham plastic
(c) Pseudo plastic (d) Dilatant

[ESE : 2008]

1.19 Consider the following statements related to the fluid properties:

1. Vapour pressure of water at 373 K is 101.5×10^3 N/m^2.
2. Capillary height in cm for water in contact with glass tube and air is (tube diameter)/0.268.
3. Blood is a Newtonian fluid.

Which of these statements is/are correct?

(a) 1 only (b) 1 and 3
(c) 1 and 2 (d) 2 only

[ESE : 2008]

1.20 What is the unit of dynamic viscosity of a fluid termed 'poise' equivalent to?

(a) dyne/cm^2 (b) gm-s/cm^2
(c) dyne-s/cm^2 (d) gm-cm/s

[ESE : 2008]

1.21 What is the pressure difference between inside and outside of a droplet of water?

(a) $\frac{2\sigma}{d}$ (b) $\frac{4\sigma}{d}$
(c) $\frac{8\sigma}{d}$ (d) $\frac{12\sigma}{d}$

where σ is surface tension and d is the diameter of the droplet. [ESE : 2008]

1.22 What is the pressure inside a soap bubble, over the atmospheric pressure if its diameter is 2 cm and the surface tension is 0.1 N/m?

(a) 0.4 N/m^2 (b) 4.0 N/m^2
(c) 40.0 N/m^2 (d) 400.0 N/m^2

[ESE : 2008]

1.23 The capillary rise or depression in a small diameter tube is

(a) directly proportional to the specific weight of the fluid
(b) inversely proportional to the surface tension
(c) inversely proportional to the diameter
(d) directly proportional to the surface area

[ESE : 2008]

1.24 Match **List-I** (Variable) with **List-II** (Dimensional expression) and select the correct answer using the codes given below the lists :

List-I
A. Dynamic viscosity
B. Moment of momentum
C. Power
D. Volume modulus of elasticity

List-II
1. ML^2T^{-3}
2. $ML^{-1}T^{-2}$
3. $ML^{-1}T^{-1}$
4. ML^2T^{-2}
5. ML^2T^{-1}

Codes:

	A	B	C	D
(a)	1	4	2	3
(b)	3	5	1	2
(c)	1	5	2	3
(d)	3	4	1	2

[ESE : 2008]

1.25 What is the capillary rise in a narrow two-dimensional slit of width *w*?

(a) Half of that in a capillary tube of diameter *w*

(b) Two-third of that in a capillary tube of diameter *w*

(c) One-third of that in a capillary tube of diameter *w*

(d) One-fourth of that in a capillary tube of diameter *w*

[ESE : 2009]

Direction: Each of the next questions consists of two statements, one labelled as the **'Assertion (A)'** and the other as **'Reason (R)'**. You are to examine these two statements carefully and select the answers to these items using the codes given below:

Codes:

(a) Both **A** and **R** are individually true and **R** is the correct explanation of **A**

(b) Both **A** and **R** are individually true but **R** is not the correct explanation of **A**

(c) **A** is true but **R** is false

(d) **A** is false but **R** is true

1.26 Assertion (A) : In a fluid, the rate of deformation is far more important than the total deformation itself.

Reason (R) : A fluid continues to deform so long as the external forces are applied.

[ESE : 2009]

1.27 Assertion (A) : A narrow glass tube when immersed into mercury causes capillary depression, and when immersed into water causes capillary rise.

Reason (R) : Mercury is denser than water.

[ESE : 2009]

1.28 Match **List-I** with **List-II** and select the correct answer using the codes given below the lists:

List-I

A. Lubrication

B. Rise of sap in trees

C. Formation of droplets

D. Cavitation

List-II

1. Capillary

2. Vapour pressure

3. Viscosity

4. Surface tension

Codes:

	A	B	C	D
(a)	2	4	1	3
(b)	3	4	1	2
(c)	2	1	4	3
(d)	3	1	4	2

[ESE : 2010]

1.29 Consider the following statements:

1. A small bubble of one fluid immersed in another fluid has a spherical shape.
2. The droplets of a fluid move upward or downward in another fluid due to unbalance between gravitational and buoyant forces.
3. Droplets of bubbles attached to a solid surface can remain stationary in a gravitational fluid if the surface tension exceeds buoyant forces.
4. Surface tension of a bubble is proportional to its radius while buoyant force is proportional to the cube of its radius.

Which of these statements are correct?

(a) 1, 2, 3 and 4 (b) 1, 2 and 4 only

(c) 1 and 3 only (d) 2, 3 and 4 only

[ESE : 2010]

1.30 The annular space between two coaxial vertical cylinders, of equal length, is filled with an incompressible liquid of constant viscosity. The outer cylinder is held fixed and the inner cylinder is slowly rotated about its axis at a uniform rotational speed. Assuming that Newton's law of viscosity holds good

(a) the tangential velocity of liquid varies linearly across the gap

(b) viscous shear stress in liquid is uniform across the gap

(c) the tangential velocity of liquid varies non-linearly across the gap

(d) viscous shear stress in liquid varies linearly across the gap

[ESE : 2010]

1.31 A thin plane lamina, of area *A* and weight *W*, slides down a fixed plane inclined to the vertical at an angle α and maintains a uniform gap ε from the surface of the plane, the gap being filled with oil of constant viscosity μ. The terminal velocity of the plane is

(a) $\dfrac{\varepsilon \cos\alpha}{\mu W A}$ (b) $\dfrac{\varepsilon \mu W}{A \sin\alpha}$

(c) $\dfrac{\varepsilon W \cos\alpha}{A \mu}$ (d) $\dfrac{\mu W \sin\alpha}{\varepsilon A}$

[ESE : 2010]

1.32 In a quiescent sea, density of water at free surface is ρ_0 and at a point much below the surface density is ρ. Neglecting variation in gravitational acceleration g and assuming a constant value of bulk modulus K, the depth h of the point from the free surface is

(a) $\dfrac{K}{g}\left(\dfrac{1}{\rho_0}+\dfrac{1}{\rho}\right)$ (b) $\dfrac{K}{g}\dfrac{(\rho-\rho_0)}{(\rho+\rho_0)^2}$

(c) $\dfrac{K}{g}\left(\dfrac{1}{\rho_0}-\dfrac{1}{\rho}\right)$ (d) $\dfrac{K}{g}\left(\dfrac{\rho\rho_0}{\rho+\rho_0}\right)$

[ESE : 2010]

1.33 Pseudo plastic is a fluid for which

(a) dynamic viscosity decreases as the rate of shear increases
(b) Newton's law of viscosity holds good
(c) dynamic viscosity increases as the rate of shear increases
(d) dynamic viscosity increases with the time for which shearing forces are applied

[ESE : 2010]

1.34 If angle of contact of a drop of liquid is acute then

(a) adhesion is more than cohesion
(b) cohesion is more than adhesion
(c) cohesion is equal to adhesion
(d) adhesion and cohesion have no bearing with angle of contact

[ESE : 2010]

1.35 With increase in pressure, the bulk modulus of elasticity

(a) Decreases (b) Increases
(c) Remains constant (d) None of the above

[ESE : 2011]

1.36 Match **List-I** with **List-II** and select the correct answer using the code given below the lists:

List-I	List-II
A. Capillarity	1. Cavitation
B. Vapour pressure	2. Density of water
C. Viscosity	3. Shear forces
D. Specific gravity	4. Surface tension

Codes:

	A	B	C	D
(a)	2	3	1	4
(b)	4	3	1	2
(c)	2	1	3	4
(d)	4	1	3	2

[ESE : 2011]

1.37 Newton's law of viscosity relates

(a) Velocity gradient and rate of shear strain
(b) Rate of shear deformation and shear stress
(c) Shear deformation and shear stress
(d) Pressure and volumetric strain

[ESE : 2011]

1.38 In an experiment to determine the rheological behaviour of a material, the observed relation between shear stress, τ, and rate of shear strain, $\dfrac{du}{dy}$, is $\tau = \tau_0 + c\left(\dfrac{du}{dy}\right)^{0.5}$. The material is

(a) a Newtonian fluid
(b) a thixotropic substance
(c) a Bingham plastic
(d) an ideal plastic

[ESE : 2011]

1.39 Match **List-I** with **List-II** and select the correct answer using the code given below the lists :

List-I
A. Coaxial cylinder viscometer
B. Capillary tube viscometer
C. Saybolt viscometer
D. Falling sphere viscometer

List-II
1. Hagen Poiseuille equation
2. Stokes law
3. Newton's law of viscosity
4. Efflux viscometer

Codes:

	A	B	C	D
(a)	2	1	4	3
(b)	3	1	4	2
(c)	2	4	1	3
(d)	3	4	1	2

[ESE : 2011]

Direction: Each of the next questions consists of two statements, one labelled as the **'Assertion (A)'** and the other as **'Reason (R)'**. You are to examine these two statements carefully and select the answers to these items using the codes given below:

Codes:
(a) Both A and R are individually true and R is the correct explanation of A
(b) Both A and R are individually true but R is not the correct explanation of A
(c) A is true but R is false
(d) A is false but R is true

1.40 Assertion (A) : In non-Newtonian fluids, the shear stress at any point is not a linear function of normal distance from the surface.
Reason (R) : This behaviour usually arises because the fluid molecules are very large, like polymers or proteins.
[ESE : 2011]

1.41 Assertion (A) : The mercury level inside the tube shall rise above the level of mercury outside.
Reason (R) : The cohesive force between the molecules of mercury is greater than the adhesive force between mercury and glass.
[ESE : 2011]

1.42 The vapour pressure is the characteristic fluid property involved in the phenomenon of
(a) water hammer in a pipe flow
(b) cavitation
(c) rise of sap in a tree
(d) spherical shape of rainwater drop
[ESE : 2012]

1.43 Match **List-I** with **List-II** and select the correct answer using the code given below the lists:

List-I (Fluids)
A. Ideal fluid
B. Newtonian fluid
C. Inviscid fluid
D. Real fluid

List-II (Viscosity equal to)
1. Zero
2. Non-zero
3. $\mu \dfrac{du}{dy}$

Code:

	A	B	C	D
(a)	1	2	3	2
(b)	2	3	3	1
(c)	1	3	1	2
(d)	2	3	1	2

[ESE : 2012]

1.44 The unit of the following property is not m^2/s
(a) thermal diffusivity
(b) kinematic viscosity
(c) dynamic viscosity
(d) mass diffusivity
[ESE : 2013]

1.45 Statement (I): In a fluid, the rate of deformation is far more important than the total deformation itself.
Statement (II): A fluid continues to deform so long as the external forces are applied.
(a) Both Statement (I) and Statement (II) are individually true and Statement (II) is the correct explanation of Statement (I).
(b) Both Statement (I) and Statement (II) are individually true but Statement (II) is **NOT** the correct explanation of Statement (I).
(c) Statement (I) is true but Statement (II) is false
(d) Statement (I) is false but Statement (II) is true.
[ESE : 2013]

1.46 Unlike the viscosity of liquids, the viscosity of gases increases with increasing temperature. This is due to
(a) Increased cohesive force between the molecules
(b) Increased momentum transfer in the molecules
(c) Decreased momentum transfer in the molecules
(d) Increases in both cohesive force and momentum transfer
[ESE : 2014]

1.47 The pressure inside a soap bubble of 50 mm diameter is 25 N/m^2 above the atmospheric pressure. The surface tension in soap film would be
(a) 0.156 N/m (b) 0.312 N/m
(c) 0.624 N/m (d) 0.078 N/m
[ESE : 2014]

1.48 Statement (I) : A small insect can sit on the free surface of a liquid though insect's density is higher than that of the liquid.
Statement (II): Liquids have viscosity.
(a) Both Statement (I) and Statement (II) are individually true and Statement (II) is the correct explanation of Statement (I).

(b) Both Statement (I) and Statement (II) are individually true but Statement (II) is **NOT** the correct explanation of Statement (I).
(c) Statement (I) is true but Statement (II) is false
(d) Statement (I) is false but Statement (II) is true.
[ESE : 2014]

1.49 Which of the following fluids exhibit a certain shear stress at zero shear strain rate followed by a straight line relationship between shear stress and shear strain rate?
(a) Newtonian fluids
(b) Ideal Bingham plastic fluids
(c) Pseudo-plastic fluids
(d) Dilatant fluids [ESE : 2015]

1.50 If angle of contact of a drop of liquid is acute, then
(a) cohesion is equal to adhesion
(b) cohesion is more than adhesion
(c) adhesion is more than cohesion
(d) both adhesion and cohesion have no connection with angle of contact
[ESE : 2016]

1.51 A spherical waterdrop of 1 mm in diameter splits up in air into 64 smaller drops of equal size. The surface tension coefficient of water in air is 0.073 N/m. The work required in splitting up the drop is
(a) 0.96×10^{-6} J (b) 0.69×10^{-6} J
(c) 0.32×10^{-6} J (d) 0.23×10^{-6} J
[ESE : 2017]

1.52 A 150 mm diameter shaft rotates at 1500 rpm within a 200 mm long journal bearing with 150.5 mm internal diameter. The uniform annular space between the shaft and the bearing is filled with oil of dynamic viscosity 0.8 poise. The shear stress on the shaft will be
(a) 1.77 kN/m^2 (b) 2.77 kN/m^2
(c) 3.77 kN/m^2 (d) 4.77 kN/m^2
[ESE : 2018]

1.53 The normal stresses within an isotropic Newtonian fluid are related to
1. Pressure
2. Viscosity of fluid
3. Velocity gradient
Which of the above are correct?
(a) 1 and 2 only (b) 1 and 3 only
(c) 2 and 3 only (d) 1, 2 and 3 [ESE : 2018]

1.54 A plate weighing 150 N and measuring 0.8 m × 0.8 m just slides down an inclined plane over an oil film of 1.2 mm thickness for an inclination of 30° and velocity of 0.2 m/s. Then the viscosity of the oil used is
(a) 0.3 Ns/m^2 (b) 0.4 Ns/m^2
(c) 0.5 Ns/m^2 (d) 0.7 Ns/m^2
[ESE : 2019]

1.55 A flat plate 0.1 m^2 area is pulled at 30 cm/s relative to another plate located at a distance of 0.01 cm from it, the fluid separating them being water with viscosity of 0.001 Ns/m^2, The power required to maintain velocity will be
(a) 0.05 W (b) 0.07 W
(c) 0.09 W (d) 0.11 W
[ESE : 2020]

1.56 When the pressure of liquid is increased from 3 MN/m^2 to 6 MN/m^2, its volume is decreased by 0.1%. The bulk modulus of elasticity of the liquid will be
(a) 3×10^{12} N/m^2 (b) 3×10^{9} N/m^2
(c) 3×10^{8} N/m^2 (d) 3×10^{4} N/m^2
[ESE : 2020]

1.57 **Statement (I):** The viscosity of liquids decreases with the increase of temperature while the viscosity of gases increases with the increase of temperature.

Statement (II): The viscous forces in a fluid are due to cohesive forces and molecular momentum transfer.
[ESE : 2021]

1.58 A liquid has a specific gravity of 1.9 and a kinematic viscosity of 6 stokes. What is the dynamic viscosity?
(a) 1.14 Ns/m^2 (b) 2.44 Ns/m^2
(c) 3.40 Ns/m^2 (d) 11.40 Ns/m^2
[ESE : 2021]

1.59 A flat plate of area 1.5×10^6 mm^2 is pulled with a speed of 0.4 m/s relative to another plate located at a distance of 0.15 mm from it. What is the power required to maintain this speed, if the fluid separating them is having viscosity as 1 poise?

(a) 160 W (b) 158 W
(c) 145 W (d) 130 W

[ESE : 2022]

1.60 A shaft of diameter 0.35 m rotates at 200 rpm inside a sleeve 100 mm long. The dynamic viscosity of lubricating oil in the 2 mm gap between sleeve and shaft is 8 poises. What is the power lost in the bearing?
(a) 0.59 kW (b) 0.69 kW
(c) 0.88 kW (d) 0.91 kW

[ESE : 2022]

1.61 A spherical water drop of 1 mm in diameter splits up in air into 64 smaller drops of equal size. The surface tension coefficient of water in air = 0.073 N/m. What is the work required in splitting up the drop?
(a) 0.12×10^{-3} J (b) 0.36×10^{-3} J
(c) 0.69×10^{-6} J (d) 0.89×10^{-3} J

[ESE : 2023]

1.62 Consider the following statements:
1. Viscosity of liquid increases with increase in temperature.
2. Viscosity of gas decreases with increase in temperature.
3. In liquids, cohesive forces predominate the molecular momentum transfer.
4. In gases, cohesive forces are small and the molecular momentum transfer predominates.

Which of the above statements are correct?
(a) 1 and 2 only (b) 2 and 3 only
(c) 3 and 4 only (d) 1, 2, 3 and 4

[ESE : 2024]

1.63 A fluid in which shear stress is more than the yield value and shear stress is proportional to the rate of shear strain is known as
(a) non-ideal plastic fluid (b) Newtonian fluid
(c) non-Newtonian fluid (d) ideal plastic fluid

[ESE : 2024]

1.64 Two horizontal plates are placed 1.25 cm apart, the space between them being filled with an oil of viscosity 15 poise. What is the shear stress in the oil if the upper plate is moved with a velocity of 2.5 m/s?
(a) 280 N/m^2 (b) 260 N/m^2
(c) 300 N/m^2 (d) 250 N/m^2

[ESE : 2024]

1.65 What is the capillary rise in a glass tube of 2.0 mm diameter when immersed vertically in mercury? (Take surface tension of mercury as 0.52 N/m. The specific gravity of mercury is 13.6 and the angle of contact is 130°)
(a) –0.25 cm (b) –0.65 cm
(c) –0.8 cm (d) –0.5 cm

[ESE : 2024]

1.66 Consider the following statements :
1. The apparent viscosity of a pseudoplastic fluid decreases with decrease in the shear rate.
2. The apparent viscosity of a dilatant fluid increases with increase in the shear rate.
3. Bingham plastic fluid requires a finite yield stress before beginning to flow.

Which of the above statements are correct?
(a) 1 and 2 only (b) 1 and 3 only
(c) 2 and 3 only (d) 1, 2 and 3

[ESE : 2024]

■■■■

Answers Fluid Properties

1.1 (d)	1.2 (b)	1.3 (d)	1.4 (d)	1.5 (d)	1.6 (d)	1.7 (b)	1.8 (d)	1.9 (c)
1.10 (c)	1.11 (c)	1.12 (b)	1.13 (b)	1.14 (d)	1.15 (b)	1.16 (b)	1.17 (d)	1.18 (d)
1.19 (a)	1.20 (c)	1.21 (b)	1.22 (c)	1.23 (c)	1.24 (b)	1.25 (a)	1.26 (a)	1.27 (b)
1.28 (d)	1.29 (a)	1.30 (c)	1.31 (c)	1.32 (c)	1.33 (a)	1.34 (a)	1.35 (b)	1.36 (d)
1.37 (b)	1.38 (b)	1.39 (b)	1.40 (c)	1.41 (d)	1.42 (b)	1.43 (c)	1.44 (c)	1.45 (a)
1.46 (b)	1.47 (a)	1.48 (b)	1.49 (b)	1.50 (c)	1.51 (b)	1.52 (c)	1.53 (d)	1.54 (d)
1.55 (c)	1.56 (b)	1.57 (a)	1.58 (a)	1.59 (a)	1.60 (a)	1.61 (c)	1.62 (c)	1.63 (d)
1.64 (c)	1.65 (d)	1.66 (c)						

Explanations Fluid Properties

1.1 (d)

Pressure on the two surfaces which are parallel to free surface are not same.

1.2 (b)

Given: μ = 9.81 poise

$$= \frac{9.81}{10}\ \text{Ns/m}^2 = 0.981\ \text{Ns/m}^2$$

dy = 1 cm = 0.01 m

U = 2 m/s

Shear stress,

$$\tau = \mu\frac{du}{dy} = \frac{\mu U}{dy}$$

$$= 0.981 \times \frac{2}{10^{-2}} = \mathbf{196.2\ N/m^2}$$

1.3 (d)

Assertion is false but Reason is true.

1.4 (d)

Vapour pressure plays very important role in cavitation.

1.5 (d)

Surface tension of water in contact with air at 20°C, σ = 0.0736 N/m

Diameter, d = 1 mm = 0.001 m

$$\therefore \text{Capillary rise, } h = \frac{4\sigma}{\rho g d} = \frac{4\times 0.0736}{1000\times 9.81\times 0.001}$$

$$= 0.030 = 30\ \text{mm}$$

1.6 (d)

$$\tau = \mu\frac{du}{dy}, \text{ Newtonian fluid}$$

$$\tau = \mu\left(\frac{du}{dy}\right)^n, \text{ Non-newtonian fluid}$$

$$\tau = 0, \text{ Ideal fluid}$$

$$\tau = \text{Ideal} + \mu\left(\frac{du}{dy}\right), \text{ Bingham plastic}$$

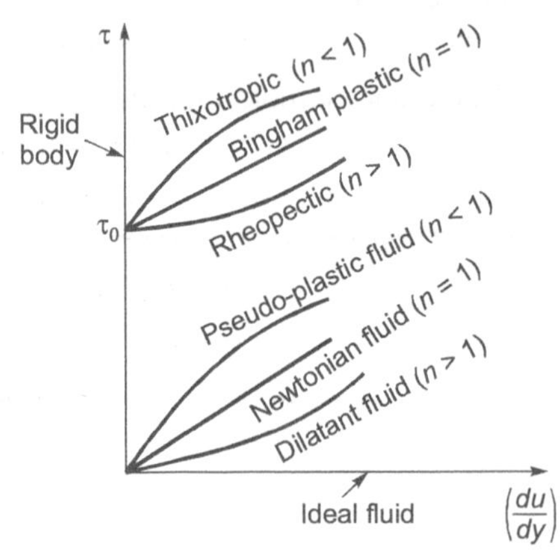

1.7 (b)

Given: $u = 2y - y^2$

$$\frac{du}{dy} = 2 - 2y$$

At y = 15 cm = 0.15 m

$$\left(\frac{du}{dy}\right)_{y=0.15\,\text{m}} = 2 - 2\times 0.15 = 1.7\ \text{s}^{-1}$$

$$\mu = 8.60 \text{ poise}$$

$$= \frac{8.60}{10} \text{ Ns/m}^2 = 0.86 \text{ Ns/m}^2$$

Shear stress,

$$\tau = \mu\left(\frac{du}{dy}\right) = 0.86 \times 1.7 = \mathbf{1.462 \ N/m^2}$$

1.8 (d)

In liquid, viscosity is due to cohesion with rise in temperature, volume of liquid increases, the distance between molecules increases, thus decreasing the cohesion. Therefore the viscosity of liquid decreases with rise in temperature. In case of gases, viscosity is due to molecular momentum exchange with rise in temperature of gas, kinetic energy of molecules increases, thus increasing the molecular momentum exchange. Therefore, the viscosity of gases increases with rise in temperature.

1.9 (c)

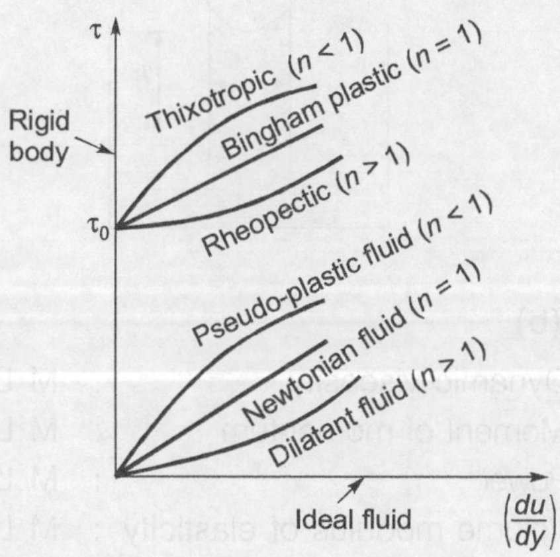

1.10 (c)

Given, $S = 0.9$

$\therefore \quad \rho = 0.9 \times 1000$

$= 900 \text{ kg/m}^3$

$\nu = 0.28$ stakes

$= 0.28 \times 10^{-4} \text{ m}^2/\text{s}$

Kinematic viscosity,

$$\gamma = \frac{\mu}{\rho}$$

$$0.28 \times 10^{-4} = \frac{\mu}{900}$$

or, $\mu = \mathbf{0.0252 \ Ns/m^2}$

1.13 (b)

Given: $w = 15 \text{ kN/m}^3$

$= 15000 \text{ N/m}^3$

$\sigma = 0.06 \text{ N/m}$

$h = 2 \text{ mm}$

$= 0.002 \text{ m}$

Capillary rise,

$$h = \frac{4\sigma}{wd} \quad \text{(for clean glass tube)}$$

$$0.002 = \frac{4 \times 0.06}{15000 \times d}$$

or $d = 0.008 \text{ m} = 8 \text{ mm}$

1.14 (d)

Bulk modulus,

$$K = \frac{\rho dp}{d\rho} = \frac{500 \times (3.5 - 3.0)}{501 - 500} = \mathbf{250 \ MPa}$$

1.15 (b)

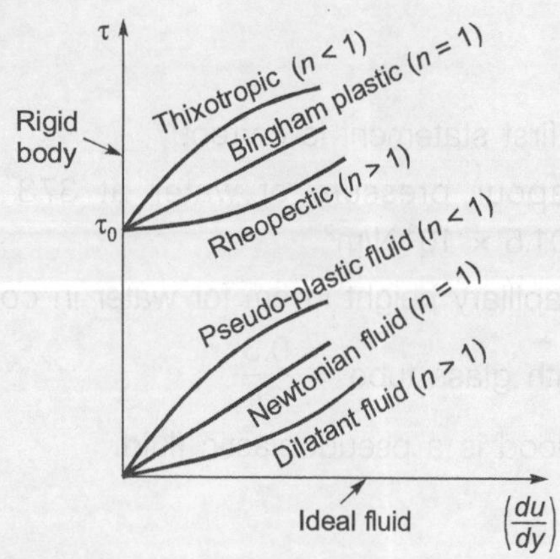

1.16 (b)

Kinematic viscosity,

$$\nu = \frac{\text{Dynamic viscosity } (\mu)}{\text{Mass density } (\rho)}$$

$$= \frac{\mu}{\rho} = \frac{ML^{-1}T^{-1}}{ML^{-3}}$$

$$= L^2T^{-1}$$

1.17 (d)

$$\tau = A\left(\frac{du}{dy}\right) + B$$

(i) $B = 0$ and $n < 1$
$\Rightarrow$ Pseudo Plastic e.g. Blood and Milk
(ii) $B = 0$ and $n > 1$
$\Rightarrow$ Dilatant Fluid e.g. Rich starch, sugar in water
(iii) $B \neq 0$ and $n = 1$
$\Rightarrow$ Bingham plastic e.g. Tooth paste
(iv) $B = 0$ and $n = 1$
$\Rightarrow$ Newtonian Fluid e.g. water.

1.18 (d)

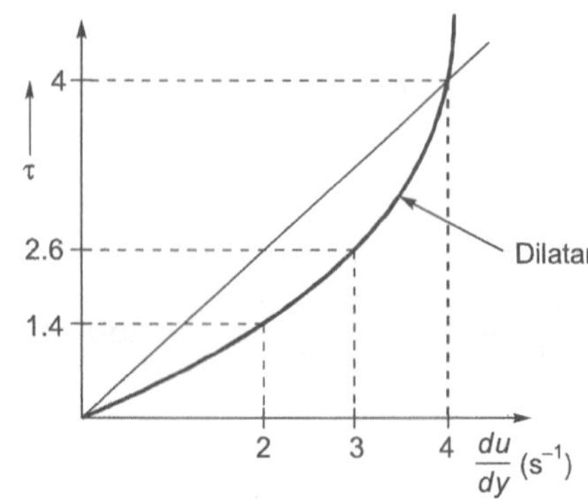

1.19 (a)

Only first statement is correct.

(i) Vapour pressure of water at 373 K is 101.5×10^3 N/m^2
(ii) Capillary height in cm for water in contact with glass tube $= \frac{0.3}{d}$
(iii) Blood is a pseudoplastic fluid.

1.20 (c)

C.G.S. unit of dynamic viscosity,

1 poise = 1 dyne s/cm^2 = $10^{-5} \times 10^4$ Ns/m^2
= 10^{-1} Ns/m^2 = **0.1 Ns/m^2**

1.21 (b)

Pressure difference between inside and outside of a droplet of water $= \frac{4\sigma}{d}$

$\because$ Bursting force $= p \times \frac{\pi d^2}{4}$

Resisting force = $\sigma \times \pi d$

Bursting force = Resisting force

$$p \times \frac{\pi d^2}{4} = \sigma \times \pi d$$

or $$p = \frac{4\sigma}{d}$$

1.22 (c)

Given: d = 2 cm = 0.02 m; σ = 0.1 N/m

$$\text{Pressure} = \frac{8\sigma}{d} = \frac{8 \times 0.1}{2 \times 10^{-2}}$$
$$= 0.4 \times 100 = \mathbf{40\ N/m^2}$$

1.23 (c)

Capillary rise is given by

$$h = \frac{4\sigma\cos\theta}{wd} = \frac{4\sigma\cos\theta}{\rho g d}$$

$$h \propto \frac{1}{d}$$

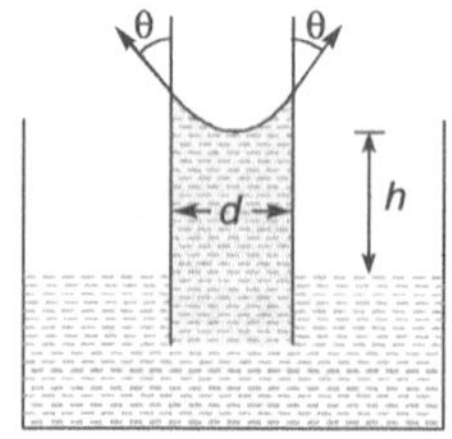

1.24 (b)

Dynamic viscosity	:	$M L^{-1} T^{-1}$
Moment of momentum	:	$M L^2 T^{-1}$
Power	:	$M L^2 T^{-3}$
Volume modulus of elasticity	:	$M L^{-1} T^{-2}$

1.25 (a)

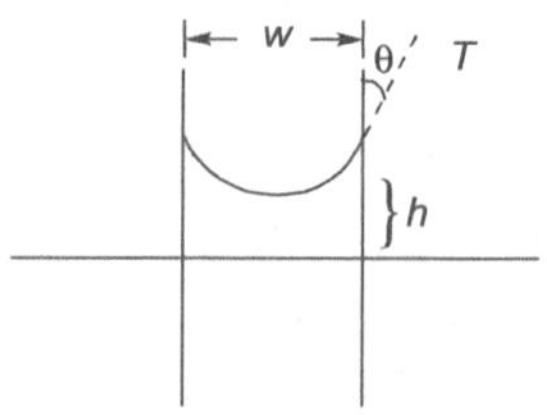

For narrow slit $\rightarrow$ weight of fluid = surface tension force

$$(w \times h \times l) \times \rho g = T\cos\theta \times 2l$$

$$\therefore \quad h = \frac{2T\cos\theta}{w\rho g}$$

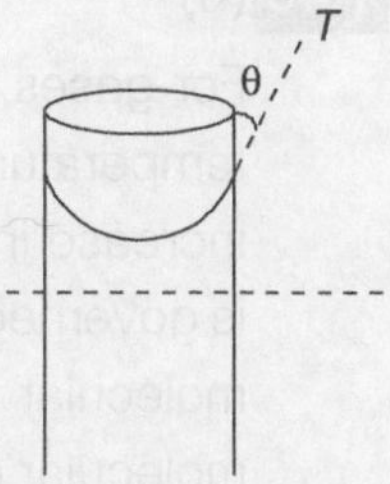

For capillary tube

$$T\cos\theta \times (w\pi) = \frac{\pi w^2}{4} h\rho g$$

$$h = \frac{4T\cos\theta}{w\rho g}$$

1.28 (d)

- Viscosity is a vital property of the lubricant.
- Rise of sap in trees is due to capillary action.
- Formation of droplets due to surface tension.
- Cavitation is related to the vapour pressure.

1.31 (c)

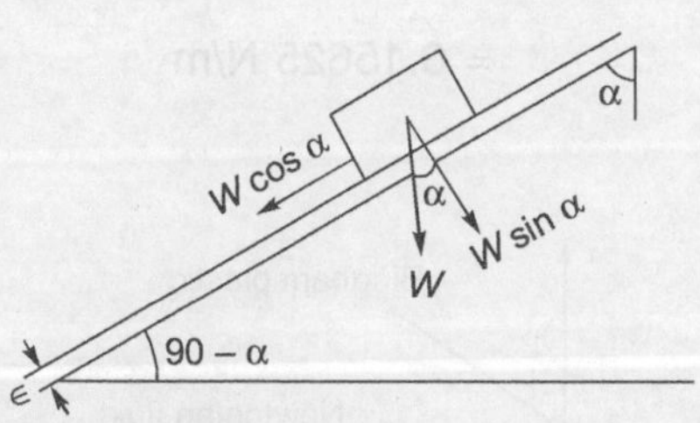

For zero acceleration

$$W \sin\alpha = \text{drag force}$$

Here

$$\text{Drag force} = \text{shear force} = \mu A \frac{V}{\epsilon}$$

$$W\cos\alpha = \mu A \frac{V}{\epsilon}$$

$$\therefore \quad V = \frac{\epsilon W \cos\alpha}{\mu A}$$

1.32 (c)

Bulk modulus,

$$K = \frac{\rho\, dp}{d\rho}$$

$$\frac{d\rho}{\rho} = \frac{\rho g dh}{K}$$

$$\left[-\frac{1}{\rho}\right]_{\rho}^{\rho_0} = \frac{gh}{K}$$

$$h = \frac{K}{g}\left[\frac{1}{\rho_0} - \frac{1}{\rho}\right]$$

1.33 (a)

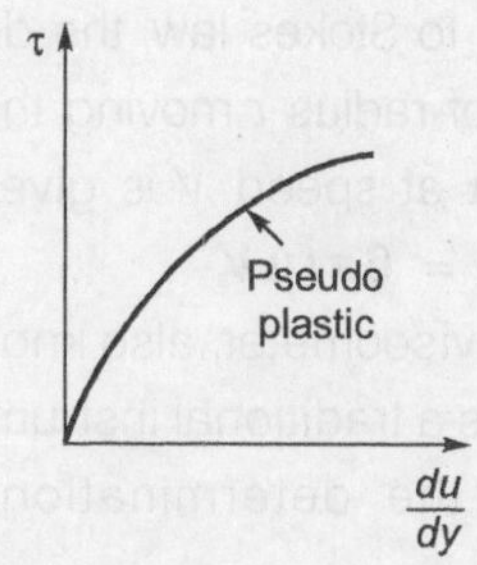

1.34 (a)

When adhesion force is more than cohesion force then fluid will wet the surface and angle of contact between fluid and surface will be less than 90° (acute).

1.35 (b)

$$\text{Bulk modulus: } K = \frac{dp}{\left(-\frac{dV}{V}\right)}$$

∴ Bulk modulus increase with increases in pressure.

1.36 (d)

Capillarity	–	Surface tension
Vapour pressure	–	Cavitation
Viscosity	–	Shear forces
Specific gravity	–	Density of water.

1.37 (b)

$$\tau \propto \left(\frac{du}{dy}\right) \propto \frac{d\theta}{dt}$$

∴ Above equation shows the relation between rate of angular deformation and shear stress.

1.38 (b)

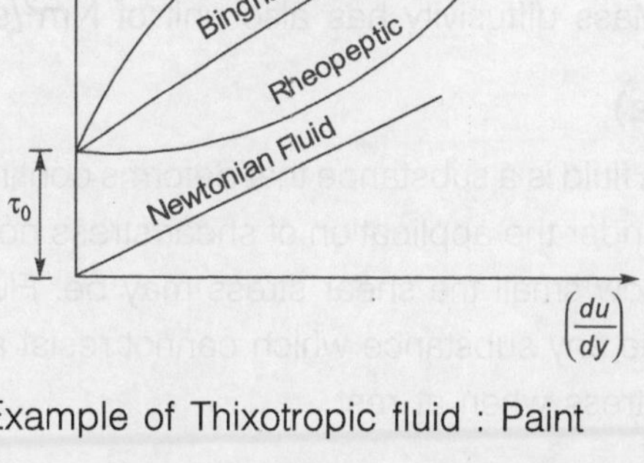

Example of Thixotropic fluid : Paint

1.39 (b)

According to Stokes law, the drag force, F, on a sphere of radius r moving through a fluid of viscosity μ at speed V is given by

$$F = 6\pi r \mu V$$

The efflux viscometer, also known as flow cup, type ford is a traditional instrument used world wide for the determination of kinematic viscosity.

1.41 (d)

Cohesive force is the action or property of like molecules sticking together, being mutually attractive. Mercury has large adhesion force with most container materials and strong cohesive forces. This causes the depression in mercury level inside the tube.

1.42 (b)

Water hammer — Sudden closure of valve
Cavitation — Vapour pressure
Rise of sap in tree — Capillarity
Spherical shape of rainwater drop — Surface tension

1.43 (c)

Ideal fluid is inviscid *i.e.*, it has zero viscosity. For Newtonian fluid,

$$\text{Shear stress: } \tau = \mu \frac{du}{dy}$$

For real fluid viscosity is non zero.

1.44 (c)

$$\text{Thermal diffusivity: } \alpha = \frac{k}{\rho c_p} \ \text{m}^2/\text{s}$$

$$\text{Kinematic viscosity: } \nu = \frac{\mu}{\rho} \ \text{m}^2/\text{s}$$

Dynamic viscosity, μ has unit **Ns/m^2**
Mass diffusivity has also unit of **Nm^2/s**

1.45 (a)

A fluid is a substance that deforms continuously under the application of shear stress no matter how small the shear stress may be. Fluid can be any substance which cannot resist a shear stress when at rest.

1.46 (b)

For gases, viscosity increases with increase in temperature while for liquids it decreases with increase in temperature. In liquid, the viscosity is governed by the cohesive forces between the molecular of the liquid, whereas in gases, the molecular activity plays a dominant role.

1.47 (a)

Pressure intensity inside a soap bubble,

$$p = \frac{4\sigma}{r}$$

$$\therefore \quad \sigma = \frac{pr}{4} = \frac{25 \times 25 \times 10^{-3}}{4}$$

$$= \mathbf{0.15625 \ N/m}$$

1.49 (b)

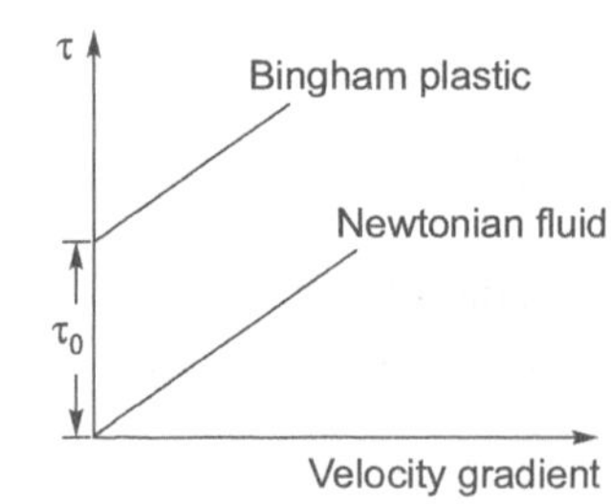

1.51 (b)

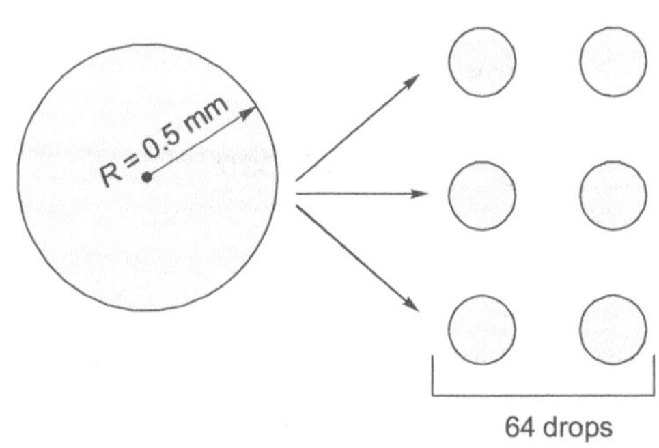

Volume before splitting = Volume after splitting

$$\frac{4}{3}\pi R^3 = n \times \frac{4}{3}\pi r^3$$

$$r = \frac{R}{n^{1/3}} = \frac{0.5}{(64)^{1/3}} = \frac{0.5}{4} = 0.125 \text{ mm}$$

$$W = \sigma (\Delta A)$$

$$= 0.073 \times (n \times 4\pi r^2 - 4\pi R^2)$$

$$= 0.073 \times 4\pi \, [64 \times (0.125)^2 - (0.5)^2]$$

$$= 0.073 \times 4\pi \, [64 \times (0.125)^2 - (0.5)^2] \times 10^{-6}$$

$$= 0.69 \times 10^{-6} \text{ J}$$

1.52 (c)

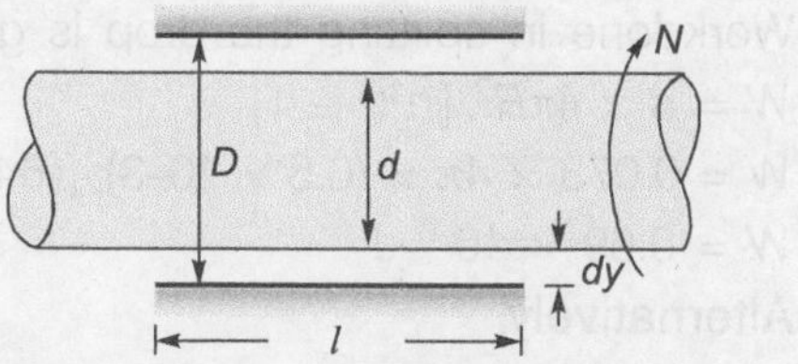

Given: d = 150 mm, N = 1500 rpm,

l = 200 mm, D = 150.5 mm

μ = 0.8 poise = 0.08 Ns/m^2

Clearance between the bearing and the journal,

$$dy = \frac{D-d}{2} = \frac{150.5-150}{2} = 0.25 \text{ mm}$$

$$= 0.25 \times 10^{-3} \text{ m}$$

The peripheral speed of the shaft,

$$u = \frac{\pi DN}{60} = \frac{\pi \times 0.150 \times 1500}{60}$$

$$= 11.78 \text{ m/s}$$

$$du = u - 0 = u = 11.78 \text{ m/s}$$

The Shear stress on the shaft,

$$\tau = \mu\frac{du}{dy} = \frac{0.08 \times 11.78}{0.25 \times 10^{-3}}$$

$$= 3769.6 \text{ N/m}^2 \approx 3.77 \text{ kN/m}^2$$

1.54 (d)

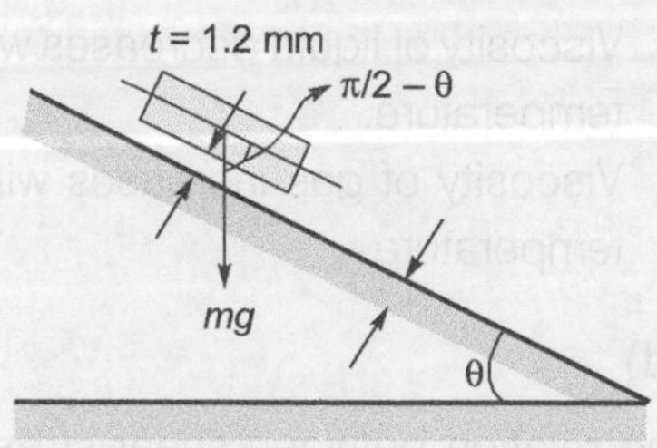

$$mg\cos\left(\frac{\pi}{2} - \theta\right) = \frac{\mu v}{t} \times A$$

$$\text{mg} \sin\theta = \frac{\mu v}{t} \times A$$

$$150 \times \sin 30 = \frac{\mu \times 0.2}{1.2 \times 10^{-3}} \times (0.8 \times 0.8)$$

$$\mu = 0.703 \ \frac{\text{N}\cdot\text{s}}{\text{m}^2}$$

1.55 (c)

Area of plate, A = 0.1 m^2

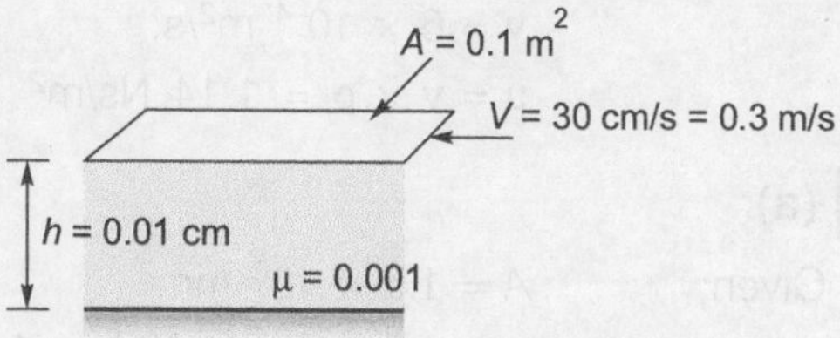

$$\text{Power} = F \times V$$

$$= (\tau A)V = \left(\mu\frac{V}{h}\right) \times AV$$

$$= 0.001\frac{0.3}{0.01 \times 10^{-2}} \times 0.1 \times 0.3$$

$$= 0.09 \text{ W}$$

1.56 (b)

Change in pressure,

$$dp = 6 - 3 = 3 \text{ MN/m}^2$$

$$= 3 \times 10^6 \text{ N/m}^2$$

%change in volume,

$$\frac{dV}{V} = \frac{-0.1}{100}$$

Bulk modulus,

$$K = \frac{dp}{\left(-\frac{dV}{V}\right)} = \frac{(6-3)}{(0.001)}$$

$$= 3 \times 10^9 \text{ N/m}^2$$

1.57 (a)

- In liquids, main cause of viscosity is cohesion between the molecules. With increase in temperature, this cohesive force decreases as the energy of particles become more, hence movement of particles become free and easy. Therefore, viscosity in liquid decreases with temperature.
- In gases, the important cause of viscosity is randomness/molecular collision due to RMS velocity. Due to rise in temperature, K.E. of molecules increases which makes RMS to increase, hence randomness and collision of molecules increased. This makes the flow difficult, therefore viscosity in gases increases with temperature.

1.58 (a)

$S = 1.9,\ \rho = 1900\ \text{kg/m}^3,$
$\nu = 6 \times 10^{-4}\ \text{m}^2/\text{s},$
$\mu = \nu \times \rho = 1.14\ \text{Ns/m}^2$

1.59 (a)

Given:
$$A = 1.5 \times 10^6\ \text{mm}^2$$
$$= 1.5 \times 10^6 \times 10^{-6} = 1.5\ \text{m}^2$$
$$V = 0.4\ \text{m/s}$$
$$y = 0.15\ \text{mm} = 0.15 \times 10^{-3}\ \text{m}$$
$$\mu = 1\ \text{poise} = 0.1\ \text{kg/m-s}$$
$$P = F \times V = \frac{\mu AV}{y} \times V = \frac{\mu AV^2}{y}$$
$$P = \frac{0.1 \times 1.5 \times (0.4)^2}{(0.15 \times 10^{-3})} = 160\ \text{W}$$

1.60 (a)

Given: $d = 0.35$ m, $N = 200$ rpm, $L = 100$ mm, $y = 2$ mm, $\mu = 0.8\ \text{Ns/m}^2$

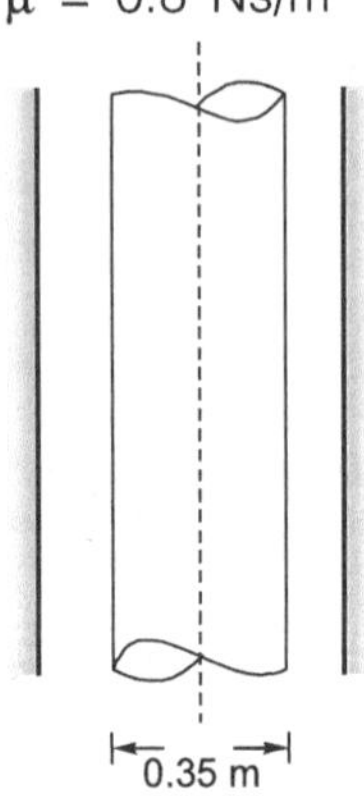

$$\omega = \frac{2\pi \times 200}{60}$$
$$T = F \times R = \frac{\mu AV \times R}{y}$$
$$= \frac{\mu 2\pi RLR\omega}{y} \times R$$
$$= \frac{2\pi\omega LR^3}{y}$$
$$= \frac{2\pi(0.8)(20.44) \times 0.1 \times 5.359375 \times 10^{-3}}{2 \times 10^{-3}}$$
$$= 28.20$$
$$P = T \times \omega = 0.59\ \text{kW}$$

1.61 (c)

Workdone in splitting the drop is given by,
$W = \sigma \times 4\pi R^2\ [n^{1/3} - 1]$
$W = 0.073 \times 4\pi \times (0.5 \times 10\text{–}3)^2[(64)^{1/3} - 1]$
$W = 0.69 \times 10^{-6}$ J

Alternatively,

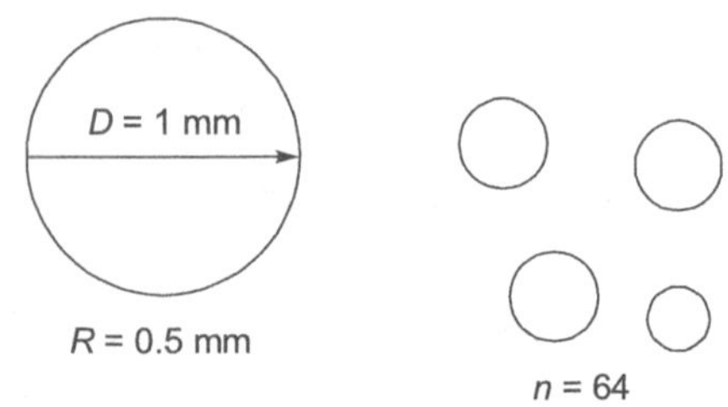

$$\text{W.D.} = (\text{Change in surface area})\sigma$$
$$= 0.073\left[n4\pi r^2 - 4\pi R^2\right]$$
$$= 0.073\left[n \times \frac{4\pi R^2}{n^{2/3}} - 4\pi R^2\right];\ r = \frac{R}{n^{1/3}}$$
$$= 0.073\left[4\pi R^2 n^{1/3} - 4\pi R^2\right]$$
$$= 0.0734\pi R^2\left[n^{1/3} - 1\right]$$
$$= 0.073 \times 4\pi\left(0.5 \times 10^{-3}\right)^2\left[(64)^{1/3} - 1\right]$$
$$= 0.69 \times 10^{-6}\text{J}$$

1.62 (c)

- Viscosity of liquid decreases with increase in temperature.
- Viscosity of gas increases with increase in temperature.

1.63 (d)

The relation between shear stress and rate of deformation for non-Newtonian fluid is given as

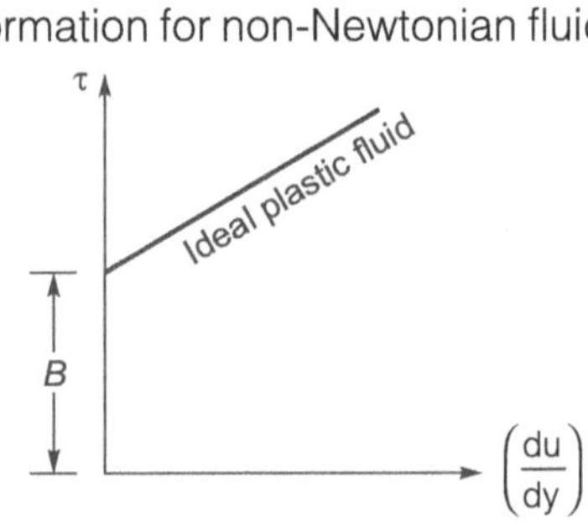

$$\tau = A\left(\frac{du}{dy}\right)^n + B$$

where, n = Flow behaviour index,
A = Consistency index, and
B = Residual strength

If $B \neq 0$ and $n = 1$.
It is ideal plastic fluid.

1.64 (c)

Given: $h = 1.25$ cm $= 0.0125$ m, $\mu = 15$ poise $= 1.5$ Pa-s, $v = 2.5$ m/s
Using Newton's laws of viscosity,

$$\tau = \mu\frac{\Delta V}{\Delta h} = \mu\left(\frac{v-0}{h-0}\right) = \mu\frac{v}{h}$$

$$= 1.5\times\frac{2.5}{0.0125} = 300\ \text{N/m}^2$$

1.65 (d)

Given: $d = 2$ m $= 0.002$ m, $\sigma = 0.52$ N/m, $\rho = 13.6 \times 10^3$ kg/m^3, $\theta = 130°$
Capillary rise is given as,

$$h = \frac{4\sigma\cos\theta}{\rho g d} = \frac{4\times0.52\times\cos(130°)}{(13.6\times10^3)\times(9.81)\times(0.002)}$$

$$= -5.010 \times 10^{-3}\ \text{m}$$
$$= -5.010\ \text{mm}$$
$$= -0.5010\ \text{cm} \simeq -0.5\ \text{cm}$$

1.66 (c)

- The apparent viscosity of a pseudoplastic fluid decreases with increase in the shear rate.

■■■■

2 Fluid Pressure and Measurement

2.1 A U-tube manometer is connected to a pipeline conveying water as shown in the figure below.

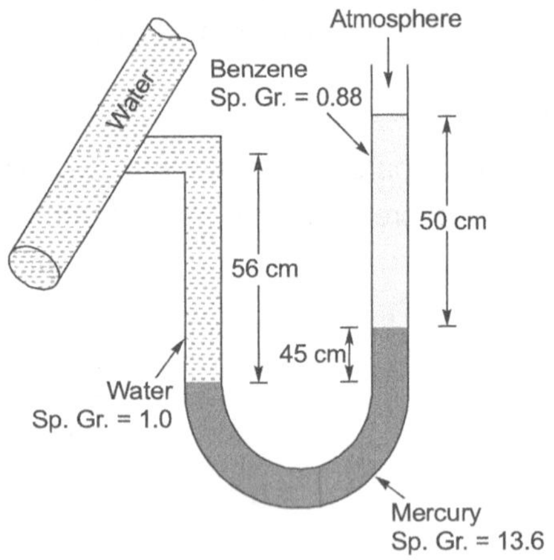

The pressure head of water in the pipeline is

(a) 7.12 m (b) 6.56 m
(c) 6.0 m (d) 5.12 m

[ESE : 2000]

2.2 A Pitot static tube is used to measure the velocity of water using a differential gauge which contains a manometric fluid of relative density 1.4. The deflection of the gauge fluid when water flows at a velocity of 1.2 m/s will be (the coefficient of the tube may be assumed to be 1)

(a) 183.5 mm (b) 52.4 mm
(c) 5.24 mm (d) 73.4 mm

[ESE : 2000]

2.3 Differential pressure head measured by mercury oil differential manometer (specific gravity of oil is 0.9) equivalent to a 600 mm difference of mercury levels will nearly be

(a) 7.62 m of oil (b) 76.2 m of oil
(c) 7.34 m of oil (d) 8.47 m of oil

[ESE : 2001]

2.4 Pressure drop of water flowing through a pipe (density 1000 kg/m^3) between two points is measured by using a vertical U-tube manometer. Manometer uses a liquid with density 2000 kg/m^3. The difference in height of manometric liquid in the two limbs of the manometer is observed to be 10 cm. The pressure drop between the two points is

(a) 98.1 N/m^2 (b) 981 N/m^2
(c) 1962 N/m^2 (d) 19620 N/m^2

[ESE : 2002]

2.5 In a pipe-flow, pressure is to be measured at a particular cross-section using the most appropriate instrument. Match **List-I** (Expected pressure range) with **List-II** (Appropriate measuring device) and select the correct answer using the codes given below the lists:

List-I

A. Steady flow with small positive gauge pressure
B. Steady flow with small negative and positive gauge pressure
C. Steady flow with high gauge pressure
D. Unsteady flow with fluctuating pressure

List-II

1. Bourdon pressure gauge
2. Pressure transducer
3. Simple piezometer
4. U-tube manometer

Codes:

	A	B	C	D
(a)	3	2	1	4
(b)	1	4	3	2
(c)	3	4	1	2
(d)	1	2	3	4

[ESE : 2002]

2.6 Hydrostatic law of pressure is given as

(a) $\frac{\partial p}{\partial z} = \rho g$ (b) $\frac{\partial p}{dz} = 0$

(c) $\frac{\partial p}{\partial z} = Z$ (d) $\frac{\partial p}{\partial z} = \text{constant}$

[ESE : 2002]

2.7 To measure the pressure head of the fluid of specific gravity S flowing through a pipeline, a simple micro-manometer containing a fluid of specific gravity S_1 is connected to it. The readings are as indicated in the diagram shown below.

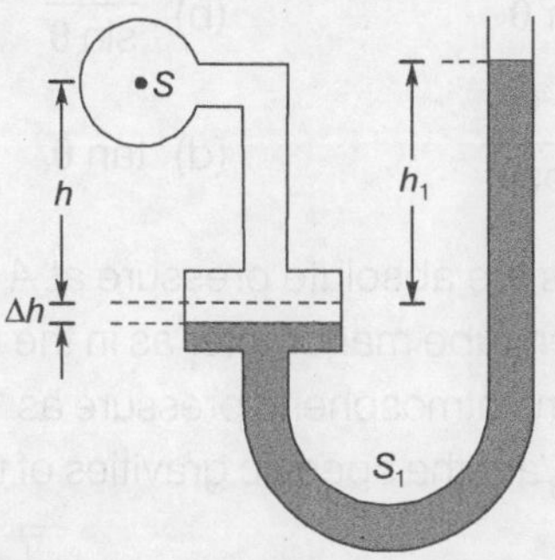

The pressure head in the pipeline is

(a) $h_1S_1 - hS - \Delta h(S_1 - S)$
(b) $h_1S_1 - hS + \Delta h(S_1 - S)$
(c) $hS - h_1S_1 - \Delta h(S_1 - S)$
(d) $hS - h_1S_1 + \Delta h(S_1 - S)$ [ESE : 2003]

2.8 A mercury-water manometer has a gauge difference of 500 mm (difference in elevation of menisci). What will be the difference in pressure?

(a) 0.5 m (b) 6.3 m
(c) 6.8 m (d) 7.3 m [ESE : 2004]

2.9 The balancing column shown in the diagram below contains 3 liquids of different densities ρ_1, ρ_2 and ρ_3. The liquid level of one limb is h_1 below the top level and there is a difference of h relative to that in the other limb.

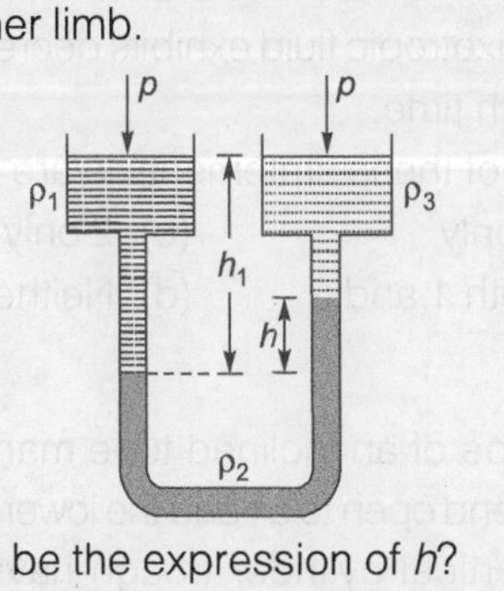

What will be the expression of h?

(a) $\left(\frac{\rho_1 - \rho_2}{\rho_1 - \rho_3}\right)h_1$ (b) $\left(\frac{\rho_2 - \rho_3}{\rho_1 - \rho_3}\right)h_1$

(c) $\left(\frac{\rho_1 - \rho_3}{\rho_2 - \rho_3}\right)h_1$ (d) $\left(\frac{\rho_1 - \rho_2}{\rho_2 - \rho_3}\right)h_1$

[ESE : 2004]

2.10 When can a piezometer be not used for pressure measurement in pipes?

(a) The pressure difference is low.
(b) The velocity is high.
(c) The fluid in the pipe is a gas.
(d) The fluid in the pipe is highly viscous.

[ESE : 2005]

2.11 A U-tube open at both ends and made of 8 mm diameter glass tube has mercury up to a height of 10 cm in both the limbs. If 19 cm^3 of water is added to one of the limbs, what is the difference in mercury levels in the two limbs at equilibrium?

(a) 4.5 mm (b) 1.0 cm
(c) 2.8 cm (d) 3.2 cm

[ESE : 2005]

2.12 The standard atmospheric pressure is 762 mm of Hg. At a specific location, the barometer reads 700 mm of Hg. At this place, what does an absolute pressure of 380 mm of Hg corresponds to?

(a) 320 mm of Hg vacuum
(b) 382 mm of Hg vacuum
(c) 62 mm of Hg vacuum
(d) 62 mm of Hg gauge

[ESE : 2006]

2.13 A manometer is made of a tube of uniform bore of 0.5 cm^2 cross-sectional area, with one limb vertical and the other limb inclined at 30° to the horizontal. Both of its limbs are open to atmosphere and, initially, it is partly filled with a manometer liquid of specific gravity 1.25. If then an additional volume of 7.5 cm^3 of water is poured in the inclined tube, what is the rise of the meniscus in the vertical tube?

(a) 4 cm (b) 7.5 cm
(c) 12 cm (d) 15 cm

[ESE : 2006]

2.14 Which property of mercury is the main reason for its use in barometers?

(a) High density
(b) Negligible capillarity effect
(c) Very low vapour pressure
(d) Low compressibility [ESE : 2007]

2.15 The pressure difference of two very light gases in two rigid vessels is being measured by a vertical U-tube water filled manometer. The reading is found to be 10 cm. What is the pressure difference?

(a) 9.81 kPa (b) 0.0981 bar
(c) 98.1 Pa (d) 981 N/m^2

[ESE : 2007]

2.16 The lower portion of a U-tube of uniform bore having both limbs vertical and open to atmosphere, is initially filled with a liquid of specific gravity 3*S*. A lighter liquid of specific gravity *S* is then poured into one of the limbs such that the length of column of lighter liquid is x. What is the resulting movement of the meniscus of the heavier liquid in the other limb?

(a) x (b) $\frac{x}{2}$

(c) $\frac{x}{3}$ (d) $\frac{x}{6}$ [ESE : 2008]

2.17 How is the difference of pressure head '*h*' measured by a mercury-oil differential manometer expressed?

(a) $h = x\left(1-\frac{S_g}{S_0}\right)$ (b) $h = x\,(S_g - S_0)$

(c) $h = x(S_0 - S_g)$ (d) $h = x\left(\frac{S_g}{S_0}-1\right)$

where x = manometer reading; S_g and S_0 are the specific gravities of mercury and oil, respectively.

[ESE : 2008]

2.18 In an open U-tube containing mercury, kerosene of specific gravity 0.8 is poured into one of its limbs so that the length of column of kerosene is about 40 cm. The level of mercury column in that limb is lowered approximately by how much?

(a) 2.4 cm (b) 1.2 cm

(c) 3.6 cm (d) 0.6 cm

[ESE : 2008]

2.19 What are the forces that influence the problem of fluid statics?

(a) Gravity and viscous forces

(b) Gravity and pressure forces

(c) Viscous and surface tension forces

(d) Gravity and surface tension forces

[ESE : 2009]

2.20 What is the difference in pressure head, measured by a mercury-oil differential manometer for a 20 cm difference of mercury level? (Sp. gravity of oil = 0.8)

(a) 2.72 m of oil (b) 2.52 m of oil

(c) 3.40 m of oil (d) 2.00 m of oil

[ESE : 2009]

2.21 In order to increase sensitivity of U-tube manometer, one leg is usually inclined by an angle θ. What is the sensitivity of inclined tube compared to sensitivity of U-tube?

(a) $\sin\theta$ (b) $\frac{1}{\sin\theta}$

(c) $\frac{1}{\cos\theta}$ (d) $\tan\theta$ [ESE : 2009]

2.22 What is the absolute pressure at *A* measured by an open tube manometer as in the figure below? (Assume atmospheric pressure as 103 kN/m^2; S_A and S_B are the specific gravities of the two fluids)

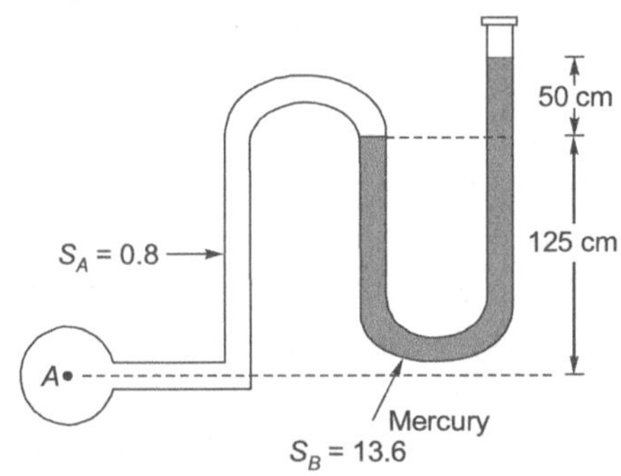

(a) 78.5 kN/m^2 (b) 180 kN/m^2

(c) 1030 kN/m^2 (d) 103 kN/m^2

[ESE : 2010]

2.23 Consider the following statements:

1. Piezometer is used to measure small variation of pressure above or below ambient pressure.
2. Thixotropic fluid exhibits decrease in viscosity with time.

Which of these statements is/are correct?

(a) 1 only (b) 2 only

(c) Both 1 and 2 (d) Neither 1 nor 2

[ESE : 2010]

2.24 The tube of an inclined tube manometer has its upper end open to air and the lower end connected to a vertical cylinder trough having a diameter 4 times that of the tube. The slope of the tube is 1 in 4 and specific gravity of manometer liquid is 0.72. The space above the liquid in the trough is connected to a gas chamber. When the movement of meniscus in the tube is 8 cm, change in gas pressure, in height of water column, is

(a) 1.35 cm (b) 1.44 cm

(c) 1.80 cm (d) 1.53 cm

[ESE : 2010]

2.25 Assertion (A): The important property for a liquid to use it as manometric fluid is its colour for visibility.

Reason (R): Inclined manometers are used to measure low pressures.

(a) Both **A** and **R** are individually true and **R** is the correct explanation of **A**
(b) Both **A** and **R** are individually true but **R** is not the correct explanation of **A**
(c) **A** is true but **R** is false
(d) **A** is false but **R** is true

[ESE : 2010]

2.26 Pascal's law states that pressure at a point is equal to all direction in a

(a) liquid at rest (b) fluid at rest
(c) laminar flow (d) turbulent flow

[ESE : 2011]

2.27 The pressure in Pascal corresponding to 3 cm column of mercury is

(a) 7988.6 (b) 3994.3
(c) 2662.8 (d) 1331.4

[ESE : 2011]

2.28 At what depth below the free surface of oil having a density of 784 kg/m^3 will the pressure be very nearly equal to 1 bar?

(a) 10 m (b) 14 m
(c) 13 m (d) 7.84 m

[ESE : 2012]

2.29 A 70 kg person walks on snow with a total foot implant area of 500 cm^2. What pressure does he exert on snow?

(a) 0.5 kPa (b) 12.5 kPa
(c) 13.73 kN/m^2 (d) 137.3 kN/m^2

[ESE : 2013]

2.30 In a differential manometer a head of 0.6 m of fluid *A* in limb 1 is found to balance a head of 0.3 m of fluid *B* in limb 2. The ratio of specific gravities of *A* to *B* is

(a) 2 (b) 0.5
(c) 0.3 (d) 0.18

[ESE : 2013]

2.31 Manometer is a device used for measuring

(a) Velocity at a point in a fluid
(b) Pressure at a point in a fluid
(c) Discharge of a fluid
(d) None of the above

[ESE : 2014]

2.32 What is the intensity of pressure in the following SI units, when specific gravity of mercury is 13.6 and the intensity of pressure is 400 kPa?

(a) 0.3 bar or 4.077 m of water or 0.299 m of Hg
(b) 4 bar or 5.077 m of water or 0.399 m of Hg
(c) 0.3 bar or 5.077 m of water or 0.599 m of Hg
(d) 4 bar or 4.077 m of water or 0.299 m of Hg

[ESE : 2015]

2.33 In a differential manometer, a head of 0.5 m of fluid *A* in limb 1 is found to balance a head of 0.3 m of fluid *B* in limb 2. The atmospheric pressure is 760 mm of mercury. The ratio of specific gravities of *A* to *B* is:

(a) 0.25 (b) 0.6
(c) 2 (d) 4

[ESE : 2016]

2.34 The pressure at a point in water column is 3.924 N/cm^2. What is the corresponding height of water?

(a) 8 m (b) 6 m
(c) 4 m (d) 2 m

[ESE : 2016]

2.35 A vacuum gauge fixed on a steam condenser reads 80 kPa vacuum. The barometer indicates 1.013 bar. The absolute pressure in terms of mercury head is, nearly

(a) 160 mm of Hg (b) 190 mm of Hg
(c) 380 mm of Hg (d) 760 mm of Hg

[ESE : 2016]

2.36 Statement (I) : A differential inverted U-tube manometer determines the difference in pressures between two points in a flow section to which it is connected.

Statement (II) : The sensitivity of an inclined gauge depends on the angle of inclination.

(a) Both Statement (I) and Statement (II) are individually true and Statement (II) is the correct explanation of Statement (I).

(b) Both Statement (I) and Statement (II) are individually true but Statement (II) is not the correct explanation of Statement (I).
(c) Statement (I) is true but Statement (II) is false.
(d) Statement (I) is false but Statement (II) is true.

[ESE : 2018]

2.37 **Statement (I) :** There exists a positive pressure difference between the inlet and throat of a venturi meter.

Statement (II) : The coefficient of discharge of a venturimeter accounts for the non-uniformity of flow at both inlet and throat.

(a) Both Statement (I) and Statement (II) are individually true and Statement (II) is the correct explanation of Statement (I).
(b) Both Statement (I) and Statement (II) are individually true but Statement (II) is not the correct explanation of Statement (I).
(c) Statement (I) is true but Statement (II) is false.
(d) Statement (I) is false but Statement (II) is true.

[ESE : 2018]

2.38 A pipe contains an oil of specific gravity 0.9. A differential manometer connected at the two points *A* and *B* shows a difference in mercury level as 15 cm. What is the difference of pressure at the two points? (Take acceleration due to gravity as 9.81 m/s^2, specific gravity of mercury as 13.6 and density of water as 1000 kg/m^3)

(a) 18688 N/m^2 (b) 19688 N/m^2
(c) 15688 N/m^2 (d) 17866 N/m^2

[ESE : 2022]

2.39 What is the intensity of pressure in the ocean at a depth of 1500 m, assuming salt water is incompressible with a specific weight of 10050 N/m^3?

(a) 15.08 MN/m^2 gauge
(b) 25.08 MN/m^2 gauge
(c) 32.06 MN/m^2 gauge
(d) 42.06 MN/m^2 gauge

[ESE : 2023]

2.40 What is the actual pressure of air in the tank if the pressure of compressed air measured by manometer is 30 cm of mercury and atmospheric pressure is 101 kPa?
(Take g = 9.78 m/s^2 and density of mercury at room temperature as 13550 kg/m^3)

(a) 140.76 GPa (b) 140.76 MPa
(c) 140.76 kPa (d) 140.76 Pa

[ESE : 2024]

2.41 Consider the following statements :

1. One end of piezometer is connected to the point where pressure is to be measured and the other end is open to the atmosphere.
2. Inverted U-tube differential manometer is used to measure high pressure.
3. U-tube differential manometer contains heavy liquids.

Which of the above statements are correct?

(a) 1 and 2 only (b) 1 and 3 only
(c) 2 and 3 only (d) 1, 2 and 3

[ESE : 2024]

■■■■

Answers Fluid Pressure and Measurement

2.1 (c)	2.2 (a)	2.3 (d)	2.4 (b)	2.5 (c)	2.6 (a)	2.7 (b)	2.8 (b)	2.9 (c)
2.10 (c)	2.11 (c)	2.12 (a)	2.13 (a)	2.14 (c)	2.15 (d)	2.16 (d)	2.17 (d)	2.18 (b)
2.19 (b)	2.20 (c)	2.21 (b)	2.22 (b)	2.23 (b)	2.24 (b)	2.25 (d)	2.26 (b)	2.27 (b)
2.28 (c)	2.29 (c)	2.30 (b)	2.31 (b)	2.32 (d)	2.33 (b)	2.34 (c)	2.35 (a)	2.36 (b)
2.37 (b)	2.38 (a)	2.39 (a)	2.40 (c)	2.41 (b)				

Explanations Fluid Pressure and Measurement

2.1 (c)

Starting from pipeline and write manometric equation

$$p + 0.56 \times 1 - 0.45 \times 13.6 - 0.5 \times 0.88 = 0$$

$\therefore \quad p =$ **6 m of water**

2.2 (a)

Pitot static tube.

Flow through pipe,

$$V = \sqrt{2gx\left(\frac{S_m}{S} - 1\right)}$$

where

x = Manometric deflection

S_m = Specific gravity of manometric fluid

S = Specific gravity of flowing fluid

$$V = \sqrt{2gx\left(\frac{S_m}{S} - 1\right)}$$

$$1.2 = \sqrt{2 \times 9.81 x\left[\frac{1.4}{1} - 1\right]}$$

$$\text{or } x = \frac{(1.2)^2}{2 \times 9.81 \times 0.4}$$

$= 0.1835$ m = **183.5 mm**

2.3 (d)

Given data:

Specific gravity of oil, $S_{oil} = 0.9$

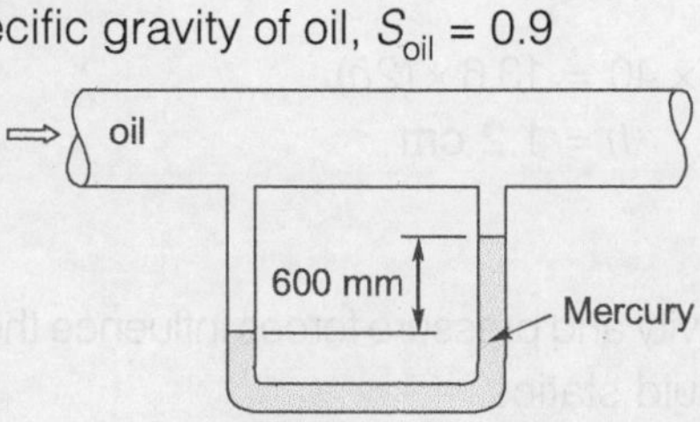

x = 600 mm of Hg = 0.60 m of Hg

$$h = x\left[\frac{S_{Hg}}{S_{oil}} - 1\right] = 0.60\left[\frac{13.6}{0.9} - 1\right]$$

= **8.47 m of oil**

2.4 (b)

$$\frac{p_1 - p_2}{\rho g} = x\left(\frac{S_m}{S} - 1\right)$$

$p_1 - p_2 = 1000 \times 9.81 \times 0.1\,(2 - 1) =$ **981 N/m²**

2.5 (c)

Bourden pressure gauge: Bourden tubes measure gauge pressure relative to ambient atmospheric pressure, as opposed to absolute pressure, vacuum is sensed as reverse motion.

Pressure transducer: It is a type of pressure sensor, that are designed to measure in a dynamic mode very high speed changes in pressure.

U-tube manometers: It can measure both negative and positive gauge pressure.

2.6 (a)

Hydrostatic law states that the pressure varies with the depth, in a static fluid.

2.7 (b)

$$\frac{p}{\rho g} + S(h + \Delta h) - S_1(h_1 + \Delta h) = 0$$

$$\therefore \quad \frac{p}{\rho g} = S_1 h_1 - hS + \Delta h(S_1 - S)$$

2.8 (b)

Difference in pressure

$$= \left(\frac{S_m}{S_0} - 1\right)x = \left(\frac{13.6}{1} - 1\right) \times 0.5 = \textbf{6.3 m}$$

2.9 (c)

$$p+\rho_1 h_1-\rho_2 h-\rho_3(h_1-h)=p$$

$$\therefore\ h=\frac{\rho_1-\rho_3}{\rho_2-\rho_3}h_1$$

2.10 (c)

Gas pressure cannot be measured by means of piezometers because a gas forms no free atmosphere surface and it can't be used when large pressures in the lighter liquid are to be measured.

2.11 (c)

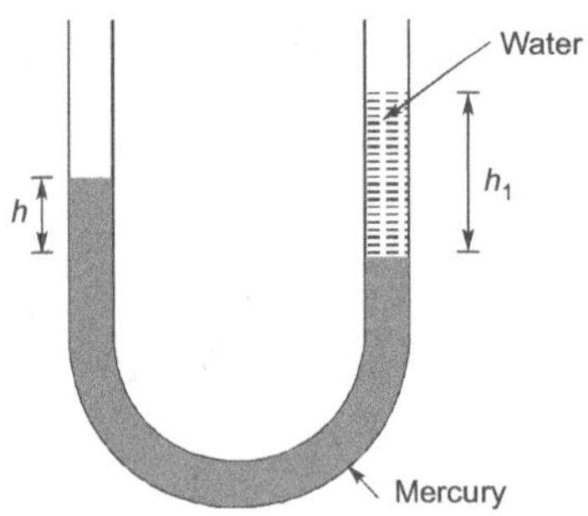

$$\frac{\pi}{4}\times(0.8)^2\times h_1=19$$

$$\therefore\ h_1=37.8\,\text{cm}$$

Now, $13.6\times h=37.8$

$$\therefore\ h=\frac{37.8}{13.6}=\mathbf{2.8\ cm}$$

2.12 (a)

Reference is always taken from atmospheric pressure which is indicated by Barometer. Barometer reads the local atmospheric pressure. So pressure = 700 – 380 = 320 mm of Hg vacuum

2.13 (a)

$$x_1+x_2=\frac{7.5}{0.5}=15\text{ cm}$$

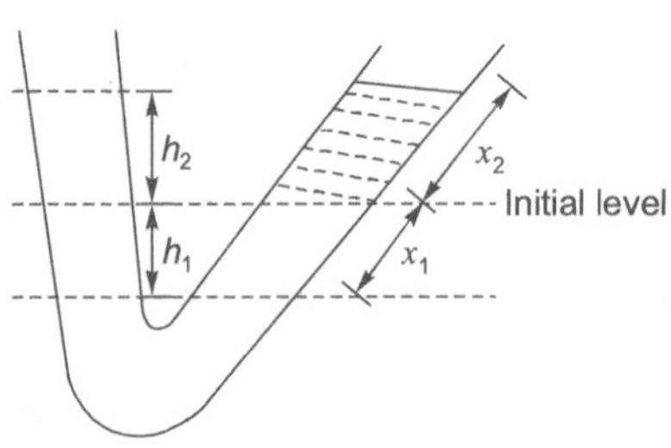

Now,

$\rho g\,(x_1+x_2)\sin 30 = 1.35\,\rho g\,(h_1+h_2)$

$\therefore\quad h_1+h_2 = 6$ cm

Total extra volume $= x_2A + h_2.A$

$7.5 = [(15-2h_1)+(6-h_1)]\,0.5$

$\therefore 15-2h_1+6-h_1=15$

$\therefore\quad h_1 = 2$ cm

and $\quad h_2 = 6-2 = \mathbf{4\ cm}$

2.14 (c)

Mercury has very low vapour pressure and hence it is an excellent fluid to be used in a barometer.

2.15 (d)

$$\frac{p_1-p_2}{\rho g}=0.1$$

$(p_1-p_2) = 0.1\times 1000\times 9.81 = \mathbf{981\ N/m^2}$.

2.16 (d)

$(s)\times(x) = (3s)\times(y)$

$$\because\quad y=\frac{x}{3}$$

Resulting movement of meniscus $=\dfrac{x}{6}$

2.17 (d)

Measurement of h using U tube manometer

Case 1. When specific gravity of manometric liquid is more than specific gravity of liquid flowing

$$h = x\left(\frac{S_g}{S_0}-1\right)$$

Case 2. When specific gravity of manometric fluid is less than the specific gravity of liquid flowing.

$$h = y\left(1-\frac{S_g}{S_0}\right)$$

2.18 (b)

$0.8\times 40 = 13.6\times(2h)$

$\Rightarrow\quad h = \mathbf{1.2\ cm}$

2.19 (b)

Gravity and pressure forces influence the problem of fluid statics.

2.20 (c)

$$h_1 s_1 = h_2 s_2$$
$$\Rightarrow 0.20 \times 13.6 = h_2 \times 0.8$$
$$\Rightarrow h_2 = \mathbf{3.4\ m}$$

2.21 (b)

Sensitivity is directly proportional to the length of fluid travel for a particular pressure difference.

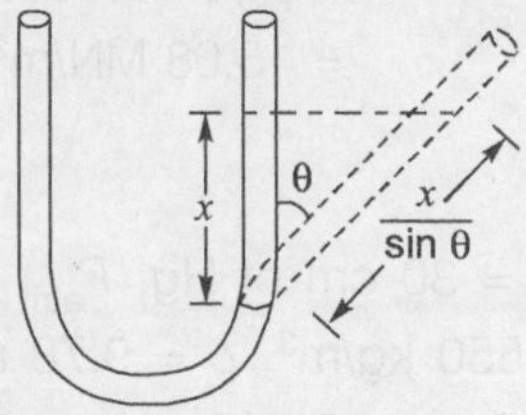

$$\therefore \quad (\text{Sensitivity})_{\text{U-tube}} \propto x$$
$$(\text{Sensitivity})_{\text{Inclined tube}} \propto \frac{x}{\sin\theta}$$

2.22 (b)

$$p_A - (\rho_A g h_A) - (\rho_m g h_m) - p_{atm} = 0$$
$$p_A - (0.8 \times 10^3 \times 9.812 \times 1.25) - (13.6 \times 10^3 \times 9.812 \times 0.5) - 103 \times 10^3 = 0$$
$$p_A = (9.812 + 66.7216 + 103)\ \text{kPa}$$
$$p_A = \mathbf{179.53\ kPa}$$

2.24 (b)

$$\frac{p}{w} = (h \sin\theta)s$$
$$\tan\theta = \frac{1}{4} = 14.036°$$
$$\sin\theta = 0.2425$$
$$\frac{p}{w} = (8 \times 0.2425) \times 0.72 = \mathbf{1.4\ cm}$$

2.25 (d)

The important property for a liquid to use it as manometric fluids is its high density and low vapour pressure. Inclined monometers are more sensitive than vertical manometers hence they are used for low pressures.

2.26 (b)

Pascal's law states that pressure at a point is equal in all directions in a static fluid. Fluid includes both liquid and gas.

2.27 (b)

$$p = \rho g h$$
$$= 13.6 \times 1000 \times 9.81 \times 0.03$$
$$= \mathbf{4002.48\ Pa}$$

Nearest match option is (b)

2.28 (c)

$$p = \rho g h$$
$$1 \times 10^5 = 784 \times 9.81 \times h$$
$$\therefore \quad h \approx \mathbf{13\ m}$$

2.29 (c)

$$\text{Pressure: } p = \frac{\text{Force}}{\text{Projected area}}$$
$$= \frac{70 \times 9.81}{500 \times 10^{-4}}$$
$$= 13734\ \text{N/m}^2 = \mathbf{13.73\ kN/m^2}$$

2.30 (b)

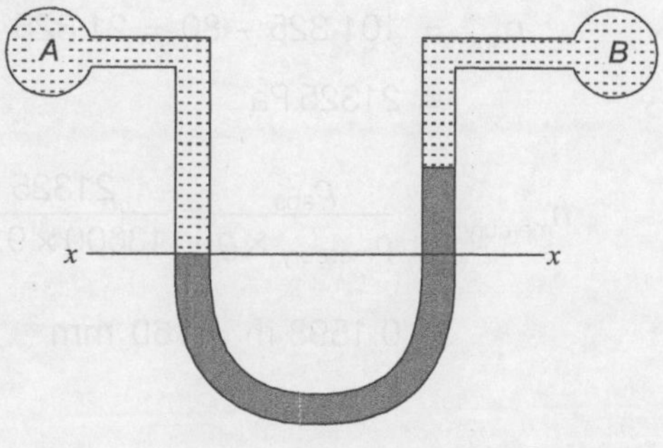

$$\rho_A g h_A = \rho_B g h_B$$
$$\frac{\rho_A}{\rho_B} = \frac{h_B}{h_A} = \frac{0.3}{0.6} = 0.5$$

Specific gravity ∝ density

$$\therefore \quad \frac{S_A}{S_B} = \frac{\rho_A}{\rho_B} = \mathbf{0.5}$$

2.31 (d)

(i) In terms of bar

$$p = \frac{400 \times 10^3}{10^5} = 4\ \text{bar}$$

(ii) In terms of meter head of water

$$h_w = \frac{p}{\rho g} = \frac{400 \times 10^3}{1000 \times 9.81} = 40.77\ \text{m}$$

(iii) In terms of meter head of mercury

$$h_{Hg} = \frac{p}{\rho g} = \frac{400 \times 10^3}{13.6 \times 10^3 \times 9.81} = \mathbf{2.99\ m}$$

2.34 (c)

$$p = 3.924\ \text{N/cm}^2 = 3.924 \times 10^4\ \text{Pa}$$

$$h = \frac{p}{\rho g} = \frac{3.924 \times 10^4}{1000 \times 9.81} = \mathbf{4\ m}$$

2.35 (a)

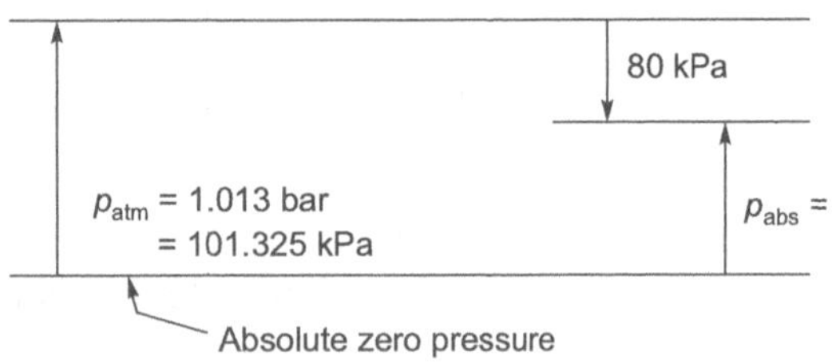

$$p_{abs} = 101.325 - 80 = 21.325\ \text{kPa}$$
$$\equiv 21325\ \text{Pa}$$

$$h_{mercury} = \frac{p_{abs}}{\rho_{mercury} \times g} = \frac{21325}{13600 \times 9.81}$$
$$= 0.1598\ \text{m} \approx \mathbf{160\ mm}$$

2.38 (a)

$$\Delta P = (\rho_{\text{manometric fluid}} - \rho_{\text{flowing fluid}}) \times g \times \text{differential head}$$
$$= (13600 - 900) \times 9.81 \times 0.15$$
$$= 18688.081\ \text{N/m}^2$$

2.39 (a)

The intensity of pressure is given by,

$$P = \rho g h = 10050 \times 1500$$
$$= 15.08\ \text{MN/m}^2$$

2.40 (c)

Given: h = 30 cm of Hg, P_{atm} = 101 kPa, ρ_{Hg} = 13550 kg/m^3, g = 9.78 m/s^2.

$$P_{actual} = P_{atm} + P = P_{atm} + \rho g h$$
$$=(101 \times 10^3) + (13550 \times 9.78 \times 0.30)$$
$$= 140755.7\ \text{Pa} = 140.7557\ \text{kPa}$$
$$\simeq 140.76\ \text{kPa}$$

2.41 (b)

- Inverted U-tube differential manometer is used to measure low pressure.

■■■■

3 Hydrostatic Forces on Surface

3.1 The height of a cylindrical container is twice that of its diameter. The ratio of the horizontal forces on the wall of the cylinder when it is completely filled to that when it is half filled with the same liquid, is

(a) 2 (b) 3
(c) 3.5 (d) 4

[ESE : 2000]

3.2 A rectangular tank of square cross-section is having its height equal to twice the length of any side at the base. If the tank is filled up with a liquid, the ratio of the total hydrostatic force on any vertical wall to that at the bottom is

(a) 2.0 (b) 1.5
(c) 1.0 (d) 0.5

[ESE : 2001]

3.3 Three immiscible liquids of specific densities ρ, 2ρ and 3ρ are kept in a jar. The height of the liquids in the jar and at the piezometer fitted to the bottom of the jar are as shown in the figure below.

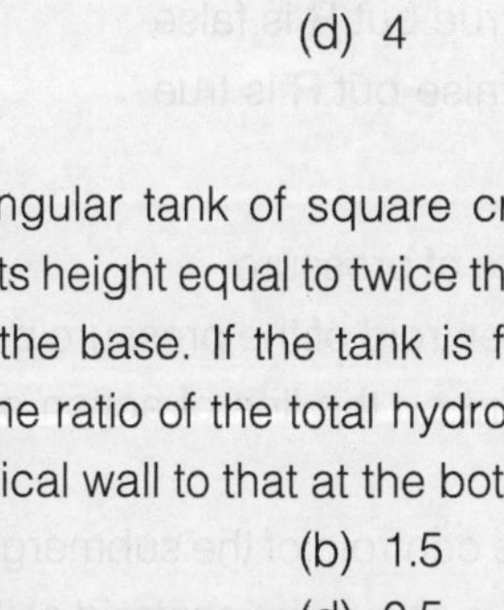

The ratio H/h is.

(a) 4 (b) 3.5
(c) 3 (d) 2.5

[ESE : 2001]

3.4 A dam is having a curved surface as shown in the figure below.

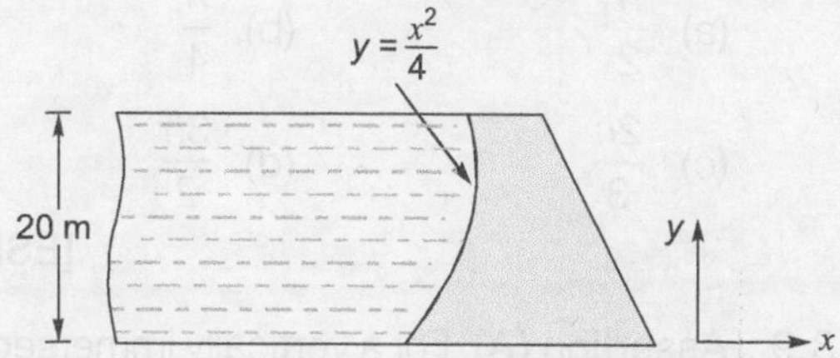

The height of the water retained by the dam is 20 m, density of water is 1000 kg/m^3. Assuming g as 9.81 m/s^2, the horizontal force acting on the dam per unit length is

(a) 1.962×10^2 N (b) 2×10^5 N
(c) 1.962×10^6 N (d) 3.924×10^6 N

[ESE : 2002]

3.5 Consider the following statements about hydrostatic force on a submerged surface:

1. It remains the same even when the surface is turned.
2. It acts vertically even when the surface is turned.

Which of these statements is/are correct?

(a) Only 1 (b) Only 2
(c) Both 1 and 2 (d) Neither 1 nor 2

[ESE : 2003]

3.6 The vertical component of force on a curved surface submerged in a static liquid is equal to the

(a) weight of liquid column above the CG of the curved surface
(b) weight of liquid above the curved surface
(c) product of pressure at CG, multiplied by the area of the curved surface
(d) product of pressure at CG, multiplied by the projected area of the curved surface

[ESE : 2003]

3.7 The point of application of a horizontal force on a curved surface submerged in liquid is (symbols have their usual meanings)

(a) $\frac{I_G}{A\bar{h}} - \bar{h}$ (b) $\frac{I_G + A\bar{h}^2}{A\bar{h}}$

(c) $\frac{A\bar{h}}{I_G}+\bar{h}$ (d) $\frac{I_G}{\bar{h}}+A\bar{h}$

[ESE : 2003]

3.8 The depth of centre of pressure for a rectangular lamina immersed vertically in water up to height h is given by

(a) $\frac{h}{2}$ (b) $\frac{h}{4}$

(c) $\frac{2h}{3}$ (d) $\frac{3h}{2}$

[ESE : 2003]

3.9 **Assertion (A):** For a vertically immersed surface, the depth of the centre of pressure is independent of the density of the liquid.

Reason (R): Centre of pressure lies above the centre of area of the immersed surface.

(a) Both **A** and **R** are individually true and **R** is the correct explanation of **A**

(b) Both **A** and **R** are individually true but **R** is not the correct explanation of **A**

(c) **A** is true but **R** is false

(d) **A** is false but **R** is true

[ESE : 2003]

3.10 A circular annular plate bounded by two concentric circles of diameter 1.2 m and 0.8 m is immersed in water with its plane making an angle of 45° with the horizontal. The centre of the circles is 1.625 m below the free surface. What will be the total pressure force on the face of the plate?

(a) 7.07 kN (b) 10.00 kN

(c) 14.14 kN (d) 18.00 kN

[ESE : 2004]

3.11 A plate of rectangular shape having the dimensions of 0.4 m × 0.6 m is immersed in water with its longer side vertical. The total hydrostatic thrust on one side of the plate is estimated as 18.3 kN. All other conditions remaining the same, the plate is turned through 90° such that its longer side remains vertical. What would be the total force on one face of the plate?

(a) 9.15 kN (b) 18.3 kN

(c) 36.6 kN (d) 12.2 kN

[ESE : 2004]

3.12 **Assertion (A):** A circular plate is immersed in a liquid with its periphery touching the free surface and the plane makes an angle θ with the free surface with different values of θ, the position of centre of pressure will be different.

Reason (R): Since the centre of pressure is dependent on second moment of area, with different values of θ, second moment of area for the circular plate will change.

(a) Both **A** and **R** are individually true and **R** is the correct explanation of **A**

(b) Both **A** and **R** are individually true but **R** is not the correct explanation of **A**

(c) **A** is true but **R** is false

(d) **A** is false but **R** is true

[ESE : 2004]

3.13 The centre of pressure

(a) the centroid of the pressure prism

(b) a point on the line of action of the resultant force

(c) at the centroid of the submerged area

(d) always above the centroid of the area

[ESE : 2005]

3.14 A semicircular plane area of diameter 1 m, is subjected to a uniform gas pressure of 420 kN/m^2. What is the moment of thrust (approximately) on the area about its straight edge?

(a) 35 kNm (b) 41 kNm

(c) 55 kNm (d) 82 kNm

[ESE : 2006]

3.15 A horizontal oil tank is in the shape of a cylinder with hemispherical ends. If it is exactly half full, what is the ratio of magnitude of the vertical component of resultant hydrostatic thrust on one hemispherical end to that of the horizontal component?

(a) $\frac{2}{\pi}$ (b) $\frac{\pi}{2}$

(b) $\frac{4}{3\pi}$ (d) $\frac{3\pi}{4}$

[ESE : 2006]

3.16 A circular plate 1.5 m diameter is submerged in water with its greatest and least depths below the surface being 2 m and 0.75 m respectively.

What is the total pressure (approximately) on one face of the plate?

(a) 12 kN (b) 16 kN
(c) 24 kN (d) None of these

[ESE : 2007]

3.17 Resultant pressure of the liquid in case of an immersed body acts through which one of the following?

(a) Centre of gravity
(b) Centre of pressure
(c) Metacentre
(d) Centre of buoyancy

[ESE : 2007]

3.18 What is the vertical component of pressure force on submerged curved surface equal to?

(a) Its horizontal component
(b) The force on a vertical projection of the curved surface
(c) The product of the pressure at centroid and surface area
(d) The gravity force of liquid vertically above the curved surface up to the free surface

[ESE : 2008]

3.19 What is the depth of centre of pressure of a vertical immersed surface from free surface of liquid equal to?

(a) $\frac{I_G}{A\bar{h}} + \bar{h}$ (b) $\frac{I_G A}{\bar{h}} + \bar{h}$
(c) $\frac{I_G \bar{h}}{A} + \bar{h}$ (d) $\frac{A\bar{h}}{I_G} + \bar{h}$

(Symbols have their usual meaning)

[ESE : 2008]

3.20 A rectangular plate 0.75 m × 2.4 m is immersed in a liquid of relative density 0.85 with its 0.75 m side horizontal and just at the water surface. If the plane of plate makes an angle of 60° with the horizontal, what is the approximate pressure force on one side of the plate?

(a) 7.80 kN (b) 15.60 kN
(c) 18.00 kN (d) 24.00 kN

[ESE : 2008]

3.21 Assertion (A): The center of pressure for a vertical surface submerged in a liquid lies above the centroid (centre of gravity) of the vertical surface.

Reason (R): The distance of the centre of pressure from the free surface of the liquid for a vertical surface submerged in a liquid is independent of the density of the liquid.

(a) Both A and R are individually true and R is the correct explanation of A
(b) Both A and R are individually true but R is not the correct explanation of A
(c) A is true but R is false
(d) A is false but R is true

[ESE : 2008]

3.22 A tank has in its side a very small horizontal cylinder fitted with a frictionless piston. The head of liquid above the piston is *h* and the piston area *a*, the liquid having a specific weight γ. What is the force that must be exerted on the piston to hold it in position against the hydrostatic pressure?

(a) $2\gamma ha$ (b) γha
(c) $\frac{2\gamma ha}{3}$ (d) $\frac{\gamma ha}{2}$

[ESE : 2009]

3.23 What is the vertical distance of the centre of pressure below the centroid of the plane area?

(a) $\frac{I_G}{A\bar{h}}$ (b) $\frac{I_G \sin\theta}{A\bar{h}}$
(c) $\frac{I_G \sin^2\theta}{A\bar{h}}$ (d) $\frac{I_G \sin^2\theta}{A\bar{h}^2}$

[ESE : 2009]

3.24 A circular area of 1.2 m diameter is immersed vertically in a liquid of unit weight 800 N/m^3 with its top edge just on the liquid surface. The depth of centre of pressure on one side, measured below the liquid surface, is

(a) 0.80 m (b) 0.75 m
(c) 0.60 m (d) 0.64 m

[ESE : 2010]

3.25 A curved surface has the shape of one-quadrant of a 3 m long 2 m diameter cylinder with two edges parallel to the axis. The curved surface is immersed in a stationary liquid of unit weight

10 kN/m^3 such that one straight edge is on the liquid surface and the other 1 m below the surface. The moment of vertical component of hydrostatic thrust on one side of the surface, about the axis of circular arc, is

(a) 7.854 kNm (b) 10.0 kNm
(c) 11.781 kNm (d) 15.708 kNm

[ESE : 2010]

3.26 The centre of pressure for an inclined surface area
(a) lies below the centroid of the surface
(b) coincides with the centroid
(c) lies above the centroid of the surface
(d) none of the above

[ESE : 2011]

3.27 **Assertion (A)** : At great depth, the vertical distance between the centre of pressure and the centre of area of immersed surface becomes negligible.

Reason (R) : The depth of centre of pressure of any immersed surface is independent of the density of the liquid.

(a) Both **A** and **R** are individually true and **R** is the correct explanation of **A**
(b) Both **A** and **R** are individually true but **R** is not the correct explanation of **A**
(c) **A** is true but **R** is false
(d) **A** is false but **R** is true

[ESE : 2011]

3.28 The depth of centre of pressure of a rectangular lamina immersed vertically in water up to a height *h* is given by

(a) $\frac{h}{3}$ (b) $\frac{h}{4}$
(c) $\frac{h}{2}$ (d) $\frac{2h}{3}$

[ESE : 2012]

3.29 The centre of pressure of a plane submerged surface
(a) is a point on the submerged area at which the resultant hydrostatic force is supposed to act
(b) should always coincide within the centre of submerged area
(c) should be at the centre of gravity of the plane surface
(d) is always below the centroid of area

[ESE : 2013]

3.30 When a dolphin glides through air, it experiences an external pressure of 0.75 m of mercury. The absolute pressure on dolphin when it is 5 m below the free surface of the water is

(a) 0.10 N/mm^2 (b) 0.5 N/mm^2
(c) 1.0 N/mm^2 (d) 0.15 N/mm^2

[ESE : 2014]

3.31 A tank of length, breadth and height in the ratio of 2 : 1 : 2 is full of water. The ratio of hydrostatic force at the bottom to that at any larger vertical surface is

(a) 1 (b) 4
(c) 2 (d) 3 [ESE : 2015]

3.32 An isosceles triangular lamina of base 1 m and height 2 m is located in the water in vertical plane and its vertex is 1 m below the free surface of the water. The position of force acting on the lamina from the free water surface is :

(a) 2.42 m (b) 2.33 m
(c) 2.00 m (d) 1.33 m

[ESE : 2016]

3.33 A house-top water tank is made of flat plates and is full to the brim. Its height is twice that of any side. The ratio of total thrust force on the bottom of the tank to that on any side will be:

(a) 4 (b) 2
(c) 1 (d) 0.5

[ESE : 2016]

3.34 The water level in a dam is 10 m. The total force acting on vertical wall per metre length is :

(a) 49.05 kN (b) 490.5 kN
(c) 981 kN (d) 490.5 N

[ESE : 2016]

3.35 **Statement (I)** : Depth of centre of pressure of any immersed surface is independent of the density of the liquid.

Statement (II) : Centre of area of the immersed body lies below the centre of pressure.

(a) Both Statement (I) and Statement (II) are individually true and Statement (II) is the correct explanation of Statement (I).

(b) Both Statement (I) and Statement (II) are individually true but Statement (II) is not the correct explanation of Statement (I).
(c) Statement (I) is true but Statement (II) is false.
(d) Statement (I) is false but Statement (II) is true.

[ESE : 2017]

3.36 A section of a dam made of concrete, $\rho = 2.6$, total height = 35 m, with top walkway width of 6 m, is shown. The upstream bottommost point is called the Heel of the dam. The sloped part on downstream side is 3 vertical on 2 horizontal. Water stands till 2 m short of the top of the dam section. The net resultant force acting on the base level of the dam is nearly

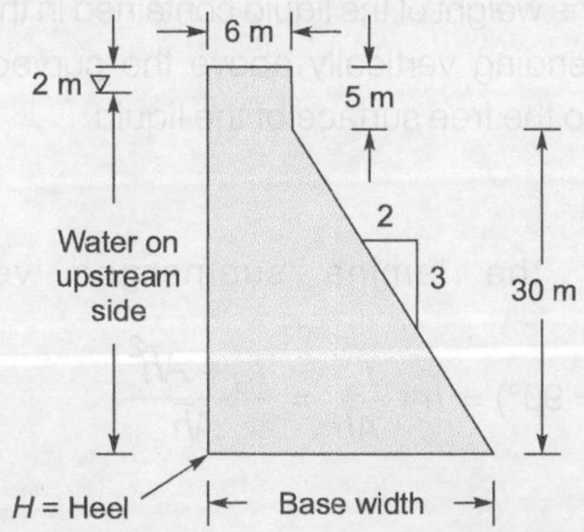

(a) 1370 k kgf (b) 1385 k kgf
(c) 1400 k kgf (d) 1433 k kgf

[ESE : 2017]

3.37 What is the force needed to apply to a piston of 2 cm radius in order to result a force of 6000 N at the working piston of radius 6 cm?

(a) 1334 N (b) 333 N
(c) 1050 N (d) 667 N

[ESE : 2021]

3.38 **Statement I :** Centre of pressure is calculated by using the "Principle of Moments".

Statement II : The moment of the resultant force about an axis is equal to the sum of moments of the components about the same axis.

(a) Both Statement (I) and Statement (II) are individually true and Statement (II) is the correct explanation of Statement (I).
(b) Both Statement (I) and Statement (II) are individually true but Statement (II) is not the correct explanation of Statement (I).
(c) Statement (I) is true but Statement (II) is false.
(d) Statement (I) is false but Statement (II) is true.

[ESE : 2022]

3.39 Oil of specific gravity 0.800 acts on a vertical triangular area whose apex is in the oil surface. The triangle is isosceles 3 m high and 4 m wide. A vertical rectangular area 2 m high is attached to the 4 m base of the triangle and is acted upon by water. What is the magnitude of the resultant hydrostatic force on the entire area? (Consider acceleration due to gravity as 9.81 m/s^2)

(a) 184 kN (b) 361 kN
(c) 421 kN (d) 520 kN

[ESE : 2023]

3.40 A circular opening, 3 m diameter, in a vertical side of a tank is closed by a disc of 3 m diameter which can rotate about a horizontal diameter. What is the force on the disc if the head of water above the horizontal diameter is 5 m?

(a) 277.4 kN (b) 324.4 kN
(c) 311.2 kN (d) 346.7 kN

[ESE : 2024]

■■■■

Answers Hydrostatic Forces on Surface

3.1 (d)	3.2 (c)	3.3 (c)	3.4 (c)	3.5 (d)	3.6 (b)	3.7 (b)	3.8 (c)	3.9 (c)
3.10 (b)	3.11 (b)	3.12 (c)	3.13 (a)	3.14 (a)	3.15 (b)	3.16 (c)	3.17 (b)	3.18 (d)
3.19 (a)	3.20 (b)	3.21 (d)	3.22 (b)	3.23 (c)	3.24 (b)	3.25 (b)	3.26 (a)	3.27 (b)
3.28 (d)	3.29 (d)	3.30 (d)	3.31 (a)	3.32 (a)	3.33 (c)	3.34 (b)	3.35 (c)	3.36 (d)
3.37 (d)	3.38 (b)	3.39 (b)	3.40 (d)					

Explanations Hydrostatic Forces on Surface

3.1 (d)

$$\text{Ratio} = \frac{\frac{1}{2}\rho g H \times \pi r H}{\frac{1}{2}\rho g \frac{H}{2} \times \pi r \frac{H}{2}} = 4$$

3.2 (c)

F_1 = Force on vertical wall

$= \gamma \times 2a \times a \times a = 2a^3\gamma$

F_2 = Force at the bottom

$= \gamma \times a \times a \times 2a = 2a^3\gamma$

$$\therefore \frac{F_1}{F_2} = 1$$

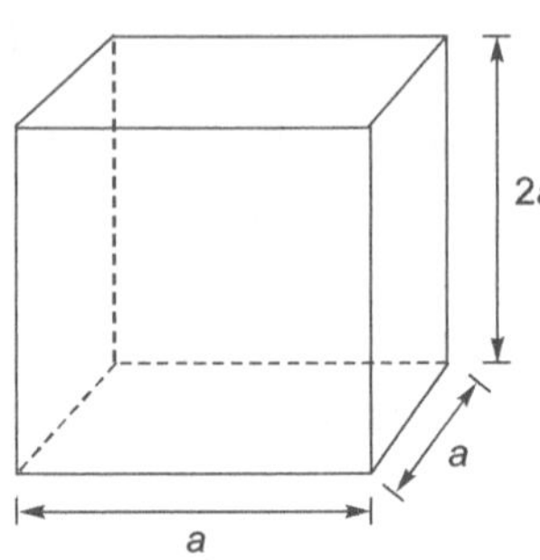

3.3)

$$3h \times \rho + 1.5h \times 2\rho + h \times 3\rho - H \times 3\rho = 0$$

$$\therefore \frac{H}{h} = 3$$

3.4 (c)

Horizontal force per unit length $= wA\bar{h}$

$= 9.81 \times 1000 \times 20 \times 1 \times 10$

$= \mathbf{1.982 \times 10^6}$ **N/m**

3.5 (d)

Magnitude of hydrostatic force on plane surface is independent of the inclination of the surface. Only when surface is rotated about its centroid.

3.6 (b)

Vertical component of a curved surface is equal to the weight of the liquid contained in that portion extending vertically above the curved surface upto the free surface of the liquid.

3.7 (b)

For the lamina submerged vertically:

$$(\theta = 90°) = \bar{h} + \frac{I_G}{A\bar{h}} = \frac{I_G + A\bar{h}^2}{A\bar{h}}$$

3.8 (c)

Depth of centre of pressure,

$$h^* = \bar{h} + \frac{I_G}{A\bar{h}}$$

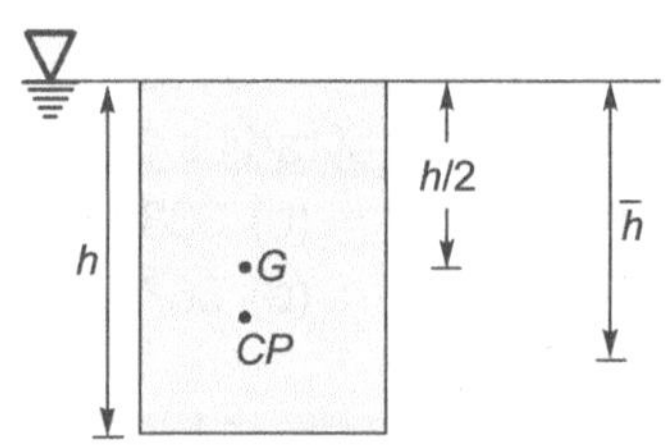

$$= \frac{h}{2} + \frac{h^3/12}{(h \times 1) \times h/2} = \frac{h}{2} + \frac{h}{6} = \frac{2h}{3}$$

3.9 (c)

$$\text{Depth of C. P. } (h^*) = \bar{h} + \frac{I_G \sin^2\theta}{A\bar{h}}$$

Which is independent of the density of the liquid.

$$h^* - \bar{h} = \frac{I_G \sin^2\theta}{A\bar{h}} > 0$$

$$\therefore \quad h^* > \bar{h}$$

Centre of pressure lies below the centre of area of the immersed surface

3.10 (b)

Total force on the face of plate = $wA\bar{h}$

$= 9.81 \times 1000 \times \frac{\pi}{4}\left[(1.2)^2 - (0.8)^2\right] \times 1.625$

= **10 kN**

3.11 (b)

The magnitude of total force is independent of the inclination of the surface.

3.12 (c)

Centre of pressure (h^*) in case of inclined surface is given by

$$h^* = \bar{h} + \frac{I_G}{A\bar{h}}\sin^2\theta$$

Here, I_G = Area moment of inertia about centroidal axis and parallel to free axis

$\bar{h}$ = Perpendicular distance of centre of gravity of body from free surface

A = Area touching fluid

3.13 (a)

The center of pressure is the point on a surface where the total sum of a pressure field acts, causing a force and no moment about that point.

3.14 (a)

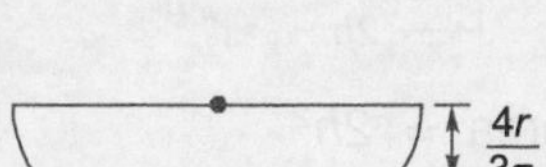

Moment of thrust = Pressure × Area × $\frac{4r}{3\pi}$

$= 420 \times \frac{\pi}{4} \times \frac{1}{2} \times 1^2 \times \frac{4 \times 0.5}{3\pi}$ = **35 kNm**

3.15 (b)

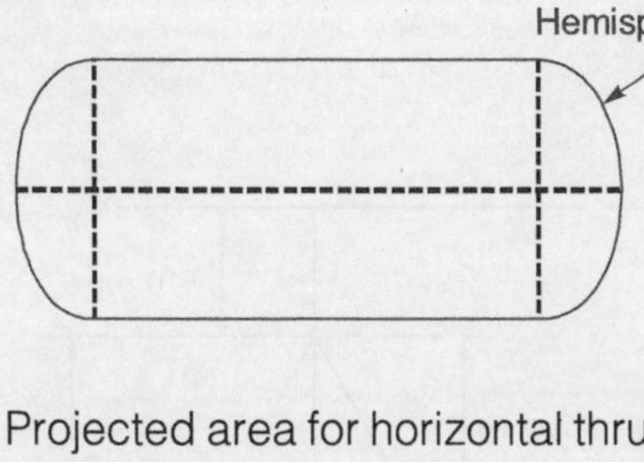

Projected area for horizontal thrust = $\frac{\pi r^2}{2}$

Volume for vertical component = $\frac{\pi r^3}{3}$

$$F_V = \rho g\left(\frac{\pi r^3}{3}\right),\ F_H = \rho g\left(\frac{\pi r^2}{2}\right)\left(\frac{4r}{3\pi}\right)$$

$$\frac{F_V}{F_H} = \frac{\pi r^3/3}{(\pi r^3)\left(\frac{2}{3\pi}\right)}$$

or $$\frac{F_V}{F_H} = \frac{3\pi}{2} \times \frac{1}{3} = \frac{\pi}{2}$$

3.16 (c)

$$\bar{h} = \frac{2 + 0.75}{2} = 1.375 \text{ m}$$

Total pressure force = $\rho g A\bar{h}$

$= 1000 \times 9.81 \times \frac{\pi}{4} \times (1.5)^2 \times 1.375$

= 23.84 kN ≈ **24 kN**

3.17 (b)

The point of application of resultant pressure is called as centre of pressure.

3.18 (d)

Vertical component of pressure force on the submerged curved surface is equal to the gravity force of liquid vertically above the curved surface upto the free surface.

3.20 (b)

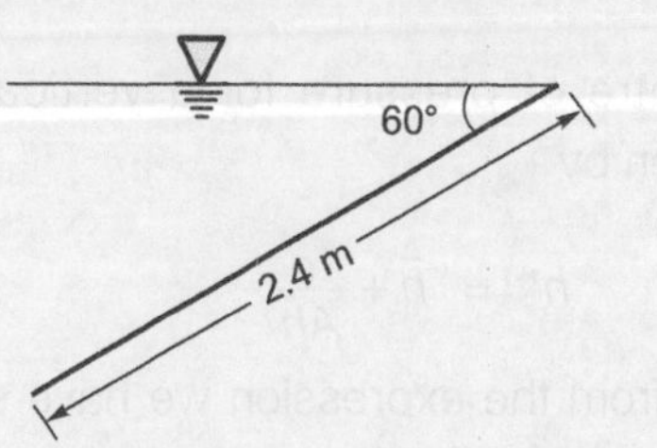

Pressure force on one side of plate = $wA\bar{h}$

$= (0.85 \times 9.81) \times (0.75 \times 2.4) \times \left(\frac{2.4 \sin 60°}{2}\right)$

= **15.60 kN**

3.21 (d)

Centre of pressure lies always below the centre of gravity of vertical surface

$$h^* = \bar{h} + \frac{I_G}{A\bar{h}}$$

Therefore the distance of centre of pressure from free surface is independent of density of liquid.

3.23 (c)

Center of = Perpendicular distance of pressure center of gravity of body from free surface + $\frac{I_G}{A\bar{x}}\cdot\sin^2\theta$

Special case: For vertical body $\theta = 90°$.

3.24 (b)

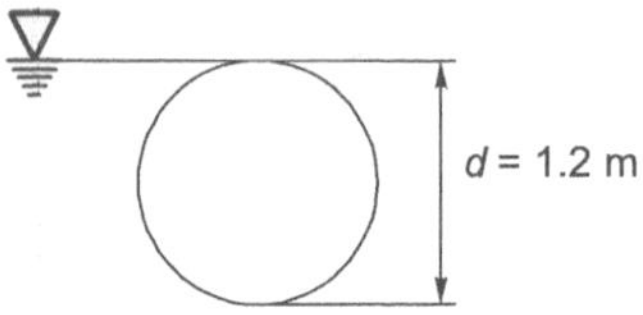

$$CP = \bar{x} + \frac{I_G}{A\bar{x}} = \frac{d}{2} + \frac{\pi}{64}d^4 \frac{1}{\frac{\pi}{4}d^2 \times \frac{d}{2}}$$

$$= \frac{5d}{8} = \frac{5\times 1.2}{8} = \mathbf{0.75\ m}$$

3.26 (a)

$$h^* = \bar{h} + \frac{I_G}{A\bar{h}}\sin^2\theta$$

$\bar{h}$ = Centre of gravity

I_G = Area moment of inertia of surface about its centroidal axis and parallel to free surface.

3.27 (b)

Centre of pressure for a vertical surface is given by

$$h^* = \bar{h} + \frac{I_G}{A\bar{h}}$$

So from the expression we have seen that for greater depth, the vertical distance between centre of pressure and centre of area of immersed surface becomes negligible.

3.28 (d)

As the pressure increases with depth centre of pressure lies below geometrical centre, its depth is $2h/3$ from surface.

3.29 (d)

Centre of pressure,

$$h^* = \frac{I_G}{A\bar{h}} + \bar{h} \quad \text{(for vertical surface)}$$

Centre of pressure,

$$h^* = \frac{I_G \sin^2\theta}{A\bar{h}} + \bar{h} \quad \text{(for inclined surface)}$$

For any plane surface $\left(\frac{I_G}{A\bar{h}}\right)$ is always positive, so h^* is always below the centroid of area.

3.30 (d)

$$p_{atm} = (\rho g h)_{Hg} = 13600 \times 9.81 \times 0.75$$
$$= 100062\ \text{N/m}^2$$
$$p_{gage} = (\rho g h)_{water} = 1000 \times 9.81 \times 5$$
$$= 49050\ \text{N/m}^2$$
$$p_{abs} = p_{atm} + p_{gage}$$
$$= 100062 + 49050 = 149112\ \text{N/m}^2$$
$$= \mathbf{0.149\ N/mm^2}$$

3.31 (a)

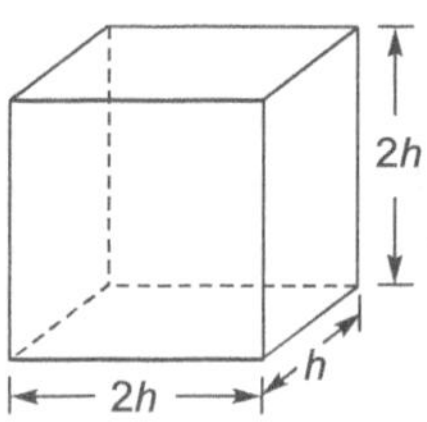

Bottom area = $2h^2$

Larger surface area = $2h \times 2h = 4h^2$

$$F_{Bottom} = \rho g \times 2h \times 2h^2 = 4\,\rho g h^3$$
$$F_{Larger\ area} = \rho g A\bar{h}$$
$$= \rho g \times 4h^2 \times h = 4\rho g A h^3$$
$$\frac{F_{Bottom}}{F_{Larger\ area}} = \frac{4\rho g h^3}{4\rho g h^3} = 1$$

3.32 (a)

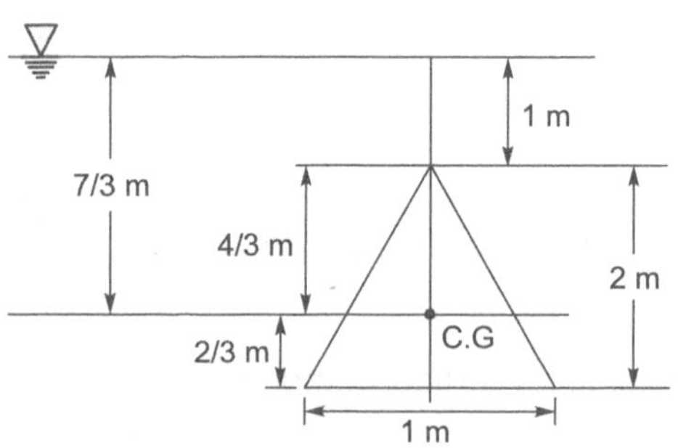

$$COP = \bar{x} + \frac{I_G}{A\bar{x}}$$

$$= \frac{7}{3} + \frac{bh^3}{36 \times \frac{bh}{2} \times \frac{7}{3}} = \frac{7}{3} + \frac{7}{42}$$

$$= 2.428$$

3.33 (c)

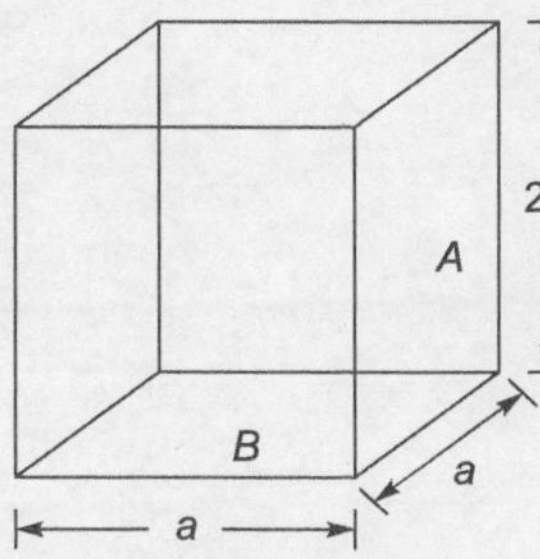

$$\frac{F_B}{F_A} = \frac{\rho g(2a) \times a^2}{\rho g(a) \times 2a^2} = 1$$

3.34 (b)

$$F = \rho g \times (5) \times (10 \times 1)$$

$$= \mathbf{490.5\ kN}$$

3.35 (c)

Statement (I) is true

Center of pressure is independent of density

$$h^* = \bar{h} + \frac{I \sin^2 \theta}{A\bar{h}}$$

Statement (II) is false

Correct → Center of area is above center of pressure.

3.36 (d)

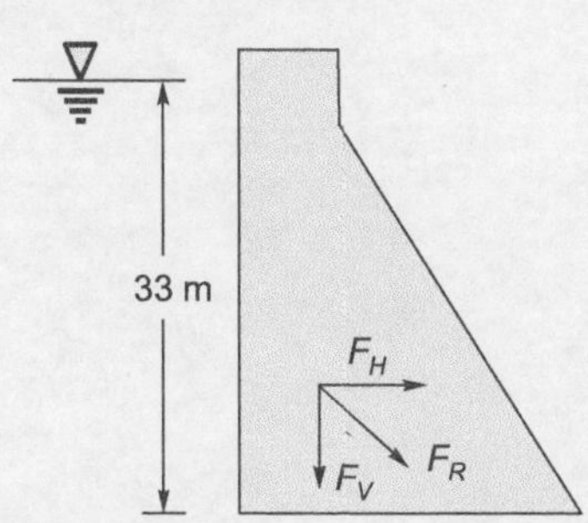

Horizontal force,

$$F_H = \rho g\bar{h}A = \rho\bar{h}A \text{ kgf}$$

$$= 1000 \times \frac{33}{2} \times 33 \times 1$$

$$= 544500 \text{ kgf} = 544.5 \text{ k kgf}$$

Vertical force,

$$F_V = mg = m \text{ kgf}$$

$$= [(6 \times 35) + (30 \times 10)] \times 1 \times 2600$$

$$= 510 \times 2600 = 1326000 \text{ kgf}$$

$$= 1326 \text{ k kgf}$$

$$F_R = \sqrt{F_H^2 + F_V^2} = \sqrt{544.5^2 + 1326^2}$$

$$= 1433.4420 \text{ k kgf}$$

3.37 (d)

$$F = PA$$

$$F \propto A$$

$$\frac{F_1}{F_2} = \frac{A_1}{A_2} = \frac{(\pi/4)D_1^2}{(\pi/4)D_2^2}$$

$$= \left(\frac{D_1}{D_2}\right)^2 = \left(\frac{R_1}{R_2}\right)^2$$

$$\frac{F_1}{6000} = \left(\frac{2}{6}\right)^2 = \frac{1}{9}$$

$$F_1 = \frac{6000}{a} = 666.66 = 667 \text{ N}$$

3.38 (b)

Both statements are correct but statement II does not support statement I.

3.39 (b)

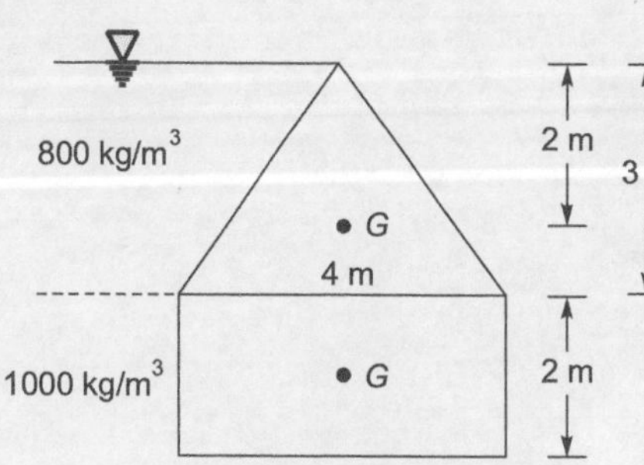

$$S = 0.800$$

$$\rho = 800 \text{ kg/m}^3$$

$$\theta = 90°$$

$$F = [(\text{Given pressure at } C.G. \text{ of triangle}) \times \text{Area}] + [(\text{Given pressure at } C.G. \text{ of rectangle}) \times \text{Area}]$$

$$F = \left[(800 \times 9.81 \times 2) \times \frac{1}{2} \times 4 \times 3\right] + \left[\{(800 \times 9.81 \times 3) + (1000 \times 9.81 \times 1)\} \times (2 \times 4)\right]$$

$$F = 94.176 + 266.832 = 361.008 \text{ kN}$$

3.40 (d)

Given: $\bar{x}$ = 5 m, d = 3 m, ρ = 1000 kg/m^3

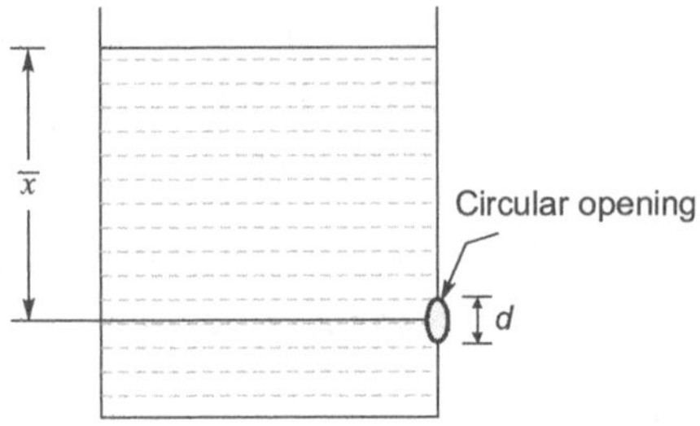

Force acting on the disc,

$$F_P = \rho A g \bar{x} = \rho\left(\frac{\pi}{4}d^2\right)g\bar{x} = 1000 \times \frac{\pi}{4} \times (3)^2 \times 9.81 \times 5$$

$$= 346714.0192 \text{ N} = 346.714 \text{ kN}$$

$$\simeq 346.7 \text{ kN}$$

■■■■

4 Buoyancy, Floatation and Liquids in Relative Equilibrium

4.1 A float of cubical shape has sides of 10 cm. The float valve just touches the valve seat to have a flow area of 0.5 cm^2 as shown in the figure below.

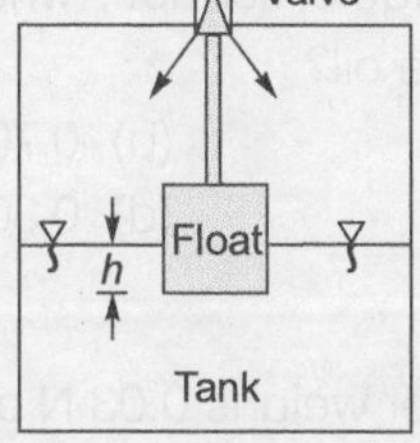

If the pressure of water in the pipeline is 1 bar, the rise of water level h in the tank to just stop the water flow will be

(a) 7.5 cm (b) 5.0 cm
(c) 2.5 cm (d) 0.5 cm

[ESE : 2000]

4.2 If a vessel containing liquid moves downward with constant acceleration g, then

(a) the pressure throughout the liquid mass is atmospheric
(b) the pressure in the liquid mass is greater than the hydrostatic pressure
(c) there will be vacuum in the liquid
(d) the pressure throughout the liquid mass is greater than atmospheric

[ESE : 2000]

4.3 A cylindrical vessel having its height equal to its diameter is filled with liquid and moved horizontally at an acceleration equal to acceleration due to gravity. The ratio of the liquid left in the vessel to the liquid at static equilibrium condition is

(a) 0.2 (b) 0.4
(c) 0.5 (d) 0.75

[ESE : 2001]

4.4 A block of aluminium having mass of 12 kg is suspended by a wire and lowered until submerged into a tank containing oil of relative density 0.8. Taking the relative density of aluminium as 2.4, the tension in the wire will be (take $g = 10$ m/s^2)

(a) 12000 N (b) 800 N
(c) 120 N (d) 80 N

[ESE : 2001]

4.5 A barge 30 m long and 10 m wide has a draft of 3 m when floating with its sides in vertical position. If its centre of gravity is 2.5 m above the bottom, the nearest value of metacentric height is

(a) 3.28 m (b) 2.78 m
(c) 1.78 m (d) zero

[ESE : 2001]

4.6 Match **List-I** (Stability) and **List-II** (Conditions) and select the correct answer using the codes given below the lists:

List-I
A. Stable equilibrium of a floating body
B. Stable equilibrium of a submerged body
C. Unstable equilibrium of a floating body
D. Unstable equilibrium of a submerged body

List-II
1. Centre of buoyancy below the centre of gravity
2. Metacentre above the centre of gravity
3. Centre of buoyancy above the centre of gravity
4. Metacentre below the centre of gravity

Codes:

	A	B	C	D
(a)	4	3	2	1
(b)	2	3	4	1
(c)	4	1	2	3
(d)	2	1	4	3

[ESE : 2002]

4.7 An open rectangular box of base 2 m × 2 m contains a liquid of specific gravity 0.80 up to a height of 2.5 m. If the box is imparted a vertically upward acceleration of 4.9 m/s^2, what will be the pressure on the base of the tank?

(a) 9.81 kPa (b) 19.62 kPa
(c) 36.80 kPa (d) 29.40 kPa

[ESE : 2004]

4.8 Which one of the following is the condition for stable equilibrium of a floating body?

(a) The metacenter coincides with the centre of gravity.
(b) The metacentre is above the centre of gravity.
(c) The metacentre is below the centre of gravity.
(d) The centre of buoyancy is below the centre of gravity.

[ESE : 2005]

4.9 **Assertion (A):** The buoyant force for a floating body passes through the centroid of the displaced volume.

Reason (R): The force of buoyancy is a vertical force and equal to the weight of the fluid displaced.

(a) Both **A** and **R** are individually true and **R** is the correct explanation of **A**
(b) Both **A** and **R** are individually true but **R** is not the correct explanation of **A**
(c) **A** is true but **R** is false
(d) **A** is false but **R** is true

[ESE : 2005]

4.10 A homogeneous solid of any arbitrary shape floats upright in a homogeneous liquid with immersed volume *V* and is in stable equilibrium. If the solid is overturned and made to float upside down in a different homogeneous liquid with exactly same volume *V* above the liquid surface, then the equilibrium

(a) would be stable
(b) would be neutral
(c) would be unstable
(d) may or may not be stable

[ESE : 2006]

4.11 A beaker of water is falling freely under the influence of gravity. Point *B* is on the surface and point *C* is vertically below *B* near the bottom of the beaker. If p_B is the pressure at point *B* and p_C the pressure of point *C*, then which one of the following is correct?

(a) $p_B = p_C$
(b) $p_B < p_C$
(c) $p_B > p_C$
(d) Insufficient information to determine any of the above

[ESE : 2006]

4.12 A 25 cm long prismatic homogeneous solid floats in water with its axis vertical and 10 cm projecting above water surface. If the same solid floats in some oil with its axis vertical and 5 cm projecting above the liquid surface, what is the specific gravity of the oil?

(a) 0.60 (b) 0.70
(c) 0.75 (d) 0.80

[ESE : 2006]

4.13 A hydrometer weighs 0.03 N and has a stem at the upper end which is cylindrical and 3 mm in diameter. It will float deeper in oil of specific gravity 0.75, than in alcohol of specific gravity 0.8 by how much amount?

(a) 10.7 mm (b) 43.3 mm
(c) 33 mm (d) 36 mm

[ESE : 2007]

4.14 A wooden rectangular block of length of *L* is made to float in water with its axis vertical. The centre of gravity of the floating body is 0.15*L* above the centre of buoyancy. What is the specific gravity of the wooden block?

(a) 0.6 (b) 0.65
(c) 0.7 (d) 0.75

[ESE : 2007]

4.15 If *B* is the centre of buoyancy, *G* is the centre of gravity and *M* is the metacentre of a floating body, the body will be in stable equilibrium if

(a) $MG = 0$ (b) *M* is below *G*
(c) $BG = 0$ (d) *M* is above *G*

[ESE : 2007]

4.16 The metacentric high of a passenger ship is kept lower than that of a naval or a cargo ship because

(a) apparent weight will increase
(b) otherwise it will be in neutral equilibrium
(c) it will decrease the frequency of rolling

(d) otherwise it will sink and be totally immersed

[ESE : 2007]

4.17 What acceleration would cause the free surface of a liquid contained in an open tank moving in a horizontal track to dip by 45°?

(a) 0.5g (b) 2g
(c) g (d) 1.5g

[ESE : 2008]

4.18 What is buoyant force?

(a) Lateral force acting on a submerged body
(b) Resultant force acting on a submerged body
(c) Resultant force due to water on a body
(d) Resultant hydrostatic force on a body due to fluid surrounding it

[ESE : 2008]

4.19 How is the metacentric height, GM expressed?

(a) $GM = BG - (I/V)$ (b) $GM = (V/I) - BG$
(c) $GM = (I/V) - BG$ (d) $GM = BG - (V/I)$

where I = Moment of inertia of the plan of the floating body at the water surface.
V = Volume of the body submerged in water.
BG = Distance between the centre of gravity (G) and the centre of buoyancy (B).

[ESE : 2008]

4.20 Stability of a floating body can be improved by which of the following?

1. Making its width large
2. Making the draft small
3. Keeping the centre of mass low
4. Reducing its density

Select the correct answer using the codes given below:

(a) 1, 2, 3 and 4 (b) 1, 2 and 3 only
(c) 1 and 2 only (d) 3 and 4 only

[ESE : 2008]

4.21 The distance from the centre of buoyancy to the metacentre is given by I/V_d, where V_d is the volume of fluid displaced.
What does I represent?

(a) Moment of inertia of a horizontal section of the body taken at the surface of the fluid
(b) Moment of inertia about its vertical centroidal axis
(c) Polar moment of inertia
(d) Moment of inertia about its horizontal centroidal axis

[ESE : 2008]

4.22 A body weighs 30 N and 15 N when weighed under submerged conditions in liquids of relative densities 0.8 and 1.2 respectively. What is the volume of the body in litres?

(a) 12.50 (b) 3.82
(c) 18.70 (d) 75.50

[ESE : 2009]

4.23 For floating bodies, how is the metacentric radius defined?

(a) The distance between centre of gravity and the metacentre.
(b) Second moment of area of plane of flotation about centroidal axis perpendicular to plane of rotation/immersed volume.
(c) The distance between centre of gravity and the centre of buoyancy.
(d) Moment of inertia of the body about its axis of rotation/immersed volume.

[ESE : 2009]

4.24 A body will be in a stable equilibrium in a floating condition if

(a) the metacentre is below the centre of gravity
(b) the metacentre is above the centre of gravity
(c) the metacentric height is zero
(d) the centre of buoyancy coincide with the centre of gravity

[ESE : 2010]

4.25 Consider the following statements:

1. Increasing the metacentric height of a ship causes increase in the period of rolling.
2. The frequency of pitching can be reduced by adding more loads along the length of a ship below its centre of gravity.
3. A ship has different metacentres in rolling and pitching.
4. Stability can be improved by increasing the width and reducing the draft.

Which of these statements are correct?

(a) 1, 2, 3 and 4
(b) 1, 2 and 4 only

(c) 1, 2 and 3 only
(d) 2, 3 and 4 only

[ESE : 2010]

4.26 Consider the following statements related to the stability of floating bodies:

1. The metacentre should be above the centre of gravity of the floating body for stable equilibrium during small oscillations.
2. For a floating body, stability is not determined simply by the relative positions of centre of gravity and centre buoyancy.
3. The position of metacentre of a floating body is fixed irrespective of the axis of oscillations.
4. Large value of metacentric height reduces the period of roll of the vessel.

Which of these statements are correct?

(a) 1 and 3 only (b) 2 and 4 only
(c) 1, 2 and 4 only (d) 1, 2, 3 and 4

[ESE : 2010]

4.27 The centre of buoyancy is

(a) centre of gravity of the body
(b) point of intersection of the buoyant force and the gravitational force
(c) point of intersection of the buoyant force and the centre line of the body
(d) centre of the volume of the displaced fluid

[ESE : 2010]

Direction: Each of the next questions consists of two statements, one labelled as the **'Assertion (A)'** and the other as **'Reason (R)'**. You are to examine these two statements carefully and select the answers to these items using the codes given below:

Codes:

(a) Both **A** and **R** are individually true and **R** is the correct explanation of **A**
(b) Both **A** and **R** are individually true but **R** is not the correct explanation of **A**
(c) **A** is true but **R** is false
(d) **A** is false but **R** is true

4.28 Assertion (A): Stability of a floating body is determined by the relative position of the centre of gravity and the centre of buoyancy.

Reason (R): If metacentre of the floating body is above the centre of gravity of the body, then the floating body will be in stable equilibrium.

[ESE : 2010]

4.29 Assertion (A): A floating body of spherical shape shall always be in neutral equilibrium condition.

Reason (R): For a spherical floating body the centre of gravity is always above the centre of buoyancy.

[ESE : 2010]

4.30 As a ship enters into a river from sea, one can expect that

(a) It rises a little (b) It sinks a little
(c) It remains at the same level
(d) Its level depends on the material used for construction [ESE : 2011]

4.31 Calculation of metacentric height of a floating body involves second moment of area. The axis about which this moment is to be calculated passes though the

(a) Top horizontal surface of the body
(b) Bottom horizontal surface of the body
(c) Centre of gravity of the body
(d) Centre of buoyancy

[ESE : 2011]

4.32 The buoyancy force is

(a) Equal to volume of liquid displace
(b) Force necessary to maintain equilibrium of a submerged body
(c) The resultant force acting on a floating body
(d) The resultant force on a body due to the fluid surrounding it [ESE : 2011]

4.33 The submerged body will be in stable equilibrium if the centre of buoyancy B

(a) Is below the centre of gravity G
(b) Coincides with G
(c) Is above the meta-centre M
(d) Is above G [ESE : 2011]

4.34 The stability of a floating body is obtained when its

(a) Centre of gravity is below the centre of buoyancy

(b) Metacentric height is negative
(c) Metacentric height is positive
(d) Metacentric height is zero

[ESE : 2011]

4.35 Assertion (A) : Increasing the metacentric height gives greater stability, but reduces the period of roll so that the vessel is less comfortable for passengers.

Reason (R) : Warships and racing yachts have larger metacentric height as their stability is more important than comfort.

(a) Both **A** and **R** are individually true and **R** is the correct explanation of **A**
(b) Both **A** and **R** are individually true but **R** is not the correct explanation of **A**
(c) **A** is true but **R** is false
(d) **A** is false but **R** is true [ESE : 2011]

4.36 For stability of a floating body (M = metacentre, G = centre of gravity and B = centre of buoyancy)

(a) M should coincide with B and G
(b) M should lie below B and G
(c) M should lie above B and G
(d) M should lie between and G

[ESE : 2012]

4.37 The resultant of all normal pressures on a body immersed in liquid acts

(a) through the centre of gravity of the body
(b) through the centre of pressure
(c) vertically upwards
(d) at metacentre [ESE : 2012]

4.38 A bucket of water is hanging from a spring balance. An iron piece is suspended into water without touching any of the side of bucket from another support. The spring balance reading will

(a) increase
(b) decrease
(c) remain the same
(d) depend on the depth of immersion

[ESE : 2012]

4.39 A piece of wood of volume V and specific gravity 0.87 floats on the surface of a liquid of specific gravity 1.31. The portion of the body which is submerged in the liquid will be

(a) $0.335V$ (b) $0.665V$
(c) $0.87V$ (d) $0.13V$

[ESE : 2012]

4.40 During floods, water entered an office having wooden tables. The position of tables, if floating, will be

(a) legs downwards (b) legs on sides
(c) legs upwards (d) Any position

[ESE : 2012]

4.41 A piece of metal of specific gravity 7 floats in mercury of specific gravity 13.6, what fraction of its volume is under mercury?

(a) 0.5 (b) 0.4
(c) 0.515 (d) Fully immersed

[ESE : 2013]

4.42 Which one of the following statements is correct?

(a) For a floating body, the stable equilibrium condition exists when position of metacentre remains higher than the centre of gravity of the body
(b) For a floating body, the stable equilibrium condition exists when position of metacentre remains lower than the centre of gravity of the body
(c) For a floating body, the neutral equilibrium condition exists when position of metacentre remains higher than the centre of gravity of the body
(d) For a floating body, the unstable equilibrium condition exists when position of metacentre remains higher than the centre of gravity of the body

[ESE : 2014]

4.43 Statement (I) : When a given body floats in different liquids, the volume displaced will decrease with increase in the specific gravity of the fluid.

Statement (II): The weight of the floating body is equal to the weight of the volume displaced.

(a) Both Statement (I) and Statement (II) are individually true and Statement (II) is the correct explanation of Statement (I).
(b) Both Statement (I) and Statement (II) are individually true but Statement (II) is not the correct explanation of Statement (I).

(c) Statement (I) is true but Statement (II) is false.
(d) Statement (I) is false but Statement (II) is true.
[ESE : 2014]

4.44 What is the specific gravity of a marble stone, which weighs 400 N in air, and 200 N in water? (g = 10 m/s^2)
(a) 8 (b) 6
(c) 4 (d) 2
[ESE : 2015]

4.45 Consider the following statements:
1. If a small upward displacement is given to a floating body, it results in the reduction of the buoyant force acting on the body
2. A slight horizontal displacement does not change either the magnitude or the location of the buoyant force

Which of the above statements is/are correct?
(a) Both 1 and 2 (b) 1 only
(c) 2 only (d) Neither 1 nor 2
[ESE : 2015]

4.46 A 25 cm long prismatic homogeneous solid floats in water with its axis vertical and 10 cm projecting above water surface. If the same solid floats in some oil with its axis vertical and 5 cm projecting above the liquid surface, then the specific gravity of the oil is
(a) 0.55 (b) 0.65
(c) 0.75 (d) 0.85 [ESE : 2016]

4.47 Consider the following statements :
The increase in metacentric height
1. Increases stability
2. Decreases stability
3. Increases comfort for passengers in a ship
4. Decreases comfort for passengers in a ship

Which of the above statements are correct ?
(a) 1 and 3 (b) 1 and 4
(c) 2 and 3 (d) 2 and 4
[ESE : 2016]

4.48 Consider the following statements pertaining to the metacentric height of ocean-going vessels :
1. Increase in the metacentric height reduces the period roll
2. Some control of period of roll is possible if Cargo is placed further from the centre line of ship
3. In warships and racing yachts, metacentric height will be larger than other categories of ships
4. For ocean-going vessels, metacentric height is of the order of 30 cm to 120 cm

Which of the above statements are correct ?
(a) 1, 2, 3 and 4 (b) 1, 2 and 3 only
(c) 1, 2 and 4 only (d) 3 and 4 only
[ESE : 2016]

4.49 A solid body of specific gravity 0.5 is 10 m long 3 m wide and 2 m high. When it floats in water with its shortest edge vertical, its metacentric height is :
(a) 0.75 m (b) 0.45 m
(c) 0.25 m (d) 0.15 m
[ESE : 2016]

4.50 Consider the following statements pertaining to stability of floating bodies:
1. A floating body will be stable when the centre of gravity is above the centre of buoyancy.
2. The positions of metacentres corresponding to different axes of rotation are generally different for the same floating object.
3. For cargo ships, the metacentric height varies with loading.

Which of the above statements are correct?
(a) 1, 2 and 3 (b) 1 and 2 only
(c) 1 and 3 only (d) 2 and 3 only

4.51 The necessary and sufficient condition for bodies in floatation to be in stable equilibrium is that the centre of gravity is located below the
(a) metacentre (b) centre of buoyancy
(c) epicentre (d) centroid
[ESE : 2018]

4.52 **Statement (I) :** If a boat, built with sheet metal on wooden frame, has an average density which is greater than that of water, then the boat can float in water with its hollow face upward but will sink once it overturns.

Statement (II) : Buoyant force always acts in the upward direction.

(a) Both Statement (I) and Statement (II) are individually true and Statement (II) is the correct explanation of Statement (I).
(b) Both Statement (I) and Statement (II) are individually true but Statement (II) is not the correct explanation of Statement (I).
(c) Statement (I) is true but Statement (II) is false.
(d) Statement (I) is false but Statement (II) is true.

[ESE : 2018]

4.53 A spherical balloon of 1.5 m diameter is completely immersed in water and chained to the bottom. If the chain has a tension of 10 kN, the weight of the balloon will be nearly

(a) 9.11 kN (b) 8.22 kN
(c) 6.44 kN (d) 7.33 kN

[ESE : 2019]

4.54 A stone weighs 400 N in air and when immersed in water it weighs 225 N. If the specific weight of water is 9810 N/m^3, the relative density of the stone will be nearly

(a) 5.9 (b) 4.7
(c) 3.5 (d) 2.3

[ESE : 2020]

4.55 If a submerged body is in unstable equilibrium, then

(a) the centre of buoyancy is below the centre of gravity.
(b) the centre of buoyancy is above the centre of gravity.
(c) meta-centre is below the centre of buoyancy.
(d) meta-centre is above the centre of buoyancy.

[ESE : 2021]

4.56 How much of concrete with $\gamma = 25$ kN/m^3 must be attached to a beam having a volume of 0.1 m^3 and specific gravity 0.6 to cause both to sink in water?
[Take g = 9.81 m/s^2]

(a) 0.825 kN (b) 0.745 kN
(c) 0.525 kN (d) 0.645 kN

[ESE : 2021]

4.57 A body of dimensions 1.5 m × 1.0 m × 2 m, weighs 1962 N in water. What is the weight of the body in air?
(Take acceleration due to gravity as 9.81 m/s^2)

(a) 31392 N (b) 23392 N
(c) 14392 N (d) 46392 N

[ESE : 2022]

■■■■

Answers Buoyancy, Floatation and Liquids in Relative Equilibrium

4.1 (b)	4.2 (a)	4.3 (c)	4.4 (d)	4.5 (c)	4.6 (b)	4.7 (d)	4.8 (b)	4.9 (a)
4.10 (d)	4.11 (a)	4.12 (c)	4.13 (d)	4.14 (c)	4.15 (d)	4.16 (c)	4.17 (c)	4.18 (d)
4.19 (c)	4.20 (b)	4.21 (a)	4.22 (b)	4.23 (b)	4.24 (b)	4.25 (d)	4.26 (c)	4.27 (d)
4.28 (d)	4.29 (b)	4.30 (b)	4.31 (a)	4.32 (d)	4.33 (d)	4.34 (c)	4.35 (b)	4.36 (c)
4.37 (c)	4.38 (a)	4.39 (b)	4.40 (c)	4.41 (c)	4.42 (a)	4.43 (a)	4.44 (d)	4.45 (a)
4.46 (c)	4.47 (b)	4.48 (a)	4.48 (c)	4.50 (d)	4.51 (a)	4.52 (b)	4.53 (d)	4.54 (d)
4.55 (a)	4.56 (d)	4.57 (a)						

Explanations Buoyancy, Floatation and Liquids in Relative Equilibrium

4.1 (b)

$$0.5\times10^{-4}\times1\times10^{5} = 1000\times\frac{10\times10}{100\times100}\times10\times h$$

$\therefore$ $h = \mathbf{5\ cm}$

4.2 (a)

A fluid at rest, pressure varies as

$$\frac{dp}{dz} = -\rho g$$

When fluid moves downward then,

$$\frac{dp}{dz} = -\rho g+\rho g = 0 \therefore p_1 = p_2 = \text{constant}$$

i.e. pressure throughout the liquid mass is atmospheric

4.3 (c)

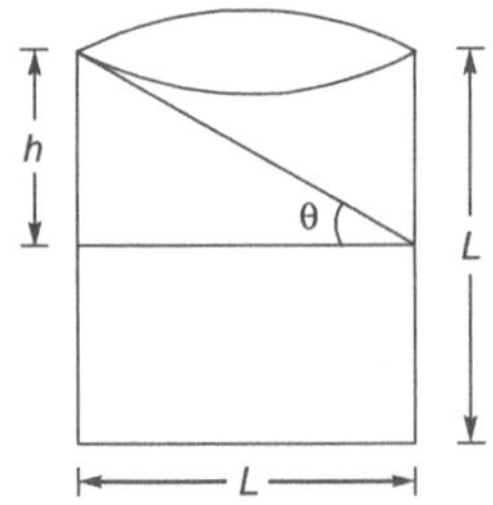

$$\tan\theta = \frac{g}{g+g} = \frac{1}{2}$$

$$\therefore \quad \frac{h}{L} = \frac{1}{2} = \mathbf{0.5}$$

4.4 (d)

Volume of aluminium,

$$V_{Al} = \frac{12}{2.4}$$

$$T + F_B = mg$$

Buoyancy force

$$= \rho V_{Al}.g = 0.8\times\frac{12}{2.4}\times10 = 40\text{ N}$$

Now,

$T + 40 = 120$

or $T = \mathbf{80\ N}$

4.5 (c)

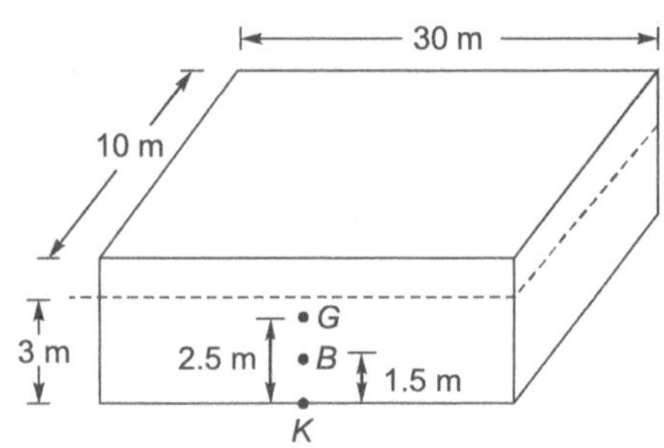

$$BM = \frac{I}{V} = \frac{(30)\frac{(10)^3}{12}}{30\times10\times3} = 2.778\text{ m}$$

where

I = Least moment of inertia

and V = Displaced volume

$KB = 3/2 = 1.5$ m

$KM = KB + BM = 2.778 + 1.5 = 4.278$ m

$GM = KM - KG$

$= 4.278 - 2.5 = 1.778\text{ m} \cong \mathbf{1.78\ m}$

4.6 (b)

- Stable equilibrium of floating body: Metacentre above the centre of gravity.
- Stable equilibrium of submerged body : Centre of buoyancy above the centre of gravity.

- Unstable equilibrium of a body floating : Metacentre below the centre of gravity.
- Unstable equilibrium of a submerged body : Centre of buoyancy below the centre of gravity.

4.7 (d)

$$\frac{\partial p}{\partial z} = \rho g\left(1+\frac{a}{g}\right)$$

$$\therefore \quad p = \rho g h\left(1+\frac{a}{g}\right)$$

$$= 9.81 \times 1000 \times 0.8 \times 2.5\left(1+\frac{4.9}{9.8}\right)$$

$$= \mathbf{29.40\ kPa}$$

4.8 (b)

The metacentre is above the centre of gravity.

4.10 (d)

The value of specific gravity in relation with object must be known.

4.11 (a)

When beaker falling freely: Then pressure

$$p = p_0 + \rho g\left(1+\frac{a}{g}\right)\cdot h$$

Here $a = -g$

$$p = p_0$$

The pressure at any point is equal to atmospheric pressure.

4.12 (c)

$$\therefore \rho_{oil} = \frac{15}{20}\rho_w \qquad (\rho_w A h_w g = \rho_{oil} A h_{oil} g)$$

$$\therefore \frac{\rho_{oil}}{\rho_w} = \mathbf{0.75}$$

4.13 (d)

When hydrometer floats is oil

$$0.75\ \rho g A\, x_1 = 0.03$$

$$0.75 \times 1000 \times 9.81 \times \frac{\pi}{4} \times (3 \times 10^{-3})^2 \times x_1 = 0.03$$

or $\quad x_1 = 0.5768$ m

When hydrometer floats in alcohol

$$0.8 \rho g A x_2 = 0.03$$

$$\Rightarrow 0.8 \times 1000 \times 9.81 \times \frac{\pi}{4} \times (3 \times 10^{-3})^2 \times x_2 = 0.03$$

$$\therefore \quad x_2 = 0.5408 \text{ m}$$

$$\therefore \quad x_1 - x_2 = 0.5968 - 0.5408$$

$$= 0.03604 \text{ m} = \mathbf{36.04\ mm}$$

4.14 (c)

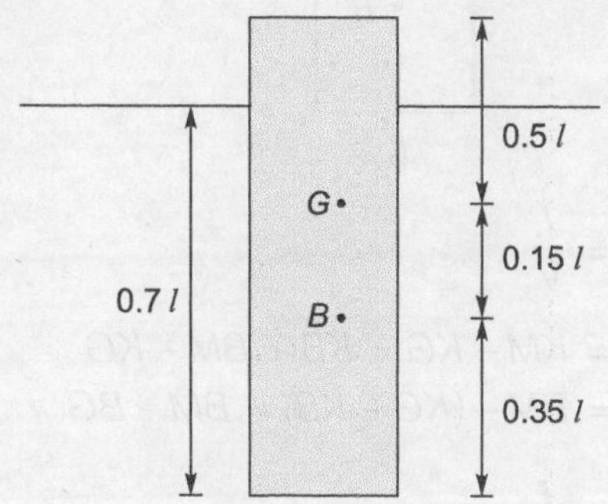

$$0.7\, l\, A\, \rho_w g = l A \rho_b g$$

$$\therefore \quad \rho_b = 0.7\, \rho_w$$

$$\therefore \quad \frac{\rho_b}{\rho_w} = \mathbf{0.7}$$

4.15 (d)

Stability of floating body

(i) If M is above G, or GM is +ve or $BM > BG$ body is in stable equilibrium.

(ii) If M is below G, or GM is –ve or $BM < BG$. The body is unstable equilibrium.

(iii) If M and G coincide, or $GM = 0$ or $BM = BG$. The body is in neutral equilibrium.

4.16 (c)

In case of ships the shifting of cargo may cause the ships to roll. As such along with the consideration of the stability of a ships, its period of roll is also required to be determined. This so because increasing the metacentric height gives greater stability to a floating body.

4.17 (c)

$$\tan 45° = \frac{a}{g}$$

$$\Rightarrow a = g$$

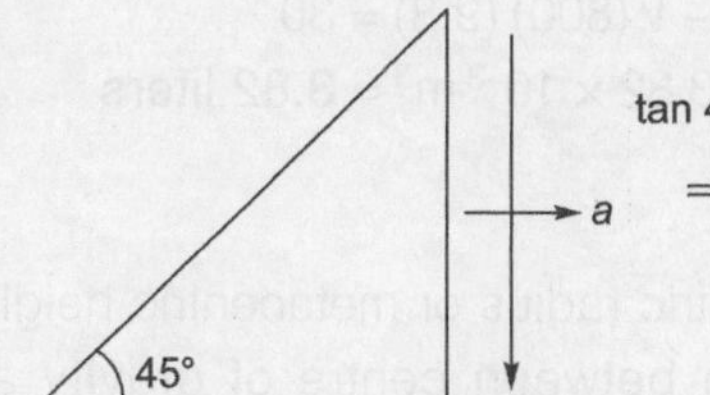

4.19 (c)

Let K be the bottom most point of the body.

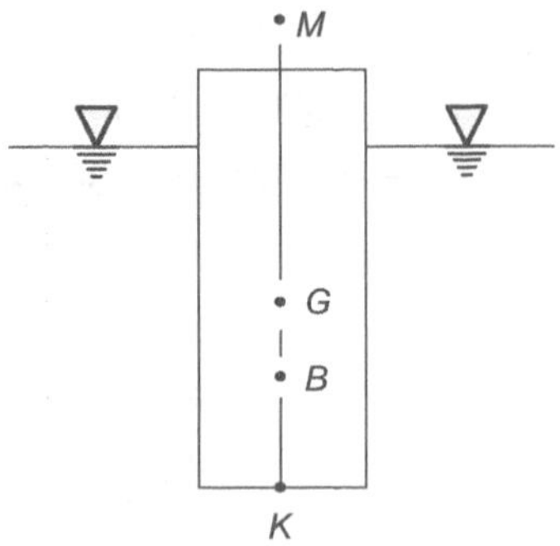

$$BM = \frac{I}{V}$$

$$GM = KM - KG = KB + BM - KG$$
$$= BM - (KG - KB) = BM - BG$$
$$= \frac{I}{V} - BG$$

4.20 (b)

Stability of a floating body can be improved by making width large which will increase I and will thus increase the metacentric height and keeping the centre of mass low and making the draft small.

4.21 (a)

BM = Distance between centre of buoyancy to metacentre is given by I/V_d.

where V_d is volume of fluid displaced.

I is moment of Inertia of a horizontal section of a body taken at the surface of the fluid.

4.22 (b)

$$W - V\rho_1 g = 30 \quad \text{...(i)}$$
$$W - V\rho_2 g = 15 \quad \text{...(ii)}$$

$$\Rightarrow \frac{W - V(800)(9.8)}{W - V(1200)(9.8)} = 2$$

$\Rightarrow$ $W - V(800)(9.8) = 2W - 2V(1200)(9.8)$

$\Rightarrow$ $W = 15680V$

Putting the value of W in Equation (i)

$15680\,V - V(800)(9.8) = 30$

$\Rightarrow$ $V = 3.82 \times 10^{-3}\,m^3$ = **3.82 liters**

4.23 (b)

Metacentric radius or metacentric height is the distance between centre of gravity and the metacentre.

4.24 (b)

For stable equilibrium, Metacentre is above the centre of gravity.

4.25 (d)

Time period of oscillations

$$T = 2\pi\sqrt{\frac{K^2}{g(GM)}}$$

(where GM = Metacentric height)

$$T \propto \frac{1}{\sqrt{GM}}$$

Hence, increase in metacentric height reduces period of oscillations.

4.26 (c)

The position of metacentre of a floating body is not fixed. As the axis oscillations change, its metacentre also changes.

4.28 (d)

Stability of a floating body is determined by relative position of centre of gravity and metacentre.

4.30 (b)

As a ship enters into a river from sea, one can expect that it sinks a little because of some buoyancy force acts. Because density of river water is lower than sea water.

4.31 (a)

$$GM = \frac{I}{V} - BG$$

where GM = Metacentric height

I = Second moment of area

V = Volume of immersed body

BG = Distance between centre of gravity & centre of buoyancy

4.32 (d)

Buoyancy force is the resultant force acting on the body due to fluid surrounding it.

4.33 (d)

Submerged body will be in stable equilibrium if the centre of buoyancy (B) is above the centre of gravity (G).

For **neutral equilibrium** : Centre of buoyancy is equal to centre of gravity.

For **unstable equilibrium** : Centre of buoyancy is less than centre of gravity.

4.34 (c)

Stability of floating body is obtained when its metacentric height is positive i.e. GM is positive.

$$GM = \frac{I}{V} - BG$$

4.35 (b)

Relation between time period of oscillation (T) and metacentric height (GM) is

$$T = 2\pi\sqrt{\frac{k^2}{g(GM)}}$$

where g = Acceleration due to gravity
k = Radius of gyration

4.36 (c)

For stability of a body, it should have positive metacentric height i.e., M should lie above B and G.

4.37 (c)

The resultant of all normal pressures always acts vertically upwards.

4.38 (a)

As the iron piece is immersed in water it will experience an upward buoyancy force by the water of bucket. The water of bucket will experience an equal force downward and hence the spring balance reading will increase.

4.39 (b)

Weight of piece of wood = Force of buoyancy

$$V \times 0.87 \times 1000 = V' \times 1.31 \times 1000$$

$$\therefore \quad V' = \mathbf{0.665}\boldsymbol{V}$$

4.40 (c)

Since there is more mass towards the top of the table.

For stable equilibrium metacentric height should be +ve.

4.41 (c)

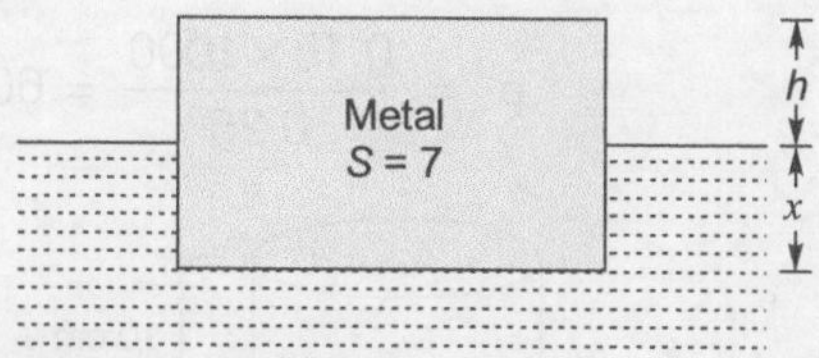

Weight of metal piece

$$= 7 \times 1000 \times V \text{ kg} = 70000V \text{ kg}$$

$$V = A(h + x)$$

Let x be the depth of metal piece under mercury be x.

$$13.6 \times A \times x = 7(h + x)A$$

$$\therefore \quad \frac{h+x}{x} = \frac{13.6}{7}$$

Fraction of volume under mercury

$$= \frac{Ax}{A(h+x)} = \frac{7}{13.6} = \mathbf{0.515}$$

4.44 (d)

$$\text{Specific gravity} = \frac{\text{Weight in air}}{\text{Weight in air} - \text{Weight in water}}$$

$$= \frac{400}{400 - 200} = 2$$

4.45 (a)

1. Buoyant force is given by volume of displaced liquid. So, small upward movement will reduce the volume of floating body is liquid and reduced displaced liquid. So Buoyant force decreases.
2. The horizontal displacement will result in motion but not change in volume so no change in direction and magnitudes.

4.46 (c)

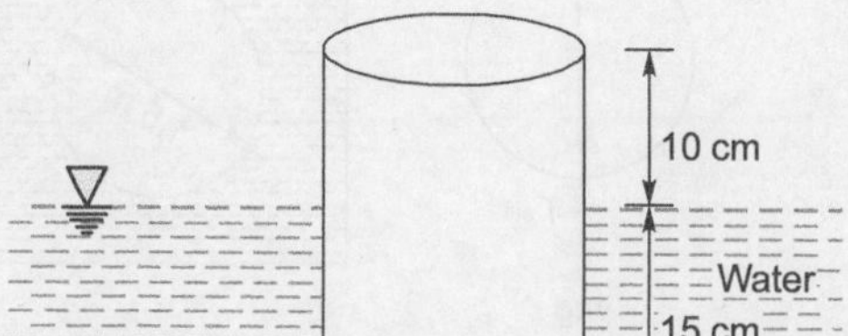

$$A \times (1.5) \times \rho_w g = \rho_s \times (0.25) \times Ag$$

$$\rho_s = \frac{0.15 \times 1000}{0.25} = 600 \text{ kg/m}^3$$

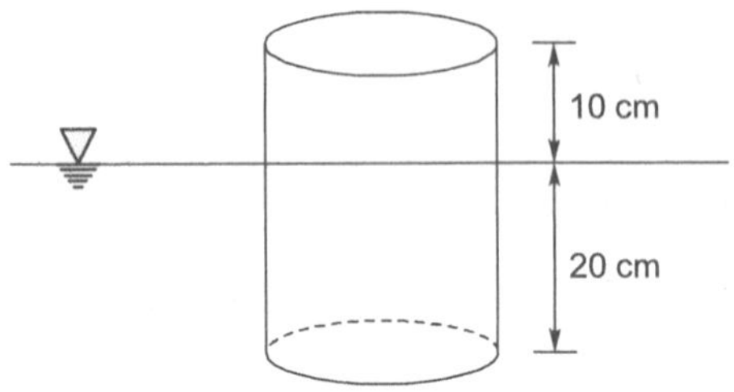

$$\rho_{oil} \times (0.2) \times A \times g = \rho_s \times (0.25) \times A \times g$$

$$\rho_{oil} = \frac{600 \times 0.25}{0.2} = 750 \text{ kg/m}^3$$

Specific gravity = **0.75**

4.49 (c)

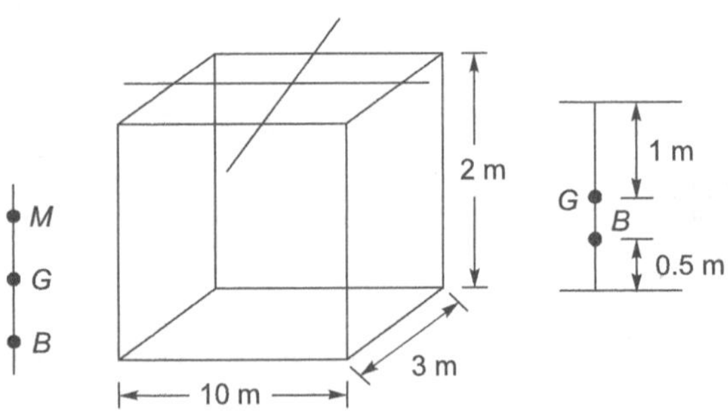

$$\rho_s = 500 \text{ kg/m}^3$$

$$BM = \frac{I_{min}}{\forall}$$

$$MG = \frac{I_{min}}{\forall} - BG$$

$$GM = \frac{10 \times 27}{12 \times 10 \times 3 \times 1} - 0.5 = \mathbf{0.25 \text{ m}}$$

4.53 (d)

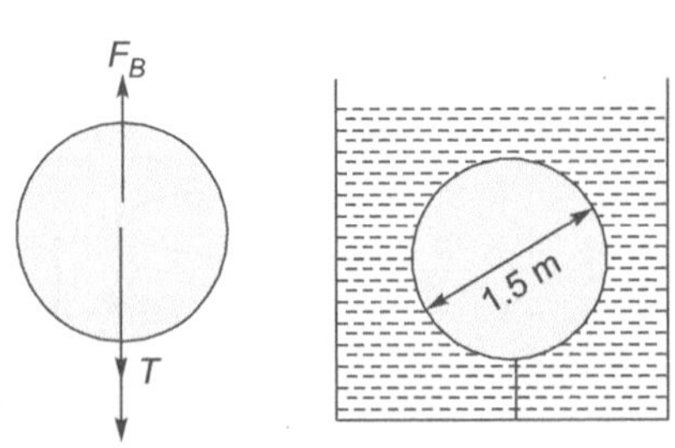

$$F_B = mg + T$$

$$mg = F_B - T = \rho g V - T$$

$$mg = 1000 \times 9.81 \times \left(\frac{4}{3}\pi \times (0.75)^3\right) - 10 \times 10^3 \text{ N}$$

$$= 7.33 \text{ kN}$$

4.54 (d)

$$W_{real} = 400 \ N = mg$$

$$= \rho_{body} \times g \times V_{body} \quad \text{...(i)}$$

$$W_{apparent} = 225 \text{ N}$$

$$W_{apparent} = W_{real} - \text{Buoyant force}$$

So, Buoyant force = 400 − 225

$$= \rho_{fluid} \times g \times V_{fd} \quad \text{...(ii)}$$

By equations (i) and (ii)

$$\frac{\rho_{body} \times g \times V_{body}}{\rho_{fluid} \times g \times V_{fluid}} = \frac{400}{175}$$

$[V_{fluid} = V_{body}, S_{body} = 2.28 \approx 2.3]$

4.55 (a)

Stable equilibrium : Centre of gravity is below the centre of buoyancy.

Unstable equilibrium : Centre of gravity is above the centre of buoyancy.

Neutral equilibrium : Centre of gravity coincides with centre of buoyancy.

4.56 (d)

$$\gamma_{concrete} = 25 \text{ kN/m}^3$$

$$S_{beam} = 0.6$$

$$V_{beam} = 0.1 \text{ m}^3$$

Let V = Volume of concrete

$$(\text{Wt})_{concrete} + (\text{Wt})_{beam} = F_B$$

$[25000 \times V] + [600 \times 9.81 \times 0.1] = 10^3 \times 9.81 [V + 0.1]$

$$V = 0.0258327 \text{ m}^3$$

Wt of concrete added $= 25 \times 10^3 \times 0.0258327$

$$= 0.645 \text{ kN}$$

4.57 (a)

$$W_{appareant\ weight} = W_{real\ weight} - F_{buoyant\ force}$$

$$W_{real\ weight} = W_{app} + F_B$$

$$= W_{app} + \rho_f g V_{fd}$$

$$= 1962 + 1000 \times 9.81 \times 1.5 \times 1 \times 2$$

$$= 31392 \text{ N}$$

■■■■

5 Fluid Kinematics

5.1 A stream function is given by $(x^2 - y^2)$. The potential function of the flow will be
(a) $2xy + f(x)$ (b) $2xy$ + constant
(c) $2(x^2 - y^2)$ (d) $2xy + f(y)$
[ESE : 2000]

5.2 The convective acceleration of fluid in the x-direction is given by
(a) $u\frac{\partial u}{\partial x} + v\frac{\partial v}{\partial y} + w\frac{\partial w}{\partial z}$
(b) $\frac{\partial u}{\partial t} + \frac{\partial v}{\partial t} + \frac{\partial w}{\partial t}$
(c) $u\frac{\partial u}{\partial x} + u\frac{\partial v}{\partial y} + u\frac{\partial w}{\partial z}$
(d) $u\frac{\partial u}{\partial x} + v\frac{\partial u}{\partial y} + w\frac{\partial u}{\partial z}$ [ESE : 2001]

5.3 The stream function in a 2-dimensional flow field is given by $\psi = xy$. The potential function is
(a) $\frac{x^2 + y^2}{2}$ (b) $\frac{x^2 - y^2}{2}$
(c) xy (d) $x^2 y + y^2 x$
[ESE : 2001]

5.4 The velocity potential of a velocity field is given by $\phi = x^2 - y^2$ + constant. Its stream function will be given by
(a) $-2xy$ + constant (b) $2xy$ + constant
(c) $-2xy + f(x)$ (d) $-2xy + f(y)$
[ESE : 2002]

5.5 A streamline is a line
(a) which is along path of the particle
(b) which is always parallel to the main direction of flow
(c) along which there is no flow
(d) on which tangent drawn at any point gives the direction of velocity [ESE : 2002]

5.6 **Assertion (A):** Stream lines are drawn in the flow field such that at a given instant of time they are perpendicular to the direction of flow at every point in the flow field.
Reason (R): Equation for a stream line in a two dimensional flow is given by
$Vxdy - V_y dx = 0$
(a) Both **A** and **R** are individually true and **R** is the correct explanation of **A**
(b) Both **A** and **R** are individually true but **R** is not the correct explanation of **A**
(c) **A** is true but **R** is false
(d) **A** is false but **R** is true [ESE : 2002]

5.7 Match **List-I** (Example) with **List-II** (Types of flow) and select the correct answer using the codes given below the lists:

List-I
A. Flow in a straight long pipe with varying flow rate
B. Flow of gas through the nozzle of a jet engine
C. Flow of water through the hose of a fire fighting pump
D. Flow in a river during tidal wave

List-II
1. Uniform, steady
2. Non-uniform, steady
3. Uniform, unsteady
4. Non-uniform, unsteady

Codes:

	A	B	C	D
(a)	1	4	3	2
(b)	3	2	1	4
(c)	1	2	3	4
(d)	3	4	1	2

[ESE : 2002]

5.8 Which one of the following stream functions is a possible irrotational flow field?
(a) $\psi = x^3y$ (b) $\psi = 2xy$
(c) $\psi = Ax^3y^2$ (d) $\psi = Ax + By^2$
[ESE : 2003]

5.9 **Assertion (A):** Streamlines can cross one another if the fluid has higher velocity.

Reason (R): At sufficiently high velocity, the Reynolds number is high and at sufficiently high Reynolds number, the structure of the flow is of turbulent type.

(a) Both **A** and **R** are individually true and **R** is the correct explanation of **A**
(b) Both **A** and **R** are individually true but **R** is not the correct explanation of **A**
(c) **A** is true but **R** is false
(d) **A** is false but **R** is true

[ESE : 2003]

5.10 The components of velocity in a two dimensional frictionless incompressible flow are $u = t^2 + 3y$ and $v = 3t + 3x$. What is the approximate resultant total acceleration at the point (3, 2) and $t = 2$?

(a) 5 (b) 49
(c) 59 (d) 54

[ESE : 2004]

5.11 The stream function $\psi = x^3 - y^3$ is observed for a two dimensional flow field. What is the magnitude of the velocity at point (1, –1)?

(a) 4.24 (b) 2.83
(c) 0 (d) – 2.83

[ESE : 2004]

5.12 Which of the following functions represent the velocity potential in a two-dimensional flow of an ideal fluid?

1. $2x + 3y$
2. $4x^2 - 3y^2$
3. $\cos(x - y)$
4. $\tan^{-1}(x/y)$

Select the correct answer using codes given below:

(a) 1 and 3 (b) 1 and 4
(c) 2 and 3 (d) 2 and 4

[ESE : 2004]

5.13 Irrotational flow is characterized as the one in which the

(a) fluid flows along a straight line
(b) fluid does not rotate as it moves along
(c) net rotation of fluid particles about their mass centres remains zero
(d) streamlines of flow are curved and closely spaced

[ESE : 2004]

5.14 If for a flow, a stream function ψ exists and satisfies the Laplace equation, then which one of the following is the correct statement?

(a) The continuity equation is satisfied and the flow is irrotational.
(b) The continuity equation is satisfied and the flow is rotational.
(c) The flow is irrotational but does not satisfy the continuity equation.
(d) The flow is rotational.

[ESE : 2005]

5.15 For irrotational and incompressible flow, the velocity potential and stream function are given by ϕ and ψ, respectively. Which one of the following sets is correct?

(a) $\nabla^2\phi = 0, \nabla^2\psi = 0$ (b) $\nabla^2\phi \neq 0, \nabla^2\psi = 0$

(c) $\nabla^2\phi = 0, \nabla^2\psi \neq 0$ (d) $\nabla^2\phi \neq 0, \nabla^2\psi \neq 0$

[ESE : 2006]

5.16 In a two-dimensional incompressible steady flow, the velocity component $u = Ae^x$ is obtained. What is the other component v of velocity?

(a) $v = Ae^{xy}$ (b) $v = Ae^y$
(c) $v = -Ae^x y + f(x)$ (d) $v = -Ae^y x + f(y)$

[ESE : 2006]

5.17 For a steady two-dimensional flow, the scalar components of the velocity field are $v_x = -2x$, $v_y = 2y$, $v_z = 0$. What are the components of acceleration?

(a) $a_x = 0, a_y = 0$ (b) $a_x = 4x, a_y = 0$
(c) $a_x = 0, a_y = 4y$ (d) $a_x = 4x, a_y = 4y$

[ESE : 2006]

5.18 Consider the following statements regarding a path line in fluid flow:

1. A path line is a line traced by a single particle over a time interval.
2. A path line shows the positions of the same particle at successive time instants.
3. A path line shows the instantaneous positions of a number of particles, passing through a common point, at some previous time instants.

Which of these statements are correct?

(a) 1 and 3 only (b) 1 and 2 only
(c) 2 and 3 only (d) 1, 2 and 3

[ESE : 2006]

5.19 Streamline, path line and streak line are identical when the

(a) flow is steady
(b) flow is uniform
(c) flow velocities do not change steadily with time
(d) flow is neither steady nor uniform

[ESE : 2007]

5.20 The relation $\frac{\partial^2\phi}{\partial x^2} + \frac{\partial^2\phi}{\partial y^2} = 0$ for an irrotational flow is known as which one of the following?

(a) Navier - Stokes equation
(b) Laplace equation
(c) Reynolds equation
(d) Euler's equation [ESE : 2007]

5.21 Which one of the following stream functions ψ is a possible irrotational flow field?

(a) $\psi = y^2 - x^2$ (b) $\psi = A \sin(xy)$
(c) $\psi = Ax^2 y^2$ (d) $\psi = Ax + By^2$

[ESE : 2007]

5.22 Match **List-I** (Condition) with **List-II** (Regulating fact) and select the correct answer using the codes given below the lists:

List-I

A. Existence of stream function
B. Existence of velocity potential
C. Absence of temporal variations
D. Constant velocity vector

List-II

1. Irrotationality of flow
2. Continuity of flow
3. Uniform flow
4. Steady flow

Codes:

	A	B	C	D
(a)	4	3	2	1
(b)	2	1	4	3
(c)	4	1	2	3
(d)	2	3	4	1

[ESE : 2007]

5.23 A velocity field is given by $u = 3xy$ and $v = \frac{3}{2}(x^2 - y^2)$. What is the relevant equation of a streamline?

(a) $\frac{dx}{dy} = \frac{x^2 - y^2}{xy}$ (b) $\frac{dx}{dy} = \frac{3xy}{x^2 - y^2}$

(c) $\frac{dx}{dy} = \frac{2xy}{x^2 - y^2}$ (d) $\frac{dx}{dy} = \frac{x^2 - y^2}{2xy}$

[ESE : 2008]

5.24 Two flows are specified as

A. $u = y$, $v = -(3/2)x$
B. $u = xy^2$, $v = x^2y$

Which one of the following can be concluded?

(a) Both flows are rotational
(b) Both flows are irrotational
(c) Flow *A* is rotational while flow *B* is irrotational
(d) Flow *A* is irrotational while flow *B* is rotational

[ESE : 2008]

5.25 A steady, incompressible flow is given by: $u = 2x^2 + y^2$ and $v = -4xy$

What is the convective acceleration along *x*-direction at point (1, 2)?

(a) $a_x = 6$ unit (b) $a_x = 24$ unit
(c) $a_x = -8$ unit (d) $a_x = -24$ unit

[ESE : 2008]

5.26 The stream function in a flow field is given by $\psi = 2xy$. In the same flow field, what is the velocity at a point (2, 1)?

(a) 4 unit (b) 5.4 unit
(c) 1.73 unit (d) 4.47 unit

[ESE : 2008]

5.27 For steady incompressible flow, if the *u*-component of velocity is $u = Ae^x$, then what is the *v*-component of velocity?

(a) Ae^y (b) $Ae^x y$
(c) $-Ae^x y$ (d) $-Ae^x$ [ESE : 2008]

5.28 Consider the following equations :

1. $A_1 V_1 = A_2 V_2$
2. $\frac{\partial u}{\partial x} + \frac{\partial v}{\partial y} = 0$
3. $\int_S \rho v dA + \frac{\partial}{\partial t}\left(\int_V \rho dV\right) = 0$
4. $\frac{1}{r}\frac{\partial}{\partial r}(rv_r) + \frac{\partial}{\partial z}(v_z) = 0$

Which of the above equations are forms of continuity equations? (where u, v are velocities and V is volume)

(a) 1 only (b) 1 and 2
(c) 2 and 3 (d) 3 and 4

[ESE : 2009]

5.29 Assertion (A) : The local acceleration is zero in a steady motion.

Reason (R) : The convective component arises due to the fact that a fluid element experiences different velocities at different locations.

(a) Both **A** and **R** are individually true and **R** is the correct explanation of **A**
(b) Both **A** and **R** are individually true but **R** is not the correct explanation of **A**
(c) **A** is true but **R** is false
(d) **A** is false but **R** is true [ESE : 2009]

5.30 The velocity potential function is given by an expression $\phi = -\frac{xy^3}{3} - x^2 + \frac{x^3 y}{3} + y^2$. The velocity component in x direction is

(a) $xy^2 - \frac{x^3}{3} - 2y$
(b) $\frac{y^3}{3} + 2x - x^2 y$
(c) $3x^2y + 2x - xy^2$
(d) $x^3y^2 + 2x^2 - y^2x$

[ESE : 2010]

5.31 The velocity potential in a flow field is $\phi = 2xy$. The corresponding value of stream function is

(a) $(y^2 - x^2)$ + constant
(b) $(x^2 - y^2)$ + constant
(c) $\frac{1}{2}(x^2 - y^2)$ + constant
(d) $2(x - y)$ + constant

[ESE : 2010]

5.32 Stream and velocity potential functions for a 2-D flow field given by $u = 2x$ and $v = -2y$ are

(a) $\psi = 2xy$; $\phi = y^2 - x^2$
(b) $\psi = x^2 - y^2$; $\phi = 2xy$
(c) $\psi = x^2y^2$; $\phi = x^2 + y^2$
(d) $\psi = x^2 + y^2$; $\phi = x^2y^2$ [ESE : 2010]

5.33 The stream function in a two-dimensional flow field is given by $\psi = x^2 - y^2$. Then the magnitude of velocity at point (1, 1) is

(a) 2 (b) $2\sqrt{2}$
(c) 4 (d) 8

[ESE : 2010]

5.34 Consider the following:

The components of velocity u and v along x and y directions in a 2D flow problem of an incompressible fluid are

1. $u = x^2 \cos y$; $v = -2x \sin y$
2. $u = x + 2$; $v = 1 - y$
3. $u = xyt$; $v = x^3 - y^2t/2$
4. $u = \ln x + y$; $v = xy - y/x$

Which of these will satisfy the continuity equation?

(a) 1, 2 and 3 only (b) 1, 2 and 4 only
(c) 2, 3 and 4 only (d) 1, 2, 3 and 4

[ESE : 2010]

5.35 The components of rotation for a three-dimensional flow field

$V = (y^2 + z^2)\hat{i} + (x^2 + z^2)\hat{j} + (x^2 + y^2)\hat{k}$ at (1, 2, 3) are

(a) $\omega_x = -1$ rad/s, $\omega_y = 2$ rad/s, $\omega_z = -1$ rad/s
(b) $\omega_x = -1$ rad/s, $\omega_y = -1$ rad/s, $\omega_z = 2$ rad/s
(c) $\omega_x = -2$ rad/s, $\omega_y = -1$ rad/s, $\omega_z = -2$ rad/s
(d) $\omega_x = 2$ rad/s, $\omega_y = -2$ rad/s, $\omega_z = -1$ rad/s

[ESE : 2010]

5.36 Assertion (A): The velocity potential provides an alternative means of expressing velocity components.

Reason (R): The existence of velocity potential in a flow field ensures that the flow must be irrotational.

[ESE : 2010]

5.37 If flow conditions satisfy 'Laplace equation' then

(a) Flow is rotational
(b) Flow does not satisfy continuity equation
(c) Flow is irrotational but does not satisfy continuity equation
(d) Flow is irrotational and satisfies continuity equation [ESE : 2011]

5.38 The stream function of a two-dimensional flow is $\psi = 2xy$. The flow is

(a) Rotational

(b) Irrotational
(c) Vortex flow
(d) Unsteady pulsating flow [ESE : 2011]

5.39 Consider the following statements pertaining to kinematics and dynamics of fluid flow :

1. For $\psi = x^2 - y^2$, velocity at (1, 1) is $2\sqrt{2}$.
2. For all types of flow, stream and potential function exist.
3. Bernoulli's equation is not valid for real fluid flow.

Which of these statements are correct?

(a) 1, 2 and 3 (b) 1 and 2 only
(c) 2 and 3 only (d) 1 and 3 only
[ESE : 2011]

5.40 For an irrotational motion
(a) The fluid element does not undergo any shear
(b) The fluid particles do not undergo a circular motion
(c) The circulation around any path is zero
(d) The viscosity may not be zero
[ESE : 2011]

5.41 The flow in a pipe whose valve is being opened or closed gradually is an example of
(a) steady flow
(b) nonsteady flow
(c) steady uniform flow
(d) steady nonuniform flow
[ESE : 2012]

5.42 A velocity potential function exists only for
(a) steady flow (b) uniform flow
(c) irrotational flow (d) compressible flow
[ESE : 2012]

5.43 A steady irrotational flow of an incompressible fluid is called
(a) streamline flow (b) creeping flow
(c) shear flow (d) potential flow
[ESE : 2012]

5.44 The streamline equation passing through the point (2, 3) will be (The two dimensional flow is given by $u = a$ and $v = a$)
(a) $x + y = 1$ (b) $y = x + 1$
(c) $2x + y = 1$ (d) None of these
[ESE : 2013]

5.45 Consider the following statements about velocity potential:

1. Velocity potential is a vector function similar to stream function.
2. It is a fully three dimensional function and not limited to two coordinates.
3. Velocity potential does not exist at stagnation points.
4. Velocity potential exists only if the flow is Irrotational.

Which of these statements are correct?

(a) 1, 2 and 3 (b) 1, 2 and 4
(c) 1, 3 and 4 (d) 2, 3 and 4
[ESE : 2013]

5.46 The velocity potential function in a two dimensional flow field is given by $\phi = x^2 - y^2$. The magnitude of velocity at point $P(1, 1)$ is
(a) zero (b) 2
(c) $2\sqrt{2}$ (d) 8 [ESE : 2013]

5.47 A 2-D flow field is defined as $\vec{V} = \vec{i}x - \vec{j}y$. The equation of streamline passing through the point (1, 1) is
(a) $xy - 1 = 0$ (b) $xy + 1 = 0$
(c) $xy + 2 = 0$ (d) $xy - 2 = 0$
[ESE : 2014]

5.48 A flownet is a graphical representation of streamlines and equipotential lines such that these lines
(a) Intersect each other at various different angles forming irregular shaped nets
(b) Intersect each other orthogonally forming curvilinear squares
(c) Indicate the direction but not magnitude of vector
(d) Indicate the direction and magnitude of vector
[ESE : 2014]

5.49 Which one of the following statements is correct for the velocity potential?
(a) Existence of velocity potential is an indication of irrotational nature of the flow
(b) The velocity potential automatically satisfies the continuity equation
(c) Velocity potential can be defined only for 2-dimensional flow
(d) All of the above [ESE : 2014]

5.50 State whether following flow field is physically possible?

$u = 3xy^2 + 2x + y^2$ and $v = x^2 - 2y - y^3$

(a) Possible for steady, incompressible flow
(b) Possible for unsteady, incompressible flow
(c) Possible for steady, compressible flow
(d) Not possible [ESE : 2015]

5.51 A steady incompressible flow field is given by $u = 2x^2 + y^2$ and $v = -4xy$. The convective acceleration along x-direction at point (1, 2) is

(a) 6 units (b) 24 units
(c) – 8 units (d) – 24 units
[ESE : 2015]

5.52 Consider the following remarks pertaining to the irrotational flow:

1. The Laplace equation of stream function $\frac{\partial^2\psi}{\partial x^2}+\frac{\partial^2\psi}{\partial y^2}=0$ must be satisfied for the flow to be potential.
2. The Laplace equation for the velocity potential $\frac{\partial^2\phi}{\partial x^2}+\frac{\partial^2\phi}{\partial y^2}=0$ must be satisfied to fulfil the criterion of mass conservation i.e. continuity equation.

Which of the above statements is/are correct?

(a) 1 only (b) Both 1 and 2
(c) 2 only (d) Neither 1 nor 2
[ESE : 2015]

5.53 In a two dimensional incompressible fluid flow field, the stream function at a point $P(2, 1)$ is given by an expression $\psi = 2xy$. The value of velocity potential at P is

(a) 3 (b) 2.5
(c) 4 (d) 5 [ESE : 2015]

5.54 In a stream line steady flow, two points A and B on a stream line are 1 m apart and the flow velocity varies uniformly from 2 m/s to 5 m/s. What is the acceleration of fluid at B?

(a) 3 m/s^2 (b) 6 m/s^2
(c) 9 m/s^2 (d) 15 m/s^2
[ESE : 2015]

5.55 The stream function is given by $\psi = 3xy$, then the velocity at the point (2, 3) is

(a) 9 (b) – 6
(c) 117 (d) 10.8 [ESE : 2015]

5.56 A two dimensional velocity field is given by
$V = (x^2 - y^2 + x)\,i - (2xy - y)j$
The convective acceleration at $(x, y) = (1, 2)$ is

(a) 0 (b) 14 units
(c) 2 units (d) None of the above
[ESE : 2015]

5.57 For a steady two-dimensional flow, the scalar components of the velocity field are $V_x = -2x$; $V_y = 2y$ and $V_z = 0$. The corresponding components of acceleration a_x and a_y respectively are :

(a) 0 and $4y$ (b) $4x$ and 0
(c) 0 and 0 (d) $4x$ and $4y$
[ESE : 2016]

5.58 The velocity of flow from a tap of 12 mm diameter is 8 m/s. What is the diameter of the jet at 1.5 m from the tap when the flow is vertically upward? Assuming that, the jet continues to be circular upto that level.

(a) 44 mm (b) 34 mm
(c) 24 mm (d) 14 mm [ESE : 2016]

5.59 **Statement (I)** : Given a flow with velocity field $\vec{V}$, $\nabla \times \vec{V} = 0$, if the flow is incompressible.

Statement (II) : Given a flow with velocity field $\vec{V}$, $\nabla \cdot (\nabla \times \vec{V}) = 0$.

Codes:

(a) Both Statement (I) and Statement (II) are individually true and Statement (II) is the correct explanation of Statement (I)
(b) Both Statement (I) and Statement (II) are individually true but Statement (II) is ***NOT*** the correct explanation of Statement (I)
(c) Statement (I) is true but Statement (II) is false
(d) Statement (I) is false but Statement (II) is true [ESE : 2016]

5.60 Uniform flow occurs when :

(a) At every point the velocity vector is identical in magnitude and direction at any given instance
(b) The follow is steady
(c) Discharge through a pipe is constant
(d) Conditions do not change with at any time
[ESE : 2016]

5.61 A two-dimensional flow field is defined as $\vec{V} = \vec{i}x - \vec{j}y$. The equation of the streamline passing through the point (1, 2) is

(a) $xy + 2 = 0$ (b) $x^2 y + 2 = 0$
(c) $xy - 2 = 0$ (d) $x^2 y - 2 = 0$

[ESE : 2017]

5.62 The stream function for a flow field is $\psi = 3x^2y + (2 + t)y^2$. The velocity at a point P for position vector $r = i + 2j - 3k$ at time $t = 2$ will be
(a) $19i - 12j$ (b) $21i - 12j$
(c) $19i + 22j$ (d) $21i + 22j$

[ESE : 2019]

5.63 A flow field satisfying $\nabla \cdot \vec{V} = 0$ as the continuity equation represents always
(a) a steady compressible flow
(b) an incompressible flow
(c) an unsteady and incompressible flow
(d) an unsteady and compressible flow

[ESE : 2019]

5.64 A duct of rectangular cross-section 600 mm x 400 mm carries 90 m^3/min of air having density of 1.2 kg/m^3. When the quantity of air in both cases is same, the equivalent diameter of a circular duct will be nearly
(a) 0.86 m (b) 0.76 m
(c) 0.64 m (d) 0.54 m [ESE : 2020]

5.65 Which of the following characteristics regarding fluid kinematics is/are correct?
1. Streamline represents an imaginary curve in the flow field so that the tangent to the curve at any point represents the direction of instantaneous velocity at that point.
2. Path lines, streamlines and streak lines are identical in steady flow.

(a) 1 only (b) 2 only
(c) Both 1 and 2 (d) Neither 1 nor 2

[ESE : 2020]

5.66 A curve that is everywhere tangent to the instantaneous local velocity vector, is
(a) Streak line (b) Path line
(c) Normal line (d) Streamline

[ESE : 2020]

5.67 Water is flowing through a pipe of diameter 200 mm with a velocity of 3 m/s. What is the head loss due to friction for a length of 5 m if the coefficient of friction is given by $f = 0.02 + \frac{0.09}{\text{Re}^{0.3}}$, where Re is Reynolds number? [Take the kinematic viscosity of water as 0.01 stoke, $g = 9.81$ m/s^2 and $(6 \times 10^5)^{0.3} = 54.13$]
(a) 0.993 m of water (b) 0.783 m of water
(c) 0.685 m of water (d) 0.552 m of water

[ESE : 2021]

5.68 The following cases represent the two velocity components, $v = 2y^2$; $w = 2xyz$. What is the third component of velocity such that they satisfy the continuity equation?
(a) $-4xy - x^2y + f(y, z)$
(b) $-3xy - x^3y + f(y, z)$
(c) $-5xy - 2x^2y + f(y, z)$
(d) $-4xy - 3x^2y + f(y, z)$

[ESE : 2022]

5.69 Consider following statements regarding dynamics of fluid flow :
1. Kinetic energy correction factor is defined as the ratio of kinetic energy of the flow per second based on actual velocity to kinetic energy of the flow per second based on average velocity.
2. Kinetic energy correction factor is defined as the ratio of kinetic energy of the flow per second based on average velocity to kinetic energy of the flow per second based on maximum velocity.
3. Momentum correction factor is defined as the ratio of momentum of the flow per second based on actual velocity to momentum of the flow per second based on average velocity.
4. Momentum correction factor is defined as the ratio of momentum of the flow per second based on average velocity to momentum of the flow per second based on maximum velocity.

Which of the above statements are correct?
(a) 1 and 2 only (b) 1 and 3 only
(c) 2 and 3 only (d) 1, 2 and 3

[ESE : 2024]

Answers Fluid Kinematics

5.1 (b)	5.2 (d)	5.3 (b)	5.4 (a)	5.5 (d)	5.6 (d)	5.7 (b)	5.8 (b)	5.9 (d)
5.10 (c)	5.11 (a)	5.12 (b)	5.13 (c)	5.14 (a)	5.15 (a)	5.16 (c)	5.17 (d)	5.18 (b)
5.19 (a)	5.20 (b)	5.21 (a)	5.22 (b)	5.23 (c)	5.24 (c)	5.25 (c)	5.26 (d)	5.27 (c)
5.28 (*)	5.29 (b)	5.30 (b)	5.31 (a)	5.32 (a)	5.33 (b)	5.34 (a)	5.35 (a)	5.36 (b)
5.37 (d)	5.38 (b)	5.39 (d)	5.40 (c)	5.41 (b)	5.42 (c)	5.43 (d)	5.44 (b)	5.45 (d)
5.46 (c)	5.47 (a)	5.48 (b)	5.49 (a)	5.50 (a)	5.51 (c)	5.52 (b)	5.53 (a)	5.54 (d)
5.55 (d)	5.56 (d)	5.57 (d)	5.58 (d)	5.59 (d)	5.60 (a)	5.61 (c)	5.62 (a)	5.63 (b)
5.64 (d)	5.65 (c)	5.66 (d)	5.67 (a)	5.68 (a)	5.69 (b)			

Explanations Fluid Kinematics

5.1 (b)

$$\frac{d\psi}{dx} = 2x = -\frac{d\phi}{dy} \qquad \therefore \frac{d\phi}{dy} = -2x$$

and $$\frac{d\psi}{dy} = -2y = \frac{d\phi}{dx} \qquad \therefore \frac{d\phi}{dx} = -2y$$

Now, $\dfrac{d\phi}{dy} = -2x$

$$\therefore \quad \phi = -2xy + f(x)$$

Differentiating w. r. to x

$$\frac{d\phi}{dx} = -2y + f'(x) = -2y$$

$$\therefore f'(x) = 0$$

$$\therefore f(x) = \text{constant}$$

$$\therefore \phi = -2xy + \text{constant}$$

5.2 (d)

$$a_x = u\frac{du}{dx} + v\frac{du}{dy} + w\frac{du}{dz} + \frac{du}{dt}$$

where $\dfrac{du}{dt}$ = Local or temporal acceleration

Remaining terms are called "convective acceleration".

5.3 (b)

$$\frac{d\psi}{dx} = y = -\frac{d\phi}{dy}, \quad \frac{d\psi}{dy} = x = \frac{d\phi}{dx}$$

$$\therefore \frac{d\phi}{dx} = x \therefore \phi = \frac{x^2}{2} + f(y)$$

Now, $\dfrac{d\phi}{dy} = f'(y) = -y$

$$\therefore f(y) = -\frac{y^2}{2} \qquad \therefore \phi = \frac{x^2 - y^2}{2}$$

5.4 (a)

Correct option is (a) i.e.,

$$\psi = -2xy + \text{constant}$$

$$\phi = x^2 - y^2 + c$$

$$u = -\frac{\partial\phi}{\partial x} = -2x$$

$$v = -\frac{\partial\phi}{\partial y} = 2y$$

also $$u = \frac{\partial\psi}{\partial y} = -2x$$

$$v = -\frac{\partial\psi}{\partial x} = 2y$$

$$\therefore \quad \partial\psi = -2x\,dy$$

$$\psi = -2xy + c \qquad \text{...(i)}$$

Differentiating (i) w.r.t. x

$$\frac{\partial\psi}{dx} = -2y + \frac{\partial c}{\partial x} = -2y$$

$$\frac{\partial c}{\partial x} = 0$$

$\therefore$ c is a constant and a neither a function of x or y

$$\psi = -2xy + c$$

5.5 (d)

A stream line is a line on which tangent drawn at any point gives the direction of velocity.

5.6 (d)

A is false but R is true. Stream lines are drawn in the flow field such that at a given instant of time they are tangent to the direction of flow at every point in the flow field.

5.7 (b)

Steady/Unsteady flow: A flow is said to be steady flow, if the fluid properties do-not-vary with respect to time at any given section, otherwise the flow is unsteady.

Uniform/Non-uniform flow: A flow is said to be uniform flow, if the velocity remains constant at different section at any given instant of time. Otherwise the flow is non-uniform.

5.8 (b)

For possible irrotational flow

$$\omega_z = \frac{\partial^2 \psi}{dx^2} + \frac{\partial^2 \psi}{dy^2} = 0$$

$\psi = 2xy$ [Only satisfied this equation].

5.10 (c)

$$a_x = v\frac{du}{dy} + \frac{du}{dt} = (3t + 3x)3 + 2t = 49 \text{ units}$$

$$a_y = u\frac{dv}{dx} + \frac{dv}{dt} = \left(t^2 + 3y\right)3 + 3 = 33 \text{ units}$$

$$a = \sqrt{a_x^2 + a_y^2} = \sqrt{(49)^2 + (33)^2}$$

$$= \mathbf{59 \text{ units}}$$

5.11 (a)

$$\frac{\partial \psi}{\partial x} = v_y = 3x^2$$

$$\frac{\partial \psi}{\partial y} = -v_x = -3y^2 \quad \therefore v_x = 3y^2$$

$$V = \sqrt{V_x^2 + V_y^2} = \mathbf{4.24 \text{ units}}$$

5.12 (b)

ϕ valid only when;

$$\frac{\partial^2 \phi}{\partial x^2} + \frac{\partial^2 \phi}{\partial y^2} = 0 \qquad \text{...(1)}$$

1. $\phi = 2x + 3y$;

$$\frac{\partial^2 \phi}{\partial x^2} = 0, \quad \frac{\partial^2 \phi}{Vy^2} = 0 \text{ Equation (1) becomes zero.}$$

i.e. valid.

2. $4x^2 - 3y^2$

$$\frac{\partial^2 \phi}{\partial x^2} = 8, \quad \frac{\partial^2 \phi}{Vy^2} = -6$$

Equation (1) becomes $8 - 6 = 2 \neq 0$ i.e. Not valid

3. $\cos(x - y)$

$$\frac{\partial \phi}{\partial x} = -\sin(x - y),$$

$$\frac{\partial \phi}{\partial y} = -\sin(x - y)(-1) = \sin(x - y)$$

$$\frac{\partial^2 \phi}{\partial x^2} = -\cos(x - y),$$

$$\frac{\partial^2 \phi}{\partial y^2} = \cos(x - y)(-1) = -\cos(x - y)$$

Equation (1) becomes

$-\cos(x - y) - \cos(x - y) = -2\cos(x - y)$

i.e. not valid

4. $\phi = \tan^{-1}\left(\dfrac{x}{y}\right)$

$$\tan \phi = \frac{x}{y}$$

$$\frac{\partial \phi}{\partial x} = \frac{y}{x^2 + y^2}, \quad \frac{\partial \phi}{\partial y} = \frac{-x}{x^2 + y^2}$$

$$\frac{\partial^2 \phi}{\partial x^2} = y\frac{(-1)}{(x^2 + y^2)} \cdot 2x = \frac{-2xy}{(x^2 + y^2)^2}$$

$$\frac{\partial^2 \phi}{dy^2} = (-x)\frac{(-1)}{(x^2 + y^2)} \cdot 2y = \frac{2xy}{(x^2 + y^2)}$$

Equation (1) becomes

$$\frac{-2xy}{(x^2 + y^2)^2} + \frac{2xy}{(x^2 + y^2)} = 0 \text{ i.e. valid}$$

5.13 (c)

Mathematically, fluid flow in which the curl of velocity function i.e. vorticity is zero everywhere, so that the circulation of the velocity about any closed curve vanishes. Irrotational motion is also known as **acyclic motion**.

5.14 (a)

Stream function always satisfies continuity equations since in this case it satisfies Laplace equation therefore it is a case of irrotational flow.

5.15 (a)

For irrotational and incompressible flow. Laplace equation must satisfied for velocity potential and stream function.

5.16 (c)

For incompressible steady flow, Laplace equation must satisfied:

$$\frac{d^2\phi}{dx^2}+\frac{d^2\phi}{dy^2} = 0$$

$$\frac{d^2\phi}{dy^2} - Ae^x = 0$$

$$\therefore \quad \frac{d^2\phi}{dy^2} = Ae^x$$

$$\frac{d\phi}{dy} = Ae^x y + f(x)$$

$$-V = Ae^x.y + f(x)$$

$$\therefore \quad V = -Ae^x.y + f(x)$$

5.17 (d)

$$a_x = u\frac{du}{dx}+v\frac{du}{dy}+w\frac{du}{dz}+\frac{du}{dt}$$

$$= (-2x) \times (2) = 4x$$

$$a_y = u\frac{dv}{dx}+v\frac{dv}{dy}+w\frac{dv}{dz}+\frac{dv}{dt}$$

$$= 0 + 2y \times 2 + 0 + 0 = \mathbf{4y}$$

5.18 (b)

Path Line: It is the line traced by a single particle as it moves over a period of time.

5.19 (a)

For steady flow, path lines, streamlines and streaklines coincide with each other.

5.20 (b)

For three dimensional steady incompressible flow the continuity equation is

$$\frac{\partial u}{\partial x}+\frac{\partial v}{\partial y}+\frac{\partial w}{\partial z} = 0$$

or $$\frac{\partial}{\partial x}\left(-\frac{\partial \phi}{\partial x}\right)+\frac{\partial}{\partial y}\left(-\frac{\partial \phi}{\partial y}\right)+\frac{\partial}{\partial z}\left(-\frac{\partial \phi}{\partial z}\right) = 0$$

or $$\frac{\partial^2\phi}{\partial x^2}+\frac{\partial^2\phi}{\partial y^2}+\frac{\partial^2\phi}{\partial z^2} = 0$$

Which is nothing but $\Delta^2\phi = 0$. The "Laplace equation". Thus any function that satisfied Laplace equation is a possible case of fluid flow.

5.21 (a)

For irrotational flow

$$\omega_z = \frac{1}{2}\left(\frac{\partial v}{\partial x}-\frac{\partial u}{\partial y}\right) = \frac{1}{2}\left[\frac{\partial^2\psi}{\partial x^2}+\frac{\partial^2\psi}{\partial y^2}\right] = 0$$

Only equation $\psi = y^2 - x^2$ is satisfied the above equation

5.22 (b)

Velocity potential function: It is a function of space and time such that its negative derivative with respect to space gives velocity in that direction. Rotational component of it will be zero. Thus flow is irrotational i.e. ideal flow.

Stream function: It is the function of space and time such that this function is defined in such a manner that it satisfies the two dimensional continuity equation

The acceleration due to change of velocity with respect to time is known as temporal or local acceleration i.e. why for steady flow it value should be zero.

5.23 (c)

$$u = 3xy$$

$$v = \frac{3}{2}\left(x^2 - y^2\right)$$

Equation of streamline is given by

$$vdx = udy$$

$$\Rightarrow \quad \frac{dx}{dy} = \frac{u}{v} = \frac{3xy}{\frac{3}{2}\left(x^2 - y^2\right)}$$

$$\Rightarrow \quad \frac{dx}{dy} = \frac{2xy}{\left(x^2 - y^2\right)}$$

5.24 (c)

(A) $\omega_z = \frac{1}{2}\left(\frac{\partial v}{\partial x} - \frac{\partial u}{\partial y}\right)$

$= \frac{1}{2}\left(\frac{\partial}{\partial x}\left(-\frac{3}{2}x\right) - \frac{\partial}{\partial y}(y)\right)$

$= \frac{1}{2}\left(-\frac{3}{2} - 1\right) = \frac{1}{2} \times \left(-\frac{5}{2}\right) = -\frac{5}{4} \neq 0$

$\therefore$ Flow A is **rotational**

(B) $\omega_z = \frac{1}{2}\left(\frac{\partial v}{\partial x} - \frac{\partial u}{\partial y}\right)$

$= \frac{1}{2}\left(\frac{\partial}{\partial x}(x^2 y) - \frac{\partial}{\partial y}(xy^2)\right)$

$= \frac{1}{2}(2xy - 2xy) = 0$

$\therefore$ Flow B is **irrotational**

5.25 (c)

Convective acceleration along x direction at point (1, 2)

$$a_x = u\frac{\partial u}{\partial x} + v\frac{\partial u}{\partial y}$$

$= (2x^2 + y^2)(4x) + (-4xy)(2y)$

$= (2 + 4)(4) + (-4)(1)(2)(2)(2)$

$= 24 - 32 =$ **−8 unit**

5.26 (d)

$$u = -\frac{\partial \psi}{\partial y} = -\frac{\partial}{\partial y}(2xy) = -2x$$

$$v = -\frac{\partial \psi}{\partial x} = \frac{\partial}{\partial x}(2xy) = 2y$$

$u(2, 1) = -4$

$v(2, 1) = 2$

$\therefore$ Velocity at point (2, 1)

$= \sqrt{(-4)^2 + 2^2} = \sqrt{20} =$ **4.47 unit**

5.27 (c)

$\Rightarrow \quad \frac{du}{dx} + \frac{dv}{dy} = 0$

$\Rightarrow \quad \frac{d(Ae^x)}{dx} + \frac{dv}{dy} = 0$

$\Rightarrow \quad Ae^x + \frac{dv}{dy} = 0$

$\Rightarrow \quad v = -\boldsymbol{Ae^x y}$

5.28 (*)

All are correct.

5.30 (b)

$$\phi = \frac{-xy^3}{3} - x^2 + \frac{x^3 y}{3} + y^3$$

$$u = \frac{-\partial\phi}{\partial x} = -\left[\frac{\partial}{\partial x}\left(\frac{-xy^3}{3} - x^2 + \frac{x^3 y}{3} + y^2\right)\right]$$

$$u = \frac{y^3}{3} + 2x - x^2 y$$

5.31 (a)

$\phi = 2xy$

$u = -\frac{\partial \phi}{\partial x}$

$v = -\frac{\partial \phi}{\partial y}$

$u = -2y \qquad v = -2x$

$u = -\frac{\partial \phi}{\partial y} \qquad v = \frac{\partial \phi}{\partial x}$

$2y = \frac{\partial \phi}{\partial y} \qquad \frac{\partial \phi}{\partial x} = -2x$

$2y\,\partial y = \partial\phi$

$y^2 + f(x) = \psi$

$f'(x) = \frac{\partial \phi}{\partial x}$

$f'(x) = -2x$

$\therefore \quad f(x) = -x^2 + c$

$\phi = y^2 - x^2 + c'$

5.32 (a)

$u = 2x \quad v = -2y$

$-\frac{\partial \phi}{\partial x} = u$

$\frac{\partial \phi}{\partial x} = -2x$

$\phi = -x^2 + f(y)$

$-\frac{\partial \phi}{\partial y} = v, \quad \frac{\partial \phi}{\partial y} = -2y$

$\frac{\partial \phi}{\partial y} = f'(y)$

Now $\quad f'(y) = -2y$

$$f(y) = y^2 + c$$

$$\therefore \quad \phi = y^2 - x^2 + c$$

5.33 (b)

$$\psi = x^2 - y^2$$

$$u = \frac{\partial \psi}{\partial y} = \frac{\partial}{\partial y}\left[x^2 - y^2\right] = -2y$$

$$v = -\frac{\partial \psi}{\partial x} = -\frac{\partial}{\partial x}\left[x^2 - y^2\right] = -2x$$

Resultant velocity at (1, 1)

$$= \sqrt{u^2 + v^2} = \sqrt{(-2y)^2 + (-2x)^2}$$

$$= \sqrt{4y^2 + 4x^2} = \sqrt{4+4}$$

$$= 2\sqrt{2} \text{ units}$$

5.34 (a)

(1) $u = x^2 \cos y$

$v = -2x \sin y$

Continuity equation

$$= \frac{\partial u}{\partial x}\left(x^2 \cos y\right) + \frac{\partial}{\partial y}(-2x \sin y)$$

$(2x)\cos y - (2x)\cos y = 0$

hence satisfy the continuity equation.

(2) $u = x + 2$

$v = 1 - y$

Continuity equation

$$\frac{\partial u}{\partial x} + \frac{\partial y}{\partial y} = 0 = \frac{\partial}{\partial x}(x+2) + \frac{\partial}{\partial y}(1-y)$$

$$= 1 - 1 = 0$$

hence satisfy the continuity equation.

(3) $u = xyt$

$$v = x^3 - y^2 \frac{t}{2}$$

Continuity equation

$$\frac{\partial u}{\partial x} + \frac{\partial v}{\partial y} = 0 = \frac{\partial}{\partial x}(xyt) + \frac{\partial}{\partial y}\left(x^3 - \frac{y^2 t}{2}\right)$$

$$= yt - yt = 0$$

(4) $u = \ln(x + y)$

$$v = xy - \frac{y}{x}$$

Continuity equation

$$\frac{\partial u}{\partial x} + \frac{\partial v}{\partial y} = 0$$

$$= \frac{\partial}{\partial x}\ln(x+y) + \frac{\partial}{\partial y}\left(xy - \frac{y}{x}\right)$$

$$= \frac{1}{x+y} + x - y \neq 0$$

Hence, does not satisfy continuity equation.

5.35 (a)

$$V = \left(y^2 + z^2\right)\hat{i} + \left(x^2 + z^2\right)\hat{j} + \left(x^2 + y^2\right)\hat{k}$$

Rotational component in x direction

$$\omega_x = \frac{1}{2}\left(\frac{\partial w}{\partial y} - \frac{\partial v}{\partial z}\right)$$

$$= \frac{1}{2}\left[\frac{\partial}{\partial y}\left(x^2 + y^2\right) - \frac{\partial}{\partial z}\left(x^2 + z^2\right)\right]$$

$$= \frac{1}{2}[2y - 2z] = y - z$$

At (1, 2, 3)

$$\omega_x = 2 - 3 = -1 \text{ rad/s}$$

Rotational component is y direction

$$\omega_y = \frac{1}{2}\left(\frac{\partial u}{\partial z} - \frac{\partial w}{\partial x}\right)$$

$$= \frac{1}{2}\left[\frac{\partial}{\partial z}\left(y^2 + z^2\right) - \frac{\partial}{\partial x}\left(x^2 + y^2\right)\right]$$

$$= \frac{1}{2}[2z - 2x] = z - x$$

At (1, 2, 3)

$$\omega_y = 3 - 1 = 2 \text{ rad/s}$$

Rotational component is z direction

$$\omega_z = \frac{1}{2}\left(\frac{\partial v}{\partial x} - \frac{\partial u}{\partial y}\right)$$

$$= \frac{1}{2}\left[\frac{\partial}{\partial x}\left(x^2 + z^2\right) - \frac{\partial}{\partial y}\left(y^2 + z^2\right)\right]$$

$$= \frac{1}{2}[2x - 2y] = x - y$$

At (1, 2, 3)

$$\omega_z = 1 - 2 = -1 \text{ rad/s}$$

5.37 (d)

If flow condition satisfies Laplace equation i.e.

$$\left(\frac{\partial^2 \phi}{\partial x^2} + \frac{\partial^2 \phi}{\partial y^2} = 0\right) \;\&\; \left(\frac{\partial^2 \psi}{\partial x^2} + \frac{\partial^2 \psi}{\partial y^2} = 0\right)$$ than it has

to satisfy continuity equation the flow is irrotational.

5.38 (b)

$$\psi = 2xy$$

$$\frac{\partial \psi}{\partial x} = 2y = v$$

$$\frac{\partial \psi}{\partial y} = -2x = -u$$

Rotational component along z-axis

$$\omega_z = \frac{1}{2}\left(\frac{\partial v}{\partial x} - \frac{\partial u}{\partial y}\right)$$

$$\therefore \quad \frac{\partial v}{\partial x} = 0; \quad \frac{\partial u}{\partial y} = 0$$

$\therefore \quad \omega_z = 0$ means irrotational.

5.39 (d)

Statement 2 is wrong because potential function exist for irrotational flow only and stream function implies that flow may be rotational or irrotational.

5.40 (c)

For irrotational motion, the circulation around any path is zero because circulation is the product of vorticity and area. Vorticity is twice of rotation. Hence for an irrotational motion, vorticity is zero and hence circulation is zero.

5.41 (b)

As the fluid velocity and flow will change with time the flow will be non-steady.

5.42 (c)

For velocity potential function to exist the flow must be irrotational since the continuity equation has to be satisfied.

5.44 (b)

$$u = -\frac{\partial \psi}{\partial y} \qquad (\partial \psi = -u dy)$$

$$\psi = -uy + f(x)$$

$$v = \frac{\partial \psi}{\partial x} = 0 + f'(x) \qquad \therefore f'(x) = a$$

$$f(x) = ax$$

So, $\psi = -uy + ax$

$$\psi|_{(2,3)} = -3a + 2a = -a$$

$$\therefore \quad 0 - a = -ay + ax$$

Stream line equation is

$$x - y = -1$$

i.e., $\mathbf{y = x + 1}$

5.45 (d)

Velocity potential function: It is defined as a scalar function of space and time such that its negative derivative w.r.t. any direction gives the fluid velocity in the direction.

Stream function: It is defined as the scalar function of space and time, such that its derivative w.r.t. and direction gives the velocity component at right angles to that direction.

From the above definition , statement (1) is wrong. All other statements except (1) is correct.

5.46 (c)

$$\phi = x^2 - y^2$$

$$u = -\frac{\partial \phi}{\partial x} = -2x, \qquad u|_{(1,1)} = -2$$

$$v = -\frac{\partial \phi}{\partial y} = +2y, \qquad v|_{(1,1)} = 2$$

Magnitude,

$$V = \sqrt{u^2 + v^2} = \sqrt{(-2)^2 + (2)^2}$$

$$= \sqrt{8} = 2\sqrt{2}$$

5.47 (a)

Equation of streamline: $\dfrac{dx}{u} = \dfrac{dy}{v}$

Here, $u = x$, $v = -y$

$$\therefore \quad \frac{dx}{x} = -\frac{dy}{y}$$

$$\Rightarrow \quad \log_e x + \log_e y = c$$

$$\therefore \quad xy = e^c$$

At (1, 1)

$$\Rightarrow \quad c = 0 \text{ or } \mathbf{xy - 1 = 0}$$

5.49 (a)

(i) Velocity potential exists 2D or 3D flows.

(ii) Velocity potential exists for irrotational flow.

5.50 (a)

$\therefore$ For Incompressible flow,

$$\frac{\partial u}{\partial x} + \frac{\partial v}{\partial y} + \frac{\partial w}{\partial z} = 0$$

$\therefore$ Given, $u = 3xy^2 + 2x + y^2$

$$v = x^2 - 2y - y^3$$

$$\therefore \frac{\partial}{\partial x}(3xy^2 + 2x + y^2) + \frac{\partial}{\partial y}(x^2 - 2y - y^3) = 0$$

$$3y^2 + 2 + 0 + 0 - 2 - 3y^2 = 0$$

$\therefore$ There is no time component is the given expression, so it is steady.

5.51 (c)

$$a_x = u\frac{\partial u}{\partial x} + v\frac{\partial u}{\partial y} + w\frac{\partial u}{\partial z}$$

$$\therefore \quad u = 2x^2 + y^2,$$

$$v = -4xy,\ w = 0$$

$$a_x = (2x^2 + y^2)(4x + 0) - 4xy(0 + 2y)$$

$$4x(2x^2 + y^2) - 8xy^2$$

$$\Rightarrow a_{x(1,2)} = 4 \times 1(2 \times 1^2 + 2^2) - 8 \times 1 \times 2^2$$

$$a_{x(1,2)} = \mathbf{-8\ m/s^2}$$

5.53 (a)

$$\Psi = 2xy$$

$$\therefore \quad v = \frac{-d\Psi}{\partial x}$$

$$v = -2y$$

but, $$v = \frac{\partial \phi}{\partial y}$$

$$\therefore \quad \frac{\partial \phi}{\partial y} = -2y$$

On Integrating it,

$$\phi = -y^2 + f(x)$$

$$\frac{\partial \phi}{\partial x} = 0 + f'(x)$$

$$\therefore \quad 4 = f'(x)$$

$$\Rightarrow \quad \frac{\partial \Psi}{\partial y} = f'(x)$$

$$f'(x) = 2x$$

$$f(x) = x^2$$

$$\therefore \quad \phi = -y^2 + x^2$$

$$\therefore \quad \phi_{(2,1)} = -1 + 2^2 = \mathbf{3}$$

5.54 (d)

$$a_B = u_B \cdot \frac{\Delta U}{\Delta x} = 5 \times \frac{(5-2)}{1} = 5 \times 3 = \mathbf{15\ m/s^2}$$

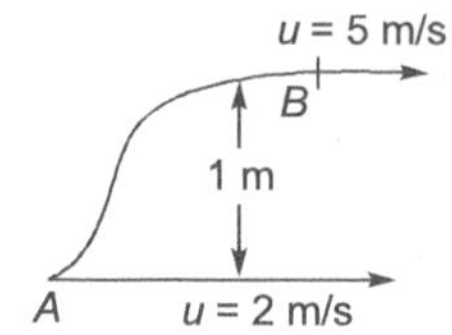

5.55 (d)

$$\Psi = 3xy$$

$$u = \frac{-\partial \Psi}{\partial y} = -3x$$

$$= -3 \times 2 = -6\ \text{m/s}$$

$$\therefore \quad v = \frac{\partial \Psi}{\partial x} = 3y = 9\ \text{m/s}$$

$$V = \sqrt{u^2 + v^2}$$

$$= \sqrt{6^2 + 9^2} = \sqrt{36 + 81}$$

$$= \sqrt{117} = \mathbf{10.817\ m/s}$$

5.56 (d)

$$V = (x^2 - y^2 + x)\hat{i}(2xy^2 - y)\hat{j}$$

$\therefore$ Convective acceleration at $(x, y) = (1, 2)$

$$\therefore a_x = u\frac{\partial u}{\partial x} + v\frac{\partial u}{\partial y}$$

$$\therefore a_x = (x^2 - y^2 + x)(2x - 0 + 1) - (2xy - y)(-2y)$$

$$\therefore a_x = (2x + 1)(x^2 - y^2 + x) + 2y(2xy - y)$$

At point (1, 2)

$$a_x = (2 + 1)(1 - 4 + 1) + 2 \times 2(2 \times 1 \times 2 - 2)$$

$$\therefore \quad a_x = 2 \text{ units}$$

$$a_y = u\frac{\partial v}{\partial y} + v\frac{\partial v}{\partial y}$$

$$a_y = -(x^2 - y^2 + x)(2y - 0) + (2xy - y)(2x - 1)$$

$$\therefore \quad a_{y(1,2)} = 10 \text{ units}$$

$$\therefore \quad a = \sqrt{a_x^2 + a_y^2} = \sqrt{2^2 + 10^2}$$

$$= \mathbf{10.2\ units}$$

$\therefore$ Answer is (d).

5.57 (d)

$$a_x = u\frac{\partial u}{\partial x} + v\frac{du}{dy}$$

$$a_x = (-2x) \times -2 + (2y) \times 0$$

$$a_x = 4x \text{ m/s}^2$$

$$a_y = u\frac{\partial v}{\partial x} + v\frac{\partial v}{\partial y}$$

$$a_y = (-2x) \times 0 + (2y) \times 2$$

$$a_y = \mathbf{4y\ m/s^2}$$

5.58 (d)

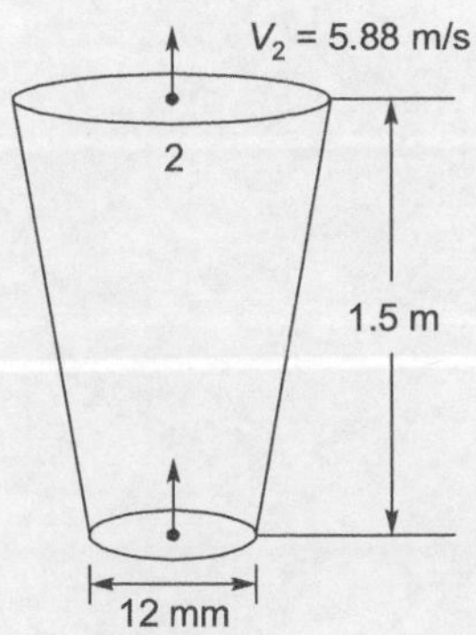

$$v^2 - u^2 = 2as$$

$$v^2 - 64 = -2 \times 9.81 \times 1.5$$

$$v = 5.88 \text{ m/s}$$

$$A_1v_1 = A_2v_2$$

$$(12)^2 \times 8 = d_2^2 \times 5.88$$

$$d_2 = \mathbf{14\ mm}$$

5.61 (c)

$$\vec{V} = \vec{i}x - \vec{j}y$$

$$u = x; \qquad v = -y$$

$$\frac{dx}{u} = \frac{dy}{V}$$

$$\int\frac{dx}{x} = \int\frac{dy}{-y}$$

$$\ln x = -\ln y + \ln c$$

$$xy = c$$

At point (1, 2) $\Rightarrow 1 \times 2 = c$

$$c = 2$$

$$\therefore \qquad xy = 2$$

$$xy - 2 = 0$$

5.62 (a)

$$\psi = 3x^2y + (2 + t)y^2$$

$$r = i + 2j - 3\hat{k}$$

At $\quad t = 2$

$$\psi = 3x^2y + 4y^2$$

$$u = +\frac{\partial\psi}{\partial y} = 3x^2 + 8y$$

$$= 3(1)^2 + 8(2) = 19$$

$$v = -\frac{\partial\psi}{\partial x} = -6xy$$

$$= -6 \times 1 \times 2 = -12$$

$$\vec{V} = u\hat{i} + v\hat{j}$$

$$\vec{V} = 19\hat{i} - 12\hat{j}$$

5.64 (d)

A_1 = Cross-section area of rectangle

A_2 = Cross-section area of circle

$$A_1 = A_2$$

$$0.6 \times 0.4 = \frac{\pi}{4} \times D^2$$

$$D = 0.55 \text{ m}$$

5.66 (d)

Tangent to streamline defines velocity vector.

5.67 (a)

D = 200 mm; $\quad \nu$ = 0.1 stokes

V = 3 m/s; $\quad g$ = 9.81

L = 5 m $\quad (6 \times 105)^{0.3} = 54.13$

Friction coefficient,

$$f = 0.02 + \frac{0.09}{(\text{Re})^{0.3}} = 0.02166$$

$$\text{Re} = \frac{\rho VD}{\mu} = \frac{10^3 \times 3 \times 0.2}{0.1 \times 10^{-4} \times 10^3} = 6 \times 10^5$$

$$h_L = \frac{4fLV^2}{2gD} = \frac{4 \times 0.02166 \times 5 \times (3)^2}{2 \times 9.81 \times 0.2}$$

$$= 0.993 \text{ m of water}$$

5.68 (a)

$$\nabla V = 0 \text{ for continuity equation}$$

$$\frac{\partial u}{\partial x}+\frac{\partial v}{\partial y}+\frac{\partial w}{\partial z} = 0$$

$$\frac{\partial u}{\partial x}+\frac{\partial}{\partial y}\{2y^2\}+\frac{\partial}{\partial z}\{2xyz\} = 0$$

$$\frac{\partial u}{\partial x}+4y+2x = 0$$

$$\frac{\partial u}{\partial x} = -2xy - 4y$$

On integrating,

$$u = -x^2y - 4xy + f(y, z)$$

5.69 (b)

Kinetic energy correction factor is given as,

$$\alpha = \frac{(\text{Kinetic energy of the flow per second})_{actual}}{(\text{Kinetic energy of the flow per second})_{average}}$$

Momentum correction factor is given as,

$$\beta = \frac{(\text{Momentum of the flow per second})_{actual}}{(\text{Momentum of the flow per second})_{average}}$$

■■■■

6 Fluid Dynamics and Flow Measurement

6.1 A glass tube with a 90° bend is open at both the ends. It is inserted into a flowing stream of oil, $S = 0.90$, so that one opening is directed upstream and the other is directed upward. Oil inside the tube is 50 mm higher than the surface of flowing oil. The velocity measured by the tube is, nearly,

(a) 0.89 m/s (b) 0.99 m/s
(c) 1.40 m/s (d) 1.90 m/s

[ESE : 2001]

6.2 Which of the following assumptions are made for deriving Bernoulli's equation?

1. Flow is steady and incompressible
2. Flow is unsteady and compressible
3. Effect of friction is neglected and flow is along a stream line
4. Effect of friction is taken into consideration and flow is along a stream line

Select the correct answer using the codes given below:

(a) 1 and 3 (b) 2 and 3
(c) 1 and 4 (d) 2 and 4

[ESE : 2002]

6.3 While measuring the velocity of air ($\rho = 1.2$ kg/m^3), the difference in the stagnation and static pressure of a Pitot static tube was found to be 380 Pa. The velocity at that location is

(a) 24.03 m/s (b) 4.02 m/s
(c) 17.8 m/s (d) 25.17 m/s

[ESE : 2002]

6.4 The expression

$$\frac{\partial \phi}{\partial t} + \int \frac{\partial p}{\rho} + \frac{1}{2}[\nabla \phi]^2 + gz = \text{constant}$$

represents

(a) steady flow energy equation
(b) unsteady irrotational Bernoulli's equation
(c) steady rotational Bernoulli's equation
(d) unsteady rotational Bernoulli's equation

[ESE : 2003]

6.5 An orifice meter with $C_d = 0.61$ is substituted by venturimeter with $C_d = 0.98$ in a pipeline carrying crude oil, having the same throat diameter as that of the orifice. For the same flow rate, the ratio of the pressure drops for the venturimeter and the orifice meter is

(a) $\frac{0.61}{0.98}$ (b) $\left(\frac{0.61}{0.98}\right)^2$
(c) $\frac{0.98}{0.61}$ (d) $\left(\frac{0.98}{0.61}\right)^2$

[ESE : 2003]

6.6 The instrument preferred in the measurement of highly fluctuating velocities in air flows is

(a) Pitot static tube
(b) propeller type anemometer
(c) three cup anemometer
(d) hot wire anemometer [ESE : 2003]

6.7 **Assertion (A):** After the fluid has re-established its flow pattern downstream of an orifice plate, it will return to same pressure that it had upstream of the orifice plate.

Reason (R): Bernoulli's equation when applied between two points having the same elevation and same velocity gives the same pressure at these points.

(a) Both **A** and **R** are individually true and **R** is the correct explanation of **A**
(b) Both **A** and **R** are individually true but **R** is not the correct explanation of **A**
(c) **A** is true but **R** is false
(d) **A** is false but **R** is true [ESE : 2003]

6.8 A constant-head water tank has, on one of its vertical sides two identical small orifices issuing two horizontal jets in the same vertical plane. The vertical distance between the centres of orifices is 1.5 m and the jet trajectories intersect at a point 0.5 m below the lower orifice. What is the

approximate height of water level in the tank above the point of intersection of trajectories?
(a) 1.0 m (b) 2.5 m
(c) 0.5 m (d) 2.0 m [ESE : 2004]

6.9 A Pitot static tube (C = 1) is used to measure air flow. With water in the differential manometer and a gauge difference of 75 mm, what is the value of air speed if ρ = 1.16 kg/m^3?
(a) 1.21 m/s (b) 16.2 m/s
(c) 35.6 m/s (d) 71.2 m/s
[ESE : 2004]

6.10 Match **List-I** (Measuring devices) with **List-II** (Measured parameter) and select the correct answer using the codes given below the lists:

List-I
A. Pitot tube
B. Micro-manometer
C. Pipe bend meter
D. Wall pressure tap

List-II
1. Flow static pressure
2. Rate of flow (indirect)
3. Differential pressure
4. Flow stagnation pressure

Codes:

	A	B	C	D
(a)	1	3	2	4
(b)	4	3	2	1
(c)	1	2	3	4
(d)	4	2	3	1

[ESE : 2004]

6.11 The Euler's equation of motion is a statement of
(a) energy balance
(b) conservation of momentum for an inviscid fluid
(c) conservation of momentum for an incompressible flow
(d) conservation of momentum for a real fluid
[ESE : 2005]

6.12 A venturimeter in an oil (sp. gr. 0.8) pipe is connected to a differential manometer in which the gauge liquid is mercury (sp. gr. 13.6). For a flow rate of 0.16 m^3/s, the manometer registers a gauge differential of 20 cm. The oil-mercury manometer being unavailable, an air-oil differential manometer is connected to the same venturimeter. Neglecting variation of discharge coefficient for the venturimeter, what is the new gauge differential for a flow rate of 0.08 m^3/s?
(a) 64 cm (b) 68 cm
(c) 80 cm (d) 85 cm [ESE : 2006]

6.13 How are the velocity coefficient C_v, the discharge coefficient C_d, and the contraction coefficient C_c of an orifice related?
(a) $C_v = C_c C_d$ (b) $C_c = C_v C_d$
(c) $C_d = C_c C_v$ (d) $C_c C_v C_d = 1$
[ESE : 2006]

6.14 While using boundary layer equations, Bernoulli's equation
(a) can be used anywhere
(b) can be used only outside the boundary layer
(c) can be used only inside the boundary layer
(d) cannot be used either inside or outside the boundary layer [ESE : 2006]

6.15 It is recommended that the diffuser angle should be kept less than 18° because
(a) pressure decreases in the flow direction and flow separation may occur
(b) pressure decreases in the flow direction and flow may become turbulent
(c) pressure increases in the flow direction and flow separation may occur
(d) pressure increases in the flow direction and flow may become turbulent [ESE : 2006]

6.16 Which one of the following is measured by a rotameter?
(a) Velocity of fluids
(b) Discharge of fluids
(c) Viscosity of fluids
(d) Rotational speed of solid shafts
[ESE : 2006]

6.17 Match **List-I** (Measuring instrument) with **List-II** (Variable to be measured) and select the correct answer using the codes given below the lists:

List-I
A. Hot wire anemometer
B. Pitot tube
C. V-notch weir
D. Tachometer

List-II
1. Discharge
2. Rotational speed

3. Velocity fluctuations
4. Stagnation pressure

Codes:

	A	B	C	D
(a)	4	3	2	1
(b)	3	4	2	1
(c)	4	3	1	2
(d)	3	4	1	2

[ESE : 2007]

6.18 An orifice meter, having an orifice of a diameter '*d*' is fitted in a pipe of diameter *D*. For this orifice meter, what is the coefficient of discharge C_d?

(a) A function of Reynolds number only
(b) A function of *d*/*D* only
(c) A function of *d*/*D* and Reynolds number
(d) Independent of *d*/*D* and Reynolds number

[ESE : 2007]

6.19 The velocity of a water stream is being measured by a L-shaped Pitot tube and the reading is 20 cm. Then, what is the approximate value of velocity?

(a) 19.6 m/s (b) 2.0 m/s
(c) 9.8 m/s (d) 20 m/s

[ESE : 2007]

6.20 Which one of the following conditions will linearize the Navier-Stokes equations to make it amenable for analytical solutions?

(a) Low Reynolds number (Re << 1)
(b) High Reynolds number (Re >> 1)
(c) Low Mach number (M << 1)
(d) High Mach number (M >> 1) [ESE : 2007]

6.21 When is Bernoulli's equation applicable between any two points in a flow field?

(a) The flow is steady, incompressible and rotational
(b) The flow is steady, compressible and irrotational
(c) The flow is unsteady, incompressible and irrotational
(d) The flow is steady, incompressible and irrotational [ESE : 2009]

6.22 In a submerged orifice flow, the discharge is proportional to which one of the following parameters?

(a) Square root of the downstream head
(b) Square root of the upstream head
(c) Square of the upstream head
(d) Square root of the difference between upstream and downstream heads

[ESE : 2009]

6.23 Consider the following statements regarding Bernoulli's theorem for fluid flow:

1. Conservation of energy
2. Steady flow
3. Viscous flow
4. Incompressible flow

Which of these statements is/are correct?

(a) 1, 2 and 4 only (b) 1 only
(c) 2, 3 and 4 only (d) 1, 2, 3 and 4

[ESE : 2010]

6.24 Assertion (A): A Pitot tube works on the principle of converting kinetic energy into potential energy.
Reason (R): The oncoming upstream velocity of fluid decelerates and comes to rest at the stagnation point at the tip of the Pitot tube.

(a) Both **A** and **R** are individually true and **R** is the correct explanation of **A**
(b) Both **A** and **R** are individually true but **R** is not the correct explanation of **A**
(c) **A** is true but **R** is false
(d) **A** is false but **R** is true

[ESE : 2010]

6.25 The condition in which Bernoulli's equation applies are

1. The flow must be inviscid.
2. The fluid may be compressible or incompressible.
3. The flow must be steady.
4. There should be only one stream line in the flow.

(a) 1, 2, 3 and 4 (b) 1, 2 and 3 only
(c) 1, 3 and 4 only (d) 2, 3 and 4 only

[ESE : 2011]

6.26 Match **List-I** with **List-II** and select the correct answer using the code given below the Lists:

	List-I		List-II
A.	Rotameter	1.	Vena contracta
B.	Venturimeter	2.	Tapering tube
C.	Orifice-meter	3.	Convergent divergent
D.	Flow nozzle	4.	Bell mouth entry

Codes:

	A	B	C	D
(a)	4	3	1	2
(b)	2	3	1	4

(c) 4 1 3 2
(d) 2 1 3 4 [ESE : 2011]

6.27 Which one of the following instruments is a rate meter?
(a) Disk meter
(b) Hot wire anemometer
(c) Pitot tube
(d) Venturimeter [ESE : 2011]

6.28 Navier-Stokes equations are useful in the analysis of
(a) turbulent flows (b) vortex flows
(c) viscous flows (d) rotational flows
[ESE : 2011]

6.29 The measurement of flow rate in a pipe is done by a combination of venturimeter and
(a) V-notch (b) pitot tube
(c) orifice-meter (d) manometer
[ESE : 2011]

6.30 **Assertion (A) :** After the fluid downstream of the orifice plate has reestablished, it will return to the same pressure that it had upstream of the orifice plate.
Reason (R) : According to Bernoulli's theorem with usual assumptions, for flow between two points, the sum of kinetic, potential and pressure energies remain constant.
(a) Both **A** and **R** are individually true and **R** is the correct explanation of **A**
(b) Both **A** and **R** are individually true but **R** is not the correct explanation of **A**
(c) **A** is true but **R** is false
(d) **A** is false but **R** is true [ESE : 2011]

6.31 The Lagrangian description of fluid motion is analogous to
(a) control volume analysis
(b) transient analysis
(c) system analysis
(d) None of the above
[ESE : 2012]

6.32 Bernoulli's equation is applicable between any two points in
(a) rotational flow of an incompressible fluid
(b) irrotational flow of compressible or incompressible fluid
(c) steady, rotational flow of an incompressible fluid
(d) steady, irrotational flow of an incompressible fluid [ESE : 2012]

6.33 A venturi meter is a device based on the Bernoulli principle and is used for measuring
(a) velocity (b) pressure drop
(c) total head (d) flow rate
[ESE : 2012]

6.34 The velocity at the exit of the pipe as shown in the below figure will be

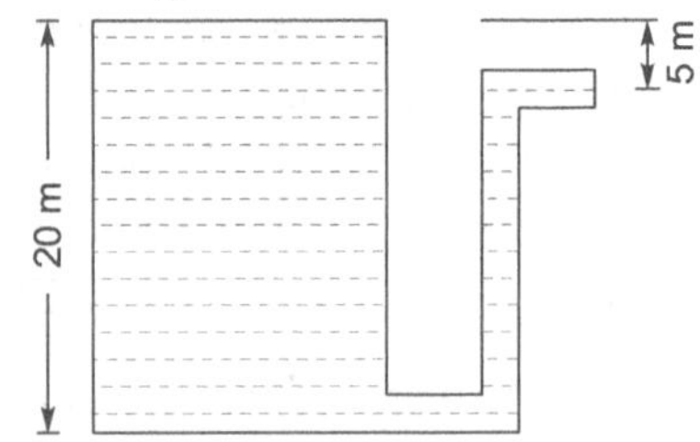

(a) 9.9 m/s (b) 14.0 m/s
(c) 17.1 m/s (d) 19.8 m/s
[ESE : 2013]

6.35 Bernoulli's equation is applicable between any two points located in
(a) rotational flow of an incompressible fluid
(b) irrotational flow of compressible or incompressible fluid
(c) steady, rotational flow of an incompressible fluid
(d) steady, irrotational flow of an incompressible fluid [ESE : 2014]

6.36 A Newtonian fluid is one which
(a) is viscous but incompressible
(b) has a linear relationship between the shear stress and the rate of angular deflection
(c) exhibits an increase in viscosity with increasing rate of deformation
(d) exhibits a decreases in viscosity with increasing rate of deformation
[ESE : 2014]

6.37 Angle of diverging portion of the venturimeter is limited to 7°, because
1. Flow decelerates in the diverging portion and pressure increases in the downstream direction. Hence, the fluid experience an adverse pressure gradient, if the divergence angle is large.

2. Flow separation takes place due to adverse pressure gradient when divergence angle is large.
3. If the divergence angle is large, a negative pressure is created at the throat which obstructs the flow of fluid.

Which of the above reasons are correct?

(a) 1, 2 and 3 (b) 1 and 2 only
(c) 1 and 3 only (d) 2 and 3 only

[ESE : 2014]

6.38

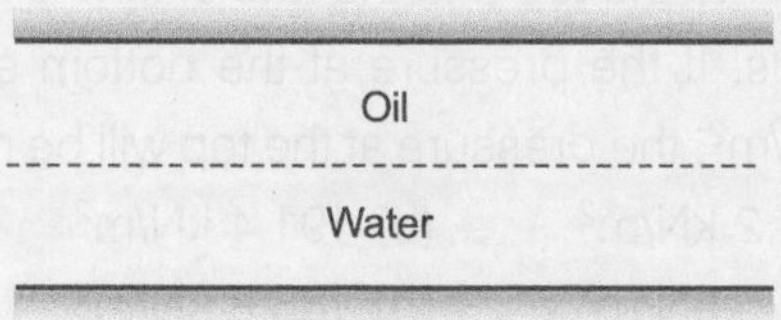

Consider flow of oil and water through a channel; the boundary conditions at the interface are

(a) velocity and shear stress are continuous
(b) shear stress is continuous and velocity is discontinuous
(c) shear stress is zero and velocity is continuous
(d) shear stress is zero

[ESE : 2014]

6.39 The head loss in a sudden expansion from 8 cm diameter pipe to 16 cm diameter pipe in terms of velocity V_1 in the smaller pipe is

(a) $\frac{1}{4}\left(\frac{V_1^2}{2g}\right)$ (b) $\frac{3}{16}\left(\frac{V_1^2}{2g}\right)$

(c) $\frac{1}{64}\left(\frac{V_1^2}{2g}\right)$ (d) $\frac{9}{16}\left(\frac{V_1^2}{2g}\right)$

[ESE : 2015]

6.40 A conical diffuser 3 m long is placed vertically. The velocity at the top (entry) is 4 m/s and at the lower end is 2 m/s. The pressure head at the top is 2 m of the oil flowing through the diffuser. The head loss in the diffuser is 0.4 m of the oil. The pressure head at the exit is :

(a) 3.18 m of oil (b) 5.21 m of oil
(c) 7.18 m of oil (d) 9.21 m of oil

[ESE : 2016]

6.41 Bernoulli's equation $\frac{p}{\rho}+\frac{v^2}{2}+gZ$ = Constant, is valid for

1. Steady flow
2. Viscous flow
3. Incompressible flow
4. Flow along a streamline

Which of the above are correct ?

(a) 1, 2 and 3 (b) 1, 2 and 4
(c) 1, 3 and 4 (d) 2, 3 and 4

[ESE : 2016]

6.42 A fluid jet is discharging from a 100 mm nozzle and the vena contracta formed has a diameter of 90 mm. If the coefficient of velocity is 0.98, then the coefficient of discharge for the nozzle is :

(a) 0.673 (b) 0.794
(c) 0.872 (d) 0.971

[ESE : 2016]

6.43 **Statement (I) :** In a venturimeter, the divergent section is much longer as compared to the convergent section.

Statement (II) : Flow separation occurs only in the diverging section of the venturimeter.

(a) Both Statement (I) and Statement (II) are individually true and Statement (II) is the correct explanation of Statement (I)
(b) Both Statement (I) and Statement (II) are individually true but Statement (II) is **NOT** the correct explanation of Statement (I)
(c) Statement (I) is true but Statement (II) is false
(d) Statement (I) is false but Statement (II) is true

[ESE : 2016]

6.44 Water is coming out from a tap and falls vertically downwards. At the tap opening, the stream diameter is 20 mm with uniform velocity of 2 m/s. Assuming steady inviscid flow, constant pressure atmosphere everywhere, and neglecting curvature and surface tension effects, the diameter of the stream 0.5 m below the tap opening is nearly

(a) 11.7 mm (b) 14.6 mm
(c) 17.5 mm (d) 20.4 mm

[ESE : 2017]

6.45 Consider the following statements regarding Bernoulli's equation :

1. It is assumed that no energy has been supplied.
2. The velocity of a steady stream of fluid flow will depend on the cross-sectional area of the stream.

3. Consider two sections 1 and 2 along a flow stream. In this reach, if q is work done by a pump, w is work absorbed by turbine, ρ is density of water and g is acceleration of gravity, with p, v and z carrying standard meanings, Bernoulli's equation will read

$$\frac{p_1}{\rho}+\frac{v_1^2}{2g}+z_1=\frac{p_2}{\rho}+\frac{v_2^2}{2g}+z_2+w+q$$

Which of the above statements are correct?
(a) 1 and 2 only (b) 1 and 3 only
(c) 2 and 3 (d) 1, 2 and 3
[ESE : 2017]

6.46 A jet of water issues from a sharp-edged vertical orifice under a constant head of 0.51 m. At a certain point of the issuing jet, the horizontal and vertical coordinates measured from vena contracta are 0.406 m and 0.085 m, respectively. What is the value of the coefficient of velocity?
(a) 0.975 (b) 0.925
(c) 0.875 (d) 0.825
[ESE : 2017]

6.47 The total energy of each particle at various places in the case of a perfect incompressible fluid flowing in a continuous stream
(a) keeps on increasing
(b) keeps on decreasing
(c) remains constant
(d) may increase or decrease [ESE : 2018]

6.48 A 40 mm diameter water jet strikes a hinged vertical plate of 800 N weight normally at its surface at its centre of gravity as shown in the figure below:

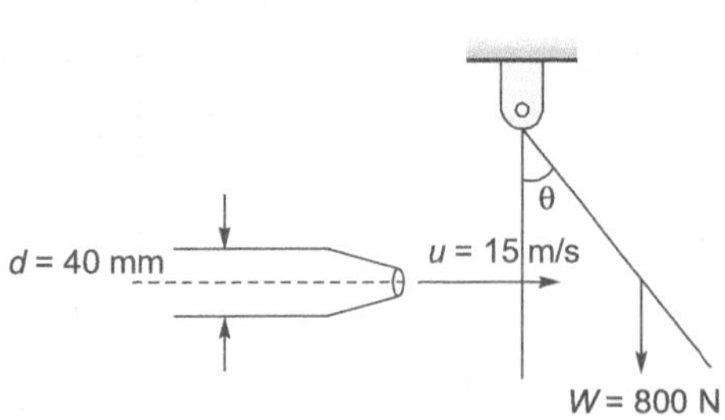

The angle of deflection is nearly
(a) $\sin^{-1} 0.353$ (b) $\sin^{-1} 0.321$
(c) $\tan^{-1} 0.353$ (d) $\tan^{-1} 0.321$
[ESE : 2018]

6.49 A nozzle at the end of an 80 mm hosepipe produces a jet 40 mm in diameter. When it is discharging the water 1200 Lpm, the force on the joint at the base of the nozzle will be
(a) 180 N (b) 200 N
(c) 220 N (d) 240 N
[ESE : 2019]

6.50 A vertical water pipe, 1.5 m long, tapers from 75 mm diameter at the bottom to 150 mm diameter at the top and the rate of flow is 50 L/s upwards. If the pressure at the bottom end is 150 kN/m^2, the pressure at the top will be nearly
(a) 195.2 kN/m^2 (b) 191.4 kN/m^2
(c) 187.6 kN/m^2 (d) 183.8 kN/m^2
[ESE : 2019]

6.51 Which of the following applications regarding Navier-Stokes equations are correct ?
1. Laminar unidirectional flow between stationary parallel plates.
2. Laminar unidirectional flow between parallel plates having no relative motion.
3. Laminar flow in circuit pipes.
4. Laminar flow between concentric rotating cylinders.

(a) 1, 2 and 3 only (b) 1, 3 and 4 only
(c) 1, 2 and 4 only (d) 2, 3 and 4 only
[ESE : 2020]

6.52 Water is flowing through a horizontal pipe of diameter 200 mm at a velocity of 3 m/s. A circular solid plate of diameter 150 mm is placed in the pipe to obstruct the flow. What is the loss of head due to obstruction in the pipe if C_c = 0.62? [Take g = 9.81 m/s^2]
(a) 3.311 m (b) 4.211 m
(c) 5.211 m (d) 6.211 m
[ESE : 2021]

6.53 Bernoulli's equation is obtained by
(a) integrating the Euler's equation of motion.
(b) differentiating the Euler's equation of motion.
(c) differentiating the Navier-Stokes equations.
(d) integrating energy equation.
[ESE : 2021]

6.54 A fire engines raises water at the rate of 6000 litres per minute through a height of 2 m and discharges it at 10 m/s. The H.P. of fire engine is nearly equal to (Take acceleration due to gravity as 9.81 m/s^2)

(a) 6.5 (b) 7.2
(c) 8.0 (d) 9.5

[ESE : 2022]

6.55 A pitot-static tube is used to measure the velocity of water in a pipe. The stagnation pressure head is 6 m and static pressure head is 5 m. What is the velocity of flow assuming the coefficient of tube equal to 0.98? (Take acceleration due to gravity as 9.81 m/s^2)

(a) 1.24 m/s (b) 2.68 m/s
(c) 3.56 m/s (d) 4.34 m/s

[ESE : 2022]

6.56 Air flow through a duct, and the Pitot-static tube measuring the velocity is attached to a differential manometer containing water. The deflection of the manometer is 100 mm, assuming the density of air is constant and equals to 1.22 kg/m^3, and that the coefficient of the tube is 0.98. What is the air velocity? [Consider acceleration due to gravity as 9.81 m/s^2]

(a) 19.3 m/s (b) 29.3 m/s
(c) 39.3 m/s (d) 49.3 m/s

[ESE : 2023]

6.57 Water flows in a circular pipe. At one section, the diameter is 0.3 m, the static pressure is 260 kPa gauge, the velocity is 3 m/s and the elevation is 10 m above ground level. The elevation at a section downstream is 0 m, and the pipe diameter is 0.15 m. Frictional effects may be neglected. Assume density of water to be 999 kg/m^3. What is the gauge pressure at the downstream section? [Consider acceleration due to gravity as 9.81 m/s^2]

(a) 180.25 kPa gauge
(b) 290.57 kPa gauge
(c) 320.25 kPa gauge
(d) 380.57 kPa gauge

[ESE : 2023]

6.58 Consider the following statements regarding dynamics of fluid flow :

1. Pitot tube is a device used to measure rate of flow of fluid through a pipe.
2. Orifice meter is a device used for measuring velocity of flow at any point in a pipe.
3. Venturi meter consists of converging part, throat and diverging part.

Which of the above statements is/are correct?

(a) 1 and 3 only (b) 2 only
(c) 3 only (d) 1 and 2 only

[ESE : 2024]

■■■■

Answers Fluid Dynamics and Flow Measurement

6.1 (b)	6.2 (a)	6.3 (d)	6.4 (b)	6.5 (b)	6.6 (d)	6.7 (d)	6.8 (b)	6.9 (c)
6.10 (b)	6.11 (b)	6.12 (c)	6.13 (c)	6.14 (b)	6.15 (c)	6.16 (b)	6.17 (d)	6.18 (b)
6.19 (b)	6.20 (a)	6.21 (d)	6.22 (d)	6.23 (a)	6.24 (a)	6.25 (c)	6.26 (b)	6.27 (d)
6.28 (c)	6.29 (d)	6.30 (d)	6.31 (c)	6.32 (d)	6.33 (d)	6.34 (a)	6.35 (d)	6.36 (b)
6.37 (b)	6.38 (a)	6.39 (d)	6.40 (b)	6.41 (c)	6.42 (b)	6.43 (a)	6.44 (b)	6.45 (a)
6.46 (a)	6.47 (c)	6.48 (a)	6.49 (d)	6.50 (a)	6.51 (b)	6.52 (a)	6.53 (a)	6.54 (d)
6.55 (d)	6.56 (c)	6.57 (b)	6.58 (c)					

Explanations Fluid Dynamics and Flow Measurement

6.1 (b)

$$S = 0.90$$
$$h = 50 \text{ mm} = 0.05 \text{ m}$$

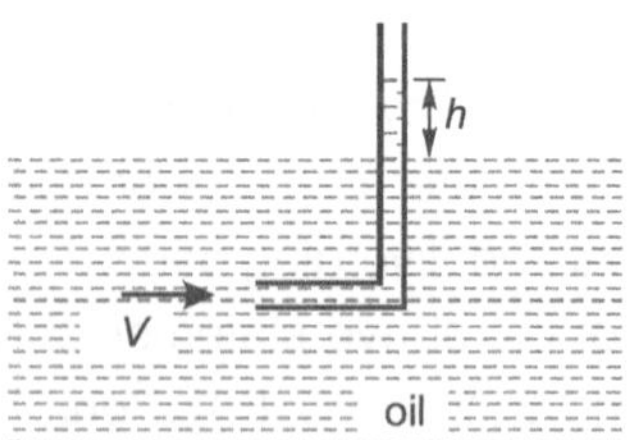

$$V = \sqrt{2gh} = \sqrt{2\times 9.81\times 0.05}$$
$$= \mathbf{0.99\ m/s}$$

6.2 (a)

Following assumptions are made in deriving the Bernoulli's equation

(1) flow in steady flow
(2) fluid is incompressible
(3) fluid is non-viscous (inviscid)
(4) equation is applicable along a stream line.
(5) effect of friction is neglected.
(6) only pressure and gravity forces are taken into account.
(7) velocity is uniform over the cross section.

6.3 (d)

$$\rho = 1.2 \text{ kg/m}^3$$
$$p_{dyn} = p_{stagnation} - p_{st} = 380 \text{ Pa}$$

also $p_{dyn} = \rho g h$

$\therefore$ $380 = 1.2 \times 9.81 \times h$

or $h = 32.279 \text{ m}$

Velocity, $V = \sqrt{2gh} = \sqrt{2\times 9.81\times 32.279}$

$= \mathbf{25.17\ m/s}$

6.4 (b)

$$\frac{\partial\phi}{\partial t} + \int\frac{dp}{\rho} + \frac{1}{2}[\nabla\phi]^2 + gz = \text{constant}$$

↓ ↓

Unsteady irrotational

6.5 (b)

$$C_d = \frac{C_d A a\sqrt{2gh}}{\sqrt{A^2 - a^2}}$$

$$C_d\sqrt{h} = C$$

where Q, A and a are constant

$$C_{d,v}\sqrt{h_v} = C_{d,o}\sqrt{h_o}$$

or $$\sqrt{\frac{h_v}{h_o}} = \frac{C_{d,o}}{C_{d,v}} = \frac{0.98}{0.61}$$

or $$\frac{h_v}{h_o} = \left(\frac{0.98}{0.61}\right)^2$$

6.6 (d)

Hot wire anemometer have extremely high frequency response and fine spatial resolution compare to other measurement method, and as such are universally employed for the detailed study of turbulent flows, or any flow in which rapid velocity fluctuation are of interest.

6.8 (b)

$$C_{v1} = \sqrt{\frac{x^2}{4HY}}$$

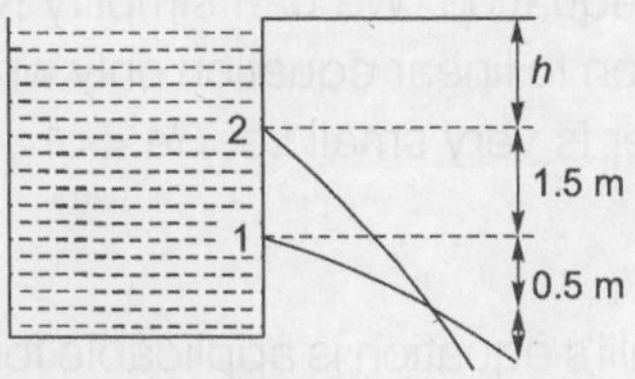

$$\therefore \quad \frac{x^2}{4\times 0.5\times(h+1.5)} = \frac{x^2}{4\times 2\times h}$$

$$\therefore \quad h = 0.5 \text{ m}$$

$$\text{Total height} = 0.5 + 1.5 + 0.5 = \mathbf{2.5\ m}$$

6.9 (c)

$$\rho_{air} g \times h_{air} = \rho_w \times g h_w$$

$$\therefore \quad h_{air} = \frac{9810\times 0.075}{1.16\times 9.81} = 64.65 \text{ m of air}$$

$$\therefore \quad V = \sqrt{2gh_{air}} = \sqrt{2\times 9.81\times 64.65} = \mathbf{35.6\ m/s}$$

6.10 (b)

The basic pitot tube consists of a tube pointing directly into the fluid flow. As this tube contains fluid, a pressure can be measured, the moving fluid is brought to rest (stagnates) as there is no outlet to allow flow to continue. This pressure is the stagnation pressure of the fluid.

Stagnation pressure = static pressure + dynamic pressure.

Micro-Manometer: It is a fluid-filled tube where an applied pressure causes the fluid height to shift proportionality. A manometer is fundamentally an instrument of differential pressure measurement, indicating the difference between two pressure by a shift in liquid column height.

6.11 (b)

Euler's equation of motion is conservation of momentum and is applicable to steady, compressible as well as incompressible flow of non-viscous fluid.

6.12 (c)

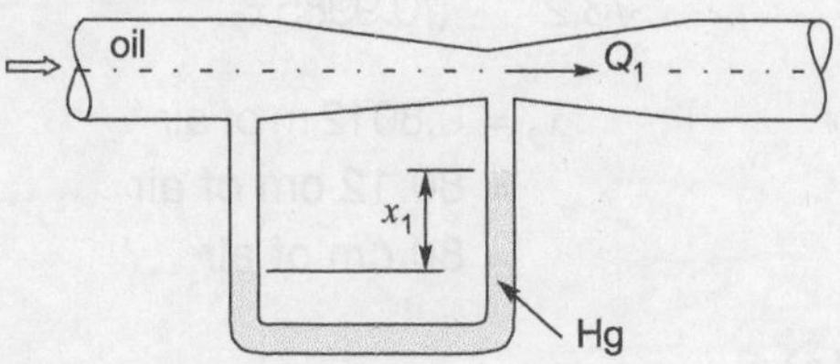

$$S_{oil} = 0.8$$

$$\therefore \quad \rho_{oil} = 0.8\times 1000 = 800 \text{ kg/m}^3$$

$$S_{Hg} = 13.6$$

$$\therefore \quad \rho_{Hg} = 13.6\times 1000 = 13600 \text{ kg/m}^3$$

$$Q_1 = 0.16 \text{ m}^3/\text{s}$$

$$x_1 = 20 \text{ cm} = 0.20 \text{ m of Hg}$$

$$h_1 = x_1\left[\frac{\rho_{Hg}}{\rho_{oil}} - 1\right]$$

$$= 0.20\left[\frac{13600}{800} - 1\right]$$

$$= 3.2 \text{ m of oil}$$

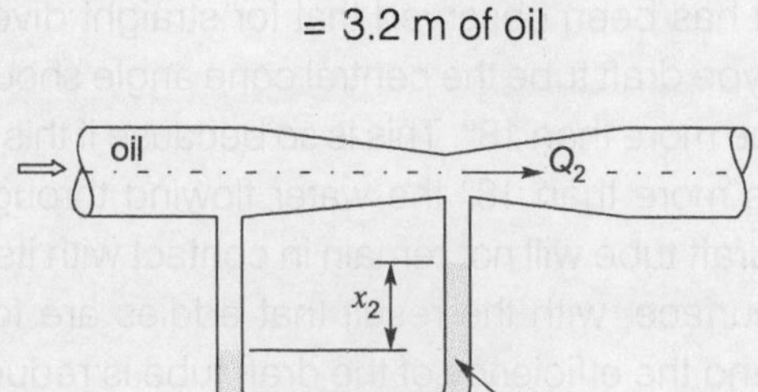

$$Q_2 = 0.08 \text{ m}^3/\text{s}$$

$$h_2 = x_2\left[1 - \frac{\rho_{air}}{\rho_{oil}}\right]$$

$$h_2 = x_2\left[1 - \frac{1.2}{800}\right]$$

$$= 0.9985\, x_2 \text{ m of oil}$$

$$Q = \frac{C_d A a\sqrt{2gh}}{\sqrt{A^2 - a^2}}$$

$$Q \propto \sqrt{h}$$

$$\text{or} \quad \frac{Q}{\sqrt{h}} = C$$

$$\frac{Q_1}{\sqrt{h_1}} = \frac{Q_2}{\sqrt{h_2}}$$

$$\frac{0.16}{\sqrt{3.2}} = \frac{0.08}{\sqrt{0.9985 x_2}}$$

or $x_2 = 0.8012$ m of air

= **80.12 cm of air**

≈ **80 cm of air**

6.13 (c)

$$C_d = \frac{\text{Actual discharge}}{\text{Theoretical discharge}} = \frac{a_c \times V_c}{a \times \sqrt{2gh}}$$

$$C_d = C_C \times C_V$$

where, $C_C = \frac{a_c}{a}$ and $C_V = \frac{V_C}{\sqrt{2gh}}$

6.14 (b)

Bernoulli's equation is applicable for non-viscous flow. At boundary layer viscous force acts. Due to this Bernoulli's equation can be used only inside the boundary layer.

6.15 (c)

It has been observed that for straight divergent type draft tube the central cone angle should not be more than 18°. This is so because if this angle is more than 18° the water flowing through the draft tube will not remain in contact with its inner surface, with the result that eddies are formed and the efficiency of the draft tube is reduced.

6.16 (b)

Rotameter and elbow meter measure discharge through pipes.

6.17 (d)

(A) Hot-wire anemometer : Velocity fluctuations
(B) Pitot-tube : Stagnation pressure
(C) V-notch weir : Discharge
(D) Tachometer : Rotational speed

6.18 (b)

For orifice meter the coefficient of discharge is a function of d/D.

6.19 (b)

$$\frac{V^2}{2g} = 0.2$$

$$\therefore \quad V = \sqrt{0.2 \times 2 \times 10} = \mathbf{2\ m/s}$$

6.20 (a)

The Navier-Stokes equation are non-linear partial differential equation in almost every real situation. In some cases such as one-dimensional flow and stokes flow the equation can be simplified to linear equation. We can simplify Navier-Stokes equation to linear equation only when Reynolds number is very small i.e. $Re << 1$.

6.21 (d)

Bernoulii's equation is applicable for steady, Non viscous, incompressible and irrotational flow.

6.22 (d)

$$Q = C_d \frac{a_1 a_2}{\sqrt{a_1^2 - a_2^2}} \sqrt{2g(h - h_L)}$$

∴In submerged orifice flow discharge is proportional to square root of the difference between upstream and downstream heads

6.23 (a)

The assumptions made in the derivation of Bernoulli's equation

1. The fluid is ideal i.e. viscosity is zero.
2. The flow is steady
3. The flow is incompressible
4. The flow is irrotational

6.25 (c)

Statement (2) is wrong.
Assumptions of Bernoulli is equation are
(a) Steady flow
(b) In viscous flow
(c) Incompressible flow
(d) These should be only one stream line in the flow.

6.26 (b)

Venturimeter is a convergent divergent shape type and orifice meter has use of vena-contracta.

6.27 (d)

Venturimeter is known as rate meter. It is a device used for finding out the discharge.

6.28 (c)

Navier-Stokes equation taken pressure force, gravity force and viscous force into consideration.

6.29 (d)

Measurement of a flow rate in a pipe is down by combination of venturimeter and manometer.

6.32 (d)

For Bernoulli's equation to be applicable the flow must be

(i) Steady
(ii) Incompressible
(iii) Non-viscous
(iv) Irrotational
(v) Along a streamline

6.33 (d)

Venturimeter is a flow measuring device.

6.34 (a)

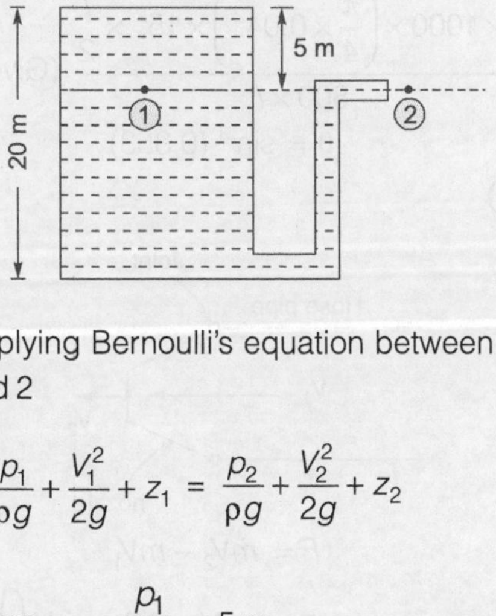

Applying Bernoulli's equation between points 1 and 2

$$\frac{p_1}{\rho g} + \frac{V_1^2}{2g} + z_1 = \frac{p_2}{\rho g} + \frac{V_2^2}{2g} + z_2$$

where $\frac{p_1}{\rho g} = 5$ m

$V_1 = 0$; $z_1 = z_2$; $p_2 = 0$

$$\therefore \quad 5 = \frac{V_2^2}{2 \times 9.81}$$

or $V_2^2 = 98.1$

or $V_2 =$ **9.9 m/s**

6.35 (d)

Bernoulli's Theorem:

In an ideal and incompressible fluid, when the flow is steady and continuous, the sum of the pressure head, velocity head and elevation head of fluid is constant at all points in the flow system.

6.37 (b)

Statement (3) is wrong because pressure decrease in the divergent part of venturi.

6.39 (d)

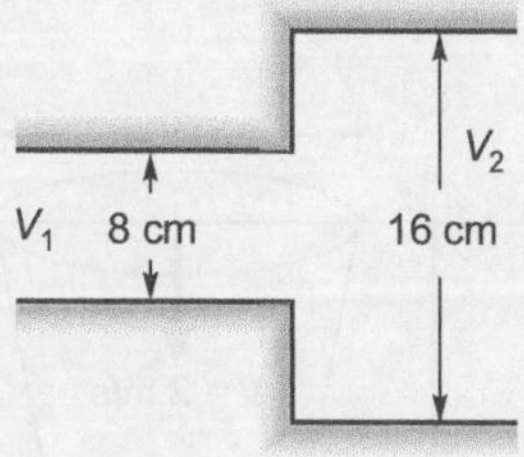

$$\therefore \quad A_1 V_1 = A_2 V_2$$

$$\therefore \quad V_2 = \frac{V_1}{4}$$

$$h_e = \frac{(V_1 - V_2)^2}{2g} = \frac{\left(V_1 - \frac{V_1}{4}\right)^2}{2g}$$

$$= \frac{V_1^2}{2g} \times \frac{3^2}{4^2} = \frac{9}{16}\left(\frac{V_1^2}{2g}\right)$$

6.40 (b)

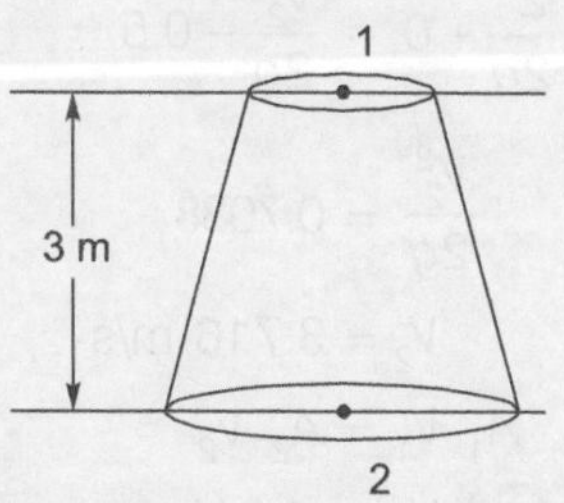

$V_1 = 4$ m/s; $V_2 = 2$ m/s

$$\frac{p_1}{\rho g} = 2 \text{ m}$$

$$h_L = 0.4 \text{ m}$$

$$\frac{p_1}{\rho g} + \frac{V_1^2}{2g} + z_1 = \frac{p_2}{\rho g} + \frac{V_2^2}{2g} + z_2 + h_L$$

$$2 + \frac{16}{2g} + 3 = \frac{p_2}{\rho g} + \frac{4}{2g} + 0 + 0.4$$

$$\frac{p_2}{\rho g} = \mathbf{5.21\ m}$$

6.42 (b)

$$C_d = C_v \times C_c$$

$$C_d = 0.98 \times \frac{(0.90)^2}{(0.100)^2} = \mathbf{0.7938}$$

6.43 (a)

A in diverging section, flow separation may occur due to adverse pressure gradient, to avoid this it is kept much longer then converging section.

6.44 (b)

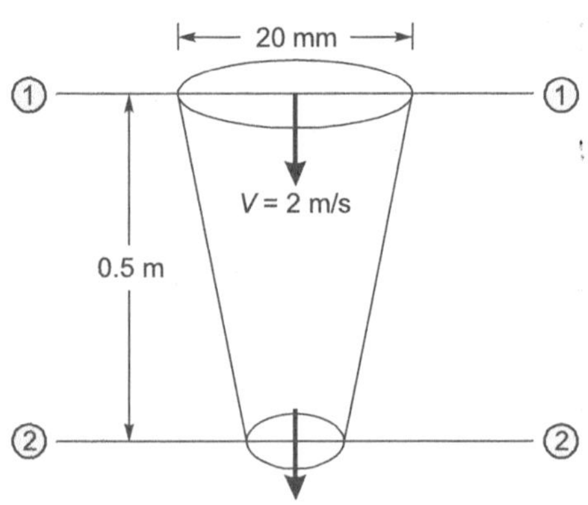

$$\frac{P_1}{\rho g} + \frac{V_1^2}{2g} + z_1 = \frac{P_2}{\rho g} + \frac{V_2^2}{2g} + z_2$$

$$\frac{P_1}{\rho g} = \frac{P_2}{\rho g} = \frac{P_{atm}}{\rho g}$$

$$\frac{2^2}{2g} + 0 = \frac{V_2^2}{2g} - 0.5$$

$$\frac{V_2^2}{2g} = 0.7038$$

$$V_2 = 3.716\ \text{m/s}$$

$$A_1 V_1 = A_2 V_2$$

$$d_1^2 V_1 = d_2^2 V_2$$

$$(20)^2 \times 2 = d_2^2 \times 3.716$$

$$d_2 = 14.67\ \text{mm}$$

Alternate method:

$$V^2 - u^2 = 2as$$

$$V^2 - 2^2 = 2g \times 0.5$$

$$V^2 = 4 + 2 \times 10 \times 0.5$$

$$V = \sqrt{14}\ \text{m/s}$$

$$A_1 u = A_2 V$$

$$d_1^2\, u = d_2^2\, V \Rightarrow (20)^2 \times 2 = \sqrt{14} \times d_2^2$$

$$800 = \sqrt{14} \times d_2^2$$

$$d_2 = 14.6\ \text{mm}$$

6.45 (a)

Statement 3 is wrong because general energy equation is

$$\frac{p_1}{\rho g} + \frac{v_1^2}{2g} + z_1 + h_P = \frac{p_2}{\rho g} + \frac{v_2^2}{2g} + z_2 + h_T$$

where h_P = Pump head

h_T = Turbine head

6.46 (a)

$$C_V = \frac{x}{2\sqrt{yH}}$$

$$C_V = \frac{0.406}{2 \times \sqrt{.085 \times .51}} = 0.975$$

6.48 (a)

$$\sin\theta = \frac{2\rho a v^2 x}{w.l.} =$$

$$\frac{2 \times 1000 \times \left(\frac{\pi}{4} \times 0.04^2\right) \times 15^2 \times \frac{l}{2}}{800 \times l} \quad (\text{Given } x = \frac{l}{2})$$

$$\theta = \sin^{-1}(0.353)$$

6.49 (d)

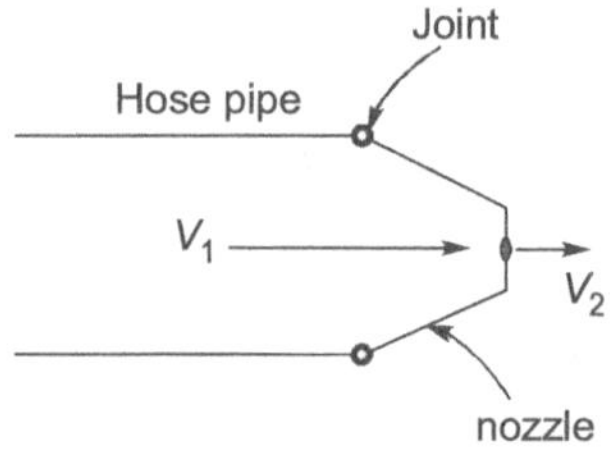

$$F = \dot{m}V_2 - \dot{m}V_1$$

$$F = \dot{m}(V_2 - V_1) = \dot{m}V_1\left(\frac{V_2}{V_1} - 1\right)$$

$$F = \rho Q V_1\left(\frac{V_2}{V_1} - 1\right)$$

$$\therefore \quad A_1V_1 = A_2V_2$$

$$\frac{V_2}{V_1} = \frac{A_1}{A_2} = \frac{d_1^2}{d_2^2}$$

$$\therefore \quad \frac{V_2}{V_1} = \frac{0.08^2}{0.04^2} = 4$$

$$Q = A_1V_1$$
$$V_1 = 3.97 \text{ m/s}$$
$$F = 1000 \times 0.02 \times 3.97\left(\frac{4}{1}-1\right)$$
$$F = 238.2 \text{ N} \approx 240 \text{ N}$$

6.50 (a)

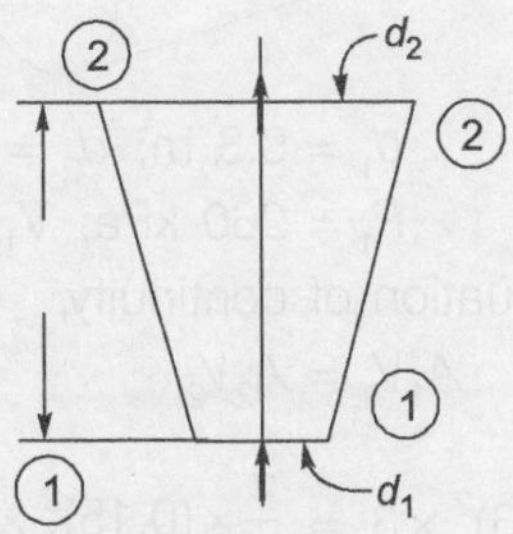

$$Q = 50 \ l/s$$
$$P_1 = 150 \text{ kN/m}^2$$
$$V_1 = \frac{4Q}{\pi d_1^2} = 11.32 \text{ m/s}$$
$$V_2 = \frac{4Q}{\pi d_2^2} = 2.83 \text{ m/s}$$
$$\frac{P_1}{\rho g} + \frac{V_1^2}{2g} + z_1 = \frac{P_2}{\rho g} + \frac{V_2^2}{2g} + z_2$$
$$\frac{150 \times 10^3}{10^3 \times 9.81} + \frac{(11.32)^2}{2 \times 9.81} + 0$$
$$= \frac{P_2}{10^3 \times 9.81} + \frac{(2.83)^2}{2 \times 9.81} + 1.5$$
$$P_2 = 195.351 \text{ kN/m}^2$$

6.52 (a)

Given: D = 200 mm, V = 3 m/s, d = 150 mm, C_C = 0.62, g = 9.81

$A \rightarrow$ Area of pipe

$a \rightarrow$ Area of obstruction

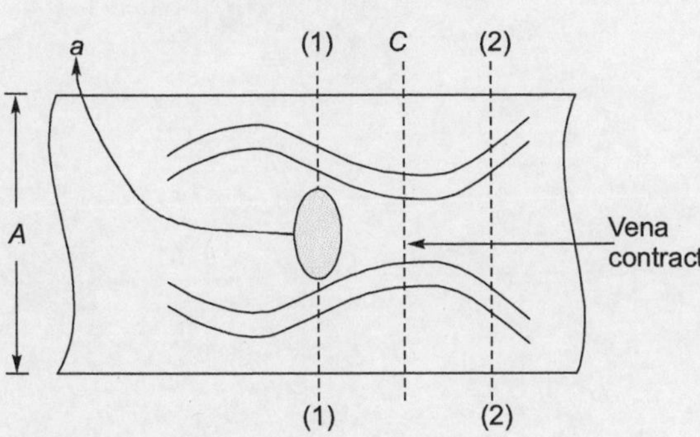

$$A_1 = (A - a)$$

Main losses are after expansion

$$h_{obs} = \frac{(V_C - V_2)^2}{2g} \quad \text{... (i)}$$
$$C_C = \frac{A_C}{A_1}$$
$$C_C = \frac{A_C}{(A-a)}$$
$$A_C = C_C(A - a)$$
$$A_CV_C = A_2V_2$$
$$V_C = \frac{A_2V_2}{A_C} = \frac{A_2V_2}{C_C(A-a)} \quad \text{... (ii)}$$
$$(h_L)_{obstruction} = \frac{1}{2g}\left[\frac{A_2V_2}{C_C(A-a)} - V_2\right]^2$$
$$(h_L)_{obstruction} = \frac{V^2}{2g}\left[\frac{A}{C_C(A-a)} - 1\right]^2$$
$$= \frac{(3)^2}{2 \times 9.81}\left[\frac{\frac{\pi}{4} \times (0.2)^2}{0.62 \times \frac{\pi}{4}\left[(0.2)^2 - (0.15)^2\right]} - 1\right]^2$$
$$= 3.311 \text{ m}$$

6.53 (a)

Bernoulli's equation is known as conservation of energy and is obtained by integrating the Euler's equation of motion.

$$\int \frac{P}{\rho} + \int gdz + \int VdV = \text{Constant}$$

6.54 (d)

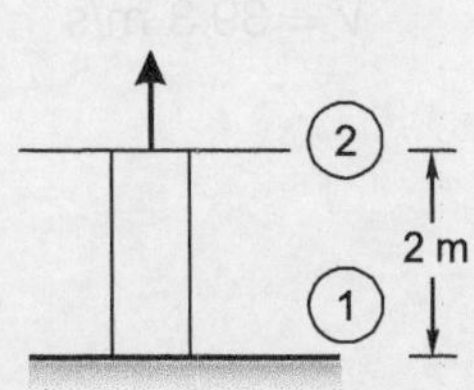

Let '2' be the point from where water exits from the engine.

Volume flow rate,

$$Q = 6000 \text{ litres/minute}$$
$$= \frac{6000 \times 10^{-3}}{60} \text{m}^3/\text{sec}$$
$$= 0.1 \text{ m}^3/\text{sec}$$

Mass flow rate,

$$\dot{m} = 0.1\,m^3/sec \times 1000\ kg/m^3$$
$$= 100\ kg/sec$$

Energy required by fire engine to raise water through a height of 2 meters with a velocity of 10 m/sec.

$$e = \dot{m}gh + \frac{1}{2}\dot{m}V_2^2$$
$$= 100 \times 9.81 \times 2 + \frac{1}{2} \times 100 \times 10^2$$
$$= 1962 + 5000 = 6962\ \text{Watts}$$

$$\therefore \text{Horse power of engine} = \frac{6962}{736} = 9.46$$

6.55 (d)

Velocity of flow,

$$V = C_V\sqrt{2g(h_{stag} - h_{static})}$$
$$= 0.98\sqrt{2g \times (6-5)}$$
$$= 4.34\ m/s$$

6.56 (c)

Air velocity in Pitot tube,

$$h = x\left[\frac{\rho_m}{\rho} - 1\right] = 100\left[\frac{1000}{1.22} - 1\right]$$
$$= 81.86\ m$$
$$\therefore \quad V = C_V \times \sqrt{2gh}$$
$$V = 0.98 \times \sqrt{2 \times 9.81 \times 81.86}$$
$$V = 39.3\ m/s$$

6.57 (b)

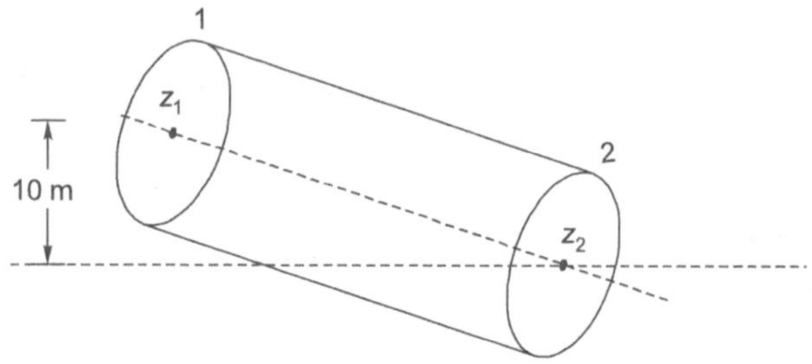

Given: $d_1 = 0.3$ m; $d_2 = 0.15$ m

$P_1 = 260$ kPa; $V_1 = 3$ m/s

Using equation of continuity,

$$A_1V_1 = A_2V_2$$
$$\frac{\pi}{4} \times (0.3)^2 \times 3 = \frac{\pi}{4} \times (0.15)^2 \times V_2$$
$$V_2 = 12\ m/s$$

Applying Bernoulli's equation between (1) and (2),

$$\frac{p_1}{\rho g} + \frac{V_1^2}{2g} + z_1 = \frac{P_2}{\rho g} + \frac{V_2^2}{2g} + z_2^{\,0}$$

$$\frac{260 \times 10^3}{9810} + \frac{(3)^2}{2 \times 9.81} + 10$$
$$= \frac{P_2}{9810} + \frac{(12)^2}{2 \times 9.81}$$
$$\therefore \quad P_2 = 290.57\ kPa$$

6.58 (c)

- Pitot tube is a device used to measure velocity of flow at any point in a pipe.
- Orifice meter is a device used to measure rate of flow of fluid through a pipe.

■■■■

7 Flow Through Pipes

7.1 A tank containing water has two orifices of the same size at depths of 40 cm and 90 cm below the free surface of water. The ratio of discharge through these orifices is
(a) 1 : 1 (b) 2 : 3
(c) 4 : 9 (d) 16 : 81
[ESE : 2000]

7.2 The velocities and corresponding flow areas of the branches labelled 1, 2, 3, 4 and 5 for a

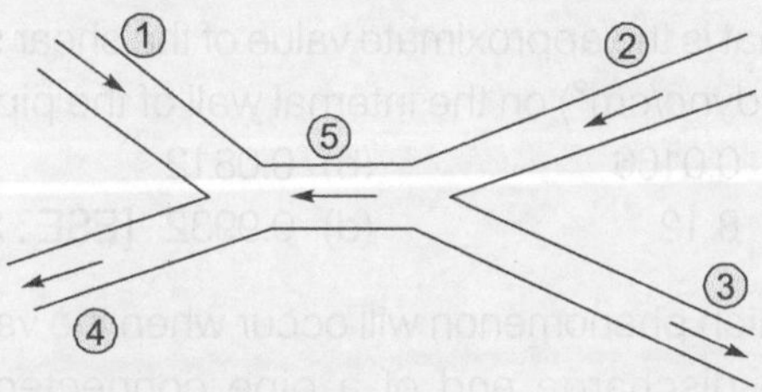

pipe system shown in the given figure are given in the following table:

Pipe Label	Velocity	Area
1.	5 cm/s	4 cm^2
2.	6 cm/s	5 cm^2
3.	V_3 cm/s	2 cm^2
4.	4 cm/s	10 cm^2
5.	V_5 cm/s	8 cm^2

The velocity V_5 would be
(a) 2.5 cm/s (b) 5 cm/s
(c) 7.5 cm/s (d) 10 cm/s
[ESE : 2000]

7.3 A pipe is connected in series to another pipe whose diameter is twice and length is 32 times that of the first pipe. The ratio of frictional head losses for the first pipe to those for the second pipe is (both the pipes have the same frictional constant)
(a) 8 (b) 4
(c) 2 (d) 1 [ESE : 2000]

7.4 A pipeline connecting two reservoirs has its diameter reduced by 20% due to deposition of chemicals. For a given head difference in the reservoirs with unaltered friction factor, this would cause a reduction in discharge of
(a) 42.8% (b) 20%
(c) 17.8% (d) 10.6%
[ESE : 2000]

7.5 Which one of the following statements is correct?
(a) Hydraulic grade line and energy grade line are the same in fluid flow problems.
(b) Energy grade line lies above the hydraulic grade line and is always parallel to it.
(c) Energy grade line lies above the hydraulic grade line and they are separated from each other by a vertical distance equal to the velocity head.
(d) The hydraulic grade line slopes upwards meeting the energy grade line only at the exit of flow. [ESE : 2000]

7.6 At location-I of a horizontal line, the fluid pressure head is 32 cm and velocity head is 4 cm. The reduction in area at location-II is such that the pressure head drops down to zero.

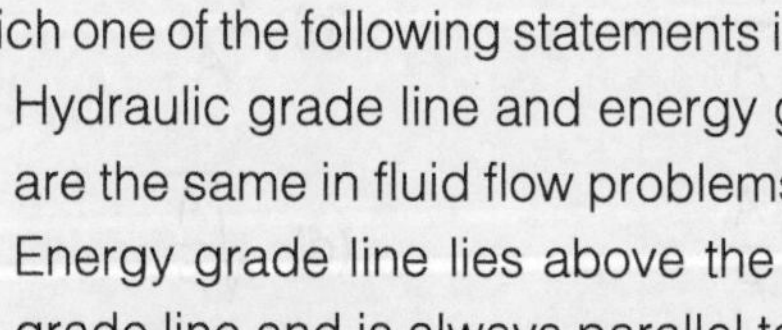

The ratio of velocities at location-II to that at location-I is
(a) 3 (b) 2.5
(c) 2 (d) 1.5
[ESE : 2001]

7.7 For maximum transmission of power through a pipeline with total head H, the head lost due to friction h_f is given by
(a) $0.1H$ (b) $H/3$

(c) $H/2$ (d) $2H/3$

[ESE : 2001]

7.8 Two pipelines of equal length and with diameters of 15 cm and 10 cm are in parallel and connect two reservoirs. The difference in water levels in the reservoirs is 3 m. If the friction is assumed to be equal, the ratio of the discharge due to the larger diameter pipe to that of the smaller diameter pipe is, nearly,

(a) 3.375 (b) 2.756

(c) 2.25 (d) 1.5

[ESE : 2001]

7.9 The hydraulic mean depth (where A = area and P = wetted perimeter) is given by

(a) $\frac{P}{A}$ (b) $\frac{P^2}{A}$

(c) $\frac{A}{P}$ (d) $\sqrt{\frac{A}{P}}$

[ESE : 2002]

7.10 Match **List-I** (Fluid parameters) with **List-II** (Basic dimensions) and select the correct answer using the codes given below the lists:

List-I

A. Dynamic viscosity

B. Chezy's roughness coefficient

C. Bulk modulus of elasticity

D. Surface tension (σ)

List-II

1. MT^{-2}

2. $ML^{-1}T^{-2}$

3. $ML^{-1}T^{-1}$

4. $L^{1/2}T^{-1}$

Codes:

	A	B	C	D
(a)	3	2	4	1
(b)	1	4	2	3
(c)	3	4	2	1
(d)	1	2	4	3

[ESE : 2002]

7.11 Velocity of pressure waves due to pressure disturbances imposed in a liquid is equal to

(a) $\left(\frac{E}{\rho}\right)^{1/2}$ (b) $(E\rho)^{1/2}$

(c) $\left(\frac{\rho}{E}\right)^{1/2}$ (d) $\left(\frac{1}{E\rho}\right)^{1/2}$

[ESE : 2003]

7.12 A 20 cm diameter 500 m long water pipe with friction factor f = 0.025, leads from a constant-head reservoir and terminates at the delivery end into a nozzle discharging into air. (Neglect all energy losses other than those due to pipe friction). What is the approximate diameter of the jet for maximum power?

(a) 6.67 mm (b) 5.98 mm

(c) 66.7 mm (d) 59.8 mm

[ESE : 2004]

7.13 A 40 mm diameter 2 m long straight uniform pipe carries a steady flow of water (viscosity 1.02 centipoise) at the rate of 3.0 litres per minute. What is the approximate value of the shear stress (in dyne/cm^2) on the internal wall of the pipe?

(a) 0.0166 (b) 0.0812

(c) 8.12 (d) 0.9932 [ESE : 2004]

7.14 Which phenomenon will occur when the valve at the discharge end of a pipe connected to a reservoir is suddenly closed?

(a) Cavitation (b) Erosion

(c) Hammering (d) Surging

[ESE : 2005]

7.15 A 12 cm diameter straight pipe is laid at a uniform downgrade and flow rate is maintained such that velocity head in the pipe is 0.5 m. If the pressure in the pipe is observed to be uniform along the length when the down slope of the pipe is 1 in 10, what is the friction factor for the pipe?

(a) 0.012 (b) 0.024

(c) 0.042 (d) 0.050 [ESE : 2006]

7.16 A compound pipeline consists of two pieces of identical pipes. The equivalent length of same diameter and same friction factor, for the compound pipeline is L_1 when pipes are connected in series, and is L_2 when connected in parallel. What is the ratio of equivalent lengths L_1 to L_2?

(a) 32 : 1 (b) 8 : 1

(c) 2 : 1 (d) $\sqrt{2}:1$

[ESE : 2006]

7.17 The energy grade line (EGL) for steady flow in a uniform diameter pipe is shown below.

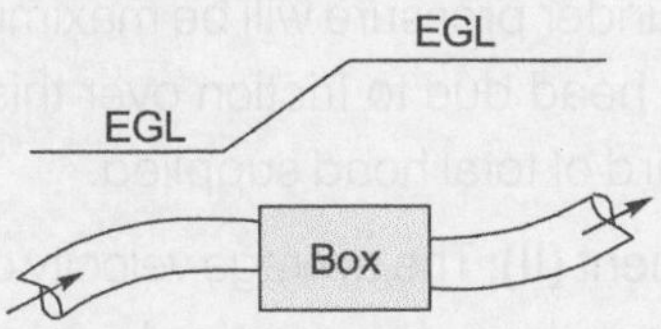

Which of the following items is contained in the box?

(a) A pump
(b) A turbine
(c) A partially closed valve
(d) An abrupt expansion **[ESE : 2006]**

7.18 The head loss in turbulent flow in pipe varies

(a) directly as the velocity
(b) inversely as the square of the velocity
(c) inversely as the square of the diameter
(d) approximately as the square of the velocity

[ESE : 2007]

7.19 Assertion (A): The power transmitted through a pipe is maximum when the loss of head due to friction is equal to one-third of total head at the inlet.

Reason (R): Velocity is maximum when the friction loss is one-third of the total head at the inlet.

(a) Both **A** and **R** are individually true and **R** is the correct explanation of **A**
(b) Both **A** and **R** are individually true but **R** is not the correct explanation of **A**
(c) **A** is true but **R** is false
(d) **A** is false but **R** is true

[ESE : 2007]

7.20 Why are surge tanks used in pipe line?

(a) To reduce frictional loss in pipe
(b) To ensure uniform flow in pipe
(c) To relieve the pressure due to water hammer
(d) To reduce cavitation

[ESE : 2008]

7.21 A penstock pipe of 10 m diameter carries water under a pressure head of 100 m. If the wall thickness is 9 mm, what is the tensile stress in the pipe wall in MPa?

(a) 2725 (b) 545.0
(c) 272.5 (d) 1090

[ESE : 2009]

7.22 Three identical pipes of length L, diameter d and friction factor f are connected in parallel between two reservoirs. What is the size of a pipe of length L and of the same friction factor f equivalent to the above pipe?

(a) $1.55d$ (b) $1.4d$
(c) $3d$ (d) $1.732d$

[ESE : 2009]

7.23 Which one of the following statements is appropriate for the free surface, the hydraulic gradient line and energy gradient line in an open channel flow?

(a) Parallel to each other but they are different lines
(b) All coinciding
(c) Such that only the first two coincide
(d) Such that they are all inclined to each other

[ESE : 2009]

7.24 The head loss in a sudden expansion from 6 cm diameter pipe to 12 cm diameter pipe in terms of velocity V_1 in the smaller diameter pipe is

(a) $\frac{3}{16} \cdot \frac{V_1^2}{2g}$ (b) $\frac{5}{16} \cdot \frac{V_1^2}{2g}$

(c) $\frac{7}{16} \cdot \frac{V_1^2}{2g}$ (d) $\frac{9}{16} \cdot \frac{V_1^2}{2g}$

[ESE : 2010]

7.25 If a fluid flows through a capillary tube of length L and diameter D, and the mass flow rate and the pressure drop are measured, the viscosity of the fluid can be estimated from the

(a) Euler equation
(b) Bernoulli's equation
(c) Hagen-Poiseuille equation
(d) Dittus-Boelter equation

[ESE : 2010]

7.26 Match **List-I** (Losses) with **List-II** (Parameters responsible) and select the correct answer using the codes given below the lists:

List-I

A. Leakage loss
B. Friction loss
C. Entrance loss

List-II

1. Zero at design point

2. Proportional to head
3. Proportional to half of relative velocity square

Codes:

	A	B	C
(a)	1	2	3
(b)	2	3	1
(c)	1	3	2
(d)	2	1	3

[ESE : 2010]

7.27 The loss of head due to sudden enlargement is attributed to

(a) Viscosity of fluid
(b) Generation of heat
(c) Roughness of pipe
(d) Production and dissipation of turbulent energy

[ESE : 2011]

7.28 The piezometric head is the summation of

(a) velocity head and pressure head
(b) pressure head and elevation head
(c) elevation head and velocity head
(d) velocity head, pressure head and elevation head

[ESE : 2012]

Directions: Each of the next questions consists of two statements, one labelled as the 'Statement (I)' and the other as 'Statement (II)'. You are to examine these two statements carefully and select the answers to these items using the codes given below:

Codes:

(a) Both Statement (I) and Statement (II) are individually true and Statement (II) is the correct explanation of Statement (I).
(b) Both Statement (I) and Statement (II) are individually true but Statement (II) is **NOT** the correct explanation of Statement (I).
(c) Statement (I) is true but Statement (II) is false
(d) Statement (I) is false but Statement (II) is true.

7.29 Statement (I): The pressure drop in circular ducts is lesser when compared to that in equivalent rectangular ducts.

Statement (II): The mean velocity in a rectangular duct will be less than that in its circular equivalent.

[ESE : 2012]

7.30 Statement (I): The hydraulic power transmitted by a pipe through certain distance by means of water under pressure will be maximum when the loss of head due to friction over this distance is one-third of total head supplied.

Statement (II): The average velocity of flow should be less than the critical velocity which corresponds to the laminar flow.

[ESE : 2012]

7.31 A tube is used (as shown in the below figure) as a siphon to discharge an oil of specific gravity 0.8 from a large open vessel into a drain at atmospheric pressure. The velocity of oil through the siphon will be

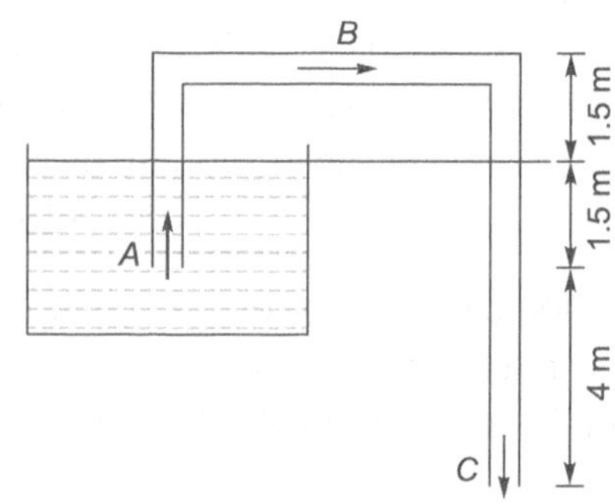

(a) 1.5 m/s
(b) 6.7 m/s
(c) 8.8 m/s
(d) 10.4 m/s

[ESE : 2013]

7.32 Water flows through a vertical contraction from a pipe of diameter d to another of diameter of *d*/2. Inlet velocity at inlet to contraction is 2 m/s and pressure 200 kN/m^2. If height of contraction measures 2 m, the pressure at exit of contraction will be nearly to

(a) 192 kN/m^2
(b) 150 kN/m^2
(c) 165 kN/m^2
(d) 175 kN/m^2

[ESE : 2013]

7.33 In flow through a pipe, the transition from laminar to turbulent flow does not depend on

(a) density of fluid
(b) length of pipe
(c) diameter of pipe
(d) velocity of the fluid

[ESE : 2013]

7.34 In the below layout of piping, what is the velocity in 200 mm diameter pipe?

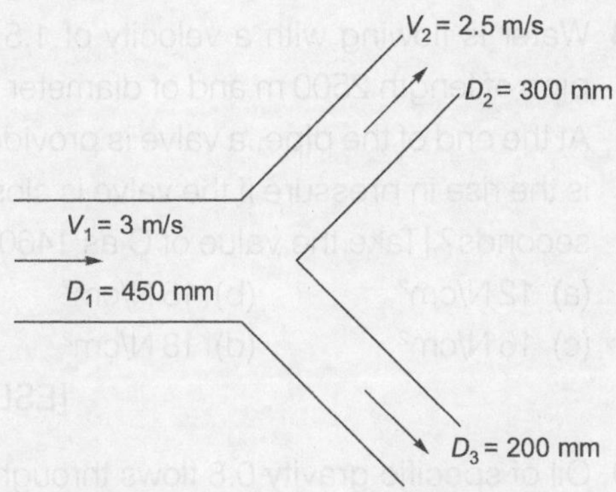

(a) 2.5 m/s (b) 5.55 m/s
(c) 7.25 m/s (d) 9.56 m/s

[ESE : 2014]

7.35 For laminar flow through a round pipe, the shear stress
(a) Remains constant over the cross-section
(b) Varies linearly with the radial distance
(c) Must be zero at all points
(d) Varies parabolically with radial distance

[ESE : 2014]

7.36 Which of the following statements is NOT correct in the context of laminar flow through with pipeline?
(a) Shear stress is zero at the centre and varies linearly with pipe radius
(b) Head loss is proportional to square of the average flow velocity
(c) The friction factor varies inversely with flow Reynolds number
(d) No dispersion of die injected into the flow stream

[ESE : 2014]

7.37 Two reservoirs connected by two pipe lines in parallel of the same diameter D and length. It is proposed to replace the two pipe lines by a single pipeline of the same length without affecting the total discharge and loss of head due to friction. The diameter of the equivalent pipe D_e in terms of the diameter of the existing pipe line, $\frac{D_e}{D}$, is :
(a) 4.0 (b) $(2)^{1/5}$
(c) $(4)^{1/4}$ (d) $(4)^{1/5}$

[ESE : 2016]

7.38 Consider fully developed laminar flow in a circular pipe of a fixed length :
1. The friction factor is inversely proportional to Reynolds number
2. The pressure drop in the pipe is proportional to the average velocity of the flow in the pipe
3. The friction factor is higher for a rough pipe as compared to a smooth pipe
4. The pressure drop in the pipe is proportional to the square of average of flow in the pipe

Which of the above statements are correct ?
(a) 1 and 4 (b) 3 and 4
(c) 2 and 3 (d) 1 and 2

[ESE : 2016]

7.39 The oil with specific gravity 0.8, dynamic viscosity of 8×10^{-3} Ns/m^2 flows through smooth pipe of 100 mm diameter and with Reynolds number 2100. The average velocity in the pipe is :
(a) 0.21 m/s (b) 0.42 m/s
(c) 0.168 m/s (d) 0.105 m/s

[ESE : 2016]

7.40 Statement (I) : In flow through a pipeline, the nature of the fluid flow depends on the velocity of the fluid.

Statement (II) : Reynolds number of the flow depends on the velocity, the diameter of the pipe and the kinematic viscosity of the fluid.
(a) Both Statement (I) and Statement (II) are individually true and Statement (II) is the correct explanation of Statement (I).
(b) Both Statement (I) and Statement (II) are individually true but Statement (II) is not the correct explanation of Statement (I).
(c) Statement (I) is true but Statement (II) is false.
(d) Statement (I) is false but Statement (II) is true.

[ESE : 2017]

7.41 The service pump in a water supply system has to maintain a net static head lift of 5 m at the tank to which it delivers freely through a 4 km long pipe, wherein all minor losses can be neglected. The diameter of the pipe is 0.2 m and its friction factor f = 0.01. The pumped water is discharged at 2 m/s. The absolute pressure differential developed by the pump is nearly (taking atmospheric pressure as 10.3 m of water)
(a) 4.5 bar (b) 5.5 bar
(c) 45 bar (d) 55 bar

[ESE : 2018]

7.42 Water is discharged from a tank maintained at a constant head of 5 m above the exit of a straight pipe 100 m long and 15 cm in diameter.

If the friction coefficient for the pipe is 0.01, the rate of flow will be nearly

(a) 0.04 m^3/s (b) 0.05 m^3/s
(c) 0.06 m^3/s (d) 0.07 m^3/s

[ESE : 2019]

7.43 In case of transmission of hydraulic power by a pipeline to a turbine in a hydroelectric power station, the maximum power transmission efficiency through the pipeline is

(a) 76% (b) 67%
(c) 54% (d) 42%

[ESE : 2019]

7.44 A pipe, having a length 200 m and 200 mm diameter with friction factor 0.015, is to be replaced by a 400 mm diameter pipe of friction factor 0.012 to convey the same quantity of flow. The equivalent length of the new pipe for the same head loss will be

(a) 8300 m (b) 8240 m
(c) 8110 m (d) 8000 m

[ESE : 2019]

7.45 Two pipes of lengths 2500 m each and diameters 80 cm and 60 cm respectively, are connected in parallel. The coefficient of friction for each pipe is 0.006 and the total flow is 250 litres/s. The rates of flow in the pipes are nearly

(a) 0.17 m^3/s and 0.1 m^3/s
(b) 0.23 m^3/s and 0.1 m^3/s
(c) 0.17 m^3/s and 0.4 m^3/s
(d) 0.23 m^3/s and 0.4 m^3/s [ESE : 2020]

7.46 The diameter of a nozzle d for maximum transmission of power through it, is

(a) $\left[\frac{D^5}{8fL}\right]^{1/4}$ (b) $\left[\frac{D^5}{8fL}\right]^{1/2}$

(c) $\left[\frac{8D^5}{fL}\right]^{1/4}$ (d) $\left[\frac{8D^5}{fL}\right]^{1/2}$

Where: D = Diameter of pipe; f = Coefficient of friction; L = Length of pipe [ESE : 2020]

7.47 Three pipes of length 800 m, 500 m and 400 m and of diameters 500 mm, 400 mm and 300 mm respectively are connected in series. These pipes are to be replaced by a single pipe of length 1700 m. What is the diameter of the single pipe?

(a) $(0.007118)^{0.2}$ m (b) $(0.003609)^{0.3}$ m
(c) $(0.003609)^{0.2}$ m (d) $(0.007118)^{0.3}$ m

[ESE : 2021]

7.48 Water is flowing with a velocity of 1.5 m/s in a pipe of length 2500 m and of diameter 500 mm. At the end of the pipe, a valve is provided. What is the rise in pressure if the valve is closed in 25 seconds? [Take the value of C as 1460 m/s]

(a) 12 N/cm^2 (b) 15 N/cm^2
(c) 16 N/cm^2 (d) 18 N/cm^2

[ESE : 2021]

7.49 Oil of specific gravity 0.8 flows through a 0.2 m diameter pipe under a pressure of 100 kN/m^2. If the datum is 5 m below the centerline of the pipe and the total energy with respect to the datum is 35 Nm/N, the discharge is [Take g = 9.81 m/s^2]

(a) 0.58 m^3/sec (b) 0.47 m^3/sec
(c) 0.31 m^3/sec (d) 0.22 m^3/sec

[ESE : 2021]

7.50 The head of water at the inlet of a pipe 2000 m long and 500 mm diameter is 60 m. A nozzle of diameter 100 mm at its outlet is fitted to the pipe. What is the velocity of water at the outlet of the nozzle if f = 0.01 for the pipe?
[Take g = 9.81 m/s^2]

(a) 30.61 m/s (b) 34.81 m/s
(c) 36.52 m/s (d) 38.36 m/s

[ESE : 2021]

7.51 A fluid of viscosity 0.7 Ns/m^2 and specific gravity 1.3 is flowing through a circular pipe of diameter 100 mm. The maximum shear stress at the pipe wall is given as 196.2 N/m^2. What is the pressure gradient of the flow?

(a) 7848 N/m^2 per m (b) –7848 N/m^2 per m
(c) –9848 N/m^2 per m (d) 9848 N/m^2 per m

[ESE : 2022]

7.52 A oil of viscosity 10 poise flows between two parallel fixed plates which are kept at distance of 50 mm apart. What is the rate of flow of oil between the plates if the drop of pressure in a length of 1.2 m be 0.3 N/cm^2 and the width of the plate is 200 mm?

(a) 6.2 litre/s (b) 3.2 litre/s
(c) 8.2 litre/s (d) 5.2 litre/s

[ESE : 2022]

Answers Flow Through Pipes

7.1 (b)	7.2 (a)	7.3 (d)	7.4 (a)	7.5 (c)	7.6 (a)	7.7 (b)	7.8 (b)	7.9 (c)
7.10 (c)	7.11 (a)	7.12 (d)	7.13 (b)	7.14 (c)	7.15 (b)	7.16 (b)	7.17 (a)	7.18 (d)
7.19 (c)	7.20 (c)	7.21 (b)	7.22 (a)	7.23 (c)	7.24 (d)	7.25 (c)	7.26 (a)	7.27 (d)
7.28 (b)	7.29 (b)	7.30 (b)	7.31 (d)	7.32 (b)	7.33 (b)	7.34 (d)	7.35 (b)	7.36 (b)
7.37 (d)	7.38 (d)	7.39 (a)	7.40 (b)	7.41 (a)	7.42 (c)	7.43 (b)	7.44 (d)	7.45 (a)
7.46 (a)	7.47 (a)	7.48 (b)	7.49 (a)	7.50 (a)	7.51 (b)	7.52 (d)		

Explanations Flow Through Pipes

7.1 (b)

$$\frac{Q_1}{Q_2} = \frac{A_1V_1}{A_2V_2} \propto \sqrt{\frac{h_1}{h_2}} = \sqrt{\frac{40}{90}} = \frac{2}{3}$$

7.2 (a)

$$Q_1 + Q_5 = Q_4$$

$$5 \times 4 + 8V_5 = 4 \times 10$$

$$\therefore \quad V_5 = \mathbf{2.5\ m/s}$$

7.3 (d)

$$\frac{h_1}{h_2} = \frac{\dfrac{fLQ^2}{12.1D^5}}{\dfrac{f \times 32L \times Q^2}{12.1 \times (2D)^5}} = 1$$

7.4 (a)

$$h_f = \frac{flQ^2}{12.1d^5}$$

$$\Rightarrow \quad Q^2 \propto d^5$$

$$\Rightarrow \quad Q \propto d^{5/2}$$

$$\therefore \quad \frac{Q_2}{Q_1} = (0.8)^{5/2} = 0.5724$$

$$\therefore \quad \%\text{ reduction} = \left(1 - \frac{Q_2}{Q_1}\right) \times 100$$

$$= (1 - 0.5724) \times 100 = \mathbf{42.8}$$

7.6 (a)

By Bernoulli's equation

$$\frac{p_1}{\rho g} + \frac{v_1^2}{2g} = \frac{p_2}{\rho g} + \frac{v_2^2}{28}$$

$$\Rightarrow \quad 32 + 4 = 0 + \frac{v_2^2}{2g}$$

$$\text{or} \quad \frac{v_2^2}{2g} = 36$$

$$\text{Now,} \quad \frac{v_2^2/2g}{v_1^2/2g} = \frac{36}{4}$$

$$\text{or,} \quad \frac{v_2^2}{v_1^2} = 9$$

$$\frac{v_2}{v_1} = \mathbf{3}$$

7.7 (b)

Power transmitted by the pipe,

$$P = \rho g Q(H - h_f)$$

$$P = \rho g A V\left(H - \frac{fL}{D}.\frac{V^2}{2g}\right)$$

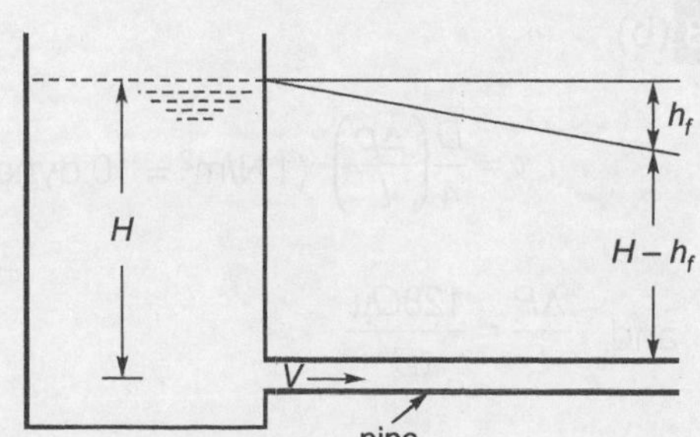

$$= \rho g A\left(HV - \frac{fL}{D}.\frac{V^3}{2g}\right)$$

For maximum power: $\dfrac{dP}{dV} = 0$

$$\therefore \quad H - 3\frac{fL}{D}\cdot\frac{V^2}{2g} = 0$$

$$h_f = \frac{H}{3}$$

7.8 (b)

$$h_f = \frac{fLV^2}{2gD}$$

$$\therefore \quad V \propto \sqrt{D}$$

$$\text{Now,} \quad \frac{Q_1}{Q_2} = \frac{A_1V_1}{A_2V_2} = \left(\frac{D_1}{D_2}\right)^{5/2} = (1.5)^{5/2}$$

$$= \mathbf{2.756}$$

7.9 (c)

$$\text{Hydraulic depth} = \frac{\text{Surface area}}{\text{Wetted perimeter}}$$

7.10 (c)

(a) Dynamic viscosity : Ns/m^2 = ML^{-1} T^{-1}

(b) Chezy's roughness coefficient: L$^{-1/2}$ T^{-1}

(c) Bulk modulus of elasticity : N/m^2 = ML^{-1} T^{-2}

(d) Surface tension $(\sigma): \frac{\text{N}}{\text{m}} = \text{MT}^{-2}$

7.12 (d)

For maximum power transmission

$$d = \left(\frac{D^5}{2fL}\right)^{1/4} = \left(\frac{(0.5)^5}{2\times 0.025\times 500}\right)^{1/4}$$

$$= \mathbf{59.8\ mm}$$

7.13 (b)

$$\tau = \frac{D}{4}\left(\frac{\Delta p}{L}\right) \quad (1\ \text{N/m}^2 = 10\ \text{dyne/cm}^2)$$

$$\text{and} \quad \frac{\Delta P}{L} = \frac{128Q\mu}{\pi D^4}$$

$$\therefore \quad \tau = \frac{128Q.\mu}{4\pi D^3} = \frac{32\times 3000\times 1.02\times 10^{-3}}{60\times 4\times \pi\times (4)^2}$$

$$= \mathbf{0.0812\ dyne/cm^2}$$

7.15 (b)

$$d = 12\ \text{cm} = 0.12\ \text{m}$$

$$\frac{V^2}{2g} = 0.5\ \text{m}$$

$$\frac{h_f}{l} = \frac{1}{10}$$

$$h_f = \frac{flV^2}{2gd} \quad \text{Darcy's formula}$$

$$\frac{h_f}{l} = f\times\frac{V^2}{2g}\times\frac{1}{d}$$

$$\frac{1}{10} = f\times 0.5\times\frac{1}{0.12}$$

$$\text{or} \quad f = \mathbf{0.024}$$

7.16 (b)

Series connection

$$\frac{L_1}{D^2} = \frac{l}{d^5} + \frac{l}{d^5} = \frac{2l}{d^5}$$

$$\therefore \quad L_1 = 2l$$

Parallel connection

$$h_{f2} = h_{f1}$$

$$\frac{fL_2Q^2}{12.1D^5} = \frac{fl(Q/2)^2}{12.1D^5}$$

$$L_2 = \frac{l}{4}$$

$$\therefore \quad \frac{L_1}{L_2} = \mathbf{8:1}$$

7.17 (a)

Pump always increases the pressure head.

7.18 (d)

The losses vary as U^n, n varying from 1.72 to 2.00. Thus for the same velocity. The losses in turbulent flow are much more than those in case of laminar flow.

7.19 (c)

The power transmitted through a pipe is maximum when the loss of head due to friction is equal to one-third of total head at inlet. Velocity is optimum for maximum power transmission.

7.20 (c)

Surge tank is an open reservoir provided in between penstock to

(i) absorb the pressure wave due to sudden blockage of flow in the turbine which will reduce water pressure hammer problem

(ii) To serve for emergency supply and to maintain constant head at inlet of turbine

7.21 (b)

Tensile stress in the pipe wall

$$= \text{Circumferential stress in pipe wall} = \frac{pd}{2t}$$

where, $p = \rho g H = 980000\ \text{N/m}^2$

$$\therefore \text{Tensile stress} = \frac{980000 \times 10}{2 \times 9 \times 10^{-3}}$$

$$= 544.44 \times 10^6\ \text{N/m}^2$$

$$= 544.44\ \text{MN/m}^2 = \mathbf{544.44\ MPa}$$

7.22 (a)

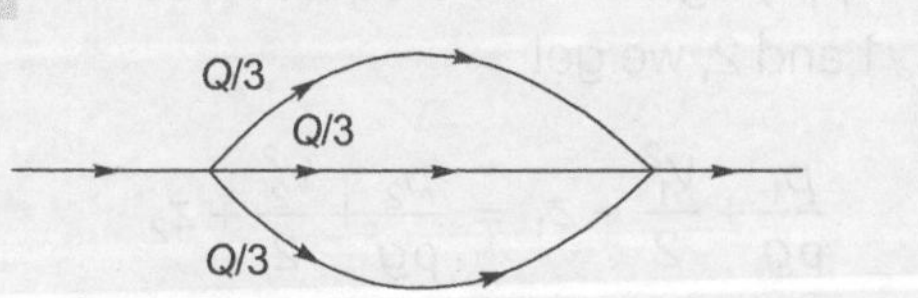

$$\frac{fL\left(\frac{Q}{3}\right)^2}{12\,d^5} = \frac{fLQ^2}{12\,d_{eq}^{\;5}}$$

$$\Rightarrow \frac{1}{9d^5} = \frac{1}{d_{eq}^{\;5}} \Rightarrow d_{eq} = (9)^{\frac{1}{5}} d = \mathbf{1.55}\boldsymbol{d}$$

7.24 (d)

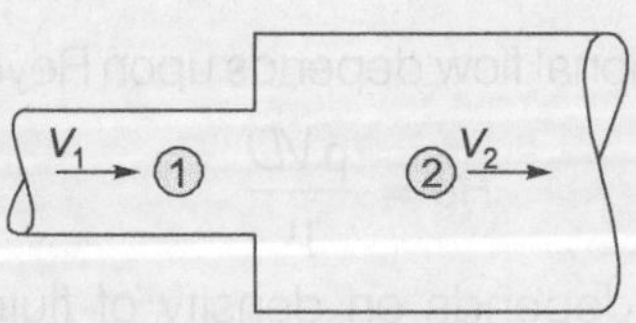

Losses due to sudden enlargement,

$$h_e = \frac{1}{2g}(V_1 - V_2)^2$$

From the continuity equation

$$A_1 V_1 = A_2 V_2$$

$$V_2 = \frac{A_1}{A_2} V_1 = \frac{36}{144} V_1 = \frac{V_1}{4}$$

we get, $$h_e = \frac{1}{2g}\left(V_1 - \frac{V_1}{4}\right)$$

$$= \frac{9}{16}\frac{V_1^2}{2g}$$

7.26 (a)

A–1: Leakage loss considered to be zero at design point.

B–2:

Frictional loss is directly proportional to the velocity head.

$$\text{Friction loss, } h_f = \frac{flV^2}{2gd} = \frac{fl}{2gd} \times 2gh$$

$$\because V = \sqrt{2gh} \text{ or } V^2 = 2gh$$

$$h_f = \frac{flh}{d}$$

i.e., $h_f \propto h$

C–3:

Entrance head loss is directly proportional to the half of the square relative velocity.

$$\text{Entrance loss, } h_i = 0.5\frac{V^2}{2g}$$

i.e. $h_i \propto \frac{V^2}{2}$

7.27 (d)

Loss of head due to sudden enlargement is

$$\frac{(V_1 - V_2)^2}{2g}$$

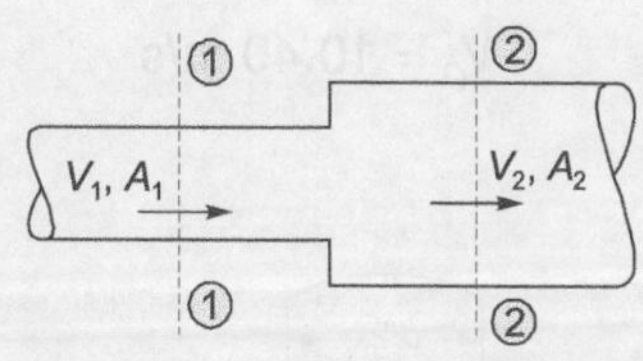

7.28 (b)

Bernoulli's eq.

$$\frac{p}{\rho g} + z + \frac{V^2}{2g} = C$$

Piezometric head + Velocity head = C

$$\text{where Piezometric head} = \frac{p}{\rho g} + z$$

Piezometric head is the summation of pressure head and elevation head.

7.29 (b)

Since the contact area of a rectangular duct is more and hence more boundary and friction, so pressure drop is more compared to circular duct, and also the mean velocity is less.

7.31 (d)

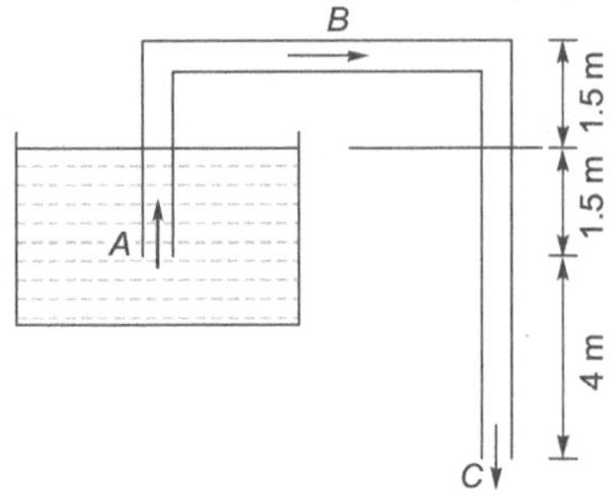

Applying the Bernoulli's equation between points A and C

$$\frac{p_A}{\rho g}+\frac{V_A^2}{2g}+z_A = \frac{p_C}{\rho g}+\frac{V_C^2}{2g}+z_C$$

where $\frac{p_A}{\rho g}$ = 1.5 m; V_A = 0; z_A = 4 m

$\frac{p_C}{\rho g}$ = 0; z_C = 0

$$\therefore \quad 1.5+0+4 = 0+\frac{V_C^2}{2g}+0$$

$$5.5 = \frac{V_C^2}{2\times 9.81}$$

or V_C = **10.40 m/s**

7.32 (b)

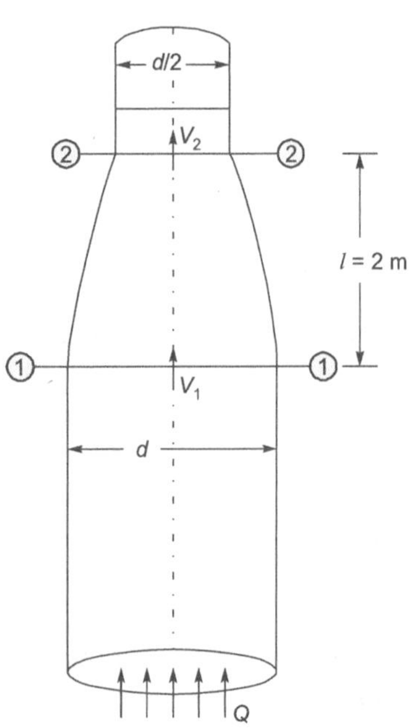

V_1 = 2 m/s; l = 2 m;

$$p_1 = 200 \text{ kN/m}^2 = 200\times 10^3 \text{ N/m}^2$$

$$d_1 = d$$

$$d_2 = \frac{d}{2}$$

Using continuity equation

$$Q = A_1V_1 = A_2V_2$$

or $$A_1V_1 = A_2V_2$$

$$\frac{\pi}{4}d_1^2V_1 = \frac{\pi}{4}d_2^2V_2$$

or $$d_1^2V_1 = d_2^2V_2$$

$$(\text{d})^2\times 2 = \left(\frac{d}{2}\right)^2\times V_2$$

or $$V_2 = 2\times 4 = 8 \text{ m/s}$$

Applying Bernoulli's equation between sections 1 and 2, we get

$$\frac{p_1}{\rho g}+\frac{V_1^2}{2}+z_1 = \frac{p_2}{\rho g}+\frac{V_2^2}{2}+z_2$$

$$\frac{200\times 10^3}{1000\times 9.81}+\frac{(2)^2}{2\times 9.81}+0=\frac{p_2}{1000\times 9.81}+\frac{(8)^2}{2\times 9.81}+2$$

$$20.38+0.20 = \frac{p_2}{9810}+3.26+2$$

or p = **150289.2 Pa ≈ 150 kPa**

7.33 (b)

Transitional flow depends upon Reynolds number

$$Re = \frac{\rho VD}{\mu}$$

∴ Re depends on density of fluid, velocity of fluid, viscosity of fluid and diameter of pipe

7.34 (d)

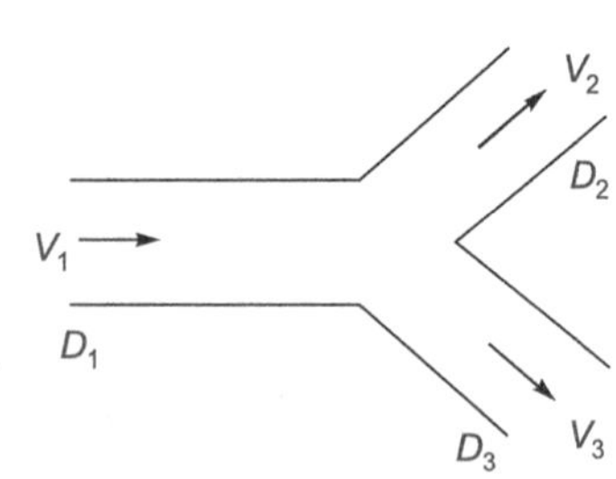

Given, V_1 = 3 m/s; V_2 = 2.5 m/s; V_3 = ?

D_1 = 450 mm; D_2 = 300 mm; D_3 = 200 mm

As per continuity equation,

$$A_1V_1 = A_2\,V_2 + A_3\,V_3$$

or $D_1^2 V_1 = D_2^2 V_2 + D_3^2 V_3$

or $\left(\frac{D_1}{D_3}\right)^2 V_1 - \left(\frac{D_2}{D_3}\right)^2 V_2 = V_3$

or $\left(\frac{450}{200}\right)^2 \times 3 - \left(\frac{300}{200}\right)^2 \times 2.5 = V_3$

or, $5.0625 \times 3 - 2.25 \times 2.5 = V_3$

or $V_3 =$ **9.5625 m/s**

7.35 (b)

Shear stress distribution is given by

$$\tau = \left(-\frac{\partial p}{\partial x}\right)\frac{r}{2}$$

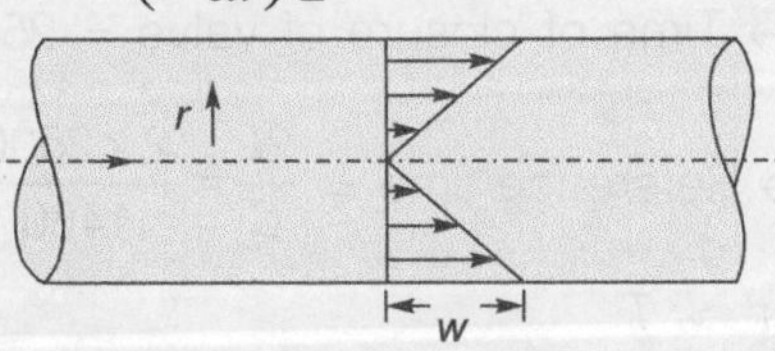

7.36 (b)

Laws of fluid friction for laminar flow as follows:

Head loss is

(i) proportional to the velocity of flow
(ii) independent of the pressure
(iii) proportional to the area of surface in contact
(iv) independent of the nature of the surface in contact
(v) greatly influenced by the variation of the temperature of the flowing fluid.

So, option (b) is wrong and valid for turbulent flow.

7.37 (d)

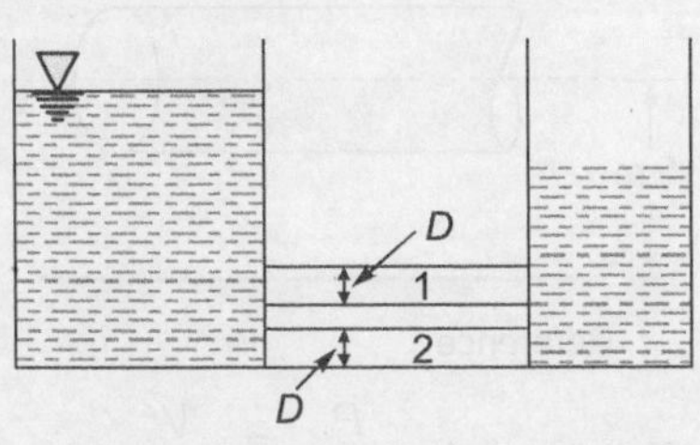

$$Q_1 = Q_2 = Q$$

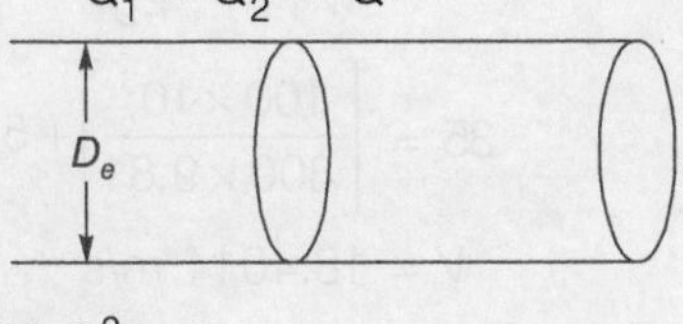

$$\frac{16fLQ^2}{\pi^2 D^5} = h_f$$

$$V = \frac{Q}{\frac{\pi}{4}D^2}$$

$$V^2 = \frac{4Q}{\pi D^2}$$

$$\frac{D_e}{D} = 4^{1/5}$$

7.39 (a)

$\rho = 800$ kg/m^3; $\mu = 800 \times 10^{-3}$ Ns/m^2

$D = 100$ mm; $Re = 2100$

$$Re = \frac{\rho \overline{V} D}{\mu}$$

$$\overline{V} = \frac{Re \times \mu}{\rho D} = \frac{2100 \times 8 \times 10^{-3}}{800 \times 1} = \mathbf{0.21\ m/s}$$

7.40 (b)

Both the statements are individually correct.

7.41 (a)

$$H_m = \frac{P_d - P_s}{\rho g} = H_s + H_f = H_s + \frac{fLV^2}{2gD}$$

$$= 5 + \frac{0.01 \times 4000 \times 2^2}{2 \times 9.81 \times 0.2}$$

$$H_m = \frac{P_d - P_s}{\rho g} = 45.77 \text{ m}$$

$$P_d - P_s = 4.49 \text{ bar} \simeq 4.5 \text{ bar}$$

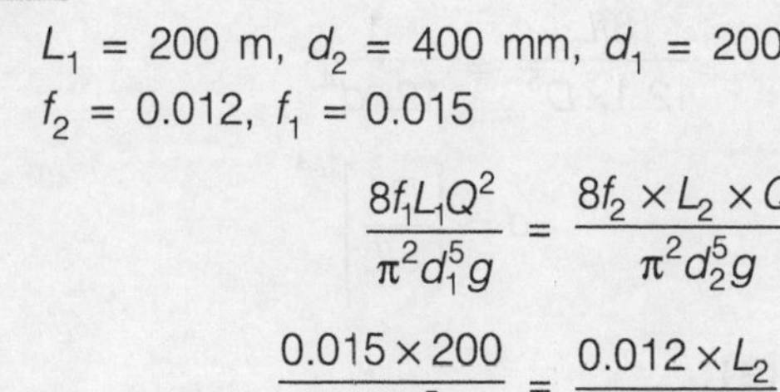

7.44 (d)

$L_1 = 200$ m, $d_2 = 400$ mm, $d_1 = 200$ mm, $f_2 = 0.012$, $f_1 = 0.015$

$$\frac{8f_1L_1Q^2}{\pi^2 d_1^5 g} = \frac{8f_2 \times L_2 \times Q^2}{\pi^2 d_2^5 g}$$

$$\frac{0.015 \times 200}{(0.2)^5} = \frac{0.012 \times L_2}{(0.4)^5}$$

$$L_2 = 8000 \text{ m}$$

7.45 (a)

For parallel connection in pipe,

Total discharge, $Q = 250$ lt/s $= 0.25$ m^3/s

$$Q = Q_1 + Q_2$$

$$Q_1 + Q_2 = 0.25 \text{ m}^3/\text{s} \quad \text{... (i)}$$

For parallel connection head loss in both pipes will be same so,

$$h_{f1} = h_{f2}$$

$$\frac{fL_1Q_1^2}{12.1\times d_1^5} = \frac{fL_2Q_2^2}{12.1\times d_2^5} \quad [\because L_1 = L_2]$$

$$\frac{Q_1^2}{0.8^5} = \frac{Q_2^2}{0.6^5}$$

$$\frac{Q_1}{Q_2} = \left(\frac{0.8}{0.6}\right)^{5/2} = 2.05$$

$$Q_1 = 2.05\ Q_2 \quad \text{... (ii)}$$

By equations (i) and (ii),

$$2.05Q_2 + Q_2 = 0.25,\ Q_2 = 0.0814 \text{ m}^3/\text{s}$$

$$Q_1 = 0.25 - 0.082 = 0.168 \text{ m}^3/\text{s}$$

7.46 (a)

Available head at nozzle = Dynamic head at exit of nozzle

$$H - h_f = \frac{V^2}{2g}$$

where, h_f is head loss and H is total head available.

For maximum power, $H = 3h_f$

$$3h_f - h_f = \frac{V^2}{2g}$$

$$2h_f = \frac{V^2}{2g}$$

$$\frac{2\times(4\times fLQ^2)}{12.1\times D^5} = \frac{Q^2}{2\times 9.81\times\frac{\pi^2}{16}\times(d^2)^2}$$

$$\frac{8fL}{12.1\times D^5} = \frac{1}{12.1d^4}$$

$$d = \left[\frac{D^5}{8fL}\right]^{1/4}$$

7.47 (a)

$$\frac{l_{eq}}{d_{eq}^5} = \frac{l_1}{d_1^5}+\frac{l_2}{d_2^5}+\frac{l_3}{d_3^5}$$

$$\frac{1700}{d_{eq}^5} = \frac{800}{0.5^5}+\frac{500}{0.4^5}+\frac{400}{0.3^5}$$

$$\frac{1700}{d_{eq}^5} = (25600 + 48828.12 + 164609.05)$$

$$\frac{1700}{d_{eq}^5} = 239037.17$$

$$d_{eq} = (0.007118)^{0.2} \text{ m}$$

7.48 (b)

Given : $V = 1.5$ m/s, $L = 2500$ m, $D = 500$ mm, $C = 1460$ m/s

$$t' = 25\text{s}$$

Reference time,

$$T = \frac{2L}{C} = \frac{2\times 2500}{1460} = 3.42$$

$t' \rightarrow$ Time of closure of valve = 25s

$$T \rightarrow \text{Reference time} = \frac{2L}{C} = \frac{2\times 2500}{1460} = 3.425$$

As $t' > T$

$\therefore$ Gradual closure

For gradual closure

$$p = \frac{\rho VL}{t'} = \frac{10^3\times 1.5\times 2500}{25}$$

$$= 15\times 10^4 \text{ N/m}^2$$

$$= 15 \text{ N/cm}^2$$

7.49 (a)

$\rho = 800$ kg/m³

$D = 0.2$ m

$p = 100$ kN/m²

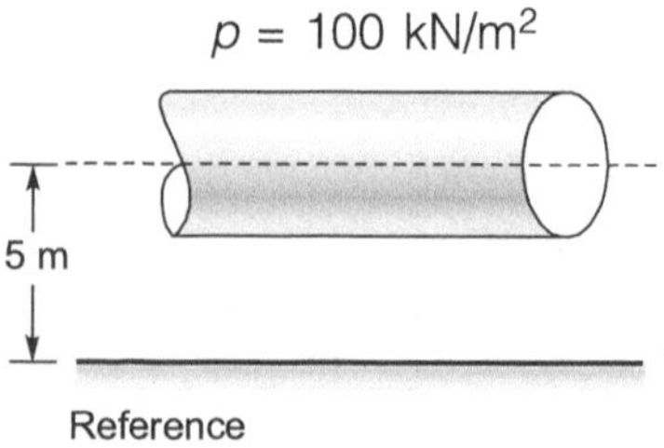

$$\text{Total head} = \frac{P}{w}+Z+\frac{V^2}{2g}$$

$$35 = \left[\frac{100\times 10^3}{800\times 9.81}\right]+5+\frac{V^2}{2g}$$

$$V = 18.4011 \text{ m/s}$$

$$Q = 0.5780 \text{ m}^3/\text{s}$$

7.50 (a)

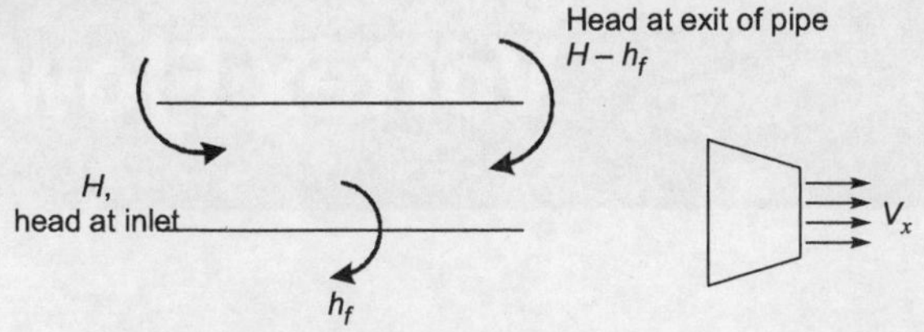

For nozzle (it converts total heat to dynamic head)

$$H - h_f = \frac{V_x^2}{2g}$$

$$\text{Area of nozzle} = \frac{\pi}{4}d_x^2$$

$$A_x = 0.00785$$

$$H - \frac{fLQ^2}{12.1d^5} = \frac{V_x^2}{2g}$$

$$60 - \frac{4 \times 0.01 \times 2000}{12.1 \times (0.5)^5} \times Q^2 = \frac{V_x^2}{2g}$$

$$2 \times g[60 - 211.57\ Q^2] = V_x^2$$

$$1177.2 - 4151\ Q^2 = V_x^2$$

$$1177.2 - 4151\ A_x^2 V_x^2 = V_x^2$$

$$1177.2 - 0.256\,V_x^2 = V_x^2$$

$$0.256\,V_x^2 = 1177.2$$

$$V_x^2 = 937.26$$

$$V_x = 30.61 \text{ m/s}$$

7.51 (b)

Given: $\mu = 0.7 \text{ Ns/m}^2$, $s = 1.2$, $d = 100$ mm,

$\tau_o = 196.2 \text{ N/m}^2$, $\frac{\partial P}{\partial x} = ?$

$$\tau = -\frac{\partial P}{\partial x} \times \frac{r}{2}$$

$$\tau_o = -\left(\frac{\partial P}{\partial x}\right) \times \frac{R}{2}$$

$$-\left(\frac{196.2 \times 2}{50 \times 10^{-3}}\right) = \frac{\partial P}{\partial x}$$

$$\frac{\partial P}{\partial x} = -7848 \text{ N/m}^2$$

7.52 (d)

Given: μ = 10 poise = 1 Ns/m^2,

y = 50 mm = 0.050 m, Q = ?

$$-\frac{\partial P}{\partial x} = \frac{0.3}{10^{-4} \times 1.2} = 2500$$

$$\text{Width} = 200 \text{ mm} = 200 \times 10^{-3} \text{ m}$$

$$Q = -\frac{1}{12\mu}\left(\frac{\partial P}{\partial x}\right) \times t^3 \times b$$

$$= \frac{1}{12 \times 1} \times 2500 \times \left(50 \times 10^{-3}\right)^3 \times 200 \times 10^{-3}$$

$$= 5.20\ l/\text{s}$$

■■■■

8 Vortex Flow

8.1 The eye of a tornado has a radius of 40 m. If the maximum wind velocity is 50 m/s, the velocity at a distance of 80 m radius is

(a) 100 m/s (b) 2500 m/s
(c) 31.25 m/s (d) 25 m/s

[ESE : 2000]

8.2 A right circular cylinder is filled with a liquid upto its top level. It is rotated about its vertical axis at such a speed that half the liquid spills out, then the pressure at the point of intersection of the axis and bottom surface

(a) is same as before rotation
(b) is half of the value before rotation
(c) is quarter of the value before rotation
(d) is equal to the atmospheric pressure

[ESE : 2001]

8.3 When a vertical cylindrical vessel containing water is rotated about its axis, then the free surface of water becomes

(a) a cycloid of revolution
(b) an ellipsoid of revolution
(c) a hyperboloid of revolution
(d) a paraboloid of revolution

[ESE : 2005]

8.4 An open circular cylinder 1.2 m high is filled with a liquid to its top. The liquid is given a rigid body rotation about the axis of the cylinder and the pressure at the centre line at the bottom surface is found to be 0.6 m of liquid. What is the ratio of volume of liquid spilled out of the cylinder to the original volume?

(a) 1/4 (b) 3/8
(c) 1/2 (d) 3/4

[ESE : 2007]

8.5 In a cylindrical vortex motion about a vertical axis, r_1 and r_2 are the radial distances of two points on the horizontal plane ($r_2 > r_1$). If for a given tangential fluid velocity at r_1, the pressure difference between the points in free vortex is one-half of that when the vortex is a forced one, then what is the value of the ratio (r_2/r_1)?

(a) $\sqrt{\frac{3}{2}}$ (b) $\sqrt{2}$
(c) 3/2 (d) $\sqrt{3}$

[ESE : 2007]

8.6 An incompressible fluid flows radially outward from a line source in a steady manner. How does velocity in any radial direction vary?

(a) r (b) r^2
(c) $1/r^2$ (d) $1/r$

[ESE : 2008]

8.7 Which one of the statements is correct for a forced vortex?

(a) Turns in an opposite direction to a free vortex
(b) Always occurs in conjunction with a free vortex
(c) Has the linear velocity directly proportional to the radius
(d) Has the linear velocity inversely proportional to the radius

[ESE : 2009]

8.8 A 50 cm diameter spherical vessel is completely filled up with a liquid of specific gravity 0.8; the vessel and the liquid are then rotated about the vertical axis, without relative motion, at a rotational speed of 20 rad/s. The points of maximum pressure lie on a horizontal plane below the centre of sphere by

(a) 12.5 cm (b) 6.96 cm
(c) 2.45 cm (d) 1.96 cm

[ESE : 2010]

8.9 **Assertion (A):** In free cylindrical vortex about a vertical axis, pressure increases with distance from axis and this increase is proportional to square of the radial distance.

Reason (R): tangential velocity is inversely proportional to radial distance and energy of a fluid element is conserved in a free cylindrical vortex.

[ESE : 2010]

8.10 The flow past the cylinder with the vortex motion causes a net transverse motion. This transverse force is known as
(a) Magnus effect
(b) Robin effect
(c) D'Alembert's Paradox
(d) Rankine doublet

[ESE : 2011]

8.11 A right circular cylinder, open at the top, is filled with a liquid of specific gravity 1.2, and rotated about its vertical axis at such a speed that half of the liquid spills out. The pressure at the centre of the bottom is
(a) zero
(b) one-fourth its value when the cylinder was full
(c) one-third its value when the cylinder was full
(d) three-fourth its value when the cylinder was full

[ESE : 2013]

8.12 An open circular tank of 20 cm diameter and 100 cm long contains water upto a height of 60 cm. The tank is rotated about its vertical axis at 300 rpm. What is the depth of parabola formed at the free surface of water? (Take acceleration due to gravity as 9.81 m/s^2)
(a) 50.28 cm (b) 55.28 cm
(c) 65.36 cm (d) 69.36 cm

[ESE : 2022]

8.13 What is the velocity of flow at radius of 0.8 m, if the water is flowing radially outward in a horizontal plane from a source at a strength of 12 m^2/s?
(a) 9.55 m/s (b) 4.77 m/s
(c) 2.38 m/s (d) 5.62 m/s

[ESE : 2022]

8.14 A hollow cylinder of 0.6 m diameter, open at the top, contains some liquid and spins about its vertical axis, producing a forced vortex motion. What is the height of the vessel so that the liquid just reaches the top of the vessel and begins to uncover the base at 100 rpm? [Consider acceleration due to gravity as 9.81 m/s^2]
(a) 3.203 m (b) 2.303 m
(c) 1.403 m (d) 0.503 m

[ESE : 2023]

8.15 The velocity of air at the outer edge of a tornado, where the pressure is 750 mm of Hg and diameter 30 meters, is 12 m/s. Consider the density of air to be constant and equal to 1.2 kg/m^3 (specific gravity of mercury = 13.6). What is the velocity of air at a radius of 2 meters from its axis?
(a) 60 m/s (b) 70 m/s
(c) 80 m/s (d) 90 m/s

[ESE : 2023]

■■■■

Answers Vortex Flow

8.1 (d) 8.2 (d) 8.3 (d) 8.4 (a) 8.5 (b) 8.6 (d) 8.7 (c) 8.8 (c) 8.9 (d)

8.10 (a) 8.11 (a) 8.12 (a) 8.13 (c) 8.14 (d) 8.15 (d)

Explanations Vortex Flow

8.1 (d)

Angular momentum is conserved.

$\therefore\ mv_1r_1 = mv_2r_2$

$\Rightarrow\ v_1r_1 = v_2r_2$

$\Rightarrow\ 50 \times 40 = v_2 \times 80$

$\Rightarrow\ v_2 = 25$ m/s

8.2 (d)

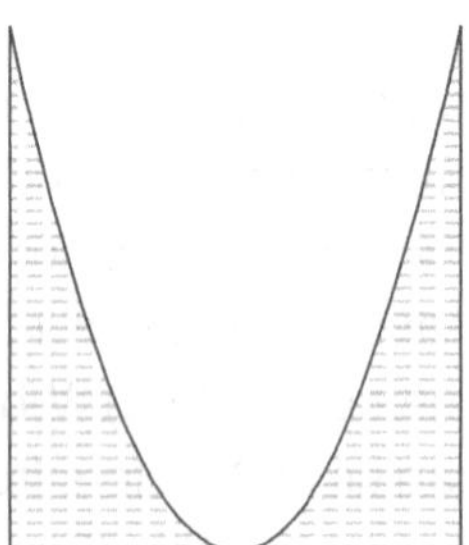

8.3 (d)

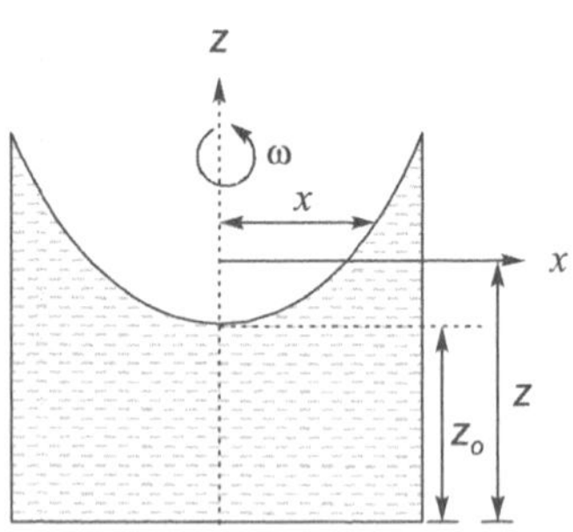

When a vertical cylindrical vessel containing water is rotated about its axis, then free surface of water becomes concave (as shown in figure)

where, $Z = Z_0 + \dfrac{\omega^2 x^2}{2g}$

This is the equation of parabola.

8.4 (a)

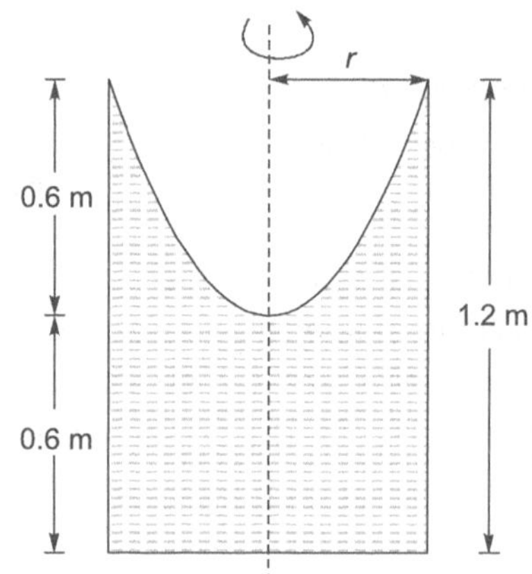

Original volume of cylinder,

$$V_1 = \pi r^2 h = \pi r^2 \times 1.2$$

Volume of liquid spilled out,

$$V_2 = \frac{1}{2}\pi r^2 \times h = \frac{1}{2}\pi r^2 \times 0.6$$

$$\therefore\ \frac{V_2}{V_1} = \frac{\frac{1}{2}\times 0.6\pi r^2}{\pi r^2 \times 1.2} = \frac{1}{4}$$

8.6 (d)

For an incompressible fluid flow radially outward from a line source in a steady manner.

Angular momentum is conserved.

$\Rightarrow\ mvr = \text{constant}$

$\Rightarrow\ vr = \text{constant}$

$$\Rightarrow\ v = \frac{\text{constant}}{r}$$

$$\therefore\ v \propto \frac{1}{r}$$

8.7 (c)

For the forced vortex, $v = \omega r$

8.8 (c)

Given data : $D = 50$ cm $= 0.50$ m; $S = 0.8$

$\omega = 2.0$ rad/s

For the sphere, the circular line of maximum

pressure is a horizontal circle, at a distance $h = g/\omega^2$ below the centre of sphere.

$$h = \frac{g}{\omega^2} = \frac{9.81}{(20)^2} = 0.0245 \text{ m} = \mathbf{2.45\ cm}$$

8.10 (a)

The flow past the cylinder with the vortex motion causes net transverse motion. This transverse force is known Magnus effect. The spinning of ball also occurs due to Magnus effect.

8.11 (a)

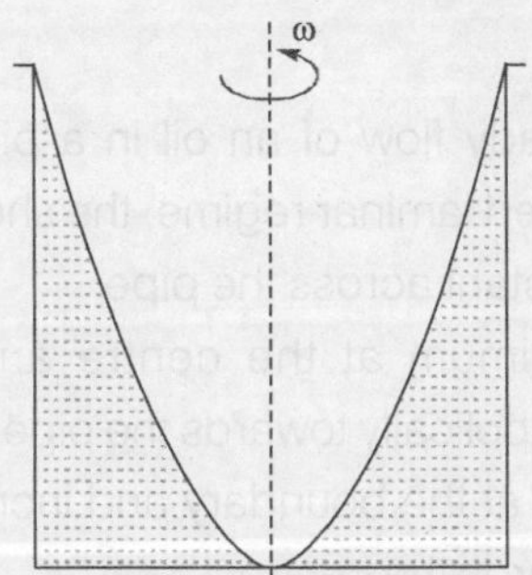

Volume of water spilled is equal to volume of paraboloid.

When half of liquid spills out, vertex of parabola lie at point P which is open to atmosphere.

So, pressure at P is zero.

8.12 (a)

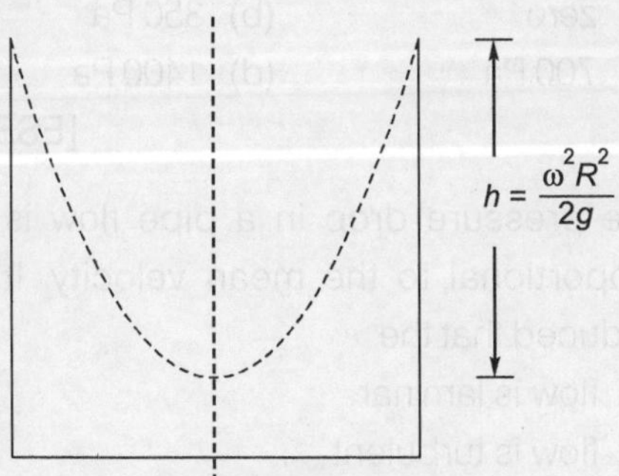

$$\omega = \frac{2\pi N}{60} = \frac{2\pi \times 300}{60} = 10\pi$$

$$h = \frac{(10\pi)^2 \times 0.1^2}{2 \times 9.81}$$

$$= 0.503 \text{ meters}$$

$$= 50.3 \text{ cm}$$

8.13 (c)

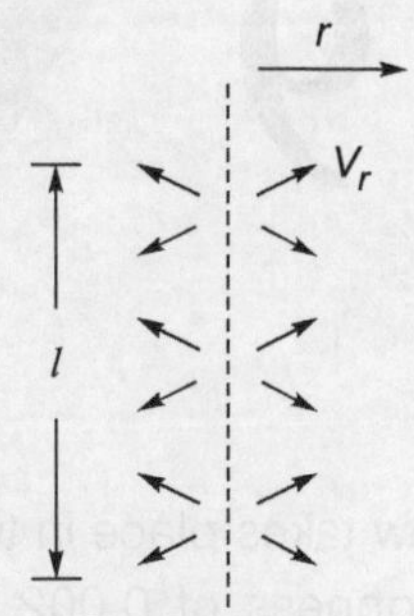

$$Q = V_r \times 2\pi r l$$

$$V_r = \frac{(Q/l)}{2\pi r}$$

$$V_r = \frac{q}{2\pi r} \quad (q = \text{source strength})$$

$$V_r = \frac{12}{2 \times \pi \times 0.8} = 2.387 \text{ m/s}$$

8.14 (d)

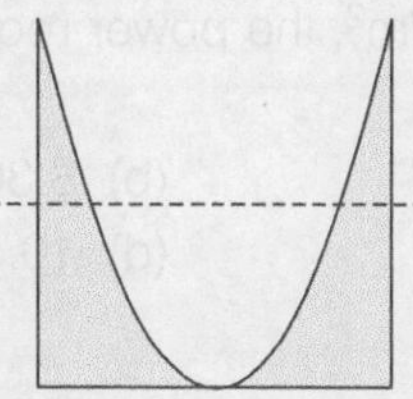

Angular velocity, ω

$$= \frac{2\pi N}{60} = \frac{2\pi \times 100}{60}$$

$$\omega = 10.47 \text{ rad/sec}$$

Height of parabola, z

$$= \frac{\omega^2 R^2}{2g} = \frac{10.47^2 \times 0.3^2}{2 \times 9.81}$$

$$z = 0.503 \text{ m}$$

8.15 (d)

$$V_1 r_1 = V_2 r_2$$

$$12 \times 15 = 2 \times V_2$$

$$V_2 = 90 \text{ m/s}$$

■■■■

9 Laminar Flow

9.1 If laminar flow takes place in two pipes, having relative roughness of 0.002 and 0.003, at a Reynolds number of 1815, then

(a) the pipe of relative roughness of 0.003 has a higher friction factor
(b) the pipe of relative roughness of 0.003 has a lower friction factor
(c) both pipes have the same friction factor
(d) no comparison is possible due to inadequate data **[ESE : 2000]**

9.2 A pipe of 20 cm diameter and 30 km length transports oil from a tanker to the shore with a velocity of 0.318 m/s. The flow is laminar. If μ = 0.1 N-s/m^2, the power required for the flow would be

(a) 9.25 kW (b) 8.36 kW
(c) 7.63 kW (d) 10.13 kW

[ESE : 2000]

9.3 The pressure drop in a 100 mm diameter horizontal pipe is 50 kPa over a length of 10 m. The shear stress at the pipe wall is

(a) 0.25 kPa (b) 0.125 kPa
(c) 0.50 kPa (d) 25.0 kPa

[ESE : 2001]

9.4 Laminar developed flow at an average velocity of 5 m/s occurs in a pipe of 10 cm radius. The velocity at 5 cm radius is

(a) 7.5 m/s (b) 10 m/s
(c) 2.5 m/s (d) 5 m/s

[ESE : 2001]

9.5 **Assertion (A):** In a pipe line, the nature of the fluid flow depends entirely on the velocity.
Reason (R): Reynolds number depends on the velocity, diameter of the pipe and kinematic viscosity of the fluid.

(a) Both **A** and **R** are individually true and **R** is the correct explanation of **A**
(b) Both **A** and **R** are individually true but **R** is not the correct explanation of **A**
(c) **A** is true but **R** is false
(d) **A** is false but **R** is true

[ESE : 2001]

9.6 In a steady flow of an oil in a pipe in the fully developed laminar regime, the shear stress is

(a) constant across the pipe
(b) maximum at the centre and decreases parabolically towards the pipe wall boundary
(c) zero at the boundary and increases linearly towards the centre
(d) zero at the centre and increases towards the pipe wall

[ESE : 2003]

9.7 The pressure drop for a relatively low Reynolds number flow in a 600 mm, 30 m long pipeline is 70 kPa. What is the wall shear stress?

(a) zero (b) 350 Pa
(c) 700 Pa (d) 1400 Pa

[ESE : 2004]

9.8 The pressure drop in a pipe flow is directly proportional to the mean velocity. It can be deduced that the

(a) flow is laminar
(b) flow is turbulent
(c) pipe is smooth
(d) pipe is rough **[ESE : 2006]**

9.9 Velocity for flow through a pipe, measured at the centre is found to be 2 m/s. Reynolds number is around 800. What is the average velocity in the pipe?

(a) 2 m/s (b) 1.7 m/s
(c) 1 m/s (d) 0.5 m/s

[ESE : 2007]

9.10 The power consumed per unit length in laminar flow for the same discharge, varies directly as D^n where D is the diameter of the pipe. What is the value of 'n'?
(a) 1/2 (b) –1/2
(c) –2 (d) –4
[ESE : 2008]

9.11 Consider the following statements:
1. The friction factor in laminar flow through pipes is independent of roughness.
2. The friction factor in laminar flow through pipes is directly proportional to Reynolds number.
3. In fully turbulent flow through pipes, friction factor is independent of Reynolds number.

Which of these statements are correct?
(a) 1, 2 and 3 (b) 1 and 3 only
(c) 2 and 3 only (d) 1 and 2 only
[ESE : 2008]

9.12 In a fully developed region of the pipe flow,
(a) the velocity profile continuously changes from linear to parabolic shape
(b) the pressure gradient remains constant in the downstream direction
(c) the pressure gradient continuously changes exceeding the wall shear stress in the downstream direction
(d) the pipe is not running full
[ESE : 2008]

9.13 Which one of the following statements is correct for a fully developed pipe flow?
(a) Pressure gradient balances the wall shear stress only and has a constant value.
(b) Pressure gradient is greater than the wall shear stress.
(c) The velocity profile is changing continuously.
(d) Inertia force balances the wall shear stress.
[ESE : 2009]

9.14 An oil of kinematic viscosity 0.5 stoke is flowing through a pipe of 5 cm diameter. The flow through the pipe becomes critical at velocity of
(a) 0.2 m/s (b) 2 m/s
(c) 2.5 m/s (d) 4 m/s
[ESE : 2010]

9.15 **Assertion (A):** For a fully developed laminar flow though a pipe, velocity distribution across any section is parabolic in shape.
Reason (R): The shear stress distribution from the centre line of the pipe upto the pipe surface increases linearly.
(a) Both **A** and **R** are individually true and **R** is the correct explanation of **A**
(b) Both **A** and **R** are individually true but **R** is not the correct explanation of **A**
(c) **A** is true but **R** is false
(d) **A** is false but **R** is true
[ESE : 2010]

9.16 For laminar flow through a pipe, the discharge varies
(a) linearly as the diameter
(b) inversely as the square of diameter
(c) as the inverse of viscosity
(d) inversely as the pressure gradient
[ESE : 2011]

9.17 Flow takes place at Reynolds number of 1500 in two different pipes with relative roughness of 0.001 and 0.002. The friction factor
(a) will be higher in case of pipe with relative roughness of 0.001
(b) will be higher in case of pipe with relative roughness of 0.002
(c) will be same in both pipes
(d) in the two pipes cannot be compared on the basis of data given
[ESE : 2011]

9.18 Flow commences between two parallel plates with the upper plate moving in the direction of flow, while the other plate is stationary. The resulting flow between the plates is called
(a) creep flow (b) Couette flow
(c) plug flow (d) Stokes flow
[ESE : 2012]

9.19 A 0.20 m diameter pipe 20 km long transports oil at a flow rate of 0.01 m^3/s. Calculate power required to maintain flow if dynamic viscosity and density of oil is 0.08 Pa.s and 900 kg/m^3 respectively
(a) 4.064 kW (b) 3.074 kW
(c) 5.064 kW (d) 4.074 kW
[ESE : 2013]

9.20 Pressure loss for laminar flow through pipeline is dependent
(a) inversely on flow of velocity
(b) directly on square of pipe radius
(c) directly on length of pipe
(d) inversely on viscosity of flowing medium
[ESE : 2013]

9.21 Laminar flow between closely spaced parallel plates is governed by the consideration of which one of the following pair of forces?
(a) Pressure and inertial forces
(b) Gravity and inertial forces
(c) Viscous and inertial forces
(d) Pressure and viscous forces
[ESE : 2014]

9.22 For fully developed laminar flow through a circular pipe with Reynolds number *Re* the friction factor is
(a) inversely proportional to *Re*
(b) proportional to *Re*
(c) proportional to square of *Re*
(d) independent of *Re*
[ESE : 2014]

9.23 The laminar flow is characterized by Reynolds number which is
(a) equal to critical value
(b) less than the critical value
(c) more than the critical value
(d) zero critical value
[ESE : 2015]

9.24 In laminar flow through a circular pipe, the discharge varies
(a) linearly with fluid density
(b) inversely with pressure drop
(c) directly as square of pipe radius
(d) inversely with fluid viscosity
[ESE : 2015]

9.25 Air is flowing over a flat plate with a free steam velocity of 24 m/s, and its kinematic viscosity is 72×10^{-6} m²/s. If a particular point, the Reynolds number is 30000, its location from the leading edge is
(a) 0.05 m (b) 0.07 m
(c) 0.08 m (d) 0.09 m
[ESE : 2015]

9.26 An oil flows through a pipe at a velocity of 1.0 m/s. The pipe is 45 m long and has 150 mm diameter. What is the head loss due to friction, if $\rho = 869$ kg/m³ and $\mu = 0.0814$ kg/ms?
(a) 0.61 m (b) 0.51 m
(c) 0.41 m (d) 0.31 m
[ESE : 2017]

9.27 Consider the following statements :
1. At low Reynolds numbers of any flow, viscous forces dominate over inertial forces.
2. Transition from laminar to turbulent flow occurs over a range of Reynolds numbers depending on the surface presented to the flow.

Which of the above statements is/are correct?
(a) 1 only (b) 2 only
(c) Both 1 and 2 (d) Neither 1 nor 2
[ESE : 2017]

9.28 In a steady laminar flow of a given discharge through a circular pipe of diameter *D*, the head loss is proportional to
(a) D^{-1} (b) D^{-2}
(c) D^{-3} (d) D^{-4} [ESE : 2017]

9.29 The centre-line velocity in a pipe flow is 2 m/s. What is the average flow velocity in the pipe if the Reynolds number of the flow is 800?
(a) 2 m/s (b) 1.5 m/s
(c) 1 m/s (d) 0.5 m/s
[ESE : 2017]

9.30 The shear stress τ_0 for steady, fully developed flow inside a uniform horizontal pipe with coefficient of friction *f*, density ρ and velocity *v*, is given by
(a) $\frac{f\rho v^2}{2}$ (b) $\frac{f\rho^2 v}{2}$
(c) $\frac{\rho^2 v}{2f}$ (d) $\frac{\rho v^2}{2f}$
[ESE : 2018]

9.31 In a laminar flow through pipe, the point of maximum instability exists at a distance of *y* from the wall which is
(a) $\frac{3}{2}$ of pipe radius *R*
(b) $\frac{2}{3}$ of pipe radius *R*

(c) $\frac{1}{2}$ of pipe radius R

(d) $\frac{1}{3}$ of pipe radius R

[ESE : 2019]

9.32 An oil of viscosity 8 poise flows between two parallel fixed plates, which are kept at a distance of 30 mm apart. If the drop of pressure for a length of 1 m is 0.3×10^4 N/m^2 and width of the plates is 500 mm, the rate of oil flow between the plates will be

(a) 4.2×10^{-3} m^3/s (b) 5.4×10^{-3} m^3/s
(c) 6.6×10^{-3} m^3/s (d) 7.8×10^{-3} m^3/s

[ESE : 2019]

9.33 To maintain 0.08 m^3/s flow of petrol with a specific gravity of 0.7, through a steel pipe of 0.3 m diameter and 800 m length, with coefficient of friction of 0.0025 in the Darcy relation, the power required will be nearly

(a) 0.6 kW (b) 1.0 kW
(c) 2.6 kW (d) 3.0 kW

[ESE : 2020]

9.34 A fluid of mass density 1790 kg/m^3 and viscosity 2.1 Ns/m^2 flows at a velocity of 3 m/s in a 6 cm diameter pipe. The head loss over a length of 12 m pipe will be nearly

(a) 62.0 m (b) 54.0 m
(c) 46.5 m (d) 38.5 m

[ESE : 2020]

9.35 For fully developed laminar pipe flow, the average velocity is

(a) One-half of the maximum velocity
(b) One-third of the maximum velocity
(c) One-fourth of the maximum velocity
(d) Two-third of the maximum velocity

[ESE : 2020]

9.36 A crude oil having a specific gravity of 0.9 flows through a pipe of diameter 0.15 m at the rate of 8 lps. If the value of μ is 0.3 Ns/m^2, the Reynolds number will be nearly

(a) 295 (b) 235
(c) 205 (d) 165

[ESE : 2020]

9.37 Air moves over a 10 m long flat plate. The transition from laminar to turbulent flow takes place between Reynolds numbers of 2.5×10^6 and 3.6×10^6. The free stream velocity is 30 m/s and $\nu = 1.5 \times 10^{-5}$ m^2/s. What is the maximum distance from the front edge of the plate along which one expects laminar flow in the boundary layer?

(a) 0.9 m (b) 1.2 m
(c) 1.5 m (d) 1.8 m

[ESE : 2023]

Directions for the following **two (02)** items:

Read the following information and answer the **two** items that follows:

Air ($\rho = 1.23$ kg/m^3 and $\nu = 1.5 \times 10^{-5}$ m^2/s) is flowing over a flat plate. The free stream speed is 15 m/s at a distance of 1 m from the leading edge.

9.38 What is the boundary layer thickness for completely laminar flow?

(a) 5.48 mm (b) 7.21 mm
(c) 2.83 mm (d) 8.35 mm

[ESE : 2023]

9.39 What is the wall shear stress for completely laminar flow?

(a) 0.101 N/m^2 (b) 1.201 N/m^2
(c) 2.301 N/m^2 (d) 3.401 N/m^2

[ESE : 2023]

9.40 What is the pressure gradient per metre width for the laminar flow of oil with a maximum velocity of 2 m/s between two horizontal parallel fixed plates which are 100 mm apart? (Take dynamic viscosity = 2.5 N-s/m^2)

(a) –4000 N/m^2 per m
(b) –3700 N/m^2 per m
(c) –3500 N/m^2 per m
(d) –3200 N/m^2 per m

[ESE : 2024]

9.41 An oil is flowing through a pipe of diameter 300 mm with velocity of 5 m/s. What is the head loss to maintain the flow for a length of 1000 m? (Take coefficient of friction = 0.005)

(a) 60.2 m (b) 72.4 m
(c) 76.7 m (d) 84.9 m

[ESE : 2024]

■■■■

Answers Laminar Flow

9.1 (c)	9.2 (c)	9.3 (b)	9.4 (a)	9.5 (d)	9.6 (d)	9.7 (b)	9.8 (a)	9.9 (c)
9.10 (d)	9.11 (b)	9.12 (b)	9.13 (a)	9.14 (b)	9.15 (a)	9.16 (c)	9.17 (c)	9.18 (b)
9.19 (d)	9.20 (c)	9.21 (d)	9.22 (a)	9.23 (b)	9.24 (d)	9.25 (d)	9.26 (a)	9.27 (c)
9.28 (d)	9.29 (c)	9.30 (a)	9.31 (b)	9.32 (a)	9.33 (b)	9.34 (d)	9.35 (a)	9.36 (c)
9.37 (d)	9.38 (c)	9.39 (a)	9.40 (a)	9.41 (d)				

Explanations Laminar Flow

9.1 (c)

In a laminar flow friction factor is independent of relative roughness.

$$f = \frac{64}{Re}$$

9.2 (c)

$$p_1 - p_2 = \frac{32\mu uL}{D^2}$$

$$\therefore \text{ Power} = \Delta p \times Q = \frac{32\mu uL}{D^2} \times \frac{\pi}{4} D^2 \times u$$

$$= 8\pi\mu v^2 L = 8 \times \pi \times 0.1 \times (0.318)^2 \times 30$$

$$= \mathbf{7.63\ kW}$$

9.3 (b)

$$\tau = \left(\frac{dp}{dx}\right)\left(\frac{R}{2}\right) = \frac{50 \times 1000 \times 50 \times 10^{-3}}{10 \times 2}$$

$$= 0.125 \text{ kPa}$$

9.4 (a)

$$U = 2U_{av}\left[1 - \left(\frac{r}{R}\right)^2\right] = 2 \times 5\left[1 - \left(\frac{5}{10}\right)^2\right]$$

$$= \mathbf{7.5\ m/s}$$

9.5 (d)

In a pipe line, nature of the fluid flow depends entirely on the Reynold number. Which is depends on velocity, diameter of pipe and kinematic viscosity of the fluid.

9.6 (d)

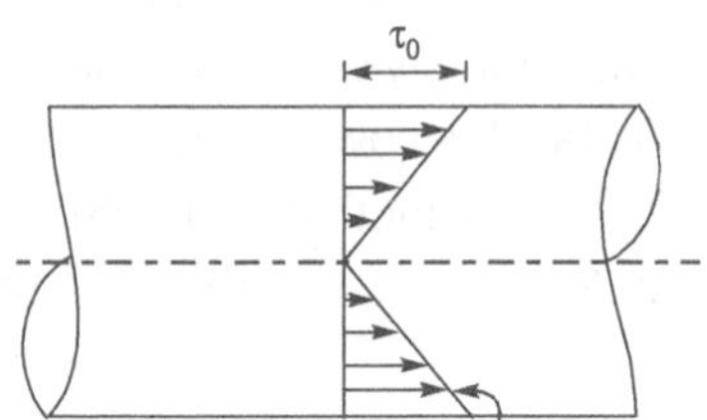

Shear stress distribution

9.7 (b)

$$\tau = \left(\frac{dp}{dx}\right) \times \frac{R}{2} = \frac{70 \times 1000}{30} \times \frac{0.3}{2} = 350 \text{ Pa}$$

9.8 (a)

In laminar flow pressure drop in a pipe flow is directly proportional to the mean velocity.

$$U = \frac{1}{4\mu}\left(-\frac{dp}{dx}\right)\left(R^2 - r^2\right);\ U \propto \frac{dp}{dx}$$

9.9 (c)

Average velocity is the pipe

$$= \frac{\text{Maximum velocity}}{2} = \frac{2}{2} = \mathbf{1\ m/s}$$

9.11 (b)

The friction factor in laminar flow through pipes is given by the expression $\frac{64}{Re}$ which is independent of roughness.

The friction factor for laminar flow through pipes is inversely proportional to Reynolds Number.

In fully turbulent flow through pipes friction factor is independent of Reynolds Number.

9.12 (b)

In a fully developed flow region of the pipe flow the pressure gradient remains constant in the downstream direction.

9.13 (a)

In case of open-channel flow, the water surface itself is hydraulic grade line. The total energy line lies at a distance of $V^2/2g$ above hydraulic gradient line at every section.

9.14 (b)

$\nu = 0.5 \text{ stoke} = 0.5 \times 10^{-4} \text{ m}^2/\text{s}$

$D = 5 \times 10^{-5}$ m

Critical value of Reynolds number = 2000

$$Re = \frac{VD}{\nu}$$

$$2000 = \frac{V \times 5 \times 10^{-2}}{0.5 \times 10^{-4}}$$

or $V =$ **2 m/s**

9.16 (c)

For Laminar flow through pipe

$$\text{Discharge: } Q = \frac{\pi D^4}{128\mu}\left(\frac{\partial P}{dx}\right)$$

$$\therefore \quad Q \propto \frac{1}{\mu}$$

9.17 (c)

As Reynold number is less then 2000 so flow is laminar therefore the friction will same for both pipe.

Friction factor: $f = \frac{64}{Re}$ for laminar flow

9.19 (d)

Pressure drop,

$$\Delta p = \frac{32\mu LV}{D^2}$$

$$\text{where, } V = \frac{4Q}{\pi d^2} = \frac{4 \times 0.01}{3.14 \times 0.04} = \frac{1}{3.14} \text{m/s}$$

$$\therefore \quad \Delta p = \frac{32 \times 0.08 \times 20 \times 10^3 \times 1}{3.14 \times 0.04} \text{ Pa}$$

Power required = $Q.\Delta p$

$$= \frac{0.01 \times 32 \times 0.08 \times 20}{3.14 \times 0.04} \times 10^3 \text{ Nm/s}$$

= **4.0764 kW**

9.20 (c)

Pressure drop for laminar flow,

$$\Delta p = \frac{32\mu LV}{D^2}$$

$$\therefore \quad \Delta p \propto \text{Length}$$

$$\propto \frac{1}{(\text{Diameter})^2}$$

$\propto$ Dynamic viscosity

$\propto$ Velocity

9.21 (d)

Simplified form of Navier-Stokes equation in rectangular coordinate system is given by

$$\frac{d\tau_{yx}}{dy} = \frac{\partial p}{\partial x} = \text{constant}$$

So, pressure force and viscous forces are considered.

9.22 (a)

$$f = \frac{64}{Re} \quad \text{for laminar flow}$$

9.23 (b)

Laminar flow is characterized by Reynolds number if it is less than critical value, above which the flow will be turbulent.

9.24 (d)

By Hagen-Poiseuille formula,

$$Q = \frac{\pi D^4}{128\mu}\frac{(p_1 - p_2)}{L}$$

$$\therefore \quad Q \propto \frac{D^4}{\mu}\frac{\Delta p}{L}$$

∴ So it varies inversely proportional to viscosity of fluid.

9.25 (d)

$$\nu = \frac{\mu}{\rho} = 72 \times 10^{-6} \text{ m}^2/\text{s}$$

$$R_{ex} = \rho\frac{Ux}{\mu}$$

$$\text{or} \quad 30000 = \frac{Ux}{\nu}$$

$$x = \frac{30000\,\nu}{u} = \frac{30000 \times 72 \times 10^{-6}}{24}$$

= 0.09 m = **9 cm**

9.26 (a)

$$\text{Re} = \frac{\rho \bar{u} D}{\mu} = \frac{869 \times 1 \times 0.15}{0.0814}$$

$= 1601.35 < 2000$ Flow is laminar

$$h_L = \frac{32\mu \bar{u} L}{\rho g D^2} = \frac{32 \times 0.0814 \times 1 \times 45}{869 \times 9.81 \times (0.15)^2}$$

$h_L = 0.611$ m

9.28 (d)

$$h_L = \frac{32\mu \times \bar{u} L}{\rho g D^2} = \frac{32u \times 4Q \times L}{\rho g \times \pi D^2 \times D^2}$$

$$= \frac{128\mu Q L}{\rho g \pi D^4}$$

$h_L \propto D^{-4}$

9.29 (c)

$$u_{avg} = \frac{u_{max}}{2} = 1 \text{ m/s}$$

9.32 (a)

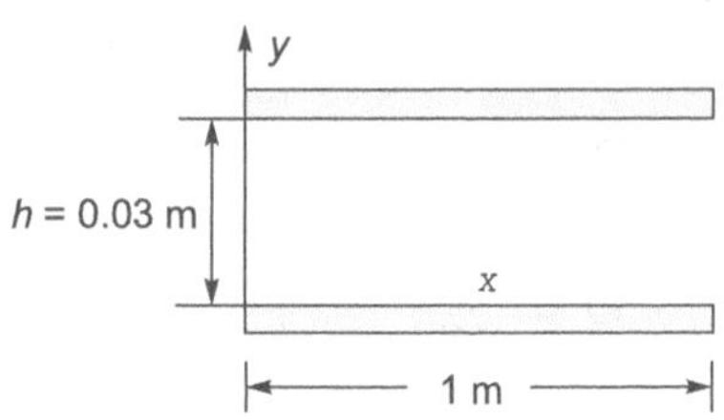

$$u = \frac{1}{2\mu}\left(\frac{-\partial P}{\partial x}\right)(hy - y^2)$$

$$\bar{u} = \frac{h^2}{12\mu}\left(\frac{-\partial P}{\partial x}\right)$$

$$\bar{u} = \frac{(0.03)^2}{12 \times 0.8} \times \left(\frac{0.3 \times 10^4}{1}\right)$$

$\bar{u} = 0.281$ m/s

$Q = \bar{u} \times (h \times b) = 4.2 \times 10^{-3}$ m^3/s

9.33 (b)

Given, $Q = 0.08$ m^3/sec;

$s = 0.7$; $d = 0.3$ m

$L = 800$ m; $f' = 0.0025$

Power $= wp = \rho g Q H_m$

$H_m = h_f =$ Head loss in pipe

$$= \left.\frac{fLV^2}{2gD}\right|_{pipe}$$

$$Q_{pump} = Q_{pipe}$$

$$0.08 = \frac{\pi}{4} d^2 \cdot V_{pipe} = \frac{\pi}{4} \times 0.3^3 \times V_{pipe}$$

$V_{pipe} = 1.13$ m/s

$$h_f = \left.\frac{(4 \times 0.0025) \times 800 \times 1.13^2}{2 \times 9.81 \times 0.3}\right|_{pipe}$$

$= 1.74$ m

$WP = (1000 \times 0.7) \times 9.81 \times 0.08 \times 1.74$

$= 1$ kW

9.34 (d)

$\rho = 1790$ kg/m^3, $\mu = 2.1$ Ns/m

$v = 3$ m/s, $D = 6$ cm $= 0.06$ m

Reynolds number,

$$\text{Re} = \frac{\rho V D}{\mu} = \frac{1790 \times 3 \times 0.06}{2.1}$$

$= 153.42 < 2300$

hence flow is laminar, so,

$$h_f = \frac{32\mu V L}{\omega d^2} = \frac{32 \times 2.1 \times 3 \times 12}{1790 \times 9.81 \times 0.06^2}$$

$= 38.26$ m

9.36 (c)

$\rho_{oil} = 0.9 \times 1000 = 900$ kg/m^3

$D = 0.15$ m, $\mu = 0.3$ Ns-m^2

$Q = 8$ lt/s $= 8 \times 10^{-3}$ m^3/s

Reynold's number,

$$\text{Re} = \frac{\rho V D}{\mu} = \frac{\rho Q D}{\mu \frac{\pi}{4} D^2} = \frac{\rho Q \times 4}{\mu \pi D}$$

$$= \frac{900 \times 8 \times 10^{-3} \times 4}{0.3 \times \pi \times 0.15}$$

$= 203.7 \approx 205$

9.37 (d)

$L = 10$ m, $U_\infty = 30$ m/s, $\nu = 1.5 \times 10^{-5}$ m²/s

$$\text{Reynolds number} = \frac{U_\infty \cdot x}{\nu}$$

$$3.6 \times 10^6 = \frac{30 \times x}{1.5 \times 10^{-5}}$$

$$\therefore \quad x = 1.8 \text{ m}$$

9.38 (c)

$$\rho = 1.23 \text{ kg/m}^3$$
$$\nu = 1.5 \times 10^{-5} \text{ m}^2\text{/s}$$
$$U_\infty = 15 \text{ m/s at } x = 1 \text{ m}$$

Reynold's number,

$$3.2 \times 10^5 = \frac{15 \times x}{1.5 \times 10^{-5}}$$

$$\Rightarrow \quad x = 0.32 \text{ m}$$

$$\therefore \quad \delta = \frac{5x}{\sqrt{Re_x}} = \frac{5 \times 0.32}{\sqrt{3.2 \times 10^5}} = 2.83 \text{ mm}$$

9.39 (a)

For laminar flow,

$$C_{f_x} = \frac{0.664}{\sqrt{Re_x}} = \frac{0.664}{\sqrt{3.2 \times 10^5}}$$

$$C_{f_x} = \frac{0.664}{565.68} = 1.17 \times 10^{-3}$$

$\therefore$ Wall shear stress,

$$\tau_w = C_{f_x} \times \rho \times \frac{V^2}{2}$$

$$= 1.17 \times 10^{-3} \times 1.23 \times \frac{15^2}{2}$$

$$= 0.162 \text{ N/m}^2$$

9.40 (a)

Given: $u_{max} = 2$ m/s, $h = 100$ mm, $\mu = 2.5$ N-s/m²

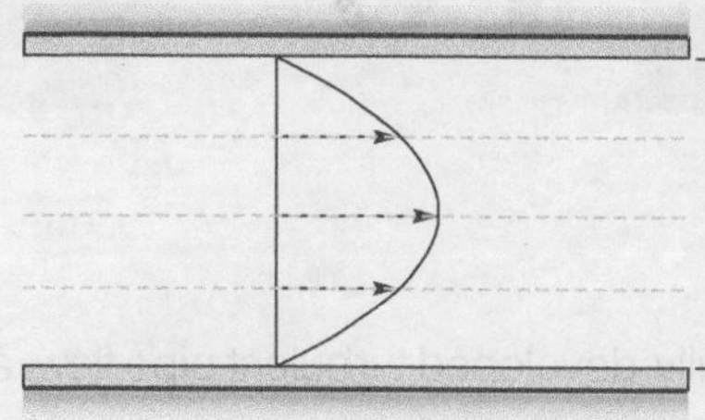

For fluid flow between two fixed parallel plate,

$$u_{max} = \frac{1}{8\mu}\left(-\frac{dP}{dx}\right)h^2$$

$$\Rightarrow \quad 2 = \frac{1}{8 \times 2.5} \times \left(-\frac{dP}{dx}\right) \times (0.1)^2$$

$$\Rightarrow \quad \frac{dP}{dx} = \frac{-2 \times 8 \times 2.5}{(0.1)^2} = -4000 \text{ N/m}^2 \text{ per m}$$

9.41 (d)

Given: $d = 300$ mm, $v = 5$ m/s, $L = 1000$ m, $f' = 0.005$

Friction head loss is given as,

$$h_f = \frac{4f'Lv^2}{2gd} = \frac{4 \times 0.005 \times 1000 \times 5^2}{2 \times 9.81 \times 0.300}$$

$$= 84.947 \text{ m} \simeq 84.9 \text{ m}$$

■■■■

10 Turbulent Flow

10.1 In a fully-developed turbulent pipe flow, assuming 1/7th power law, the ratio of time mean velocity at the centre of the pipe to the average velocity of the flow is

(a) 2.0 (b) 1.5
(c) 1.22 (d) 0.817

[ESE : 2001]

10.2 The critical value of Reynolds number for transition from laminar to turbulent boundary layer in external flows is taken as

(a) 2300 (b) 4000
(c) 5×10^5 (d) 3×10^6

[ESE : 2002]

10.3 The friction factor for fully developed turbulent flow through a rough circular pipe depends on

(a) Reynolds number
(b) relative roughness
(c) Reynolds number and relative roughness
(d) Reynolds number, relative roughness and diameter to length ratio

[ESE : 2003]

10.4 Consider the following statements regarding laminar sublayer of boundary layer flow:

1. The laminar sublayer exists only in a region that occurs before the formation of laminar boundary layer.
2. The laminar sublayer is a region next to the wall where the viscous force is predominate while the rest of the flow is turbulent.
3. The laminar sublayer occurs only in turbulent flow past a smooth plate.

Which of these statements is/are correct?

(a) 1, 2 and 3 (b) 1 and 2
(c) Only 2 (d) 1 and 3

[ESE : 2004]

10.5 How does the head loss in turbulent flow in pipe vary?

(a) Directly as velocity
(b) Inversely as square of velocity
(c) Approximately as square of velocity
(d) Inversely as velocity

[ESE : 2009]

10.6 The parameters which determine the friction for turbulent flow in a rough pipe are

(a) Froude number and relative roughness
(b) Froude number and Mach number
(c) Reynolds number and relative roughness
(d) Mach number and relative roughness

[ESE : 2010]

10.7 A flat plate of 4 m length is kept parallel to air flowing at 5 m/s at 15°C. Assuming the density of air to be 1.2 kg/m^3 and dynamic viscosity to be 1.76×10^{-5} kg/ms, and knowing that the flow becomes turbulent when Reynolds number exceeds 5×10^5, the flow will become turbulent at

(a) 1.25 m (b) 1.47 m
(c) 1.52 m (d) 1.74 m

[ESE : 2012]

10.8 Which of the following factors determine the friction for turbulent flow of incompressible fluids in a rough pipe?

(a) Froude number and Mach number
(b) Mach number and relative roughness
(c) Reynolds number and relative roughness
(d) Froude number and relative roughness

[ESE : 2012]

10.9 Water flows through a smooth circular pipe of diameter *D* and length *L* because of a pressure difference ΔP across the length. The volume flow rate is *Q* and the flow is turbulent with Reynolds number 10^5. If the pressure difference is increased to 4 ΔP the volume flow rate will be

(a) 2*Q*
(b) A little more than 2*Q*
(c) A little less than 2*Q*
(d) 4*Q*

[ESE : 2014]

10.10 $Q = \dfrac{\partial u'}{\partial x} = -\dfrac{\partial v'}{\partial y}$ for a turbulent flow signifies

(a) conservation of bulk momentum transport
(b) increase in u' in *x*-direction followed by increase in v' in negative *y*-direction
(c) turbulence is anisotropic
(d) turbulence is isotropic

[ESE : 2019]

■■■■

Answers Turbulent Flow

10.1 (d) 10.2 (c) 10.3 (b) 10.4 (c) 10.5 (c) 10.6 (c) 10.7 (b) 10.8 (c) 10.9 (b) 10.10 (b)

Explanations Turbulent Flow

10.2 (c)

For flow over flat plate in the boundary layer, in case of laminar flow the Reynold number must be less-than 5×10^5.

10.4 (c)

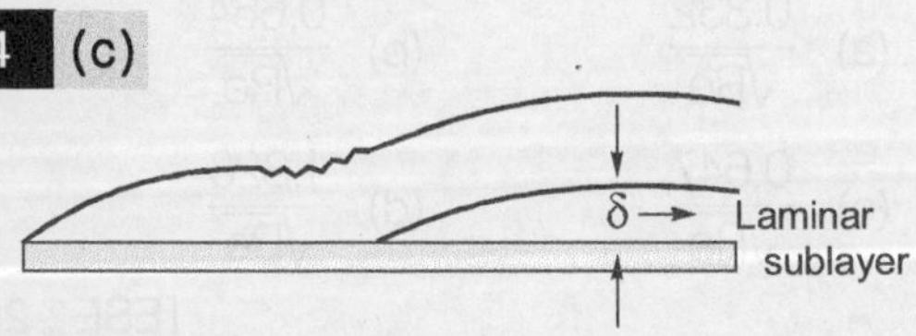

10.5 (c)

Head loss in a turbulent flow

$$H_L = V^{1.7 \text{ to } 2}$$

10.6 (c)

In laminar flow friction factor depend only on Reynolds number but in case of turbulent flow, the friction factor will depend upon Reynolds number and Roughness factor both.

10.7 (b)

$$Re = \frac{\rho V x}{\mu}$$

For flow to become turbulent

$$Re = 5 \times 10^5$$

$$\therefore \quad 5 \times 10^5 = \frac{1.2 \times 5 \times x}{1.76 \times 10^{-5}}$$

or $$x = 1.47 \text{ m}$$

10.9 (b)

$$\Delta h = \frac{flv^2}{2gd}$$

$$f \propto \frac{0.316}{(\text{Re})^{1/4}} \Rightarrow f \propto \frac{1}{(v)^{0.25}}$$

$$\Delta h \propto v^{7/4}$$

$$\Delta p \propto (Q_1)^{7/4}$$

$$4\Delta p \propto (Q_2)^{7/4}$$

$$Q_2 = 2.208\, Q_1$$

■■■■

11 Boundary Layer Theory

11.1 The development of boundary layer zones labelled *P*, *Q*, *R* and *S* over a flat plate is shown in the figure below.

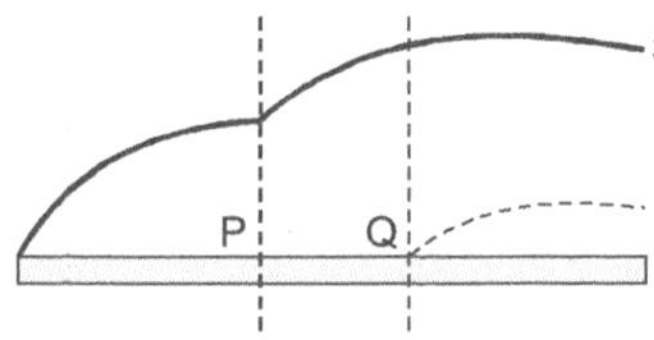

Based on this figure, match **List-I** (Boundary layer zones) with **List-II** (Types of boundary layer) and select the correct answer using the codes given below the lists:

List-I	List-II
A. *P*	1. Transitional
B. *Q*	2. Laminar viscous sub-layer
C. *R*	3. Laminar
D. *S*	4. Turbulent

Codes:

	A	B	C	D
(a)	3	1	2	4
(b)	3	2	1	4
(c)	4	2	1	3
(d)	4	1	2	3

[ESE : 2000]

11.2 In a turbulent boundary layer over the entire length of a plate, the boundary layer thickness increases with its distance x from the leading edge as

(a) $x^{1/2}$ (b) $x^{1/5}$
(c) $x^{2/5}$ (d) $x^{4/5}$ **[ESE : 2000]**

11.3 Separation of fluid flow is caused by

(a) reduction of pressure in the direction of flow
(b) reduction of the boundary layer thickness
(c) presence of adverse pressure gradient
(d) presence of favourable pressure gradient

[ESE : 2000]

11.4 The velocity distribution in the boundary layer is given as $\frac{u}{U_\infty} = \frac{y}{\delta}$, where u is the velocity at a distance y from the boundary, U_∞ is the free stream velocity and δ is the boundary layer thickness at a certain distance from the leading edge of plate. The ratio of displacement thickness to momentum thickness is

(a) 5 (b) 4
(c) 3 (d) 2

[ESE : 2001]

11.5 For the velocity profile $\frac{u}{U_\infty} = \eta$, the momentum thickness of a laminar boundary layer on a flat plate at a distance of 1 m from leading edge for air (kinematic viscosity = 2×10^{-5} m^2/s) flowing at a free stream velocity of 2 m/s is given by

(a) 3.16 mm (b) 2.1 mm
(c) 3.16 m (d) 2.1 m

[ESE : 2001]

11.6 According to Blasius law, the local skin friction coefficient in the boundary layer over a flat plate is given by

(a) $\frac{0.332}{\sqrt{Re}}$ (b) $\frac{0.664}{\sqrt{Re}}$
(c) $\frac{0.647}{\sqrt{Re}}$ (d) $\frac{1.328}{\sqrt{Re}}$

[ESE : 2001]

11.7 The boundary layer flow separates from the surface if

(a) $\frac{du}{dy} = 0$ and $\frac{dp}{dx} = 0$
(b) $\frac{du}{dy} = 0$ and $\frac{dp}{dx} > 0$
(c) $\frac{du}{dy} = 0$ and $\frac{dp}{dx} < 0$
(d) the boundary layer thickness is zero

[ESE : 2002]

11.8 The laminar boundary layer thickness, δ at any point x for flow over a flat plate is given by $\frac{\delta}{x}$ =

(a) $\frac{0.664}{\sqrt{Re_x}}$ (b) $\frac{1.328}{\sqrt{Re_x}}$

(c) $\frac{1.75}{\sqrt{Re_x}}$ (d) $\frac{5.0}{\sqrt{Re_x}}$

[ESE : 2002]

11.9 Match **List-I** (Phenomena) with **List-II** (Causes) and select the correct answer using the codes given below the lists:

List-I

A. Shock wave
B. Flow separation
C. Capillary rise
D. Cavitation

List-II

1. Surface tension
2. Vapour pressure
3. Compressibility
4. Adverse pressure gradient

Codes:

	A	B	C	D
(a)	3	1	2	4
(b)	4	2	1	3
(c)	3	4	1	2
(d)	4	1	2	3

[ESE : 2003]

11.10 Consider the following statements pertaining to boundary layer:

1. Boundary layer is a thin layer adjacent to the boundary where maximum viscous energy dissipation takes place.
2. Boundary layer thickness is a thickness by which the ideal flow is shifted.
3. Separation of boundary layer is caused by presence of adverse pressure gradient.

Which of these statements are correct?

(a) 1, 2 and 3 (b) 1 and 2
(c) 1 and 3 (d) 2 and 3

[ESE : 2003]

11.11 The velocity profile for turbulent layer over a flat plate is

(a) $\frac{u}{U} = \sin\left(\frac{\pi}{2} - \frac{y}{\delta}\right)$

(b) $\frac{u}{U} = \left(\frac{y}{\delta}\right)^{1/7}$

(c) $\frac{u}{U} = 2\left(\frac{y}{\delta}\right) - \left(\frac{y}{\delta}\right)^2$

(d) $\frac{u}{U} = \frac{3}{2}\left(\frac{y}{\delta}\right) - \frac{1}{2}\left(\frac{y}{\delta}\right)^3$

[ESE : 2003]

11.12 **Assertion (A):** Aircraft wings are slotted to control separation of boundary layer especially at large angles of attack.

Reason (A): This helps to increase the lift and the aircraft can take off from, and land on, short runways.

(a) Both **A** and **R** are individually true and **R** is the correct explanation of **A**
(b) Both **A** and **R** are individually true but **R** is not the correct explanation of **A**
(c) **A** is true but **R** is false
(d) **A** is false but **R** is true

[ESE : 2003]

11.13 A flat plate, 2 m × 0.4 m is set parallel to a uniform stream of air (density 1.2 kg/m^3 and viscosity 16 centistoke) with its shorter edges along the flow. The air velocity is 30 km/h. What is the approximate estimated thickness of boundary layer at the downstream end of the plate?

(a) 1.96 mm (b) 4.38 mm
(c) 13.12 mm (d) 9.51 mm

[ESE : 2004]

11.14 What is the ratio of displacement thickness to momentum thickness for linear velocity distribution over a flat plate?

(a) 1.5 (b) 2.0
(c) 2.5 (d) 3.0

[ESE : 2004]

11.15 Match **List-I** (Variables in laminar boundary layer flow over a flat plate set parallel to the stream) with **List-II** (Related expression with usual notations) and select the correct answer using the codes given below the lists:

List-I

A. Boundary layer thickness
B. Average skin friction coefficient
C. Shear stress at boundary
D. Displacement thickness

List-II

1. $1.729 / \sqrt{Ux / \nu}$
2. $0.332 \rho U^2 / \sqrt{Ux / \nu}$
3. $5\sqrt{\nu / Ux}$
4. $0.664\sqrt{\nu / Ux}$
5. $1.328 / \sqrt{UL / \nu}$

Codes:

	A	B	C	D
(a)	3	5	4	2
(b)	2	4	1	3
(c)	3	5	2	1
(d)	5	4	1	2

[ESE : 2004]

11.16 For linear distribution of velocity in the boundary layer on a flat plate, what is the ratio of displacement thickness (δ^*) to the boundary layer thickness (δ)?

(a) $\frac{1}{4}$ (b) $\frac{1}{3}$

(c) $\frac{1}{2}$ (d) $\frac{1}{5}$

[ESE : 2005]

11.17 The velocity distribution in the boundary layer over a flat plate, set parallel to the direction of an incompressible free stream, is given by one-sixth power law of the $\frac{u}{U} = \left(\frac{y}{\delta}\right)^{1/6}$. What is the ratio of displacement thickness to boundary layer thickness?

(a) 1 : 8 (b) 1 : 7

(c) 7 : 72 (d) 3 : 8

[ESE : 2006]

11.18 In the boundary layer, the flow is

(a) viscous and rotational
(b) inviscid and irrotational
(c) inviscid and rotational
(d) viscous and irrotational

[ESE : 2006]

11.19 The velocity profile in a laminar boundary layer is given by $\frac{u}{U} = \frac{y}{\delta}$. The ratio of momentum thickness to displacement thickness for the boundary is given by which one of the following?

(a) 2 : 3 (b) 1 : 2

(c) 1 : 6 (d) 1 : 3

[ESE : 2008]

11.20 The velocity distribution in a turbulent boundary layer is given by $\frac{u}{U} = \left(\frac{y}{\delta}\right)^{1/7}$

What is the displacement thickness δ^*?

(a) δ (b) $\frac{\delta}{7}$

(c) $\frac{7\delta}{8}$ (d) $\frac{\delta}{8}$

[ESE : 2008]

11.21 Assertion (A): In the boundary layer concept the shear stress at the outer edge of the layer is considered to be zero.

Reason (R): Local velocity is almost equal to velocity in potential flow.

(a) Both **A** and **R** are individually true and **R** is the correct explanation of **A**
(b) Both **A** and **R** are individually true but **R** is not the correct explanation of **A**
(c) **A** is true but **R** is false
(d) **A** is false but **R** is true

[ESE : 2008]

11.22 What is the commonly used boundary layer control method to prevent separation?

(a) Use of smooth boundaries
(b) Using large divergence angle in the boundary
(c) Suction of accelerating fluid within the boundary layer
(d) Suction of retarded fluid within the boundary layer

[ESE : 2009]

11.23 Assertion (A) : The 'dimples' on a golf ball are intentionally provided.

Reason (R) : A turbulent boundary layer, since it has more momentum than a laminar boundary layer, can better resist an adverse pressure gradient.

(a) Both **A** and **R** are individually true and **R** is the correct explanation of **A**
(b) Both **A** and **R** are individually true but **R** is not the correct explanation of **A**

(c) A is true but R is false
(d) A is false but R is true [ESE : 2009]

11.24 Consider the following statements regarding turbulent boundary layer on a flat plate with zero pressure gradient:
1. The thickness of the boundary layer increases approximately as the power of the distance from the leading edge.
2. The shear stress at the boundary is approximately inversely proportional to the square root of the local Reynolds number.

Which of these statements is/are correct?
(a) Both 1 and 2 (b) Neither 1 nor 2
(c) 1 only (d) 2 only
[ESE : 2010]

11.25 Air (density 1.2 kg/m^3 and kinematic viscosity 15 centistokes) flows over a flat plate, at zero angle of incidence, at a velocity of 20 m/s. If Reynolds number at transition is taken as 2.5×10^5, maximum distance, from leading edge up to which the boundary layer remains laminar, is
(a) 375 mm (b) 93.75 mm
(c) 187.5 mm (d) 250 mm
[ESE : 2010]

11.26 The boundary layer separation occurs when
(a) Pressure gradient is positive
(b) Pressure gradient is zero
(c) Pressure gradient is negative
(d) None of the above
[ESE : 2011]

11.27 The thickness of boundary layer in a turbulent flow is
(a) $\frac{5x}{\sqrt{R_{Nx}}}$ (b) $\frac{5.835x}{\sqrt{R_{Nx}}}$
(c) $\frac{0.377x}{(R_{Nx})^{\frac{1}{5}}}$ (d) $\frac{5.377x}{(R_{Nx})^{\frac{1}{5}}}$
[ESE : 2012]

11.28 The Reynolds number of a fluid flowing over a flat plate at a distance of 16 cm from the leading edge is 25600. The thickness of the boundary layer at this point will be
(a) 5 mm (b) 7.5 mm
(c) 10 mm (d) 10.5 mm
[ESE : 2012]

11.29 Boundary layer separation is caused due to
(a) laminar flow changing to turbulent flow
(b) velocity gradient being zero at the wall
(c) reduction in pressure to vapour pressure
(d) decrease in boundary layer thickness to a negligible value
[ESE : 2012]

11.30 Consider the following statements pertaining to boundary layer on solid surfaces:
1. The boundary layer separation takes place if the pressure gradient is zero.
2. The condition of boundary layer separation is $(\partial u/\partial y)_{y=0} = 0$.
3. Boundary layer on a flat plate is laminar if the Reynolds number is less than 5×10^5.

Which of the above statements is/are correct?
(a) 1, 2 and 3 (b) 1 and 2 only
(c) 2 and 3 only (d) 1 and 3 only
[ESE : 2014]

11.31 What is the ratio of momentum thickness to the boundary layer thickness δ when the layer velocity profile is given by $\frac{u}{U_\infty} = \left(\frac{y}{\delta}\right)^{\frac{1}{2}}$?
(a) 0.133 (b) 0.333
(c) 0.166 (d) 0.136

where u is velocity at height y above surface and U_∞ is free stream velocity of flow.
[ESE : 2015]

11.32 The boundary layer thickness at a given distance from the leading edge of a flat plate is
(a) More for lighter fluid
(b) More for denser fluid
(c) Less for denser fluid
(d) Less for lighter fluid
[ESE : 2015]

11.33 A fluid is flowing over a flat plate. At distance of 8 cm from the leading edge, the Reynolds number is found to be 25600. The thickness of the boundary layer at this point is
(a) 1.5 mm (b) 2.5 mm
(c) 4.0 mm (d) 5.0 mm
[ESE : 2015]

11.34 Consider the following statements pertaining to boundary layer on a flat plate:

1. The thickness of laminar boundary layer at a distance x from the leading edge varies as $x^{1/2}$.
2. The thickness of turbulent boundary layer at a distance x from the leading edge varies as $x^{4/5}$.
3. Boundary layer is laminar when Reynolds number is less than 5×10^5

Which of the above statements are correct?

(a) 1, 2 and 3 (b) 1 and 2 only
(c) 1 and 3 only (d) 2 and 3 only

[ESE : 2015]

11.35 The thickness of the boundary layer for a fluid flowing over a flat plate at a point 20 cm from the leading edge is found to be 4 mm. The Reynolds number at the point (adopting 5 as the relevant constant) is :

(a) 48400 (b) 57600
(c) 62500 (d) 77600

[ESE : 2016]

11.36 What is the ratio of displacement thickness to boundary layer thickness for a linear distribution of velocity $\frac{u}{u_\infty} = \frac{y}{\delta}$ in the boundary layer on a flat plate, where δ is the boundary layer thickness and u_∞ is the free steam velocity ?

(a) 0.5 (b) 0.67
(c) 0.75 (d) 0.8 [ESE : 2016]

11.37 Statement (I) : In subsonic flow in a diverging channel, it is possible that the flow may separate.

Statement (II) : In subsonic flow in a diverging channel, there is adverse pressure gradient in the channel.

Codes:

(a) Both Statement (I) and Statement (II) are individually true and Statement (II) is the correct explanation of Statement (I)
(b) Both Statement (I) and Statement (II) are individually true but Statement (II) is ***NOT*** the correct explanation of Statement (I)
(c) Statement (I) is true but Statement (II) is false
(d) Statement (I) is false but Statement (II) is true

[ESE : 2016]

11.38 Statement (I) : In a boundary layer formed by uniform flow past a flat plate, the pressure gradient in the x direction is zero.

Statement (II) : In a boundary layer formed by uniform flow past a flat plate, the pressure gradient in the y direction is negligible.

Codes:

(a) Both Statement (I) and Statement (II) are individually true and Statement (II) is the correct explanation of Statement (I)
(b) Both Statement (I) and Statement (II) are individually true but Statement (II) is ***NOT*** the correct explanation of Statement (I)
(c) Statement (I) is true but Statement (II) is false
(d) Statement (I) is false but Statement (II) is true

[ESE : 2016]

11.39 Which one of the following is not the methodology of control separation of flow from boundary in the application of aerofoils?

(a) Streamlining of blunt body shapes
(b) Fluid ejection from the boundary layer
(c) Suction of fluid from the boundary layer
(d) Creating a motion of the boundary wall

[ESE : 2021]

11.40 The region from the tube inlet to the point at which the boundary layer merges at the centreline, is called

(a) Hydrodynamic entry length
(b) Thermal entrance region
(c) Hydrodynamic entrance region
(d) Thermal entry length

[ESE : 2022]

■■■■

Answers Boundary Layer Theory

11.1 (a)	11.2 (d)	11.3 (c)	11.4 (c)	11.5 (b)	11.6 (b)	11.7 (b)	11.8 (d)	11.9 (c)
11.10 (a)	11.11 (b)	11.12 (c)	11.13 (b)	11.14 (d)	11.15 (a)	11.16 (c)	11.17 (b)	11.18 (a)
11.19 (d)	11.20 (d)	11.21 (a)	11.22 (c)	11.23 (b)	11.24 (c)	11.25 (c)	11.26 (a)	11.27 (c)
11.28 (a)	11.29 (b)	11.30 (c)	11.31 (c)	11.32 (a)	11.33 (b)	11.34 (a)	11.35 (c)	11.36 (a)
11.37 (a)	11.38 (b)	11.39 (b)	11.40 (c)					

Explanations Boundary Layer Theory

11.1 (a)

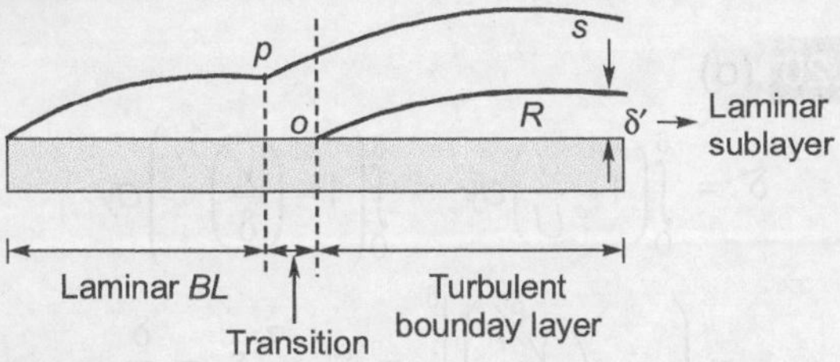

11.2 (d)

$$\delta \propto \left(\frac{x^4}{u}\right)^{1/5}$$

11.4 (c)

$$\delta = \int_0^{\delta}\left(1-\frac{y}{\delta}\right)dy = \left[y-\frac{y^2}{2\delta}\right]_0^{\delta} = \frac{\delta}{2}$$

$$\theta = \int_0^{\delta}\left(1-\frac{y}{\delta}\right)\left(\frac{y}{\delta}\right)dy = \int_0^{\delta}\left[\frac{y}{\delta}-\left(\frac{y}{\delta}\right)^2\right]dy$$

$$= \left[\frac{y^2}{2\delta}-\frac{y^2}{3\delta^2}\right]_0^{\delta} = \frac{\delta}{6}$$

$$\therefore\ H = \frac{\delta^*}{\theta} = \mathbf{3}$$

11.5 (b)

$$Re = \frac{2\times 1}{2\times 10^{-5}} = 10^5$$

$$\delta = \frac{5\times 1}{\sqrt{10^5}} = 0.0158$$

$$\theta = \int_0^{\delta}\left(1-\frac{y}{\delta}\right)\frac{y}{\delta}.dy = \frac{\delta}{6}$$

$\therefore\ d = \mathbf{2.1\ mm}$

11.6 (b)

$$C_{f l} = \frac{0.664}{\sqrt{Re_x}}$$

11.8 (d)

$$\delta = \frac{5x}{\sqrt{Re_x}}$$

11.9 (c)

Shock wave	:	Compressibility
Flow separation	:	Adverse pressure gradient
Capillary rise	:	Surface tension
Cavitation	:	Vapour pressure

11.11 (b)

In a turbulent boundary layer, near the boundary large velocity change occurs in a relatively small vertical distance, and hence at the boundary the velocity gradient $\left(\frac{du}{dy}\right)$ is steeper in turbulent boundary layer than in a laminar boundary layer. The velocity distribution in a turbulent boundary layer follows a logarithm law, which can also be represented by a power law of the type.

$$\frac{u}{U} = \left(\frac{y}{\delta}\right)^n$$

The value of the exponent n is approximately (1/7) for the moderate Reynolds number.

11.13 (b)

Boundary layer thickness δ for flat plate

$$\delta = 5\sqrt{\frac{x\nu}{V}} = 5\sqrt{\frac{16\times10^{-4}\times0.4}{30\times\left(\frac{5}{18}\right)}}$$

$= 0.00438 \text{ m} = \mathbf{4.38\ mm}$

11.14 (d)

Let velocity distribution is

$$\frac{u}{U} = \frac{y}{\delta}$$

Displacement thickness

$$(\delta^*) = \int_0^\delta\left(1-\frac{u}{U}\right)dy = \int_0^\delta\left(1-\frac{y}{\delta}\right)dy = \frac{\delta}{2}$$

Momentum thickness

$$(\theta) = \int_0^\delta\left(1-\frac{u}{U}\right)\left(\frac{u}{U}\right)dy = \int_0^\delta\left(1-\frac{y}{\delta}\right)\left(\frac{y}{\delta}\right)dy = \frac{\delta}{6}$$

$$\therefore\ \frac{\delta^*}{\theta} = \mathbf{3}$$

11.16 (c)

$$\delta^* = \int_0^\delta\left(1-\frac{y}{\delta}\right)dy = \frac{\delta}{2}$$

$$\therefore\ \frac{\delta^*}{\delta} = \frac{1}{2}$$

11.17 (b)

$$\delta^* = \int_0^\delta\left(1-\frac{u}{U}\right)dy = \int_0^\delta\left(1-\left(\frac{y}{\delta}\right)^{1/6}\right)dy$$

$$\delta^* = \left[y - \frac{6y^{7/6}}{7\delta^{1/6}}\right]_0^\delta = \delta - \frac{6\delta}{7} = \frac{\delta}{7}$$

$$\therefore\ \frac{\delta^*}{\delta} = \frac{1}{7}$$

11.18 (a)

In boundary layer, the flow is viscous and rotational.

11.19 (d)

Velocity profile of laminar boundary layer is given

$$\frac{u}{U} = \frac{y}{\delta}$$

Displacement thickness,

$$\delta^* = \int_0^\delta\left(1-\frac{u}{U}\right)dy = \delta^* = \int_0^\delta\left(1-\frac{u}{U}\right)dy$$

$$= \int_0^\delta\left(1-\frac{y}{\delta}\right)dy = \frac{\delta}{2}$$

Momentum thickness,

$$\theta = \int_0^\delta\left(1-\frac{u}{U}\right)\frac{u}{U}dy = \int_0^\delta\left(\frac{y}{\delta}\right)\left(1-\frac{y}{\delta}\right)dy = \frac{\delta}{6}$$

$$\therefore\ \frac{\theta}{\delta^*} = \frac{1}{3}$$

11.20 (d)

$$\delta^* = \int_0^\delta\left(1-\frac{u}{U}\right)dy = \int_0^\delta\left[1-\left(\frac{y}{\delta}\right)^{1/7}\right]dy$$

$$= \left(y - \frac{7}{8}\frac{y^{8/7}}{\delta^{1/7}}\right)\Bigg|_0^\delta = \delta - \frac{7}{8}\delta = \frac{\delta}{8}$$

11.21 (a)

In the boundary layer concept the shear stress at the outer edge of the layer is considered to be zero because

$\frac{du}{dy} = 0$ and $\tau = \frac{\mu du}{dy}$

because local velocity is equal to velocity in potential flow.

∴ **R** is the correct explanation of **A** and answer is (a).

11.23 (b)

Both **A** and **R** is true but **R** is not the correct explanation of **A**.

The correct reason is to tripping the boundary at low Reynold number. This low Reynold number is a main advantage as if the Reynold number is high it may actually increase the drag coefficient.

11.24 (c)

For turbulent flow:

Boundary layer thickness,

$$\delta = \frac{0.37x}{(Re_x)^{1/5}}$$

Hence, $\delta \propto x^{4/5}$

Shear stress,

$$\tau = \frac{\text{Drag force}}{\text{Area}} = \frac{1}{2}\frac{\rho A U^2}{A} \times C_D$$

$$= \frac{1}{2}\rho U^2 \times \frac{0.072}{(Re)^{1/5}}$$

Hence shear stress,

$$\tau \propto \frac{1}{(Re)^{1/5}}$$

11.25 (c)

$\rho = 1.2$ kg/m^3, $\nu = 15 \times 10^{-6}$ m^2/s

$U_\infty = 20$ m/s

If $(Re)_{Cr} = 2.5 \times 10^5$

$$2.5 \times 10^5 = \frac{U_\infty x}{\nu} = \frac{20 \times x}{15 \times 10^{-16}}$$

or $$x = \frac{2.5 \times 10^5 \times 15 \times 10^{-6}}{20}$$

$= 0.1875$ m $=$ **187.5 mm**

i.e. up to 187.5 mm from leading edge flow will be laminar.

11.28 (a)

$$\delta = \frac{5x}{\sqrt{R_{e_x}}} = \frac{5.16}{\sqrt{25600}}$$

$= 0.5$ cm $=$ **5 mm**

11.30 (c)

The boundary layer separation takes place in the region of adverse pressure gradient. So statement (1) is wrong.

11.31 (c)

$$\therefore \quad Q = \int_0^\delta \frac{U}{U_\infty}\left(1 - \frac{v}{U_\infty}\right)\cdot dy$$

$$\because \quad \frac{U}{U_\infty} = \left(\frac{y}{\delta}\right)^{\frac{1}{2}}$$

$$\therefore \quad Q = \int_0^\delta \left(\frac{y}{\delta}\right)^{\frac{1}{2}}\left(1 - \frac{y^{1/2}}{\delta^{1/2}}\right)\cdot dy$$

$$\therefore \quad Q = \frac{1}{\sqrt{8}}\left[\frac{2}{3}y^{3/2}\right]_0^\delta - \frac{1}{28}\left[y^2\right]_0^\delta$$

$$\therefore \quad Q = \frac{1}{\sqrt{8}} \times \frac{2}{3} \times 8^{3/2} - \frac{\delta^2}{2\delta}$$

$$\therefore \quad Q = \frac{2}{3}\delta - \frac{\delta}{2} = \frac{4-3}{6}\delta = \frac{\delta}{6}$$

$$\therefore \quad \frac{Q}{\delta} = \frac{1}{6} = \mathbf{0.1667}$$

11.32 (a)

$$\delta = \frac{5x}{\sqrt{R_{ex}}} \Rightarrow \delta = \frac{5x}{\sqrt{\frac{\rho U x}{\mu}}}$$

$$\therefore \quad \delta \propto \frac{1}{\sqrt{\rho}}$$

⇒ Boundary layer thickness is more for lighter fluid.

11.33 (b)

$$\delta = \frac{5x}{\sqrt{R_{ex}}} = \frac{5 \times 8}{\sqrt{25600}} = \frac{40}{160}\text{ cm}$$

$\delta = 0.25$ cm $=$ **2.5 mm**

11.34 (a)

1. Laminar flow,

$$\delta = \frac{5x}{\sqrt{R_{ex}}} \Rightarrow \delta = \frac{5x}{\sqrt{\rho\frac{Ux}{\mu}}}$$

$$\delta \propto x^{1/2}$$

2. Turbulent flow,

$$\delta = \frac{0.377x}{(R_{ex})^{1/5}} = \frac{0.377x}{\left(\frac{\rho U x}{u}\right)^{1/5}}$$

$$\therefore \quad \delta \propto x^{4/5}$$

$$\Rightarrow \quad \delta \propto x^{0.8}$$

3. Statement is true.

11.35 (c)

$$\delta = \frac{5x}{\sqrt{R_{ex}}} = \frac{5 \times 0.2}{\sqrt{R_{ex}}}$$

$R_{ex} =$ **62500**

11.36 (a)

$$\delta^* = \int_0^\delta \left(1 - \frac{u}{u_\infty}\right) dr = \int_0^\delta \left(1 - \frac{y}{\delta}\right) dy$$

$$\delta^* = \delta - \frac{\delta}{2} = \frac{\delta}{2}$$

$$\frac{\delta^*}{\delta} = \frac{1}{2}$$

■■■■

12 Open Channel Flow

(Not in Syllabus)

12.1 The critical depth of a rectangular channel of width 4.0 m for a discharge of 12 m^3/s is nearly

(a) 300 mm (b) 30 mm
(c) 0.972 m (d) 0.674 m

[ESE : 2001]

12.2 An open channel flow encounters a hydraulic jump as shown in the figure below.

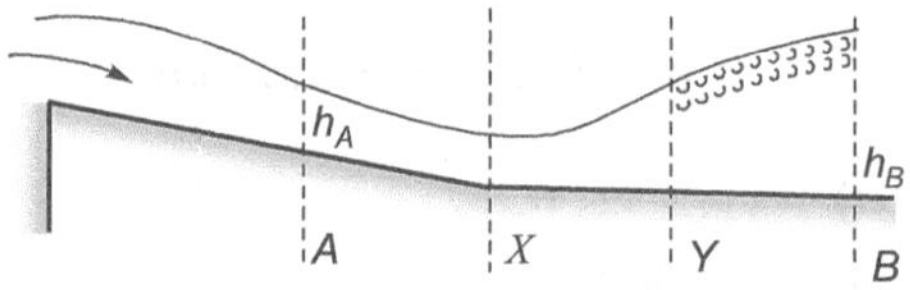

The following fluid flow conditions are observed between A and B:

1. Critical depth
2. Steady non-uniform flow
3. Unsteady non-uniform flow
4. Steady uniform flow

The correct sequence of the flow conditions in the direction of flow is

(a) 1, 2, 3, 4 (b) 1, 4, 2, 3
(c) 2, 1, 4, 3 (d) 4, 2, 3, 1

[ESE : 2001]

12.3 Consider the following statements:

A hydraulic jumps occurs in an open channel

1. when the Froude number is equal to or less than one
2. at the toe of a spillway
3. downstream of a sluice gate in a canal
4. when the bed slope suddenly changes

Which of these statements are correct?

(a) 1, 2, 3 and 4 (b) 1, 2 and 3
(c) 2, 3 and 4 (d) 1 and 4 **[ESE : 2003]**

12.4 For a circular channel, the wetted perimeter (where R = radius of circular channel, θ = half the angle subtended by the water surface at the centre) is given by

(a) $R\theta/2$ (b) $3R\theta$
(c) $2R\theta$ (d) $R\theta$

[ESE : 2003]

12.5 Which one of the following is the correct expression for the area of flow for a circular channel? (where θ = half the angle subtended by water surface at the centre and R = radius of the circular channel)

(a) $R^2\left(2\theta - \frac{\sin 2\theta}{2}\right)$ (b) $R^2\left(\theta - \frac{\sin 2\theta}{2}\right)$

(c) $R^2(2\theta - \sin 2\theta)$ (d) $2R^2(\theta - \sin 2\theta)$

[ESE : 2004]

12.6 Match **List-I** (Flow depth) with **List-II** (Basic hydraulic condition associated therewith) and select the correct answer using the codes given below the lists:

List-I

A. Conjugate depth
B. Critical depth
C. Alternate depth
D. Normal depth

List-II

1. Uniform flow
2. Same specific energy
3. Minimum specific energy
4. Same specific force
5. Same bed slope

Codes:

	A	B	C	D
(a)	3	5	4	2
(b)	2	4	1	3
(c)	4	3	2	1
(d)	5	4	1	2

[ESE : 2004]

12.7 How is the best hydraulic channel cross-section defined?

(a) The section with minimum roughness coefficient.
(b) The section that has a maximum area for a given flow.
(c) The section that has a minimum wetted perimeter.
(d) The section that has maximum wetted area.

[ESE : 2005]

12.8 An open channel of symmetric right angled triangular cross-section is conveying a discharge *Q*. Taking *g* as the acceleration due to gravity, what is the critical depth?

(a) $\left(\frac{Q^2}{g}\right)^{1/3}$ (b) $\left(\frac{2Q^2}{g}\right)^{1/3}$

(c) $\left(\frac{Q^2}{g}\right)^{1/5}$ (d) $\left(\frac{2Q^2}{g}\right)^{1/5}$

[ESE : 2006]

12.9 For hydraulically most efficient symmetric trapezoidal section of an open channel, which one of the following is the false characterization?

(a) Wetted perimeter is minimum for a given area of flow section
(b) Hydraulic radius is half the flow depth
(c) Width at top of liquid is twice the hydraulic depth
(d) Discharge is maximum for given area of flow, bed slope and roughness [ESE : 2008]

12.10 A sluice gate discharges water into a horizontal rectangular channel with a velocity of 12 m/s and depth of flow of 1 m. What is the depth of flow after the hydraulic jump?

(a) 6.42 m (b) 5.84 m
(c) 4.94 m (d) 2.21 m

[ESE : 2009]

12.11 For a given discharge in a channel at critical depth

(a) The specific energy is minimum
(b) The specific energy is maximum
(c) The total energy is minimum
(d) The total energy is maximum

[ESE : 2011]

12.12 **Statement (I):** Hydraulic jump is similar to normal shock in gases where supersonic flow becomes subsonic across the shock plane resulting an increase in pressure.

Statement (II): In hydraulic jump, the Froude number is less than one and the subcritical flow changes to supercritical across the jump.

[ESE : 2012]

12.13 **Statement (I):** Gravity is the driving force behind flows through open channels.

Statement (II): Gravity stands to reason that the ratio of inertial to gravitational forces will play a major role in open channel flow analysis.

(a) Both Statement (I) and Statement (II) are individually true and Statement (II) is the correct explanation of Statement (I).
(b) Both Statement (I) and Statement (II) are individually true but Statement (II) is **NOT** the correct explanation of Statement (I).
(c) Statement (I) is true but Statement (II) is false
(d) Statement (I) is false but Statement (II) is true.

[ESE : 2021]

■■■■

Answers Open Channel Flow

12.1 (c) 12.2 (b) 12.3 (c) 12.4 (c) 12.5 (b) 12.6 (c) 12.7 (c) 12.8 (d) 12.9 (c)

12.10 (c) 12.11 (a) 12.12 (c) 12.13 (a)

Explanations Open Channel Flow

12.1 (c)

Critical depth (y_c) for a constant unit flow q in a rectangular channel occur when the specific energy is minimum and it is given by

$$y_c = \sqrt[3]{\frac{q^2}{g}}$$

$$q = \frac{Q}{L} = \frac{12}{4} = 3 \text{ m}^2/\text{s}$$

$$y_c = \sqrt[3]{\frac{(3)^2}{9.81}} = 0.972$$

12.3 (c)

It occurs frequently in a canal below a regulating sluice, at the foot of a spillway, or at the place where a step channel bottom slope suddenly changes to a flat slope.

12.4 (c)

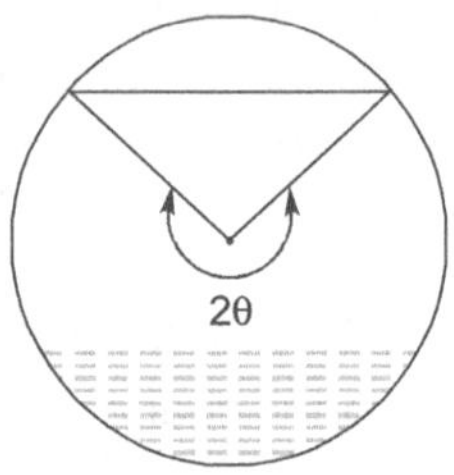

When $\theta = \pi$, then wetted perimeter will $2\pi R$

$\therefore \quad P = 2\theta R$

12.5 (b)

Area of flow for a circular channel

$$= R^2\left(\theta - \frac{\sin 2\theta}{2}\right)$$

12.7 (c)

When wetted perimeter is minimum.

12.8 (d)

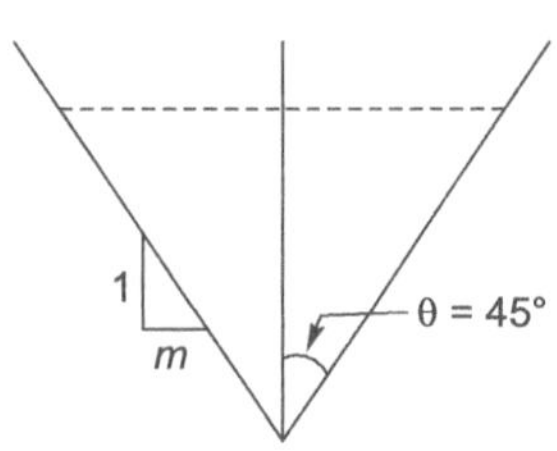

For triangular cross-section

critical depth: $y_c = \left(\frac{2Q^2}{gm^2}\right)^{1/5}$

where m = 1 for triangular

$$\therefore \quad y_c = \left(\frac{2Q^2}{g}\right)^{1/5}$$

12.9 (c)

Hydraulically most efficient is symmetric trapezoidal section of an open channel.

The wetted perimeter is minimum for trapezoidal half hexagon, hydraulic radius is half the flow depth and discharge is maximum for given area of flow , bed slope and roughness.

The assumption width at top is twice the hydraulic depth is wrong and is valid for rectangular channel.

12.10 (c)

$$\frac{V}{\sqrt{gy}} = F$$

$$\Rightarrow \quad F = \frac{12}{\sqrt{9.8 \times 1}} = 3.83$$

$$\because \quad y_1 = 1 \text{ m}$$

$$\frac{y_2}{y_1} = \frac{1}{2}\left[-1 + \sqrt{1 + 8F^2}\right]$$

$$y_2 = \mathbf{4.94 \text{ m}}$$

12.11 (a)

For a given value of specific energy, the critical depth gives the greatest discharge, or conversely, for a given discharge, the specific energy is a minimum for the critical depth.

For rectangular channels, the critical depth, d_c (m), is given by

$$d_c = [Q^2/b^2g]^{1/3}$$

where d_c = critical depth, (m)

Q = quantity of flow or discharge, (m^3/s)

b = width of channel, (m)

12.12 (c)

In hydraulic jump the flow changes from supercritical to subcritical.

■■■■

13 Drag and Lift

13.1 A circular cylinder of 400 mm diameter is rotated about its axis in a stream of water having a uniform velocity of 4 m/s. When both the stagnation points coincide, the lift force experienced by the cylinder is
(a) 160 kN/m (b) 10.05 kN/m
(c) 80 kN/m (d) 40.2 kN/m
[ESE : 2000]

13.2 An automobile moving at a velocity of 40 km/hr is experiencing a wind resistance of 2 kN. If the automobile is moving at a velocity of 50 km/hr, the power required to overcome the wind resistance is
(a) 43.4 kW (b) 3.125 kW
(c) 2.5 kW (d) 27.776 kW
[ESE : 2000]

13.3 When a cylinder is placed in an ideal fluid and the flow is uniform, the pressure coefficient C_p is equal to
(a) $1 - \sin^2\theta$ (b) $1 - 2\sin^2\theta$
(c) $1 - 4\sin^2\theta$ (d) $1 - 8\sin^2\theta$
[ESE : 2000]

13.4 Which one of the following sets of standard flows is superimposed to represent the flow around a rotating cylinder?
(a) Doublet, vortex and uniform flow
(b) Source, vortex and uniform flow
(c) Sink, vortex and uniform flow
(d) Vortex and uniform flow [ESE : 2000]

13.5 Improved streamlining produces 25% reduction in the drag coefficient of a torpedo. When it is travelling fully submerged and assuming the driving power to remain the same, the increase in speed will be
(a) 10% (b) 20%
(c) 25% (d) 30% [ESE : 2000]

13.6 When pressure drag over a body is large as compared to the friction drag, then the shape of the body is that of
(a) an aerofoil
(b) a streamlined body
(c) a two-dimensional body
(d) a bluff body
[ESE : 2000]

13.7 Match **List-I** (Types of flow) with **List-II** (Basic ideal flows) and select the correct answer using the codes given below the lists:

List-I
A. Flow over a stationary cylinder
B. Flow over a half Rankine body
C. Flow over a rotating body
D. Flow over a Rankine oval

List-II
1. Source + sink + uniform flow
2. Doublet + uniform flow
3. Source + uniform flow
4. Doublet + free vortex + uniform flow

Codes:

	A	B	C	D
(a)	1	4	3	2
(b)	2	4	3	1
(c)	1	3	4	2
(d)	2	3	4	1

[ESE : 2001]

13.8 Match **List-I** with **List-II** and select the correct answer using the codes given below the lists:

List-I
A. Stoke's law
B. Bluff body
C. Streamline body
D. Karman Vortex Street

List-II
1. Strouhal number
2. Creeping motion

3. Pressure drag
4. Skin friction drag

Codes:

	A	B	C	D
(a)	2	3	1	4
(b)	3	2	4	1
(c)	2	3	4	1
(d)	3	2	1	4

[ESE : 2001]

13.9 The drag force exerted by a fluid on a body immersed in the fluid is due to

(a) pressure and viscous forces
(b) pressure and gravity forces
(c) pressure and surface tension forces
(d) viscous and gravity forces

[ESE : 2002]

13.10 Which one of the following causes lift on an immersed body in a fluid stream?

(a) Buoyant forces
(b) Resultant fluid force on the body
(c) Dynamic fluid force component exerted on the body parallel to the approach velocity
(d) Dynamic fluid force component exerted on the body normal to the approach velocity

[ESE : 2005]

13.11 A cylindrical object is rotated with constant angular velocity about its symmetry axis in a uniform flow field of an ideal fluid producing streamlines as shown in the figure given below.

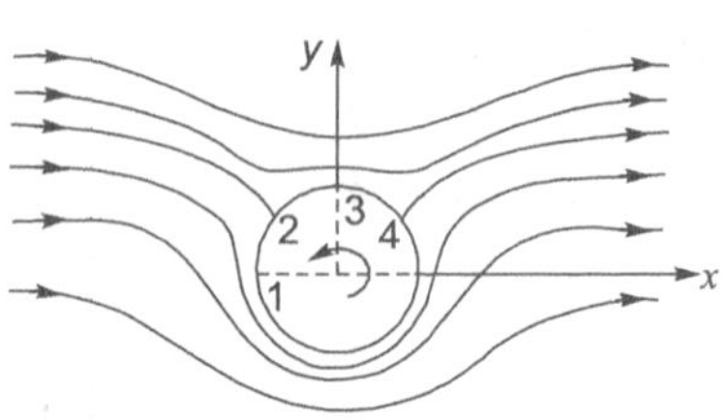

At which point(s), is the pressure on the cylinder surface maximum?

(a) Only at point 3
(b) Only at point 2
(c) At points 1 and 3
(d) At points 2 and 4

[ESE : 2007]

13.12 Consider the following statements:

1. The phenomenon of lift produced by imposing circulation over a doublet in a uniform flow is known as Magnus effect.
2. The path deviation of a cricket ball from its original trajectory is due to the Magnus effect.

Which of these statements is/are correct?

(a) 1 only (b) 2 only
(c) Both 1 and 2 (d) Neither 1 nor 2

[ESE : 2007]

13.13 Whenever a plate is submerged at an angle with the direction of flow of liquid, it is subjected to some pressure. What is the component of this pressure in the direction of flow of liquid, known as?

(a) Stagnation pressure
(b) Lift
(c) Drag
(d) Bulk modulus

[ESE : 2007]

13.14 For flow of the ideal fluid over a cylinder from the front stagnation point,

(a) pressure first decreases then increases
(b) velocity first decreases then increases
(c) pressure remains the same
(d) velocity remains the same

[ESE : 2008]

13.15 In the flow past bluff bodies

(a) pressure drag is smaller than friction drag
(b) friction drag occupies the major part of total drag
(c) pressure drag occupies the major part of total drag
(d) pressure drag is less than that of streamlined body

[ESE : 2008]

13.16 Match **List-I** with **List-II** and select the correct answer using the codes given below the lists :

List-I

A. Singing of telephone wires
B. Velocity profile in a pipe is initially parabolic and then flattens
C. Formation of cyclones
D. Shape of rotameter tube

List-II

1. Vortex flow
2. Drag
3. Vortex shedding
4. Turbulence

Codes :

	A	B	C	D
(a)	3	1	4	2
(b)	2	1	4	3
(c)	3	4	1	2
(d)	2	4	1	3

[ESE : 2009]

13.17 How could 'Magnus effect' be simulated as a combination?

(a) Uniform flow and doublet
(b) Uniform flow, irrotational vortex and doublet
(c) Uniform flow and vortex
(d) Uniform flow and line source

[ESE : 2009]

Direction: Each of the next questions consists of two statements, one labelled as the **'Assertion (A)'** and the other as **'Reason (R)'**. You are to examine these two statements carefully and select the answers to these items using the codes given below:

Codes:

(a) Both **A** and **R** are individually true and **R** is the correct explanation of **A**
(b) Both **A** and **R** are individually true but **R** is not the correct explanation of **A**
(c) **A** is true but **R** is false
(d) **A** is false but **R** is true

13.18 Assertion (A): When a circular cylinder is placed normal to the direction of flow, drag force is essentially a function of the Reynolds number of the flow.

Reason (R): As Reynolds number is about 100 and above, eddies formed break away from either side in periodic fashion, forming a meandering street called the Karman Vortex street.

[ESE : 2010]

13.19 Assertion (A): When a thin flat plate is held parallel to a fluid stream, both pressure drag a well as viscous drag will be negligible.

Reason (R): The total drag will be entirely due to shear stress and pressure drag will be negligible.

[ESE : 2010]

13.20 A thin plate has been placed parallel to flow direction. The relative magnitude of friction and pressure drags will be

(a) negligible friction as well as pressure drag
(b) negligible pressure drag and maximum friction drag
(c) maximum pressure drag and negligible friction drag
(d) pressure drag equals the friction drag

[ESE : 2011]

13.21 A sphere of diameter 2 mm falls 150 mm in 20 seconds in a viscous liquid. The density of the sphere is 7500 kg/m^3 and of liquid is 900 kg/m^3. What is the coefficient of viscosity of the liquid? (Take acceleration due to gravity as 9.81 m/s^2)

(a) 22.31 poise (b) 25.62 poise
(c) 19.17 poise (d) 32.21 poise

[ESE : 2022]

■■■■

Answers Drag and Lift

13.1 (b)	13.2 (a)	13.3 (c)	13.4 (a)	13.5 (a)	13.6 (d)	13.7 (d)	13.8 (c)	13.9 (a)
13.10 (d)	13.11 (d)	13.12 (c)	13.13 (c)	13.14 (a)	13.15 (c)	13.16 (c)	13.17 (b)	13.18 (a)
13.19 (d)	13.20 (b)	13.21 (c)						

Explanations Drag and Lift

13.1 (b)

$$\text{Lift force} = \frac{1}{2}\rho A v^2$$

$$F_L = \frac{1}{2}\rho \times \pi DL \times v^2$$

$$\frac{F_L}{L} = \frac{1}{2}\rho\pi D v^2 = \frac{1000 \times \pi \times 0.4 \times 16}{2}$$

$$\frac{F_L}{L} = \mathbf{10.5\ kN/m}$$

13.2 (a)

$$\frac{F}{\rho v^2 d^2} = \text{Constant}$$

$$\Rightarrow \frac{F_2}{F_1} = \left(\frac{V_2}{V_1}\right)^2$$

$$\Rightarrow F_2 = 2\left(\frac{50}{40}\right)^2$$

$$= 2 \times (1.25)^2 = 3.125 \text{ kN}$$

$$\text{Power} = 3.125 \times \left(50 \times \frac{5}{18}\right) \text{kW}$$

$$= \mathbf{43.4\ kW}$$

13.3 (c)

For rotating cylinder

$$P_S = P + \frac{\omega . u^2}{2g}\left[1 - 4\sin^2\theta - \frac{\tau^2}{4\pi^2 R^2 v^2} - \frac{4\sin\theta}{u . 2\pi R}\right]$$

For ideal flow

$$\tau = 0$$

$$\therefore P_S - P = P = \frac{\omega v^2}{2g} C_P$$

where

$$C_P = 1 - 4\sin^2\theta$$

13.6 (d)

A bluff body is one in which the length in the flow direction is close to or equal to the length perpendicular to the flow direction. This results in a unique characteristics, that skin friction drag is much lower than pressure drag.

13.7 (d)

(a) Flow over a stationary cylinder → doublet + uniform flow

(b) Flow over a half Rankine body → source + uniform flow

(c) Flow over a rotating body → Doublet + free vortex + uniform flow

(d) Flow over a Rankine oval → source + sink + uniform flow

13.8 (c)

Bluff body: A bluff body is one in which the length in the flow direction is close to or equal to the length perpendicular to the flow direction. This results in a unique characteristics, that skin friction drag is much lower then pressure drag.

Stoke's law: Creeping flow also known as stokes flow is a type of fluid flow where advective inertial forces are small compared with viscous forces. For this Reynolds number is very low i.e., $R_e << 1$. For this type of flow the inertial forces are assumed to be negligible and the Navier-stokes equation simplify to give the stokes equation.

13.9 (a)

Viscous forces and pressure forces are responsible for drag forces.

13.12 (c)

The lift produced by circulation around a circular cross-section placed in a uniform stream of fluid is commonly known as Magnus effect.

13.13 (c)

The component of total force F_R in the direction of motion is called as Drag. The component of total force in the direction perpendicular to the direction of motion is known as Lift.

13.14 (a)

At the stagnation point pressure is maximum

$\therefore$ for flow past ideal fluid over a cylinder from front stagnation point pressure first decreases then increases.

13.15 (c)

In the flow past bluff bodies the pressure drag occupies the major part of total drag.

13.16 (c)

Vortex shedding is caused when a fluid flows past a blunt object. The fluid flow past the object creates alternating low pressure vortices on the down stream of object. The object will tend to move towards low pressure zone. This causes the telephone wire to swing which is called swinging of telephone lines.

In cyclone, the fluid moves in a circular/spiral motion due to the pressure difference caused. This is a type of vortex flow.

13.17 (b)

Magnus Effect could be simulated as combination of uniform flow, irrotational vortex and doublet.

13.19 (d)

When a thin flat plate is hold parallel to a fluid stream, viscous drag will be maximum.

13.20 (b)

A thin plate has been placed parallel to flow direction. That body is known as streamlined body and the relative magnitude of friction drag will be maximum and that of pressure drag will be minimum.

13.21 (c)

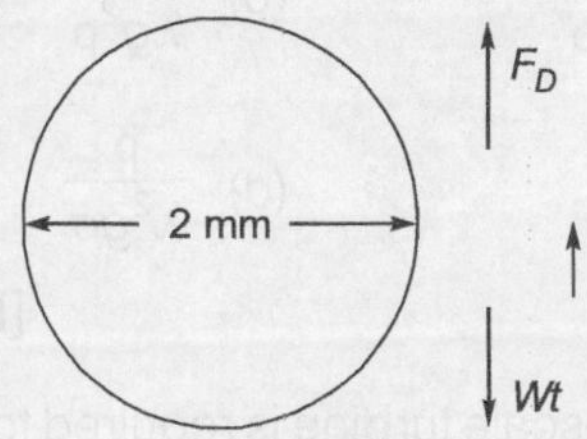

$$U_\infty = \frac{150\times10^{-3}}{20} = 7.5\times10^{-3}$$

$$R_o = \frac{\rho U_\infty D}{\mu}$$

Given: d = 2 mm, t = 20s, ρ_s = 7500 kg/m^3, ρ_f = 900 kg/m^3

$$F_D + F_B = Wt$$

Terminal velocity,

$$U_\infty = \frac{d^2 g(\rho_s - \rho_f)}{18\mu}$$

Coefficient of viscosity,

$$\mu = \frac{d^2 g(\rho_s - \rho_f)}{18 U_\infty}$$

$$= \frac{(0.002)^2 \times 9.81 \times [7500 - 900]}{18 \times 7.5 \times 10^{-3}}$$

$$= 19.18 \text{ poise}$$

■■■■

14 Dimensional Analysis

14.1 The dimensionless group formed by wavelength λ, density of fluid ρ, acceleration due to gravity g and surface tension σ, is

(a) $\frac{\sigma}{\lambda^2 g\rho}$ (b) $\frac{\sigma}{\lambda g^2 \rho}$

(c) $\frac{\sigma g}{\lambda^2 \rho}$ (d) $\frac{\rho}{\lambda^2 g\sigma}$

[ESE : 2000]

14.2 If the full-scale turbine is required to work under a head of 30 m and to run at 428 rpm, then a quarter-scale turbine model tested under a head of 10 m must run at

(a) 143 rpm (b) 341 rpm
(c) 428 rpm (d) 988 rpm

[ESE : 2000]

14.3 Match **List-I** (Dimensionless numbers) with **List-II** (Definition as the ratio of) and select the correct answer using the codes given below the lists:

List-I
A. Reynolds number
B. Froude number
C. Weber number
D. Mach number

List-II
1. Inertia force and elastic force
2. Inertia force and surface tension force
3. Inertia force and gravity force
4. Inertia force and viscous force

Codes:

	A	B	C	D
(a)	1	2	3	4
(b)	4	3	2	1
(c)	1	3	2	4
(d)	4	2	3	1

[ESE : 2001]

14.4 A ship with hull length of 100 m is to run with a speed of 10 m/s. For dynamic similarity, the velocity for a 1 : 25 model of the ship in a towing tank should be

(a) 2 m/s (b) 10 m/s
(c) 20 m/s (d) 25 m/s

[ESE : 2001]

14.5 A model test is to be conducted in a water tunnel using a 1 : 20 model of a submarine, which is to travel at a speed of 12 km/h deep under sea surface. The water temperature in the tunnel is maintained, so that its kinematic viscosity is half that of sea water. At what speed is the model test to be conducted to produce useful data for the prototype?

(a) 12 km/h (b) 240 km/h
(c) 24 km/h (d) 120 km/h

[ESE : 2002]

14.6 A model test is to be conducted for an under water structure, which is likely to be exposed to strong water currents. The significant forces are known to be dependent on structure geometry, fluid velocity, fluid density, fluid viscosity, fluid depth and acceleration due to gravity. Which of the following numbers must match for the model with that of the prototype?

1. Mach number
2. Weber number
3. Froude number
4. Reynolds number

Select the correct answer using the codes given below:

(a) 3 only (b) 1, 2, 3 and 4
(c) 1 and 2 (d) 3 and 4

[ESE : 2002]

14.7 Match **List-I** (Flow/Wave) with **List-II** (Dimensionless Number) and select the correct answer using the codes given below the lists:

List-I
A. Capillary waves in channel
B. Testing of aerofoils
C. Flow around bridge piers
D. Turbulent flow through pipes

List-II
1. Reynolds number
2. Froude number
3. Weber number
4. Euler number
5. Mach number

Codes:

	A	B	C	D
(a)	5	4	3	2
(b)	3	5	4	1
(c)	5	4	2	1
(d)	3	5	2	1

[ESE : 2003]

14.8 Consider the following statements:
1. Dimensional analysis is used to determine the number of variables involved in a certain phenomenon.
2. The group of repeating variables in dimensional analysis should include all the fundamental units.
3. Buckingham's π theorem stipulates the number of dimensionless groups for a given phenomenon.
4. The coefficient in Chezy's equation has no dimension.

Which of these statements are correct?
(a) 1, 2, 3 and 4 (b) 2, 3 and 4
(c) 1 and 4 (d) 2 and 3

[ESE : 2003]

14.9 Consider the following statements:
1. Complete similarity between model and prototype envisages geometric and dynamic similarities only.
2. Distorted models are necessary where geometric similarity is not possible due to practical reasons.
3. In testing of model of a ship, the surface tension forces are generally neglected.
4. The scale effect takes care of the effect of dissimilarity between model and prototype.

Which of these statements are correct?
(a) 1 and 3 (b) 1, 2 and 4
(c) 2 and 3 (d) 2 and 4

[ESE : 2003]

14.10 The power ratio of a pump and its one-fourth scale model, if the ratio of the heads is 5 : 1, will be
(a) 100 (b) 3.2
(c) 179 (d) 12.8

[ESE : 2003]

14.11 **Assertion (A):** Reynolds number must be same for the model and prototype immersed in subsonic flows.

Reason (R): Equality of Reynolds number for the model and prototype satisfies the dynamic similarity criteria.

(a) Both **A** and **R** are individually true and **R** is the correct explanation of **A**
(b) Both **A** and **R** are individually true but **R** is not the correct explanation of **A**
(c) **A** is true but **R** is false
(d) **A** is false but **R** is true

[ESE : 2003]

14.12 Match **List-I** (Type of model) with **List-II** (Transference ratio for velocity) and select the correct answer using the codes given below the lists:

List-I	List-II
A. Reynolds model	1. $\sqrt{\dfrac{K_r}{\rho_r}}$
B. Froude model	2. $\sqrt{\dfrac{\sigma_r}{\rho_r L_r}}$
C. Weber model	3. $\dfrac{\mu_r}{\rho_r L_r}$
D. Mach model	4. $\sqrt{g_r L_r}$

(Where symbols g, μ, ρ, σ and K have their usual meanings and subscript 'r' refers to the ratio)

Codes:

	A	B	C	D
(a)	3	1	2	4
(b)	3	4	2	1
(c)	2	1	3	4
(d)	2	4	3	1

[ESE : 2004]

14.13 In a flow condition where both viscous and gravity forces dominate and both the Froude number and the Reynolds number are the same in model and prototype; and the ratio of kinematic viscosity of model to that of the prototype is 0.0894. What is the model scale?

(a) 1 : 3.3 (b) 3.3 : 1
(c) 5 : 1 (d) 1 : 5

[ESE : 2004]

14.14 A ship whose full length is 100 m is to travel at 10 m/s. For dynamic similarity, with what velocity should a 1 : 25 model of the ship be towed?

(a) 2 m/s (b) 10 m/s
(c) 25 m/s (d) 250 m/s

[ESE : 2004]

14.15 A 1.0 m long model of a ship is towed at a speed of 81 cm/s in a towing tank. To what speed of the ship, 64 m long does this correspond to?

(a) 7.20 m/s (b) 6.48 m/s
(c) 5.76 m/s (d) 3.60 m/s

[ESE : 2004]

14.16 Consider the following statements:

1. For achieving dynamic similarity in model studies on ships, Froude numbers are equated.
2. Reynolds numbers should be equated for studies on aerofoil for dynamic similarity.
3. In model studies on a spillway, the ratio of width to height is equated for kinematic similarity.

Which of these statements are correct?

(a) 1, 2 and 3 (b) 1 and 2
(c) 2 and 3 (d) 1 and 3

[ESE : 2005]

14.17 Which one of the dimensionless numbers identifies the compressibility effect of a fluid?

(a) Euler number (b) Froude number
(c) Mach number (d) Weber number

[ESE : 2005]

14.18 It is observed in a flow problem that total pressure, inertia and gravity forces are important. Then, similarity requires that

(a) Reynolds and Weber numbers be equal
(b) Mach and Froude numbers be equal
(c) Euler and Froude numbers be equal
(d) Reynolds and Mach numbers be equal

14.19 A ship's model with scale 1 : 100, has a wave resistance of 10 N at its design speed. What is the corresponding prototype wave resistance in kN?

(a) 100
(b) 1000
(c) 10000
(d) Cannot be determined because of insufficient data

[ESE : 2007]

14.20 In a fluid machine, the relevant parameters are volume flow rate, density, viscosity, bulk modulus, pressure difference, power consumption, rotational speed and characteristic dimension. Using the Buckingham pi (π) theorem, what would be the number of independent non-dimensional groups?

(a) 3 (b) 4
(c) 5 (d) None of these

[ESE : 2007]

14.21 Match **List-I** (Dimensionless number) with **List-II** (Nature of forces involved) and select the correct answer using the codes given below the lists :

List-I	List-II
A. Euler number	1. Surface tension
B. Weber number	2. Gravity
C. Mach number	3. Pressure
D. Froude number	4. Elastic

Codes:

	A	B	C	D
(a)	3	1	4	2
(b)	3	4	1	2
(c)	4	1	2	3
(d)	4	2	1	3

[ESE : 2008]

14.22 The time period of a simple pendulum depends on its effective length L and the local acceleration due to gravity g. What is the number of dimensionless parameter involved?

(a) Two (b) One
(c) Three (d) Zero

[ESE : 2009]

14.23 What is the correct dimensionless group formed with the variable ρ-density, *N*-rotational speed, d-diameter and μ-coefficient of viscosity?

(a) $\frac{\rho N d^2}{\mu}$ (b) $\frac{\rho N d}{\mu}$

(c) $\frac{N d}{\rho \mu}$ (d) $\frac{N d^2}{\rho \mu}$

[ESE : 2009]

14.24 In similitude with gravity force, where equality of Froude number exists, the velocity ratio becomes

(a) 1.0 (b) $(L_r)^{1/2}$
(c) $1/L_r$ (d) $(L_r)^{3/2}$

where L_r = ratio of model to prototype linear dimension.

[ESE : 2010]

14.25 Weber number is ratio of square root of inertia force to

(a) Surface tension force
(b) Gravity force
(c) Pressure force
(d) Viscous force

[ESE : 2011]

14.26 **Assertion (A)** : In fluid system model studies, a simple scaling-up of measurements made on the model may not yield results accurately corresponding to the prototype.

Reason (R) : Surface tension forces may be relatively much more significant in the model than in the prototype.

(a) Both **A** and **R** are individually true and **R** is the correct explanation of **A**
(b) Both **A** and **R** are individually true but **R** is not the correct explanation of **A**
(c) **A** is true but **R** is false
(d) **A** is false but **R** is true

[ESE : 2011]

14.27 Principles of similitude form the basis of

(a) performing acceptance tests
(b) comparing two identical equipments
(c) comparing similarity between design and actual equipment
(d) design and testing models of prototype based on results of models

[ESE : 2012]

14.28 The dimensional formula of bulk modulus of elasticity in MLT notation system is

(a) MLT^{-1} (b) $M^{-1} L T^2$
(c) $M^{-1}LT^{-1}$ (d) $M L^{-1} T^{-2}$

[ESE : 2024]

■■■■

Answers Dimensional Analysis

14.1 (a)	14.2 (d)	14.3 (b)	14.4 (a)	14.5 (d)	14.6 (d)	14.7 (d)	14.8 (d)	14.9 (c)
14.10 (c)	14.11 (a)	14.12 (b)	14.13 (c)	14.14 (a)	14.15 (b)	14.16 (a)	14.17 (c)	14.18 (c)
14.19 (c)	14.20 (c)	14.21 (a)	14.22 (b)	14.23 (a)	14.24 (b)	14.25 (a)	14.26 (b)	14.27 (d)
14.28 (d)								

Explanations Dimensional Analysis

14.1 (a)

Total number of constant = 4 – 3 = 1

$\therefore \quad \pi_1 = \nu^a.\rho^b.g^C.\sigma$

$M^0 L^0 T^0 = L^{a-3b+c}.M^{b+1}.T^{-2c-2}$

$\therefore \; b = -1, \; c = -1, \; a = -2$

$$\pi_1 = \frac{\sigma}{\lambda^2 g\rho}$$

14.2 (d)

$$\left(\frac{gH}{N^2d^2}\right)_P = \left(\frac{gH}{N^2d^2}\right)_m$$

$$\therefore \quad N_M = N_P\left(\frac{d_P}{d_m}\right)\sqrt{\frac{H_m}{H_P}}$$

$$= 428 \times 4 \times \sqrt{\frac{10}{30}} = \textbf{988 rpm}$$

14.3 (b)

Inertia-Viscous force ratio : Reynold number
Inertia-Gravity force ratio : Froude number
Inertia-Surface tension ratio : Weber number
Inertia-Elastic force ratio : Mach number

14.4 (a)

Since the model is tested at a velocity hence the similarity is based on Froude model law, according to which

$$\left(\frac{V}{\sqrt{gL}}\right)_m = \left(\frac{V}{\sqrt{gL}}\right)_P$$

$$V_m = V_P\sqrt{\left(\frac{L_m}{L_P}\right)} = 10\sqrt{\frac{1}{25}} = \textbf{2 m/s}$$

14.5 (d)

$(Re)_{\text{model}} = (Re)_{\text{prototype}}$

$$\frac{V_mD_m}{\nu_m} = \frac{V_PD_P}{\nu_P}$$

$$V_m = V_P\left(\frac{D_P}{D_m}\right)\left(\frac{\nu_m}{\nu_P}\right)$$

$$V_m = 12(20)\left(\frac{1}{2}\right) = \textbf{120 km/h.}$$

14.6 (d)

$$Re = \frac{\rho VD}{\mu}$$

$$\text{Froude number} = \frac{V}{\sqrt{gL}}$$

14.7 (d)

(A) Capillary waves in channel : Weber number
(B) Testing of aerofoils : Mach number
(C) Flow around bridge : Froude number
(D) Turbulent flow through pipes : Reynolds number

14.8 (d)

Chezy's constant has dimension

$= L^{1/2}\, T^{-1}$

14.9 (c)

2nd and 3rd statement are correct, therefore correct option will be (c).

In a general flow field, complete similarity between a model and prototype is achieved only when there is geometric, kinematic and dynamic similarity.

14.10 (c)

$$\left(\frac{P}{N^3 D^5}\right)_m = \left(\frac{P}{N^3 D^5}\right)_P \quad ...(i)$$

also

$$\left(\frac{gH}{N^2 D^2}\right)_m = \left(\frac{gH}{N^2 D^2}\right)_P \quad ...(ii)$$

$$(i) \Rightarrow \quad \frac{P_m}{N_m^3 D_m^5} = \frac{P_p}{N_P^3 (4D_m)^5}$$

$$\frac{P_P}{P_m} = \frac{N_P^3}{N_m^3}(4)^5$$

$$(ii) \Rightarrow \quad \frac{H_m}{N_m^2 D_m^2} = \frac{5H_m}{N_P^2 (4D_m)^2}$$

$$\left(\frac{N_P}{N_m}\right)^2 = \frac{5}{16}$$

$$\frac{N_P}{N_m} = \frac{\sqrt{5}}{4}$$

$$\frac{P_P}{P_m} = \mathbf{179}$$

14.13 (c)

Given, $(Fr)_m = (Fr)_P$

Also, $(Re)_m = (Re)_P$

$$\therefore \quad \left(\frac{V}{\sqrt{gL}}\right)_m = \left(\frac{V}{\sqrt{gL}}\right)_P \quad ...(i)$$

$$\text{and} \quad \left(\frac{\rho VL}{\mu}\right)_m = \left(\frac{\rho VL}{\mu}\right)_P \quad ...(ii)$$

Given $\frac{\mu_m}{\mu_P} = 0.0894$

From equations (i) and (ii), we have

Model scale

$$= \frac{L_P}{L_m} = \left(\frac{\mu_P}{\mu_m}\right)^{2/3} = \left(\frac{1}{0.0894}\right)^{2/3} = \mathbf{5 : 1}$$

14.14 (a)

Since the model is tested at a velocity hence the similarity is based on Froude model law, according to which

$$\left(\frac{V}{\sqrt{gL}}\right)_m = \left(\frac{V}{\sqrt{gL}}\right)_P$$

$$V_m = V_P\sqrt{\left(\frac{L_m}{L_P}\right)} = 10\sqrt{\frac{1}{25}}$$

$$\therefore \quad V_m = \mathbf{2\ m/s}$$

14.17 (c)

A higher Mach number signifies the predominance of the effect of compressibility of the fluid. However, when Mach number is relatively small say $M < 0.4$, the effect of compressibility of the fluid can be altogether neglected.

14.18 (c)

$$\text{Euler number : } E_u = \left(\frac{\text{Inertia force}}{\text{Pressure force}}\right)^{1/2}$$

$$\text{Froude number: } Fr = \left(\frac{\text{Inertia force}}{\text{Gravity force}}\right)^{1/2}$$

14.19 (c)

Froude model law

$$\frac{V_p}{V_m} \propto \sqrt{\frac{L_p}{L_m}}$$

$$\frac{F_D}{F_M} = \frac{\rho V_p^2 D_p^2}{\rho V_m^2 D_m^2}$$

$$R_p = \left(\frac{L_p}{L_m}\right)^3 r_m$$

$$= (100)^3 \times 10 = \mathbf{10000\ kN}$$

14.20 (c)

No. of variables: $n = 8$

No. of fundamental dimension: $m = 3$

No. of π terms $= n - m = 8 - 3 = \mathbf{5}$

14.21 (a)

$$\text{Euler number} = \frac{\text{Inertia force}}{\text{Pressure force}}$$

$$\text{Weber number} = \frac{\text{Inertia force}}{\text{Surface tension force}}$$

$$\text{Froude number} = \frac{\text{Inertia force}}{\text{Gravity force}}$$

$$\text{Mach number} = \frac{\text{Inertia force}}{\text{Elastic force}}$$

14.22 (b)

$$T = 2\pi\sqrt{\frac{l}{g}}$$

The only dimensionless parameter is 'π'.

14.23 (a)

Among the four groups

$$\frac{\rho N d^2}{\mu} = \frac{(ML^{-3})(T^{-1})(L^2)}{(ML^{-1}T^{-1})}$$

$$= M^0L^0T^0$$

$\therefore \frac{\rho N d^2}{\mu}$ is a dimensionless group.

14.24 (b)

Froude number for model and prototype is given as

$$(Fr)_m = (Fr)_p$$

$$\frac{V_m}{\sqrt{g_m L_m}} = \frac{V_p}{\sqrt{g_p L_p}}$$

Since $g_m = g_p$

$$\frac{V_m}{\sqrt{L_m}} = \frac{V_p}{\sqrt{L_p}}$$

$$\frac{V_m}{V_p} \times \frac{1}{\sqrt{\frac{L_m}{L_p}}} = 1$$

$$\frac{V_p}{V_m} = \sqrt{\frac{L_p}{L_m}} = \sqrt{L_r}$$

where L_r = Scale ratio for length

14.25 (a)

Weber number,

$$We = \sqrt{\frac{\text{Inertia force}}{\text{Surface tension force}}}$$

$$= \sqrt{\frac{\rho L^2 v^2}{\sigma L}} = \sqrt{\frac{\rho L V^2}{\sigma}}$$

Weber number is used in case of capillarity.

14.28 (d)

Dimension of bulk modulus

$$= \frac{\text{Dimension of hydrostatic stress}}{\text{Dimension of volumetric strain}}$$

$$= \frac{[ML^{-1}L^{-2}]}{[M^0L^0T^0]} = [ML^{-1}T^{-2}]$$

■■■■

15 Flow over Notches and Weirs

15.1 A standard 90° V-notch weir is used to measure discharge. The discharge is Q_1 for a height H_1 above the sill and Q_2 is the discharge for a height H_2. If $\frac{H_2}{H_1}$ is 4, then $\frac{Q_2}{Q_1}$ is

(a) 32 (b) $16\sqrt{2}$
(c) 16 (d) 8

[ESE : 2001]

15.2 Which of the following is/are related to measure the discharge by rectangular notch?

1. $\frac{2}{3}C_d b\sqrt{2g}H^2$
2. $\frac{2}{3}C_d b\sqrt{2g}H^{3/2}$
3. $\frac{2}{3}C_d b\sqrt{2g}H^{5/2}$
4. $\frac{2}{3}C_d b\sqrt{2g}H^{1/2}$

Select the correct answer using the codes given below:

(a) 1 and 3 (b) 2 and 3
(c) 2 only (d) 4 only

[ESE : 2002]

15.3 The discharge, over a rectangular weir of width 100 m when the head of water over the weir is 1.5 m and the head due to velocity of approach is 0.02 m, is given by

(a) 842.5 m^3/s (b) 653.4 m^3/s
(c) 814.5 m^3/s (d) 552.5 m^3/s

[ESE : 2010]

15.4 **Statement (I) :** The coefficient of discharge for a a mouthpiece is higher than of an orifice.

Statement (II): The discharge through an orifice varies as $H^{1/2}$ whereas the discharge through a mouthpiece varies as $H^{2/3}$ (where H is the head causing the flow in both cases).

(a) Both Statement (I) and Statement (II) are individually true and Statement (II) is the correct explanation of Statement (I).
(b) Both Statement (I) and Statement (II) are individually true but Statement (II) is **NOT** the correct explanation of Statement (I).
(c) Statement (I) is true but Statement (II) is false
(d) Statement (I) is false but Statement (II) is true.

[ESE : 2014]

15.5 The discharge through an orifice fitted in a tank can be increased by :

(a) Fitting a short length of pipe to the outside
(b) Sharpening the edge of orifice
(c) Fitting a long length of pipe to the outside
(d) Fitting a long length of pipe to the inside

[ESE : 2016]

15.6 Water at 20°C is flowing between a two-dimensional channel in which the top and bottom walls are 1.5 mm apart. If the average velocity is 2 m/s, what is the maximum velocity?

(a) 1 m/s (b) 1.5 m/s
(c) 2 m/s (d) 3 m/s

[ESE : 2023]

■■■■

Answers Flow over Notches and Weirs

15.1 (a) 15.2 (c) 15.3 (d) 15.4 (c) 15.5 (a) 15.6 (d)

Explanations Flow over Notches and Weirs

15.1 (a)

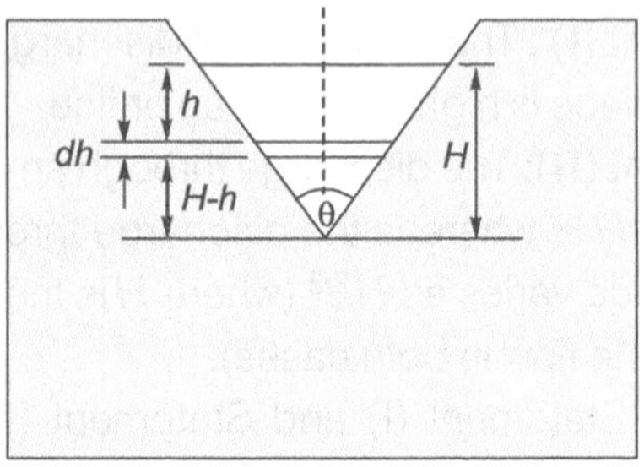

$$dQ = \sqrt{2gh}.dA = \sqrt{2gh}.(b.dh)$$

$$\tan\frac{\theta}{2} = \frac{b/2}{(H-h)}$$

$$\Rightarrow \quad b = 2(H-h)\tan\frac{\theta}{2}$$

$$dQ = 2(H-h)\tan\frac{\theta}{2}\sqrt{2gh}\cdot dh$$

$$Q = \int dQ = \frac{8}{15}\sqrt{2g}\tan\frac{\theta}{2}H^{5/2}$$

$$\frac{Q_2}{Q_1} = \left(\frac{H_2}{H_1}\right)^{5/2} = 4^{5/2} = \mathbf{32}$$

15.2 (c)

Discharge by rectangular notch,

$$Q = \frac{2}{3}C_d\, b\sqrt{2g}\cdot H^{3/2}$$

15.3 (d)

$$Q = \frac{2}{3}C_d \times L \times \sqrt{2g}\quad\left[(H_1+h_a)^{3/2} - h_a^{3/2}\right]$$

$C_d = 1$; $L = 100$ m; $H_1 = 1.5$ m; $h_a = 0.02$ m

$$Q = \frac{2}{3}\times 100\times\sqrt{2\times 9.812}\left[(1.5+0.02)^{3/2} - (0.02)^{3/2}\right]$$

$$= \frac{2}{3}\times 442.9898[1.87115] = \mathbf{552.6\ m^3/s}$$

15.4 (c)

The discharge through an orifice varies as $H^{1/2}$ whereas the discharge through a mouthpiece varies also as $H^{1/2}$ (where H is the head causing the flow in both cases).

So, statement (II) is wrong.

15.6 (d)

Maximum velocity,

$$V_{max} = \frac{3}{2}V_{avg}$$

$$V_{max} = 1.5 \times 2$$

$$V_{max} = 3 \text{ m/s}$$

■■■■

UNIT

II

Thermodynamics

Syllabus

Thermodynamic systems and processes; properties of pure substance; Zeroth, First and Second Laws of Thermodynamics; Entropy, Irreversibility and availability; analysis of thermodynamic cycles related to energy conversion, ideal and real gases; compressibility factor; Gas mixtures.

Contents

Sl.	Topic	Page No.
1.	Thermodynamic System, Processes and Zeroth Law	132
2.	First Law, Heat and Work, Energy	140
3.	Second Law, Carnot Cycle and Entropy	162
4.	Irreversibility and Availability	188
5.	Thermodynamic Relations	193
6.	Pure Substances	199

OOOO

1 Thermodynamic System, Processes and Zeroth Law

1.1 Number of components (C), phases (P) and degrees of freedom (F) are related by Gibbs phase rule as

(a) $C - P - F = 2$ (b) $F - C - P = 2$
(c) $C + F - P = 2$ (d) $P + F - C = 2$

[ESE : 2001]

1.2 In new temperature scale say °ρ, the boiling and freezing points of water at one atmosphere are 100°ρ and 300°ρ respectively. Correlate this scale with the Centigrade scale. The reading of 0°ρ on the Centigrade scale is

(a) 0°C (b) 50°C
(c) 100°C (d) 150°C

[ESE : 2001]

1.3 Pressure reaches a value of absolute zero

(a) at a temperature of –273 K
(b) under vacuum condition
(c) at the earth's centre
(d) when molecular momentum of system becomes zero

[ESE : 2002]

1.4 Which one of the following correctly defines 1 K, as per the internationally accepted definition of temperature scale?

(a) 1/100th of the difference between normal boiling point and normal freezing point of water
(b) 1/273.15th of the normal freezing point of water
(c) 100 times the difference between the triple point of water and the normal freezing point of water
(d) 1/273.16th of the triple point of water

[ESE : 2004]

1.5 Which of the following are intensive properties?

1. Kinetic energy 2. Specific enthalpy
3. Pressure 4. Entropy

Select the correct answer using the code given below:

(a) 1 and 3 (b) 2 and 3
(c) 1, 3 and 4 (d) 2 and 4

[ESE : 2005]

1.6 Which one of the following is the characteristic equation of a real gas?

(a) $(p + a/v^2)(v - b) = RT$
(b) $(p - a/v^2)\,(v + b) = RT$
(c) $pv = RT$
(d) $pv = nRT$

[ESE : 2006]

1.7 **Assertion (A):** At very high densities, compressibility of a real gas is less than one.

Reasons (R): As the temperature is considerably reduced, the molecules are brought closer together and thermonuclear attractive forces become greater at pressure around 4 MPa.

(a) Both **A** and **R** are individually true and **R** is the correct explanation of **A**
(b) Both **A** and **R** are individually true but **R** is not the correct explanation of **A**
(c) **A** is true but **R** is false
(d) **A** is false but **R** is true

[ESE : 2006]

1.8 Which of the following are intensive properties?

1. Kinetic energy
2. Thermal conductivity
3. Pressure
4. Entropy

Select the correct answer using the code given below:

(a) 1 and 2 (b) 2 and 3 only
(c) 2, 3 and 4 (d) 1, 3 and 4

[ESE : 2007]

1.9 Which one of the following thermodynamic processes approximates the steaming of food in a pressure cooker?

(a) Isenthalpic (b) Isobaric
(c) Isochoric (d) Isothermal

[ESE : 2007]

1.10 Match **List-I** (Type of thermometer) with **List-II** (Thermometric property) and select the correct answer using the code given below the lists:

List-I

A. Mercury in glass
B. Thermocouple
C. Thermistor
D. Constant volume gas

List-II

1. Pressure
2. Electrical resistance
3. Volume
4. Induced electric voltage

Codes:

	A	B	C	D
(a)	1	4	2	3
(b)	3	2	4	1
(c)	1	2	4	3
(d)	3	4	2	1

[ESE : 2007]

1.11 An ideal gas at 27°C is heated at constant pressure till its volume becomes three times. What would be then the temperature of gas?
(a) 81°C (b) 627°C
(c) 543°C (d) 327°C
[ESE : 2008]

1.12 Ice kept in a well insulated thermoflask is an example of which system?
(a) Closed system
(b) Isolated system
(c) Open system
(d) Non-flow adiabatic system
[ESE : 2009]

1.13 Measurement of temperature is based on which law of thermodynamics?
(a) Zeroth law of thermodynamics
(b) First law of thermodynamics
(c) Second law of thermodynamics
(d) Third law of thermodynamics
[ESE : 2009]

1.14 Consider the following properties:
1. Temperature 2. Viscosity
3. Specific entropy 4. Thermal conductivity
Which of the above properties of a system is/are intensive?
(a) 1 only (b) 2 and 3
(c) 2, 3 and 4 (d) 1, 2, 3 and 4
[ESE : 2009]

1.15 Consider the following properties:
1. Entropy
2. Viscosity
3. Temperature
4. Specific heat at constant volume
Which of the above properties of a system is/are extensive?
(a) 1 only (b) 1 and 2
(c) 2, 3 and 4 (d) 1, 2 and 4
[ESE : 2009]

1.16 Consider the following:
1. Kinetic energy 2. Entropy
3. Thermal conductivity 4. Pressure
Which of these are intensive properties?
(a) 1, 2 and 3 only (b) 2 and 4 only
(c) 3 and 4 only (d) 1, 2, 3 and 4
[ESE : 2010]

1.17 Molar specific heats of an ideal gas depend on
(a) its pressure
(b) its temperature
(c) both its pressure and temperature
(d) the number of atoms in a molecule
[ESE : 2010]

1.18 Consider the following:
1. Temperature 2. Viscosity
3. Internal energy 4. Entropy
Which of these are extensive properties?
(a) 1, 2, 3 and 4 (b) 2 and 4 only
(c) 2 and 3 only (d) 3 and 4 only
[ESE : 2010]

1.19 The Zeroth law of thermodynamics states that
(a) two thermodynamic systems are always in thermal equilibrium with each other
(b) if two systems are in thermal equilibrium, then the third system will also be in thermal equilibrium
(c) two systems not in thermal equilibrium with a third system are also not in thermal equilibrium with each other
(d) when two systems are in thermal equilibrium with a third system, they are in thermal equilibrium with each other **[ESE : 2010]**

1.20 Hot coffee stored in a well insulated thermos flask is an example of
(a) isolated system
(b) closed system

(c) open system
(d) non-flow adiabatic system [ESE : 2010]

1.21 A control volume is
(a) an isolated system
(b) a closed system but heat and work can cross the boundary
(c) a specific amount of mass in space
(d) a fixed region in space where mass, heat and work can cross the boundary of that region [ESE : 2010]

1.22 A closed thermodynamic system manifests when
(a) matter is not allowed to cross the boundary but energy transfer occurs across the boundary
(b) there is transfer of both mass and energy across the system boundaries
(c) there is only transfer of mass but no heat and work energy are transferred
(d) there is absolutely no interaction of the system with surroundings [ESE : 2010]

Direction: Each of the next questions consists of two statements, one labelled as the **'Assertion (A)'** and the other as **'Reason (R)'**. You are to examine these two statements carefully and select the answers to these items using the codes given below:

Codes:
(a) Both **A** and **R** are individually true and **R** is the correct explanation of **A**
(b) Both **A** and **R** are individually true but **R** is not the correct explanation of **A**
(c) **A** is true but **R** is false
(d) **A** is false but **R** is true

1.23 **Assertion (A):** When the surface of a lake is cooled during winter, the cooled surface water descends to the bottom as long as its temperature exceeds 4°C, but when cooled below 4°C it floats at the top.
Reason (R): Ice forms at the top of a pond in winter while bottom fluid stays warmer at more than 4°C.
[ESE : 2010]

1.24 **Assertion (A):** An air filled balloon released from the ground goes up and up till it reaches a certain elevation and floats in the air.
Reason (R): With increase in elevation the temperature of the atmospheric air increases and so the density decreases.
[ESE : 2010]

1.25 A closed system is one in which
(a) Mass does not cross boundaries of the system, though energy may do so
(b) Mass crosses the boundary but not the energy
(c) Neither mass nor energy cross the boundary of the system
(d) Both energy and mass cross the boundaries of the system [ESE : 2011]

1.26 Match **List-I** with **List-II** and select the correct answer using the code given below the lists:

List-I
A. Interchange of matter is not possible in a system
B. Any processes in which the system returns to its original condition or state is called
C. Interchange of matter is possible in a
D. The quantity of matter under consideration in thermodynamic is called

List-II
1. Open
2. System
3. Closed system
4. Cycle

Codes:

	A	B	C	D
(a)	2	1	4	3
(b)	3	1	4	2
(c)	2	4	1	3
(d)	3	4	1	2

[ESE : 2011]

1.27 A thermodynamic system is considered to be an isolated one if
(a) Mass transfer and entropy change are zero
(b) Entropy change and energy transfer are zero
(c) Energy transfer and mass transfer are zero
(d) Mass transfer and volume change are zero
[ESE : 2011]

1.28 Which one of the following represents open thermodynamic system?
(a) Manual ice cream freezer
(b) Centrifugal pump
(c) Pressure cooker
(d) Bomb calorimeter [ESE : 2011]

1.29 In highly rarefied gases, the concept of this loses validity
(a) Thermodynamic equilibrium
(b) Continuum
(c) Stability
(d) Macroscopic viewpoint [ESE : 2012]

1.30 The constant volume gas thermometer works on the principle that
(a) at low pressure, the temperature of the gas is independent of its pressure at constant volume
(b) at high pressure, the temperature of the gas is independent of its pressure at constant volume
(c) at low pressure, the temperature of the gas is proportional to its pressure at constant volume
(d) at high pressure, the temperature of the gas is proportional to its pressure at constant volume [ESE : 2012]

1.31 An ideal gas at 27°C is heated at constant pressure till the volume becomes three times. The temperature of the gas will then be
(a) 81°C (b) 900°C
(c) 627°C (d) 927°C
[ESE : 2012]

1.32 The value of an extensive property is essentially dependent on
(a) mass of the system
(b) interaction of the system with its surroundings
(c) path followed by the system in going from one state to another
(d) nature of boundaries, rigid or flexible
[ESE : 2013]

1.33 If the time taken by a system to execute a process through a finite gradient is infinitely large, the process
(a) becomes reversible
(b) is irreversible
(c) becomes isothermal
(d) is adiabatic [ESE : 2013]

1.34 Internal energy of a system is dependent on the following aspects:
1. Molecular weight
2. Molecular structure
3. Degree of molecular activity

Which of the above are correct?
(a) 1 and 2 only (b) 1 and 3 only
(c) 2 and 3 only (d) 1, 2 and 3
[ESE : 2013]

1.35 Statement (I) : Negative temperature are impossible on the Kelvin scale.
Statement (II): The Kelvin scale is thermodynamic temperature scale.
(a) Both Statement (I) and Statement (II) are individually true and Statement (II) is the correct explanation of Statement (I).
(b) Both Statement (I) and Statement (II) are individually true but Statement (II) is **NOT** the correct explanation of Statement (I).
(c) Statement (I) is true but Statement (II) is false
(d) Statement (I) is false but Statement (II) is true.
[ESE : 2013]

1.36 Which of the following thermodynamic properties are intensive properties?
1. Density
2. Entropy
3. Viscosity
(a) 1, 2 and 3 (b) 1 and 2 only
(c) 2 and 3 only (d) 1 and 3 only
[ESE : 2014]

1.37 The property of a thermodynamic system is:
(a) A path function
(b) A point function
(c) A quantity which does not change in reversible process
(d) A quantity which changes when system undergoes a cycle [ESE : 2016]

1.38 Statement (I) : A homogeneous mixture of gases that do not react within themselves can be treated as a pure substance.
Statement (II) : Flue gases can be treated as homogeneous mixture of gases.
(a) Both Statement (I) and Statement (II) are individually true and Statement (II) is the correct explanation of Statement (I).
(b) Both Statement (I) and Statement (II) are individually true but Statement (II) is not the correct explanation of Statement (I).
(c) Statement (I) is true but Statement (II) is false.
(d) Statement (I) is false but Statement (II) is true.
[ESE : 2017]

1.39 Which one of the following substances has constant specific heat at all pressures and temperatures?
(a) Mano-atomic gas (b) Di-atomic gas
(c) Tri-atomic gas (d) Poly-atomic gas
[ESE : 2018]

1.40 Certain quantities cannot be located on the graph by a point but are given by the area under the curve corresponding to the process. These quantities in concepts of thermodynamics are called as
(a) cyclic functions (b) point functions
(c) path functions (d) real functions
[ESE : 2019]

1.41 **Statement (I):** If two systems are in thermal equilibrium with a third system, then they are not in thermal equilibrium with each other.

Statement (II): Zeroth law of thermodynamics is the basis for temperature.
(a) Both Statement (I) and Statement (II) are individually true, and Statement (II) is the correct explanation of Statement (I)
(b) Both Statement (I) and Statement (II) are individually true, but Statement (II) is not the correct explanation of Statement (I)
(c) Statement (I) is true, but Statement (II) is false
(d) Statement (I) is false, but Statement (II) is true
[ESE : 2021]

1.42 1 kg of water falls from an altitude of 1000 m above ground level. What is the change in the temperature of water at the foot of the fall, if there are no losses during the fall? [Take specific heat of water as 1 $kcal.kg^{-1}.K^{-1}$ and $g = 9.81$ m/s^2]
(a) 3.35°C (b) 2.35°C
(c) 3.32°C (d) 4.12°C
[ESE : 2021]

1.43 In a thermoelectric thermometer for t°C temperature, the emf is given as: $E = 0.003t - 5 \times 10^{-7}\, t^2 + 0.5 \times 10^{-3}$ volts. Thermometer is having reference junction at ice point and is calibrated at ice point and steam point. What is the temperature shown by the thermometer for a substance at 30°C?
(a) 33.23°C (b) 36.28°C
(c) 41.23°C (d) 46.28°C
[ESE : 2021]

1.44 Which one of the following is the intensive property in thermodynamic system?
(a) Pressure (b) Enthalpy
(c) Internal energy (d) Entropy
[ESE : 2022]

1.45 100 kg of ice at –5°C is placed in a bunker to cool some vegetables. 24 hours later, the ice has melted into water at 10°C. What is the average rate of cooling in kJ/h provided by the ice? (Take specific heat of ice is 1.94 kJ/kgK; Specific heat of water is 4.1868 kJ/kgK; latent heat of fusion of ice at 0°C is 335 kJ/kg)
(a) 1611 kJ/h (b) 1811 kJ/h
(c) 1711 kJ/h (d) 1911 kJ/h
[ESE : 2022]

1.46 The work required for compressing one kg of air is minimum if the compression is carried out
(a) adiabatically (b) isothermally
(c) isentropically (d) isochorically
[ESE : 2022]

1.47 What is the heat stored in 5 m^3 of water, if specific heat is 4.19 kJ/(kg°C), temperature final = 26°C, temperature initial = 18°C?
(a) 67.6×10^3 kJ (b) 147.6×10^3 kJ
(c) 167.6×10^3 kJ (d) 47.6×10^3 kJ
[ESE : 2023]

1.48 Which one of the following properties is not an extensive property of a system?
(a) Energy (b) Enthalpy
(c) Volume (d) Temperature
[ESE : 2024]

1.49 Thermoelectric thermometer works on the principle of
(a) Wheatstone bridge (b) entropy
(c) Seebeck effect (d) ohmmeter
[ESE : 2024]

■■■■

Answers Thermodynamic System, Processes and Zeroth Law

1.1 (d)	1.2 (d)	1.3 (d)	1.4 (d)	1.5 (b)	1.6 (a)	1.7 (a)	1.8 (b)	1.9 (c)
1.10 (d)	1.11 (b)	1.12 (b)	1.13 (a)	1.14 (d)	1.15 (a)	1.16 (c)	1.17 (d)	1.18 (d)
1.19 (d)	1.20 (a)	1.21 (d)	1.22 (a)	1.23 (b)	1.24 (c)	1.25 (a)	1.26 (d)	1.27 (c)
1.28 (b)	1.29 (b)	1.30 (c)	1.31 (c)	1.32 (a)	1.33 (a)	1.34 (d)	1.35 (b)	1.36 (d)
1.37 (b)	1.38 (b)	1.39 (a)	1.40 (c)	1.41 (d)	1.42 (b)	1.43 (a)	1.44 (a)	1.45 (a)
1.46 (b)	1.47 (c)	1.48 (d)	1.49 (c)					

Explanations Thermodynamic System, Processes and Zeroth Law

1.1 (d)

Gibbs phase rule:

$$P + F = C + 2$$
$$P + F - C = 2$$

1.2 (d)

$$\frac{300-0}{300-100} = \frac{0-C}{0-100}$$

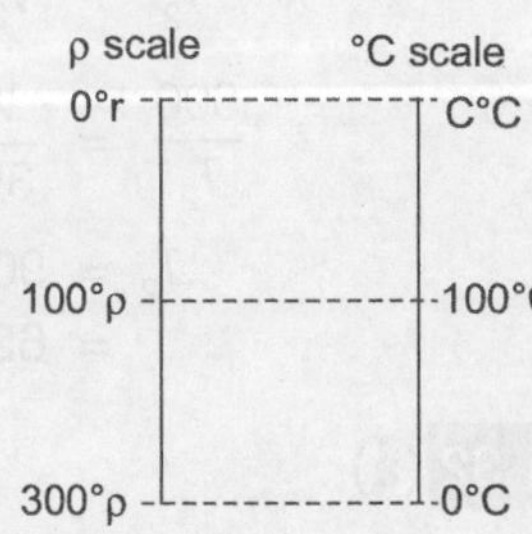

$$\frac{3}{2} = \frac{C}{100}$$

$$C = \mathbf{150°C}$$

Alternate:

Assume $\rho = a + bT$

$$300 = a + b \times 0$$
$$a = 300$$
$$100 = a + b \times 100$$
$$\frac{100-300}{100} = b$$
$$b = -2$$
$$\rho = 300 - 2T$$

So, If $\rho = 0$

$$300 - 2T = 0$$
$$T = 150°C$$

1.4 (d)

The CGPM 1967 defined kelvin as equal to $\frac{1}{273.16}$ of thermodynamic temperature of the triple point of water (0.01°C)

1.8 (b)

Intensive properties are independent of the mass of the system. Examples: Pressure, temperature, thermal conductivity etc.

1.9 (c)

Volume of pressure cooker is always constant, hence it is isochoric process.

1.10 (d)

Type of thermometer		Thermometric Property
(a) Mercury in glass	:	Volume
(b) Thermocouple	:	Induced electric voltage
(c) Thermistor	:	Electrical resistance
(d) Constant volume	:	Pressure gas

1.11 (b)

$$\frac{V_1}{T_1} = \frac{V_2}{T_2}$$

$$\frac{V_1}{(273+27)} = \frac{3V_1}{T_2}$$

or $T_2 = 300 \times 3 = 900\text{ K} = \mathbf{627°C}$

1.13 (a)

Measurement of temperature is based on Zeroth law of thermodynamics.

1.14 (d)

Intensive property is that property which is independent of mass. All the four properties are intensive properties.

1.15 (a)

Extensive property is that property which is dependent on mass. Here entropy is an extensive property

1.21 (d)

The boundary of closed system usually changes shape and no mass flow occurs, but in control volume boundary remains fixed and unaltered but through it moving substance flows.

1.25 (a)

In a closed system, mass does not cross the boundary but the energy may cross the boundary. For example: Piston cylinder without valves.

1.26 (d)

Open system – Exchange of heat and mass with surroundings. Example: Compressor
Closed system – Exchange of heat and with surrounding. Example: Green house
Isolated system – Neither exchanger of heat nor exchange of mass with surroundings.
Example: Universe

Interaction of thermodynamic system

Type of system	Mass flow	Work	Heat
Open	Yes	Yes	Yes
Closed	No	Yes	Yes
Isolated	No	No	No

Cyclic Processes: A cyclic process is a process that can be repeated indefinitely often without changing the final state of the system in which the process occurs. The only traces of the effects of a cyclic process are to be found in the surroundings of the system or in other systems.

1.27 (c)

An isolated system is one in which there is no energy and mass transfer between the system and surrounding.
Examples : Universe, coffee in thermo flask.

1.28 (b)

Examples of open thermodynamic system are centrifugal pump, boiler, turbine, evaporator, condenser.

1.29 (b)

In highly rarefied gases, since the density is very less matter cannot be assumed to be continuously distributed throughout the space. Hence the concept of continuum loses its validity.

1.30 (c)

For a constant volume gas thermometer, pressure acts as thermometric property and at low pressure the gas behaves as ideal gas and follows ideal gas equation

$$pV = mRT$$

$\therefore$ for V = constant (m, R and also constant)

$$p \propto T$$

1.31 (c)

$$p = \text{Constant}$$

$$\therefore \quad \frac{T_1}{T_2} = \frac{V_1}{V_2}$$

$$\frac{300}{T_2} = \frac{V_1}{3V_1}$$

$$T_2 = 900\text{ K} = (900 - 273)^\circ\text{C}$$

$$= \mathbf{627^\circ C}$$

1.32 (a)

Properties are of two types:
Intensive property: Independent of the mass of the system.
Extensive property: It is related to mass.

1.36 (d)

Entropy is extensive property which is independent of mass.

1.40 (c)

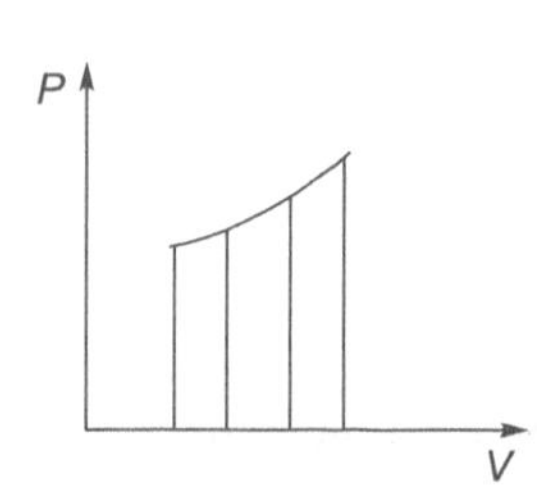

Area under curve on P-V represents work when projected on volume axis for closed system

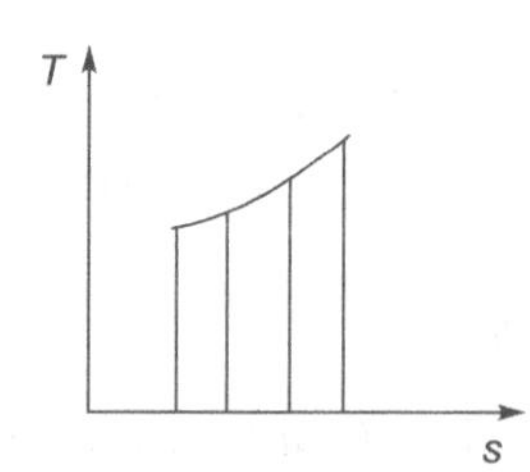

Area under T-s diagram represents heat transfer for a reversible process

1.41 (d)

The zeroth law of thermodynamics states that if two thermodynamic system are each in thermal equilibrium with a third, then they are in thermal equilibrium with each other. Therefore, statement (I) is wrong.

1.42 (b)

$$m = 1 \text{ kg},\ Z = 1000 \text{ m}$$

$$mgz = mc\ dt$$

$$9.81 \times 1000 \times 10^{-3} \text{ kJ} = 1 \times 4.186 \times dt$$

$$dt = 2.343°\text{C} \approx 2.35°\text{C}$$

1.43 (*)

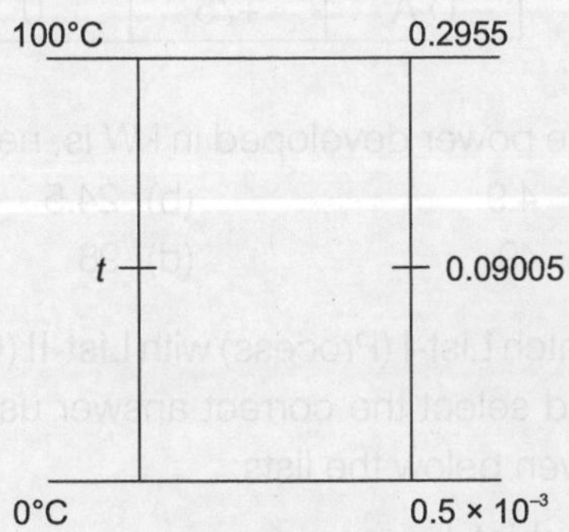

$$E = 0.003t - 5 \times 10^{-7}t^2 + 0.5 \times 10^{-3}$$

$$E_0 = 0.5 \times 10^{-3}$$

$$E_{100} = 0.2955$$

$$E_{30} = 0.09005$$

$$\frac{t-0}{100-0} = \frac{0.09 - 0.5\times10^{-3}}{0.2955 - 0.5\times10^{-3}}$$

$$t = 30.393°\text{C} \simeq 33.23°\text{C}$$

1.45 (a)

Given: $m = 100$ kg, $C_{ice} = 1.94$ kJ/kgK,

Latent heat of water = 335 kJ/kg,

$C_{water} = 4.1868$ kJ/kgK

$$-5°\text{C} \xrightarrow{\text{ice}} 0°\text{C} \longrightarrow 0°\text{C} \xrightarrow{\text{water}} 10°\text{C}$$

$$Q = m(C_{ice} \times \Delta T_{ice} + LH + C_w \times \Delta T_{water})$$

$$\therefore \quad Q = 378656.8 \text{ kJ}$$

$$\dot{Q} = 1610.7 \text{ kJ/h}$$

1.46 (b)

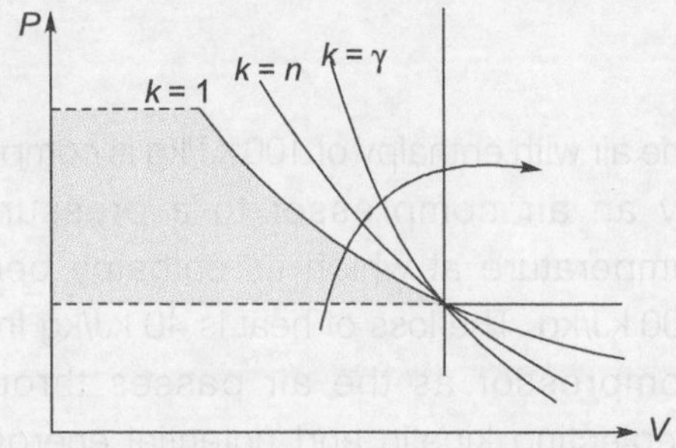

The work required for compression is minimum for compression carried isothermally.

$$\text{As,} \quad W = \int -VdP$$

1.47 (c)

Given, Volume of water,

$$V = 5 \text{ m}^3$$

Specific heat, $c = 4.19$ kJ/kg°C

Final temperature,

$$T_f = 26°\text{C}$$

Initial temperature,

$$T_i = 18°\text{C}$$

Mass of water,

$$m = \rho V$$

where, density of water, $\rho = 1000$ kg/m^3

$$\therefore \quad m = 1000 \times 5 = 5000 \text{ kg}$$

Thus, heat stored is

$$Q = mc\Delta T$$

$$= 5000 \times 4.19 \times (26 - 18)$$

$$= 167600 \text{ kJ} = 167.6 \times 10^3 \text{ kJ}$$

1.49 (c)

The Seebeck effect is s phenomenon in which a temperature difference between two dissimilar electrical conductors or semiconductors produce a voltage difference between the two substances. Thermoelectric thermometer works on Seebeck principle.

■■■■

2 First Law, Heat and Work, Energy

2.1 The air with enthalpy of 100 kJ/kg is compressed by an air compressor to a pressure and temperature at which its enthalpy becomes 200 kJ/kg. The loss of heat is 40 kJ/kg from the compressor as the air passes through it. Neglecting kinetic and potential energies the power required for an air mass flow of 0.5 kg/s is

(a) 30 kW (b) 50 kW
(c) 70 kW (d) 90 kW

[ESE : 2000]

2.2 **Assertion (A):** A thermodynamic system may be considered as a quantity of working substance with which interaction of heat and work are studied.

Reason (R): Energy in the form of work and heat are mutually convertible.

(a) both **A** and **R** are true and **R** is the correct explanation of **A**
(b) both **A** and **R** are true but **R** is not a correct explanation of **A**
(c) **A** is true but **R** is false
(d) **A** is false but **R** is true

[ESE : 2000]

2.3 The data given in the table refers to an engine based on Carnot cycle,

where Q_1 = Heat received (kJ/min)
Q_2 = Heat rejected (kJ/s)
W = Work output (kW)

S. No.	Q_1	Q_2	W
1.	1500	16.80	8.20
2.	1600	17.92	8.75
3.	1700	19.03	9.30
4.	1800	20.15	9.85

If heat received by the engine is 2000 kJ/min the work output will be nearly

(a) 9.98 (b) 10.39
(c) 11.54 (d) 10.95 [ESE : 2001]

2.4 A system while undergoing a cycle A–B–C–D–A has the values of heat and work transfers as given in the table

Process	Q(kJ/min)	W(kJ/min)
A-B	+687	+474
B-C	−269	0
C-D	−199	−180
D-A	+75	0

The power developed in kW is, nearly,

(a) 4.9 (b) 24.5
(c) 49 (d) 98 [ESE : 2001]

2.5 Match **List-I** (Process) with **List-II** (Characteristic) and select the correct answer using the codes given below the lists:

List-I

A. Throttling process
B. Isentropic process
C. Free expansion
D. Isothermal process

List-II

1. No work done
2. No change in entropy
3. Constant internal energy
4. Constant enthalpy

Codes:

	A	B	C	D
(a)	4	2	1	3
(b)	1	2	4	3
(c)	4	3	1	2
(d)	1	3	4	2

[ESE : 2001]

Direction: Each of the next questions consists of two statements, one labelled as the **'Assertion (A)'** and the other as **'Reason (R)'**. You are to examine these two statements carefully and select the answers to these items using the codes given below:

Codes:

(a) Both **A** and **R** are individually true and **R** is the correct explanation of **A**

(b) Both **A** and **R** are individually true but **R** is not the correct explanation of **A**

(c) **A** is true but **R** is false

(d) **A** is false but **R** is true

2.6 **Assertion (A):** If the enthalpy of a closed system decreased by 25 kJ while the system receives 30 kJ of energy by heat transfer, the work done by the system is 55 kJ.

Reason (R): The first law energy balance for a closed system is (notations have their usual meaning) $\Delta E = Q - W$. [ESE : 2001]

2.7 **Assertion (A):** Specific heat at constant pressure for an ideal gas is always greater than the specific heat at constant volume.

Reason (R): Heat added at constant volume is not utilized for doing any external work.

[ESE : 2002]

2.8 The heat absorbed or rejected during a polytropic process is equal to

(a) $\left(\dfrac{\gamma - n}{\gamma - 1}\right)^{1/2} \times$ work done

(b) $\left(\dfrac{\gamma - n}{n - 1}\right) \times$ work done

(c) $\left(\dfrac{\gamma - n}{\gamma - 1}\right) \times$ work done

(d) $\left(\dfrac{\gamma - n}{\gamma - 1}\right)^{2} \times$ work done [ESE : 2002]

2.9 Identify the process for which the two integrals $\int pdV$ and $-\int Vdp$, evaluated between any two given states gives the same value

(a) Isenthalpic (b) Isothermal

(c) Isentropic (d) Polytropic

[ESE : 2003]

2.10 Which one of the following phenomenon occurs when gas in a piston-in-cylinder assembly expands reversibly at constant pressure?

(a) Heat is added to the gas

(b) Heat is removed from the gas

(c) Gas does work from its own stored energy

(d) Gas undergoes adiabatic expansion

[ESE : 2003]

2.11 During steady flow compression process of a gas with mass flow rate of 2 kg/s, increase in specific enthalpy is 15 kJ/kg and decrease in kinetic energy is 2 kJ/kg. The rate of heat rejection to the environment is 3 kW. The power needed to drive the compressor is

(a) 23 kW (b) 26 kW

(c) 29 kW (d) 37 kW

[ESE : 2003]

2.12 For a simple closed system of constant composition, the difference between the net heat and work interactions is identifiable as the change in

(a) Enthalpy (b) Entropy

(c) Flow energy (d) Internal energy

[ESE : 2003]

2.13 Gas contained in a closed system consisting of piston cylinder arrangement is expanded. Work done by the gas during expansion is 50 kJ. Decreases in internal energy of the gas during expansion is 30 kJ. Heat transfer during the process is equal to

(a) –20 kJ (b) +20 kJ

(c) –80 kJ (d) +80 kJ

[ESE : 2003]

2.14 A perfect gas at 27°C was heated until its volume was doubled using the following three different processes separately:

1. Constant pressure process
2. Isothermal process
3. Isentropic process

Which one of the following is the correct sequence in the order of increasing value of the final temperature of the gas reached by using the above three different processes?

(a) 1 – 2 – 3 (b) 2 – 3 – 1

(c) 3 – 2 – 1 (d) 3 – 1 – 2

[ESE : 2004]

2.15 In a steady-flow adiabatic turbine, the changes in the internal energy, enthalpy, kinetic energy and potential energy of the working fluid, from inlet to exit, are –100 kJ/kg, –140 kJ/kg, –10 kJ/kg and 0 kJ/kg respectively. Which one of the following gives the amount of work developed by the turbine?

(a) 100 kJ/kg (b) 110 kJ/kg
(c) 140 kJ/kg (d) 150 kJ/kg

[ESE : 2004]

2.16 An ideal gas undergoes an isothermal expansion from state R to state S in a turbine as shown in the diagram given below:

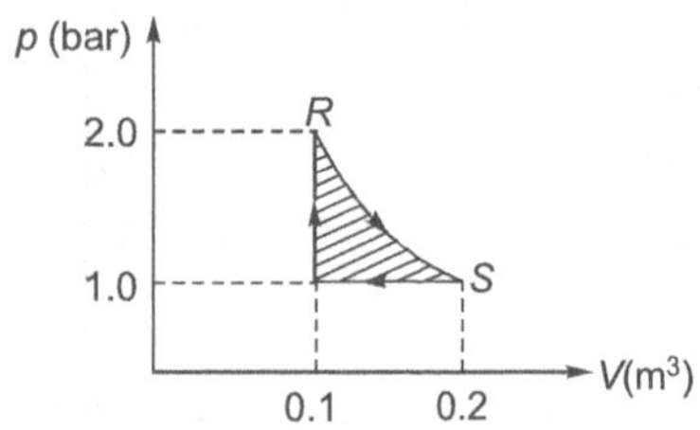

The area of shaded region is 1000 Nm. The amount is turbine work done during the process is

(a) 14,000 Nm (b) 12,000 Nm
(c) 11,000 Nm (d) 10,000 Nm

[ESE : 2004]

2.17 170 kJ of heat is supplied to a system at constant volume. Then the system rejects 180 kJ of heat at constant pressure and 40 kJ of work is done on it. The system is finally brought to its original state by adiabatic process. If the initial value of internal energy is 100 kJ, then which one of the following statements is correct?

(a) The highest value of internal energy occurs at the end of the constant volume process
(b) The highest value of internal energy occurs at the end of constant pressure process
(c) The highest value of internal energy occurs after adiabatic expansion
(d) Internal energy is equal at all points

[ESE : 2004]

2.18 A system undergoes a process during which the heat transfer to the system per degree increase in temperature is given by the equation:

$\delta Q/dT = 2.0$ kJ/°C

The work done by the system per degree increase in temperature is given by the equation

$\delta W/dT = 2 - 0.1T$

where T is in °C. If during the process the temperature of water varies from 100°C to 150°C, the change in internal energy is

(a) 125 kJ (b) –250 kJ
(c) 625 kJ (d) –1250 kJ

[ESE : 2004]

2.19 An insulated box containing 0.5 kg of a gas having c_v = 0.98 kJ/kgK falls from a balloon 4 km above the earth's surface. The temperature rise of the gas when the box hits the ground is

(a) 0 K (b) 20 K
(c) 40 K (d) 60 K

[ESE : 2004]

2.20 Which one of the following is the steady flow energy equation for a boiler?

(a) $h_1 + \dfrac{V_1^2}{2gJ} = h_2 + \dfrac{V_2^2}{2gJ}$

(b) $Q = h_2 - h_1$

(c) $h_1 + \dfrac{V_1^2}{2gJ} + Q = h_2 + \dfrac{V_2^2}{2gJ}$

(d) $W_s = (h_2 - h_1) + Q$

[ESE : 2005]

2.21 Change in internal energy in a reversible process occurring in a closed system is equal to the heat transferred if the process occurs at constant:

(a) Pressure (b) Volume
(c) Temperature (d) Enthalpy

[ESE : 2005]

2.22 In a polytropic process, the term

$\left[\left(\dfrac{\gamma - n}{\gamma - 1}\right)\left(\dfrac{p_1V_1 - p_2V_2}{n-1}\right)\right]$ is equal to:

(a) Heat absorbed or rejected
(b) Change in internal energy
(c) Ratio of T_1/T_2
(d) Work done during polytropic expansion

[ESE : 2005]

2.23 A closed system undergoes a process 1-2 for which the values of Q_{1-2} and W_{1-2} are +20 kJ and + 50 kJ, respectively. If the system is returned to state 1 and Q_{2-1} is –10 kJ, what is the value of the work W_{2-1}?

(a) +20 kJ (b) –40 kJ
(c) –80 kJ (d) +40 kJ [ESE : 2005]

2.24 Change in enthalpy in a closed system is equal to the heat transferred, if the reversible process takes place at constant

(a) Temperature (b) Internal energy
(c) Pressure (d) Entropy

[ESE : 2005]

2.25 A gas is compressed in a cylinder by a movable piston to a volume one-half of its original volume. During the process, 300 kJ heat left the gas and the internal energy remained same. The work done on the gas is

(a) 100 kNm (b) 150 kNm
(c) 200 kNm (d) 300 kNm

[ESE : 2005]

2.26 Assertion (A): Thermodynamic work is path-dependent except for an adiabatic process.

Reason (R): It is always possible to take a system from a given initial state to any final state by performing adiabatic work only.

(a) Both **A** and **R** are individually true and **R** is the correct explanation of **A**
(b) Both **A** and **R** are individually true but **R** is not the correct explanation of **A**
(c) **A** is true but **R** is false
(d) **A** is false but **R** is true [ESE : 2005]

2.27 Which one of the following is the correct sequence of the three processes *A*, *B* and *C* in the increasing order of the amount of work done by a gas following ideal-gas expansions by these processes?

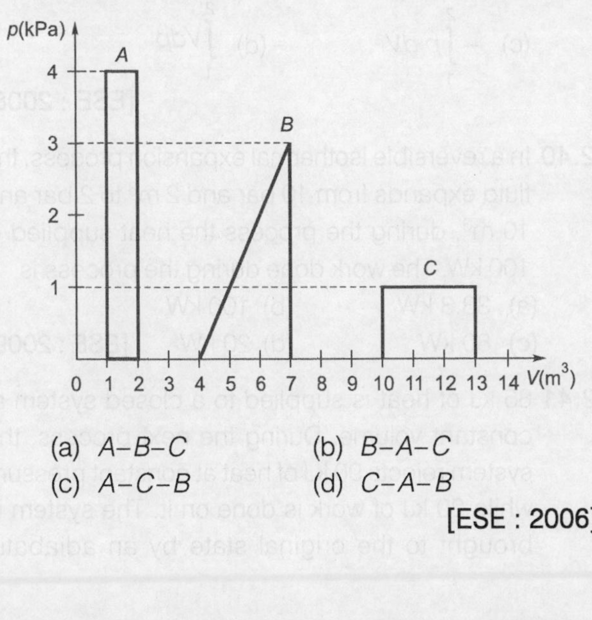

(a) *A–B–C* (b) *B–A–C*
(c) *A–C–B* (d) *C–A–B*

[ESE : 2006]

2.28 A thermoelectric engine which consists of two dissimilar electric conductors connected at two junctions maintained at different temperatures, converts

(a) Electric energy into heat energy
(b) Heat energy into electric energy
(c) Mechanical work into electric energy
(d) Electric energy into mechanical work

[ESE : 2006]

2.29 Assertion (A): The internal energy depends on the internal state of a body, as determined by its temperature, pressure and composition.

Reason (R): Internal energy of a substance does not include any energy that it may possess as a result of its macroscopic position or movement.

(a) Both **A** and **R** are individually true and **R** is the correct explanation of **A**
(b) Both **A** and **R** are individually true but **R** is not the correct explanation of **A**
(c) **A** is true but **R** is false
(d) **A** is false but **R** is true

[ESE : 2006]

2.30 A fluid flowing along a pipe line undergoes a throttling process from 10 bar to 1 bar is passing through a partially open valve. Before throttling, the specific volume of the fluid is 0.5 m^3/kg and after throttling is 2.0 m^3/kg. The change in specific internal energy during the throttling process is

(a) Zero (b) 100 kJ/kg
(c) 200 kJ/kg (d) 300 kJ/kg

[ESE : 2007]

2.31 An insulated tank initially contains 0.25 kg of a gas with an internal energy of 200 kJ/kg. Additional gas with an internal energy of 300 kJ/kg and an enthalpy of 400 kJ/kg enters the tank until the total mass of gas contained is 1 kg. The final internal energy (in kJ/kg) of the gas in the tank is

(a) 250 (b) 275
(c) 350 (d) None of these

[ESE : 2007]

2.32 Which one of the following is correct?

The cyclic integral of $(\delta Q - \delta W)$ for a process is

(a) positive (b) negative
(c) zero (d) unpredictable

[ESE : 2007]

2.33 One kg of ice at 0°C is completely melted into water at 0°C at 1 bar pressure. The latent heat of fusion of water is 333 kJ/kg and the densities of water and ice at 0°C are 999.0 kg/m^3 and 916.0 kg/m^3, respectively. What are the approximate values of the work done and energy transferred as heat for the process, respectively?
(a) –9.4 J and 333.0 kJ
(b) 9.4 J and 333.0 kJ
(c) 333.0 kJ and – 9.4 J
(d) None of the above

[ESE : 2007]

2.34 Assertion (A): Though heat is added during a polytropic expansion process for which $\gamma > n > 1$, the temperature of the gas decreases during the process.

Reason (R): The work done by the system exceeds the heat added to the system.
(a) both **A** and **R** are true and **R** is the correct explanation of **A**
(b) both **A** and **R** are true but **R** is not a correct explanation of **A**
(c) **A** is true but **R** is false
(d) **A** is false but **R** is true **[ESE : 2007]**

2.35 Consider the four processes *A*, *B*, *C* and *D* shown in the graph given below:

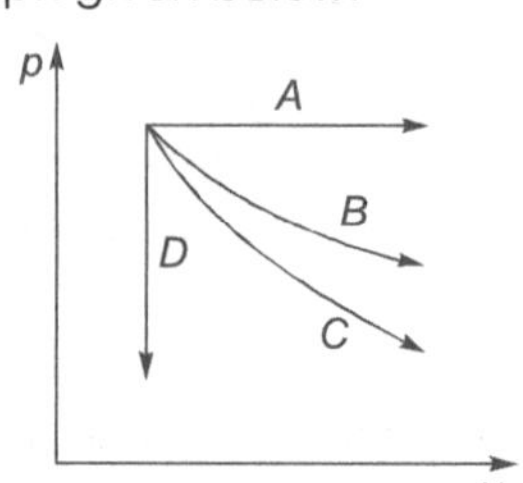

Match **List-I** (Processes shown in the graph) with **List-II** (Index *n* in the equation $pv^n = C$) and select the correct answer using the code given below the lists:

	List-I		List-II
A.	A	**1.**	0
B.	B	**2.**	1
C.	C	**3.**	1.4
D.	D	**4.**	∞

Codes:

	A	B	C	D
(a)	4	2	3	1
(b)	1	2	3	4
(c)	1	3	2	4
(d)	4	3	2	1

[ESE : 2007]

2.36 A balloon which is initially collapsed and flat is slowly filled with a gas at 100 kPa so as to form it into a sphere of 1 m radius. The work done by the gas in the balloon during the filling process is
(a) 428.9 kJ (b) 418.9 kJ
(c) 420.9 kJ (d) 416.9 kJ

[ESE : 2008]

2.37 A 4 kW, 20 litre water heater is switched on for 10 minutes. The heat capacity c_p for water is 4 kJ/kgK. Assuming all the electrical energy has gone into heating the water. The increase of the water temperature is
(a) 15°C (b) 20°C
(c) 26°C (d) 30°C **[ESE : 2008]**

2.38 In which one of the following processes, in a closed system the thermal energy transferred to a gas is completely converted to internal energy resulting in an increase in gas temperature?
(a) Isochoric process (b) Adiabatic process
(c) Isothermal process (d) Free expansion

[ESE : 2008]

2.39 Which one of the following expresses the reversible work done by the system (steady flow) between states 1 and 2?
(a) $\int_1^2 p\,dV$ (b) $-\int_1^2 Vdp$
(c) $-\int_1^2 p\,dV$ (d) $\int_1^2 Vdp$

[ESE : 2008]

2.40 In a reversible isothermal expansion process, the fluid expands from 10 bar and 2 m^3 to 2 bar and 10 m^3, during the process the heat supplied is 100 kW. The work done during the process is
(a) 33.3 kW (b) 100 kW
(c) 80 kW (d) 20 kW **[ESE : 2009]**

2.41 85 kJ of heat is supplied to a closed system at constant volume. During the next process, the system rejects 90 kJ of heat at constant pressure while 20 kJ of work is done on it. The system is brought to the original state by an adiabatic

process. The initial internal energy is 100 kJ. Then the quantity of work transfer during the process is
(a) 30 kJ (b) 25 kJ
(c) 20 kJ (d) 15 kJ [ESE : 2009]

2.42 A thermal electric power plant produces 1000 MW of power. If the coal releases 900×10^7 kJ/h of energy, then the rate at which heat is rejected from the power plant is
(a) 500 MW (b) 1000 MW
(c) 1500 MW (d) 2000 MW
[ESE : 2009]

2.43 Match **List-I** with **List-II** and select the correct answer using the code given below the lists :

List-I
A. Isolated system
B. Nozzle
C. Throttling device
D. Centrifugal compressor

List-II
1. Energy is always constant
2. Increase in velocity at the expense of its pressure drop
3. Appreciable drop in pressure without any change in energy
4. Enthalpy of the fluid increases by the amount of work input

Code :

	A	B	C	D
(a)	4	3	2	1
(b)	1	3	2	4
(c)	4	2	3	1
(d)	1	2	3	4

[ESE : 2009]

2.44 0.70 kg/s of air enters with a specific enthalpy of 290 kJ and leaves it with 450 kJ of specific enthalpy. Velocities at inlet and exit are 6 m/s and 2 m/s respectively. Assuming adiabatic process, what is power input to the compressor?
(a) 120 kW (b) 118 kW
(c) 115 kW (d) 112 kW
[ESE : 2009]

2.45 In a throttling process, which one of the following parameters remains constant?
(a) temperature (b) pressure
(c) enthalpy (d) entropy [ESE : 2009]

2.46 A closed system receives 60 kJ heat but its internal energy decreases by 30 kJ. Then the work done by the system is
(a) 90 kJ (b) 30 kJ
(c) –30 kJ (d) –90 kJ
[ESE : 2010]

2.47 Match **List-I** (Process index 'n') with **List-II** (T-s traces) and select the correct answer using the code given below the lists:

List-I
A. 0
B. 1.0
C. 1.4
D. ∞

List-II

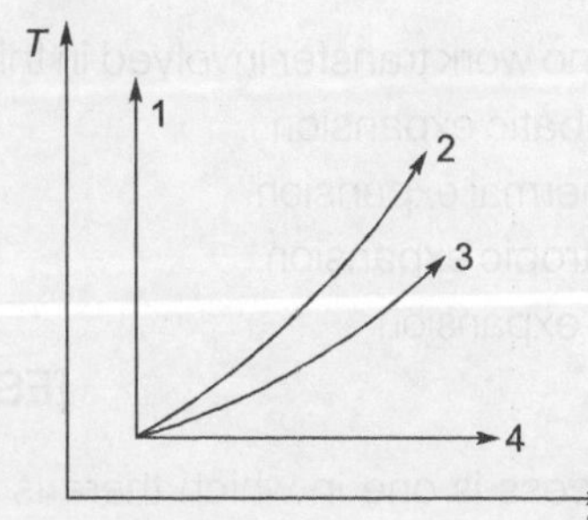

Codes:

	A	B	C	D
(a)	3	1	4	2
(b)	2	1	4	3
(c)	3	4	1	2
(d)	2	4	1	3

[ESE : 2010]

2.48 Match **List-I** with **List-II** and select the correct answer using the code given below the Lists :

List-I
A. $n = \infty$
B. $n = 1.4$
C. $n = 1.0$
D. $n = 0$

List-II

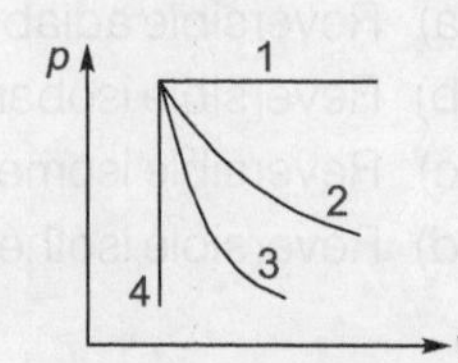

Codes:

	A	B	C	D
(a)	4	3	2	1
(b)	1	3	2	4
(c)	4	2	3	1
(d)	1	2	3	4

[ESE : 2011]

2.49 Air is being forced by the bicycle pump into a tyre against a pressure of 4.5 bars. A slow downward movement of the piston can be approximated as

(a) Isobaric process (b) Adiabatic process
(c) Throttling process (d) Isothermal process

[ESE : 2011]

2.50 Work transfer between the system and the surroundings

(a) Is a point function
(b) Is always given by $\int pdv$
(c) Is a function of pressure only
(d) Depends on the path followed by the system

[ESE : 2011]

2.51 There is no work transfer involved in this process

(a) Adiabatic expansion
(b) Isothermal expansion
(c) Polytropic expansion
(d) Free expansion

[ESE : 2012]

2.52 This process is one in which there is only work interaction between the system and the surroundings

(a) Diabetic process
(b) Adiabatic process
(c) Isothermal process
(d) Quasi-static process

[ESE : 2012]

2.53 In which of the following processes, the heat is fully converted into work?

(a) Reversible adiabatic process
(b) Reversible isobaric process
(c) Reversible isometric process
(d) Reversible isothermal process

[ESE : 2012]

2.54 Consider the following statements for a throttling process:

1. It is an adiabatic process
2. There is no work transfer in the process
3. Entropy increases in throttling process.

Which of these statements are correct?

(a) 1, 2 and 3 (b) 1 and 2 only
(c) 2 and 3 only (d) 1 and 3 only

[ESE : 2012]

2.55 If the work done on a closed system is 20 kJ/kg, and 40 kJ/kg heat is rejected from the system, its internal energy decreases by

(a) 20 kJ/kg (b) 60 kJ/kg
(c) –20 kJ/kg (d) –60 kJ/kg

[ESE : 2012]

2.56 The enthalpy drop for flow through convergent horizontal nozzles is 100 kJ/kg. If the velocity of approach at inlet to the nozzle is negligible, the exit velocity of the fluid is

(a) 20 m/s (b) 400 m/s
(c) 447.2 m/s (d) 520.8 m/s

[ESE : 2012]

2.57 By integrating Euler equation between two sections 1 and 2 for flow of an incompressible inviscid fluid through a pipe, we get

(a) steady flow energy equation
(b) Bernoulli equation
(c) continuity equation
(d) variable flow equation

[ESE : 2012]

2.58 **Statement (I):** Thermodynamic work is given as the product of an intensive thermodynamic property and the differential of an extensive property.

Statement (II): Work is not a thermodynamic property.

(a) Both Statement (I) and Statement (II) are individually true and Statement (II) is the correct explanation of Statement (I).
(b) Both Statement (I) and Statement (II) are individually true but Statement (II) is **NOT** the correct explanation of Statement (I).
(c) Statement (I) is true but Statement (II) is false
(d) Statement (I) is false but Statement (II) is true.

[ESE : 2012]

2.59 For steady flow through an insulated horizontal constant diameter pipe, this property remains constant

(a) Enthalpy (b) Internal energy
(c) Entropy (d) Volume

[ESE : 2012]

2.60 Joule's experiment states that for a cycle
(a) change of pressure is proportional to temperature change
(b) change of volume is proportional to temperature change
(c) change of internal energy is proportional to temperature change
(d) sum of all heat transfer is proportional to sum of all work transfer
[ESE : 2013]

2.61 An insulated container is divided into two compartments *A* and *B* by a thin diaphragm. While *A* contains a mass of gas at pressure *P* and temperature *T*, *B* is evacuated. The diaphragm is punctured and the gas in *A* rushes into *B*. The process is called
(a) adiabatic
(b) isentropic
(c) constant internal energy
(d) free expansion [ESE : 2013]

2.62 The expression $\int_1^2 pdV$ gives the measure of work during
(a) non flow reversible process
(b) steady flow reversible process
(c) open system and any process
(d) any system and any process
[ESE : 2013]

2.63 The displacement work done by a system is given by $\int pdV$. It is valid for
(a) any process
(b) a quasi-static process
(c) an isentropic process
(d) a non quasi-static process
[ESE : 2013]

2.64 A non flow quasi-static (reversible) process occurs for which
$P = (-3\,V + 16)$ bar,
where *V* is volume in m^3. What is work done when *V* changes from 2 to 6 m^3?
(a) 16×10^5 J (b) 16.5×10^5 J
(c) 16×10^3 J (d) 16.5×10^2 J
[ESE : 2013]

2.65 Statement (I): Steady flow means that the rates of flow of mass and energy through the control volume are constant.
Statement (II): At the steady state of a system, any thermodynamic property will have a fixed value at a particular location and will not alter with time.
(a) Both Statement (I) and Statement (II) are individually true and Statement (II) is the correct explanation of Statement (I).
(b) Both Statement (I) and Statement (II) are individually true but Statement (II) is **NOT** the correct explanation of Statement (I).
(c) Statement (I) is true but Statement (II) is false
(d) Statement (I) is false but Statement (II) is true.
[ESE : 2013]

2.66 In a cyclic process, the heat transfer are +30 J, –50 J and –10 J and +60 J. The network for the cyclic process is
(a) 30 Nm (b) 40 Nm
(c) 50 Nm (d) 60 Nm
[ESE : 2014]

2.67 A system executes a cyclic process during which there are two processes as given below:
${}_1Q_2 = 460$ kJ, ${}_2Q_1 = -100$ kJ and ${}_1W_2 = 210$ kJ,
What will be work interaction in process ${}_2W_1$?
(a) 100 kJ (b) –210 kJ
(c) 150 kJ (d) –150 kJ
[ESE : 2014]

2.68 A body of mass 20 kg falls freely in vacuum. It has fallen through a vertical distance of 50 m. The gravitational acceleration may be assumed as 10 m/s^2. What is the thermodynamic work done by the body?
(a) 1000 Nm (b) 10 kJ
(c) 0 (d) 1 kNm
[ESE : 2015]

2.69 When a system is taken from state *x* to state *y*, 30 kJ of heat flows into the system and the system does 10 kJ of work. When the system is returned from *y* to *x* along another path, work done on the system is 8 kJ. What is the amount of heat liberated or absorbed?

(a) 12 kJ of heat liberated
(b) 28 kJ of heat liberated
(c) 12 kJ of heat absorbed
(d) 28 kJ of heat absorbed [ESE : 2015]

2.70 A closed gaseous system undergoes a reversible constant pressure process at 2 bar in which 100 kJ of heat is rejected and the volume changes from 0.2 m^3 to 0.1 m^3. The change in the internal energy of the system is

(a) – 100 kJ (b) – 80 kJ
(c) – 60 kJ (d) – 40 kJ [ESE : 2015]

2.71 The values of heat transfer and work transfer for the processes of a thermodynamic cycle are given below:

Process	Heat transfer (kJ)	Work transfer (kJ)
1	300	300
2	00	250
3	–100	–100
4	00	–250

The thermal efficiency of the cycle and the work ratio will be respectively:

(a) 33% and 0.66 (b) 66% and 0.36
(c) 36% and 0.66 (d) 33% and 0.36
[ESE : 2015]

2.72 The performance of reciprocating compressors with provision of cooling cylinder is compared with

(a) Mechanical efficiency
(b) Isothermal efficiency
(c) Adiabatic efficiency
(d) Isentropic efficiency [ESE : 2015]

2.73 Which of the following statements are correct for a throttling process ?

1. It is an adiabatic steady flow process
2. The enthalpy before and after throttling is same
3. In the process, due to fall in pressure, the fluid velocity at outlet is always more than inlet velocity

(a) 1 and 2 only (b) 1 and 3 only
(c) 2 and 3 only (d) 1, 2 and 3
[ESE : 2016]

2.74 Consider the following statements for isothermal process :

1. Change in internal energy is zero
2. Heat transfer is zero

Which of the above statements is/are correct ?

(a) 1 only (b) 2 only
(c) Both 1 and 2 (d) Neither 1 nor 2
[ESE : 2016]

2.75 Which one of the following statements is correct during adiabatic charging of an ideal gas into an empty cylinder from a supply main?

(a) The specific enthalpy of the gas in the supply main is equal to the specific enthalpy of the gas in the cylinder
(b) The specific enthalpy of the gas in the supply main is equal to the specific internal energy of the gas in the cylinder
(c) The specific internal energy of the gas in the supply main is equal to the specific enthalpy of the gas in the cylinder
(d) The specific internal energy of the gas in the supply main is equal to the specific internal energy of the gas in the cylinder

[ESE : 2016]

2.76 Consider the following systems :

1. An electric heater
2. A gas turbine
3. A reciprocating compressor

The steady flow energy equation can be applied to which of the above systems?

(a) 1 and 2 only (b) 1 and 3 only
(c) 1, 2 and 3 (d) 2 and 3 only
[ESE : 2016]

2.77 Consider the following conditions for the reversibility of a cycle :

1. The *P* and *T* of the working substance must not differ appreciably, from those of the surroundings at any state in the process
2. All the processes, taking place in the cycle, must be extremely slow
3. The working parts of the engine must be friction-free

Which of the above conditions are correct ?

(a) 1, 2 and 3 (b) 1 and 2 only
(c) 1 and 3 only (d) 2 and 3 only
[ESE : 2016]

Directions: Each of the next consists of two statements, one labelled as the 'Statement (I)' and the other as 'Statement (II)'. You are to examine these two statements carefully and select the answers to these items using the codes given below:

Codes:

(a) Both Statement (I) and Statement (II) are individually true and Statement (II) is the correct explanation of Statement (I).

(b) Both Statement (I) and Statement (II) are individually true but Statement (II) is **NOT** the correct explanation of Statement (I).

(c) Statement (I) is true but Statement (II) is false

(d) Statement (I) is false but Statement (II) is true.

2.78 **Statement (I):** Thermometers using different thermometric property substance may give different readings except at two fixed points.

Statement (II) : Thermodynamic temperature scale is independent of any particular thermometric substance.

[ESE : 2016]

2.79 **Statement (I) :** First law of thermodynamics analyses the problem quantitatively whereas second law of thermodynamics analyses the problem qualitatively.

Statement (II) : Throttling process is reversible process.

[ESE : 2016]

2.80 **Statement (I) :** The specific heat at constant pressure for an ideal gas is always greater than the specific heat at constant volume.

Statement (II) : Heat added at constant volume is not utilized for doing any external work.

[ESE : 2017]

2.81 During a constant pressure expansion of a gas, 33.3% heat is converted into work while the temperature rises by 20 K. The specific heat of the gas at constant pressure as proportion of work, *W* is

(a) 8% (b) 10%
(c) 12% (d) 15%

[ESE : 2017]

2.82 A cylinder contains 10 m^3 of an ideal gas at a pressure of 2 bar. This gas is compressed in a reversible isothermal process till its pressure increases to 15 bar. What quantum of work will be required for this process? (You can use the table given herewith.)

Number	2	2.5	3	5	7
$\log_{10}$	0.301	0.397	0.475	0.698	0.845

(a) 4500 kJ (b) 4030 kJ
(c) 450 kJ (d) 403 kJ

[ESE : 2017]

2.83 When the valve of an evacuated bottle is opened, the atmospheric air rushes into it. If the atmospheric pressure is 101.325 kPa and 0.6 m^3 of air enters into the bottle, then the work done by the air will be

(a) 80.8 kJ (b) 70.8 kJ
(c) 60.8 kJ (d) 50.8 kJ

[ESE : 2018]

2.84 A thermodynamic cycle is composed of four processes. The heat added and the work done in each process are as follows:

Process	Heat transfer (J)	Work done (J)
1–2	0	50 (by the gas)
2–3	50 (from the gas)	0
3–4	0	20 (on the gas)
4–1	80 (to the gas)	0

The thermal efficiency of the cycle is

(a) 20.3% (b) 37.5%
(c) 40.3% (d) 62.5%

[ESE : 2018]

2.85 A steel tank placed in hot environment contains 5 kg of air at 4 atm at 30°C. A portion of the air is released till the pressure becomes 2 atm. Later, the temperature of the air in the tank is found to be 150°C. The quantity of air allowed to escape is

(a) 4.72 kg (b) 4.12 kg
(c) 3.71 kg (d) 3.21 kg

[ESE : 2018]

2.86 An engine works on the basis of Carnot cycle operating between temperatures of 800 K and 400 K. If the heat supplied is 100 kW, the output is

(a) 50 kW (b) 60 kW
(c) 70 kW (d) 80 kW **[ESE : 2018]**

2.87 A system absorbs 100 kJ as heat and does 60 kJ work along the path 1–2–3. The same system does 20 kJ work along the path 1–4–3. The heat absorbed during the path 1–4–3 is

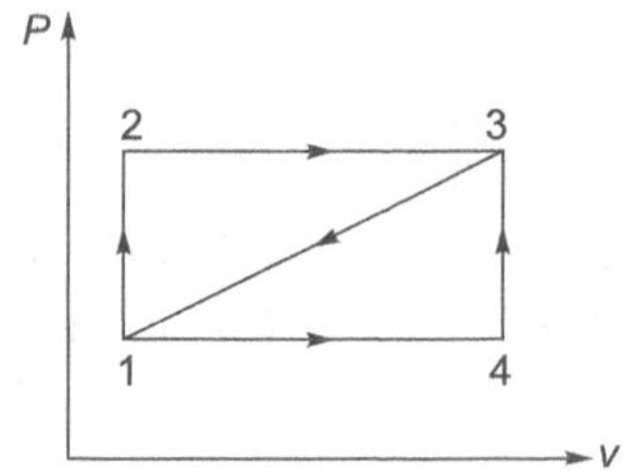

(a) – 140 kJ (b) – 80 kJ
(c) 80 kJ (d) 60 kJ [ESE : 2018]

2.88 What is the ratio of the efficiencies $\left(\frac{\eta_I}{\eta_{II}}\right)$ for the two cycles as shown in the T-s diagrams?

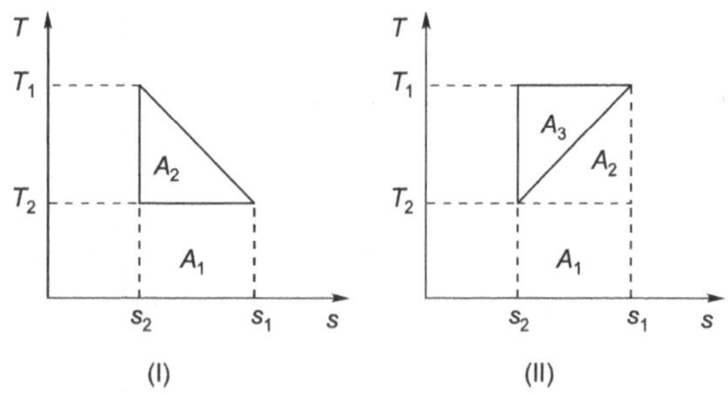

(a) $\frac{T_1+T_2}{2T_1}$ (b) $\frac{T_1+T_2}{2T_2}$
(c) $\frac{2T_2}{T_1+T_2}$ (d) $\frac{2T_1}{T_1+T_2}$

[ESE : 2018]

2.89 In a steam turbine with steam flow rate of 1 kg/s, inlet velocity of steam of 100 m/s, exit velocity of steam of 150 m/s, enthalpy at inlet of 2900 kJ/kg, enthalpy at outlet of 1600 kJ/kg, the power available from the turbine will be nearly
(a) 1575.5 kW (b) 1481.6 kW
(c) 1387.7 kW (d) 1293.8 kW
[ESE : 2019]

2.90 In an isentropic flow through a nozzle, air flows at the rate of 600 kg/hr. At inlet to nozzle, the pressure is 2 MPa and the temperature is 127 °C. The exit pressure is of 0.5 MPa. If the initial velocity of air is 300 m/s, the exit velocity will be
(a) 867 m/s (b) 776 m/s
(c) 685 m/s (d) 594 m/s
[ESE : 2019]

2.91 A 2 kg of steam occupying 0.3 m^3 at 15 bar is expanded according to the law $PV^{1.3}$ = constant to a pressure of 1.5 bar. The work done during the expansion will be
(a) 602.9 kJ (b) 606.7 kJ
(c) 612.45 kJ (d) 618.3 kJ
[ESE : 2019]

2.92 Water vapour at 90 kPa and 150°C enters a subsonic diffuser with a velocity of 150 m/s and leaves the diffuser at 190 kPa with a velocity of 55 m/s, and during the process, 1.5 kJ/kg of heat is lost to the surrounding. For water vapour, c_p is 2.1 kJ/kgK. The final temperature of water vapour will be
(a) 154°C (b) 158°C
(c) 162°C (d) 166°C [ESE : 2019]

2.93 A piston-cylinder device with air at an initial temperature of 30°C undergoes an expansion process for which pressure and volume are related as given below:

p(kPa)	100	37.9	14.4
V(m^3)	0.1	0.2	0.4

The work done by the system for n = 1.4 will be
(a) 4.8 kJ (b) 6.8 kJ
(c) 8.4 kJ (d) 10.6 kJ [ESE : 2020]

2.94 0.8 kg of air flows through a compressor under steady state conditions. The properties of air at entry are: pressure 1 bar, velocity 10 m/s, specific volume 0.95 m^3/kg and internal energy 30 kJ/kg. The corresponding values at exit are 8 bar, 6 m/s, 0.2 m^3/kg and 124 kJ/kg. Neglecting change in potential energy, the power input will be
(a) 117 kW (b) 127 kW
(c) 137 kW (d) 147 kW [ESE : 2020]

2.95 A stationary mass of gas is compressed without friction from an initial state of 0.3 m^3 and 0.105 MPa to a final state of 0.15 m^3 and 0.105 MPa, the pressure remaining constant during the process. There is a transfer of 40 kJ of heat from the gas during the process. How much does the internal energy of the gas change?
(a) –24.25 kJ (b) –19.62 kJ
(c) –15.91 kJ (d) –12.72 kJ
[ESE : 2021]

2.96 Consider the following statements for comparison of heat and work:

1. Both heat and work are transient phenomena.
2. Both heat and work are boundary phenomena.
3. Both heat and work are path functions and inexact differentials.

Which of the above statements are correct?

(a) 1 and 2 only (b) 1, 2 and 3
(c) 2 and 3 only (d) 1 and 3 only

[ESE : 2021]

2.97 A tank containing a fluid is stirred by a paddle wheel. The work input to the paddle wheel is 5090 kJ. The heat transfer from the tank is 1500 kJ. What is the change in internal energy of this control mass? (Consider the tank and the fluid inside a control surface)

(a) –3590 kJ (b) +3590 kJ
(c) +4590 kJ (d) –4590 kJ

[ESE : 2021]

2.98 During the charging of a storage battery, the current is 20 A and the voltage is 12.8 V. The rate of heat transfer from the battery is 10W. At what rate is the internal energy increasing?

(a) –256 J/s (b) +246 J/s
(c) +256 J/s (d) –246 J/s

[ESE : 2021]

2.99 Statement I : Heat and work are boundary phenomenon and recognized only when they cross the boundary of a system.

Statement II : Heat and work depend on the path followed by the system during a process.

(a) Both Statement (I) and Statement (II) are individually true and Statement (II) is the correct explanation of Statement (I).
(b) Both Statement (I) and Statement (II) are individually true but Statement (II) is **NOT** the correct explanation of Statement (I).
(c) Statement (I) is true but Statement (II) is false
(d) Statement (I) is false but Statement (II) is true.

[ESE : 2022]

2.100 In the cylinder of an air motor, the compressed air has a specific internal energy of 420 kJ/kg at the beginning of the expansion and a specific internal energy of 200 kJ/kg after expansion. What is the heat flow to or from the cylinder when the work done by the air during the expansion is 100 kJ/kg?

(a) –120 kJ/kg (b) +200 kJ/kg
(c) –80 kJ/kg (d) +100 kJ/kg

[ESE : 2023]

2.101 In the turbine of a gas turbine unit the gases flow through the turbine at 17 kg/s and the power developed by the turbine is 14000 kW. The specific enthalpies of the gases at inlet and outlet are 1200 kJ/kg and 360 kJ/kg respectively, and the velocities of the gases at inlet and outlet 60 m/s and 150 m/s respectively. What is the rate at which heat is rejected from the turbine?

(a) –89.5 kW (b) 96.2 kW
(c) –121.5 kW (d) 119.3 kW

[ESE : 2023]

2.102 If 1 m^3 of a gas is compressed adiabatically (the ratio of specific heats = 1.4) from 1 bar to 5 bar in a reciprocating compressor, what is the work of compression?

(a) 192543 N-m (b) 245361 N-m
(c) 158643 N-m (d) 204050 N-m

[ESE : 2024]

2.103 Two cylindrical vessels of 2 m^3 each are inter-connected through a pipe with a valve in-between. Initially, the valve is closed and one vessel has 20 kg air while 4 kg of air is there in second vessel. Assuming the system to be at 27 °C temperature initially and perfectly insulated, what is the final pressure in vessels after the valve is opened to attain equilibrium?

(a) 516.6 kPa (b) 51.66 kPa
(c) 561.6 kPa (d) 56.16 kPa

[ESE : 2024]

■■■■

Answers First Law, Heat and Work, Energy

2.1 (c)	2.2 (b)	2.3 (d)	2.4 (a)	2.5 (a)	2.6 (a)	2.7 (a)	2.8 (c)	2.9 (b)
2.10 (a)	2.11 (c)	2.12 (d)	2.13 (b).	2.14 (c)	2.15 (d)	2.16 (c)	2.17 (a)	2.18 (c)
2.19 (c)	2.20 (b)	2.21 (b)	2.22 (a)	2.23 (b)	2.24 (c)	2.25 (d)	2.26 (c)	2.27 (d)
2.28 (b)	2.29 (a)	2.30 (d)	2.31 (c)	2.32 (c)	2.33 (a)	2.34 (a)	2.35 (b)	2.36 (b)
2.37 (d)	2.38 (a)	2.39 (b)	2.40 (b)	2.41 (d)	2.42 (c)	2.43 (d)	2.44 (d)	2.45 (c)
2.46 (a)	2.47 (c)	2.48 (a)	2.49 (d)	2.50 (d)	2.51 (d)	2.52 (b)	2.53 (d)	2.54 (a)
2.55 (a)	2.56 (c)	2.57 (b)	2.58 (b)	2.59 (a)	2.60 (d)	2.61 (d)	2.62 (a)	2.63 (b)
2.64 (a)	2.65 (a)	2.66 (a)	2.67 (c)	2.68 (c)	2.69 (b)	2.70 (b)	2.71 (b)	2.72 (b)
2.73 (a)	2.74 (a)	2.75 (b)	2.76 (c)	2.77 (d)	2.78 (b)	2.79 (c)	2.80 (a)	2.81 (d)
2.82 (b)	2.83 (c)	2.84 (b)	2.85 (d)	2.86 (a)	2.87 (d)	2.88 (d)	2.89 (d)	2.90 (d)
2.91 (d)	2.92 (a)	2.93 (d)	2.94 (b)	2.95 (a)	2.96 (b)	2.97 (b)	2.98 (b)	2.99 (b)
2.100 (a)	2.101 (d)	2.102 (d)	2.103 (a)					

Explanations First Law, Heat and Work, Energy

2.1 (c)

Given data :

$h_1 = 100$ kJ/kg; $h_2 = 200$ kJ/kg

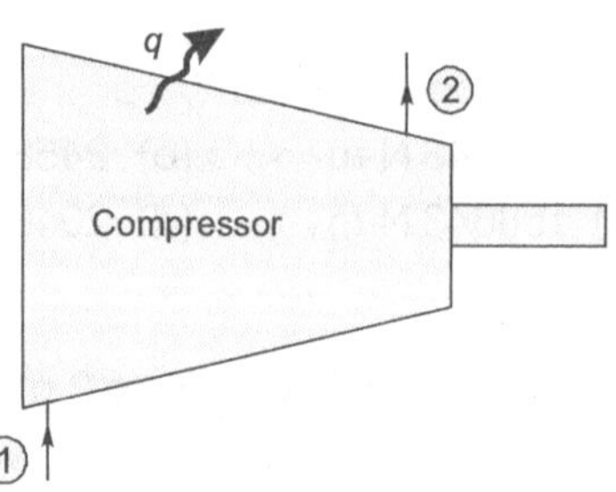

$q = -40$ kJ/kg, $m = 0.5$ kg/s

Applying SFEE at inlet and exit of the compressor

$$m\left[h_1 + \frac{V_1^2}{2} + gz_1\right] + mq = m\left[h_2 + \frac{V_2^2}{2} + gz_2\right] + W$$

Neglecting kinetic and potential energies

$\therefore \quad mh_1 + mq = mh_2 + W$

$0.5 \times 100 - 0.5 \times 40 = 0.5 \times 200 + W$

or $\quad W = -70$ kW

The power required to run the compressor is 70 kW.

2.2 (b)

A definite quantity of matter which is enclosed by a boundary that separates it from the rest of universe is called thermodynamic system. Generally it is sometimes called control mass.

Energy in the form of work and heat are mutually convertible and upto the degree to which it is governed depends upon the second law of thermodynamics.

2.3 (d)

S. No.	Q_1 (kJ/min)	Q_2 (kJ/s)	W(kW)	η(%)
1.	1500	16.8	8.2	32.8
2.	1600	17.92	8.75	32.8
3.	1700	19.03	9.30	32.8
4.	1800	20.15	9.05	32.8
5.	2000	–	–	32.8

$$W = Q_1 - Q_2$$

$$\eta = \frac{W}{Q_1}$$

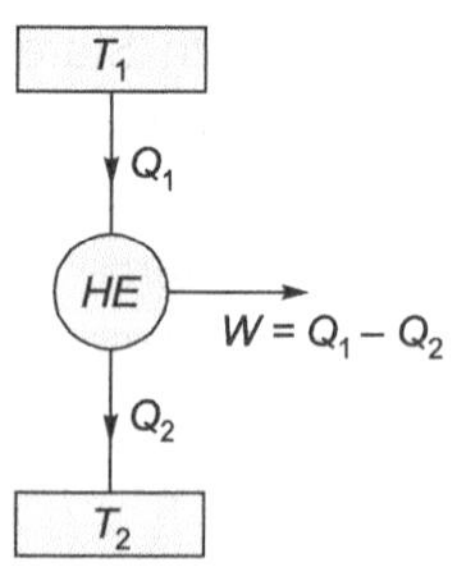

$$0.328 = \frac{W}{\left(\frac{2000}{60}\right)}$$

or $\quad W = 10.93$ kW

2.4 (a)

Under the cyclic work

$$\oint Q = \oint W$$

$$W = 474 - 180 = 294 \text{ kJ/min}$$

$$W = \frac{294}{60} = \textbf{4.9 kW}$$

2.5 (a)

Throttling process: Enthalpy is constant (to measure the quality of steam)
Isentropic process: Entropy is constant.
Free expansion: No work done
Isothermal process: Constant internal energy.

2.6 (a)

$$H_1 + Q = H_2 + W$$
$$25 = W - 30 \text{ or } W = \textbf{55 kJ}$$

2.8 (c)

$$\text{Heat rejected, } Q = \left(\frac{\gamma - n}{\gamma - 1}\right) \times W$$

2.9 (b)

For isothermal process

$$pV = C$$
$$pdV + Vdp = 0$$
$$pdV = -Vdp$$
$$\int pdV = -\int Vdp$$

2.11 (c)

From *SFEE* for compressor

$$m\left(h_1 + \frac{V_1^2}{2}\right) + Q = m\left(h_2 + \frac{V_2^2}{2}\right) + W$$

$$m(h_1 - h_2) + m\left(\frac{V_1^2 - V_2^2}{2}\right) + Q = W$$

$$2(-15 + 2) - 3 = W$$
$$-26 - 3 = W$$
$$-29 = W$$

Work required: $W = \textbf{29 kW}$

2.12 (d)

The given statement is for first law of Thermodynamics *i.e.*

$$\delta Q = dU + \delta W$$
$$\delta Q - \delta W = dU$$

2.13 (b)

$$Q = \Delta U + \delta W$$

$\Delta U = -30$ kJ (decrease in internal energy)
$\delta W = +50$ kJ (Work done by the system)
$\therefore\ Q = -30 + 50 = \textbf{+ 20 kJ}$

2.14 (c)

For perfect gas

$T_1 = 27°C = 300$ K
V_1 = Initial volume
V_2 = Final volume, $V_2 = 2V_1$

(i) For constant pressure process:

$$V \propto T$$
$$\frac{V_1}{T_1} = \frac{V_2}{T_2}$$
$$\frac{V_1}{T_1} = \frac{2V_1}{T_2}$$
$$\Rightarrow\quad T_2 = 2T_1 = 600 \text{ K}$$

(ii) For isothermal process:

$$T = \text{Constant}$$
$$T_1 = T_2 = 300 \text{ K}$$

(iii) For isentropic process:

$$T_1 V_1^{\gamma-1} = T_2 V_2^{\gamma-1}$$
$$T_2 = T_1\left(\frac{V_1}{V_2}\right)^{\gamma-1} = T_1\left(\frac{V_1}{2V_1}\right)^{1.4-1}$$
$$T_2 = 0.757\,T_1 = \textbf{227.35 K}$$

2.15 (d)

This is the case of *SFEE* neglect internal energy

$$m\left(h_1 + \frac{V_1^2}{2}\right) + Q = m\left(h_2 + \frac{V_2^2}{2}\right) + W$$

($Q = 0\quad \because$ adiabatic process)

$\therefore\ W = 140 + 10 = \textbf{150 kJ/kg}$

2.16 (c)

For steady flow device,

$$W = -V\int dp = (2-1) \times 0.1 \times 10^5 + 1000$$
$$= \textbf{11000 Nm}$$

2.17 (a)

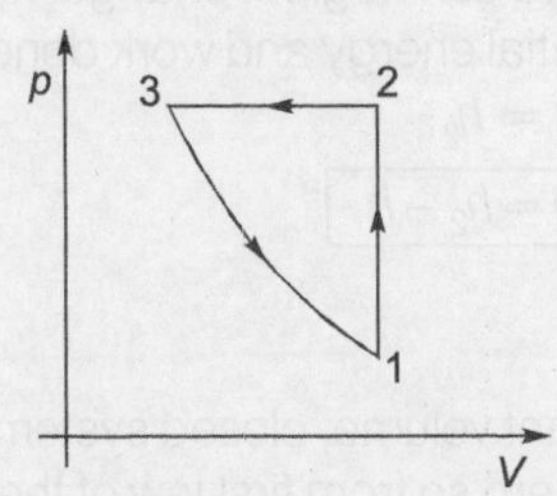

Given $U_1 = 100$ kJ

For constant volume process 1-2

From first law of thermodynamic

$$U_1 + Q = U_2 + W$$

$$100 + 170 = U_2 \quad (W = 0)$$

$$U_2 = 270 \text{ kJ}$$

For constant pressure process 2-3

$$U_2 + Q = U_3 + W$$

$$270 - 180 = U_3 - 40$$

$$U_3 = 130 \text{ kJ}$$

$$U_2 > U_3 > U_1$$

hence at the end of constant volume process, the internal energy is highest.

2.18 (c)

$$\frac{dQ}{dT} = \frac{dU}{dT} + \frac{dW}{dT}$$

Change in internal energy

$$\frac{dU}{dT} = \frac{dQ}{dT} - \frac{dW}{dT} = 2 - (2 - 0.1T)$$

$$U_2 - U_1 = \int_{T_1}^{T_2} 0.1\,TdT = \left|\frac{0.1T^2}{2}\right|_{T_1}^{T_2}$$

$$= \frac{0.1}{2}[150^2 - 100^2] = \mathbf{625\ kJ}$$

2.19 (c)

Change is internal energy

= Change is potential energy

$$mc_V dT = mgh$$

$$dT = \frac{9.81 \times 4000}{0.98 \times 10^3} = \mathbf{40\ K}$$

2.20 (b)

According to steady flow energy equation, for per unit mass between states '1' and '2'

$$h_1 + \frac{1}{2}V_1^2 + gh_1 + q = h_2 + \frac{1}{2}V_2^2 + gh_2 + w.$$

In boiler, we can neglect change in kinetic energy and potential energy and work done is zero, Thus,

$$h_1 + q = h_2$$

$$\Rightarrow \quad \boxed{q = h_2 - h_1}$$

2.21 (b)

At constant volume, closed system external work done is zero so from first law of thermodynamics

$$\delta Q = dU + \delta W$$

since $\delta W = 0$

$$\delta Q = dU$$

2.22 (a)

Heat added or rejected during polytropic,

$$Q = \left(\frac{\gamma - n}{\gamma - 1}\right) \times W$$

where W = Polytropic work done

2.23 (b)

For process 1-2

$$U_1 + Q = U_2 + W$$

$$U_1 - U_2 = W - Q = 50 - 20 = 30 \text{ kJ}$$

For process 2-1

$$U_2 + Q = U_1 + W$$

$$Q + U_2 - U_1 = W$$

$$-10 - 30 = W$$

$$W = \mathbf{-40\ kJ}$$

2.24 (c)

$\delta Q = dU + pdV$

at constant pressure

$pdV = d(pV)$

$(\delta Q)_p = d(U + pV) = dh$

Hence, heat transferred at constant pressure increases the enthalpy of a system.

2.25 (d)

$$\delta Q = dU + \delta W$$

since $dU = 0$

$\therefore \quad \delta Q = \delta W = \mathbf{300\ kNm}$

2.27 (d)

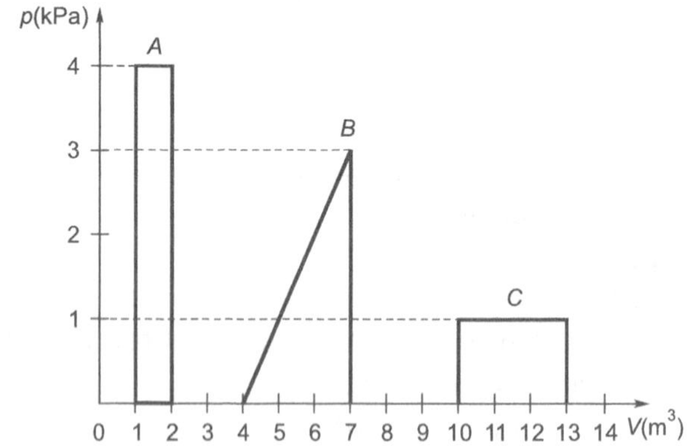

For process A,

$$W_A = 4(2 - 1) = 4 \text{ kJ}$$

For process B,

$$W_B = \int pdV = \frac{1}{2} \times 3 \times (7 - 4)$$

$$= 4.5 \text{ kJ}$$

For process C

$$W_C = \int pdV = 1(13-10) = 3 \text{ kJ}$$

$$W_c < W_A < W_B$$

2.30 (d)

Enthalpy of fluid before throttling is equal to the enthalpy of fluid after throttling

$$h_1 = h_2$$

$$\Rightarrow u_1 + p_1V_1 = u_2 + p_2V_2$$

$$u_2 - u_1 = p_1V_1 - p_2V_2$$
$$= 10 \times 10^5 \times 0.5 - 1 \times 10^5 \times 2$$
$$= (5 \times 10^5 - 2 \times 10^5) \text{ J/kg}$$
$$= \mathbf{300\ kJ/kg}$$

2.32 (c)

$$\delta W - \delta Q = 0$$

This is the first law for a closed system undergoing a cycle.

2.33 (a)

$$\text{Work done} = \int_{V_1}^{V_2} pdv$$

$$= p(v_2 - v_1) = p\left(\frac{m}{\rho_2} - \frac{m}{\rho_1}\right)$$

$$= 1\times10^5 \times 1\left(\frac{1}{999} - \frac{1}{916}\right) \approx -9.4 \text{ J}$$

Heat transfer = $m\,h_{fg} = 1 \times 333 =$ **333 kJ**

2.34 (a)

For polytropic process

Heat added or rejected,

$$Q = \frac{\gamma - n}{\gamma - 1} \times \text{Work done in polytropic process}$$

$$Q = \frac{\gamma - n}{\gamma - 1}\left(\frac{p_1V_1 - p_2V_2}{n-1}\right)$$

if $\gamma = 1.4,\ n = 1.3$

$$Q = \frac{1.4 - 1.3}{1.4 - 1} \times W$$

$$Q = 0.25W$$

Hence $Q < W$

2.36 (b)

Work done by the gas during filling process

$$= -\int Vdp$$

$$= \left(\frac{4}{3}\pi\, 1^3\right)(100) = 418.9 \text{ kJ}$$

2.37 (d)

$$mc_p\Delta T = (4 \times 10 \times 60)$$

$$20 \times 4 \times \Delta T = 2400$$

or $\Delta T = \mathbf{30\ °C}$

2.38 (a)

$$\delta Q = dU + \delta W$$

For $\delta W = 0, \delta Q = dU$

δW is zero for isochoric process i.e. constant volume process.

2.39 (b)

For steady flow process, reversible work is given by $-\int_1^2 Vdp$.

2.40 (b)

A reversible isothermal expansion process is constant internal energy process i.e.

$$dU = 0$$

$\because \quad \delta Q = dU + \delta W$

$\therefore \quad \delta Q = \delta W \quad (\because dU = 0)$

$\therefore$ Work done during the process = 100 kW

2.41 (d)

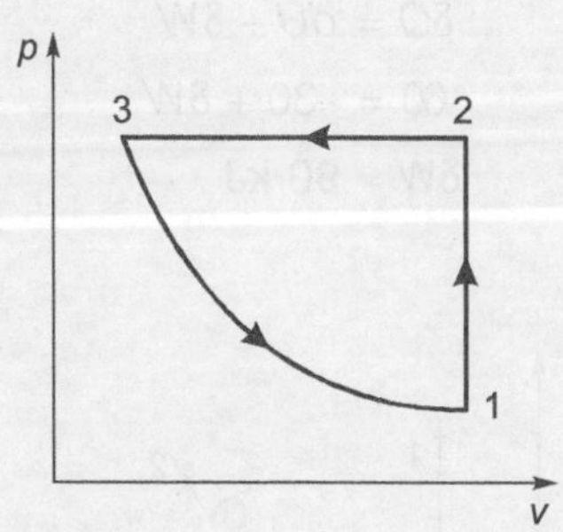

For the process 1-2,

$$\delta Q = +85,\ dW = 0$$

$$U_2 - U_1 = 85 \text{ kJ}$$

For the process 2-3,

$$\delta Q = -90 \text{ kJ}$$

$$\delta W = -20 \text{ kJ}$$

$$U_3 - U_2 = -70 \text{ kJ}$$

For the process 3-1,

$$\delta Q = 0$$

$$\delta W = U_3 - U_1$$

We have $U_2 - U_1 = 85$ kJ

and $U_3 - U_2 = -70$ kJ

by solving these two equations, we get

$$U_3 - U_1 = \mathbf{15\ kJ}$$

2.42 (c)

Energy released by the coal

$= 900 \times 10^7$ kJ/hr = 2500 MW

Heat rejected from the power plant

2500 – 1000 = **1500 MW**

2.44 (d)

Power input to compressor

$$= m\left[\left(h_2 + \frac{V_2^2}{2g}\right) - \left(h_1 + \frac{V_1^2}{2g}\right)\right]$$

$$= 0.7\left[\left(450 + \frac{2^2}{2 \times 9.81 \times 1000}\right) - \left(290 + \frac{6^2}{2 \times 9.81 \times 1000}\right)\right] = \mathbf{112\ kW}$$

2.46 (a)

$$Q_{\text{supplied}} = 60 \text{ kJ}$$

$$dU = -30 \text{ kJ}$$

From the first law of thermodynamics

$$\delta Q = dU + \delta W$$

$$60 = -30 + \delta W$$

or $\delta W = \mathbf{90\ kJ}$

2.47 (c)

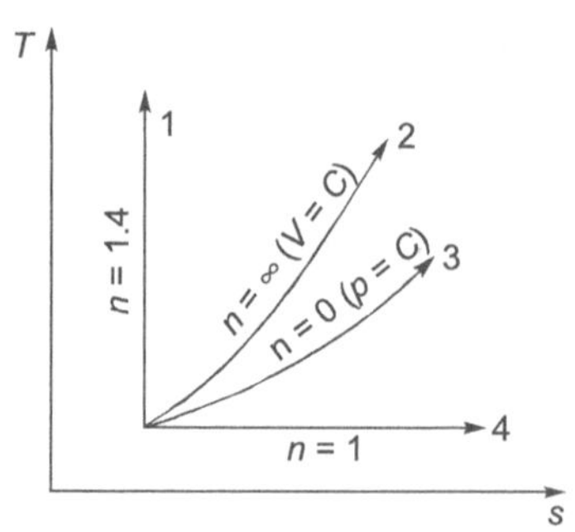

2.48 (a)

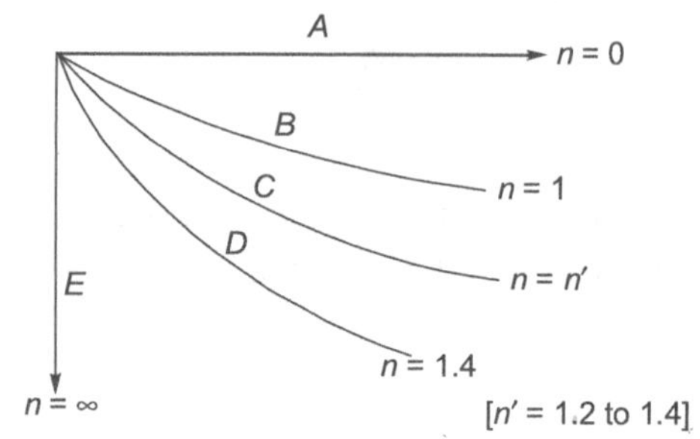

A = Constant pressure process
B = Isothermal process
C = Polytropic process
D = Isentropic process
E = Isochoric process

2.49 (d)

The slow downward movement of piston, can be treated as isothermal process.

2.50 (d)

Both work and heat transfer are path function.

2.51 (d)

Since free expansion occurs against vacuum hence no work is done by the expanding fluid.

2.52 (b)

For adiabatic process there is no heat interaction and hence energy transfer takes place only through work interaction.

2.53 (d)

For isothermal process, T = constant. So any heat added is spent by the fluid as work.

2.54 (a)

During a throttling process, there is not heat or work transfer. The pressure decreases and since the process is inherently irreversible, the entropy increases.

2.55 (a)

According to first law of thermodynamics

$$\delta q = du + \delta w$$

$$du = \delta q - \delta w$$

$$\delta q = -40 \text{ kJ/kg}$$

(–ve since heat rejected by system)

$$\delta w = -20 \text{ kJ/kg}$$

(–ve since work done on the system)

$\therefore \quad du = -40 + 20$

$du = -20$ kJ/kg

$\therefore$ Decrease in internal energy of system

$= \textbf{20 kJ/kg.}$

2.56 (c)

Applying steady flow energy equation at inlet and exit of nozzle and neglecting change in datum and assuming no heat and work transfer.

$$h_1 + \frac{V_1^2}{2} = h_2 + \frac{V_2^2}{2}$$

$$V_1 \approx 0$$

$$V_2 = \sqrt{2(h_1 - h_2)}$$

$$V_2 = \sqrt{2(100 \times 1000)} = \mathbf{447.2\ m/s}$$

2.57 (b)

Euler's equation

$$\frac{dp}{\rho} + gdz + VdV = 0$$

Integrating

$$\int \frac{dp}{\rho} + g\int dz + \int VdV = C\ (\rho = \text{constant})$$

$$\frac{p}{\rho} + gz + \frac{V^2}{2} = \text{Constant} \quad \text{...(i)}$$

Equation (i) is called Bernoulli's equation.

2.59 (a)

Applying steady flow energy equation

$$q + h_1 + \frac{V_1^2}{2} + z_1 g = h_2 + \frac{V_2^2}{2} + z_2 g + w$$

Insulated, $q = 0$

No work interaction, $w = 0$

Horizontal pipe, $z_1 = z_2$

Since constant diameter

$\therefore$ At any sections (i) and (ii),

$$A_1 V_1 \rho_1 = A_2 V_2 \rho_2$$

$$\Rightarrow \quad V_1 \rho_1 = V_2 \rho_2 \quad (\because A_1 = A_2)$$

Since any thing is not mentioned about compressibility, taking the fluid as incompressible.

$$\therefore \quad V_1 = V_2$$

$\therefore$ SFEE reduces to

$$h_1 = h_2$$

$\Rightarrow$ Enthalpy is constant

2.60 (d)

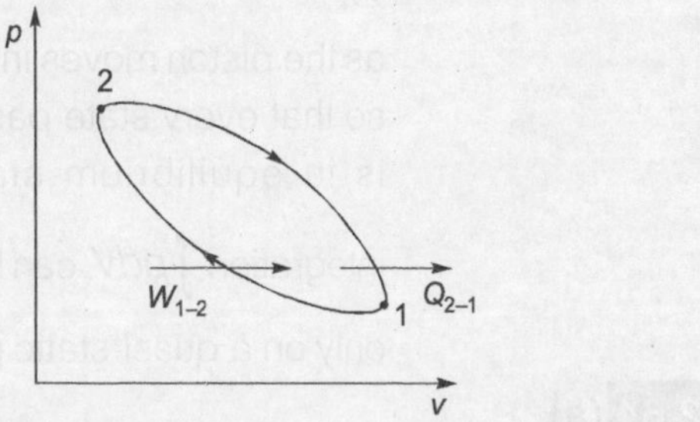

Definite amount of work input W_{1-2} to the system followed by the transfer of an amount of heat Q_{2-1}. Joule's experiment found that W_{1-2} is always proportional to Q_{2-1}.

$$(\Sigma W)_{cycle} = J(\Sigma Q)_{cycle}$$

where J is Joule's equivalent or mechanical equivalent of heat.

2.61 (d)

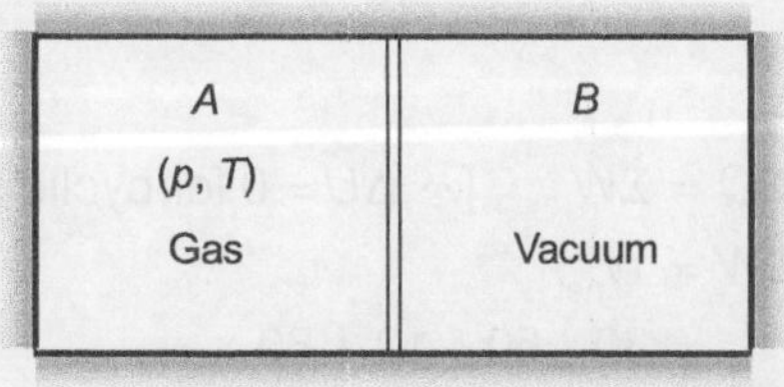

When the partition is removed, the gas rushes to fill the entire volume. The expansion of a gas against vacuum is called **free expansion**.

2.62 (a)

$$W = \int_1^2 pdV$$

This expression is work done during a quasi-equilibrium expansion or compression process of a closed system. So, it is the measure of work during non-flow (*i.e.*, closed system) reversible (quasi-equilibrium) process.

2.63 (b)

$\int pdV$ work is also known as moving boundary or displacement work.

$dW = F.dL = pAdL = pdV$

dL = infinitesimal distance moved by piston

$\therefore \quad dV = AdL$ is infinitesimal displacement volume.

$\Rightarrow \quad W_{1-2} = \int_1^2 pdV$ is the amount of work done as the piston moves infinitely slowly so that every state passed through is in equilibrium state. So, the integration $\int pdV$ can be performed only on a quasi-static path.

2.64 (a)

$$W_{1-2} = \int_1^2 pdV = \int_1^2 (-3V+16)dV$$

$$= -3\left(\frac{V_2^2 - V_1^2}{2}\right) + 16(V_2 - V_1)$$

$$= -3\left(\frac{6^2 - 2^2}{2}\right) + 16(6-2)$$

$$= \frac{-3\times 8\times 4}{2} + 64 = \mathbf{16 \times 10^5\ J}$$

2.66 (a)

$\Sigma Q = \Sigma W \quad [\because \Delta U = 0$ for cyclic process$]$

or, $\Sigma W = W_{net}$

$= 30 - 50 - 10 + 60$

$= 90 - 60 =$ **30 J**

2.67 (c)

${}_1Q_2 = {}_1W_2 + {}_1U_2$

${}_2Q_1 = {}_2W_1 + {}_2U_1$

For cyclic process,

${}_1U_2 + {}_2U_1 = 0$

$$\oint dQ = \oint dW$$

$460 - 100 = 210 + x$

$x = 360 - 210 =$ **150 kJ**

2.68 (c)

Since vacuum does not offer any resistance, there is no work transfer involved here.

So 'W' is zero and option (c) is correct.

2.69 (b)

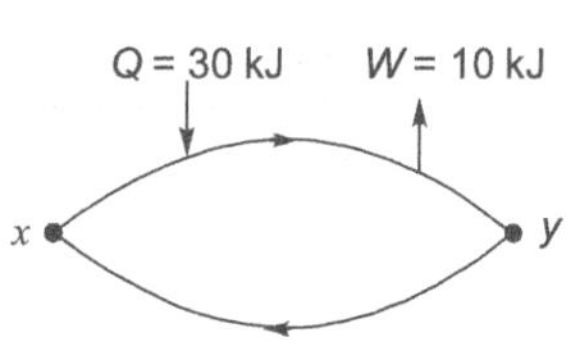

Process $(x - y)$:

$Q_{xy} = (U_y - U_x) + W_{xy}$

$30 = (U_y - U_x) + 10$

$\Rightarrow (U_y - U_x) = 20$ kJ

$\Rightarrow (U_x - U_y) = -20$ kJ

Process $(y - x)$:

$Q_{y-x} = \Delta U_{x-y} + W_{y-x}$

$= -20 - 8$

$= -28$ kJ

That is, 28 kJ of heat is liberated.

2.70 (b)

$$W = \int_{V_1}^{V_2} pdV = \int_{0.2}^{0.1} (2.0\times 10^5)\times dV$$

$= 2 \times 10^5 \times [0.1 - 0.2] = -20$ kJ

$Q = \Delta U + W$

or $\quad \Delta U = Q - W = -100 + 20 =$ **– 80 kJ**

Option (b) is correct.

2.71 (b)

$$\eta_{th} = \frac{\text{Net work done}}{\text{Heat supplied}}$$

$$= \frac{300 + 250 - 100 - 250}{300}$$

$$= \frac{200}{300} = 0.66 = 66\%$$

$$\text{Work ratio} = \frac{W_{net}}{W_T} = \frac{200}{550} = \mathbf{0.36}$$

Option (b) is correct.

2.72 (b)

Isothermal efficiency,

$$\eta_{iso} = \frac{\text{Isothermal work input}}{\text{Actual work input}}$$

2.81 (d)

$Q = mc_p\Delta T$

$W = 0.333\ Q$

$$\Rightarrow \quad \frac{W}{0.333} = mc_p \times 20$$

$$\Rightarrow \quad c_p = \frac{W}{0.333 \times 20 \times m} = \frac{0.15\,W}{m}$$

$$\Rightarrow \quad c_p \propto 0.15\ W$$

2.82 (b)

Reversible isothermal process

$$W = P_i V_i \ln\frac{P_i}{P_h} = 10 \times 200 \times \ln\frac{2}{15}$$

$$= 10 \times 200 \times \left[2.303 \times \ln_{10}\left(\frac{2}{15}\right)\right]$$

$$= 10 \times 200 \times \left[2.303 \times [\ln_{10} 2 - \ln_{10} 15]\right]$$

$$= 10 \times 200 \times \left[2.303 \times [\ln_{10} 2 (\ln_{10} 3 + \ln_{10} 5)]\right]$$

$$= 10{\times}200{\times}\{2.303{\times}[0.301-(0.475+0.698)]\}$$

$$= -4016.432 \text{ kJ} = 4016.42 \text{ kJ}$$

2.83 (c)

Work done by air,

$$W = P \times \Delta V$$

$$= 101.325 \text{ kPa} \times 0.6 \text{ m}^3$$

$$W = 60.8 \text{ kJ}$$

2.84 (b)

$$Q_{in} = 80 \text{ J} \quad \text{(for process 4-1)}$$

$$W_{net} = 50 \text{ J} - 20 \text{ J} = 30 \text{ J}$$

$$\eta = \frac{W_{net}}{Q_{in}} = \frac{30}{80} = 0.375 = 37.5\%$$

2.85 (d)

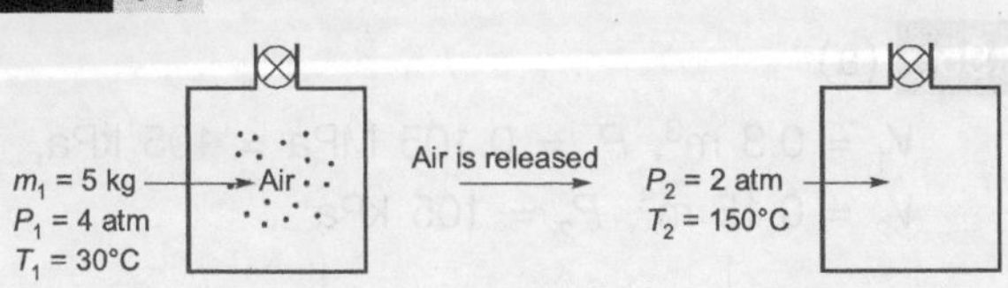

$$m_1 = \frac{P_1 V}{RT_1}$$

$$m_2 = \frac{P_2 V}{RT_2}$$

$$5 \text{ kg} = \frac{4 \times 101.325 \times V}{0.287 \times 303}$$

$$m_2 = \frac{2 \times (101.325)\, V}{0.287 \times 423}$$

$$m_2 = \frac{2 \times (101.325) \times 5 \times 0.287 \times 303)}{0.287 \times 423(4 \times 101.325)} = 1.79 \text{ kg}$$

Quantity of air escape = $m_1 - m_2$ = 3.2 kg

2.86 (a)

$$\eta = 1 - \frac{400}{800} = \frac{\text{Power output}}{100}$$

Power output = 50 kW

2.87 (d)

$$\delta Q = dU + \delta W$$

$$100 = dU + 60 \qquad (1\text{–}2\text{–}3)$$

$$dU = 40 \text{ kJ}$$

$$\delta Q = dU + \delta W \qquad (1\text{–}4\text{–}3)$$

$$= 20 + 40 = 60 \text{ kJ}$$

2.88 (d)

$$\eta_I = \frac{\frac{1}{2}(T_1 - T_2)(S_1 - S_2)}{\frac{1}{2}[T_1 + T_2](S_1 - S_2)}$$

$$\eta_{II} = \frac{\frac{1}{2}(T_1 - T_2)(S_1 - S_2)}{T_1(S_1 - S_2)}$$

$$\frac{\eta_I}{\eta_{II}} = \frac{T_1(S_1 - S_2)}{\frac{1}{2}(T_1 + T_2)(S_1 - S_2)} = \frac{2T_1}{(T_1 + T_2)}$$

2.89 (d)

$$\dot{m} = 1 \text{ kg/s}$$

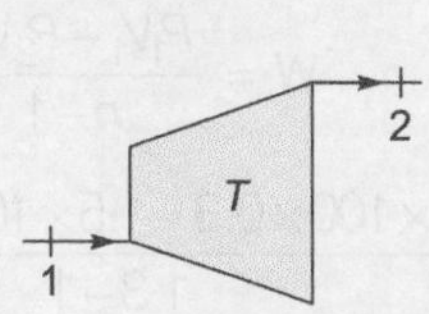

$$\dot{m}\left\{h_1 + \frac{C_1^2}{2000}\right\} = \dot{m}\left\{h_2 + \frac{C_2^2}{2000}\right\} + \dot{W}_{output}$$

$$1\left\{2900 + \frac{100^2}{2000}\right\} = 1\left\{1600 + \frac{150^2}{2000}\right\} + \dot{W}_{output}$$

$$2905 = 1600 + 11.25 + \dot{W}_{output}$$

$$\dot{W}_{output} = 1293.8 \text{ kW}$$

2.90 (d)

$\dot{m}$ = 600 kg/hr, = 0.167 kg/s

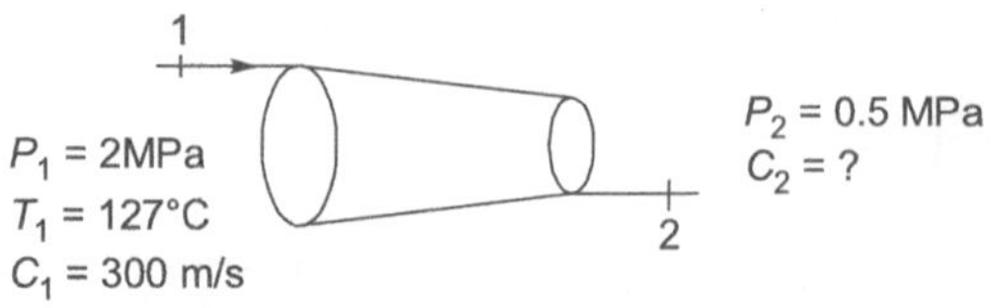

$$\frac{T_2}{T_1} = \left(\frac{P_2}{P_1}\right)^{\frac{\gamma-1}{\gamma}}$$

$$\frac{T_2}{273+127} = \left(\frac{0.5}{2}\right)^{\frac{1.4-1}{1.4}}$$

$$T_2 = 269.18 \text{ K}$$

S.F.E.E.,

$$\dot{m}\left(h_1 + \frac{C_1^2}{2000}\right) = \dot{m}\left(h_2 + \frac{C_2^2}{2000}\right)$$

$$c_pT_1 + \frac{300^2}{2000} = c_pT_2 + \frac{C_2^2}{2000}$$

$$1.005 \times 400 + 45 = 1.005 \times 269.18 + \frac{C_2^2}{2000}$$

$$C_2 = 594.094 \text{ m/s}$$

2.91 (d)

m = 2 kg; V_1 = 0.3 m^3;

P_1 = 15 bar; P_2 = 1.5 bar

$$P_1V_1^{1.3} = P_2V_2^{1.3}$$

$$V_2 = \left(\frac{P_1}{P_2}\right)^{1/1.3} \times V_1$$

$$V_2 = 5.878 \times 0.3 = 1.7634 \text{ m}^3$$

$$W = \frac{P_1V_1 - P_2V_2}{n-1}$$

$$W = \frac{15\times100\times0.3 - 1.5\times100\times1.7634}{1.3-1}$$

$$= 618.3 \text{ kJ}$$

2.92 (a)

P_1 = 90 kPa, P_2 = 190 kPa

T_1 = 150°C

C_1 = 150 m/s, C_2 = 55 m/s

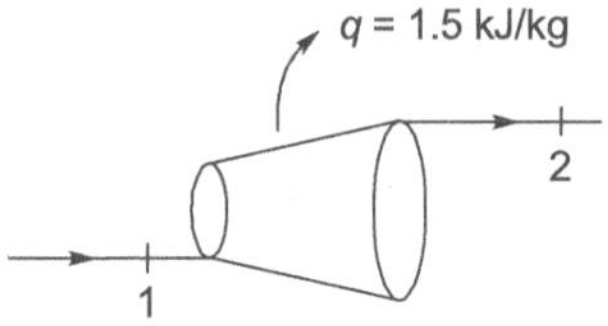

S.F.E.E.

$$h_1 + \frac{C_1^2}{2000} + q = h_2 + \frac{C_2^2}{2000}$$

$$c_PT_1 + \frac{150^2}{2000} + (-1.5) = c_PT_2 + \frac{55^2}{2000}$$

$$2.1\times423 + \frac{150^2}{2000} - 1.5 = 2.1\times T_2 + \frac{55^2}{2000}$$

$$888.3 + 11.25 - 1.5 = 2.1 \times T_2 + 1.5125$$

$$T_2 = 426.922 \text{ K} = 153.922\text{°C}$$

2.93 (d)

Initial state, P_1 = 100 kPa, V_1 = 0.1 m^3

Final state, P_2 = 14.4 kPa, V_4 = 0.4 m^3

Work done, W =

$$\frac{P_1V_1 - P_2V_2}{n-1} = \frac{100\times0.1 - 14.4\times0.4}{1.4-1} = 10.6 \text{ kJ}$$

2.94 (b)

Power input can be found from steady flow energy equation,

$$H_1 + \frac{1}{2}mV_1^2 + W = H_2 + \frac{1}{2}mV_2^2 + Q$$

∵ PE is zero.

$$U_1 + P_1v_1 + \frac{1}{2}mV_1^2 + W = U_2 + P_2v_2 + \frac{1}{2}mV_2^2 + Q$$

$$\dot{m}\left\{30 + 100\times0.95 + \frac{10^2}{2\times1000}\right\} + W = \dot{m}\left\{124 + 800\times0.2 + \frac{6^2}{2000}\right\}$$

$$W = \dot{m}\left[(U_2 - U_1) + P_2\cdot v_2 - P_1\cdot v_1 + \frac{V_2^2 - V_1^2}{2000}\right]$$

$$= 127.17 \text{ kW}$$

2.95 (a)

V_1 = 0.3 m^3, P_1 = 0.105 MPa = 105 kPa,

V_2 = 0.15 m^3, P_2 = 105 kPa

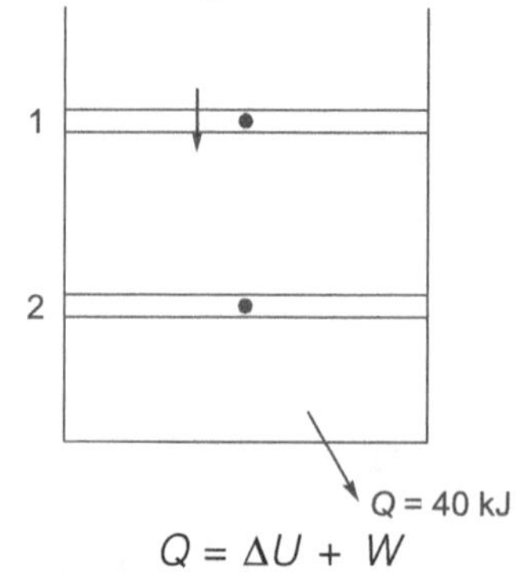

$$Q = \Delta U + W$$

$$-40 = \Delta U + [+105\,(0.15 - 0.3)]$$

$$\Delta U = -24.25 \text{ kJ}$$

2.97 (b)

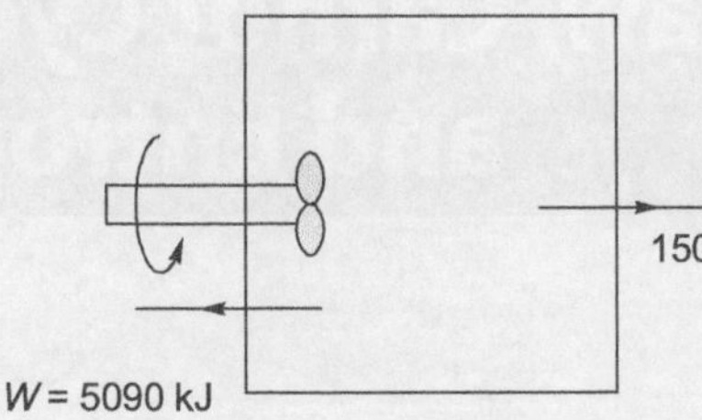

$$Q = \Delta U + W$$
$$-1500 = \Delta U + (-5090)$$
$$\Delta U = +3590 \text{ kJ}$$

2.98 (b)

$$W = VI = 12.8 \times 20 = 256 \text{ W}$$
$$Q = \Delta U + W$$
$$-10 = \Delta U + (-256)$$
$$\Delta U = +246 \text{ kJ}$$

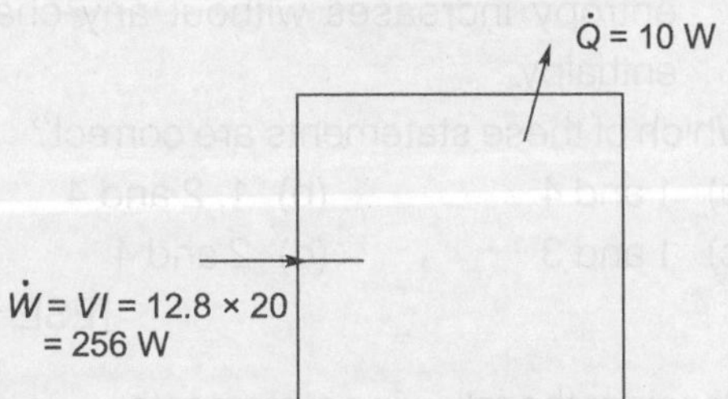

2.100 (a)

Given: u_1 = 420 kJ/kg, u_2 = 200 kJ/kg

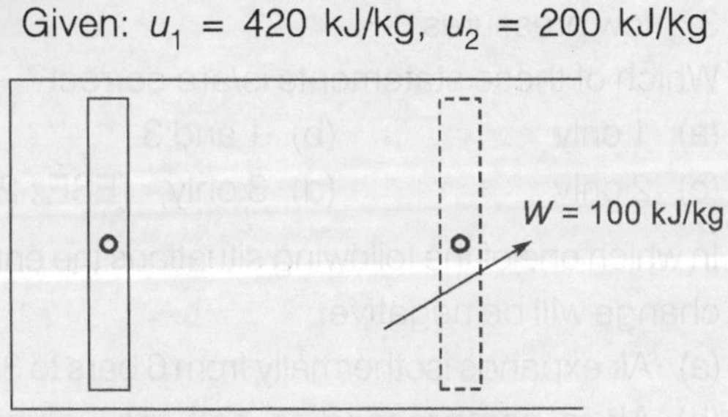

From first law, $Q = \Delta u + W$

$$Q = (200 - 420) + 100$$
$$= -120 \text{ kJ/kg}$$

2.101 (d)

Given: mass flow rate, $\dot{m}$ = 17 kg/s
Power developed, P = 14000 kW
Enthalpy at inlet, h_i = 1200 kJ/kg
Enthalpy at outlet, h_e = 360 kJ/kg
Velocity at inlet, V_i = 60 m/s
Velocity at exit, V_e= 60 m/s
Using SFEE,

$$\dot{m}\left[h_i - h_e + \frac{V_i^2}{2000} - \frac{V_e^2}{2000}\right] - P = -Q$$

$$17\left[1200 - 360 + \frac{60^2}{2000} - \frac{150^2}{2000}\right] - 14000 = -Q$$

$$Q = -119.35 \text{ kW}$$

2.102 (d)

Given: P_1 = 1 bar, P_2 = 5 bar, V_1 = 1 m^3

For adiabatic compression,

$$P_1V_1^{\gamma} = P_2V_2^{\gamma}$$
$$1 \times (1)^{1.4} = 5 \times V_2^{1.4}$$
$$\Rightarrow \quad V_2 = 0.3167 \text{ m}^3$$

Now, work of compression is

$$W =$$
$$\frac{P_1V_1 - P_2V_2}{\gamma - 1} = \frac{1 \times 100 \times 1 - 5 \times 100 \times 0.3167}{0.4}$$
$$= -145.875 \text{ kJ} = -145875 \text{ N/m}$$

This is not given in any options.

Now, considering compressor work (i.e. open system)

$$W_C = \frac{\gamma}{\gamma - 1}(P_1V_1 - P_2V_2) = \gamma \times W$$
$$= 1.4 \times (-145875)$$
$$= -204224 \text{ Nm}$$

Thus, among the options, the most appropriate is (d).

2.103 (a)

After attainment of equilibrium, total mass of air,

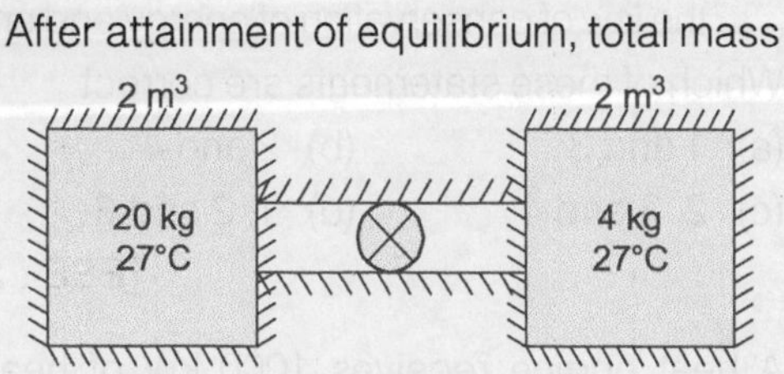

$$m = 24 \text{ kg}$$

Final total volume, V = 2 + 2 = 4 m^3

Assuming air to be an ideal gas,

$$PV = mRT$$

$$P = \frac{24 \times 287 \times 300}{4} = 516600 \text{ N/m}^2$$

$$P = 516.6 \text{ kPa}$$

■■■■

3 Second Law, Carnot Cycle and Entropy

3.1 Which one of the following sets of thermodynamic laws/relations is directly involved in determining the final properties during an adiabatic mixing process?

(a) The first and second laws of thermodynamics
(b) The second law of thermodynamics and steady flow relations
(c) Perfect gas relationship and steady flow relations
(d) The first law of thermodynamics and perfect gas relationship

[ESE : 2000]

3.2 Consider the following statements:

1. The first law of thermodynamics is a law of conservation of energy.
2. Perpetual motion machine of the first kind converts energy into equivalent work.
3. A closed system does not exchange work or energy with its surroundings.
4. The second law of thermodynamics stipulates the law of conservation of energy and entropy.

Which of these statements are correct

(a) 1 and 3 (b) 2 and 4
(c) 2, 3 and 4 (d) 1, 2 and 3

[ESE : 2000]

3.3 A heat engine receives 1000 kW of heat at a constant temperature of 285°C and rejects 492 kW of heat at 5°C. Consider the following thermodynamic cycles in this regard:

1. Carnot cycle
2. Reversible cycle
3. Irreversible cycle

Which of these cycles could possibly be executed by the engine?

(a) 1 only (b) 3 only
(c) 1 and 2 (d) None of 1, 2 & 3

[ESE : 2000]

3.4 Consider the following statements:
When dry saturated steam is throttled from a higher pressure to a lower pressure, the

1. pressure decreases and the volume increases.
2. temperature decreases and the steam becomes superheated.
3. temperature and the dryness fraction increase.
4. entropy increases without any change in enthalpy.

Which of these statements are correct?

(a) 1 and 4 (b) 1, 2 and 4
(c) 1 and 3 (d) 2 and 4

[ESE : 2000]

3.5 Consider the following statements:
A real gas obeys perfect gas law at very

1. high temperatures 2. high pressures
3. low pressures

Which of these statements is/are correct?

(a) 1 only (b) 1 and 3
(c) 2 only (d) 3 only **[ESE : 2000]**

3.6 In which one of the following situations the entropy change will be negative:

(a) Air expands isothermally from 6 bars to 3 bars
(b) Air is compressed to half the volume at constant pressure
(c) Heat is supplied to air at constant volume till the pressure becomes three folds
(d) Air expands isentropically from 6 bars to 3 bars

[ESE : 2000]

3.7 The heat added to a closed system during a reversible process is given by $Q = \alpha T + \beta T^2$, where α and β are constants. The entropy change of the system as its temperature changes from T_1 to T_2 is equal to

(a) $\alpha + \beta(T_2 - T_1)$
(b) $[\alpha(T_2 - T_1) + \beta/2(T_2^2 - T_1^2]/T_1$

(c) $\dfrac{\left[\dfrac{\alpha}{2}\left(T_2^2 - T_1^2\right) + \dfrac{\beta}{3}\left(T_2^3 - T_1^3\right)\right]}{T_1^3}$

(d) $\alpha \ln\left(\dfrac{T_2}{T_1}\right) + 2\beta(T_2 - T_1)$

[ESE : 2000]

3.8 Assertion (A): In thermodynamic analysis, the concept of reversibility is that, a reversible process is the most efficient process.

Reasons (R): The energy transfer as heat and work during the forward process, as always identically equal to the energy transfer is heat and work, during the reversal of the process.

[ESE : 2001]

3.9 A reversible heat engine receives 6 kJ of heat from thermal reservoir at temperature 800 K, and 8 kJ of heat from another thermal reservoir at temperature 600 K. If it rejects heat to a third thermal reservoir at temperature 100 K, then the thermal efficiency of the engine is approximately equal to:

(a) 65% (b) 75%
(c) 80% (d) 85%

[ESE : 2002]

3.10 A system comprising of a pure substance executes reversibly a cycle 1–2–3–4–1 consisting of two isentropic and two isochoric processes as shown in the figure

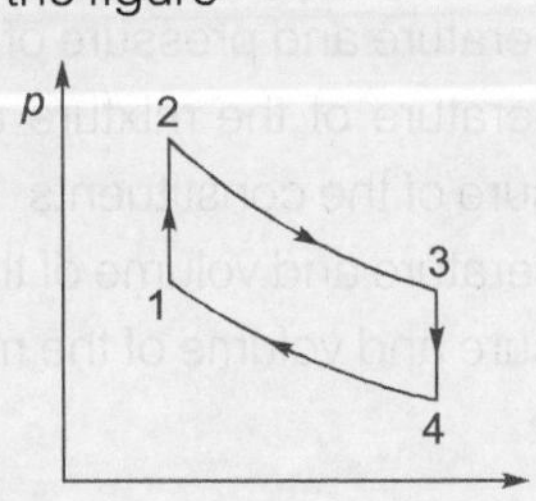

Which one of the following is the correct representation of this cycle on the temperature-entropy coordinates?

(a)
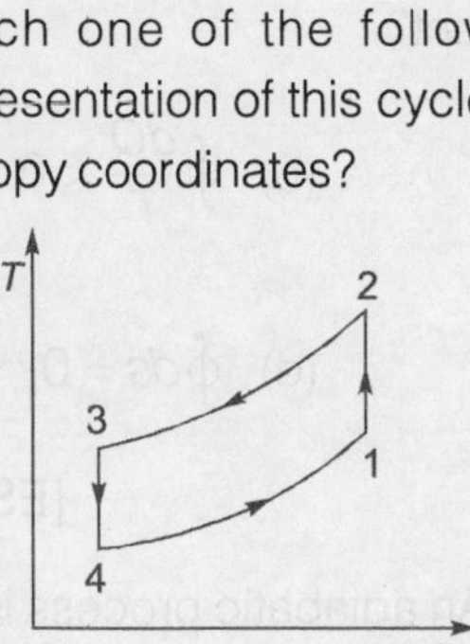

(b)
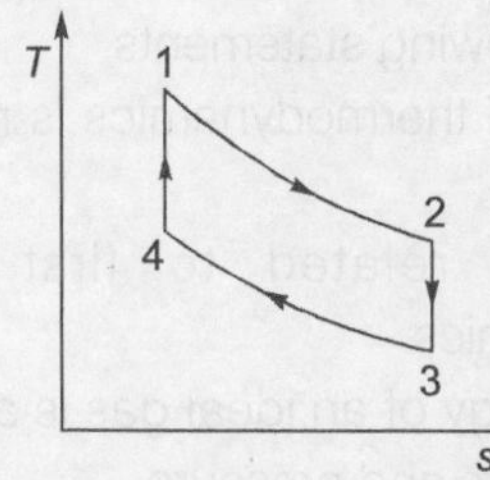

(c)
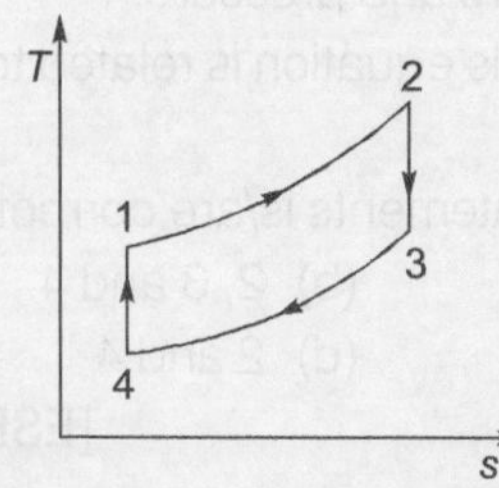

(d)
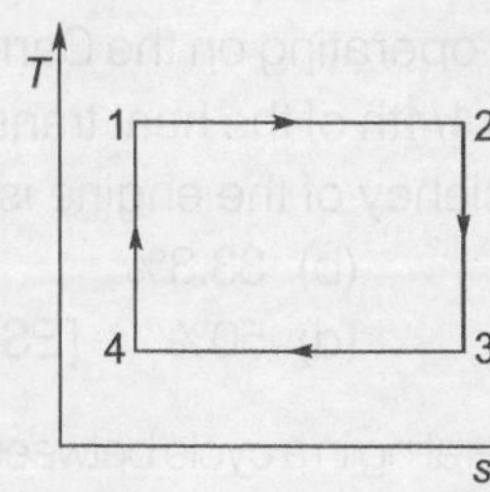

[ESE : 2002]

3.11 A reversible engine operates between temperatures T_1 and T_2. The energy rejected by this engine is received by a second reversible engine at temperature T_2 and rejected to a reservoir at temperature T_3. If the efficiencies of the engines are same then the relationship between T_1, T_2 and T_3 is given by

(a) $T_2 = \dfrac{T_1 + T_3}{2}$ (b) $T_2 = \sqrt{T_1^2 + T_3^2}$

(c) $T_2 = \sqrt{T_1 T_3}$ (d) $T_2 = \dfrac{T_1 + 2T_3}{2}$

[ESE : 2002]

3.12 Assertion (A): Efficiency of reversible engine operating between temperature limits T_1 and T_2 is maximum.

Reason (R): Efficiency of a reversible engine is greater than that of an irreversible engine.

(a) Both A and R are individually true and R is the correct explanation of A
(b) Both A and R are individually true but R is not the correct explanation of A
(c) A is true but R is false
(d) A is false but R is true

[ESE : 2002]

3.13 Consider the following statements:
1. Zeroth law of thermodynamics is related to temperature.
2. Entropy is related to first law of thermodynamics.
3. Internal energy of an ideal gas is a function of temperature and pressure.
4. Van der Waals equation is related to an ideal gas.

Which of thes statements is/are correct?
(a) 1 only (b) 2, 3 and 4
(c) 1 and 3 (d) 2 and 4

[ESE : 2003]

3.14 For a heat engine operating on the Carnot cycle, the work output is 1/4th of the heat transferred to the sink. The efficiency of the engine is
(a) 20% (b) 33.3%
(c) 40% (d) 50% [ESE : 2003]

3.15 In a heat engine operating in a cycle between a source temperature of 606°C and a sink temperature of 20°C, what will be the least rate of heat rejection per kW net output of the engine?
(a) 0.50 kW (b) 0.667 kW
(c) 1.5 kW (d) 0.0341 kW

[ESE : 2004]

3.16 One kg of air is subjected to the following processes:
1. Air expands isothermally from 6 bar to 3 bar.
2. Air is compressed to half the volume at constant pressure.
3. Heat is supplied to air at constant volume till the pressure becomes three fold.

In which of the above processes, the change in entropy will be positive.
(a) 1 and 2 (b) 2 and 3
(c) 1 and 3 (d) 1, 2 and 3

[ESE : 2004]

3.17 Two reversible engines operate between thermal reservoirs at 1200 K, T_2 K and 300 K such that first engine receives heat from 1200 K reservoir and rejects heat to thermal reservoir at T_2 K, while the second engine at 300 K. The efficiency of both the engines is equal then, the value of temperature T_2 is
(a) 400 K (b) 500 K
(c) 600 K (d) 700 K

[ESE : 2004]

3.18 Which of the following is/are reversible process (es)?
1. Isentropic expansion.
2. Slow heating of water from a hot source.
3. Constant pressure heating of an ideal gas from a constant temperature source.
4. Evaporation of a liquid at constant temperature.

Select the correct answer using the code given below:
(a) 1 only (b) 1 and 2
(c) 2 and 3 (d) 1 and 4

[ESE : 2005]

3.19 A reversible engine operates between temperatures 900 K and T_2 (T_2 < 900 K) and another reversible engine between T_2 and 400 K (T_2 > 400 K) in series. The value of T_2 if work outputs of both the engines are equal is
(a) 600 K (b) 625 K
(c) 650 K (d) 675 K

[ESE : 2005]

3.20 A Carnot engine operates between 27°C and 327°C. If the engine produces 300 kJ of work, what is the entropy change during heat addition?
(a) 0.5 kJ/K (b) 1.0 kJ/K
(c) 1.5 kJ/K (d) 2.0 kJ/K

[ESE : 2005]

3.21 The entropy of a mixture of ideal gases is the sum of the entropies of constituents evaluated at
(a) Temperature and pressure of the mixture
(b) Temperature of the mixture and the partial pressure of the constituents
(c) Temperature and volume of the mixture
(d) Pressure and volume of the mixture

[ESE : 2005]

3.22 For a real thermodynamic cycle, which one of the following is correct?
(a) $\oint \frac{dQ}{T} > 0$ (b) $\oint \frac{dQ}{T} < 0$
(c) $\oint \frac{dQ}{T} = 0$ (d) $\oint ds = 0$

[ESE : 2005]

3.23 Assertion (A): An adiabatic process is always a constant entropy process.

Reason (R): In an adiabatic process there is no heat transfer.
(a) Both A and R are individually true and R is the correct explanation of A
(b) Both A and R are individually true but R is not the correct explanation of A
(c) A is true but R is false
(d) A is false but R is true **[ESE : 2005]**

3.24 A heat source H_1 can supply 6000 kJ/min at 300°C and another heat source H_2 can supply 60000 kJ/min at 100°C. Which one of the following statements is correct if the surroundings are at 27°C?
(a) Both the heat sources have the same efficiency
(b) The first heat source has lower efficiency
(c) The second heat source has lower efficiency
(d) The first heat source produces higher power
[ESE : 2006]

3.25 Assertion (A): All constant entropy processes are adiabatic, but all adiabatic processes are not isentropic.
Reason (R): An adiabatic process which resists the exchange of energy to the surroundings may have irreversibility due to friction and heat conduction.
(a) Both **A** and **R** are individually true and **R** is the correct explanation of **A**
(b) Both **A** and **R** are individually true but **R** is not the correct explanation of **A**
(c) **A** is true but **R** is false
(d) **A** is false but **R** is true
[ESE : 2006]

3.26 A Carnot engine uses nitrogen as the working fluid (γ = 1.4). The heat supplied is 52 kJ and adiabatic expansion ratio is 32 : 1. The receiver temperature is 295 K. The amount of heat rejected is
(a) 11 kJ (b) 13 kJ
(c) 26 kJ (d) 28 kJ
[ESE : 2007]

3.27 Which one of the following is correct on the basis of the second law of thermodynamics?
(a) For any spontaneous process, the entropy of the universe increases.
(b) $\Delta S = Q_{rev}/T$ at constant temperature
(c) Efficiency of the Stirling cycle is more than that of a Carnot cycle
(d) $\Delta E = Q + W$
(the symbols have their usual meaning)
[ESE : 2007]

3.28 Assigning the basic dimensions to mass, length, time and temperature respectively as *M*, *L*, *T* and θ (Temperature), what are the dimensions of entropy?
(a) $MLT^{-2}\theta$ (b) $ML^2T^{-1}\theta^{-1}$
(c) $ML^2T^{-2}\theta^{-1}$ (d) $ML^3T^{-2}\theta^{-1}$
[ESE : 2007]

3.29 Two polytropic processes undergone by a perfect gas are shown below in the pressure volume co-ordinates.

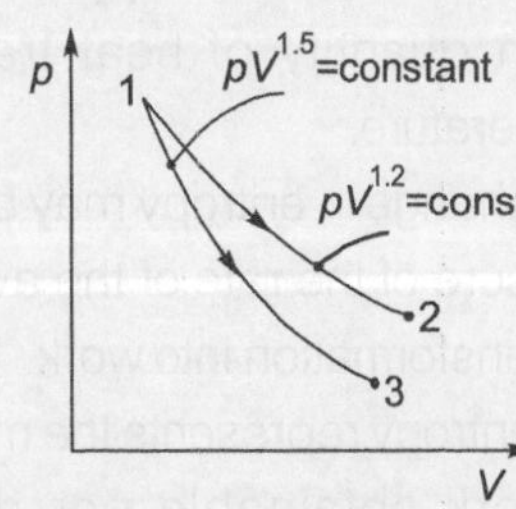

Which representation shows correctly the above processes on the temperature-entropy coordinates?

(a)
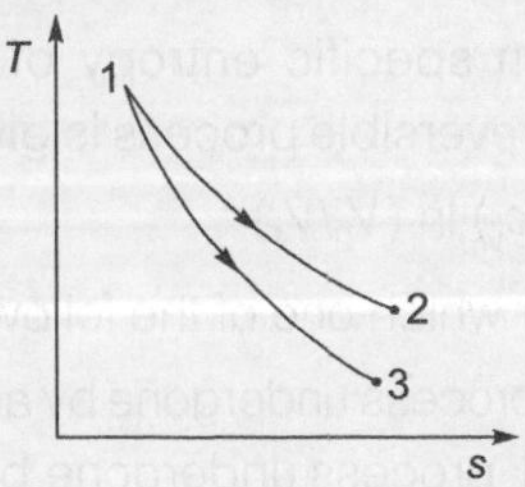

(b)
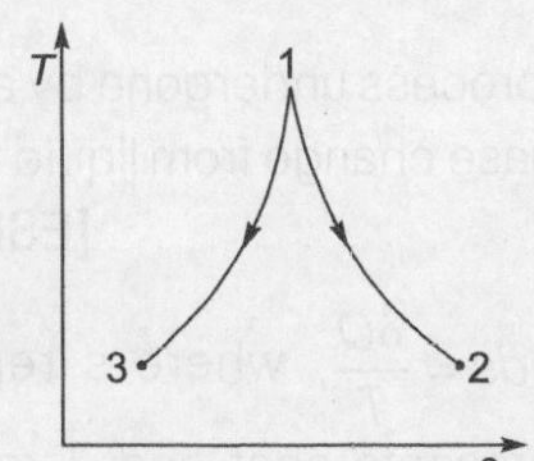

(c)
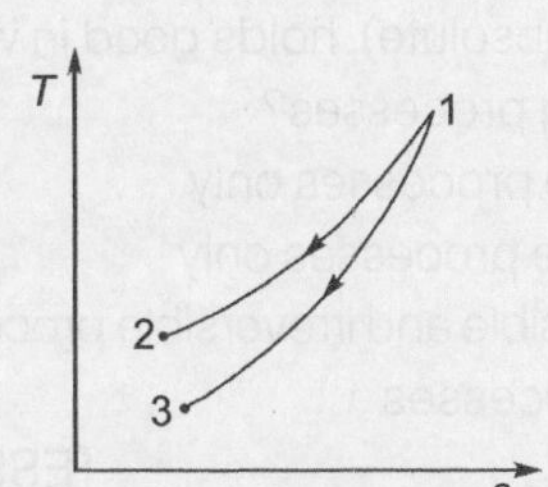

(d) 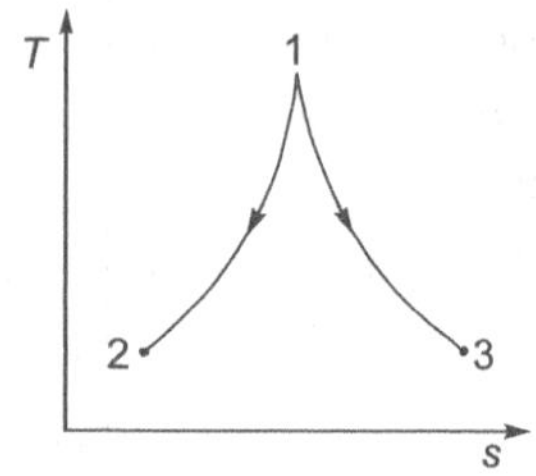

[ESE : 2008]

3.30 A Carnot engine operates between 327°C and 27°C. If the engine produces 300 kJ of work, the entropy change during heat addition is

(a) 0.5 kJ/K (b) 1.0 kJ/K
(c) 1.5 kJ/K (d) 2.0 kJ/K

[ESE : 2008]

3.31 Which of the following statements is correct?

(a) The increase in entropy is obtained from a given quantity of heat transfer at a low temperature.
(b) The change in entropy may be regarded as a measure of the rate of the availability of heat for transformation into work.
(c) The entropy represents the maximum amount of work obtainable per degree drop in temperature
(d) All of the above

[ESE : 2008]

3.32 The change in specific entropy of a system undergoing a reversible process is given by

$s_2 - s_1 = (c_p - c_v) \ln (V_2/V_1)$

This is valid for which one of the following?

(a) Adiabatic process undergone by an ideal gas
(b) Isothermal process undergone by an ideal gas
(c) Polytropic process undergone by a real gas
(d) Isobaric phase change from liquid to vapour

[ESE : 2008]

3.33 The relation $ds = \frac{\delta Q}{T}$, where s represents entropy, Q represents heat and T represents temperature (absolute), holds good in which one of the following processes?

(a) Reversible processes only
(b) Irreversible processes only
(c) Both reversible and irreversible processes
(d) All real processes

[ESE : 2009]

3.34 An inventor says that his new concept of an engine, while working between temperature limits of 27°C and 327°C rejects 45% of heat absorbed from the source. His engine is then equivalent to which one of the following engines?

(a) Carnot engine
(b) Diesel engine
(c) An impossible engine
(d) Ericsson engine

[ESE : 2009]

3.35 Three engines A, B and C operating on Carnot cycle use working substances as Argon, Oxygen and Air respectively. Which engine will have higher efficiency?

(a) Engine A
(b) Engine B
(c) Engine C
(d) All engines have same efficiency

[ESE : 2009]

3.36 A series combination of two Carnot's engines operate between the temperatures of 180°C and 20°C. If the engines produce equal amount of work, then the intermediate temperature is

(a) 80°C (b) 90°C
(c) 100°C (d) 110°C

[ESE : 2009]

3.37 An engine working on Carnot cycle rejects 40% of absorbed heat from the source, while the sink temperature is maintained at 27°C, then the source temperature is

(a) 750°C (b) 477°C
(c) 203°C (d) 67.5°C

[ESE : 2009]

3.38 A heat engine is supplied with 250 kJ/s of heat at a constant fixed temperature of 227°C; the heat is rejected at 27°C, the cycle is reversible, then the amount of heat rejected is

(a) 250 kJ/s (b) 200 kJ/s
(c) 180 kJ/s (d) 150 kJ/s

[ESE : 2009]

3.39 For a given value of T_H (Source temperature) for a reversed Carnot cycle, the variation of T_L (Sink temperature) for different values of COP is represented by which one of the following graphs?

(a)
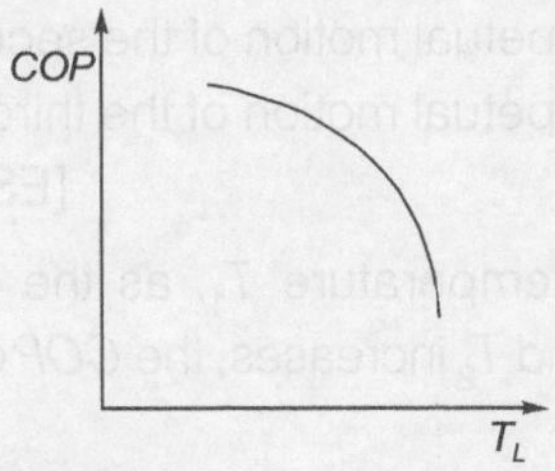

(b)
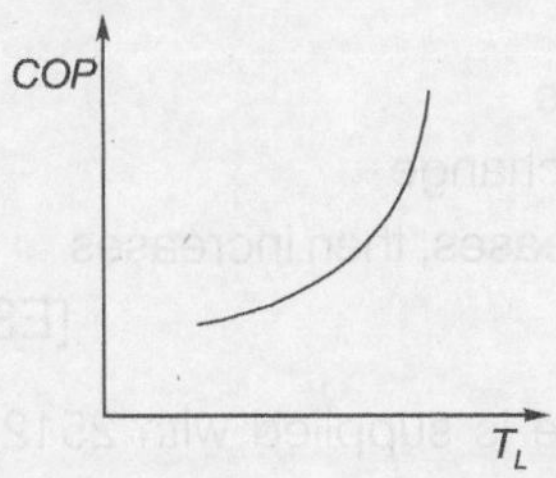

(c)
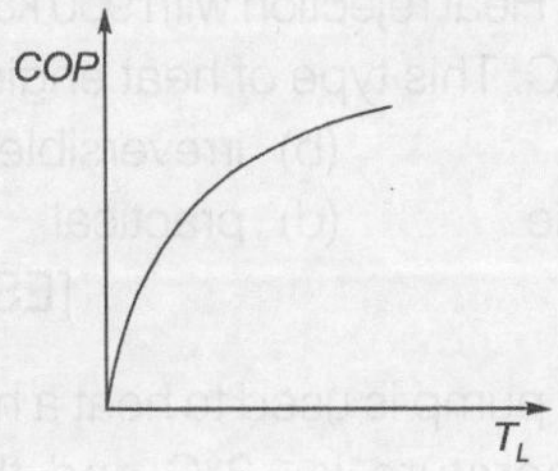

(d)
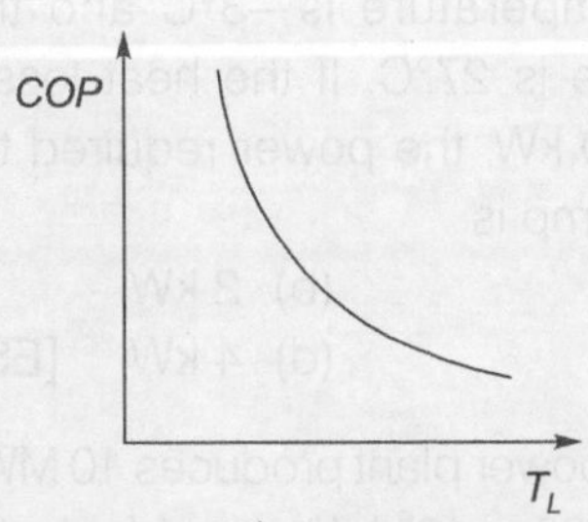

[ESE : 2009]

3.40 A reversible heat engine rejects 50 percent of the heat supplied during a cycle of operation. If this engine is reversed and operates as a heat pump, then what is its coefficient of performance?

(a) 1.0 (b) 1.5
(c) 2.0 (d) 2.5

[ESE : 2009]

3.41 The thermal efficiency of a Carnot heat engine is 30%. If the engine is reversed in operation to work as a heat pump with operating conditions unchanged, then what will be the COP for heat pump?

(a) 0.30
(b) 2.33
(c) 3.33
(d) Cannot be calculated

[ESE : 2009]

3.42 Efficiency of a Carnot engine is 75%. If the cycle direction is reversed, *COP* of the reversed Carnot cycle is

(a) 1.33 (b) 0.75
(c) 0.33 (d) 1.75

[ESE : 2010]

3.43 A Carnot engine rejects 30% of absorbed heat to a sink at 30°C. The temperature of the heat source is

(a) 100°C (b) 433°C
(c) 737°C (d) 1010°C

[ESE : 2010]

3.44 An engine operates between temperature limits of 900 K and T_2 and another between T_2 and 400 K. For both to be equally efficient, the value of T_2 will be

(a) 700 K (b) 600 K
(c) 750 K (d) 650 K

[ESE : 2010]

3.45 An inventor states that his new engine rejects to the sink 40% of heat absorbed from the source while the source and sink temperatures are 327°C and 27°C respectively. His engine is therefore equivalent to

(a) Joule engine (b) Striling engine
(c) Impossible engine (d) Carnot engine

[ESE : 2010]

3.46 Assertion (A): The constant pressure lines are steeper than the constant volume lines for a perfect gas on the T-s plane.

Reason (R): The specific heat at constant pressure is more than the specific heat at constant volume for a perfect gas.

(a) Both A and R are individually true and R is the correct explanation of A
(b) Both A and R are individually true but R is not the correct explanation of A
(c) A is true but R is false
(d) A is false but R is true [ESE : 2010]

3.47 Consider the following statements:

1. Boiling of water from a heat source at the same boiling temperature.
2. Theoretical isothermal compression of a gas.
3. Theoretical polytropic compression process with heat rejection to atmosphere.

4. Diffusion of two ideal gases into each other at constant pressure and temperature.

Which of these processes are irreversible?

(a) 1, 2, 3 and 4 (b) 1 and 4 only
(c) 2, 3 and 4 only (d) 3 and 4 only

[ESE : 2010]

3.48 Consider the following statements:

1. Slow heating of water from an electric heater.
2. Isentropic expansion of air.
3. Evaporation of a liquid from a heat source at the evaporation temperature.
4. Constant pressure heating of a gas by a constant temperature source.

Which of these processes is/are reversible?

(a) 3 only (b) 2 and 3 only
(c) 2 and 4 only (d) 1, 2, 3 and 4

[ESE : 2010]

3.49 The value of $\oint \frac{dQ}{T}$ for an irreversible cycle is

(a) Equal to zero (b) Greater than zero
(c) Less than zero (d) Unity

[ESE : 2011]

3.50 Increase in entropy of a system represents

(a) Increase in availability of energy
(b) Increase in temperature
(c) Decreases in pressure
(d) Degradation of energy [ESE : 2011]

3.51 Isentropic flow is

(a) Irreversible adiabatic flow
(b) Reversible adiabatic flow
(c) Ideal fluid flow
(d) Frictionless reversible flow [ESE : 2011]

3.52 An inventor states that his new conceptual engine, while operating between temperature limits of 377°C and 27°C, will reject 50% of heat absorbed from the source. What type of cycle will this engine have?

(a) Carnot cycle (b) Stirling cycle
(c) Impossible cycle (d) Possible cycle

[ESE : 2012]

3.53 The continual motion of a movable device in absence of friction

(a) violates the first law of thermodynamics
(b) violates the second law of thermodynamics
(c) is the perpetual motion of the second kind
(d) is the perpetual motion of the third kind

[ESE : 2012]

3.54 For a given temperature T_1, as the difference between T_1 and T_2 increases, the *COP* of a Carnot heat pump

(a) increases
(b) decreases
(c) does not change
(d) first decreases, then increases

[ESE : 2012]

3.55 A heat engine is supplied with 2512 kJ/min of heat at 650°C. Heat rejection with 900 kJ/min takes place at 100°C. This type of heat engine is

(a) ideal (b) irreversible
(c) impossible (d) practical

[ESE : 2012]

3.56 A Carnot heat pump is used to heat a house. The outside temperature is –3°C and the indoor temperature is 27°C. If the heat loss from the house is 40 kW, the power required to operate the heat pump is

(a) 1 kW (b) 2 kW
(c) 3 kW (d) 4 kW [ESE : 2012]

3.57 An electric power plant produces 10 MW of power consuming 90×10^6 kJ/hour of fuel energy. The heat rejected by the power plant is

(a) 5 MW (b) 10 MW
(c) 15 MW (d) 20 MW [ESE : 2013]

3.58 Which thermodynamics law predicts correctly, the degree of completion of chemical reaction?

(a) Zeroth law (b) First law
(c) Second law (d) Third law

[ESE : 2013]

3.59 In a cyclic heat engine operating between a source temperature of 600°C and sink temperature of 20°C, the least rate of heat rejection per kW net output of engine is

(a) 0.588 kW (b) 0.650 kW
(c) 0.505 kW (d) 0.460 kW

[ESE : 2013]

3.60 A Carnot engine rejects 50% of heat absorbed from a source to a sink at 27°C. What is the source temperature?

(a) 54°C (b) 150°C
(c) 327°C (d) 600°C

[ESE : 2013]

3.61 A refrigeration unit has a *COP* of 4 and extracts 10 kJ of heat from the cold reservoir. If this machine works as a heat pump, how much heat will it deliver to the environment?

(a) 2 kJ (b) 2.5 kJ
(c) 12.5 kJ (d) 25 kJ [ESE : 2013]

3.62 The entropy of universe tends to

(a) become zero
(b) remain constant
(c) be maximum
(d) attain a certain finite minimum value

[ESE : 2013]

3.63 A reversible engine has ideal thermal efficiency of 30%. When it is used as a refrigerating machine with all conditions unchanged, the coefficient of performance will be

(a) 3.33 (b) 3.00
(c) 2.33 (d) 1.33 [ESE : 2013]

Directions: Each of the next consists of two statements, one labelled as the 'Statement (I)' and the other as 'Statement (II)'. You are to examine these two statements carefully and select the answers to these items using the codes given below:

Codes:

(a) Both Statement (I) and Statement (II) are individually true and Statement (II) is the correct explanation of Statement (I).
(b) Both Statement (I) and Statement (II) are individually true but Statement (II) is **NOT** the correct explanation of Statement (I).
(c) Statement (I) is true but Statement (II) is false
(d) Statement (I) is false but Statement (II) is true.

3.64 Statement (I): There is entropy transfer both in heat transfer and work transfer.

Statement (II): Both heat and work are energy in transition. [ESE : 2013]

3.65 Statement (I): The 'Inequality of Clausius' provides the criterion of the reversibility of a cycle.

Statement (II): $\oint \frac{dQ}{T} > 0$, the cycle is irreversible and possible.

[ESE : 2013]

3.66 Which of the following processes are thermodynamically irreversible?

1. Gradual heating of water from a hot source
2. Isothermal expansion in a Stirling cycle
3. Mixing of two ideal gases at constant pressure

(a) 1, 2 and 3 (b) 1 and 2 only
(c) 2 and 3 only (d) 1 and 3 only

[ESE : 2013]

3.67 A researcher claims that he has developed an engine, which while working between source and sink temperatures of 377°C and 27° C rejects only 50% of absorbed heat. What will his engine be?

(a) An impossible engine
(b) A stirling engine
(c) A reversible engine
(d) A practical engine

[ESE : 2014]

3.68 A reversible engine works between temperature limits 260°C and 60°C. To improve the performance we have to

(a) Raise the source temperature to 300°C
(b) Lower the sink temperature to 30°C
(c) Insulate the engine
(d) None of these

[ESE : 2014]

3.69 A Carnot engine operates between 37°C and 347°C. If the engine produces 620 kJ of work, the entropy change during heat addition is

(a) 1 kJ/K (b) 2 kJ/K
(c) 3 kJ/K (d) 4 kJ/K

[ESE : 2014]

3.70 An amount of 1000 kJ of heat is added to a system during a constant pressure vapourization process at a temperature of 227°C. The available energy added to the system, if the temperature of the surroundings is 27°C, is

(a) 600 kJ (b) 500 kJ
(c) 400 kJ (d) 300 kJ

[ESE : 2014]

3.71 An inventor claims to have developed a refrigeration unit which maintains −10°C in the refrigerator which is kept in a room where the surrounding temperature is 25°C and which has COP 8.5. His claims is

(a) Valid (b) Marginally correct
(c) Invalid (d) None of the above

[ESE : 2014]

3.72 Practically it is not feasible to design an engine which closely follows the 'Carnot cycle' for the following reasons:

1. Transfer of heat energy at constant temperature is very difficult to achieve
2. Isentropic processes are very fast processes
3. It makes uses of smaller pressure ratios
4. Thermal efficiency is not a function of source and sink temperatures

Which of the above reasons are correct?

(a) 1 and 2 (b) 2 and 3
(c) 3 and 4 (d) 4 and 1

[ESE : 2014]

3.73 Two identical finite bodies of constant heat capacity at temperatures T_1 and T_2 are available to do work in a heat engine. The final temperature T_f reached by the bodies on delivery of maximum work is

(a) $T_f = \frac{T_1 + T_2}{2}$ (b) $T_f = \sqrt{T_1 T_2}$
(c) $T_f = T_1 - T_2$ (d) $T_f = \sqrt{T_1^2 + T_2^2}$

[ESE : 2014]

3.74 In the figure shown below, E is the heat engine with efficiency of 0.4 and R is the refrigerator. If $Q_2 + Q_4 = 3Q_1$, the COP of the refrigerator will be

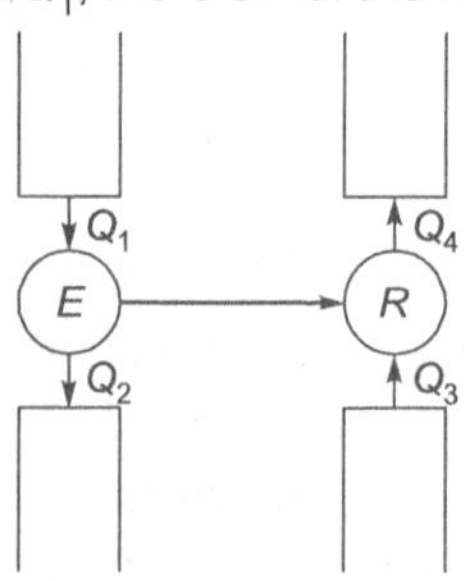

(a) 3.0 (b) 4.5
(c) 5.0 (d) 5.5

[ESE : 2014]

3.75 The *COP* of an ideal refrigerator is *N*. If the machine is operated as a heat pump between the same temperature limits, its *COP* will be

(a) $N - 1$ (b) N
(c) $N + 1$ (d) $2N$

[ESE : 2014]

3.76 An ideal refrigerator based on reversed Carnot cycle works between –23°C and +27°C. What will be the required power in kW, if a cooling rate of 1.5 kW is desired?

(a) 0.25 kW (b) 0.3 kW
(c) 3.25 kW (d) 7.5 kW

[ESE : 2014]

3.77 A liquid of heat capacity 5 J/K in an insulated container is heated electrically from 300 K to 600 K. If ln 2 = 0.693, entropy generation of the universe would be

(a) 6.93 J/K (b) 3.465 J/K
(c) 34.65 J/K (d) 10.65 J/K

[ESE : 2015]

3.78 Which of the following relationships represents the change of entropy of a perfect gas?

1. $C_p \frac{dT}{T} + \frac{R}{V} dV$ 2. $C_v \frac{dP}{P} - \frac{R}{P} dP$
3. $C_v \frac{dP}{P} + C_p \frac{dV}{V}$ 4. $C_p \frac{dP}{P} - C_v \frac{dV}{V}$

(a) 1, 2 and 4 only (b) 1, 2 and 3 only
(c) 2, 3 and 4 only (d) 1, 2, 3 and 4

[ESE : 2015]

Direction: The following items consists of two statements, one labelled as **Assertion 'A'** and the other labelled as **Reason 'R'**. You are to examine these two statements carefully and select your answers to these items using the codes given below:

Codes:

(a) Both **A** and **R** are true and **R** is the correct explanation of **A**
(b) Both **A** and **R** are true but **R** is not a correct explanation of **A**
(c) **A** is true but **R** is false
(d) **A** is false but **R** is true

3.79 **Assertion (A):** In an isolated system, the heat transfer δQ and the work transfer δW are always zero.

Reason (R): In an isolated system, the entropy always remains constant. [ESE : 2015]

3.80 **Assertion (A):** $\oint \frac{\delta Q}{T} = 0$, the cycle is reversible.

Reason (R): $\oint \frac{\delta Q}{T} > 0$, the cycle is irreversible and possible. [ESE : 2015]

3.81 **Assertion (A):** Entropy across the normal shock increases.

Reason (R): Stagnation temperature across the normal shock remains constant. **[ESE : 2015]**

3.82 Consider the following statements:

1. Free expansion of a gas
2. Slow heating of oil from a constant temperature source
3. Evaporation of water at its saturation temperature by a source at the same temperature
4. Isentropic compression of an ideal gas

Which of these processes are irreversible?

(a) 1 and 2 (b) 2 and 3
(c) 3 and 4 (d) 1 and 4

[ESE : 2015]

3.83 A Reversed Carnot Engine removes 50 kW from a heat sink. The temperature of the heat sink is 250 K and the temperature of the heat reservoir is 300 K. The power required of the engine is :

(a) 10 kW (b) 20 kW
(c) 30 kW (d) 50 kW

[ESE : 2016]

3.84 A heat engine receives heat at the rate of 2500 kJ/min and gives an output of 12.4 kW. Its thermal efficiency is, nearly :

(a) 18% (b) 23%
(c) 26% (d) 30% **[ESE : 2016]**

3.85 One reversible heat engine operates between 1000 K and T_2 K. and another reversible heat engine operates between T_2 K and 400 K. If both the engines have the same heat input and output, then the temperature T_2 must be equal to :

(a) 582.7 K (b) 632.5 K
(c) 682.8 K (d) 732.5 K

[ESE : 2016]

3.86 A system of 100 kg mass undergoes a process in which its specific entropy increases from 0.3 kJ/kg K to 0.4 kJ/kg K. At the same time, the entropy of the surroundings decreases from 80 kJ/K to 75 kJ/K. The process is :

(a) Reversible and isothermal
(b) Irreversible
(c) Reversible only
(d) Isothermal only **[ESE : 2016]**

3.87 A Carnot engine operates between 300 K and 600 K. If the entropy change during heat addition is 1 kJ/K, the work produced by the engine is:

(a) 100 kJ (b) 200 kJ
(c) 300 kJ (d) 400 kJ

[ESE : 2016]

3.88 Which of the following statements pertaining to entropy are correct ?

1. The entropy of a system reaches its minimum value when it is in a state of equilibrium with its surroundings
2. Entropy is conserved in all reversible processes
3. Entropy of a substance is least in solid phase
4. Entropy of a solid solution is not zero at absolute zero temperature

(a) 1, 2 and 3 only (b) 2, 3 and 4 only
(c) 3 and 4 only (d) 1, 2, 3 and 4

[ESE : 2016]

3.89 An ideal heat engine, operating on a reversible cycle, produces 9 kW. The engine operates between 27°C and 927°C. What is the fuel consumption given that the calorific value of the fuel is 40000 kJ/kg?

(a) 0.8 kg/hr (b) 1.02 kg/hr
(c) 1.08 kg/hr (d) 1.28 kg/hr

[ESE : 2016]

3.90 The Carnot cycle is impracticable because :

(a) Isothermal process is very fast; and isentropic process is very slow
(b) Isothermal process is very slow; and isentropic process is very fast
(c) Isothermal process and isentropic process are both very slow
(d) Isothermal process and isentropic process are both very fast

[ESE : 2016]

3.91 A solar collector receiving solar radiation at the rate of 0.6 kW/m^2 transforms it to the internal energy of a fluid at an overall efficiency of 50%. The fluid heated to 350 K is used to run a heat engine which rejects heat at 313 K. If the heat engine is to deliver 2.5 kW power, the minimum area of the solar collector required would be, nearly :

(a) 8 m^2 (b) 17 m^2
(c) 39 m^2 (d) 79 m^2
[ESE : 2016]

3.92 A reversible heat engine, operating on Carnot cycle, between the temperature limits of 300 K and 1000 K produces 14 kW of power. If the calorific value of the fuel is 40,000 kJ/kg. The fuel consumption will be :
(a) 1.4 kg/hr (b) 1.8 kg/hr
(c) 2.0 kg/hr (d) 2.2 kg/hr
[ESE : 2016]

3.93 Statement (I): Clausius inequality is valid for all cycles, reversible or irreversible including refrigeration cycles.
Statement (II): Clausius statement is a negative statement which has no proof.
[ESE : 2016]

3.94 Consider the following processes :
1. Extension of a spring
2. Plastic deformation of a material
3. Magnetization of a material exhibiting hysteresis

Which of the above processes are irreversible?
(a) 1 and 2 only (b) 1 and 3 only
(c) 2 and 3 only (d) 1, 2 and 3
[ESE : 2016]

3.95 A system of 100 kg mass undergoes a process in which its specific entropy increases from 0.3 kJ/kgK to 0.4 kJ/kgK. At the same time, the entropy of the surroundings decreases from 80 kJ/K to 75 kJ/K. The process is
(a) reversible and isothermal
(b) irreversible
(c) reversible only
(d) isothermal only [ESE : 2017]

3.96 A reversible heat engine rejects 80% of the heat supplied during a cycle of operation. If the engine is reversed and operates as a refrigerator, then its coefficient of performance shall be
(a) 6 (b) 5
(c) 4 (d) 3 [ESE : 2017]

3.97 A heat engine working on the Carnot cycle receives heat at the rate of 50 kW from a source at 1300 K and rejects it to a sink at 400 K. The heat rejected is
(a) 20.3 kW (b) 15.4 kW
(c) 12.4 kW (d) 10.8 kW
[ESE : 2017]

3.98 Consider the following statements:
On heating an elastomer under tensile load, its shrinkage
1. maximizes the enthalpy
2. maximizes the entropy
3. minimizes the free energy
4. avoids breaking

Which of the above statements are correct?
(a) 1 and 2 (b) 2 and 3
(c) 3 and 4 (d) 1 and 4
[ESE : 2017]

3.99 The thermal efficiency of the hypothetical cycle shown is

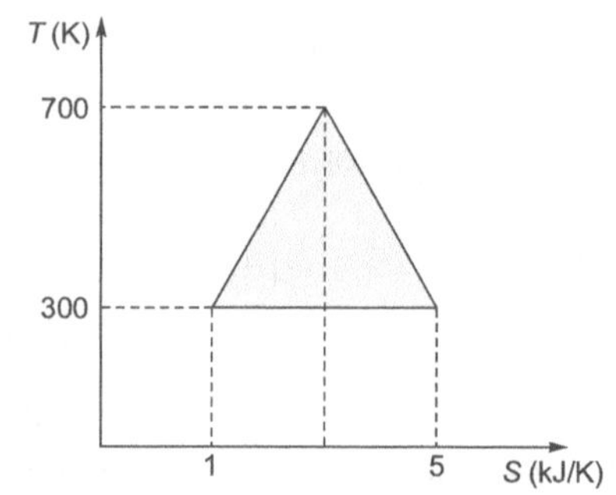

(a) 0.6 (b) 0.5
(c) 0.4 (d) 0.3 [ESE : 2017]

3.100 Two reversible engines are connected in series between a heat source and a sink. The efficiencies of these engines are 60% and 50%, respectively. If these two engines are replaced by a single reversible engine, the efficiency of this engine will be
(a) 60% (b) 70%
(c) 80% (d) 90% [ESE : 2018]

3.101 Consider the following statements:
1. Entropy is related to the first law of thermodynamics.
2. The internal energy of an ideal gas is a function of temperature and pressure.
3. The zeroth law of thermodynamics is the basis for measurement of temperature.

Which of the above statements are correct?
(a) 1 and 2 only (b) 1 and 3 only
(c) 2 and 3 only (d) 1, 2 and 3
[ESE : 2018]

3.102 Consider the following statements:

1. The entropy of a pure crystalline substance at absolute zero temperature is zero.
2. The efficiency of a reversible heat engine is independent of the nature of the working substance and depends only on the temperature of the reservoirs between which it operates.
3. Carnot's theorem states that of all heat engines operating between a given constant temperature source and a given constant temperature sink, none has a higher efficiency than a reversible engine

Which of the above statement are correct?

(a) 1 and 2 only (b) 1 and 3 only
(c) 2 and 3 only (d) 1, 2 and 3

[ESE : 2018]

3.103 Which of the following devices complies with the Clausius statement of the second law of thermodynamics?

(a) Closed-cycle gas turbine
(b) Internal combustion engine
(c) Steam power plant
(d) Domestic refrigerator [ESE : 2018]

3.104 A reversible Carnot engine operates between 27°C and 1527°C, and produces 400 kW of net power. The change of entropy of the working fluid during the heat addition process is

(a) 0.222 kW/K (b) 0.266 kW/K
(c) 0.288 kW/K (d) 0.299 kW/K

[ESE : 2018]

3.105 An ideal gas is flowing through an insulated pipe at the rate of 3 kg/s. There is a 10% pressure drop from an inlet to exit of the pipe. The values of $R = 0.287$ kJ/kg.K and $T_o = 300$ K. The rate of energy loss for the pressure drop due to friction, will be nearly

(a) 34 kW (b) 30 kW
(c) 26 kW (d) 22 kW

[ESE : 2020]

3.106 A cyclic heat engine operates between a source temperature of 800°C and a sink temperature of 30°C. The least rate of heat rejection per kW net output of engine will be nearly

(a) 0.2 kW (b) 0.4 kW
(c) 0.6 kW (d) 0.8 kW

[ESE : 2020]

3.107 **Statement (I):** A thermal energy reservoir is a system that always remains at constant temperature even though the heat is added to or removed from it.

Statement (II): A thermal reservoir that supplies heat energy is called sink and one that absorbs the heat energy is called source.

[ESE : 2021]

3.108 Entropy generated (S_{gen}) can be taken as a criterion to indicate feasibility of process. Which of the following conditions are correct?

1. If $S_{gen} = 0$, then the process is a reversible process.
2. If $S_{gen} > 0$, then the process is an irreversible process.
3. If $S_{gen} < 0$, then the process is impossible.

Select the correct answer using the code given below

(a) 1 and 2 only (b) 2 and 3 only
(c) 1 and 3 only (d) 1, 2 and 3

[ESE : 2021]

3.109 What is the highest possible theoretical efficiency of a heat engine operating with a hot reservoir of furnace gases at 2000°C when the cooling water available is at 10°C?

(a) 47.37% (b) 57.37%
(c) 77.54% (d) 87.54%

[ESE : 2023]

3.110 The performance test of an air- conditioning unit rated as 140.7 kW (40 TR) seems to be indicating poor cooling. The test on heat rejection to atmosphere in its condenser shows the following:

Cooling water flow rate : 4 L/s

Water temperatures: In 30 °C; Out 40 °C

Power input to motor: 48 kW (95% efficiency)

What is the actual refrigeration capacity of the unit? (The specific heat capacity of water is 4.1868 kJ/kg-K)

(a) 101.5 kW (b) 167.4 kW
(c) 121.9 kW (d) 189.3 kW

[ESE : 2024]

3.111 A machine working on a Carnot cycle operates between 305 K and 260 K. What is the COP when it is operated as a refrigerating machine?

(a) 5.78 (b) 4.35
(c) 1.17 (d) 6.78

[ESE : 2024]

3.112 A small water-cooled condenser uses mains water at 13 °C and heats this to 24 °C before it goes to waste. The evaporator duty is 4.2 kW and the motor output is 1.7 kW. What is the water mass flow?

(a) 0.09 kg/s (b) 0.19 kg/s
(c) 0.21 kg/s (d) 0.13 kg/s

[ESE : 2024]

■■■■

Answers Second Law, Carnot Cycle and Entropy

3.1 (d)	3.2 (*)	3.3 (d)	3.4 (b)	3.5 (b)	3.6 (b)	3.7 (d)	3.8 (b)	3.9 (d)
3.10 (c)	3.11 (c)	3.12 (b)	3.13 (a)	3.14 (a)	3.15 (a)	3.16 (c)	3.17 (c)	3.18 (d)
3.19 (c)	3.20 (b)	3.21 (b)	3.22 (b)	3.23 (d)	3.24 (c)	3.25 (d)	3.26 (b)	3.27 (a)
3.28 (c)	3.29 (b)	3.30 (b)	3.31 (a)	3.32 (b)	3.33 (a)	3.34 (c)	3.35 (d)	3.36 (c)
3.37 (b)	3.38 (d)	3.39 (b)	3.40 (c)	3.41 (c)	3.42 (c)	3.43 (c)	3.44 (b)	3.45 (c)
3.46 (d)	3.47 (d)	3.48 (a)	3.49 (c)	3.50 (d)	3.51 (b)	3.52 (d)	3.53 (d)	3.54 (b)
3.55 (c)	3.56 (d)	3.57 (c)	3.58 (c)	3.59 (c)	3.60 (c)	3.61 (c)	3.62 (c)	3.63 (c)
3.64 (d)	3.65 (c)	3.66 (d)	3.67 (d)	3.68 (b)	3.69 (b)	3.70 (c)	3.71 (c)	3.72 (a)
3.73 (b)	3.74 (c)	3.75 (c)	3.76 (b)	3.77 (b)	3.78 (*)	3.79 (c)	3.80 (c)	3.81 (b)
3.82 (d)	3.83 (a)	3.84 (d)	3.85 (b)	3.86 (b)	3.87 (c)	3.88 (b)	3.89 (c)	3.90 (b)
3.91 (d)	3.92 (b)	3.93 (c)	3.94 (c)	3.95 (b)	3.96 (c)	3.97 (b)	3.98 (b)	3.99 (c)
3.100 (c)	3.101 (*)	3.102 (d)	3.103 (d)	3.104 (b)	3.105 (c)	3.106 (b)	3.107 (c)	3.108 (d)
3.109 (d)	3.110 (c)	3.111 (a)	3.112 (d)					

Explanations Second Law, Carnot Cycle and Entropy

3.2 (*)

Only statements 1 is right. statements 2, 3 and 4 are wrong.

3.3 (d)

$Q_1 = 1000$ kW

$T_1 = 285°C = (285 + 273)$ K $= 558$ K

$Q_2 = 492$ kW

$T_2 = 5°C = (5 + 273)$ K $= 278$ K

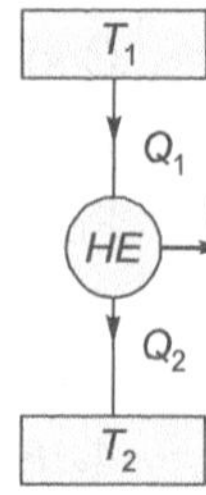

$$\eta_{Carnot} = 1-\frac{T_2}{T_1}$$

$$= 1-\frac{278}{558} = 1-0.4982$$

$$= 0.5017 = \mathbf{50.17\%}$$

$$\eta_{act} = 1-\frac{Q_2}{Q_1} = 1-\frac{492}{1000}$$

$$= 1-0.492 = 0.508 = \mathbf{50.8\%}$$

Thus, $\eta_{act} > \eta_{Carnot}$, heat engine is impossible.

3.4 (b)

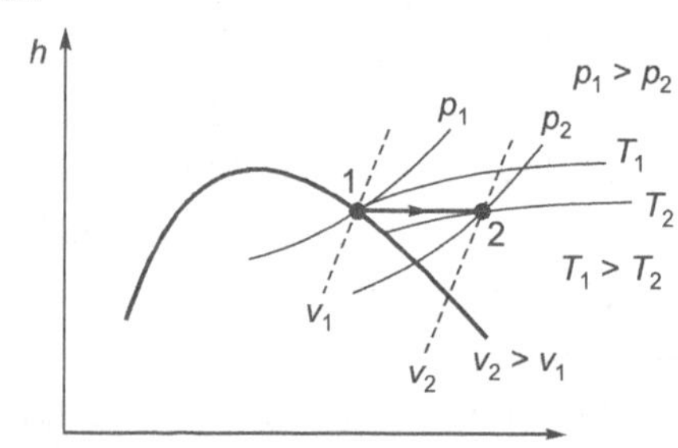

When dry saturated steam is throttled from a higher pressure to a low pressure, the

- pressure decreases and the volume increases

- temperature decreases and the steam becomes superheated.
- entropy increases without any change in enthalpy.

3.5 (b)

A real gas starts behaving as ideal gas at low pressure and high temperature.

3.6 (b)

$$\Delta s = c_v \ln\frac{T_2}{T_1} + R\ln\frac{V_2}{V_1}$$

$$\Delta s = c_p \ln\frac{T_2}{T_1} - R\ln\frac{p_2}{p_1}$$

3.7 (d)

$$Q = \alpha T + \beta T^2$$

$$\frac{dQ}{dT} = \alpha + 2\beta T$$

$$dQ = (\alpha + 2\beta T)dT$$

$$\Delta s = \int_{T_1}^{T_2}\frac{dQ}{T} = \int_{T_1}^{T_2}\frac{(\alpha+2\beta T)}{T}dT$$

$$= \alpha\ln\left(\frac{T_2}{T_1}\right) + 2\beta(T_2 - T_1)$$

3.9 (d)

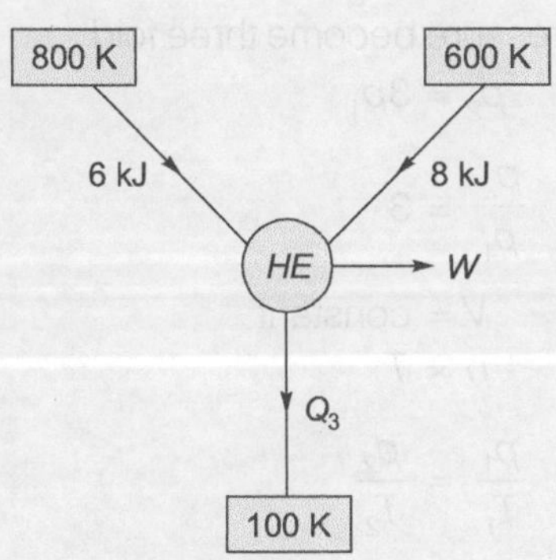

$$\oint\frac{\delta Q}{T} = 0 \text{ for reversible heat engine}$$

$$\frac{Q_1}{T_1} + \frac{Q_2}{T_2} - \frac{Q_3}{T_3} = 0$$

$$\frac{6}{800} + \frac{8}{600} - \frac{Q_3}{100} = 0$$

$$Q_3 = 2 \text{ kJ}$$

$$\eta = \frac{\text{Work done}}{\text{Heat supplied}} = \frac{Q_1 + Q_2 - Q_3}{Q_1 + Q_2}\times 100$$

$$= \frac{12}{14}\times 100 = \mathbf{85\%}$$

3.10 (c)

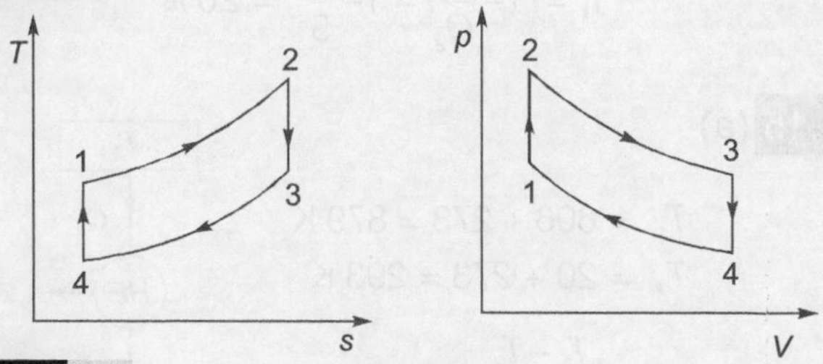

3.11 (c)

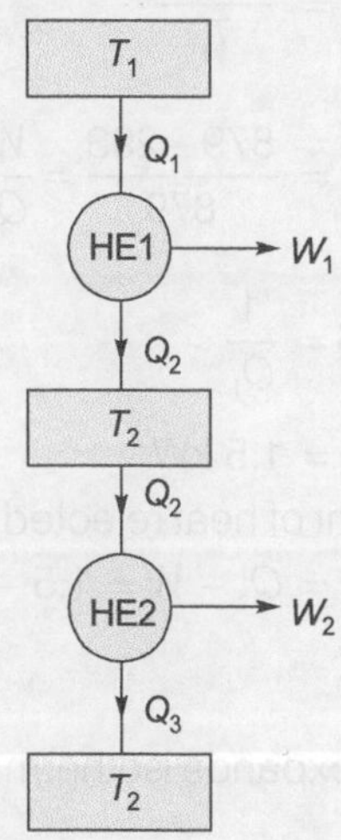

If $\eta_1 = \eta_2$

$$1 - \frac{T_2}{T_1} = 1 - \frac{T_3}{T_2}$$

$$\Rightarrow \quad T_2^2 = T_3 T_1$$

or $\quad T_2 = \sqrt{T_3 T_1}$

3.12 (b)

Carnot Theorem: The efficiency of a reversible engine is greater than the efficiency of irreversible engine working between same temperature limit.

3.13 (a)

Entropy is related with second law of Thermodynamics and for ideal gas internal energy is function of temperature only. Van der Waal's equation is characteristic equation for real gases.

3.14 (a)

$$W = \frac{1}{4}Q_2$$

$$Q_1 - Q_2 = \frac{1}{4}Q_2$$

$$Q_1 = \left(1 + \frac{1}{4}\right)Q_2 = \frac{5}{4}Q_2$$

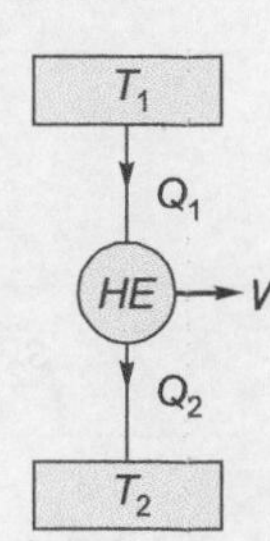

$$\eta = 1-\frac{Q_2}{Q_2} = 1-\frac{4}{5} = \mathbf{20\%}$$

3.15 (a)

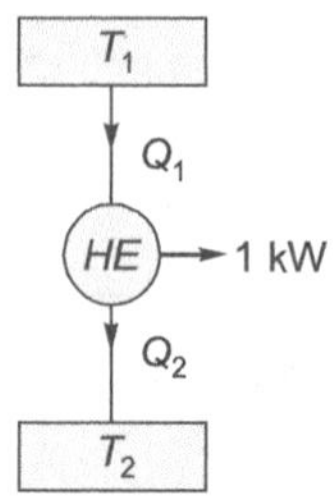

$$T_1 = 606 + 273 = 879 \text{ K}$$
$$T_2 = 20 + 273 = 293 \text{ K}$$
$$\eta = \frac{T_1 - T_2}{T_1}$$
$$= \frac{879-293}{879} = \frac{W}{Q_1} \quad \text{(since } W = 1 \text{ kW)}$$
$$0.6666 = \frac{1}{Q_1}$$
$$Q_1 = 1.5 \text{ kW}$$

Amount of heat rejected

$$Q_2 = Q_1 - W = 1.5 - 1.0 = \mathbf{0.5\ kW}$$

3.16 (c)

(i) Air expands isothermally from 6 bar to 3 bar

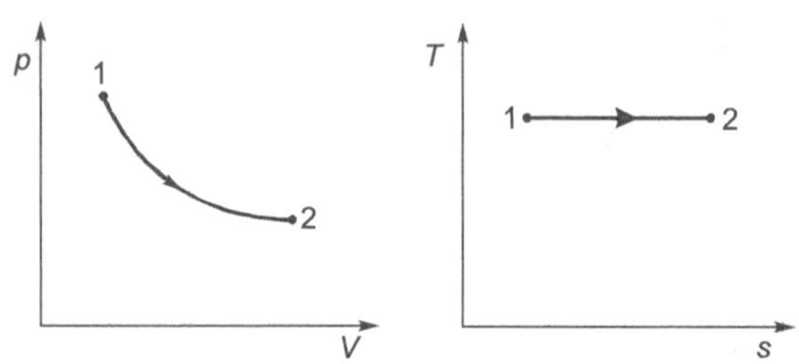

$$s_2 - s_1 = c_p \ln\left(\frac{V_2}{V_1}\right) + c_v \ln\left(\frac{p_2}{p_1}\right)$$
$$p_1V_1 = p_2V_2$$
$$(6)V_1 = (3)V_2$$
$$V_1 = 0.5\ V_2$$
$$\frac{V_2}{V_1} = 2$$
$$\frac{p_2}{p_1} = \frac{3}{6} = \frac{1}{2}$$
$$c_p > c_v$$
$$s_2 - s_1 = c_p \ln(2) + c_v \ln\frac{1}{2}$$
$$= (c_p - c_v) \ln 2 > 0$$

(ii) Air is compressed to half the volume at constant pressure

$$V_2 = \frac{1}{2}V_1$$

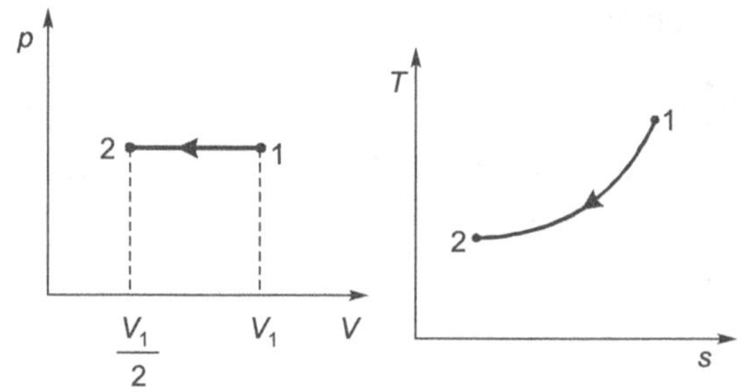

$$s_2 - s_1 = c_v \ln\left(\frac{T_2}{T_1}\right) + R\ln\left(\frac{V_2}{V_1}\right)$$
$$V \propto T$$
$$\frac{V_1}{T_1} = \frac{V_2}{T_2}$$
$$\frac{T_2}{T_1} = \frac{V_2}{V_1} = \frac{V_1/2}{V_1} = \frac{1}{2}$$
$$s_2 - s_1 = c_v \ln\left(\frac{1}{2}\right) + R\ln\left(\frac{1}{2}\right)$$
$$= -(c_v + R) \ln 2 < 0$$

(iii) Heat is supplied to air at constant volume till pressure become three fold

$$p_2 = 3p_1$$
$$\frac{p_2}{p_1} = 3$$
$$V = \text{constant}$$
$$p \propto T$$
$$\frac{p_1}{T_1} = \frac{p_2}{T_2}$$
$$\frac{T_2}{T_1} = \frac{p_2}{p_1} \Rightarrow \frac{T_2}{T_1} = \frac{3p_1}{p_1} = 3$$

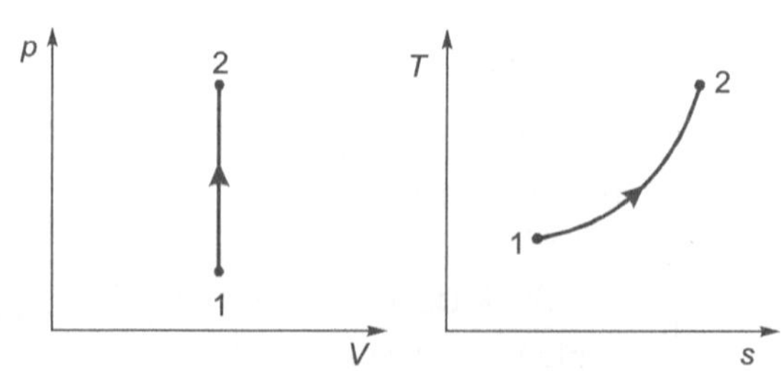

$$s_2 - s_1 = c_v \ln\left(\frac{T_2}{T_1}\right) - R\ln\left(\frac{p_2}{p_1}\right)$$

$$= c_p \ln 3 - R \ln 3$$

$$= (c_p - R) \ln 3 > 0 \qquad (\because c_p > R)$$

3.17 (c)

When the efficiency of both the engines are equal then intermediate temperature

$$T_2 = \sqrt{T_1 T_3} = \sqrt{1200 \times 300} = \mathbf{600\ K}$$

3.19 (c)

When power output of both the engines are same then

$$Q_1 - Q_2 = Q_2 - Q_3 \left(\frac{Q}{T} = \text{constant}\right)$$

$$\Rightarrow T_1 - T_2 = T_2 - T_3$$

$$\Rightarrow \quad T_2 = \frac{T_1 + T_3}{2} = \frac{900 + 400}{2} = 650\ K$$

3.20 (b)

Given data

$$T_2 = 27°C = (27 + 273)\ K = 300\ K$$

$$T_1 = 327°C = (327 + 273)\ K = 600\ K$$

$$W = 300\ kJ$$

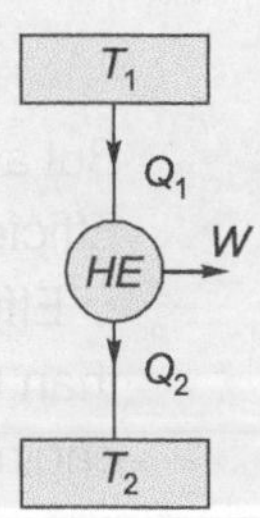

$$\eta_{Carnot} = 1 - \frac{T_2}{T_1} = 1 - \frac{300}{600} = 1\ 0.5 = 0.5$$

$$\text{also } \eta_{Carnot} = \frac{W}{Q_1}$$

$$\therefore \quad 0.5 = \frac{300}{Q_1}$$

$$\text{or} \quad Q_1 = 600\ kJ$$

Entropy change during heat addition :

$$\Delta S_1 = \frac{Q_1}{T_1} = \frac{600}{600} = \mathbf{1\ kJ/kg}$$

3.22 (b)

From the Clausius inequality for real thermodynamic cycle

$$\oint \frac{\delta Q}{T} < 0$$

3.23 (d)

Reversible adiabatic process is an isentropic process but if adiabatic process is not reversible, it will not be isentropic.

3.24 (c)

Heat source-1

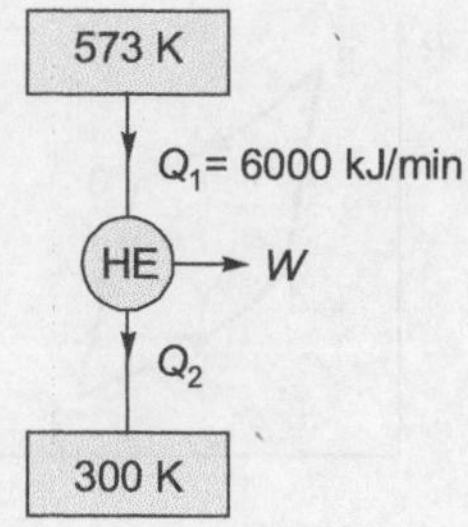

$$\eta = \frac{W}{Q_1} = \frac{T_1 - T_2}{T_1}$$

$$\eta = \frac{T_1 - T_2}{T_1}$$

$$= 1 - \frac{300}{573}$$

$$= 0.476$$

$$\eta = \frac{W}{Q_1}$$

$$0.476 = \frac{W_1}{6000}$$

$$\Rightarrow \quad W_1 = 2856\ kJ/min$$

Heat source-2

$$\eta = 1 - \frac{T_2}{T_1}$$

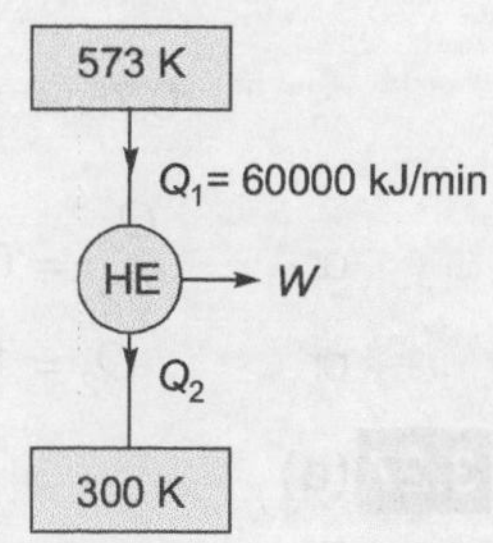

$$= 1 - \frac{300}{373}$$

$$= 0.1957$$

$$\eta = \frac{W}{Q_1}$$

$$0.1957 = \frac{W}{60000}$$

$$\text{or} \quad W = \mathbf{11742.62\ kJ/min.}$$

3.25 (d)

For irreversible process

$$ds > \left(\frac{\delta Q}{T}\right)_{irr.}$$

$$\Rightarrow \quad ds = \left(\frac{\delta Q}{T}\right)_{irr} + (\delta s)_{generation}$$

Consider heat rejection from system

Let us take any arbitrary value.

$$ds = -4 + 4 = 0$$

Thus isentropic process.

Thus an isentropic process may not be reversible adiabatic.

3.26 (b)

Given: $\gamma = 1.4$, $Q_1 = 52$ kJ

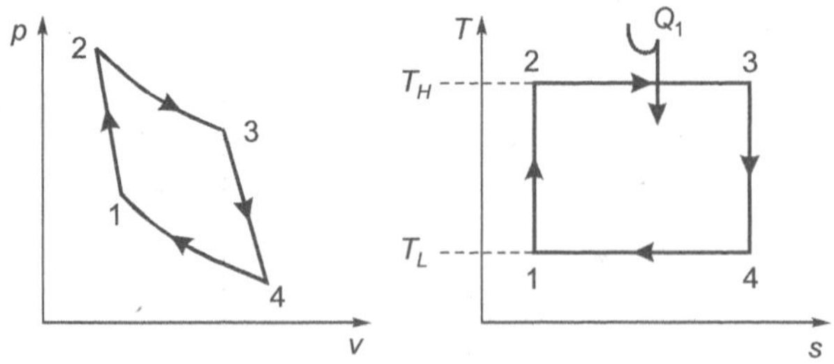

Expansion ratio,

$$\frac{V_4}{V_3} = 32$$

For expansion power 3 – 4,

$$T_H V_3^{\gamma-1} = T_L V_4^{\gamma-1}$$

or $$\frac{T_L}{T_H} = \left(\frac{V_3}{V_4}\right)^{\gamma-1} = \left(\frac{1}{32}\right)^{1.4-1} = 0.25$$

$$\eta_{\text{Carnot}} = 1 - \frac{T_L}{T_H} = 1 - 0.25 = 0.75$$

also $$\eta_{\text{Carnot}} = 1 - \frac{Q_2}{Q_1}$$

$\therefore$ $$0.75 = 1 - \frac{Q_2}{52}$$

or $$\frac{Q_2}{52} = 0.25$$

or $Q_2 =$ **13 kJ**

3.27 (a)

For any spontaneous process, the entropy of the universe increases. It is called principle of increase of entropy.

3.28 (c)

$$dS = \frac{\delta Q}{T} = \frac{ML^2T^{-2}}{\theta} = ML^2T^{-2}\theta^{-1}$$

3.30 (b)

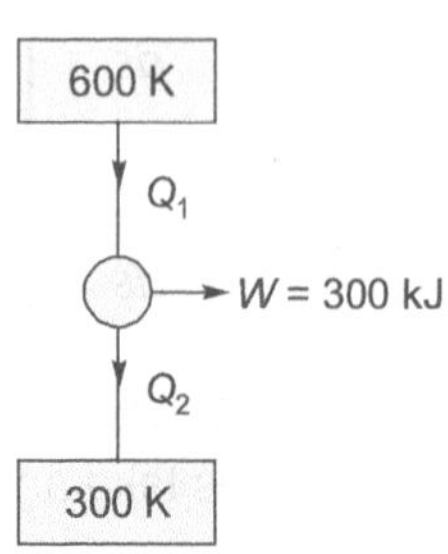

$$\eta = \frac{T_1 - T_2}{T_1} = \frac{W}{Q_1}$$

$$\frac{W}{Q_1} = 1 - \frac{300}{600} = 0.5$$

$$Q_1 = \frac{300}{0.5} = 600 \text{ kJ}$$

$\therefore$ Change in entropy during heat addition

$$= \frac{Q_1}{T_1} = \frac{600}{600} = 1 \text{ kJ/K}$$

3.31 (a)

The increase in entropy is obtained from a given quantity of heat transfer at a low temperature

$$\Delta S = \frac{\delta Q}{T}$$

3.32 (b)

$$Tds = du + pdV$$

$$\Rightarrow \quad Tds = c_v dT + pdV$$

$$\Rightarrow \quad ds = c_v \frac{dT}{T} + \frac{p}{T} dV$$

$$\Rightarrow \quad ds = c_v \frac{dT}{T} + \frac{R}{V} dV$$

Integrating the above expression

$$\Rightarrow s_2 - s_1 = c_v \ln\frac{T_2}{T_1} + R\ln\frac{V_2}{V_1}$$

For isothermal process undergone by ideal gas.

$$T_2 = T_1 \text{ thus } \ln\frac{T_2}{T_1} = 0$$

$$\Rightarrow s_2 - s_1 = (c_p - c_v)\ln\frac{V_2}{V_1}$$

3.34 (c)

Carnot efficiency of engine,

$$\eta = 1 - \frac{T_2}{T_1} = 1 - \frac{300}{600} = \mathbf{0.5}$$

But according to the inventor's claim

Efficiency of engine = 1 – 0.45 = 0.55

$\because$ Efficiency of actual engine cannot be greater than Carnot efficiency. So this is an impossible engine.

3.35 (d)

Since Efficiency of the engine purely depends upon source and sink temperatures and independent of the working substances hence all the engines will have same efficiency.

3.36 (c)

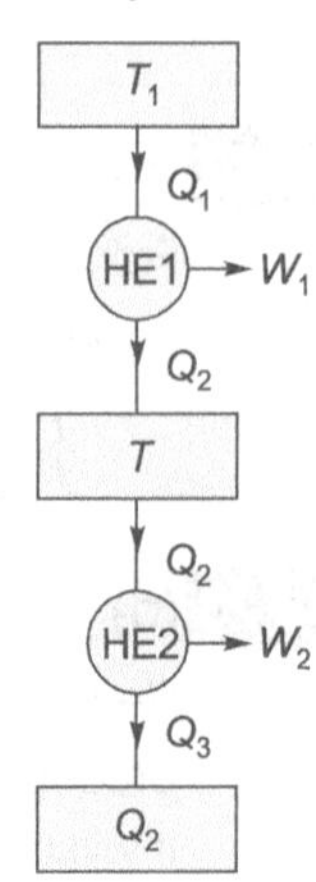

Source temperature = T_1

Intermediate temperature = T

Sink Temperature = T_2

$\because$ $$W_1 = W_2$$

$$Q_1\left(1 - \frac{T}{T_1}\right) = Q_2\left(1 - \frac{T_2}{T}\right)$$

$$\Rightarrow \quad \frac{T_1}{T_2}\left(1 - \frac{T}{T_1}\right) = \left(1 - \frac{T_2}{T}\right)$$

$$\Rightarrow \quad \frac{T_1}{T} - 1 = 1 - \frac{T_2}{T}$$

$$\Rightarrow \quad \frac{T_1 + T_2}{T} = 2$$

$$\Rightarrow \quad T = \frac{T_1 + T_2}{2}$$

$$= \frac{180 + 20}{2} = 100°C$$

3.37 (b)

Sink temperature: $T_2 = 27°C = 300$ K.

It is given that engine rejects 40% of absorbed heat from the source

For a Carnot cycle engine

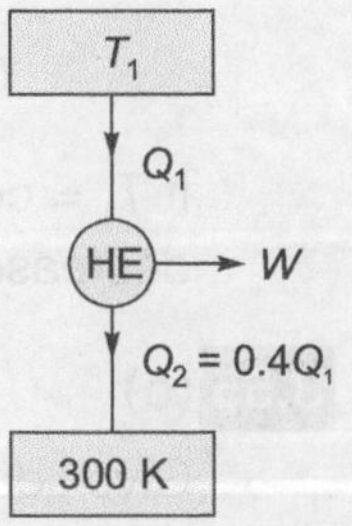

$$\frac{Q_1}{T_1} = \frac{Q_2}{T_2}$$

$$\frac{Q_1}{T_1} = \frac{0.4\, Q_1}{300}$$

$$\Rightarrow \quad T = \frac{300}{0.4} = 750 \text{ K}$$

$$= \mathbf{477°C}$$

3.38 (d)

Heat supplied by the heat engine,

$$Q_1 = 250 \text{ kJ/s}$$

Source temperature,

$$T_1 = 227°C = 500 \text{ K}$$

Sink temperature,

$$T_2 = 27°C = 300 \text{ K}$$

$$\frac{250}{500} = \frac{Q_2}{300}$$

$$\Rightarrow \quad Q_2 = 250 \times 0.6 = \mathbf{150 \text{ kJ/s}}$$

3.40 (c)

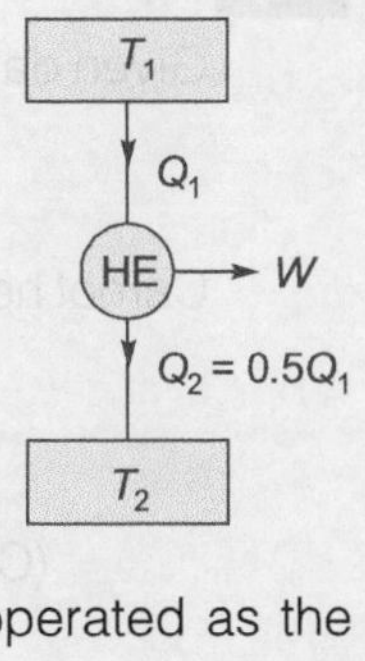

$$\frac{Q_1}{T_1} = \frac{Q_2}{T_2}$$

$$\Rightarrow \quad \frac{Q_1}{T_1} = \frac{0.5\, Q_1}{T_2}$$

$$\Rightarrow \quad \frac{T_2}{T_1} = 0.5$$

If the engine is reversed and operated as the heat pump.

Then $(COP)_{HP}$ coefficient of performance,

$$= \frac{T_1}{T_1 - T_2} = \frac{1}{1 - \frac{T_2}{T_1}} = \frac{1}{1 - 0.5} = 2$$

3.41 (c)

$$\eta_{H\cdot E} = 1 - \frac{Q_2}{Q_1}$$

$$\Rightarrow \quad \frac{Q_2}{Q_1} = 0.70$$

$$(COP)_{H.P.} = \frac{1}{1 - \frac{Q_2}{Q_1}} = \frac{1}{1 - 0.70} = \frac{1}{0.33}$$

$$= \mathbf{3.33}$$

3.42 (c)

$$\eta_{Carnot} = 75\%$$

COP of reversed Carnot cycle is

$$= \frac{1}{\eta_{Carnot}} - 1 = \frac{1}{0.75} - 1 = \mathbf{0.33}$$

3.43 (c)

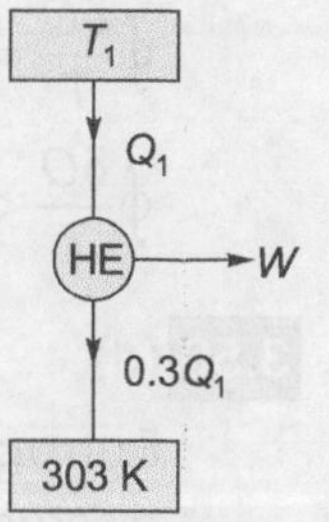

$$\eta = \frac{Q_1 - 0.3Q_1}{Q_1} = \frac{T_1 - T_2}{T_1}$$

$$\eta = 0.7 = \frac{T_1 - 303}{T_1}$$

$$0.7\, T_1 = T_1 - 303$$

$$303 = 0.3\, T_1$$

$$T_1 = \frac{303}{0.3} = 1010 \text{ K} = \mathbf{737°C}$$

3.44 (b)

Intermediate temperature for both engines to be equally efficient

$$T_2 = \sqrt{T_1 T_3} = \sqrt{900 \times 400} = \mathbf{600 \text{ K}}$$

3.45 (c)

Maximum possible efficiency of the engine or Carnot efficiency of the engine

$$\eta = \frac{T_1 - T_2}{T_1}, T_1 = 600 \text{ K}, T_2 = 300 \text{ K}$$

$$\eta = \frac{600 - 300}{600} = 0.5$$

$$\eta_{inventor} = \frac{Q_1 - 0.4Q_1}{Q_1} = 0.6$$

Since no engine is possible which is more efficient than carnot hence inventor's claims is wrong and engine is impossible.

3.46 (d)

On *T-s* plane

$$\text{Slope of constant pressure line} = \left(\frac{dT}{ds}\right)_p = \frac{T}{c_p}$$

$$\text{Slope of constant volume line} = \left(\frac{dT}{ds}\right)_v = \frac{T}{c_v}$$

$$\because \quad c_p > c_v \quad \Rightarrow \left(\frac{dT}{ds}\right)_p < \left(\frac{dT}{ds}\right)_v$$

∴ Constant volume line is more steeper.

3.49 (c)

From Clausius inequality

$$\oint \frac{\delta Q}{T} \le 0$$

$$\oint \frac{\delta Q}{T} = 0 \qquad \text{for reversible cycle}$$

$$\oint \frac{\delta Q}{T} < 0 \qquad \text{for irreversible cycle}$$

3.50 (d)

Entropy is the measure of degree of disorderness. Increase in entropy of a system represents degradation of energy.

3.51 (b)

The second law of thermodynamics states that,

$$\frac{\delta Q}{T} \le dS$$

where δQ is the amount energy the system gains by heating, *T* is the temperature of the system, and dS is the change in entropy. The equality sign will hold for a reversible process. For a reversible isentropic process, there is no transfer of heat energy and therefore the process is also adiabatic. For an irreversible process, the entropy will increase. Hence removal of heat from the system (cooling) is necessary to maintain a constant internal entropy for an irreversible process in order to make it isentropic. Thus an irreversible isentropic process is not adiabatic.

3.52 (d)

∴ Carnot efficiency,

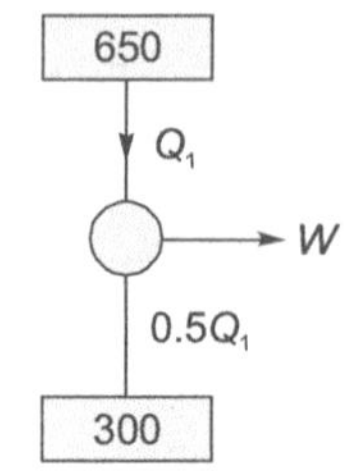

$$\eta_c = 1 - \frac{300}{650} = 0.538$$

Claimed efficiency,

$$\eta = \frac{W}{Q_1} = 0.5$$

$$\because \quad \eta < \eta_c$$

The cycle is possible

3.54 (b)

$$COP = \frac{T_1}{T_1 - T_2} \propto \frac{1}{T_1 - T_2}$$

If T_1 = constant and $T_1 - T_2$ increase then *COP* decreases.

3.55 (c)

Given data

$$Q_1 = 2512 \text{ kJ/min}$$
$$Q_2 = 900 \text{ KJ/min}$$
$$T_1 = 650°C = (650 + 273) K = 923 K$$
$$T_2 = 100°C = (100 + 273) K = 373 K$$
$$W = Q_1 - Q_2 = 2512 - 900 = 1612 \text{ kJ/min}$$

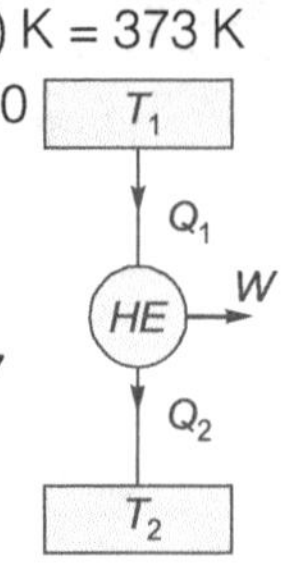

Actual efficiency,

$$\eta = \frac{W}{Q_1} = \frac{1612}{2512} = 0.6417$$

Carnot efficiency,

$$\eta_{carnot} = 1 - \frac{T_2}{T_1} = 1 - \frac{373}{923} = 0.5958$$

Thus, $\eta > \eta_{Carnot}$

Therefore, the heat engine is impossible.

3.56 (d)

Given data

$$T_2 = -3°C = (-3 + 273) K = 270 K$$

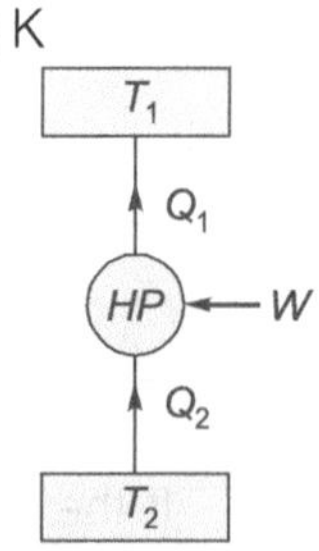

Carnot heat pump

$$T_1 = 27°C = 300 K$$
$$Q_1 = 40 \text{ kW}$$

$$(COP)_{HP} = \frac{T_1}{T_1 - T_2} = \frac{300}{300 - 270} = \frac{300}{30} = 10$$

also $(COP)_{HP} = \frac{Q_1}{W}$

$\therefore \quad 10 = \frac{40}{W}$

or $W = \frac{40}{10} = \mathbf{4\ kW}$

3.57 (c)

Heat supplied = 90×10^6 kJ/hour

$$= \frac{90 \times 10^6}{3600} \times 10^3 \text{ W} = 25 \text{ MW}$$

$\therefore$ Heat supplied – Heat rejected = Power produced

$\therefore \quad Q_R = Q_S - W_{net} = 25 - 10 = \mathbf{15\ MW}$

3.58 (c)

It is second law of thermodynamics which provides the criterion as to the probability of various process. So, the degree of completion of chemical reaction is predicted correctly by second law of thermodynamics.

3.59 (c)

Given data :

$T_1 = 600°C = 873$ K

$T_2 = 20°C = 293$ K

Least rate of heat rejection means reversible heat engine.

$W = 1$ kW

$$\eta = 1 - \frac{T_2}{T_1} = 1 - \frac{293}{873}$$

also $\eta = \frac{W}{Q_1}$

$\therefore \quad 0.6643 = \frac{1}{Q_1}$

or $Q_1 = 1\,.\,505$ kW

also $Q_1 = W + Q_2$

$\therefore \quad 1.505 = 1 + Q_2$

or $Q_2 = \mathbf{0.505\ kW}$

3.60 (c)

$$Q_S - Q_R = W_{net}$$

$$Q_R = 0.5Q_S$$

$\therefore \quad 0.5Q_S = W_{net}$

or $\frac{W_{net}}{Q_S} = 1 - \frac{T_{sink}}{T_{source}}$

$\therefore \quad \frac{T_{sink}}{T_{source}} = 0.5$

$\therefore \quad T_{source} = 300 \times 2 = 600$ K

$\Rightarrow \quad T_{source} = 327°C$

3.61 (c)

$$(COP)_{HP} - (COP)_{Refrigerator} = 1$$

$\therefore \quad (COP)_{HP} = 5 \qquad \therefore \frac{q_R}{W} = 4$

$\Rightarrow \quad \frac{q_R}{W} = 5 \quad$ or, $W = 2.5$ kJ

$q_R = 5 \times W = 5 \times 2.5 = 12.5$ kJ

3.62 (c)

Universe is formed by system and surrounding and considered as isolated system.

$$dS_{univ} \geq 0$$

or $dS_{sys} + dS_{surr} \geq 0$

(i) The energy of the world (universe) is constant.

(ii) The entropy of the world tends towards a maximum.

(iii) The entropy increase of an isolated system is a measure of the extent of irreversibility of the process undergone by the system.

3.63 (c)

$$\eta_E = 1 - \frac{T_L}{T_H}$$

$\therefore \quad \frac{T_L}{T_H} = 0.7$

$$COP = \frac{T_L}{T_H - T_L}$$

$$= \frac{0.7}{1 - 0.7} = \frac{0.7}{0.3} = 2.33$$

3.64 (d)

$$\Delta S = \frac{\delta Q}{T}$$

Statement (I) is not correct.

3.65 (c)

$\oint \frac{\delta Q}{T} = 0$, reversible cycle.

$\oint \frac{\delta Q}{T} < 0$, irreversible and possible cycle.

$\oint \frac{\delta Q}{T} > 0$, impossible cycle.

So, statement (II) is wrong.

3.66 (d)

Isothermal expansion in a Stirling cycle is a reversible process, So, option 2 is not correct. Other options 1 and 3 are the example of thermodynamically irreversible process.

3.67 (d)

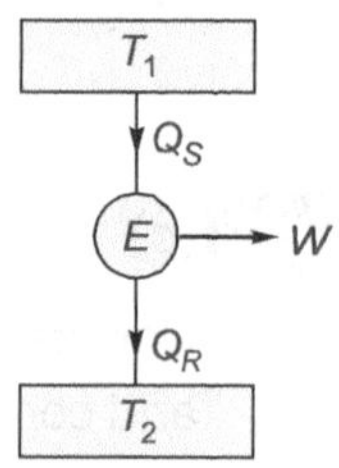

$$T_1 = 377°C = 650\text{ K}$$
$$T_2 = 27°C = 300\text{ K}$$
$$\eta_{Carnot} = 1 - \frac{T_2}{T_1}$$
$$= 1 - \frac{300}{650} = 0.5385$$
$$Q_R = 0.5\, Q_S$$
$$W = Q_S - Q_R = 0.5 Q_S$$
$$\eta_{Real} = \frac{W}{Q_S} = 0.5 < \eta_{Carnot}$$

Hence, it will be a practical engine.

3.68 (b)

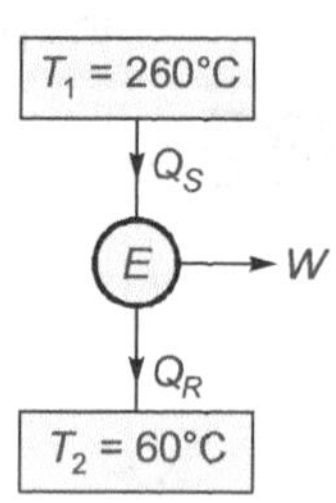

$$T_1 = 260°C = 533\text{ K}$$
$$T_2 = 60°C = 333\text{ K}$$
$$\eta_{ideal} = 1 - \frac{333}{533} = 0.3752$$

Case-I

$T_1 = 300°C = 573$ K

$T_2 = 60°C = 333$ K

$$\eta_I = 1 - \frac{333}{573} = \mathbf{0.4188}$$

Case-II

$T_1 = 260°C = 533$ K

$T_2 = 30°C = 303$ K

$$\eta_{II} = 1 - \frac{303}{533} = \mathbf{0.4315}$$

$$\boldsymbol{\eta_{II} > \eta_I}$$

3.69 (b)

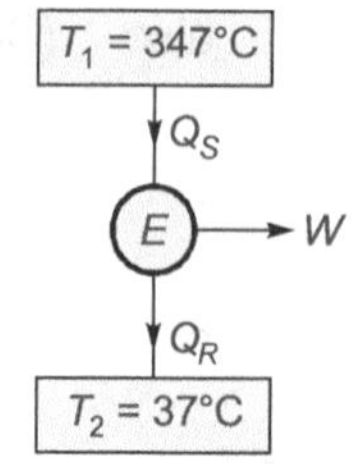

$$\eta = \frac{W}{Q_s}$$
$$= 1 - \frac{(273+37)}{(273+347)} = 0.5$$
$$\therefore\ Q_s = \frac{W}{0.5} = \frac{620}{0.5}$$
$$= 1240\text{ kJ}$$
$$\therefore\ \Delta S = -\frac{Q_s}{T_1} = -\frac{1240}{620} = \mathbf{-2\text{ kJ/K}}$$

3.70 (c)

Given data : $Q = 1000$ kJ

$T = 227°C = 500$ K

$T_0 = 27°C = 300$ K

Available energy,

$$AE = Q\left[1 - \frac{T_0}{T}\right] = 1000\left[1 - \frac{300}{500}\right]$$
$$= \mathbf{400\text{ kJ}}$$

3.71 (c)

$$(COP)_{Carnot} = \frac{273-10}{35} = \frac{263}{35} = 7.5143$$
$$(COP)_{inventor} = 8 > (COP)_{Carnot}$$

which is impractical

So, Inventor's claim is invalid.

3.73 (b)

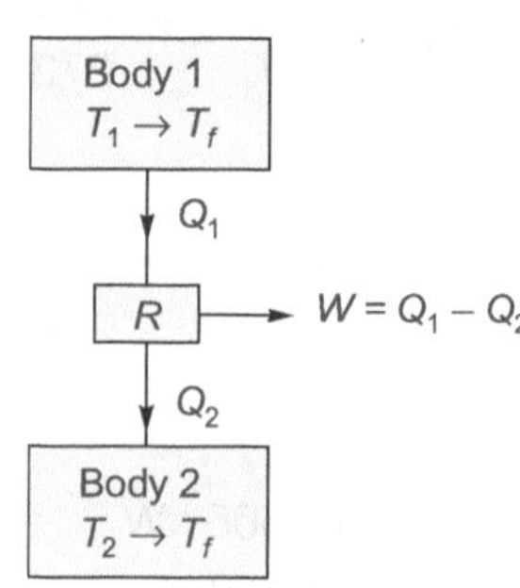

Work obtainable, W, will be maximum when engine is reversible, i.e. $\Delta S = 0$.

$$\Delta S_1 = \int_{T_1}^{T_f} c_p \frac{dT}{T} = c_p \ln \frac{T_f}{T_1}$$

$$\Delta S_2 = \int_{T_2}^{T_f} c_p \frac{dT}{T} = c_p \ln \frac{T_f}{T_2}$$

$\because\ \Delta S_{univ} \geq 0$

$$\therefore\ c_p \ln \frac{T_f}{T_1} + c_p \ln \frac{T_f}{T_2} \geq 0$$

$$\text{or } c_p \ln \left[\frac{T_f^2}{T_1 T_2}\right] \geq 0$$

For maximum work,

$$\ln\left[\frac{T_f^2}{T_1 T_2}\right] = 0 = \ln 1$$

$$\therefore\quad T_f = \sqrt{T_1 T_2}$$

3.74 (c)

$$\eta_E = 0.4 = \frac{W}{Q_1}$$

$$COP = \frac{Q_3}{W} = \frac{Q_3}{0.4Q_1} = \frac{Q_4 - W}{0.4Q_1}$$

$$= \frac{Q_4 - (Q_1 - Q_2)}{0.4Q_1} = \frac{2Q_1}{0.4Q_1} = 5$$

3.75 (c)

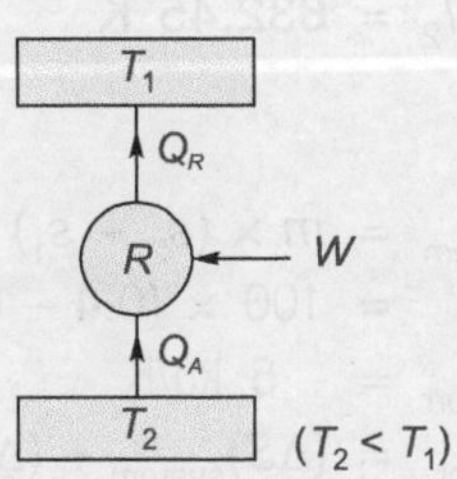

$$(COP)_R = \frac{T_2}{T_1 - T_2} = N$$

$$(COP)_{HP} = \frac{T_1}{T_1 - T_2}$$

$$(COP)_{HP} - (COP)_R = 1$$

$$\therefore\quad (COP)_{HP} = N + 1$$

3.76 (b)

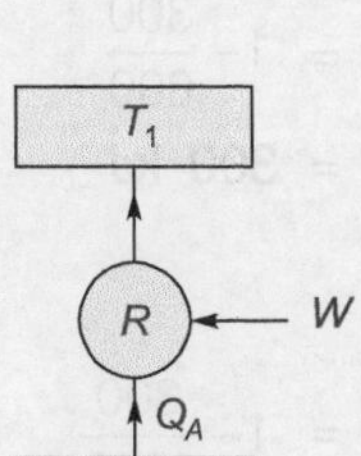

$$T_1 = 27°C = 300\text{ K}$$

$$T_2 = 23°C = 250\text{ K}$$

$$Q_A = 1.5\text{ kW}$$

$$(COP)_R = \frac{250}{300 - 250} = 5$$

$$(COP)_R = \frac{Q_A}{W}$$

$$\therefore\quad W = \frac{Q_A}{(COP)_R} = \frac{1.5}{5} = \mathbf{0.3\ kW}$$

3.77 (b)

$$\text{Entropy generated} = c_p \ln \frac{T_2}{T_1}$$

$$= 5 \times \ln \frac{600}{300}$$

$$= 5 \times 0.693 = \mathbf{3.465\ J/K}$$

Option (b) is correct.

3.78 (*)

For perfect gas

$$TdS = dU + PdV$$

$$\Rightarrow\quad dS = \frac{dU}{T} + \frac{P}{T}dV = \frac{C_V \cdot dT}{T} + \frac{R \cdot dV}{V}$$

$$\because\quad PV = RT$$

$$\Rightarrow\quad \frac{P}{T} = \frac{R}{V}$$

Hence (1) is correct.

Also, we know:

$$TdS = dH - VdP$$

$$= \frac{C_p dT}{T} - \frac{R \cdot dP}{P}$$

Hence, (2) is incorrect.

and (3) i.e.

$$dS = \frac{dV}{V} + C_V \cdot \frac{dP}{P} \text{ is also true.}$$

Hence (4) is also incorrect.

3.79 (c)

The isolated system is one in which there is no interaction between the system and surroundings either in terms of heat or work.

3.80 (c)

By Clausius inequality

$$\oint \frac{\delta Q}{T} = 0, \text{ the cycle is reversible.}$$

$\oint \frac{\delta Q}{T} < 0$, the cycle is irreversible and possible.

$\oint \frac{\delta Q}{T} > 0$, the cycle is impossible.

3.81 (b)

Entropy across the normal shock increases stagnation temperature remains constant across the normal shock.

3.82 (d)

(i) In isentropic compression process:
Increase in entropy due to irreversibility is compensated by decrease in entropy due to heat loss and process becomes isentropic but not reversible.

(ii) Free expansion is always irreversible process due to absence of resistance against the expansion.

(iii) Slow heating is a quasi-static process which is reversible in nature.

(iv) Evaporation process is reversible.

3.83 (a)

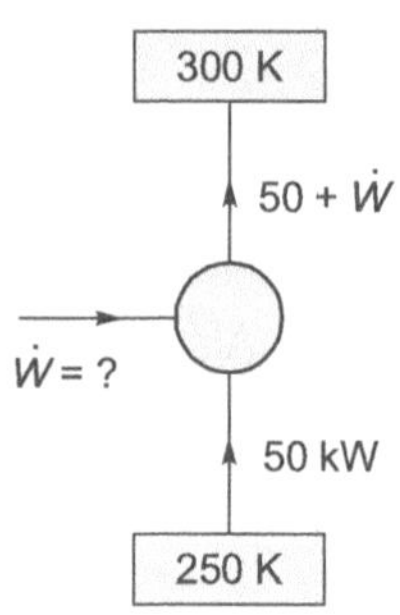

$$\frac{50+\dot{W}}{50} = \frac{300}{250}$$

$$1+\frac{\dot{W}}{50} = 1.2$$

$$\dot{W} = \mathbf{10\ kW}$$

3.84 (d)

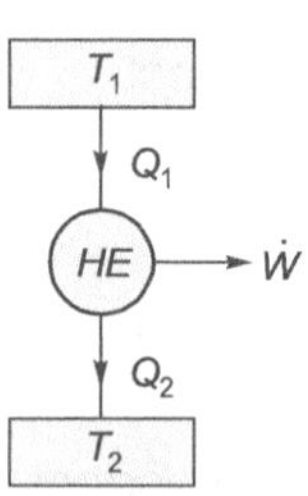

Given data:

$$Q_1 = 2500 \text{ kJ/min} = 41.67 \text{ kW}$$

$$W = 12.4 \text{ kW}$$

$$\eta = \frac{W}{Q_1} = \frac{12.4}{41.67} = \mathbf{29.76\%}$$

3.85 (b)

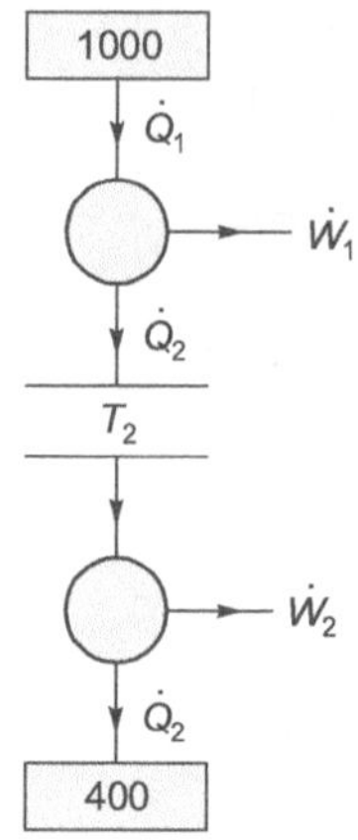

$$\Rightarrow \quad \dot{W}_1 = \dot{W}_2$$

$$\therefore \dot{Q}_1 \times \left(1-\frac{T_2}{1000}\right) = \dot{Q}_1 \times \left(1-\frac{400}{T_2}\right)$$

$$T_2^2 = 1000 \times 400$$

$$T_2 = \mathbf{632.45\ K}$$

3.86 (b)

$$(\Delta S)_{system} = m \times (s_2 - s_1)$$
$$= 100 \times (0.4 - 0.3) = 10 \text{ kJ/K}$$
$$(\Delta S)_{surr} = -5 \text{ kJ/K}$$
$$(\Delta S)_{univ} = (\Delta S)_{system} + (\Delta S)_{surr}$$
$$= 10 - 5 = 5 \text{ kJ/K}$$

$\therefore$ Irreversible

3.87 (c)

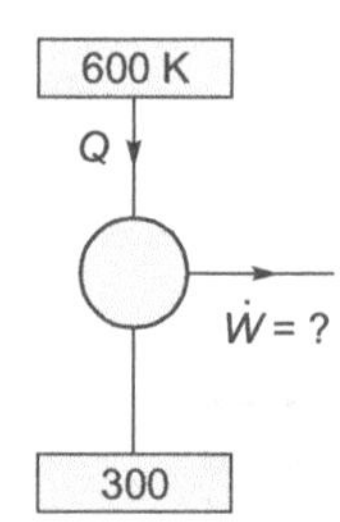

$$(\Delta S)_{heat\ addition} = 1$$

$$\frac{Q}{600} = 1$$

$$Q = 600$$

$$\therefore \quad \frac{W}{Q} = 1-\frac{300}{600}$$

$$W = \mathbf{300\ kJ}$$

3.89 (c)

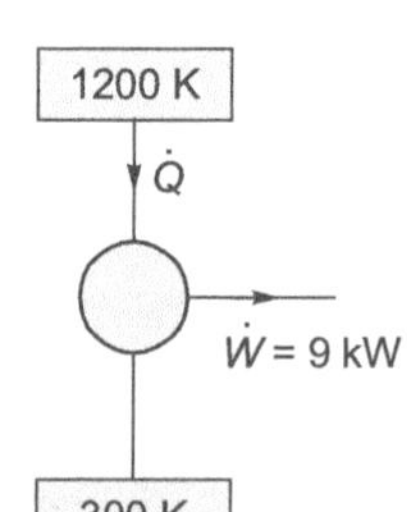

$$\frac{9}{\dot{Q}} = 1-\frac{300}{1200}$$

$$\dot{Q} = 12 \text{ kW}$$

also $\quad \dot{Q} = m_f \times \text{C.V.}$

$\therefore \quad 12 = m_f \times 40000$

or $\quad m_f = 3 \times 10^{-4} \text{ kg/s} = \mathbf{1.08\ kg/hr}$

3.91 (d)

$$\eta_{Th} = \frac{P}{Q_s} = \frac{P}{A \times 0.6 \times 0.5}$$

$$\Rightarrow \quad = \frac{2.5}{A \times 0.6 \times 0.5} = 1 - \frac{313}{350}$$

$$\Rightarrow \quad A = \mathbf{78.89\ m^2}$$

3.92 (b)

$$\eta_{Th} = 1 - \frac{300}{1000} = \frac{14}{\dot{m}_f \times 40{,}000}$$

$$\Rightarrow \quad \dot{m}_f = \mathbf{1.8\ kg/hr}$$

3.95 (b)

$$(\Delta S)_{univ} = (\Delta S)_{sys} + (\Delta S)_{surr}$$

$$(\Delta S)_{sys} = m(s_f - s_i)$$

$$= 100 \times (0.4 - 0.3) = 10\ \text{kJ/K}$$

$$(\Delta S)_{surr} = 75 - 80 = -5\ \text{kJ/K}$$

[Unit for surrounding entropy is error]

$$(\Delta S)_{univ} = 10 - 5 = 5\ \text{kJ/K}$$

$(\Delta S)_{univ} > 0 \Rightarrow$ **Irreversible**

3.96 (c)

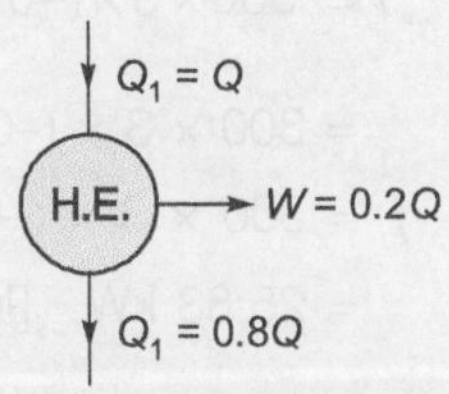

$$\therefore \quad \eta = \frac{W}{Q_1} = 0.2$$

$$(COP)_{HP} = 1 + COP_R = \frac{1}{\eta_{HE}}$$

$$(COP)_R = \frac{1}{\eta_{HE}} - 1 = \frac{1}{0.2} - 1$$

$$= 5 - 1 = 4$$

3.97 (b)

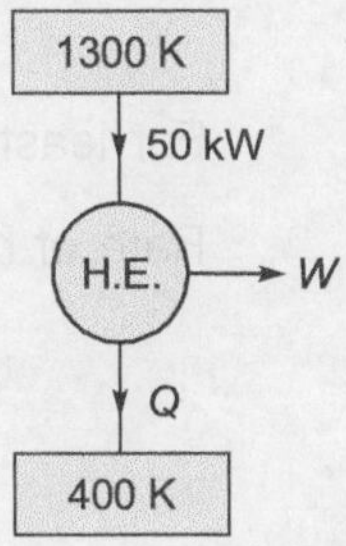

$$\frac{50}{1300} = \frac{Q}{400}$$

$$Q = 15.4\ \text{kW}$$

3.98 (b)

An elastic rubberband is an example of elastomer so lets consider that any process that is spontaneous (i.e. which proceeds on its own) will cause decrease in Gibbs free energy, so when we heat an elastic rubber band its shrinkage happens on its own. Hence it is spontaneous process so ΔG must be negative. i.e. gibbs free energy decreases.

Also the shrinkage of the rubber band causes the molecule to be more randomly coiled into each other, hence it increases the randomness of the rubber band molecules therefore the entropy increases.

3.99 (c)

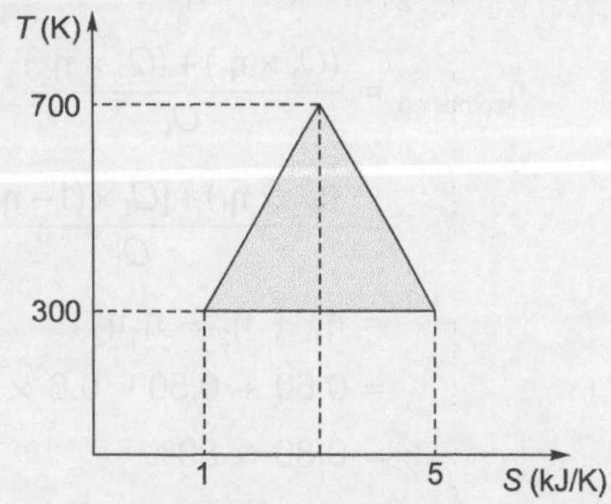

$$\eta = \frac{\text{Work output}}{\text{Heat supplied}}$$

$$= \frac{\frac{1}{2}(5-1)(700-300)}{\frac{1}{2}(5-1)(700-300) + 300(5-1)}$$

$$= \mathbf{0.4}$$

3.100 (c)

$$\eta_1 = \frac{W_1}{Q_1}, \quad \eta_2 = \frac{W_2}{Q_2}$$

$$\eta_{combined} = \frac{W_1 + W_2}{Q_1}$$

Also, $$\eta = 1 - \frac{Q_2}{Q_1}$$

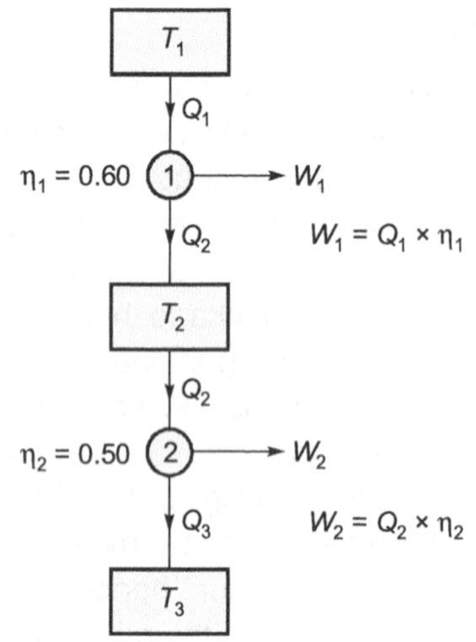

$$\Rightarrow \quad \frac{Q_2}{Q_1} = 1 - \eta_1$$

$$Q_2 = Q_1(1 - \eta_1)$$

$$\eta_{combined} = \frac{(Q_1 \times \eta_1) + (Q_2 \times \eta_2)}{Q_1}$$

$$= \frac{(Q_1 \times \eta_1) + [Q_1 \times (1 - \eta_1)\eta_2]}{Q_1}$$

$$= \eta_1 + \eta_2 - \eta_1\eta_2$$

$$= 0.60 + 0.50 - 0.6 \times 0.50$$

$$= 0.80 = 80\%$$

3.101 (*)

1. Entropy is related to the second law of thermodynamics.
2. The internal energy of an ideal gas is function of temperature only.
   Therefore only 3rd statement is correct.

3.104 (b)

$$\eta = 1 - \frac{300}{1800} = 0.833 = \frac{400}{Q_s}$$

$$\Delta S = \frac{480}{1800} = 0.266 \text{ kW/K},$$

$$Q_s = 480 \text{ kW}$$

3.105 (c)

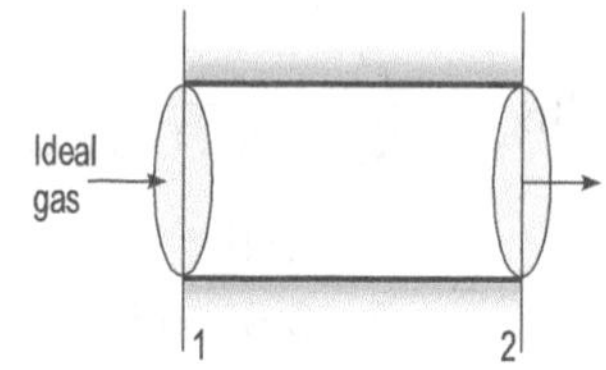

$\dot{m} = 3$ kg/s; $P_2 = 0.9P_1$

$T_o = 300$ K; $R = 0.287$ kJ/kgK

Applying SFEE between section 1 and 2.

$$h_1 + \frac{C_1^2}{2} + gz_1 + q = h_2 + \frac{C_2^2}{2} + gz_2 + w_{cv}$$

$$[q = 0, \ w_{cv} = 0]$$

Neglecting change in KE and PE

$$\Rightarrow \quad h_1 = h_2$$

$$C_pT_1 = C_pT_2$$

$$T_1 = T_2$$

The rate of energy loss,

$$I = T_o \times (\Delta S)$$

$$= 300 \times \dot{m}\left[c_p \ln\frac{T_2}{T_1} - R\ln\frac{P_2}{P_1}\right]$$

$$[T_2 = T_1]$$

$$I = 300 \times 3 \times (-0.287)\ln\frac{0.9P_1}{P_1}$$

$$= 300 \times 3 \times (-0.287)\ln(1 - 0.1)$$

$$\dot{I} = 300 \times 3 \times (-0.287) \times (-0.1)$$

$$= 25.83 \text{ kW} \quad [\ln(1 - x) \approx (-x)]$$

3.106 (b)

$$T_H = 800° + 273 = 1073 \text{ K}$$

$$T_L = 30 + 273 = 303 \text{ K}$$

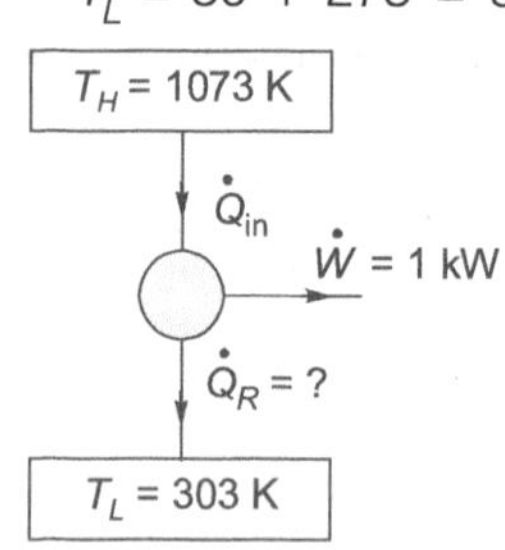

For least $\dot{Q}_R$ efficiency should be maximum

Rate of heat rejection ($\dot{Q}_R$)

$$\therefore \quad \eta_{max} = \frac{\dot{W}}{\dot{Q}_{in}}$$

$$1 - \frac{303}{1073} = \frac{\dot{W}}{\dot{W} + \dot{Q}_R} = \frac{1}{1 + \dot{Q}_R}$$

According to 1st law, $[Q_{in} = \dot{W} + \dot{Q}_R]$

$$0.7176 = \frac{1}{1 + \dot{Q}_R}$$

$$1 + \dot{Q}_R = 1.393$$

$$\dot{Q}_R = 0.393 \text{ kW} \approx 0.4 \text{ kW}$$

3.109 (d)

Given: Temperature of hot reservoir,

$$T_H = 2273 \text{ K}$$

Temperature of cooling water, $T_L = 283$ K

$$\eta = 1 - \frac{283}{2273} = 0.8754$$

3.110 (c)

Given : $P_{motor} = 48$ kW (95% efficiency);

$T_{w0} = 40°C$; $T_{wi} = 30°C$; $\dot{V}_w = 4 L/s$;

$C_{PW} = 4.1868$ kJ/kgK

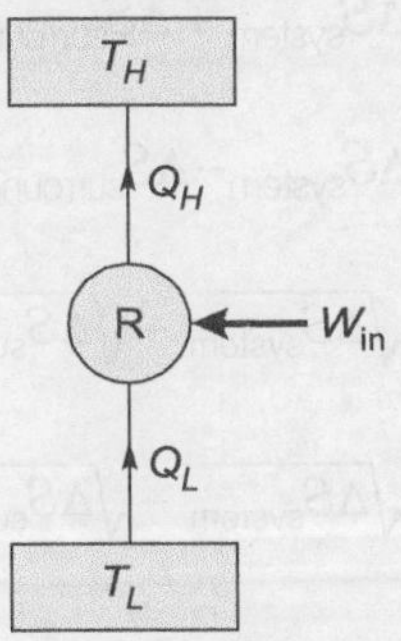

$$Q_H = \dot{m} C_p \Delta T$$

$$= \rho V C_{PW} \Delta T - \rho V C_{PW} (T_{wo} - T_{wi})$$

$$= 1000 \times 4 \times 10^{-3} \times 4.1868 \times (40 - 30)$$

$$= 167.472 \text{ kW}$$

$$\dot{W}_{in} = 48 \times 0.95 = 45.6 \text{ kW}$$

$$\therefore \quad Q_L = Q_H - W_{in} = 167.472 - 45.6$$

$$= 121.872 \text{ kW} \simeq 121.9 \text{ kW}$$

3.111 (a)

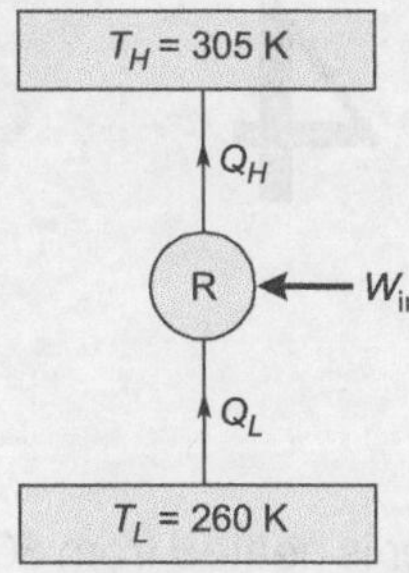

Given : $T_L = 260$K; $T_H = 305$K

$$\text{COP}_{\text{Ref}} = \frac{T_L}{T_H - T_L} = \frac{260}{305 - 260}$$

$$= 5.778$$

3.112 (d)

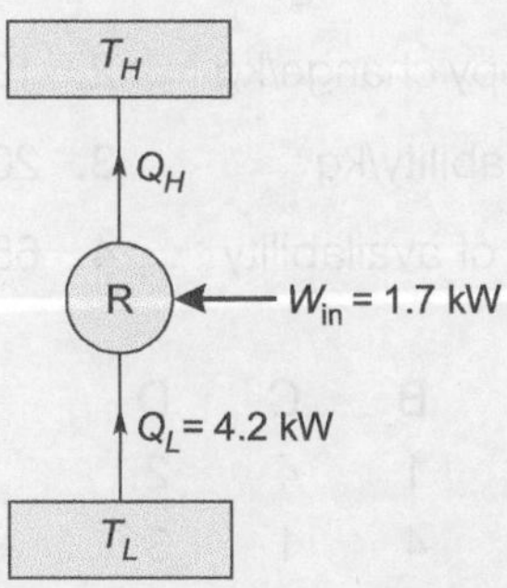

$$Q_H = Q_L + W_{in}$$

$$= 4.2 + 1.7 = 5.9 \text{ kW}$$

$$\therefore \quad Q_H = \dot{m}_w C_{pw} (T_0 - T_i)$$

$$5.9 = \dot{m}_w \times 4.187 \times (24 - 13)$$

$$\Rightarrow \quad \dot{m}_w = \frac{5.9}{4.187 \times 11}$$

$$= 0.128 \text{ kg/s} \simeq 0.13 \text{ kg/s}$$

Note: There is an error in question given by UPSC; the term "motor input" should be used instead of "motor output".

■■■■

4 Irreversibility and Availability

4.1 10 kg of water is heated from 300 K to 350 K in an insulated tank due to churning action by a stirrer. The ambient temperature is 300 K. In this context, match **List-I** with **List-II** and select the correct answer using the codes given below the lists:

List-I	List-II
A. Enthalpy change	1. 12.2 kJ/kg
B. Entropy change/kg	2. 1968 kJ
C. Availability/kg	3. 2090 kJ
D. Loss of availability	4. 656 J/kg-K

Codes:

	A	B	C	D
(a)	3	1	4	2
(b)	2	4	1	3
(c)	3	4	1	2
(d)	2	1	4	3

[ESE : 2000]

4.2 Availability function for a closed system is expressed as:

(a) $\phi = u + p_0V - T_0s$

(b) $\phi = du + p_0dV + T_0dS$

(c) $\phi = du + p_0\, dV - T_0ds$

(d) $\phi = u + p_0V + T_0s$

[ESE : 2002]

4.3 For a steady flow process from state 1 to 2, enthalpy changes from h_1 = 400 kJ/kg to h_2=100 kJ/kg and entropy changes from s_1 = 1.1 kJ/kgK to s_2 = 0.7 kJ/kgK. Surrounding environmental temperature is 300 K Neglect changes in kinetic and potential energy. The change in availability of the system is

(a) 420 kJ/kg (b) 300 kJ/kg
(c) 180 kJ/kg (d) 90 kJ/kg

[ESE : 2003]

4.4 What will be the loss of available energy associated with the transfer of 1000 kJ of heat from constant temperature system at 600 K to another at 400 K when the environment temperature is 300 K?

(a) 150 kJ (b) 250 kJ
(c) 500 kJ (d) 700 kJ

[ESE : 2004]

4.5 The irreversibility is defined as the difference of the maximum useful work and actual work: $I = W_{\text{max useful}} - W_{\text{actual}}$. How can this be alternatively expressed?

(a) $I = T_0(\Delta S_{\text{system}} + \Delta S_{\text{surrounding}})$

(b) $I = T_0(\Delta S_{\text{system}} - \Delta S_{\text{surrounding}})$

(c) $I = T_0\left(\sqrt{\Delta S_{\text{system}}} + \sqrt{\Delta S_{\text{surrounding}}}\right)$

(d) $I = T_0\left(\sqrt{\Delta S_{\text{system}}} - \sqrt{\Delta S_{\text{surrounding}}}\right)$

[ESE : 2005]

4.6 **Assertion (A):** The change in availability of a system is equal to the change in the Gibbs function of the system at constant temperature and pressure.

Reason (R): The Gibbs function is useful when evaluating the availability of systems in which chemical reactions occur.

(a) Both **A** and **R** are individually true and **R** is the correct explanation of **A**
(b) Both **A** and **R** are individually true but **R** is not the correct explanation of **A**
(c) **A** is true but **R** is false
(d) **A** is false but **R** is true

[ESE : 2006]

4.7 The loss of available energy associated with the transfer of 1000 kJ of heat from a constant temperature system at 600 K to another at 400 K when the environmental temperature is 300 K is

(a) 150 kJ (b) 250 kJ
(c) 166.67 kJ (d) 180 kJ

[ESE : 2008]

4.8 100 kJ of energy is transferred from a heat reservoir at 1000 K to a heat reservoir at 500 K. The ambient temperature is 300 K. The loss of available energy due to heat transfer process is

(a) 20 kJ (b) 30 kJ
(c) 40 kJ (d) 50 kJ

[ESE : 2012]

4.9 Which of the following processes are irreversible?

1. Heat transfer through a finite temp. difference
2. Free expansion
3. Transfer of electricity through a resistor

(a) 1, 2 and 3 (b) 1 and 2 only
(c) 2 and 3 only (d) 1 and 3 only

[ESE : 2013]

4.10 A body of mass 2 kg and c_p = 1.00 kJ/kgK is available at 600 K. If the atmosphere is 300 K and ln 2 = 0.693, the maximum work obtainable from the body till it comes to equilibrium with the atmosphere is

(a) 150 kJ (b) 142 kJ
(c) 184.2 kJ (d) 190.5 kJ

[ESE : 2015]

4.11 Consider the following statements regarding availability:

1. It is generally conserved
2. It can either be negative or positive
3. It is the maximum theoretical worked obtainable
4. It can be destroyed in irreversibility

Which of the above statements are correct?

(a) 3 and 4 (b) 1 and 2
(c) 1 and 3 (d) 2 and 4

[ESE : 2015]

4.12 During a thermodynamic process, 100 kJ of heat is transferred from a reservoir at 800 K to a sink at 400 K. The ambient temperature is 300 K. The loss of available energy is

(a) 27.5 kJ (b) 32.5 kJ
(c) 37.5 kJ (d) 62.5 kJ

[ESE : 2015]

4.13 1000 kJ/s of heat is transferred from a constant temperature heat reservoir maintained at 1000 K to a system at a constant temperature of 500 K. The temperature of the surroundings is 300 K. The net loss of available energy as a result of this heat transfer is :

(a) 450 kJ/s (b) 400 kJ/s
(c) 350 kJ/s (d) 300 kJ/s

[ESE : 2016]

4.14 The effects of heat transfer from a high temperature body to a low temperature body are :

1. The energy is conserved
2. The entropy is not conserved
3. The availability is not conserved

Which of the above statements are correct?

(a) 1 and 2 only (b) 1 and 3 only
(c) 2 and 3 only (d) 1, 2 and 3

[ESE : 2016]

4.15 An ideal gas is flowing through an insulated pipe at the rate of 3.3 kg/s. There is a pressure drop of 15% from the inlet to the outlet of the pipe. What is the rate of exergy loss because of this pressure drop due to friction, given that R_{gas} = 0.287 kJ/kgK and the reference temperature T_0 is 300 K?

(a) 42.62 kW (b) 40.26 kW
(c) 38.14 kW (d) 35.13 kW

[ESE : 2017]

4.16 For a steady process, the conditions at stage 1 and stage 2 are, respectively, h_1 = 300 kJ/kg, h_2 = 150 kJ/kg, s_1 = 1.25 kJ/kgK and s_2 = 0.8 kJ/kgK. The 'availability' at the ambient temperature 300 K will be

(a) 15 kJ (b) 20 kJ
(c) 25 kJ (d) 35 kJ

[ESE : 2017]

4.17 A heat reservoir is maintained at 927°C. If the ambient temperature is 27°C, the availability of heat from the reservoir is limited to

(a) 57% (b) 66%
(c) 75% (d) 88%

[ESE : 2018]

4.18 When 25 kg of water at 95°C is mixed with 35 kg of water at 35°C, the pressure being taken as constant at surrounding temperature of 15°C, and c_p of water is 4.2 kJ/kgK, the decrease in available energy due to mixing will be nearly

(a) 270.5 kJ (b) 277.6 kJ
(c) 281.8 kJ (d) 288.7 kJ

[ESE : 2019]

■■■■

Answers Irreversibility and Availability

4.1 (c)	4.2 (a)	4.3 (c)	4.4 (b)	4.5 (a)	4.6 (b)	4.7 (b)	4.8 (b)	4.9 (a)
4.10 (c)	4.11 (a)	4.12 (c)	4.13 (d)	4.14 (d)	4.15 (a)	4.16 (a)	4.17 (c)	4.18 (c)

Explanations Irreversibility and Availability

4.1 (c)

Given data:

$m = 10$ kg, $T_1 = 300$ K, $T_2 = 350$ K, $T_0 = 300$ K

Enthalpy change,

$$dH = mc_p(T_2 - T_1)$$
$$= 10 \times 4.18(350 - 300) = 2090 \text{ kJ}$$

Entropy change per kg,

$$\Delta s = c_p \log_e \frac{T_2}{T_1} = 4.18 \log_e \frac{350}{300}$$
$$= 0.64435 \text{ kJ/kgK} = 644.35 \text{ J/kgK}$$

$$\text{Availability/kg} = Q - T_0\,\Delta s$$
$$= c_p(T_2 - T) - 300 \times 644.35$$
$$= 4.18(350 - 300) - 300 \times 0.64435$$
$$= 15.69 \text{ kJ/kg}$$

Loss of availability,

$$I = T_0 \Delta S_{uni} = T_0 \times m \times \Delta s$$
$$= 300 \times 10 \times 0.64435 = \mathbf{1933\ kJ}$$

4.2 (a)

$$\phi = u + p_o V - T_o s$$

4.3 (c)

Availability for a flow process

$= (h_1 - h_2) - T_0(s_1 - s_2)$

$= (400 - 100) - 300\,(1.1 - 0.7)$

$= 300 - 120 = \mathbf{180\ kJ/kg}$

4.4 (b)

Given data:

$Q = 1000$ kJ, $T_1 = 600$ K, $T_2 = 400$ K, $T_0 = 300$ K

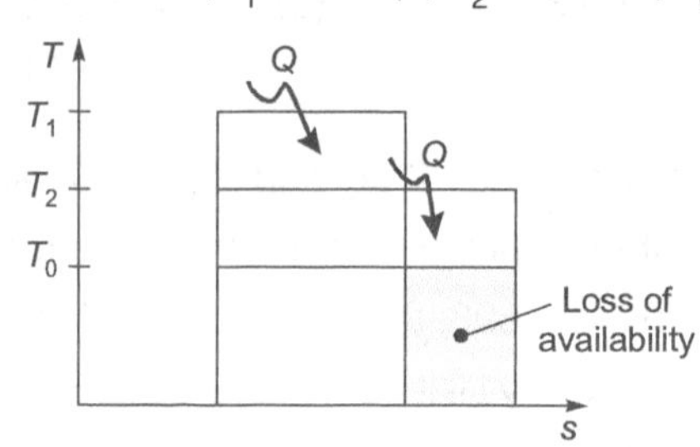

Loss of available energy,

$$I = T_0[\Delta S_{uni}] = T_0[\Delta S_{sys} + \Delta S_{sur}]$$
$$= T_0\,\Delta S_{sys} \qquad \because \Delta S_{sur} = 0$$
$$= T_0\left[-\frac{Q}{T_1} + \frac{Q}{12}\right]$$
$$= 300\left[-\frac{1000}{600} + \frac{1000}{400}\right]$$
$$= 300[-1.667 + 2.5] = 249.9 \text{ kJ}$$
$$\simeq \mathbf{250\ kJ}$$

4.5 (a)

Irreversibility,

$$I = T_0\,(\Delta s)_{universe}$$
$$= T_0\,(\Delta s_{system} + \Delta s_{surrounding})$$

4.7 (b)

Given data:

$Q = 1000$ kJ, $T_1 = 600$ K, $T_2 = 400$ K, $T_0 = 300$ K

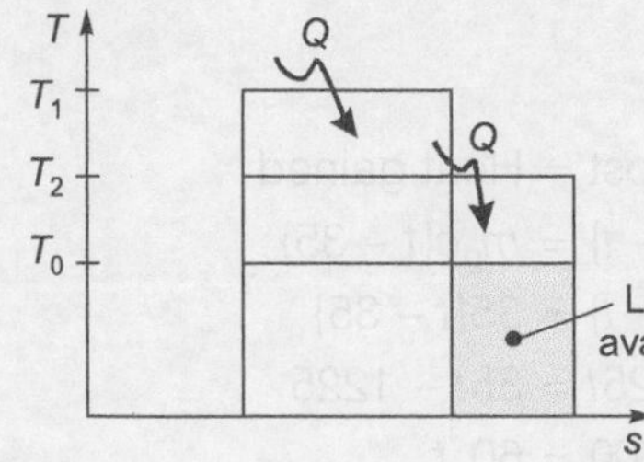

Loss of available energy,

$$I = T_0[\Delta S_{uni}] = T_0[\Delta S_{sys} + \Delta S_{sur}]$$

$$= T_0\,\Delta S_{sys} \qquad \because \Delta S_{sur} = 0$$

$$= T_0\left[-\frac{Q}{T_1}+\frac{Q}{12}\right]$$

$$= 300\left[-\frac{1000}{600}+\frac{1000}{400}\right]$$

$$= 300[-1.667 + 2.5] = 249.9 \text{ kJ}$$

$$\simeq \mathbf{250\ kJ}$$

4.8 (b)

Given data:

$Q = 100$ kJ, $T_1 = 1000$ K, $T_2 = 500$ K, $T_0 = 300$ K

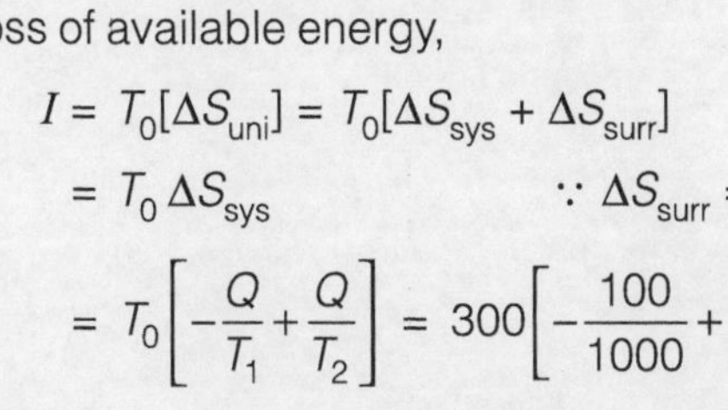

Loss of available energy,

$$I = T_0[\Delta S_{uni}] = T_0[\Delta S_{sys} + \Delta S_{surr}]$$

$$= T_0\,\Delta S_{sys} \qquad \because \Delta S_{surr} = 0$$

$$= T_0\left[-\frac{Q}{T_1}+\frac{Q}{T_2}\right] = 300\left[-\frac{100}{1000}+\frac{100}{500}\right]$$

$$= 300\,[-0.10 + 0.2] = \mathbf{30\ kJ}$$

4.9 (a)

The irreversibility of a process may be due to either one or both of the following:

(a) Lack of equilibrium during the process

(b) Involvement of dissipative effect.

So, all processes (1), (2) and (3) are irreversible.

4.10 (c)

$$W_{max} = Q - UE$$

$$= m[c_p\cdot\Delta T - T_0\cdot\Delta s]$$

$$= 2\left[1\times 300 - 300\times 1\times \ln\frac{600}{300}\right]$$

$$= \mathbf{184.2\ kJ}$$

$$\left(\because \Delta s = c_p \ln\frac{T_2}{T_1}\right)$$

4.11 (a)

- Availability is the maximum theoretical work obtainable, when a system passes from a given state to dead state, while interacting with the atmosphere only.
- Availability associated with a given amount of energy is higher at higher temperature and lower at lower temperature. If the temperature at which energy is stored decreases, the availability in that given amount of energy will also decrease. It means availability can be destroyed in irreversibility.
- Availability cannot be negative and also cannot be conserved. It is an extensive property.

4.12 (c)

Given data:

$Q = 100$ kJ, $T_1 = 800$ K, $T_2 = 400$ K, $T_0 = 300$ K

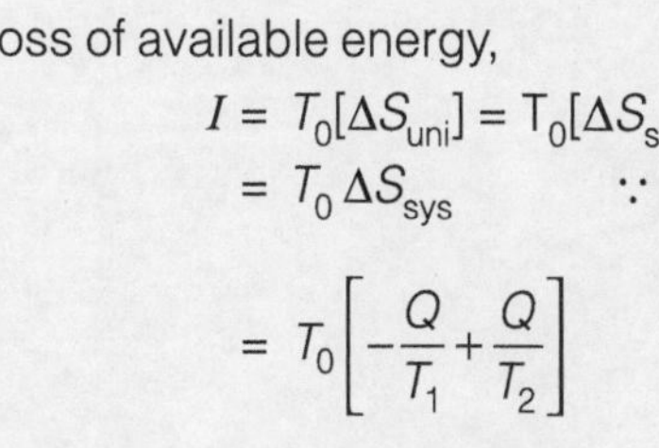

Loss of available energy,

$$I = T_0[\Delta S_{uni}] = T_0[\Delta S_{sys} + \Delta S_{surr}]$$

$$= T_0\,\Delta S_{sys} \qquad \because \Delta S_{surr} = 0$$

$$= T_0\left[-\frac{Q}{T_1}+\frac{Q}{T_2}\right]$$

$$= 300\left[-\frac{100}{800}+\frac{100}{400}\right]$$

$$= 300\,[-0.125 + 0.25] = \mathbf{37.5\ kJ}$$

4.13 (d)

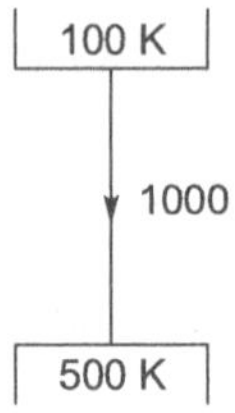

$T_0 = 300$ K

Loss of *A.E.*

$$= \underbrace{1000\left\{1-\frac{300}{1000}\right\}}_{AE_1} - \underbrace{1000\left\{1-\frac{300}{500}\right\}}_{AE_2}$$

$= 700 - 400 =$ **300 kJ**

4.15 (a)

$$\dot{I} = T_0 \dot{S}_{gen}$$

$$= 300 \times \dot{m}(s_2 - s_1)$$

$$= 300 \times 3.3\left(C_p \ln\frac{T_2}{T_1} - R\ln\frac{P_2}{P_1}\right)$$

$T_2 = T_1$ ($\because h_1 = h_2$ from SFEE)

$P_2 = 0.85P_1$

$$\dot{I} = 300 \times 3.3(0 - R\ln 0.85)$$

$$= 300 \times 3.3(-R\ln(1 - 0.15))$$

Approximating $\ln(1 - x) \approx -x$

$\therefore \quad \dot{I} = 300 \times 3.3 \times (-0.287) \times (-0.15)$

$\dot{I} = 42.619$ kJ/sec

4.17 (c)

$$\eta = 1 - \frac{300}{1200} = 75\%$$

4.18 (c)

Heat lost = Heat gained

$m_1 c\{95 - t\} = m_2 c\{t - 35\}$

$25\{95 - t\} = 35\{t - 35\}$

$2735 - 25t = 35t - 1225$

$3600 = 60\,t$

$t = 60°C$

$T = 333$ K

$$S_{gen} = \Delta S_1 + \Delta S_2$$

$$= m_1 c \ln\frac{T}{T_1} + m_2 c \ln\frac{T}{T_2}$$

$$= 25 \times 4.2 \ln\frac{333}{273+95} + 35 \times 4.2 \ln\frac{333}{273+35}$$

$= -10.4938 + 11.472$

$= 0.9784$ kJ/K

Loss of A.E.

$=$ Irreversibility $= T_0 S_{gen}$

$= 288 \times 0.9784 = 281.80$ kJ

■■■■

5 Thermodynamic Relations

5.1 Consider the following thermodynamic relation

1. $Tds = du + pdV$
2. $Tds = du - pdV$
3. $Tds = dh + Vdp$
4. $Tds = dh - Vdp$

Which of these thermodynamic relations are correct?

(1) 1 and 3 (b) 1 and 4
(c) 2 and 3 (d) 2 and 4

[ESE : 2000]

5.2 The Clapeyron equation with usual notations is given by

(a) $\left(\frac{dT}{dp}\right)_{sat} = \frac{h_{fg}}{TV_{fg}}$ (b) $\left(\frac{dp}{dT}\right)_{sat} = \frac{h_{fg}}{TV_{fg}}$

(c) $\left(\frac{dT}{dp}\right)_{sat} = \frac{Th_{fg}}{TV_{fg}}$ (d) $\left(\frac{dp}{dT}\right)_{sat} = \frac{Th_{fg}}{V_{fg}}$

[ESE : 2000]

5.3 Match **List-I** (Name of entity) with **List-II** (Definition) and select the correct answer given below the lists:

List-I

A. Compressibility factor
B. Joule-Thomson coefficient
C. Constant pressure specific heat
D. Isothermal compressibility

List-II

1. $-\frac{1}{V}\left(\frac{\partial V}{\partial p}\right)_T$ 2. $\left(\frac{\partial h}{\partial T}\right)_p$

3. $\left(\frac{\partial T}{\partial p}\right)_h$ 4. $\left(\frac{pV}{RT}\right)$

Codes:

	A	B	C	D
(a)	2	1	4	3
(b)	4	3	2	1
(c)	2	3	4	1
(d)	4	1	2	3

[ESE : 2001]

5.4 Consider the following statements:

1. Availability is the maximum theoretical work obtainable.
2. Clapeyron's equation for dry saturated steam is given by

$$v_g - v_f = \frac{dT_s}{dp}\left(\frac{h_g - h_f}{T_s}\right).$$

3. A gas can have any temperature at a given pressure unlike a vapour which has a fixed temperature at a given pressure.
4. Joule Thomson coefficient is expressed as $\mu = (\partial s/\partial p)_h$.

Which of these statements are correct?

(a) 1, 2 and 3 (b) 1, 3 and 4
(c) 2 and 3 (d) 1, 2 and 4

[ESE : 2001]

5.5 Which one of the following functions represents the Clapeyron equation pertaining to the change of phase of a pure substance?

(a) $f(T, p, h_{fg})$ (b) $f(T, p, h_{fg}, V_{fg})$
(c) $f(T, p, h_{fg}\ s_{fg})$ (d) $f(T, p, h_{fg}, s_{fg}, V_{fg})$

[ESE : 2002]

5.6 Tds equation can be expressed as

(a) $TdS = C_v\,dT + \frac{T\beta dV}{K}$

(b) $TdS = C_v\,dT + \frac{T}{K}dV$

(c) $TdS = C_v\,dT + \frac{TK}{\beta}dV$

(d) $TdS = C_v\,dT + \frac{T\beta}{K}dp$

[ESE : 2002]

5.7 Match **List-I** (Terms) with **List-II** (Relations) and select the correct answer using the codes given below the lists:

List-I

A. Specific heat at constant volume (C_v)
B. Isothermal compressibility (K_T)
C. Volume expansivity (β)
D. Difference between specific heat at constant pressure and at constant volume ($C_p - C_v$)

List-II

1. $\frac{1}{V}\left(\frac{\partial V}{\partial T}\right)_p$
2. $-T\left(\frac{\partial p}{\partial V}\right)_T\left(\frac{\partial V}{\partial T}\right)_p^2$
3. $T\left(\frac{\partial s}{\partial T}\right)_V$
4. $-\frac{1}{V}\left(\frac{\partial V}{\partial p}\right)_T$

Codes:

	A	B	C	D
(a)	3	4	2	1
(b)	4	1	3	2
(c)	3	4	1	2
(d)	4	1	2	3

[ESE : 2003]

5.8 Van der Waals equation of state is given by $(p + a/V^2)(V - b) = RT$. The constant b in the equation in terms of specific volume at critical point v_c is equal to

(a) $V_c/3$
(b) $2\,V_c$
(c) $3\,V_c$
(d) $8a/(27\,V_cR)$

[ESE : 2003]

5.9 **Assertion (A):** Throttling process for real gases at initial temperature higher than maximum inversion temperature is accompanied by decrease in temperature of the gas .

Reason (R): Joule-Kelvin coefficient μ_j is given by $(\partial T/\partial p)_h$ and should have a positive value for decrease in temperature during throttling process.

(a) both **A** and **R** are true and **R** is the correct explanation of **A**
(b) both **A** and **R** are true but **R** is not a correct explanation of **A**
(c) **A** is true but **R** is false
(d) **A** is false but **R** is true

[ESE : 2003]

5.10 The variation of saturation pressure with saturation temperature for a liquid is 0.1 bar/K at 400 K. The specific volume of saturated liquid and dry saturated vapour at 400 K are 0.251 m^3/kg and 0.001 m^3/kg. The value of latent heat of vaporization using Clausius Clapeyron equation is

(a) 16000 kJ/kg
(b) 1600 kJ/kg
(c) 1000 kJ/kg
(d) 160 kJ/kg

[ESE : 2004]

5.11 The equation relating the following measurable properties:

(i) the slope of saturation pressure temperature line
(ii) the latent heat and
(iii) the change in volume during phase transformation is known as

(a) Maxwell relation
(b) Joules equation
(c) Clapeyron equation
(d) None of the above

[ESE : 2005]

5.12 Consider the following statements pertaining to the Clapeyron equation

1. It is useful to estimate properties like enthalpy from other measurable properties.
2. At a change of phase, it can be used to find the latent heat at a given pressure.
3. It is derived from the relationship

$$\left(\frac{\partial p}{\partial V}\right)_T = \left(\frac{\partial s}{\partial T}\right)_V$$

Which of these statements are correct?

(a) 1, 2 and 3
(b) 1 and 2
(c) 1 and 3
(d) 2 and 3

[ESE : 2006]

5.13 According to the Clapeyron equation, which one of the following gives the slope of the saturated liquid pressure-temperature curve $\frac{dp}{dT}$?
(suffix f and g denote saturated liquid and saturated vapour respectively)

(a) $\frac{(u_g - u_f)}{(v_g - v_f)}$
(b) $\frac{(h_g - h_f)}{(v_g - v_f)}$
(c) $\frac{(u_g - u_f)}{T(v_g - v_f)}$
(d) $\frac{(h_g - h_f)}{T(v_g - v_f)}$

[ESE : 2006]

5.14 The throttling of certain gases may be used for getting the refrigeration effect. The value of Joule-Thomson coefficient (μ) for such a throttling process is

(a) $\mu = 0$ (b) $\mu = 1$
(c) $\mu < 1$ (d) $\mu > 1$

[ESE : 2007]

5.15 Which one of the following statements is correct?

(a) Compressibility factor is unity for ideal gases
(b) Compressibility factor is zero for ideal gases
(c) Compressibility factor is lesser than unity for ideal gases
(d) Compressibility factor is more than unity for ideal gases.

[ESE : 2007]

5.16 Which thermodynamic property is evaluated with the help of Maxwell equations from the data of other measurable properties of a system?

(a) Enthalpy (b) Entropy
(c) Latent heat (d) Specific heat

[ESE : 2007]

5.17 Consider the following statements in respect of the Clausius - Clapeyron equation:

1. It points to one possible way of measuring thermodynamic temperature.
2. It permits latent heat of vaporization to be estimated from measurements of specific volumes of saturated liquid, saturated vapour and the saturation temperature at two nearby pressures.
3. It does not apply to changes from solid to the liquid phase and from solid to the vapour phase.

Which of these statements are correct?

(a) 1, 2 and 3 (b) 1 and 2 only
(c) 2 and 3 only (d) 1 and 3 only

[ESE : 2007]

5.18 Which one of the following relationships defines the Helmholtz function F?

(a) $F = H + TS$ (b) $F = H - TS$
(c) $F = U - TS$ (d) $F = U + TV$

[ESE : 2007]

5.19 When a real gas undergoes Joule-Thomson expansion the temperature

(a) may remain constant
(b) always increases
(c) may increases or decreases
(d) always decreases

[ESE : 2007]

5.20 At critical point the enthalpy of vaporization is

(a) dependent on temperature only
(b) maximum
(c) minimum
(d) zero

[ESE : 2008]

5.21 Which one of the following parameters is significant to ascertain chemical equilibrium of a system?

(a) Clapeyron relation
(b) Maxwell relation
(c) Gibbs function
(d) Helmholtz function

[ESE : 2009]

5.22 Clapeyron's equation is used for finding out the

(a) Dryness fraction of steam only
(b) Entropy of superheated vapour only
(c) Specific volume at any temperature and pressure
(d) Total heat of superheated steam only

[ESE : 2011]

5.23 When a system reaches the state of equilibrium, the following property assumes its maximum value

(a) Availability (b) Entropy
(c) Gibbs function (d) Helmholtz function

[ESE : 2012]

5.24 The difference between constant pressure specific heat C_p and constant volume specific heat C_v for pure substance

(a) approaches zero at triple point
(b) approaches zero as the absolute temperature approaches zero
(c) is always equal to the gas constant R
(d) approaches zero at critical point

[ESE : 2012]

5.25 Joule-Thomson coefficient is the slope of

(a) constant enthalpy lines on T-s diagram
(b) constant enthalpy lines on T-p diagram
(c) inversion curve on T-s diagram
(d) inversion curve on T-p diagram

[ESE : 2012]

5.26 Which gas shows a heating effect in the Joule-Thomson experiment, while undergoing throttling process through a porous plug of cotton wool?

(a) Carbon dioxide (b) Hydrogen
(c) Oxygen (d) Nitrogen

[ESE : 2012]

5.27 It is very useful in predicting the properties of gases for which more precise data are not available but their critical properties are known
(a) Van der Waal's equation of state
(b) Breattie-Bridgeman equation of state
(c) Redlich-Kwong equation of state
(d) Generalized compressibility chart

[ESE ; 2013]

5.28 In a real gas equation $pv = zRT$, depending on the values of pressure and temperature of the real gas, the value of z
(a) Should always be less than 1
(b) May be less than 1, may be greater than 1 or equal to 1
(c) Should always be greater than 1
(d) Should always be equal to 1 [ESE : 2014]

Direction: The following items consists of two statements, one labelled as **Assertion 'A'** and the other labelled as **Reason 'R'**. You are to examine these two statements carefully and select your answers to these items using the codes given below:

Codes:
(a) Both **A** and **R** are true and **R** is the correct explanation of **A**
(b) Both **A** and **R** are true but **R** is not a correct explanation of **A**
(c) **A** is true but **R** is false
(d) **A** is false but **R** is true

5.29 Assertion (A) : Property tables list different values of some properties for a substance at the different state as a result of using same reference states.

Reason (R): The reference state chosen is of no consequence in thermodynamic process calculations as long as we use values from the single consistent set of tables.

[ESE : 2014]

5.30 Assertion (A): The Clapeyron equation enables us to determine the enthalpy change associated with phase change.

Reason (R): Using usual notations, the Clapeyron equation is given by $\left(\frac{dT}{dP}\right)_{sat} = \frac{h_{fg}}{Tv_{fg}}$.

[ESE : 2015]

5.31 The vapour pressure of a liquid at any arbitrary temperature can be estimated approximately with the help of
(a) Gibbs equation
(b) Joule-Kelvin equation
(c) Clausius-Clapeyron equation
(d) Gibbs-Duhem equation [ESE : 2015]

5.32 Consider the following statements pertaining to Clapeyron equation :
1. It is useful in estimating properties like enthalpy from other measurable properties
2. At a change of phase, it can be used to find the latent heat at a given pressure
3. It is derived from the relationship

$$\left(\frac{\partial p}{\partial v}\right)_T = \left(\frac{\partial s}{\partial T}\right)_v$$

Which of the above statements are correct?
(a) 1 and 3 only (b) 2 and 3 only
(c) 1 and 2 only (d) 1, 2 and 3

[ESE : 2016]

■■■■

Answers Thermodynamic Relations

5.1 (b) 5.2 (b) 5.3 (b) 5.4 (a) 5.5 (b) 5.6 (a) 5.7 (c) 5.8 (a) 5.9 (d)
5.10 (c) 5.11 (c) 5.12 (b) 5.13 (d) 5.14 (d) 5.15 (a) 5.16 (b) 5.17 (b) 5.18 (c)
5.19 (c) 5.20 (d) 5.21 (c) 5.22 (c) 5.23 (b) 5.24 (b) 5.25 (b) 5.26 (b) 5.27 (d)
5.28 (b) 5.29 (a) 5.30 (c) 5.31 (c) 5.32 (c)

Explanations Thermodynamic Relations

5.1 (b)

We know that

$$\delta q = du + \delta w$$

⇒ For reversible process (closed system)

$$\Rightarrow \text{but } ds = \frac{\delta Q}{T}$$

$$\Rightarrow \boxed{T \cdot ds = du + p \cdot dv}$$

We know that

$$h = u + pV$$

$$dh = \underbrace{du + p \cdot dV}+ V \cdot dp$$

$$\Rightarrow \quad dh = \delta Q + V \cdot dp$$

$$\Rightarrow \quad \delta Q = dh - V \cdot dp$$

$$\Rightarrow \boxed{T \cdot ds = dh - V \cdot dp}$$

5.2 (b)

The Clapeyron equation is $\left(\frac{dp}{dT}\right)_{\text{sat}} = \frac{h_{fg}}{T \cdot v_{fg}}$

It enables us to determine the enthalpy of vaporization h_{fg} at a given temperature by simply measuring the slope of saturation curve on a p-T diagram and the specific volume of saturated liquid and saturated vapour at given temperature.

5.3 (b)

Compressibility factor: $Z = \frac{pV}{RT}$

Joule Thomson coefficient,

$$\mu_j = \left(\frac{\partial T}{\partial p}\right)_h; \quad c_p = \left(\frac{\partial h}{\partial T}\right)_p$$

5.4 (a)

Availability of a given system is defined as the maximum useful of the work that can be obtained in a process in which system comes to equilibrium with the surrounding or attains dead state.

5.6 (a)

$$Tds = c_v dT + T\left(\frac{\partial p}{\partial T}\right)_v dv \quad \text{...(i)}$$

$$\beta = \frac{1}{V}\left(\frac{\partial v}{\partial T}\right)_p \quad \text{...(ii)}$$

$$K_T = -\frac{1}{V}\left(\frac{\partial V}{\partial p}\right)_T \quad \text{...(iii)}$$

$$\frac{\beta}{K_T} = -\left(\frac{\partial V}{\partial T}\right)_p \times \left(\frac{\partial p}{\partial V}\right)_T = \left(\frac{\partial p}{\partial T}\right)_v \quad \text{...(iv)}$$

From Eqs. (i) and (iv)

$$\Rightarrow \quad Tds = c_v dT + T\frac{\beta}{K}dv$$

5.7 (c)

Volume expansivity: $\beta = \frac{1}{V}\left(\frac{\partial V}{\partial T}\right)_p$

Specific heat at constant volume,

$$c_v = T\left(\frac{\partial s}{\partial T}\right)_v$$

Isothermal compressibility

$$K_T = \frac{-1}{V}\left(\frac{\partial V}{\partial p}\right)_T$$

$$C_p - C_v = -T\left(\frac{\partial V}{\partial T}\right)_p^2\left(\frac{\partial p}{\partial V}\right)_T$$

5.8 (a)

In Van der Waal's equation

$$\left(p + \frac{a}{V^2}\right)(V - b) = RT$$

where constant, $a = 3p_c V_c^2$ and $b = \frac{V_C}{3}$.

5.9 (d)

The fluid temperature will decrease during throttling process that takes place on the left hand side of inversion line.

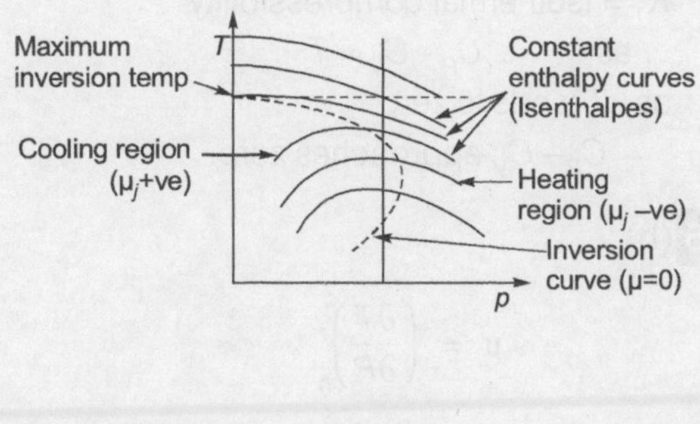

5.10 (c)

$$\frac{dp}{dT} = \frac{h_{fg}}{T(v_g - v_f)}$$

$$10 = \frac{h_{fg}}{400(0.251-0.001)}$$

$$h_{fg} = \mathbf{1000\ kJ/kg}$$

5.15 (a)

The ratio $\frac{p\bar{V}}{RT}$ is called the compressibility factor (Z).

For ideal gas $Z = 1$

5.20 (d)

At critical point enthalpy of vaporization is zero.

5.22 (c)

Clapeyron's equation,

$$\left(\frac{dp}{dT}\right)_{sat} = \frac{h_g - h_f}{T(v_g - v_f)}$$

5.23 (b)

When system reaches equilibrium, driving force that executed the process becomes zero and entropy attains its maximum value. Moreover as the process progress towards equilibrium the available energy or availability decreases continuously.

5.24 (b)

For a pure substance

$$C_P - C_V = \frac{TV\beta^2}{K_T}$$

V = Volume

β = Coefficient of volume expensivity

T = Absolute temperature

K_T = Isothermal compressibility

$\therefore$ as $C_P - C_V \propto T$

as T tends to zero

$C_P - C_V$ approaches zero.

5.25 (b)

$$\mu = \left(\frac{\partial T}{\partial P}\right)_h$$

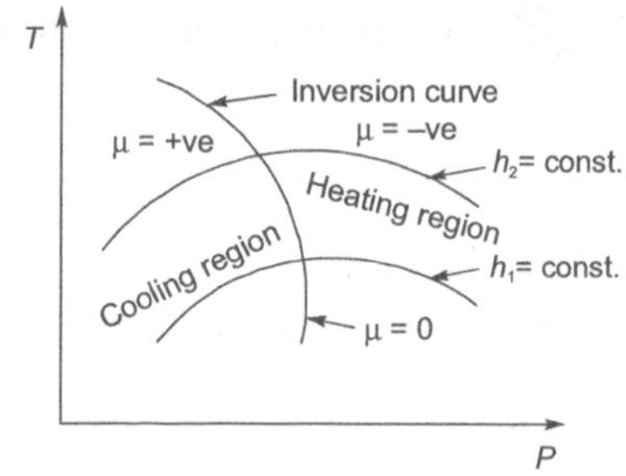

i.e., Joule Thomson coefficient μ is the slope of constant enthalpy lines on T-p diagram.

5.26 (b)

All gases except Hydrogen, Helium and Neon cool upon expansion by the Joule-Thomson process.

5.27 (d)

The compressibility factor Z is a function of P and T. The magnitude of Z for a certain gas at a particular pressure and temperature gives an indication of the extent of deviation of the gas from the ideal gas behaviour. For each substance, there is a compressibility factor chart which is found to be satisfactory for a great variety of substances. These plots are extremely useful where detailed data on a particular gas are lacking but its critical properties are available.

5.30 (c)

Clapeyron equation is given by

$$\left(\frac{dp}{dT}\right)_{sat} = \frac{h_{fg}}{Tv_{fg}}$$

Clapeyron equation is used to determine the enthalpy change associated with phase change.

5.31 (c)

Clausius-Clapeyron equation can be used to estimate approximately the vapor pressure of a liquid at any arbitrary temperature.

5.32 (c)

It should be

$$\Rightarrow \quad \left(\frac{\partial p}{\partial T}\right)_v = \left(\frac{\partial s}{\partial v}\right)_T$$

■■■■

6 Pure Substances

6.1 In which one of the following working substances, does the relation $T_2/T_1 = (p_2/p_1)^{0.286}$ hold if the process takes place with zero heat transfer?

(a) Wet steam

(b) Isentropic

(c) Petrol vapour and air mixture

(d) Air **[ESE : 2000]**

6.2 Consider the phase diagram of a certain substance as shown in the given figure. Match **List-I** (Process) with **List-II** (Curves/lines) and select the correct answer using the codes given below the lists:

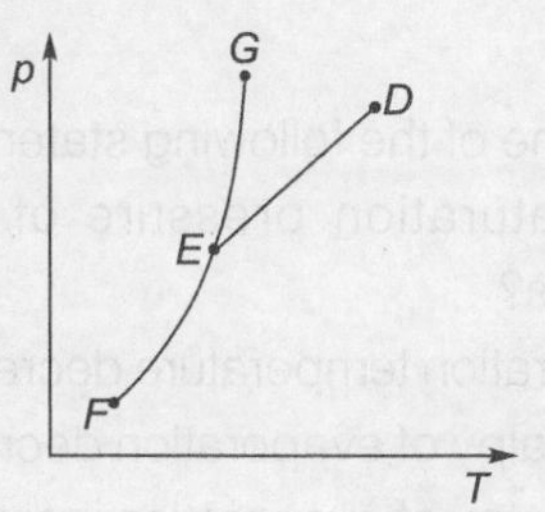

List-I	List-II
A. Vaporization	1. *EF*
B. Fusion	2. *EG*
C. Sublimation	3. *ED*

Codes:

	A	B	C
(a)	1	3	2
(b)	1	2	3
(c)	3	2	1
(d)	3	1	2

[ESE : 2001]

6.3 With increase of pressure, the latent heat of steam

(a) remains same

(b) increases

(c) decreases

(d) behaves unpredictably **[ESE : 2002]**

6.4 Consider the following statements regarding the throttling process of wet steam:

1. The steam pressure and temperature decrease but enthalpy remains constant.
2. The steam pressure decreases, the temperature increases but enthalpy remains constant.
3. The entropy, specific volume, and dryness fraction increase.
4. The entropy increases but the volume and dryness fraction decrease.

Which of these statements are correct?

(a) 1 and 4 (b) 2 and 3

(c) 1 and 3 (d) 2 and 4

[ESE : 2002]

6.5 The internal energy of a gas obeying Van der Waals equation $\left(p+\frac{a}{v^2}\right)(v-b)=RT$ depends on its

(a) temperature

(b) temperature and pressure

(c) temperature and specific volume

(d) pressure and specific volume

[ESE : 2000]

6.6 The process 1-2 for steam shown in the given figure is

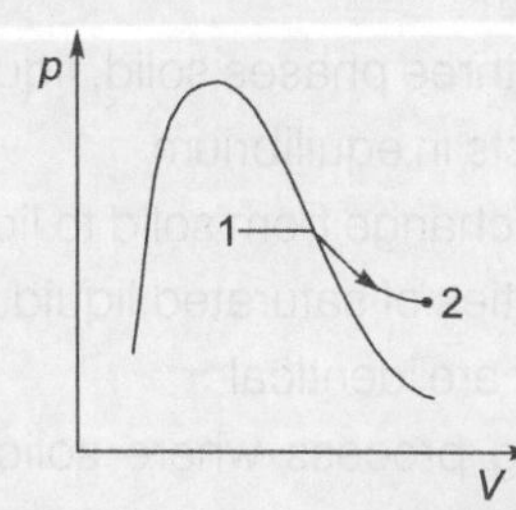

(a) isobaric (b) isentropic

(c) isenthalpic (d) isothermal

[ESE : 2000]

6.7 The value of compressibility factor for an ideal gas may be:

1. Less or more than one
2. Equal to one
3. Zero
4. Less than zero

The correct value(s) is/are given by

(a) 1 and 2 (b) 1 and 4
(c) 3 only (d) 2 only

[ESE : 2002]

6.8 Saturated liquid at a high pressure p_1 having enthalpy of saturated liquid 1000 kJ/kg is throttled to a lower pressure p_2, at pressure p_2 enthalpy of saturated liquid and that of the saturated vapour are 800 and 2800 kJ/kg respectively. The dryness fraction of vapour after throttling process is

(a) 0.1 (b) 0.5
(c) 18/28 (d) 0.8

[ESE : 2003]

6.9 Which one of the following represents the condensation of a mixture of saturated liquid and saturated vapour on the enthalpy-entropy diagram?

(a) A horizontal line
(b) An inclined line of constant slope
(c) A vertical line
(d) A curved line

[ESE : 2004]

6.10 Match **List-I** with **List-II** and select the correct answer using the code given below the lists:

List-I

A. Critical point
B. Sublimation
C. Triple point
D. Melting

List-II

1. All the three phases solid, liquid and vapour co-exists in equilibrium
2. Phase change from solid to liquid
3. Properties of saturated liquid and saturated vapour are identical
4. Heating process where solid gets directly transformed to gaseous phase

Codes:

	A	B	C	D
(a)	2	1	4	3
(b)	3	4	1	2
(c)	2	4	1	3
(d)	3	1	4	2

[ESE : 2005]

6.11 Assertion (A): For a mixture of solid, liquid and vapour phases of a pure substance in equilibrium, the number of independent intrinsic properties needed is equal to one.

Reason (R): The three phases can coexist only at one particular pressure.

[ESE : 2005]

6.12 Which one of the following is the fluid whose properties in all its three phases are made use of in thermodynamics?

(a) Ammonia (b) Freon 12
(c) Helium (d) Water

[ESE : 2007]

6.13 For a pure substance, what are the numbers of the thermodynamic degrees of freedom for saturated vapour and superheated vapour, respectively?

(a) 1 and 1 (b) 1 and 2
(c) 2 and 1 (d) 2 and 2

[ESE : 2007]

6.14 Which one of the following statements is correct when saturation pressure of water vapour increases?

(a) Saturation temperature decreases
(b) Enthalpy of evaporation decreases
(c) Enthalpy of evaporation increases
(d) Specific volume change of phase increases

[ESE : 2007]

6.15 Which p–V diagram for steam illustrates correctly the isothermal process undergone by wet steam till it becomes superheated?

(a)

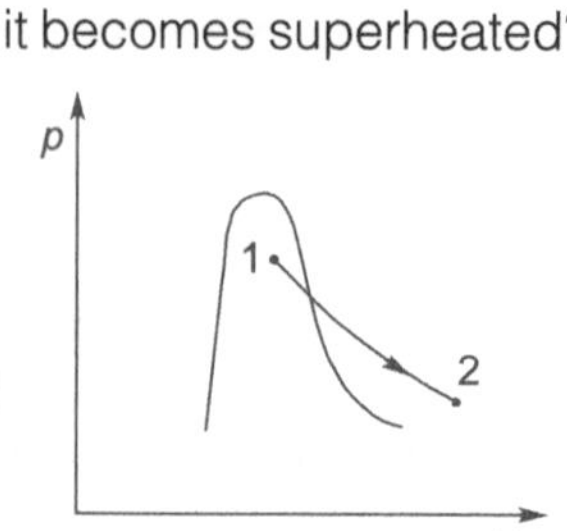

(b)

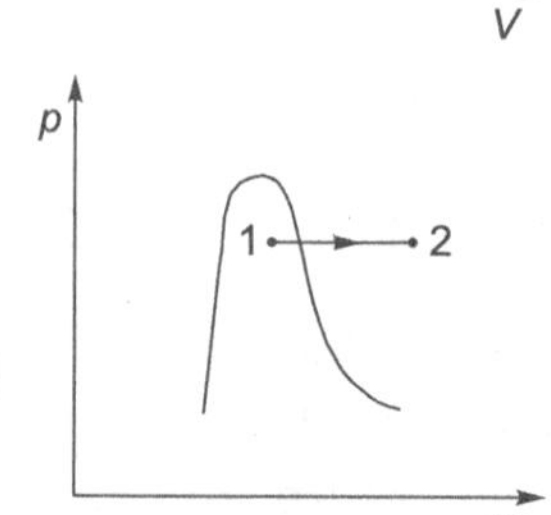

(c) 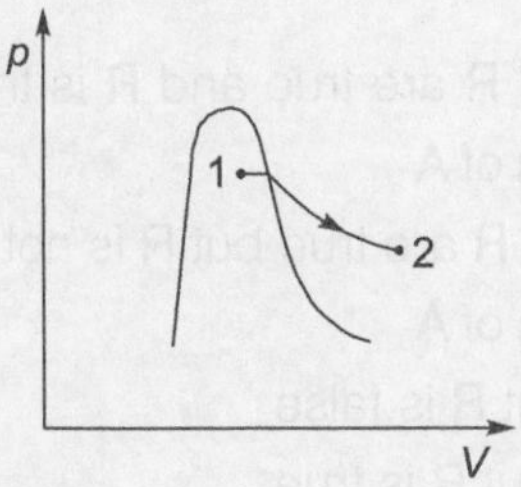

(d) 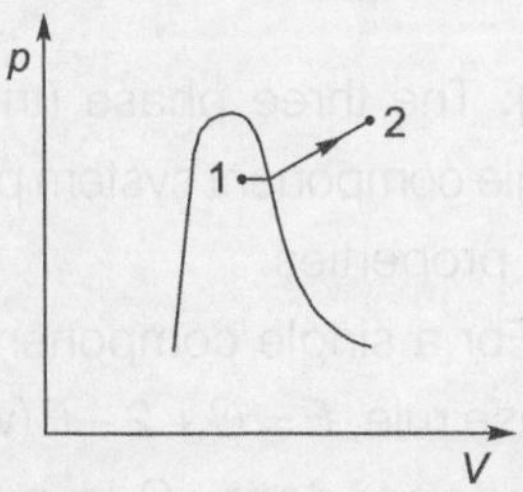

[ESE : 2007]

6.16 Consider the following statements:

1. A gas with a compressibility factor more than 1 is more compressible than a perfect gas.
2. The *x* and *y* axes of the compressibility chart are compressibility factor on *y*-axis and reduced pressure on *x*-axis.
3. The first and second derivatives of the pressure with respect to volume at critical points are zero.

Which of the statements given above are correct?

(a) 2 and 3 only (b) 1 and 3 only
(c) 1 and 2 only (d) 1, 2 and 3

[ESE : 2007]

6.17 The specific volume of water when heated from 0ºC

(a) first increases and then decreases
(b) first decreases and then increases
(c) increases steadily
(d) decreases steadily [ESE : 2008]

6.18 Consider the following properties of vapour

1. Pressure 2. Temperature
3. Dryness fraction 4. Specific volume

Which of these two properties alone are not sufficient to specify the condition of a vapour?

(a) 1 and 2 (b) 1 and 3
(c) 2 and 3 (d) 3 and 4[ESE : 2009]

6.19 Consider the following:

1. Air
2. Gaseous combustion products
3. Steam

Which of these are pure substances, assuming there is no phase change?

(a) 1 and 2 only (b) 1 and 3 only
(c) 2 and 3 only (d) 1, 2 and 3

[ESE : 2009]

6.20 State of a wet vapour cannot be specified only by

(a) Pressure and temperature
(b) Pressure and dryness fraction
(c) Temperature and dryness fraction
(d) Pressure and volume [ESE : 2011]

6.21 Reduced pressure is

(a) Always less than atmospheric pressure
(b) Always unity
(c) An index of molecular position of a gas
(d) Dimensionless [ESE : 2011]

6.22 The following remarks pertain to the properties of gases and vapours

1. The equation of state is not by itself sufficient for the calculation of properties such as *u*, *h* and *s*.
2. Throttling process is an adiabatic steady flow process.
3. Increase in entropy in wet region is directly proportional to the increase in enthalpy.

Which of these remarks are correct?

(a) 1, 2 and 3 (b) 1 and 2 only
(c) 1 and 3 only (d) 2 and 3 only

[ESE : 2012]

6.23 No substance can exist in the liquid phase in stable equilibrium

(a) at temperature below the triple point temperature
(b) at pressure below the triple point pressure
(c) at pressure above the triple point pressure
(d) at temperature above the triple point temperature

[ESE : 2012]

6.24 If a pure substance is below the triple point temperature, the solid on being heated will only

(a) solidify or freeze
(b) liquefy
(c) vaporize or sublimate
(d) have its temperature increased

[ESE : 2013]

6.25 Along the 'triple line' in a *p-v* diagram showing all three phases of water, which one of the following statements is correct?

(a) A substance has the same pressure and temperature but different specific volume
(b) A substance has same temperature and specific volume but different pressure
(c) A substance has same specific volume and pressure but different temperature
(d) A substance has same specific volume, pressure and temperature

[ESE : 2014]

6.26 Consider the following statements:

1. In an ideal gas, there are no inter molecular forces of attraction and repulsion.
2. At very low pressure, all gases and vapours approach ideal gas behaviour.
3. Enthalpy of an ideal gas depends only on temperature.

Which of the above statements are correct?

(a) 1, 2 and 3 (b) 1 and 2 only
(c) 1 and 3 only (d) 2 and 3 only

[ESE : 2014]

6.27 Consider the following statements pertaining to the properties of perfect, non reacting gas mixtures:

1. The total volume of a mixture is the sum of partial volumes at the same pressure and temperature.
2. The entropy of a mixture of gases is the same as the entropies of the constituents.
3. The total pressure of a mixture of gases is the sum of the partial pressure of the substances.
4. The mole fraction of a mixture of gases is equal to both pressure and volume fraction.

Which of the above statements is/are correct ?

(a) 1, 2, 3 and 4 (b) 1, 2 and 3 only
(c) 1, 2 and 4 only (d) 3 and 4 only

[ESE : 2014]

Direction: The following items consists of two statements, one labelled as **Assertion 'A'** and the other labelled as **Reason 'R'**. You are to examine these two statements carefully and select your answers to these items using the codes given below:

Codes:

(a) Both **A** and **R** are true and **R** is the correct explanation of **A**
(b) Both **A** and **R** are true but **R** is not a correct explanation of **A**
(c) **A** is true but **R** is false
(d) **A** is false but **R** is true

6.28 **Assertion (A):** The three phase (triple state/ point) of a single component system possesses a single set of properties.

Reason (R): For a single component system, the Gibbs phase rule, $F = C + 2 - P$ (where F is number of independent, C is number of components in the system and P is number of phases), reduces to $F = 3 - P$.

[ESE : 2015]

6.29 **Assertion (A):** A mixture of liquid air an air cannot to considered as pure substance.

Reason (R): Proportions of oxygen and nitrogen differ in liquid and gaseous states in equilibrium.

[ESE : 2015]

6.30 Consider the following statements :

1. There is no change in temperature when a liquid is being evaporated into vapour
2. Vapour is a mixed phase of liquid and gas in the zone between saturated liquid line and saturated vapour line
3. The saturated dry vapour curve is steeper as compared to saturated liquid curve on a *T-s* diagram
4. The enthalpy of vaporization decreases with increase in pressure

Which of the above statements are correct ?

(a) 1, 2 and 3 only (b) 3 and 4 only
(c) 1, 2 and 4 only (d) 1, 2, 3 and 4

[ESE : 2016]

6.31 The ordinate and abscissa of the diagram to depict the isobaric processes of an ideal gas as a hyperbola are, respectively

(a) temperature and entropy
(b) internal energy and volume

(c) temperature and density

(d) enthalpy and entropy [ESE : 2018]

6.32 A frictionless piston cylinder device contains 5 kg of steam at 400 kPa and 200°C. The heat is now transferred to the steam until the temperature reaches 250°C. If the piston is not attached to a shaft, its mass is constant, and by taking the values of specific volume v_1 as 0.53434 m^3/kg and v_2 as 0.59520 m^3/kg, the work done by the steam during this process is

(a) 121.7 kJ (b) 137.5 kJ

(c) 153.3 kJ (d) 189.1 kJ

[ESE : 2019]

6.33 A gas mixture consists of 3 kg of O_2, 5 kg of N_2 and 12 kg of CH_4. The mass fraction and mole fraction of O_2 are

(a) 0.25 and 0.125 (b) 0.15 and 0.092

(c) 0.25 and 0.092 (d) 0.15 and 0.125

[ESE : 2019]

6.34 What is the value of mass of the air in a room of size 4 m × 5 m × 6 m at 100 kPa and 25°C? [Take $R = 0.287$ kPa·m^3·kg^{-1}·K^{-1}]

(a) 150 kg (b) 180 kg

(c) 140 kg (d) 130 kg

[ESE : 2021]

6.35 The state of a simple compressible pure substance can be fixed by specifying

(a) one independent property

(b) two independent properties

(c) three independent properties

(d) four independent properties [ESE : 2021]

6.36 A gas weighs 16 N/m^3 at 25°C and at an absolute pressure 0.25 N/mm^2. What is the gas constant approximately? (Take acceleration due to gravity as 9.81 m/s^2)

(a) 514.68 Nmkg^{-1}K^{-1} (b) 542.55 Nmkg^{-1}K^{-1}

(c) 562.68 Nmkg^{-1}K^{-1} (d) 592.55 Nmkg^{-1}K^{-1}

[ESE : 2022]

6.37 A certain perfect gas of mass 0.01 kg occupies a volume of 0.003 m^3 at a pressure of 7 bar and a temperature of 131°C. The gas is allowed to expand until the pressure is 1 bar and the final volume is 0.02 m^3. What is the molar mass of the gas?

(Take universal gas constant as 8314.5 J/K. kmol)

(a) 8 kg/kmol (b) 10 kg/kmol

(c) 12 kg/kmol (d) 16 kg/kmol

[ESE : 2023]

6.38 What is the specific volume of steam at 17672 kPa and 712 K considering it as a perfect gas? [Take critical pressure = 22.09 MPa, critical temperature = 647.3 K, R_{steam} = 0.4615 kJ/kg.K]

(a) 0.0186 m^3/kg (b) 0.0986 m^3/kg

(c) 0.2146 m^3/kg (d) 0.3146 m^3/kg

[ESE : 2023]

■■■■

Answers Pure Substances

6.1 (d)	6.2 (c)	6.3 (c)	6.4 (c)	6.5 (c)	6.6 (d)	6.7 (d)	6.8 (a)	6.9 (b)
6.10 (b)	6.11 (d)	6.12 (d)	6.13 (b)	6.14 (b)	6.15 (c)	6.16 (a)	6.17 (b)	6.18 (a)
6.19 (d)	6.20 (a)	6.21 (d)	6.22 (a)	6.23 (b)	6.24 (c)	6.25 (a)	6.26 (a)	6.27 (a)
6.28 (a)	6.29 (a)	6.30 (d)	6.31 (c)	6.32 (a)	6.33 (b)	6.34 (c)	6.35 (b)	6.36 (a)
6.37 (d)	6.38 (a)							

Explanations Pure Substances

6.1 (d)

$$\left(\frac{T_2}{T_1}\right) = \left(\frac{p_2}{p_1}\right)^{(\gamma-1)/\gamma}$$

Put $\gamma = 1.4$ (For air)

$$\left(\frac{T_2}{T_1}\right) = \left(\frac{p_2}{p_1}\right)^{(\gamma-1)/\gamma} = \left(\frac{p_2}{p_1}\right)^{0.286}$$

6.2 (c)

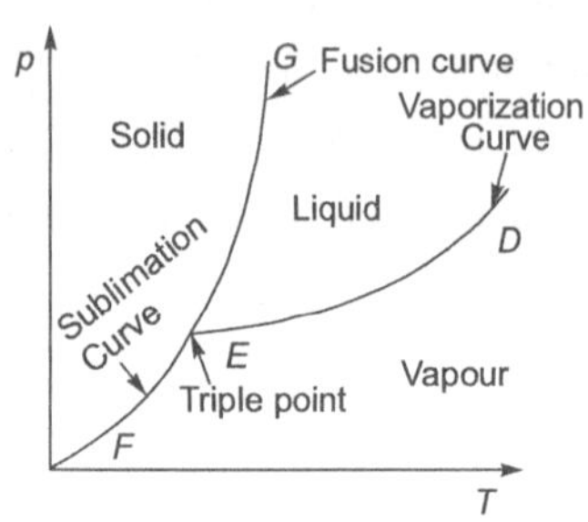

6.4 (c)

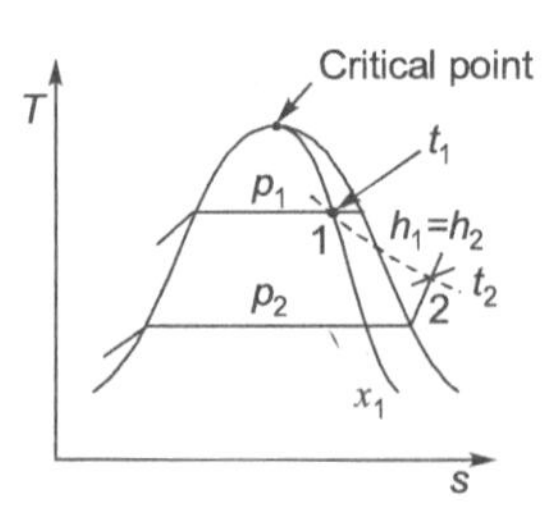

During throttling of wet steam, enthalpy remains constant, entropy increases, specific volume increases, temperature decreases.

6.5 (c)

Change in internal energy, $u = f(T, V)$

$$du = \left(\frac{\partial u}{\partial T}\right)_V dT + \left(\frac{\partial u}{\partial V}\right)_T dV$$

$$du = c_v dT + \left[T - \left(\frac{\partial p}{\partial T}\right)_V - p\right] dV$$

6.7 (d)

Compressibility factor of ideal gas is one

$Z = \frac{pV}{RT}$; $Z = 1$ for ideal gas

For real gases Z can be greater than or less than unity. The farther Z is away from unity the more the gas deviates from ideal gas behaviour.

6.8 (a)

Since enthalpy remains constantin throttling process,

$$\therefore \quad h_1 = h_2$$

$$1000 = 800 + x(2800 - 800)$$

$$x = 0.1$$

6.9 (b)

For Mollier diagram, condensation takes place at constant pressure which is along vertically inclined.

6.10 (b)

Critical point : Properties of saturated liquid and saturated vapour are identical.

Sublimation : Heating process, where solid gets directly transformed to gaseous phase.

Triple point : All the three phases-solid, liquid and vapour co-exists in equilibrium.

6.13 (b)

Saturated vapour has only one degree of freedom but superheated vapour has two degrees of freedom.

6.14 (b)

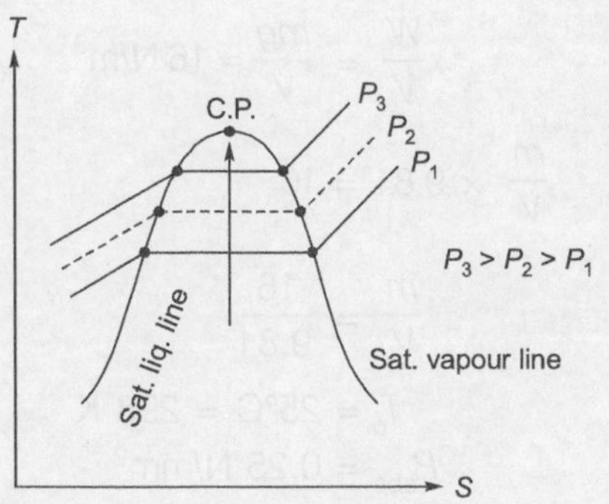

6.17 (b)

The density of water is maximum at 4°C therefore the specific volume of water is minimum at 4°C. So when water is heated at 0°C the density of water increases and as a result the specific volume of water decreases and reaches the minimum. Afterwards the specific volume of water increases once the temperature exceeds 4°C.

6.18 (a)

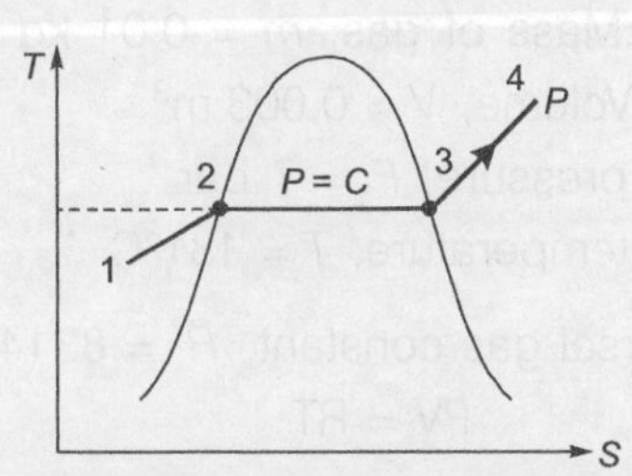

In 2-3, only pressure and temperature does not tells about quality of vapour i.e. dryness fraction.

6.20 (a)

State of wet vapour can not be specified only by pressure and temperature.

6.21 (d)

Ratio of pressure of a substance to its critical pressure is known as reduced pressure. Therefore it is dimensionless term.

6.22 (a)

The equation of state is not by itself sufficient for the calculation of properties.

6.24 (c)

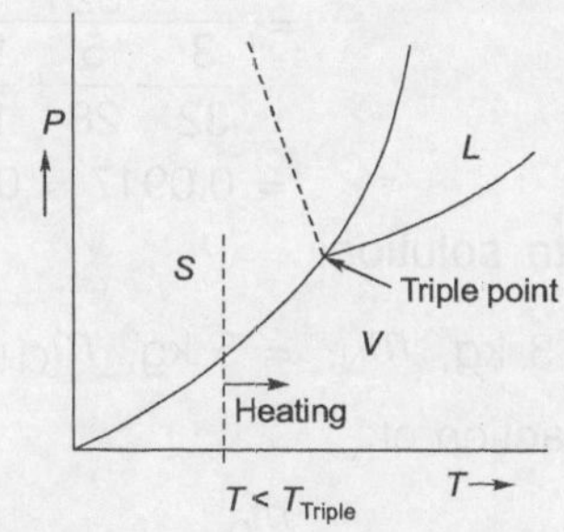

Upon heating, phase changes from S to V, so solid will vaporize or sublime.

6.26 (a)

At very low pressure or density, all gases and vapours approach ideal gas behaviour.

6.29 (a)

A pure substance is defined as one that is homogeneous and invariable in chemical composition throughout its mass.

The mixture of air and liquid air is not pure substance since the relative proportion of oxygen and nitrogen differ in gas and liquid phase.

6.31 (c)

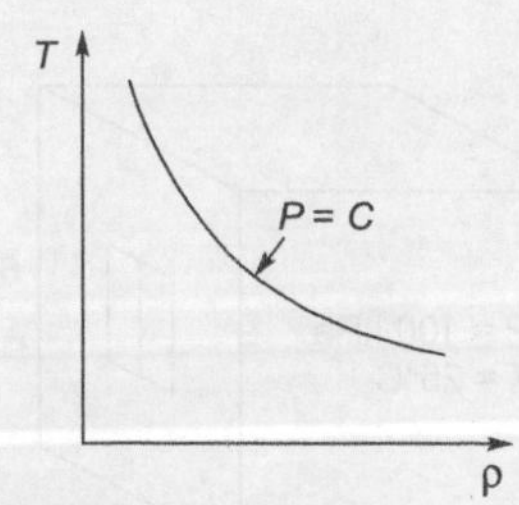

$$P = \rho RT \ ; \ xy = \text{Const.}$$

6.32 (a)

Constant pressure process

$$\text{Work} = P(V_2 - V_1) = P \times m(v_2 - v_1)$$
$$= 400 \times 5(0.59520 - 0.53434) \text{ kJ}$$
$$= 121.72 \text{ kJ}$$

6.33 (b)

$$\text{Mass fraction of } O_2 = \frac{3}{3+5+12} = 0.15$$

$$\text{Mole fraction of } O_2 = \frac{n_{O_2}}{n_{O_2} + n_{N_2} + n_{CH_4}}$$

$$= \frac{\frac{3}{32}}{\frac{3}{32}+\frac{5}{28}+\frac{12}{16}}$$

$$= 0.0917 \approx 0.092$$

Alternate solution:

$m_{O_2} = 3$ kg, $m_{N_2} = 5$ kg, $m_{CH_4} = 12$ kg

Mass fraction of

$$O_2 = \frac{m_{O_2}}{m_{O_2}+m_{N_2}+m_{CH_4}}$$

$$= \frac{3}{3+5+12} = \frac{3}{20} = 0.15$$

$$n_{O_2} = \frac{3}{32} = 0.09375$$

$$n_{N_2} = \frac{5}{28} = 0.17857$$

$$n_{CH_4} = \frac{12}{16} = 0.75$$

Mole fraction of

$$O_2 = \frac{n_{O_2}}{n_{O_2}+n_{N_2}+n_{CH_4}} = 0.0917$$

6.34 (c)

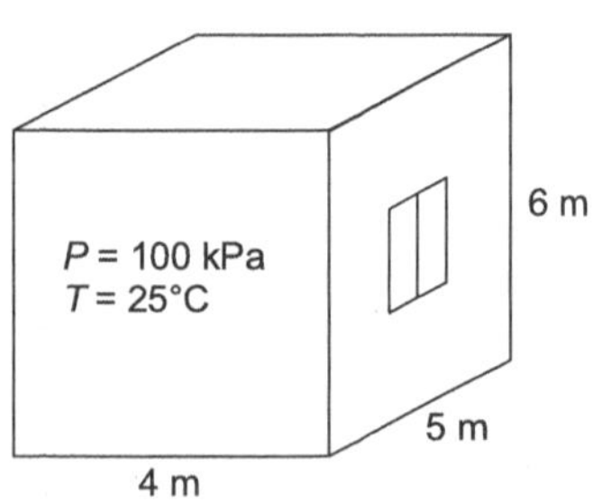

$V = 4$ m × 5 m × 6 m = 120 m^3

$P = 100$ kPa

$T = 298$ K

$$m = \frac{PV}{RT} = 140.3 \text{ kg}$$

6.36 (a)

$$\frac{W}{V} = \frac{mg}{V} = 16 \text{ N/m}^3$$

$$\frac{m}{V} \times 9.81 = 16$$

$$\frac{m}{V} = \frac{16}{9.81}$$

$T_o = 25°C = 298$ K

$P_{abs} = 0.25$ N/mm^2

$= 0.25$ N/$(10^{-3}$ m$)^2$

$PV = mRT$

$$P = \frac{m}{V}RT$$

$$0.25 \times 10^6 = \frac{16}{9.81}\times R\times 298$$

$R = 514.36$ Nm/kgK

6.37 (d)

Given: Mass of gas, $m = 0.01$ kg

Volume, $V = 0.003$ m^3

Initial pressure, $P_1 = 7$ bar

Initial temperature, $T = 131°C$

Universal gas constant, $\bar{R} = 8314.5$ J/K kmol

PV = RT

$$7 \times 10^5 \times 0.003 = \frac{0.01}{M}\times 8314.5\times(273+131)$$

$M = 15.99$ kg/kmol

6.38 (a)

Given : $P = 17672$ kPA

$T = 712$ K

$R = 0.4615$ kJ/kg.K

Using ideal gas equation,

$Pv = RT$

$$v = \frac{RT}{P} = \frac{0.4615\times 712}{17672}$$

$v = 0.01859$ m^3/kg

■■■■

UNIT

III

Heat Transfer

Syllabus

Modes of heat transfer, Steady and unsteady heat conduction, Thermal resistance, Fins, Free and forced convection, Correlations for convective heat transfer, Radiative heat transfer – Radiation heat transfer coefficient; boiling and condensation, Heat exchanger performance analysis.

Contents

Sl.	Topic	Page No.
1.	Conduction	208
2.	Unsteady State Conduction Heat Transfer	227
3.	Heat Dissipation From Extended Surfaces (Fins)	230
4.	Free and Forced Convection	234
5.	Radiation	248
6.	Heat Exchangers	261
7.	Boiling and Condensation	271

OOOO

1 Conduction

1.1 The outer surface of a long cylinder is maintained at constant temperature. The cylinder does not has any heat source. The temperature in the cylinder will
(a) increase linearly with radius
(b) decrease linearly with radius
(c) be independent of radius
(d) vary logarithmically with radius

[ESE : 2000]

1.2 A copper wire of radius 0.5 mm is insulated with a sheathing of thickness 1 mm having a thermal conductivity of 0.5 W/mK. The outside surface convective heat transfer coefficient is 10 W/m^{2}K. If the thickness of insulation sheathing is raised by 10 mm, then the electrical current-carrying capacity of the wire will
(a) increase
(b) decrease
(c) remain the same
(d) vary depending upon the electrical conductivity of the wire [ESE : 2000]

1.3 The overall heat transfer coefficient U for a plane composite wall of n layers is given by (the thickness of the ith layer is t_i, thermal conductivity of the i th layer is k_i, convective heat transfer coefficient is h)

(a) $\dfrac{1}{\dfrac{1}{h_1}+\sum_{i=1}^{n}\dfrac{t_i}{k_i}+\dfrac{1}{h_n}}$ (b) $h_1+\sum_{i=1}^{n}\dfrac{t_i}{k_i}+h_n$

(c) $\dfrac{1}{\dfrac{1}{h_1}+\sum_{i=1}^{n}\dfrac{t_i}{k_i}+h_n}$ (d) $\dfrac{1}{h_1}+\sum_{i=1}^{n}\dfrac{t_i}{k_i}+\dfrac{1}{h_n}$

[ESE : 2000]

1.4 A composite plane wall is made up of two different materials of the same thickness and having thermal conductivities of k_1 and k_2 respectively. The equivalent thermal conductivity of the slab is

(a) $k_1 + k_2$ (b) $k_1 k_2$
(c) $\dfrac{k_1 + k_2}{k_1 k_2}$ (d) $\dfrac{2k_1 k_2}{k_1 + k_2}$

[ESE : 2000]

Direction: Each of the next consists of two statements, one labelled as the **'Assertion (A)'** and the other as **'Reason (R)'**. You are to examine these two statements carefully and select the answers to these items using the codes given below:

Codes:
(a) Both **A** and **R** are individually true and **R** is the correct explanation of **A**
(b) Both **A** and **R** are individually true but **R** is not the correct explanation of **A**
(c) **A** is true but **R** is false
(d) **A** is false but **R** is true

1.5 **Assertion (A):** The leakage heat transfer from the outside surface of a steel pipe carrying hot gases is reduced to a greater extent on providing refractory brick lining on the inside of the pipe as compared to that with brick lining on the outside.
Reason (R): The refractory brick lining on the inside of the pipe offers a higher thermal resistance.

[ESE : 2000]

1.6 **Assertion (A):** Thermal conductance of heat pipe is several hundred times that of the best available metal conductor under identical conditions.
Reason (R): The value of latent heat is far greater than that of specific heat.

[ESE : 2000]

1.7 A cylinder made of a metal of conductivity 40 W/mK is to be insulated with a material of conductivity 0.1 W/mK. If the convective heat transfer coefficient with the ambient atmosphere is 5 W/m^{2}K, the critical radius of insulation is

(a) 2 cm (b) 4 cm
(c) 8 cm (d) 50 cm

[ESE : 2001]

1.8 A plane wall of thickness $2L$ has a uniform volumetric heat source q^* (W/m^3). It is exposed to local ambient temperature T_∞ at both the ends ($x = \pm L$). The surface temperature t_s of the wall under steady-state condition (where h and k have their usual meanings) is given by

(a) $T_s = T_\infty + \frac{q^* L}{h}$ (b) $T_s = T_\infty + \frac{q^* L}{2k}$

(c) $T_s = T_\infty + \frac{q^* L^2}{h}$ (d) $T_s = T_\infty + \frac{q^* L^3}{2k}$

[ESE : 2001]

1.9 A flat plate has thickness 5 cm, thermal conductivity 1 W/mK, convective heat transfer coefficients on its two flat faces of 10 W/m^{2}K and 20 W/m^{2}K. The overall heat transfer coefficient for such a flat plate is

(a) 5 W/m^{2}K (b) 6.33 W/m^{2}K
(c) 20 W/m^{2}K (d) 30 W/m^{2}K

[ESE : 2001]

1.10 A 0.5 m thick plane wall has its two surfaces kept at 300°C and 200°C. Thermal conductivity of the wall varies linearly with temperature and its values at 300°C and 200°C are 25 W/mK and 15 W/mK, respectively. Then the steady heat flux through the wall is

(a) 8 kW/m^2 (b) 5 kW/m^2
(c) 4 kW/m^2 (d) 3 kW/m^2

[ESE : 2002]

1.11 Using thermal-electrical analogy in heat transfer, match **List-I** (Electrical quantities) with **List-II** (Thermal quantities) and select the correct answer using the codes given below the lists:

List-1	List-II
A. Voltage	1. Thermal resistance
B. Current	2. Thermal capacity
C. Resistance	3. Heat flow
D. Capacitance	4. Temperature

Codes:

	A	B	C	D
(a)	2	3	1	4
(b)	4	1	3	2
(c)	2	1	3	4
(d)	4	3	1	2

[ESE : 2002]

1.12 A metal rod of 2 cm diameter has a conductivity of 40 W/mK, which is to be insulated with an insulating material of conductivity of 0.1 W/mK. If the convective heat transfer coefficient with the ambient atmosphere is 5 W/m^{2}K, the critical thickness of insulation will be

(a) 1 cm (b) 2 cm
(c) 7 cm (d) 8 cm [ESE : 2003]

1.13 A composite wall is made of two layers of thickness δ_1 and δ_2 having thermal conductivities k and $2k$ and equal surface areas normal to the direction of heat flow. The outer surfaces of the composite wall are at 100°C and 200°C respectively. The heat transfer takes place only by conduction and the required surface temperature at the junction is 150°C What will be the ratio of their thicknesses, $\delta_1 : \delta_2$?

(a) 1 : 1 (b) 2 : 1
(c) 1 : 2 (d) 2 : 3 [ESE : 2004]

1.14 A hollow pipe of 1 cm outer diameter is to be insulated by thick cylindrical insulation having thermal conductivity 1 W/mK. The surface heat transfer coefficient on the insulation surface is 5 W/m^{2}K. What is the minimum effective thickness of insulation for causing the reduction in heat leakage from the insulated pipe?

(a) 10 cm (b) 15 cm
(c) 19.5 cm (d) 20 cm

[ESE : 2004]

1.15 What will be the geometric radius of heat transfer for a hollow sphere of inner and outer radii r_1 and r_2?

(a) $\sqrt{r_1 r_2}$ (b) $r_2 r_1$
(c) r_2/r_1 (d) $r_2 - r_1$

[ESE : 2004]

1.16 Upto the critical radius of insulation:

(a) Added insulation increases heat loss
(b) Added insulation decreases heat loss
(c) Convection heat loss is less than conduction heat loss
(d) Heat flux decreases

[ESE : 2005]

1.17 Match **List-I** (Governing Equations of Heat Transfer) with **List-II** (Specific Cases of Heat Transfer) and select the correct answer using the code given below the lists:

List-I

A. $\frac{d^2T}{dr^2}+\frac{2}{r}\frac{dT}{dr}=0$ B. $\frac{\partial^2T}{\partial x^2}=\frac{1}{\alpha}\frac{\partial T}{\partial t}$

C. $\frac{d^2T}{dr^2}+\frac{1}{r}\frac{dT}{dr}=0$ D. $\frac{d^2\theta}{dx^2}-m^2\theta=0$

List-II

1. Pin fin 1-D case
2. 1-D conduction in cylinder
3. 1-D conduction in sphere
4. Plane slab

(Symbols have their usual meaning)

Codes:

	A	B	C	D
(a)	2	4	3	1
(b)	3	1	2	4
(c)	2	1	3	4
(d)	3	4	2	1

[ESE : 2005]

1.18 6.0 kJ of conduction heat transfer has to take place in 10 minutes from one end to other end of a metallic cylinder of 10 cm^2 cross-sectional area, length 1 meter and thermal conductivity as 100 W/mK. What is the temperature difference between the two ends of the cylindrical bar?

(a) 80°C (b) 100°C
(c) 120°C (d) 160°C [ESE : 2005]

1.19 Heat is conducted through a 10 cm thick wall at the rate of 30 W/m^2 when the temperature difference across the wall is 10°C. What is the thermal conductivity of the wall?

(a) 0.03 W/mK (b) 0.3 W/mK
(c) 3.0 W/mK (d) 30.0 W/mK

[ESE : 2005]

1.20 A composite hollow sphere with steady internal heating is made of 2 layers of materials of equal thickness with thermal conductivities in the ratio of 1 : 2 for inner to outer layers. Ratio of inside to outside diameter is 0.8 What is ratio of temperature drop across the inner and outer layers?

(a) 0.4 (b) 1.6
(c) 2 ln 0.8 (d) 2.5 [ESE : 2005]

1.21 A plane wall is 25 cm thick with an area of 1 m^2, and has a thermal conductivity of 0.5 W/mK. If a temperature difference of 60°C is imposed across it, what is the heat flow?

(a) 120 W (b) 140 W
(c) 160 W (d) 180 W

[ESE : 2005]

1.22 A metal plate has surface area of 2 m^2, thickness 10 mm and a thermal conductivity of 200 W/mK. What is the thermal resistance of the plate?

(a) 4×10^{-4} K/W (b) 2.5×10^{-3} K/W
(c) 1.5×10^{-5} K/W (d) 2.5×10^{-5} K/W

[ESE : 2006]

1.23 Which one of the following expresses the thermal diffusivity of a substance in terms of thermal conductivity of a substance (k), mass density (ρ) and specific heat (c)?

(a) $k^2\rho c$ (b) $1/\rho kc$
(c) $k/\rho c$ (d) $\rho c/k^2$

[ESE : 2006]

1.24 What is the ratio of thermal conductivity to electrical conductivity equal to?

(a) Prandtl number (b) Schmidt number
(c) Lorenz number (d) Lewis number

[ESE : 2006]

1.25 A copper block and an air mass block having similar dimensions are subjected to symmetrical heat transfer from one face of each block. The other face of the block will be reaching to the same temperature at a rate

(a) faster in air block
(b) faster in copper block
(c) equal in air as well as copper block
(d) cannot be predicted with the given information

[ESE : 2006]

1.26 A wall of thickness 0.6 m has a normal area 1.5 m^2 and is made up of material of thermal conductivity 0.4 W/mK. The temperatures on the two sides are 800°C and 100°C. What is the thermal resistance of the wall?

(a) 1 W/K (b) 1.8 W/K
(c) 1 K/W (d) 1.8 K/W

[ESE : 2007]

1.27 For conduction through a spherical wall with constant thermal conductivity and with inner side temperature greater than outer wall temperature,

(one dimensional heat transfer), what is the type of temperature distribution?
(a) Linear (b) Parabolic
(c) Hyperbolic (d) None of these
[ESE : 2007]

1.28 A composite wall having three layers of thickness 0.3 m, 0.2 m and 0.1 m and of thermal conductivities 0.6, 0.4 and 0.1 W/mK, respectively, is having surface area 1 m^2. If the inner and outer temperatures of the composite wall are 1840 K and 340 K, respectively, what is the rate of heat transfer?
(a) 150 W (b) 1500 W
(c) 75 W (d) 750 W [ESE : 2007]

1.29 What is the expression for the thermal conduction resistance to heat transfer through a hollow sphere of inner radius r_1 and outer radius r_2, and thermal conductivity k?
(a) $\dfrac{(r_2 - r_1) r_1 r_2}{4\pi k}$ (b) $\dfrac{4\pi k(r_2 - r_1)}{r_1 r_2}$
(c) $\dfrac{r_2 - r_1}{4\pi k r_1 r_2}$ (d) None of these
[ESE : 2007]

1.30 The value thermal conductivity of thermal insulation applied to a hollow spherical vessel containing very hot material is 0.5 W/mK. The convective heat transfer coefficient at the outer surface of insulation is 10 W/m^{2}K. What is the critical radius of the sphere?
(a) 0.1 m (b) 0.2 m
(c) 1.0 m (d) 2.0 m
[ESE : 2008]

1.31 A composite wall of a furnace has 2 layers of equal thickness having thermal conductivities in the ratio of 3 : 2. What is the ratio of the temperature drop across the two layers?
(a) 2 : 3 (b) 3 : 2
(c) 1 : 2 (d) $\log_e 2 : \log_e 3$
[ESE : 2008]

1.32 In which one of the following materials, is the heat energy propagation minimum due to conduction heat transfer?
(a) Lead (b) Copper
(c) Water (d) Air
[ESE : 2008]

1.33 Match **List-I** with **List-II** and select the correct answer using the code given below the lists:

List-I
A. Heat transfer through solid
B. Heat transfer from hot surface to surrounding fluid
C. Heat transfer in boiling liquid
D. Heat transfer from one body to another separated in space

List-II
1. Radiation heat transfer
2. Fourier's law of heat conduction
3. Convection heat transfer
4. Newton's law of cooling

Codes:

	A	B	C	D
(a)	3	1	2	4
(b)	2	4	3	1
(c)	2	1	3	4
(d)	3	4	2	1

[ESE : 2008]

1.34 A wall as shown below, is made up of two layers (*A*) and (*B*). The temperatures are also shown in the sketch. The ratio of thermal conductivity of two layers is $k_A/k_B = 2$

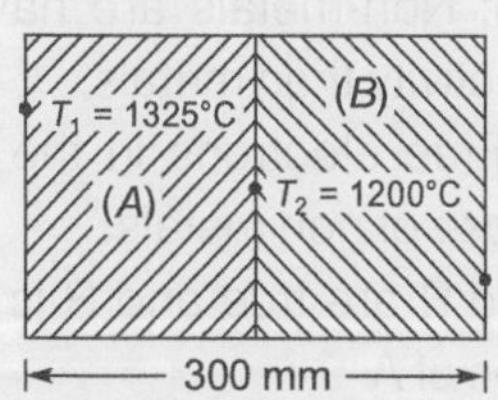

What is the ratio of thickness of two layers?
(a) 0.105 (b) 0.213
(c) 0.555 (d) 0.840 [ESE : 2008]

1.35 What is the critical radius of insulation for a sphere equal to?

$$\begin{pmatrix} k = \text{thermal conductivity in W/mK} \\ h = \text{heat transfer coefficent in W/m}^2\text{K} \end{pmatrix}$$

(a) $2kh$ (b) $2k/h$
(c) k/h (d) $\sqrt{2kh}$ [ESE : 2008]

1.36 Match **List-I** (Law) with **List-II** (Effect) and select the correct answer using the code given below the lists:

List-I
A. Fourier's law
B. Stefan Boltzmann law
C. Newton's law of cooling
D. Ficks law

List-II
1. Mass transfer
2. Conduction
3. Convection
4. Radiation

Codes:

	A	B	C	D
(a)	3	1	2	4
(b)	2	4	3	1
(c)	3	4	2	1
(d)	2	1	3	4

[ESE : 2008]

1.37 Two walls of same thickness and cross sectional area have thermal conductivities in the ratio 1 : 2. If the same temperature difference is maintained across the two faces of both the walls, what is the ratio of heat flow Q_1/Q_2?

(a) 1/2 (b) 1
(c) 2 (d) 4 [ESE : 2008]

1.38 **Assertion (A):** Non-metals are having higher thermal conductivity than metals.

Reason (R): Free electrons in the metals are higher than those of non metals.

(a) Both **A** and **R** are true and **R** is the correct explanation of **A**
(b) Both **A** and **R** are true but **R** is not a correct explanation of **A**
(c) **A** is true but **R** is false
(d) **A** is false but **R** is true [ESE : 2008]

1.39 In $MLT\theta$ system (T being time and θ temperature), what is the dimension of thermal conductivity?

(a) $ML^{-1}T^{-1}\theta^{-3}$ (b) $MLT^{-1}\theta^{-1}$
(c) $ML\theta^{-1}T^{-3}$ (d) $ML\theta^{-1}T^{-2}$

[ESE : 2009]

1.40 A steel plate of thermal conductivity 50 W/mK and thickness 10 cm passes a heat flux by conduction of 25 kW/m^2. If the temperature of the hot surface of the plate is 100°C, then what is the temperature of the cooler side of the plate?

(a) 30°C (b) 40°C
(c) 50°C (d) 60°C [ESE : 2009]

1.41 A composite slab has two layers of different materials having internal conductivities k_1 and k_2. If each layer has the same thickness, then what is the equivalent thermal conductivity of the slab?

(a) $\dfrac{k_1 k_2}{(k_1+k_2)}$ (b) $\dfrac{k_1 k_2}{2(k_1+k_2)}$

(c) $\dfrac{2k_1}{(k_1+k_2)}$ (d) $\dfrac{2k_1 k_2}{(k_1+k_2)}$

[ESE : 2009]

1.42 A composite wall of a furnace has 3 layers of equal thickness having thermal conductivities in the ratio of 1 : 2 : 4. What will be the temperature drop ratio across the three respective layers?

(a) 1 : 2 : 4 (b) 4 : 2 : 1
(b) 1 : 1 : 1 (d) log4 : log2 : log1

[ESE : 2009]

1.43 What is the heat lost per hour across a wall 4 m high, 10 m long and 115 mm thick, if the inside wall temperature is 30°C and outside ambient temperature is 10°C? Conductivity of brick wall is 1.15 W/mK, heat transfer coefficient for inside wall is 2.5 W/m^{2}K and that for outside wall is 4 W/m^{2}K.

(a) 3635 kJ (b) 3750 kJ
(c) 3840 kJ (d) 3920 kJ

[ESE : 2009]

1.44 A large concrete slab 1 m thick has one dimensional temperature distribution :

$$T = 4 - 10x + 20x^2 + 10x^3$$

where T is temperature and x is distance from one face towards other face of wall. If the slab material has thermal diffusivity of 2×10^{-3} m^2/hr, what is the rate of change of temperature at the other face of the wall?

(a) 0.1°C/h (b) 0.2°C/h
(c) 0.3°C/h (d) 0.4°C/h [ESE : 2009]

1.45 **Assertion (A):** Cork is a good insulator.

Reason (R): Good insulators are highly porous.

(a) Both **A** and **R** are individually true and **R** is the correct explanation of **A**
(b) Both **A** and **R** are individually true but **R** is not the correct explanation of **A**

(c) A is true but R is false

(d) A is false but R is true [ESE : 2009]

1.46 Upto the critical radius of insulation

(a) convection heat loss will be less than conduction heat loss

(b) heat flux will decrease

(c) added insulation will increase heat loss

(d) added insulation will decrease heat loss

[ESE : 2010]

1.47 The equivalent thermal conductivity of the wall as shown in the figure below is

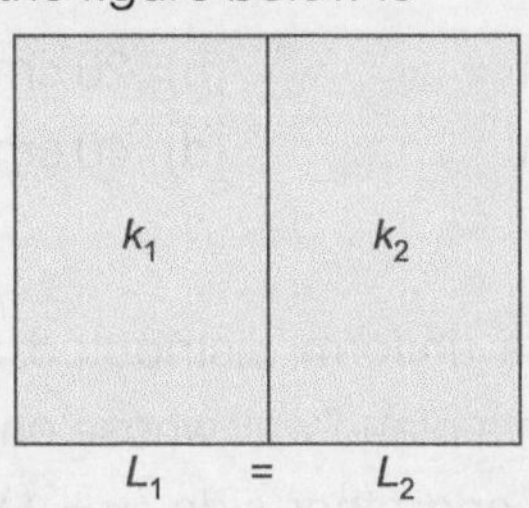

(a) $\frac{k_1 + k_2}{2}$ (b) $\frac{k_1 k_2}{k_1 + k_2}$

(c) $\frac{2k_1 k_2}{k_1 + k_2}$ (d) $\sqrt{k_1 k_2}$ [ESE : 2010]

1.48 A refrigerated van has to be maintained at −10°C, while ambient temperature is 32°C. Overall surface area of the van is 25 m^2. The walls of the van are of 16 mm plywood (k = 0.06 W/mK) at inner surface and 12 mm of glass work (k = 0.06 W/mK). The capacity of the refrigerated van in tons of refrigeration will be

(a) 0.10 tonne (b) 0.35 tonne

(c) 0.55 tonne (d) 0.65 tonne

[ESE : 2010]

1.49 Which substance has the minimum value of thermal conductivity

(a) Air (b) Water

(c) Plastic (d) Rubber [ESE : 2011]

1.50 If the radius of any current carrying conductor is less than the critical radius, then the addition of electrical insulation will enable the wire to carry a higher current because

(a) The heat loss from the wire would decrease

(b) The heat loss from the wire would increase

(c) The thermal resistance of the insulation is reduced

(d) The thermal resistance of the conductor is increased

[ESE : 2011]

1.51 If the thermal conductivity of a material of wall varies as $k_0(1 + \alpha t)$ then the temperature at the centre of the wall as compared to that in case of constant thermal conductivity will be

(a) more

(b) less

(c) same

(d) possible in all as above

[ESE : 2011]

1.52 The unit of which of the following parameter is not m^2/s

(a) thermal diffusivity (b) kinematic viscosity

(c) mass diffusivity (d) dynamic viscosity

[ESE : 2011]

1.53 Which of the following expressions gives the thermal resistance for heat conduction through a hollow sphere of radii r_1 and r_2?

(a) $\frac{4\pi k r_1 r_2}{r_2 - r_1}$ (b) $\frac{(r_2 - r_1)\ln\frac{r_2}{r_1}}{4\pi k}$

(c) $\frac{r_2 - r_1}{4\pi k r_1 r_2}$ (d) $\frac{4\pi k(r_2 - r_1)}{r_1 r_2}$

[ESE : 2012]

1.54 A furnace has a 20 cm thick wall with thermal conductivity 0.8 W/mK. For the same heat loss from the furnace, what will be the thickness of the wall if the thermal conductivity of the material is 0.16 W/mK?

(a) 4 cm (b) 6.3 cm

(c) 10 cm (d) 40 cm

[ESE : 2012]

1.55 A furnace wall is 10 cm thick and has a thermal conductivity of 0.1 kW/m-K. Inner temperature is maintained at 525°C, while the surrounding temperature outside the furnace is 25°C. If the surface area of the furnace is 20 m^2, the heat flux through the wall is

(a) 50 kW/m^2 (b) 500 kW/m^2

(c) 1000 kW/m^2 (d) 10000 kW/m^2

[ESE : 2012]

1.56 Which of the following properties of air increase with rise in temperature?

1. Specific gravity
2. Specific heat
3. Thermal conductivity
4. Kinematic viscosity

(a) 1, 2, 3 and 4 (b) 1, 2 and 3 only
(c) 2, 3 and 4 only (d) 1 and 4 only

[ESE : 2012]

1.57 The Poisson's equation of general conduction heat transfer applies to the case

(a) steady state heat conduction with heat generation
(b) steady state heat conduction without heat generation
(c) unsteady state heat conduction without heat generation
(d) unsteady state heat conduction with heat generation

[ESE : 2013]

1.58 A plane slab of 100 mm thickness generates heat. It is observed that the temperature drop between the centre and its surface to be 50°C. If the thickness is increased to 20 cm the temperature difference will be

(a) 100°C (b) 200°C
(c) 400°C (d) 600°C [ESE : 2013]

1.59 Uniform heat generation takes place in a symmetric slab so that heat flows towards both sides in contact with fluid. The zero-gradient boundary condition $\frac{\partial T}{\partial x} = 0$ occurs at

(a) left wall of slab (b) right wall of slab
(c) centreline of slab (d) nowhere in slab

[ESE : 2013]

1.60 The conduction heat diffuses in a material when the materials has:

1. High thermal conductivity
2. Low density
3. High specific heat
4. High viscosity

Which of the above are correct?

(a) 1 and 2 (b) 2 and 3
(c) 3 and 4 (d) 4 and 1 [ESE : 2014]

1.61 In an equation of Fourier law of heat conduction, heat flow through a body per unit time is $Q = -kA\frac{dT}{dx}$, the negative sign of k in this equation is to take care of

(a) decreasing temperature along the direction of increasing thickness
(b) increasing temperature along the direction of increasing thickness
(c) constant temperature along the direction with constant thickness
(d) all of the above [ESE : 2014]

1.62 A flat wall with a thermal conductivity of 0.2 kW/mK has its inner and outer surface temperatures 600°C and 200°C respectively. If the heat flux through the wall is 200 kW/m^2, what is the thickness of the wall?

(a) 10 cm (b) 20 cm
(c) 30 cm (d) 40 cm

[ESE : 2014]

1.63 There is a uniform distributed source of heat present in a plane wall whose one side ($x = 0$) is insulated and other side ($x = L$) is exposed to ambient temperature (T_∞), with heat transfer coefficient (h). Assuming constant thermal conductivity (k), steady state and one dimensional conduction, the temperature of the wall is maximum at x equal to

(a) 0 (b) L
(c) $L/2$ (d) $L/4$ [ESE : 2015]

1.64 An insulating material with a thermal conductivity, k = 0.12 W/mK is used for a pipe carrying steam. The local coefficient of heat transfer (h) to the surroundings is 4 W/m^{2}K. In order to provide effective insulation, the minimum outer diameter of the pipe should be

(a) 45 mm (b) 60 mm
(c) 75 mm (d) 90 mm

[ESE : 2015]

1.65 A plane wall is 20 cm thick with an area perpendicular to heat flow of 1 m^2 and has a thermal conductivity of 0.5 W/mK. A temperature difference of 100°C is imposed across it. The rate of heat flow is

(a) 0.10 kW (b) 0.15 kW
(c) 0.20 kW (d) 0.25 kW

[ESE : 2015]

1.66 In a wall of constant thermal conductivity, the temperature profile for heat conduction in the presence of a heat source inside the wall is
(a) linear (b) logarithmic
(c) parabolic (d) hyperbolic
[ESE : 2015]

1.67 Determine the heat transfer through a plane of length 4 m, height 3 m and thickness 0.2 m. The temperatures of inner and outer surfaces are 150°C and 90°C respectively. Thermal conductivity of the wall is 0.5 W/mK.
(a) 1800 W (b) 2000 W
(c) 2200 W (d) 2400 W
[ESE : 2015]

1.68 In a furnace the heat loss through the 150 mm thick refractory wall lining is estimated to be 50 W/m^2. If the average thermal conductivity of the refractory material is 0.05 W/mK, the temperature drop across the wall will be :
(a) 140°C (b) 150°C
(c) 160°C (d) 170°C
[ESE : 2016]

1.69 A plane wall is 20 cm thick with an area of 1 m^2 and has a thermal conductivity of 0.5 W/mK. A temperature difference of 100°C is imposed across it. The heat flow is at:
(a) 150 W (b) 180 W
(c) 220 W (d) 250 W
[ESE : 2016]

1.70 Consider the following statements about thermal conductivity :
1. Thermal conductivity decreases with increasing molecular weight
2. Thermal conductivity of non-metallic liquids generally decreases with increasing temperature
3. Thermal conductivity of gases and liquids is generally smaller than that of solids

Which of the above statements are correct ?
(a) 1 and 2 only (b) 1 and 3 only
(c) 2 and 3 only (d) 1, 2 and 3
[ESE : 2016]

1.71 A furnace is provided with an insulating refractory lining. The overall thermal conductivity of the material is 0.03 W/mK. The thickness of the lining is 100 mm. The inner and outer temperatures are 250°C and 50°C, respectively. The heat loss to the surroundings will be
(a) 30 J/m^2/s (b) 60 J/m^2/s
(c) 60 J/s (d) 30 J/s
[ESE : 2017]

1.72 A wall of 0.6 m thickness has normal area of 1.5 m^2 and is made up of material of thermal conductivity 0.4 W/mK. If the temperatures on the two sides of the wall are 800°C and 1000°C, the thermal resistance of the wall is
(a) 1.8 K/W (b) 1.8 W/K
(c) 1 K/W (d) 1 W/K
[ESE : 2017]

1.73 A wall surface of 200 mm thickness has an outside temperature of 50°C and inside temperature of 25°C with thermal conductivity of 0.51 W/mK. The heat transfer through this wall will be
(a) 63.75 W/m^2 (b) 65.75 W/m^2
(c) 70.25 W/m^2 (d) 73.25 W/m^2
[ESE : 2018]

1.74 The refractory lining of a furnace has a thickness of 200 mm. The average thermal conductivity of the refractory material is 0.04 W/mK. The heat loss is estimated to be 180 kJ/hr/m^2. The temperature difference across the lining will be
(a) 280°C (b) 250°C
(c) 240°C (d) 220°C
[ESE : 2018]

1.75 It is desired to increase the heat dissipation rate from the surface of an electronic device of spherical shape of 5 mm radius exposed to convection with h = 10 W/m^{2}K by encasing it in a spherical sheath of conductivity 0.04 W/mK. For maximum heat flow, the critical diameter of the sheath shall be
(a) 20 mm (b) 18 mm
(c) 16 mm (d) 12 mm
[ESE : 2018]

1.76 A cold storage has 23 cm brick wall on the outside and 8 cm plastic foam on the inside. The inside and outside temperatures are –2°C and 22°C, respectively. If the thermal conductivities of brick and foam are 0.98 W/m-K and 0.02 W/m-K, and the inside and outside heat transfer coefficients are 29 W/m^2-K and 12 W/m^2-K, respectively, then the rate of heat removal for a (projected) wall area of 90 m^2 will nearly be

(a) 503 W (b) 497 W
(c) 490 W (d) 481 W

[ESE : 2018]

1.77 A wire of 8 mm diameter at a temperature of 60°C is to be insulated by a material having k = 0.174 W/m K. The heat transfer coefficient on the outside h_a = 8 W/m^{2}K and ambient temperature T_a = 25°C. The maximum thickness of insulation for maximum heat loss will be:

(a) 15.25 mm (b) 16.50 mm
(c) 17.75 mm (d) 18.25 mm

[ESE : 2019]

1.78 What is the critical radius of insulation for asbestos (thermal conductivity = 0.17 W.m^{-1}°C^{-1}) surrounding a circular pipe and exposed to room air at 20°C with heat transfer coefficient 3 W.m^{-2}°C^{-1}?

(a) 7.21 cm (b) 6.37 cm
(c) 5.67 cm (d) 6.93 cm

[ESE : 2021]

1.79 The resistance wire of a 1200 W hair dryer is 80 cm long and has a diameter of 0.3 cm. What is the rate of heat generation in the wire per unit volume?

(a) 212 W/cm^3 (b) 312 W/cm^3
(c) 512 W/cm^3 (d) 412 W/cm^3

[ESE : 2022]

1.80 A 2 kW resistance heater wire with thermal conductivity 15 W/m°C, diameter 0.4 cm and length 50 cm is used to boil the water by immersing it in water. Assuming the variation of the thermal conductivity of the wire with temperature to be negligible. What is the differential equation that describes the variation of the temperature in the wire during steady operation?

(a) $\frac{1}{r}\frac{d}{dr}\left(r\frac{dT}{dr}\right)+\frac{g}{k}=0$

(b) $\frac{1}{2r}\frac{dT}{dr}\left(r\frac{dT}{dr}\right)+\frac{g}{k}=0$

(c) $\frac{1}{r}\frac{d}{dr}\left(\frac{dT}{dr}\right)+\frac{g}{k}=0$

(d) $\frac{1}{r}\frac{d}{dr}\left(r\frac{dT}{dr}\right)=0$

[ESE : 2022]

1.81 A 2 kW resistance heater wire whose thermal conductivity is 15 W/m°C has a diameter of 4 mm and a length of 0.5 m, is used to boil the water. If the outer surface temperature of resistance wire is 105°C, what is the temperature at the centre of the wire?

(a) 136°C (b) 126°C
(c) 146°C (d) 156°C

[ESE : 2022]

1.82 The thermal contact conductance at the interface of two 1 cm thick aluminium plates is measured to be 11000 W/m^2°C. What is the thickness of the aluminium plate whose thermal resistance is equal to the thermal resistance of the interface between the plates? (Take the thermal conductivity of aluminium at room temperature as 237 Wm°C)

(a) 3.15 cm (b) 5.13 cm
(c) 1.28 cm (d) 2.15 cm

[ESE : 2022]

1.83 **Statement (I):** Heat is defined as the form of energy that is transferred between the system and surroundings due to the temperature difference between them.

Statement (II): The temperature difference is the driving force or potential for heat transfer.

(a) Both statement (I) and statement (II) are individually true statement (II) is the corect explanation of statement (I).

(b) Both statement (I) and statement (II) are individually true, but statement (II) is not the corect explanation of statement (I).

(c) Statement (I) is true, but Statement (II) is false.
(d) Statement (I) is false, but Statement (II) is true.

[ESE : 2023]

1.84 The inner surface of a plane brick wall is at 40°C, the outer surface is at 20°C, the wall is 250 mm thick and the thermal conductivity of the brick is 0.52 W/mK. What is the rate of heat transfer per unit area of wall surface?

(a) 24.9 W/m² (b) 34.9 W/m²
(c) 41.6 W/m² (d) 51.6 W/m²

[ESE : 2023]

1.85 The equation for two-dimensional steady-state conduction without heat sources is

(a) $\frac{d^2T}{dr^2}+\frac{q}{k}=0$ (b) $\frac{d^2T}{dx^2}+\frac{1}{r}\frac{dT}{dr}=0$

(c) $\frac{d^2T}{dx^2}=0$ (d) $\frac{\partial^2T}{\partial x^2}+\frac{\partial^2T}{\partial y^2}=0$

[ESE : 2024]

■■■■

Answers Conduction

1.1	(d)	1.2	(a)	1.3	(a)	1.4	(d)	1.5	(a)	1.6	(a)	1.7	(a)	1.8	(a)	1.9	(a)
1.10	(c)	1.11	(d)	1.12	(a)	1.13	(c)	1.14	(c)	1.15	(a)	1.16	(a)	1.17	(d)	1.18	(b)
1.19	(b)	1.20	(d)	1.21	(a)	1.22	(d)	1.23	(c)	1.24	(c)	1.25	(b)	1.26	(c)	1.27	(c)
1.28	(d)	1.29	(c)	1.30	(a)	1.31	(a)	1.32	(d)	1.33	(b)	1.34	(b)	1.35	(b)	1.36	(b)
1.37	(a)	1.38	(d)	1.39	(c)	1.40	(c)	1.41	(d)	1.42	(b)	1.43	(c)	1.44	(b)	1.45	(a)
1.46	(c)	1.47	(c)	1.48	(d)	1.49	(a)	1.50	(b)	1.51	(a)	1.52	(d)	1.53	(c)	1.54	(a)
1.55	(b)	1.56	(c)	1.57	(a)	1.58	(b)	1.59	(c)	1.60	(a)	1.61	(a)	1.62	(d)	1.63	(a)
1.64	(b)	1.65	(d)	1.66	(c)	1.67	(a)	1.68	(b)	1.69	(d)	1.70	(d)	1.71	(b)	1.72	(c)
1.73	(a)	1.74	(b)	1.75	(c)	1.76	(b)	1.77	(c)	1.78	(c)	1.79	(a)	1.80	(a)	1.81	(b)
1.82	(d)	1.83	(b)	1.84	(c)	1.85	(d)										

Explanations Conduction

1.1 (d)

$$\frac{1}{r}\cdot\frac{d}{dr}\left\{\frac{r\cdot dT}{dr}\right\}=0$$

$$d\left(\frac{r\cdot dT}{dr}\right)=0$$

Integrating both sides

$$\frac{r\cdot dT}{dr}=C_1$$

$$dT=\frac{C_1\cdot dr}{r}$$

$$T=C_1\ln r+C_2$$

Temperature variation will be logarithmic for cylinder.

1.2 (a)

$$r_c=\frac{k}{h}=\frac{0.5}{10}=0.05\text{ m}=50\text{ mm}$$

New radius = 0.5 + 1 + 10 mm = 11.5 mm

Since, radius is less then critical radius hence insulation provided will improve the heat loss and current carring capacity will increase.

1.3 (a)

$$UA = \frac{1}{\Sigma R} = \frac{1}{\frac{1}{h} + \sum_{i=1}^{n} \frac{t_i}{k_i} + \frac{1}{h_o}}$$

1.4 (d)

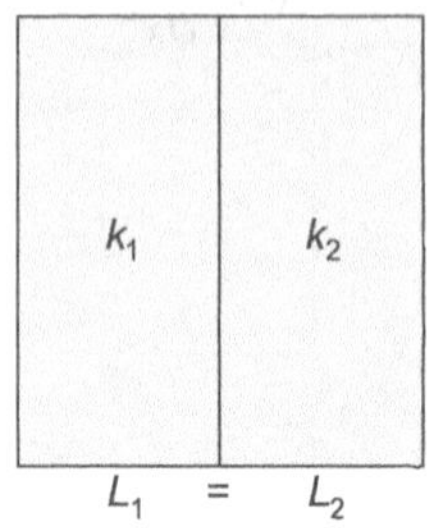

$$R_{eq} = R_1 + R_2$$

$$\frac{L_{eq}}{k_{eq}A} = \frac{L_1}{k_1 A} + \frac{L_2}{k_2 A}$$

Since $L_1 = L_2 = L$

$$\frac{2L}{k_{eq}} = \frac{L}{k_1} + \frac{L}{k_2}$$

$$\frac{k_{eq}}{2} = \frac{k_1 k_2}{k_1 + k_2}$$

$$k_{eq} = \frac{2k_1 k_2}{k_1 + k_2}$$

1.5 (a)

When the refractory brick is set inside, both convective and conductive thermal resistance will increase. Since area of contact between the gases and inside surface will decrease there by increasing convection resistance.

1.6 (a)

In heat pipe the rate of heat transfer is very high because phase change of fluid is made use of because the convection heat transfer coefficient during phase change is very high.

1.7 (a)

For cylinder

$$r_c = \frac{k}{h_o} = \frac{0.1}{5} \times 100 = \mathbf{2\ cm}$$

1.8 (a)

$$t - t_\infty = \frac{q^* L^2}{2k} = \left[1 - \left(\frac{x}{L}\right)^2\right] + \frac{q^* L}{h}$$

At $x = L,\ t = t_s$

$$t_s - t_\infty = \frac{q^* L}{h}$$

$$\Rightarrow \quad t_s = t_\infty + \frac{q^* L}{h}$$

1.9 (a)

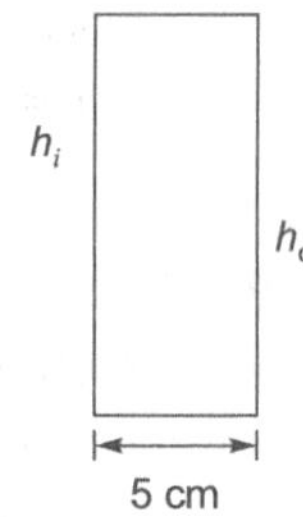

$$U = \frac{1}{\frac{1}{h_i} + \frac{l}{k} + \frac{1}{h_o}}$$

$$U = \frac{1}{\frac{1}{10} + \frac{1}{20} + \frac{5 \times 10^{-2}}{1}}$$

$$= \frac{1}{0.1 + 0.05 + 0.05} = \mathbf{5\ W/m^2K}$$

1.10 (c)

$$\frac{300 - T}{300 - 200} = \frac{25 - k}{25 - 15}$$

$$\therefore \quad k = \frac{T - 50}{10}$$

$$q = -k\frac{dT}{dx} = \int_{300}^{200} \frac{(T-50)dT}{10 \times 0.5}$$

$$= \frac{1}{5}\int_{200}^{300} (TdT - 50dT)$$

$$= \frac{1}{5}\left[\frac{T^2}{2} - 50T\right]_{200}^{300}$$

$$= 4000\ \text{W/m}^2 = \mathbf{4\ kW/m^2}$$

1.11 (d)

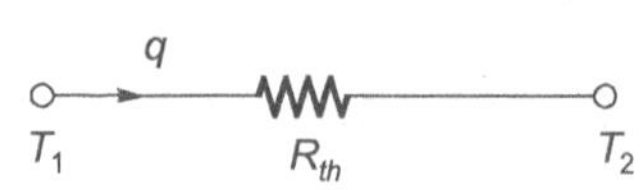

$$q = \frac{\Delta T}{R_{th}}$$

As we know that, $i = \frac{\Delta V}{R}$

A. Voltage — Temperature
B. Current — Heat flow
C. Resistance — Thermal resistance
D. Capacitance — Thermal capacity

1.12 (a)

$$r_c = \frac{k}{h_o} = \frac{0.1}{5} = 2\ \text{cm}$$

Critical thickness of insulation

$$= 2 - 1 = \mathbf{1\ cm}$$

1.13 (c)

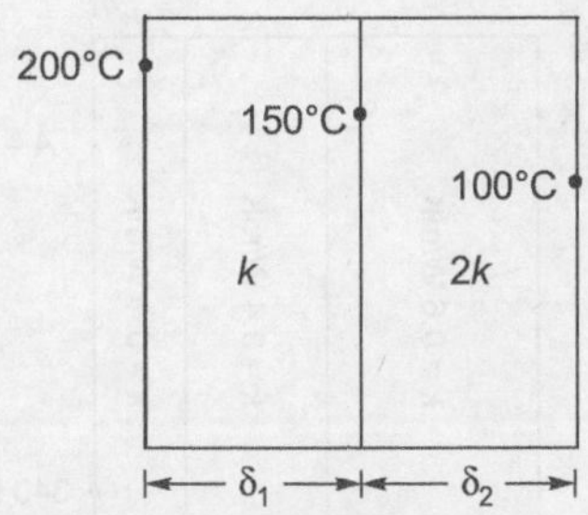

$$q = \frac{200-150}{\frac{\delta_1}{k}} = \frac{150-100}{\frac{\delta_2}{2k}}$$

$$\Rightarrow \frac{50\times k}{\delta_1} = \frac{50\times 2k}{\delta_2} = \frac{\delta_1}{\delta_2} = \frac{1}{2}$$

1.14 (c)

$$r_c = \frac{k}{h} = \frac{1}{5} = 20\text{ cm}$$

Thickness = 20 cm – 0.5 cm = **19.5 cm**

1.15 (a)

$$Q = \frac{4\pi r_1 r_2 k(T_1 - T_2)}{r_2 - r_1}$$

$$A_m = 4\pi r_m^2 = 4\pi r_1 r_2$$

$$\therefore \quad r_m = \sqrt{r_1 r_2}$$

1.16 (a)

Upto critical radius, added insulation inhances the heat transfer.

1.17 (d)

$$\frac{d^2\theta}{dx^2} - m^2\theta = 0 \text{ pin fin } m = \sqrt{\frac{HP}{kA}}$$

$$\frac{\partial^2 T}{\partial x^2} = \frac{1}{\alpha}.\frac{dT}{dt} \quad \text{For plane slab}$$

(α = Thermal diffusivity)

$$\frac{d^2T}{dr^2} + \frac{1}{r}.\frac{dT}{dr} = 0 \quad \text{For cylinder}$$

$$\frac{d^2T}{dr^2} + \frac{2}{r}.\frac{dT}{dr} = 0 \quad \text{For sphere}$$

1.18 (b)

$$Q = kA\frac{dT}{dx} \qquad (A = 10\text{ cm}^2)$$

$$\frac{6.0\times 10^3}{10\times 60} = 100\times 10\times 10^{-4} \times \frac{\Delta T}{1}$$

$$\Delta T = \mathbf{100°C}$$

1.19 (b)

$$Q = kA\frac{\Delta T}{l}$$

$$\Rightarrow \quad \frac{Q}{A} = \frac{k\Delta T}{l}$$

$$30 = \frac{k\times 10}{10\times 10^{-2}}$$

$$\Rightarrow \quad k = \frac{30}{100} = \mathbf{0.3\ W/mK}$$

1.20 (d)

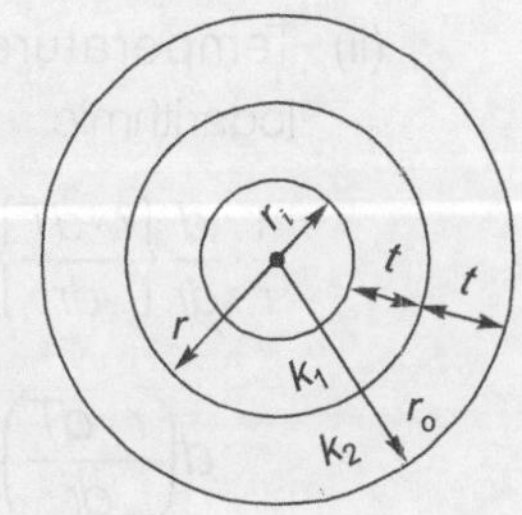

$$R_{\text{shpere}} = \frac{r_2 - r_1}{4\pi k r_1 r_2}$$

$$Q = \frac{\Delta T}{R_{\text{sphere}}}$$

$$\frac{d_i}{d_o} = 0.8$$

$$\therefore \quad \frac{r_i}{r_o} = 0.8$$

$$r_o - r_i = 2t$$

$$\therefore t = 0.1 r_o$$

$$r = 0.8 r_o + 0.1\, r_o = 0.9 r_o$$

$$\because \quad \frac{k_1}{k_2} = \frac{1}{2} \text{ (given)}$$

$$\therefore \quad \frac{\Delta T_{\text{inner}}}{\Delta T_{\text{outer}}} = \frac{R_{\text{inner}}}{R_{\text{outer}}}$$

$$= \frac{(0.9r_o - 0.8r_o)}{4\pi k_1 (0.9r_o)(0.8r_o)} \times \frac{4\pi r_o (0.9r_o) k_2}{(r_o - 0.9r_o)}$$

$$= \frac{k_2}{k_1} \times \frac{1}{0.8} = 2\times 1.25 = \mathbf{2.5}$$

1.21 (a)

$$Q = \frac{kA(T_1 - T_2)}{l} = \frac{0.5\times 1\times 60}{0.25}$$

$$= \mathbf{120\ W}$$

1.22 (d)

$$R_{th} = \frac{l}{kA} = \frac{10\times 10^{-3}}{200+2} = \mathbf{2.5\times 10^{-5}\ K/W}$$

1.23 (c)

$$\alpha = \frac{k}{\rho c}$$

1.24 (c)

$$\text{Lorenz number} = \frac{\text{Thermal conductivity}}{\text{Electrical conductivity}}$$

1.25 (b)

$$\left(\frac{k}{\rho C_p}\right)_{cu} > \left(\frac{k}{\rho C_p}\right)_{air}$$

1.26 (c)

$$R_{th} = \frac{kA}{l} = \frac{0.4 \times 1.5}{0.6} = 1 \text{ K/W}$$

1.27 (c)

(i) Temperature distribution in slab is linear

(ii) Temperature distribution in cylinder is logarithmic

$$\frac{1}{r} \cdot \frac{d}{dr}\left\{\frac{r \cdot dT}{dr}\right\} = 0$$

$$d\left(\frac{r \cdot dT}{dr}\right) = 0$$

Integrating both sides

$$\frac{r \cdot dT}{dr} = C_1$$

$$dT = \frac{C_1 \cdot dr}{r}$$

$$T = C_1 \ln r + C_2$$

Temperature variation will be logarithmic for cylinder.

(iii) Temperature distribution in sphere is hyperbolic

$$\frac{d}{dr}\left\{r^2 \frac{dT}{dr}\right\} = 0$$

$$r^2 \frac{dT}{dr} = C_1$$

$$dT = \frac{C_1}{r^2} dr$$

$$T = \frac{-C_1}{r} + C_2$$

1.28 (d)

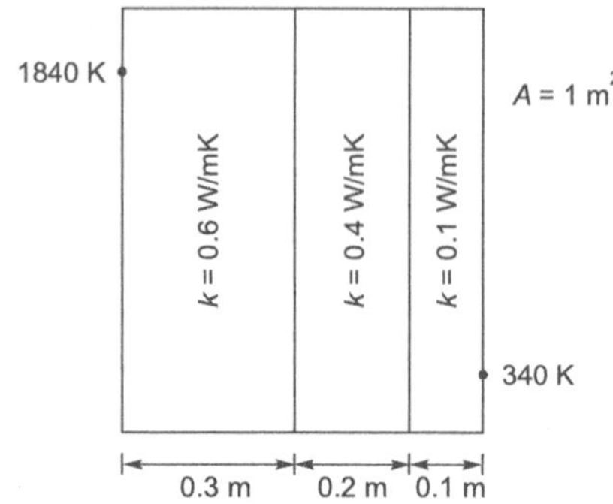

$$Q = \frac{(T_1 - T_2)}{R_{th}}$$

$$= \frac{(1840 - 340)}{\frac{0.3}{0.6 \times 1} + \frac{0.2}{0.4 \times 1} + \frac{0.1}{0.1 \times 1}} = \mathbf{750\ W}$$

1.29 (c)

For hollow sphere

$$Q = \frac{4\pi k r_1 r_2 (T_1 - T_2)}{(r_2 - r_1)}$$

Rearranging, we get

$$Q = \frac{T_1 - T_2}{\frac{(r_2 - r_1)}{4\pi k r_1 r_2}}$$

1.30 (a)

$$(r_o)_{critical} = \frac{2k}{h} \quad \text{(For sphere)}$$

$$= \frac{2 \times 0.5}{10} = \mathbf{0.1\ m}$$

1.31 (a)

$$\frac{k_1 A(\Delta T_1)}{dx} = \frac{k_2 A(\Delta T_2)}{dx}$$

$$\Rightarrow \quad k_1(\Delta T_1) = k_2(\Delta T_2)$$

$$\Rightarrow \quad \frac{\Delta T_1}{\Delta T_2} = \frac{k_2}{k_1} = \frac{2}{3}$$

1.32 (d)

Due to minimum thermal conductivity of air heat conduction is minimum in air.

1.33 (b)

Heat transfer through solid	:	Fourier's law of heat conduction
Heat transfer from	:	Newton's law of

hot surface to surrounding fluid	cooling
Heat transfer in boiling liquid	: Convection heat transfer
Heat transfer from one body to another separated in space	: Radiation heat transfer

1.34 (b)

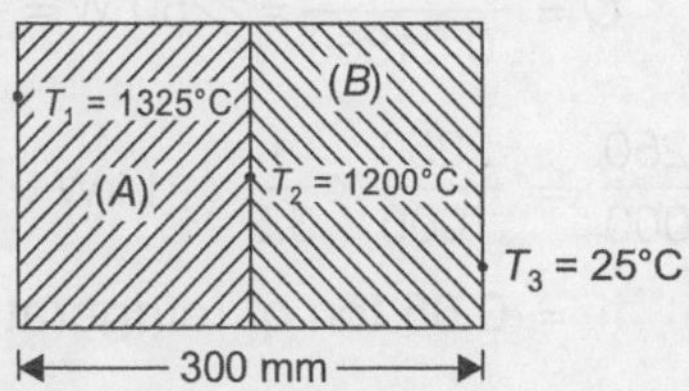

$$\frac{k_A(1325-1200)}{l_A} = \frac{k_B(1200-25)}{l_B}$$

$$\Rightarrow \frac{l_A}{l_B} = \frac{2\times 125}{1175} = 0.2127 \simeq 0.213$$

1.35 (b)

Critical radius of insulation for sphere is $\frac{2k}{h}$.

1.36 (b)

Fourier's law	: Conduction
Stefan Boltzmann law	: Radiation
Newton's law of cooling	: Convection
Fick's law	: Mass transfer

1.37 (a)

$$\frac{Q_1}{Q_2} = \frac{k_1 A\frac{dT}{dx}}{k_2 A\frac{dT}{dx}}$$

$$\Rightarrow \frac{Q_1}{Q_2} = \frac{k_1}{k_2} = \frac{1}{2}$$

1.38 (d)

Non-metals have lower thermal conductivity and free electrons in metal is higher than non metal therefore (d) is the answer

1.39 (c)

$$Q = -kA\frac{dT}{dx}$$

$$(ML^2T^{-3}) = k(L^2)\frac{(\theta)}{(L)}$$

$$\Rightarrow ML^2T^{-3} = k(L)(\theta)$$

$$\Rightarrow k = \frac{ML^2T^{-3}}{L\theta} = MLT^{-3}\theta^{-1}$$

1.40 (c)

$$Q = -kA\frac{dT}{dx}$$

$$\Rightarrow \frac{Q}{A} = -k\frac{dT}{dx}$$

$$\Rightarrow 25\times 10^3 = 50\times\frac{(100-T_2)}{(0.1)}$$

$$\Rightarrow T_2 = \mathbf{50°C}$$

1.41 (d)

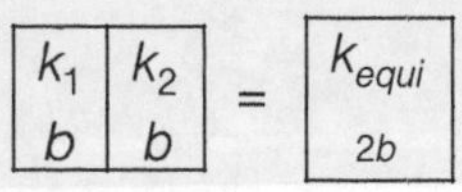

Equating R_{th},

$$\frac{b}{k_1A}+\frac{b}{k_2A} = \frac{2b}{k_{equi}\times A}$$

$$k_{equi} = \frac{2k_1k_2}{k_1+k_2}$$

1.42 (b)

$$\Rightarrow k_1\Delta T_1 = k_2\Delta T_2 = k_3\Delta T_3 = Q$$

$$\Rightarrow \Delta T_1 : \Delta T_2 : \Delta T_3 = \frac{Q}{k_1}:\frac{Q}{k_2}:\frac{Q}{k_3}$$

$$= \frac{1}{1}:\frac{1}{2}:\frac{1}{4} = \mathbf{4:2:1}$$

1.43 (c)

Heat loss per second

$$= \frac{(T_1-T_2)}{\frac{1}{h_1A}+\frac{l}{k_1A}+\frac{1}{h_2A}} = \frac{(30-10)}{\frac{1}{40}\left(\frac{0.115}{1.15}+\frac{1}{2.5}+\frac{1}{4}\right)}$$

$$= \frac{40\times 20}{(0.1+0.4+0.25)} = 1066.66 \text{ kJ/s}$$

$$= \frac{3840,000}{1000}\text{ kJ/hour} = \mathbf{3840\ kJ/hour}$$

1.44 (b)

$$\frac{\partial T}{\partial x} = -10 + 40x + 30x^2$$

$$\frac{\partial^2 T}{\partial x^2} = 40 + 60x$$

$$\left.\frac{\partial^2 T}{\partial x^2}\right|_{x=1} = \frac{1}{\alpha}\frac{\partial T}{\partial \tau}$$

$$40 + 60(1) = \left(\frac{1}{2\times10^{-3}}\right)\left(\frac{\partial T}{\partial \tau}\right)$$

$$\Rightarrow \frac{\partial T}{\partial \tau} = (2\times10^{-3})(100) = \mathbf{0.2°C/hour}$$

1.45 (a)

In porous cavities available in insulating materials the ambient air may go and gets entraped. Air being bad conductor of heat it may offer insulation.

1.46 (c)

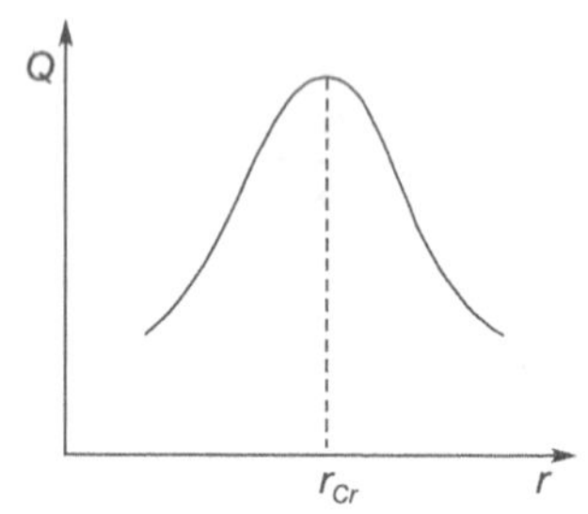

From the above plot between heat transfer rate versus r, it can be concluded that by providing insulation, up to critical radius heat transfer rate increases and become maximum at critical radius and further insulation decreases the heat transfer rate.

1.48 (d)

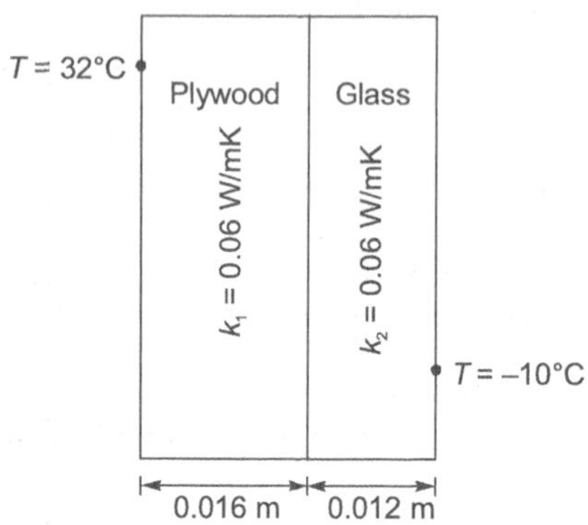

Total thermal resistance

$$R_{th} = \frac{L_1}{k_1 A} + \frac{L_2}{k_2 A}$$

$$= \frac{1}{25}\left(\frac{0.016}{0.06} + \frac{0.012}{0.06}\right) = 0.01866$$

$$Q = \frac{T_1 - T_2}{R_{th}}$$

$$Q = \frac{32 + 10}{0.01866} = 2250 \text{ W} = \frac{2250}{1000} \text{ kW}$$

$$\therefore \frac{2250}{1000} = \frac{2250}{1000} \times \frac{1}{3.5} \left(\because 1 \text{ kW} = \frac{1}{3.5} \text{ tonne}\right)$$

$= \mathbf{0.65}$ **ton of refrigeration**

1.49 (a)

Among the following options given value of thermal conductivity of air is minimum.

$$k_{Air} = \mathbf{0.024\ W/mK}$$

1.50 (b)

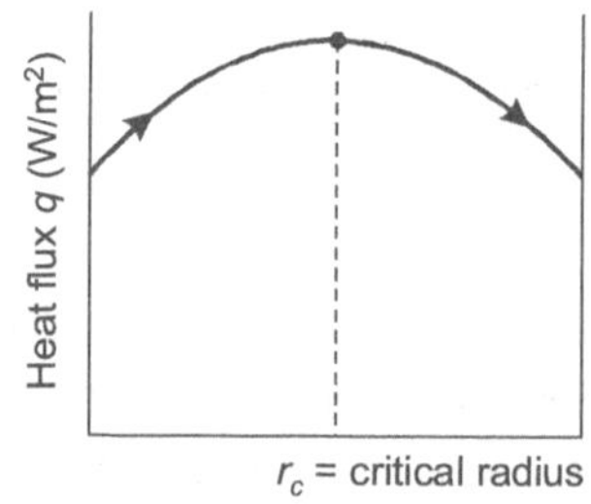

If the radius of any current carrying conductor is less than the critical radius, then addition of electrical insulation will enable to carry higher current because heat loss from the wire would increase.

1.51 (a)

$$k = k_0(1 + \alpha t)$$

where, k_0 = thermal conductivity at 0°C

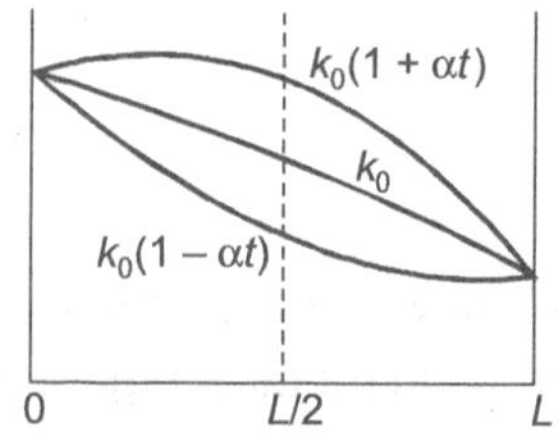

So, temperature at centre of the wall is more.

1.52 (d)

Unit of dynamic viscosity = **Ns/m² = Pa.s**

1.53 (c)

$$dQ = k(A_x)\frac{dT}{dx}$$

$$\text{Thermal resistance} = \frac{dT}{dQ}$$

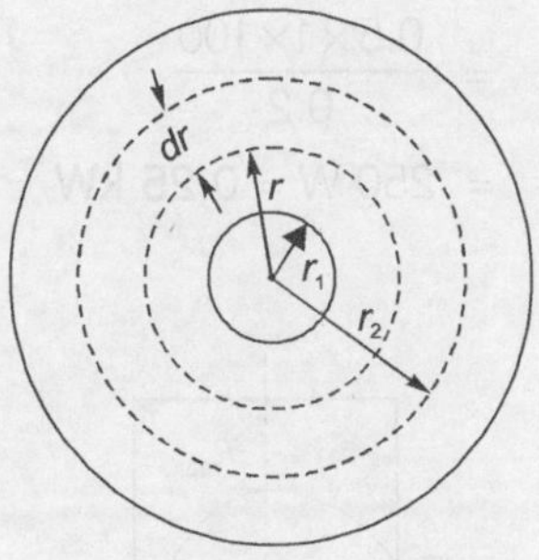

$$dR = \frac{dx}{k(A_x)}$$

Here $dx = dr$

$A_x = 4\pi r^2$

$$\therefore \quad dR = \frac{dr}{k4\pi r^2}$$

$$R = \int dR = \int_{r_1}^{r_2} \frac{dr}{4\pi k r^2}$$

$$\therefore \quad R = \frac{r_2 - r_1}{4\pi k r_1 r_2}$$

1.54 (a)

$$Q = kA\frac{dT}{dx}$$

$$\therefore \quad Q \propto \frac{k}{dx}$$

As Q is same is both cases

$$\therefore \quad \frac{k_1}{(dx)_1} = \frac{k_2}{(dx)_2}$$

$$\frac{0.8}{20} = \frac{0.16}{(dx)_2}$$

$(dx)_2$ = **4 cm**

1.55 (b)

$$\text{Heat flux, } q = \frac{Q}{A} = k\frac{dT}{dx} = 0.1\times\frac{(525-25)}{10\times10^{-2}}$$

= **500 kW/m²**

1.56 (c)

With increasing temperature, density of air decrease and hence the specific gravity decreases.

1.57 (a)

Generalized heat conduction equation is

$$\nabla^2 T + \frac{q_G}{k} = \frac{1}{\alpha}\frac{\partial T}{\partial t}$$

Laplace's equation, $\nabla^2 T = 0$

Poisson's equation, $\nabla^2 T + \frac{q_G}{k} = 0$

Here, $\frac{1}{\alpha}\frac{\partial T}{\partial t} = 0 \quad \therefore$ Steady state

So, Poisson's equation applies to steady state heat condition with heat generation.

1.58 (b)

$$T_{max} = T_1 + \frac{q_g L^2}{8K}$$

$$\Delta T = T_{max} - T_1 = \frac{q_g L^2}{8K}$$

$$\text{So,} \quad \frac{\Delta T_1}{\Delta T_2} = \left(\frac{L_1}{L_2}\right)^2$$

$$\therefore \quad \Delta T_2 = \Delta T_1\left(\frac{L_2}{L_1}\right)^2$$

$$\text{or} \quad \Delta T_2 = 50\times\left(\frac{200}{100}\right)^2$$

= 50 × 4 = **200°C**

1.59 (c)

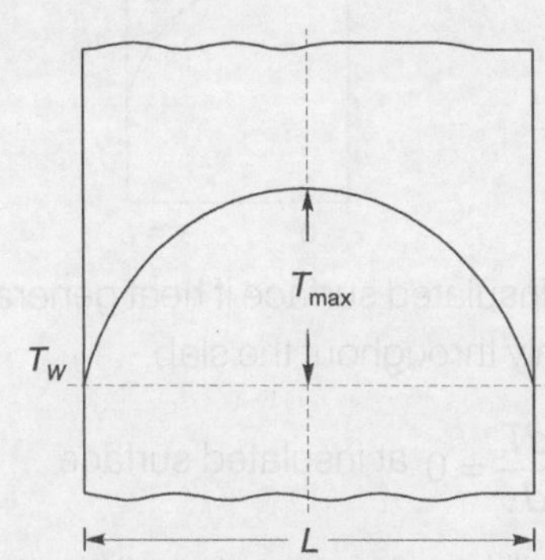

Temperature is maximum at the centreline.

$$\text{So,} \quad \left.\frac{\partial T}{\partial x}\right|_{L/2} = 0$$

1.60 (a)

Conduction heat diffuses faster in a material when its thermal diffusivity is high

i.e. $\alpha = \dfrac{k}{\rho c_p}$ is high.

For α to be high,

(i) k should be high

(ii) Density should be low

(iii) Specific heat should be low.

1.61 (a)

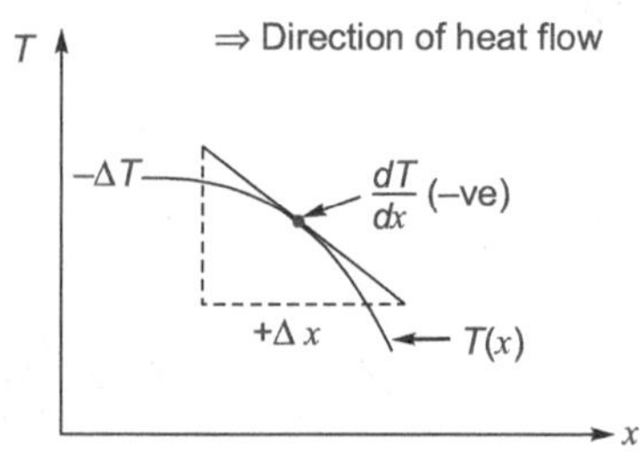

The negative sign is used because heat flows from high to a low temperature and the slope (dT/dx) is negative.

1.62 (d)

$$\frac{q}{A} = 200 \text{ w/m}^2 = \frac{0.2 \times (600-200)}{b}$$

$$\Rightarrow \quad b = 0.4 \text{ m} = 40 \text{ cm}$$

1.63 (a)

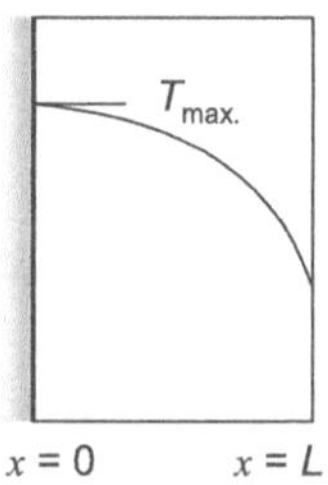

For an insulated surface if heat generated uniformly throughout the slab,

Since, $\dfrac{dT}{dx} = 0$ at insulated surface

Temp will be maximum at $x = 0$

1.64 (b)

For effective insulation

$$r_c = \frac{k}{h} = \frac{0.12}{4} = 0.03 \text{ m}$$

$\therefore$ Diameter = **60 mm**

1.65 (d)

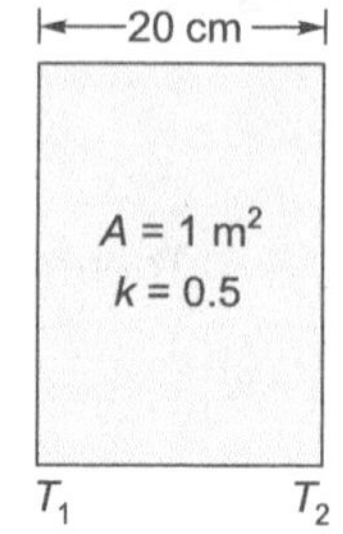

$$T_1 - T_2 = 100°\text{C}$$

$$Q = kA\frac{(T_1 - T_2)}{L}$$

$$= \frac{0.5 \times 1 \times 100}{0.2}$$

$$= 250 \text{ W} = \mathbf{0.25\ kW}$$

1.66 (c)

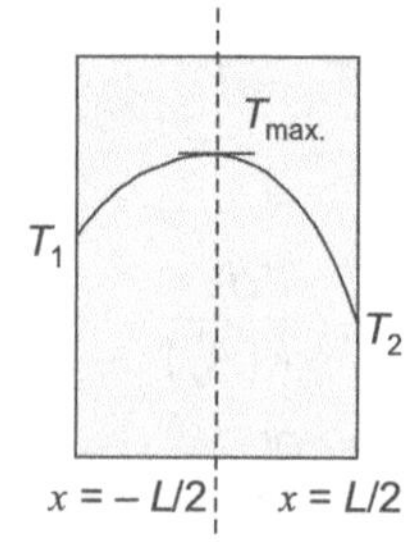

In the pressure of heat generation temperature profile is parabolic and maximum temperature is within the body and not at the walls.

1.67 (a)

$$Q = kA\frac{(T_1 - T_2)}{L} = \frac{0.5 \times 4 \times 3 \times 60}{0.2}$$

$$= \mathbf{1800\ W}$$

1.68 (b)

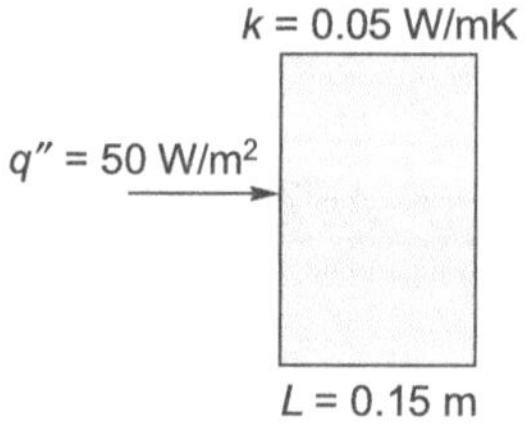

$$q'' = \frac{k\Delta T}{L}$$

$$\Delta T = \frac{50 \times 0.15}{0.05}$$

$$= \mathbf{150°C}$$

1.69 (d)

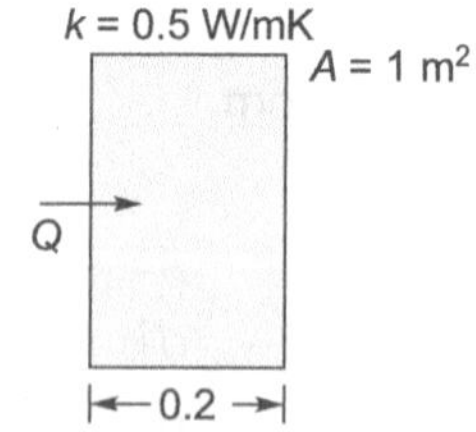

$$\Delta T = 100$$

$$Q = \frac{kA\Delta T}{L} = \frac{0.5 \times 1 \times 100}{0.2} = \mathbf{250\ W}$$

1.71 (b)

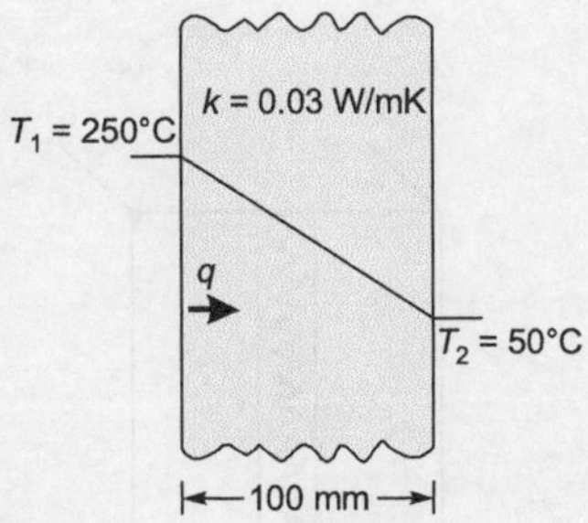

Heat flux, $q = \dfrac{Q}{A} = \dfrac{k(T_1 - T_2)}{b}$

$$= \frac{0.03(250-50)}{0.1} \quad \text{(Fourier's Law)}$$

$= 60$ J/m^2-sec

1.72 (c)

$$(R_{th})_{slab} = \frac{b}{kA} = \frac{0.6}{0.4 \times 1.5} = 1 \text{ K/W}$$

1.73 (a)

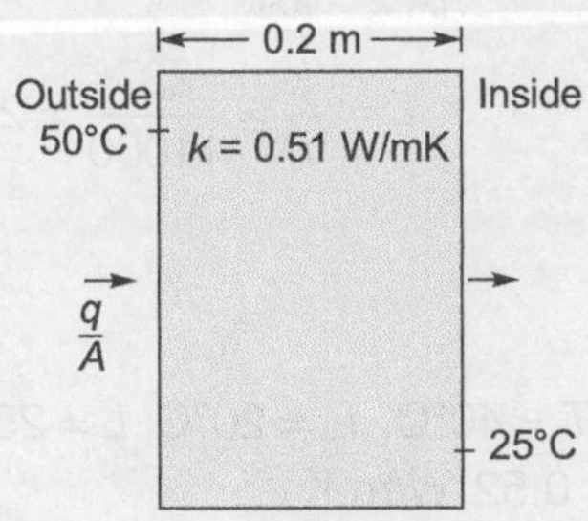

$k = 0.51$ W/m-K

$$\dot{q} = \frac{\dfrac{50-25}{0.2}}{0.51 \times A}$$

$$\Rightarrow \quad \frac{\dot{q}}{A} = 63.75 \text{ W/m}^2$$

1.74 (b)

$$\frac{\dot{Q}}{A} = k\frac{\partial T}{\partial x}$$

$$\frac{180 \times 10^3}{3600} = 0.04 \times \frac{\Delta T}{0.2}$$

$\Delta T = 250°C$

1.75 (c)

Critical radius of insulation for sphere,

$$= \frac{2k}{h} = \frac{2 \times 0.04}{10} \times 1000 \text{ mm}$$

$= 8$ mm

Critical diameter $= 2 \times 8 = 16$ mm

1.76 (b)

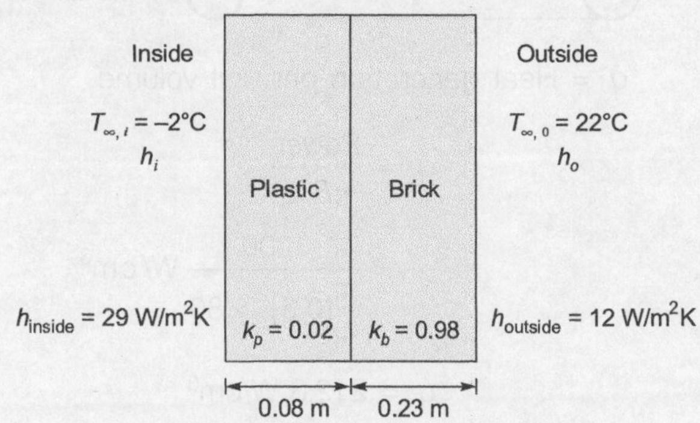

$k_b = 0.98$ W/m-K; $k_p = 0.02$ W/m-K

$A = 90$ m^2; $h_i = 29$ W/m^{2}K

$h_o = 12$ W/m^{2}K

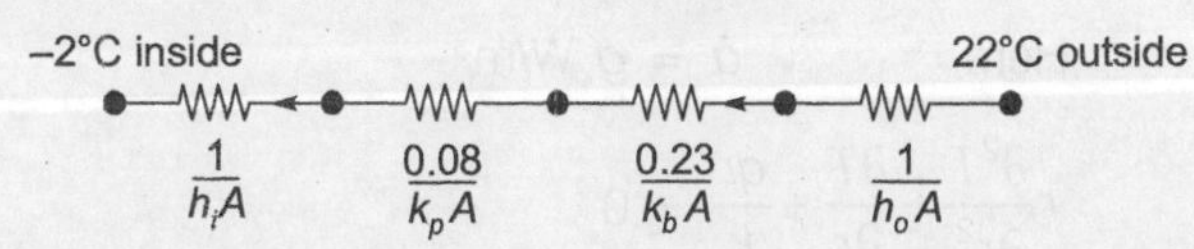

$$\dot{Q} = \frac{22-(-2)}{\dfrac{1}{29 \times 90} + \dfrac{0.08}{0.02 \times 90} + \dfrac{0.23}{0.98 \times 90} + \dfrac{1}{12 \times 90}}$$

$$\dot{Q} = 496.26 \text{ W}$$

1.77 (c)

Critical radius of insulation,

$$r_{critical} = \frac{k_{Ins}}{h} = \left(\frac{0.174}{8}\right) \text{m}$$

$= 0.02175$ m

$\therefore$ Critical thickness of insulation

$= r_{critical} - r_{wire}$

$= (0.02175 - 0.004)$ m

$= 17.75$ mm

1.78 (c)

$$r_{critical} = \frac{k_{ins}}{h} = \left(\frac{0.17}{3}\right) \text{m} = 5.67 \text{ cm}$$

1.79 (a)

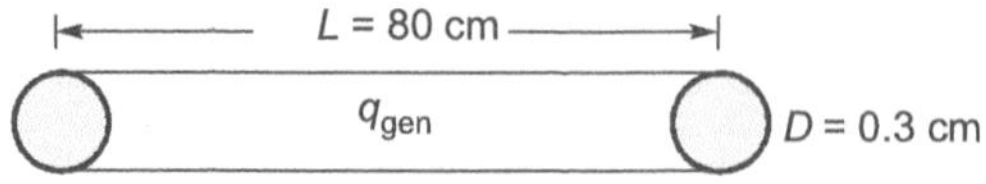

$\dot{q}$ = Heat generation per unit volume

$$\dot{q} = \frac{\text{Power}}{\pi R^2 L}$$

$$= \frac{1200}{\frac{\pi}{4}(0.3)^2 \times 80} \text{ W/cm}^3$$

$$\dot{q} = 212.3 \text{ W/cm}^3$$

1.80 (a)

$$\frac{\partial^2 T}{\partial r^2} + \frac{1}{r}\frac{\partial T}{\partial r} + \frac{1}{r^2}\frac{\partial^2 T}{\partial D^2}^{\,0} + \frac{\partial^2 T}{\partial z^2}^{\,0} + \frac{\dot{q}}{k} = \frac{1}{\alpha}\left(\frac{\partial T}{\partial T}\right)^{0}$$

Here, $\dot{q} = g \text{ W/m}^3$

$$r\frac{\partial^2 T}{\partial r^2} + \frac{\partial T}{\partial r} + \frac{\dot{q}r}{k} = 0$$

$$\frac{\partial}{\partial r}\left(r\frac{\partial T}{\partial r}\right) + \frac{\dot{q}r}{k} = 0$$

$$\frac{1}{r}\frac{\partial}{\partial r}\left(r\frac{\partial T}{\partial r}\right) + \frac{\dot{q}}{k} = 0$$

1.81 (b)

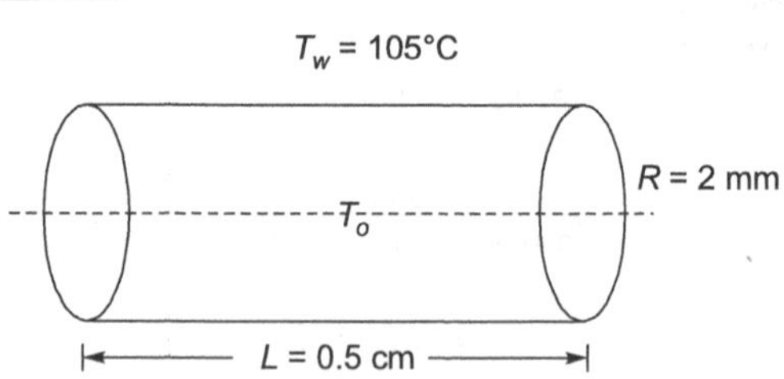

$$T_o - T_w = \frac{\dot{q}R^2}{4k}$$

$$T_o - 105 = \frac{2000}{4\times 15}\times\left(\frac{2}{1000}\right)^2 \times \frac{1}{\pi R^2 \times L}$$

$$= \frac{2000}{4\times 15}\times\frac{1}{\pi \times 0.5}$$

$$T_o = 126°\text{C}$$

1.82 (d)

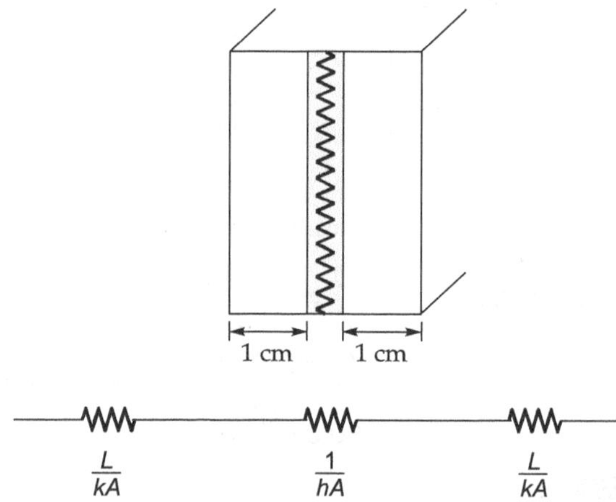

Contact resistance = Plate resistance

$$\frac{1}{hA} = \frac{L}{kA}$$

$$L = \frac{k}{h} = \frac{237}{11000} = 2.15 \text{ cm}$$

1.84 (c)

Given: T_i = 40°C, T_o = 20°C, L = 250 mm = 0.25 m, k = 0.52 W/m-K

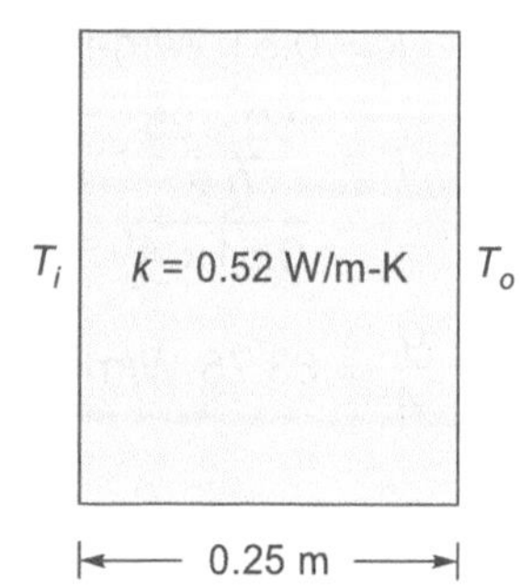

$$q = \frac{\Delta T}{R_{th}} = \frac{\Delta T}{\frac{L}{kA}} = \frac{kA\Delta T}{L}$$

$$\therefore \quad \frac{q}{A} = q'' = \frac{(\Delta T)k}{L} = \frac{(40-20)\times 0.52}{0.25}$$

$$\therefore \quad q'' = 41.6 \text{ W/m}^2$$

■■■■

2 Unsteady State Conduction Heat Transfer

2.1 Match **List-I** with **List-II** and select the correct answer using the codes given below the lists:

List-I	List-II
A. Fin	1. UA/C_{min}
B. Heat exchanger	2. $x/2\sqrt{\alpha\tau}$
C. Transient conduction	3. $\sqrt{HP/kA}$
D. Heisler chart	4. hL/k

Codes:

	A	B	C	D
(a)	3	1	2	4
(b)	3	1	4	2
(c)	3	4	2	1
(d)	2	4	3	1

[ESE : 2000]

2.2 The curve for unsteady state cooling or heating of bodies is

(a) parabolic curve asymptotic to time axis
(b) exponential curve asymptotic to time axis
(c) exponential curve asymptotic both to time and temperature axis
(d) hyperbolic curve asymptotic both to time and temperature axis

[ESE : 2004]

2.3 **Assertion (A):** In lumped heat capacity systems the temperature gradient within the system is negligible.

Reason (R): In analysis of lumped capacity systems, the thermal conductivity of the system material is considered very high irrespective of the size of the system.

(a) both **A** and **R** are true and **R** is the correct explanation of **A**
(b) both **A** and **R** are true but **R** is not a correct explanation of **A**
(c) **A** is true but **R** is false
(d) **A** is false but **R** is true [ESE : 2004]

2.4 Heisler charts are used to determine transient heat flow rate and temperature distribution when:

(a) Solids possess infinitely large thermal conductivity
(b) Internal conduction resistance is small and convective resistance is large
(c) Internal conduction resistance is large and the convective resistance is small
(d) Both conduction and convection resistance are almost of equal significance

[ESE : 2005]

2.5 A sphere, a cube and a thin circular plate, all made of same material and having same mass are initially heated to a temperature of 250°C and then left in air at room temperature for cooling. Then, which one of the following is correct?

(a) All will be cooled at the same rate
(b) Circular plate will be cooled at lowest rate
(c) Sphere will be cooled faster
(d) Cube will be cooled faster than sphere but slower than circular plate

[ESE : 2008]

2.6 The time constant of a thermocouple is

(a) the time taken to attain 100% of initial temperature difference
(b) the time taken to attain 63.2% of initial temperature difference
(c) the time taken to attain 50% of initial temperature difference
(d) the minimum time taken to record a temperature reading

[ESE : 2010]

2.7 **Assertion (A):** Lumped capacity analysis of unsteady heat conduction assumes a constant uniform temperature throughout a solid body.

Reason (R): The surface convection resistance is very large compared with the internal conduction resistance.

(a) Both **A** and **R** are true and **R** is the correct explanation of **A**
(b) Both **A** and **R** are true but **R** is not a correct explanation of **A**
(c) **A** is true but **R** is false
(d) **A** is false but **R** is true [ESE : 2010]

2.8 Dimensionless time is represented by
(a) Biot number (b) Fourier number
(c) Euler number (d) Graetz number
[ESE : 2011]

2.9 In unsteady-state conduction for bodies with negligible temperature gradients, the time temperature variation curve is
(a) linear (b) parabolic
(c) sinusoidal (d) exponential
[ESE : 2012]

2.10 For quick response of a thermocouple to observe varying temperature of fluids
(a) wire diameter must be large
(b) wire material density must be large
(c) wire material specific heat must be large
(d) wire surface heat transfer coefficients must be large
[ESE : 2012]

2.11 When is transient heat transfer problem considered as a lumped capacity problem?
(a) The internal resistance of the object is negative
(b) The internal resistance of the object is zero
(c) The internal resistance of the object is infinite
(d) The internal resistance of the object is negligible
[ESE : 2012]

2.12 Consider the following statements with regard to heat transfer:
1. The temperature variations in lumped heat capacity analysis is exponential with time
2. In situations involving simultaneous heat and mass transfer, the ratio of convective heat transfer to convective mass transfer varies with Lewis number, Le, as $(Le)^{1/3}$?

Which of the above statements is/are correct?
(a) Both 1 and 2 (b) Neither 1 nor 2
(c) 1 only (d) 2 only
[ESE : 2015]

2.13 Assertion (A): Temperature potential difference is a necessary condition for heat interaction between systems.
Reason (R): Heat transfer to a system inevitably increases the temperature of the system.
(a) Both **A** and **R** are true and **R** is the correct explanation of **A**
(b) Both **A** and **R** are true but **R** is not a correct explanation of **A**
(c) **A** is true but **R** is false
(d) **A** is false but **R** is true [ESE : 2015]

2.14 Lumped system analysis assumes a uniform temperature distribution throughout the body, which will be the case only when the thermal resistance of the body to heat conduction is
(a) 3 (b) 2
(c) 1 (d) 0
[ESE : 2022]

2.15 In the design of a fin, the error involved in one dimensional fin analysis is negligible (less than about 1%) when
(a) $\frac{h\delta}{k} < 0.2$ (b) $\frac{hk}{\delta} < 0.2$
(c) $\frac{h\delta}{k} < 2$ (d) $\frac{hk}{\delta} < 2$
[ESE : 2022]

■■■■

Answers Unsteady State Conduction Heat Transfer

2.1 (b) 2.2 (b) 2.3 (a) 2.4 (d) 2.5 (d) 2.6 (b) 2.7 (a) 2.8 (b) 2.9 (d)
2.10 (d) 2.11 (d) 2.12 (c) 2.13 (c) 2.14 (d) 2.15 (a)

Explanations Unsteady State Conduction Heat Transfer

2.1 (b)

For fin $m = \sqrt{\frac{HP}{kA}}$

For heat exchanger: $NTU = UA/C_{min}$

For transient conduction Biot number:

$$Bi = hl/k_{solid}$$

For Heisler chart : $\frac{x}{2\sqrt{\alpha\tau}}$

2.2 (b)

$$\frac{\theta}{\theta_0} = e^{-\frac{hA\tau}{\rho c_p V}}$$

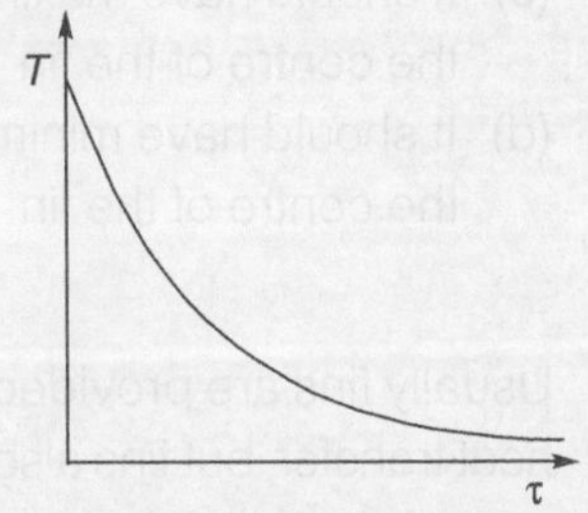

Exponential and asymptotic to time, axis

2.3 (a)

We assume internal temperature gradient is negligible in which conduction resistance offered by body is too small. The internal temperature gradient to be minimized when thermal conductivity is very high.

2.4 (d)

The Heisler charts are extensively used to determine the temperature distribution and heat flow rate when both conduction and convection resistances are almost of equal importance i.e. Biot number is nearly one.

2.5 (d)

For a given volume since the thin circular plate has a very large area of exposure the rate of heat transfer will be maximum. Sphere has got lowest area, therefore cools at the slowest rate.

2.6 (b)

Time constant of the thermocouple represents the time required for the thermo couple to record 63.2% of initial temperature difference.

2.7 (a)

The internal conduction resistance $\left(\frac{L}{kA}\right)$ can be assumed to be small or negligible in comparison with the surface convective resistance $\left(\frac{1}{hA}\right)$.

Lumped capacity analysis of unsteady heat conduction assumes uniform temperature throughout a solid body.

2.8 (b)

Dimensionless time is represented by Fourier number.

Mathematically, Fourier number = $\frac{\alpha t}{L_C^2}$

where α = Thermal diffusivity = $\left(\frac{k}{\rho C}\right)$

L_C = characteristics length; t = Time

2.10 (d)

If surface that transfer coefficient is large then the rate of heat transfer will be more and hence thermocouple will be quick to respond to varying temperature of fluids.

2.11 (d)

For considering a transient heat transfer problem as lump capacity problem. There should be no temperature gradients within the body. This is possible only when the internal resistance of the object is negligeable.

2.12 (c)

Statement 2 is incorrect as ratio of convective heat transfer to convective mass transfer varies with Lewis number Le as $(Le)^{2/3}$.

2.13 (c)

During phase change heat transfer occurs but the temperature throughout the process remain same. Statement (I) is correct but statement (II) is wrong.

2.14 (d)

Biot number (Bi) is given by

$$Bi = \frac{R_{cond}}{R_{conv}}$$

For uniform temperature distribution throughout the body, $R_{cond} = 0$.

$$\Rightarrow \quad Bi = 0$$

■■■■

3 Heat Dissipation From Extended Surfaces (Fins)

3.1 The efficiency of a pin fin with insulated tip is

(a) $\frac{\tanh mL}{\sqrt{hA/kP}}$ (b) $\frac{\tanh mL}{mL}$

(c) $\frac{mL}{\tanh mL}$ (d) $\frac{\sqrt{hA/kP}}{\tanh mL}$

[ESE : 2001]

3.2 The insulated tip temperature of a rectangular longitudinal fin having an excess (over ambient) root temperature of θ_0 is

(a) $\theta_0 \tanh(ml)$ (b) $\frac{\theta_0}{\sinh(ml)}$

(c) $\frac{\theta_0 \tanh(ml)}{(ml)}$ (d) $\frac{\theta_0}{\cosh(ml)}$

[ESE : 2002]

3.3 Consider the following statements pertaining to large heat transfer rate using fins:

1. Fins should be used on the side where heat transfer coefficient is small.
2. Long and thick fins should be used.
3. Short and thin fins should be used.
4. Thermal conductivity of fin material should be large.

Which of these statements are correct?

(a) 1, 2 and 3 (b) 1, 2 and 4
(c) 2, 3 and 4 (d) 1, 3 and 4

[ESE : 2002]

3.4 **Assertion (A):** In a liquid-to-gas heat exchanger fins are provided in the gas side.

Reason (R): The gas offers less thermal resistance than liquid.

(a) Both **A** and **R** are individually true and **R** is the correct explanation of **A**
(b) Both **A** and **R** are individually true but **R** is not the correct explanation of **A**
(c) **A** is true but **R** is false
(d) **A** is false but **R** is true [ESE : 2002]

3.5 In order to achieve maximum heat dissipation, the fin should be designed in such a way that :

(a) It should have maximum lateral surface at the root side of the fin
(b) It should have maximum lateral surface towards the tip side of the fin
(c) It should have maximum lateral surface near the centre of the fin
(d) It should have minimum lateral surface near the centre of the fin

[ESE : 2005]

3.6 Usually fins are provided to increases the rate of heat transfer, but fins also act as insulation. Which one of the following non-dimensional numbers decides this factor?

(a) Eckert number (b) Biot number
(c) Fourier number (d) Peclet number

[ESE : 2007]

3.7 Fins are used to increase the heat transfer from surface by

(a) increasing the temperature difference
(b) increasing the effective surface area
(c) increasing the convective heat transfer coefficient
(d) none of the above [ESE : 2008]

3.8 A fin length l protrudes from a surface held at temperature T_0; it being higher than the ambient temperature T_a. The heat dissipation from the free end of the fin is stated to be negligibly small. What is the temperature gradient $\left(\frac{dT}{dx}\right)_{x=l}$ at the tip of the fin?

(a) Zero (b) $\frac{T_0 - T_l}{l}$

(c) $h(T_0 - T_a)$ (d) $\frac{T_l - T_a}{T_0 - T_a}$

[ESE : 2008]

3.9 The effectiveness of a fin will be maximum in an environment with

(a) free convection
(b) forced convection

(c) radiation
(d) convection and radiation [ESE : 2008]

3.10 On heat transfer surface, fins are provided
(a) to increase temperature gradient so as to enhance heat transfer
(b) to increase turbulence in flow for inhencing heat transfer
(c) to increase surface area to promote the rate of heat transfer
(d) to decrease the pressure drop of the fluid

[ESE : 2010]

3.11 Fins are made as thin as possible to
(a) reduce the total weight
(b) accommodate more number of fins
(c) increase the width for the same profile area
(d) improve flow of coolant around the fin

[ESE : 2010]

3.12 Extended surfaces are used to increase the rate of heat transfer. When the convective heat transfer coefficient $h = mk$, the addition of extended surface will
(a) increase the rate of heat transfer
(b) decrease the rate of heat transfer
(c) not increase the rate of heat transfer
(d) increase the rate of heat transfer when the length of the fin is very large

[ESE : 2010]

3.13 A fin will be necessary and effective only when
(a) k is small and h is large
(b) k is large and h is also large
(c) k is small and h is also small
(d) k is large and h is small
Where k = thermal conductivity of fin material, h = convective heat transfer coefficient between the fin surface and environment temperature

[ESE : 2012]

3.14 Which one of the following statements is correct?
(a) Fins should be attached on the side where heat transfer coefficients are high
(b) Effectiveness of fins depends on thermal conductivity only
(c) Fins must have small thickness for better heat dissipation
(d) In boiling heat transfer appliances, fins will be very effective

[ESE : 2012]

3.15 Three fins of equal length and diameter but made of aluminium, brass and cast-iron are heated to 200°C at one end. If the fins dissipate heat to the surrounding air at 25°C the temperature at the free end will be least in case of
(a) aluminium fin
(b) brass fin
(c) cast-iron fin
(d) each fin will have the same temperature at the free end

[ESE : 2013]

3.16 Statement (I): Fins are equally effective irrespective of whether they are on the hot-side or cold-side of the fluid.
Statement (II): The temperature along the fin is variable and hence the rate of heat transfer varies along the elements of fins.
(a) Both Statement (I) and Statement (II) are individually true and Statement (II) is the correct explanation of Statement (I).
(b) Both Statement (I) and Statement (II) are individually true but Statement (II) is **NOT** the correct explanation of Statement (I).
(c) Statement (I) is true but Statement (II) is false
(d) Statement (I) is false but Statement (II) is true.

[ESE : 2013]

3.17 Consider the following statements: An increase in pin fin effectiveness is caused by high value of
1. Convective coefficient
2. Thermal conductivity
3. Sectional area
4. Circumference

Which of the above statements are correct?
(a) 1 and 3 (b) 1 and 4
(c) 2 and 3 (d) 2 and 4

[ESE : 2015]

3.18 A fin will be more effective when Biot number is

(a) Greater than 1 (b) Equal to 1

(c) Between $\frac{1}{4}$ and $\frac{3}{4}$ (d) Less than 1

[ESE : 2015]

3.19 Two rods, one of length l and the other of length $2l$, are made of the same material and have same diameter. Both ends of the longer rod are maintained at 100°C. One end of the shorter rod is maintained at 100°C while the other end is insulated. Both rods are exposed to the same environment at 40°C. The temperature at the insulated end is measured to be 55°C. The temperature at the midpoint of the longer rod would be

(a) 45°C (b) 50°C

(c) 55°C (d) 60°C [ESE : 2017]

3.20 The purpose of providing fins on heat transfer surface is to increase

(a) Temperature gradient so as to enhance heat transfer by convection

(b) Effective surface area to promote rate of heat transfer by convection

(c) Turbulence in flow for enhancing heat transfer by convection

(d) Pressure drop of the fluid

[ESE : 2020]

■■■■

Answers Heat Dissipation From Extended Surfaces (FINS)

3.1 (b)	3.2 (d)	3.3 (d)	3.4 (c)	3.5 (a)	3.6 (b)	3.7 (b)	3.8 (a)	3.9 (a)
3.10 (c)	3.11 (b)	3.12 (c)	3.13 (d)	3.14 (c)	3.15 (c)	3.16 (d)	3.17 (d)	3.18 (d)
3.19 (c)	3.20 (b)							

Explanations Heat Dissipation From Extended Surfaces (FINS)

3.1 (b)

Efficiency of infinitely long fin, $\eta = \frac{1}{mL}$

Efficiency of fin with insulated tip,

$$\eta = \frac{\tanh(ml)}{(ml)}$$

where $m = \sqrt{\frac{hP}{kA}}$

3.2 (d)

Temperature distribution

$$\frac{\theta}{\theta_0} = \frac{\cosh m(l-x)}{\cosh (ml)}$$

when $x = l;\ \frac{\theta}{\theta_0} = \frac{1}{\cosh (ml)}$

3.3 (d)

Ideally the fin material should have a large thermal conductivity to minimise the temperature variation from root to its tip. In the limit of infinite thermal conductivity, the entire fin would be at the temperature of the base surface, therby providing the maximum possible heat transfer enhancement.

$$\varepsilon = \frac{km}{h}$$

Hence, fin should be placed where heat transfer coefficient is small to have high effectiveness.

3.4 (c)

Fin should be kept on the side where heat transfer coefficient is minimum so that effectiveness is higher hence it should be provided on gas side.

3.5 (a)

Fin should be designed so that base surface is maximum: the more the heat transfer influxed through the base area, higher will be heat dissipation.

3.6 (b)

The heat flow rate will be minimum when biot number is equal to unity and will increase when Biot number < 1.

3.7 (b)

Fins are used to increase the heat transfer from surface by increasing the effective surface area.

3.8 (a)

It is mentioned that heat dissipation from the free end of fin is stated to be negligibly small.

$$\therefore \left(\frac{dT}{dx}\right)_{x=l} = 0$$

3.9 (a)

The effectiveness of fin will be maximum in an environment where h is minimum.

3.10 (c)

By providing fins, surface area exposed increases which promotes the heat transfer.

3.11 (b)

Conditions for fins to be effective are

1. Thermal conductivity (k) should be large
2. Heat transfer coefficient (h) should be small.
3. Thickness of the fin (y) should be small.

$\therefore$ Fins are made as thin as possible to accommodate more number of fins.

3.12 (c)

For the fins,

Since effeciveness, $\varepsilon_{fin} = \dfrac{km}{h}$

and when $h = mk$

effectiveness will become unity and by providing fin, heat transfer rate will be unaltered.

3.13 (c)

Fins should be attached to side where heat transfer coefficients are small and fins must have small thickness to accommodate more number of fins and have adequate between the fins for better heat dissipation.

3.15 (c)

Thermal conductivity of cast iron is least, fin material will provide more thermal resistance therefore temperature drop from the root will be higher.

3.16 (d)

Fins are more effective when they are on the hot side of the fluids.

3.17 (d)

$$\in_{fin} = \sqrt{\frac{kP}{hA}}$$

$$\in \propto k^{1/2}$$

$$\in \propto P^{1/2}$$

Thus fin effectiveness will increase if thermal conductivity and circumference increases.

3.18 (d)

A fin will be more effective when

$$Bi < 1$$

$$Bi = \frac{hL_c}{K} \qquad \therefore \boxed{k\uparrow Bi\downarrow \in\uparrow}$$

as $\in = \sqrt{\dfrac{kP}{hA}}$

3.19 (c)

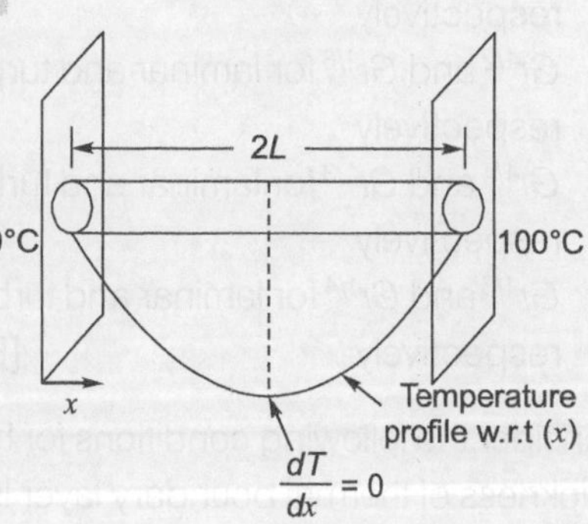

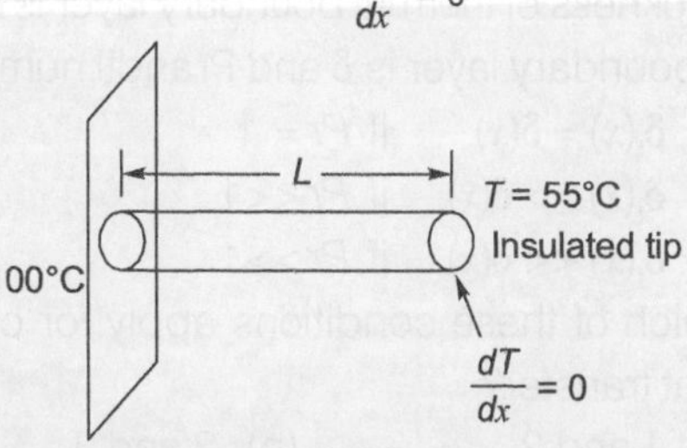

Since the boundary conditions at the mid-plane of the longer rod (in first fig.) and at the tip of the shorter rod being same. The temperature at the midplane of longer rod is 55°C.

3.20 (b)

Providing fins can increase the area of contact between solid and fluid surface. Hence increasing heat transfer from the surface.

■■■■

4 Free and Forced Convection

4.1 For the fully developed laminar flow and heat transfer in a uniformly heated long circular tube, if the flow velocity is doubled and the tube diameter is halved, the heat transfer coefficient will be
(a) double of the original value
(b) half of the original value
(c) same as before
(d) four times of the original value
[ESE : 2000]

4.2 In respect of free convection over a vertical flat plate the Nusselt number varies with Grashof number '*Gr*' as
(a) Gr and $Gr^{1/4}$ for laminar and turbulent flows respectively
(b) $Gr^{1/2}$ and $Gr^{1/3}$ for laminar and turbulent flows respectively
(c) $Gr^{1/4}$ and $Gr^{1/3}$ for laminar and turbulent flows respectively
(d) $Gr^{1/3}$ and $Gr^{1/4}$ for laminar and turbulent flows respectively [ESE : 2000]

4.3 Consider the following conditions for heat transfer (thickness of thermal boundary layer is δ_t, velocity of boundary layer is δ and Prandtl number is Pr)
1. $\delta_t(x) = \delta(x)$ if $Pr = 1$
2. $\delta_t(x) >> \delta(x)$ if $Pr << 1$
3. $\delta_t(x) << \delta(x)$ if $Pr >> 1$

Which of these conditions apply for convective heat transfer?
(a) 1 and 2 (b) 2 and 3
(c) 1 and 3 (d) 1, 2 and 3
[ESE : 2000]

4.4 The Nusselt number is related to Reynolds number in laminar and turbulent flows respectively as
(a) $Re^{-1/2}$ and $Re^{0.8}$ (b) $Re^{1/2}$ and $Re^{0.8}$
(c) $Re^{-1/2}$ and $Re^{-0.8}$ (d) $Re^{1/2}$ and $Re^{-0.8}$
[ESE : 2000]

4.5 If heat and mass transfer take place simultaneously, the ratio of heat transfer coefficient to the mass transfer coefficient is a function of the ratio of
(a) Schmidt and Reynolds numbers
(b) Schmidt and Prandtl numbers
(c) Nusselt and Lewis numbers
(d) Reynolds and Lewis numbers
[ESE : 2000]

4.6 **Assertion (A):** All analysis of heat transfer in turbulent flow must eventually rely on experimental data.
Reason (R): The eddy properties vary across the boundary layer and no adequate theory is available to predict their behaviour.
(a) Both **A** and **R** are individually true and **R** is the correct explanation of **A**
(b) Both **A** and **R** are individually true but **R** is not the correct explanation of **A**
(c) **A** is true but **R** is false
(d) **A** is false but **R** is true
[ESE : 2000]

4.7 For flow over a flat plate the hydrodynamic boundary layer thickness is 0.5 mm. The dynamic viscosity is 25 x 10^{-6} Pa-s, specific heat is 2.0 kJ/kgK and thermal conductivity is 0.05 W/mK. The thermal boundary layer thickness would be
(a) 0.1 mm (b) 0.5 mm
(c) 1 mm (d) 2 mm
[ESE : 2001]

4.8 The Nusselt number is related to Reynolds number in laminar and turbulent flows respectively as
(a) $Re^{-1/2}$ and $Re^{0.8}$ (b) $Re^{1/2}$ and $Re^{0.8}$
(c) $Re^{-1/2}$ and $^{-0.8}$ (d) $Re^{1/2}$ and $Re^{-0.8}$
[ESE : 2001]

4.9 For natural convective flow over a vertical flat plate as shown in the given figure, the governing differential equation for momentum is

$$\left(u\frac{\partial u}{\partial x}+v\frac{\partial u}{\partial y}\right)=g\beta\,(T-T_\infty)+\gamma\frac{\partial^2 u}{\partial y^2}$$

If equation is non-dimensionalized by

$$U=u/U_\infty,\ X=x/L,\ y=y/L \text{ and } \theta=\frac{T-T_\infty}{T_s-T_\infty}$$

then the term $g\beta(T-T_\infty)$ is equal to

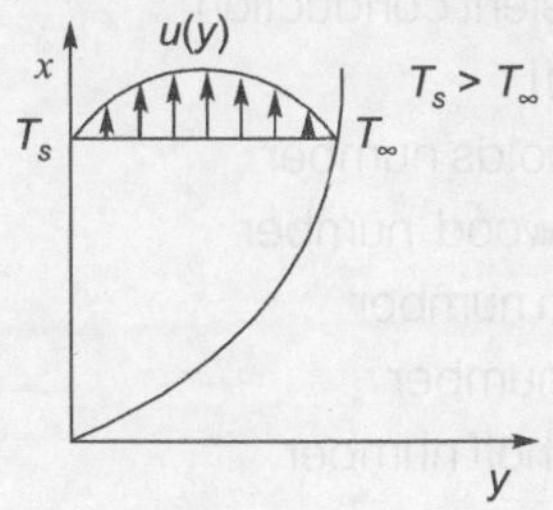

(a) Grashof number
(b) Prandtl number
(c) Rayleigh number
(d) Grashof number/(Reynolds number)^2

[ESE : 2001]

4.10 Nusselt number for fully developed turbulent flow in a pipe is given by $Nu = C \cdot Re^a \cdot Pr^b$. The values of *a* and *b* are

(a) $a = 0.5$ and $b = 0.33$ for heating and cooling both
(b) $a = 0.5$ and $b = 0.4$ for heating and $b = 0.3$ for cooling
(c) $a = 0.8$ and $b = 0.4$ for heating and $b = 0.3$ for cooling
(d) $a = 0.8$ and $b = 0.3$ for heating and $b = 0.4$ for cooling

[ESE : 2001]

Direction: Each of the next questions consists of two statements, one labelled as the **'Assertion (A)'** and the other as **'Reason (R)'**. You are to examine these two statements carefully and select the answers to these items using the codes given below:

Codes:

(a) Both **A** and **R** are individually true and **R** is the correct explanation of **A**
(b) Both **A** and **R** are individually true but **R** is not the correct explanation of **A**
(c) **A** is true but **R** is false
(d) **A** is false but **R** is true

4.11 Assertion (A): According to Reynolds analogy for Prandtl number equal to unity, Stanton number is equal to one half of the friction factor.

Reason (R): If thermal diffusivity is equal to kinematic viscosity, the velocity and the temperature distribution in the flow will be the same.

[ESE : 2001]

4.12 Assertion (A): Nusselt number is always greater than unity.

Reason (R): Nusselt number is the ratio of two thermal resistances, one the thermal resistance which would be offered by the fluid, if it was stationary and the other, the thermal resistance associated with convective heat transfer coefficient at the surface.

[ESE : 2001]

4.13 A 320 cm high vertical pipe at 150°C wall temperature is in a room with still air at 10°C. This pipe supplies heat at the rate of 8 kW into the room air by natural convection. Assuming laminar flow, the height of the pipe needed to supply 1 kW only is

(a) 10 cm (b) 20 cm
(c) 40 cm (d) 80 cm

[ESE : 2002]

4.14 The average Nusselt number in laminar natural convection from a vertical wall at 180°C with still air at 20°C is found to be 48. If the wall temperature becomes 30°C, all other parameters remaining same, the average Nusselt number will be

(a) 8 (b) 16
(c) 24 (d) 32

[ESE : 2002]

4.15 Match **List-I** (Type of heat transfer) with **List-II** (Governing dimensionless parameter) and select the correct answer using the codes given below the lists:

List-I

A. Forced convection
B. Natural convection
C. Combined free and forced convection
D. Unsteady conduction with convection at surface

List-II
1. Reynolds, Grashof and Prandtl number
2. Reynolds and Prandtl number
3. Fourier number and Biot number
4. Prandtl number and Grashof number

Codes:

	A	B	C	D
(a)	2	1	4	3
(b)	3	4	1	2
(c)	2	4	1	3
(d)	3	1	4	2

[ESE : 2002]

4.16 Prandtl number of a flowing fluid greater than unity indicates that hydrodynamic boundary layer thickness is

(a) greater than thermal boundary layer thickness
(b) equal to thermal boundary layer thickness
(c) greater than hydrodynamic boundary layer thickness
(d) independent of thermal boundary layer thickness

[ESE : 2002]

4.17 A fluid of thermal conductivity 1.0 W/mK flows in fully developed flow with Reynolds number of 1500 through a pipe of diameter 10 cm. The heat transfer coefficient for uniform heat flux and uniform wall temperature boundary conditions are, respectively

(a) 36.57 and 43.64 W/m^2K
(b) 43.64 and 36.57 W/m^2K
(c) 43.64 W/m^2K for both the cases
(d) 36.57 W/m^2K for both the cases

[ESE : 2002]

4.18 The velocity and temperature distribution in a pipe flow are given by $U(r)$ and $T(r)$. If u_m is the mean velocity at any section of the pipe, the bulk mean temperature at that section is

(a) $\int_0^{r_0} u(r)T(r)r^2 dr$
(b) $\int_0^{r_0} \frac{u(r)}{3r}\frac{T(r)}{2r} dr$
(c) $\int_0^{r_0} \frac{u(r)T(r).dr}{2\pi r_0^3}$
(d) $\frac{2}{u_m r_0^2}\int_0^{r_0} u(r)(T_r)r dr$

[ESE : 2003]

4.19 Match **List-I** (Process) with **List-II** (Predominant Parameter Associated With the Process) and select the correct answer using the codes given below the lists:

List-I
A. Mass transfer
B. Forced convection
C. Free convection
D. Transient conduction

List-II
1. Reynolds number
2. Sherwood number
3. Mach number
4. Biot number
5. Grashoff number

Codes:

	A	B	C	D
(a)	5	1	2	3
(b)	2	1	5	4
(c)	4	2	1	3
(d)	2	3	5	4

[ESE : 2003]

4.20 For fully-developed turbulent flow in a pipe with heating, the Nusselt number Nu, varies, with Reynolds number Re and Prandtl number Pr as

(a) $Re^{0.5}\, Pr^{1/3}$
(b) $Re^{0.8}Pr^{0.2}$
(c) $Re^{0.8}Pr^{0.4}$
(d) $Re^{0.8}Pr^{0.3}$

[ESE : 2003]

4.21 In a convective heat transfer situation Reynolds number is very large but the Prandtl number is so small that the product $Re\, x\, Pr$ is less than one. In such a condition which one of the following is correct?

(a) Thermal boundary layer does not exist
(b) Viscous boundary layer thickness is less than the thermal boundary layer thickness
(c) Viscous boundary layer thickness is equal to the thermal boundary layer thickness
(d) Viscous boundary layer thickness is greater than the thermal boundary layer thickness

[ESE : 2004]

4.22 Match **List-I** (Process) with **List-II** (Predominant parameter associated with the flow) and select the correct answer using the code given below the lists:

List-I
A. Transient conduction
B. Mass transfer
C. Forced convection
D. Free convection

List-II
1. Sherwood Number
2. Mach Number
3. Biot Number
4. Grashof Number
5. Reynolds number

Codes:

	A	B	C	D
(a)	2	4	3	1
(b)	3	1	5	4
(c)	2	1	3	4
(d)	3	4	2	1

[ESE : 2004]

4.23 Nusselt number for a pipe flow heat transfer coefficient is given by the equation $Nu_D = 4.36$. Which one of the following combinations of conditions do exactly apply for use of this equation?
(a) Laminar flow and constant wall temperature
(b) Turbulent flow and constant wall heat flux
(c) Turbulent flow and constant wall temperature
(d) Laminar flow and constant wall heat flux

[ESE : 2004]

4.24 The non-dimensional parameter known as Stanton number (St) is used in
(a) forced convection heat transfer in flow over flat plate
(b) condensation heat transfer with laminar film layer
(c) natural convection heat transfer over flat plate
(d) unsteady heat transfer from bodies in which internal temperature gradients cannot be neglected

[ESE : 2004]

4.25 Which one of the following numbers represents the ratio of kinematic viscosity to the thermal diffusivity?
(a) Grashof number (b) Prandtl number
(c) Mach number (d) Nusselt number

[ESE : 2005]

4.26 A thin flat plate 2 m × 2 m is hanging freely in air. The temperature of the surroundings is 25°C. Solar radiation is falling on one side of the plate at the rate of 500 W/m^2. What should be the convection heat transfer coefficient in W/m^2°C, if the temperature of the plate is to remain constant at 30°C?
(a) 25 (b) 50
(c) 100 (d) 200

[ESE : 2005]

4.27 Match List-I (Phenomenon) with List-II (Associated Dimensionless parameter) and select the correct answer using the code given below the lists:

List-I
A. Transient conduction
B. Forced convection
C. Mass transfer
D. Natural convection

List-II
1. Reynolds number 2. Grashof number
3. Biot number 4. Mach number
5. Sherwood number

Codes:

	A	B	C	D
(a)	3	2	5	1
(b)	5	1	4	2
(c)	3	1	5	2
(d)	5	2	4	1

[ESE : 2006]

4.28 Match List-I (Non-dimensional Number) with List-II (Application) and select the correct answer using the code given below the lists:

List-I
A. Grashof number
B. Stanton number
C. Sherwood number
D. Fourier number

List-II
1. Mass transfer
2. Unsteady state heat conduction
3. Free convection
4. Forced convection

Codes:

	A	B	C	D
(a)	4	3	1	2

(b) 3 4 1 2
(c) 4 3 2 1
(d) 3 4 2 1 [ESE : 2007]

4.29 For steady, uniform flow through pipes with constant heat flux supplied to the wall, what is the value of Nusselt number?
(a) 48/11 (b) 11/48
(c) 24/11 (d) 11/24 [ESE : 2007]

4.30 Which one of the following non-dimensional numbers is used for transition from laminar to turbulent flow in free convection?
(a) Reynolds number (b) Grashof number
(c) Peclet number (d) Rayleigh number
[ESE : 2007]

4.31 For calculation of heat transfer by natural convection from a horizontal cylinder, what is the characteristic length in Grashof number?
(a) Diameter of the cylinder
(b) Length of the cylinder
(c) Circumference of the base of the cylinder
(d) Half the circumference of the base of the cylinder [ESE : 2007]

4.32 For a fluid having Prandtl number equal to unity, how are the hydrodynamic boundary layer thickness δ, and the thermal boundary layer thickness δ_t related?
(a) $\delta = \delta_t$ (b) $\delta > \delta_t$
(c) $\delta < \delta_t$ (d) $\delta_t = \delta^{1/3}$
[ESE : 2007]

4.33 Natural convection heat transfer coefficients over surface of vertical pipe and vertical flat plate for same height and fluid are equal.
What is/are the possible reasons for this?
1. Same height
2. Both vertical
3. Same fluid
4. Same fluid flow pattern
Select the correct answer using the codes given below:
(a) 1 only (b) 1 and 2
(c) 3 and 4 (d) 4 only
[ESE : 2008]

4.34 Which non-dimensional number relates the thermal boundary layer and hydrodynamic boundary layer?
(a) Rayleigh number (b) Peclet number
(c) Grashof number (d) Prandtl number
[ESE : 2008]

4.35 What is hydraulic diameter used in place of diameter for non-circular ducts equal to?
(a) A/m (b) $4\,A/m$
(c) $A/(4\,m)$ (d) $4\,m/A$
where A = Area of flow and m = Perimeter.
[ESE : 2008]

4.36 Air at 20°C blows over a hot plate of 50 × 60 cm made of carbon steel maintained at 220°C. The convective heat transfer coefficient is 25 W/m^{2}K. What will be the heat loss from the plate?
(a) 1500 W (b) 2500 W
(c) 3000 W (d) 4000 W
[ESE : 2009]

4.37 Assertion (A): For the similar conditions the values of convection heat transfer coefficients are more in forced convection than in free convection.
Reason (R): In case of forced convection system the movement of fluid is by means of external agency.
(a) Both **A** and **R** are individually true and **R** is the correct explanation of **A**
(b) Both **A** and **R** are individually true but **R** is not the correct explanation of **A**
(c) **A** is true but **R** is false
(d) **A** is false but **R** is true
[ESE : 2009]

4.38 Match **List-I** with **List-II** and select the correct answer using the code given below the Lists :

	List-I		List-II
A.	Stanton number	1.	Natural convection
B.	Grashof number	2.	Mass transfer
C.	Peclet number	3.	Forced convection
D.	Schmidt number	4.	Forced convection for small Pr. nummber

Codes:

	A	B	C	D
(a)	2	1	4	3
(b)	3	1	4	2
(c)	2	4	1	3
(d)	3	4	1	2

[ESE : 2011]

4.39 For a fluid flowing over a flat plate, the Nusselt number at a point 0.5 m from the leading edge is 100. If the thermal conductivity of the fluid is 0.025 W/mK, the coefficient of convective heat transfer is

(a) 200 W/m^2K
(b) 5 W/m^2K
(c) 5×10^{-4} W/m^2K
(d) 1.25×10^{-4} W/m^2K

[ESE : 2011]

4.40 In laminar flow over a flat plate, the convective heat transfer coefficient is proportional to (x is the distance from the leading edge)

(a) $x^{1/2}$
(b) $x^{-1/2}$
(c) $x^{1/3}$
(d) $x^{-1/5}$

[ESE : 2012]

4.41 The laminar boundary layer occurs when a cold fluid flows over a hot plate. In which of the following positions, the temperature gradient assumes zero value?

(a) At bottom of boundary layer
(b) In mid free stream of fluid
(c) At top of boundary layer
(d) At the junction of laminar and turbulent boundary layer

[ESE : 2012]

4.42 The Reynolds-Colburn analogy, which is used to determine heat transfer coefficient from the measurement of frictional drag, is applicable to

(a) circumferential fins
(b) flat plates
(c) rectangular fins
(d) triangular fins

[ESE : 2012]

4.43 Hot air at 150°C flows over a flat plate maintained at 50°C. If the forced convection heat transfer coefficient is 75 W/m^2K, the heat gain rate by the plate through an area of 2 m^2 will be

(a) 15 kW
(b) 22.5 kJ/s
(c) 7.5 kJ/s
(d) None of the above

[ESE : 2013]

4.44 Dimensionless quantity that connects the link between velocity flow field and the temperature field is

(a) Nusselt number
(b) Prandtl number
(c) Reynolds number
(d) Grashof number

[ESE : 2014]

4.45 The characteristic length for computing Grashof number in the case of horizontal cylinder is

(a) the length of the cylinder
(b) the diameter of the cylinder
(c) the perimeter of the cylinder
(d) the radius of the cylinder

[ESE : 2014]

4.46 For which of these configurations is a minimum temperature difference required for natural convection to set in

(a) fluid near a heated vertical plate
(b) fluid near a heated plate inclined at 45° to the vertical
(c) fluid over a heated horizontal plate
(d) fluid near a heated cylinder

[ESE : 2014]

4.47 For a fluid with Prandtl number Pr > 1, momentum boundary layer thickness

(a) decreases rapidly compared to the thermal boundary layer thickness
(b) and thermal boundary layer thickness increase at the same rate
(c) increases rapidly compared to the thermal boundary layer thickness
(d) and thermal boundary layer thickness decrease at the same rate

[ESE : 2015]

4.48 Consider the following statements:

1. In natural convection turbulent flow over heated vertical plate, h is independent of the characteristic length
2. In turbulent flow, non-dimensional heat transfer coefficient for natural convection over a heated vertical plate is given by
$$Nu = c(Pr)^{\frac{1}{3}}$$

Which of the above statements is/are correct?

(a) 1 only
(b) Both 1 and 2
(c) 2 only
(d) Neither 1 nor 2

[ESE : 2015]

4.49 In a pipe, laminar flow in fully developed region with constant heat flux from pipe wall, bulk mean temperature of fluid

(a) and pipe wall temperature increase in flow direction
(b) and pipe wall temperature decrease in flow direction
(c) remains constant, but pipe wall temperature increases in flow direction

(d) increases but pipe wall temperature remains constant [ESE : 2015]

4.50 In a laminar developing flow through a pipe with constant wall temperature, the magnitude of the pipe wall inner surface convective heat transfer coefficient shall be maximum at the:
(a) Middle length of flow
(b) Beginning of flow
(c) End of flow
(d) None of the above [ESE : 2015]

4.51 Thermal boundary layer is a region where:
(a) Heat dissipation is negligible
(b) Inertia and convection are of the same order of magnitude
(c) Convection and dissipation terms are of the same order of magnitude
(d) Convection and conduction terms are of the same order of magnitude
[ESE : 2016]

4.52 Heat is lost from a 100 mm diameter steam pipe placed horizontally in ambient air at 30°C. If the Nusselt number is 25 and the thermal conductivity of the air is 0.03 W/mK, then the heat transfer coefficient will be
(a) 7.5 W/m²K (b) 15 W/m²K
(c) 25 W/m²K (d) 35 W/m²K
[ESE : 2017]

4.53 Air at 1 atmospheric pressure and 27°C blows across a 12 mm diameter sphere at a free stream velocity of 4 m/s. A small heater inside the sphere maintains the surface temperature at 77°C. With k = 0.026 W/mK and with (Nu) = 31.4, the heat loss by the sphere would be
(a) 1.93 J/s (b) 1.76 J/s
(c) 1.65 J/s (d) 1.54 J/s
[ESE : 2017]

4.54 In forced convection, the surface heat transfer coefficient from a heated flat plate is a function of
(a) Re and Gr (b) Pr and Gr
(c) Re and Pr (d) Re, Gr and Pr
where Re is Reynolds number, Pr is Prandtl number and Gr is Grashoff number.
[ESE : 2018]

4.55 An insulated pipe of 50 mm outside diameter with ε = 0.8 is laid in a room at 30°C. If the surface temperature is 250°C and the convective heat transfer coefficient is 10 W/m²K, the total heat loss per unit length of the pipe will be:
(a) 896.6 W/m (b) 818.8 W/m
(c) 786.4 W/m (d) 742.2 W/m
[ESE : 2019]

4.56 In liquid metals, thermal boundary layer develops much faster than velocity boundary layer due to
(a) lower value of Nusselt number
(b) higher value of Prandtl number
(c) lower value of Prandtl number
(d) higher value fo Nusselt number
[ESE : 2019]

4.57 Consider the following statements:
For the laminar condensation on a vertical plate, the Nusselt theory says that
1. inertia force in the film is negligible compared to viscosity and weight.
2. heat flow is mainly by conduction through the liquid film, convection in liquid film as well as in vapour is neglected.
3. velocity of vapour is very high.
Which of the above statements are correct?
(a) 1, 2 and 3 (b) 1 and 2 only
(c) 1 and 3 only (d) 2 and 3 only
[ESE : 2019]

4.58 Which one of the following is NOT a fluid property for convection heat transfer coefficient?
(a) Dynamic viscosity (b) Thermal conductivity
(c) Density (d) Rate of convection
[ESE : 2022]

4.59 In friction coefficient, the boundary layer thickness (δ), the local friction coefficient (C_f) at location x and Reynolds number (Re) at location x for turbulent flow over a flat plate are

(a) $\delta = \dfrac{5x}{\text{Re}_x^{0.5}}; C_f = \dfrac{0.664}{\text{Re}_x^{0.5}}$

(b) $\delta = \dfrac{0.382x}{\text{Re}_x^{0.5}}; C_f = \dfrac{0.0592}{\text{Re}_x^{0.5}}$

(c) $\delta = \dfrac{0.382x}{\text{Re}_x^{1/5}}; C_f = \dfrac{0.0592}{\text{Re}_x^{1/5}}$

(d) $\delta = \frac{5x}{Re_x^{1/5}}; C_f = \frac{0.664}{Re_x^{1/5}}$ [ESE : 2022]

4.60 What is the traditional expression for calculation of heat transfer in fully developed turbulent flow in smooth tubes that recommended by Dittus Boelter?

(a) $Nu_d = 0.023(Re_d^{0.8})(Pr)^n$

(b) $Nu_d = 0.023(Re_d^{0.4})(Pr)^{2n}$

(c) $Nu_d = 0.023(Re_d^{0.8})(Pr)^{2n}$

(d) $Nu_d = 0.023(Re_d^{0.4})(Pr)^n$ [ESE : 2022]

4.61 An electric current is passed through a wire 1 mm in diameter and 10 cm long. The wire is submerged in liquid water at atmospheric pressure and the current is increased until the water boils. For this situation, h is 5000 W/m²°C and the water temperature will be 100°C. How much electric power must be supplied to the wire to maintain the wire surface at 114°C?

(a) 21.99 W (b) 32.39 W
(c) 42.39 W (d) 52.36 W

[ESE : 2022]

Directions for the following **two (02)** items:
Read the following information and answer the **two** items that follows:
Hot water at 98°C flows through a 2-in schedule 40 horizontal steel pipe,
ID = 0.0525 m, OD = 0.06033 m [k = 54 W/m°C] and is exposed to atmospheric air at 20°C. The water velocity is 25 cm/s.

The properties of water at 98°C are
ρ = 960 kg/m³, μ = 2.82 × 10^{-4} kg/m.s, k = 0.68 W/m°C, Pr = 1.76, Nu = 151.4

4.62 What is the Reynolds number?

(a) 31340 (b) 44680
(c) 48450 (d) 51230 [ESE : 2023]

4.63 What is the convective heat transfer coefficient at inlet?

(a) 1961 W/m²°C (b) 1961 W/m²K
(c) 1348 W/m²°C (d) 1348 W/m²K

[ESE : 2023]

4.64 Which one of the following expressions is correct about the relation among Grashof number (Gr), Prandtl number (Pr) and Rayleigh number (Ra)?

(a) Ra = Gr Pr (b) Gr = Ra Pr
(c) Ra = Gr – Pr (d) Gr = Ra – Pr

[ESE : 2024]

4.65 An order-of-magnitude analysis of the free convection boundary layer equations will indicate a general criterion for determining whether free convection effects dominate. The criterion is that when

(a) Gr/Re² < 10, then free convection is of primary importance

(b) Gr/Re² = 10, then free convection is of primary importance

(c) Gr/Re² > 10, then free convection is of primary importance

(d) Gr/Re² > 10, then forced convection is of primary importance [ESE : 2024]

Answers Free and Forced Convection

4.1 (a)	4.2 (c)	4.3 (d)	4.4 (b)	4.5 (b)	4.6 (a)	4.7 (b)	4.8 (b)	4.9 (d)
4.10 (c)	4.11 (d)	4.12 (a)	4.13 (b)	4.14 (c)	4.15 (c)	4.16 (a)	4.17 (b)	4.18 (d)
4.19 (b)	4.20 (c)	4.21 (b)	4.22 (b)	4.23 (d)	4.24 (a)	4.25 (b)	4.26 (b)	4.27 (c)
4.28 (b)	4.29 (a)	4.30 (d)	4.31 (a)	4.32 (a)	4.33 (c)	4.34 (d)	4.35 (b)	4.36 (a)
4.37 (a)	4.38 (b)	4.39 (b)	4.40 (b)	4.41 (c)	4.42 (b)	4.43 (a)	4.44 (b)	4.45 (b)
4.46 (a)	4.47 (c)	4.48 (a)	4.49 (a)	4.50 (b)	4.51 (d)	4.52 (a)	4.53 (d)	4.54 (c)
4.55 (b)	4.56 (c)	4.57 (b)	4.58 (d)	4.59 (c)	4.60 (a)	4.61 (a)	4.62 (b)	4.63 (a)
4.64 (a)	4.65 (c)							

Explanations Free and Forced Convection

4.1 (a)

For fully developed flow

$$Nu = \frac{hd}{k}$$

Hence, $h \propto \frac{1}{d}$

4.2 (c)

For laminar flow:

$$Nu = 0.53(Gr.Pr)^{1/4}$$

$$\{10^4 \le Gr.Pr \le 10^9\}$$

For turbulent flow: $= 0.13(G.rPr)^{1/3}$

$$\{10^9 \le Gr.Pr \le 10^{12}\}$$

4.3 (d)

$$\delta_t = \frac{\delta}{(Pr)^{1/3}}$$

If $Pr = 1 \Rightarrow \delta_t = \delta$

If $Pr << 1 \Rightarrow \delta_t >> \delta$

If $Pr >> 1 \Rightarrow \delta_t << \delta$

4.4 (b)

For laminar flow

$$Nu = C(Re)^{1/2}(Pr)^{1/3}$$

For turbulent flow

$$Nu = C(Re)^{0.8}(Pr)^{1/3}$$

4.5 (b)

Schmidt number $(Sc) = \frac{\nu}{D}$

where ν is kinematic viscosity and D is diffusion coefficient

Prandtl number $(Pr) = \frac{\nu}{\alpha}$

where α is thermal diffusivity

$$\frac{S_c}{Pr} = \frac{\nu / D}{\nu / \alpha} = \frac{\alpha}{D}$$

4.6 (a)

Because in turbulent flow there is continuous mixing of fluid hence any correlation to satisfy it should be based on experimental data.

4.7 (b)

$$\delta_t = \frac{\delta}{(Pr)^{1/3}}$$

where δ = Boundary layer thickness

$$Pr = \frac{\mu c_p}{k} = \frac{25\times10^{-6}\times2\times10^2}{0.05}$$

$$= \frac{50\times10^{-3}\times10^2}{5} = 1$$

$$\delta_t = \delta = \mathbf{0.5\ mm}$$

4.9 (d)

$\frac{Gr}{Re^2}$; Gives dimensionless number which signifies whether it is forced or free convection.

$\frac{Gr}{Re^2} << 1$; Forced convection.

$\frac{Gr}{Re^2} >> 1$; Natural convection.

4.10 (c)

Fully developed turbulent flow in a pipe:

$Nu = 0.023\, Re^{0.8}\, Pr^n$

$n = 0.4$ for heating of fluids

$n = 0.3$ for cooling of fluids

4.11 (d)

Friction factor and stanton number are related as per Reynold analogy with the following expression

$$St_x \cdot Pr^{2/3} = \frac{f}{8}$$

If frictional drag coefficient which is related to stanton number as

$$St_x \times Pr^{2/3} = \frac{C_{fx}}{2}$$

4.12 (a)

Convection is the combined effect of conduction and bulk fluid motion. Higher the velocity of fluid, higher will be the value of 'h'

$$Nu = \frac{hL}{k} = \frac{hA.\Delta T}{\frac{k.A.\Delta T}{L}}$$

$$= \frac{\text{Rate of heat transfer by convection}}{\text{Rate of heat transfer by conduction}}$$

It is clear that bulk motion is predominant due to motion of fluids so Nusselt number is always greater than one.

Nusselt number can he interpreted as

$$\frac{\frac{L}{kA}}{\frac{1}{hA}} = \frac{\text{Thermal resistance of conduction offered by fluid}}{\text{Thermal resistance offered by convection}}$$

4.13 (b)

For vertical pipe characteristic dimension is the length of the pipe

For laminar flow

$$Nu = (Gr.Pr)^{1/4}$$

$$\frac{hL}{k} = C\left(\frac{g\beta\Delta TL^3}{\nu^2}\times \text{Pr}\right)^{1/4}$$

$$\Rightarrow \quad h \propto \frac{1}{L^{1/4}}$$

$$\frac{q_1}{q_2} = \frac{h_1.A.\Delta T}{h_2.A.\Delta T} = \frac{h_1\pi DL_1\Delta T}{h_2\pi DL_2\Delta T}$$

$$\frac{q_1}{q_2} = \left(\frac{L_2}{L_1}\right)^{1/4}\times\frac{L_1}{L_2} = \left(\frac{L_1}{L_2}\right)^{3/4}$$

$$\frac{8}{q_2} = \left(\frac{320}{L_2}\right)^{3/4}$$

Put $q_2 = 1$ kW

$$\Rightarrow \quad \frac{320}{L_2} = (8)^{4/3} = 16$$

$$\Rightarrow \quad L_2 = \frac{320}{16} = 20 \text{ cm}$$

4.14 (c)

$$Nu = (Gr.Pr)^{1/4}$$

$$Gr = \frac{g\beta L^3(\Delta T)}{\nu^2}$$

$$Nu = (\Delta T_1)^{1/4}$$

$$\Rightarrow \quad \frac{48}{x} = \frac{(180-20)^{1/4}}{(30-20)^{1/4}}$$

$$Nu = \frac{48}{2} = 24$$

4.15 (c)

In forced convection, Reynolds number and Prandtl number is significant; in free convection, Grashoff number and Prandtl number is significant; in unsteady conduction with convection Fourier number and Biot number is significant.

4.16 (a)

$$\delta = \delta_t.Pr^{1/3}, \text{ when } Pr>1$$

$$\Rightarrow \quad \nu > \alpha$$

$$\Rightarrow \quad \delta > \delta_t$$

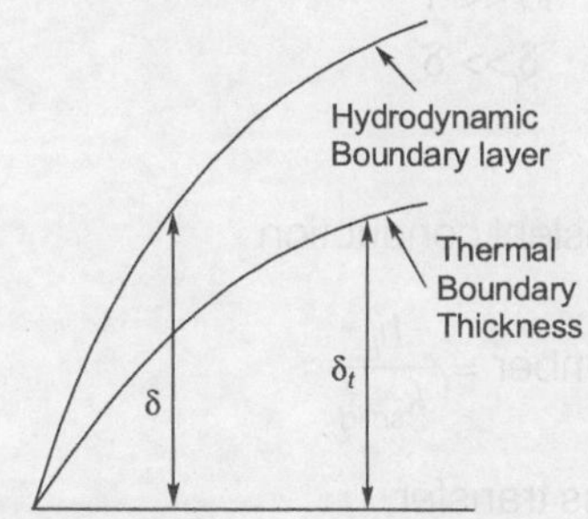

4.17 (b)

For constant heat flux

$$Nu = \frac{hD}{k} = 4.36$$

$$4.36 = \frac{h\times 10\times 10^{-2}}{1}$$

$$h = \frac{4.36}{10^{-1}} = 43.6 \text{ W/m}^2\text{K}$$

For uniform wall temperature

$$Nu = 3.66 = \frac{hD}{k}$$

$$3.66 = \frac{h\times 10\times 10^{-2}}{1}$$

or $h =$ **36.6 W/m²K**

4.18 (d)

$$T_{\text{mean}} \text{ or } T_{\text{bulk temp}} = \frac{2}{U_m r_0^2}\int_0^{r_0} uTrdr$$

4.19 (b)

Free convection	Grashof number
Force convection	Reynold number
Tansient conduction	Biot number
Mass transfer	Sherwood number

4.20 (c)

For heating

$$Nu = 0.023\, Re^{0.8}\, Pr^{0.4}$$

4.21 (b)

When Pr number is less than one. The Reynold number only signifies about the thickness of hydrodynamic boundary layer and related to thermal boundary through Prandtl number.

$$\delta_t = \frac{\delta}{(Pr)^{1/3}}$$

If $Pr << 1$

then $\delta_t >> \delta$

4.22 (b)

(*A*) Transient conduction

$$\text{Biot number} = \frac{h_L}{k_{solid}}$$

(B) Mass transfer

$$\text{Sherwood number} = \frac{h_L}{D}$$

(C) Forced convection

$$\text{Reynolds number} = \frac{\rho v L}{\mu}$$

(D) Free convection

$$\text{Grashof number} = \frac{F_{inertia} \times F_{buoyancy}}{F^2_{viscous}}$$

4.23 (d)

For laminar pipe flow

$Nu = 4.36$ (for constant heat flux)

$Nu = 3.66$ (for constant wall temperature)

4.24 (a)

$$\text{Stanton number } (St) = \frac{h}{\rho C_p u} = \frac{Nu_x}{Re_x Pr}$$

It is used for forced connection case when applied to plates. By measuring shear stress due to fluid on plate we can infer heat transfer data.

4.25 (b)

Prandtl number,

$$= \frac{\mu c_p}{k_f} = \frac{\nu}{\alpha} = \frac{\text{Kinematic viscosity}}{\text{Thermal diffusivity}}$$

It provides a measure of relative effectiveness of momentum and energy transport by diffusion in velocity and thermal boundary layer simultaneously

4.26 (b)

Solar radiation is falling on one side but convection is taking place from both the side.

$$Q = hA\Delta t$$

$$500 \times 4 = h \times 8 \times 5$$

$$\therefore \quad h = \frac{500 \times 4}{5 \times 8} = \mathbf{50\ W/m^2\,°C}$$

4.27 (c)

(a) Transient conduction : Biot number
(b) Forced convection : Reynolds number
(c) Mass transfer : Sherwood number
(d) Natural convection : Grashof number

4.28 (b)

(a) Grashof number : Free convection
(b) Stanton number : Forced convection
(c) Sherwood number : Mass transfer
(d) Fourier number : Unsteady state heat conduction

4.29 (a)

Heat transfer through pipe

$Nu = \frac{48}{11} = 4.36$ (For constant heat flux)

$Nu = 3.66$ (For constant wall temperature)

4.30 (d)

Rayleigh number is used to determine whether the flow is laminer or turbulent in natural convection.

$Ra < 10^9$ Laminar

$10^9 < Ra < 10^{10}$ Transition

$Ra > 10^{10}$ Turbulent

4.31 (a)

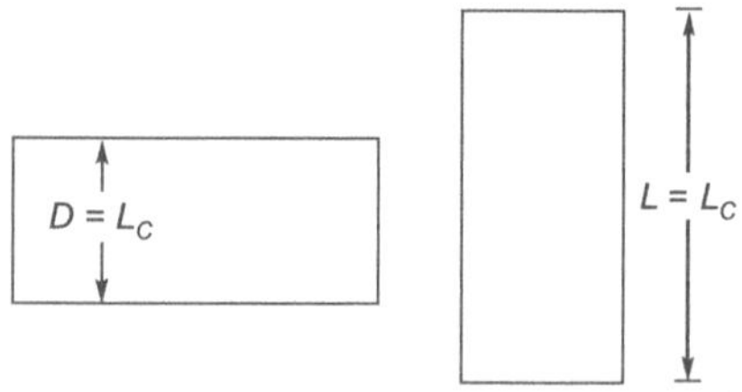

For horizontal cylinder characteristic length will be its diameter and for vertical cylinder characteristic length will be its length.

4.32 (a)

$$\delta_t = \frac{\delta}{(Pr)^{1/3}}$$

(i) When $Pr = 1$; $\delta_t = \delta$
(ii) When $Pr > 1$; $\delta_t < \delta$
(iii) When $Pr < 1$; $\delta_t > \delta$

4.34 (d)

Prandtl number relates the thermal boundary layer and hydrodynamic boundary layer.

$$\frac{\delta}{\delta_t} = \frac{\text{Hydrodynamic boundary layer}}{\text{Thermodynamic boundary layer}} = (Pr)^{1/3}$$

4.35 (b)

Hydraulic diameter for non circular duct is equal to $\frac{4A}{m}$.
where A = Area of flow and m = Perimeter.

4.36 (a)

Convective heat loss will take place from the one side of the plate since it is written that air blows over the hot plate

$$\therefore \quad Q = hA(T_1 - T_2)$$
$$= 25 \times (0.5 \times 0.6)(220 - 20)$$
$$= 25 \times (0.3)(200) = \mathbf{1500\ W}$$

4.38 (b)

Stanton number	:	Force convection
Grashoff number	:	Natural convection
Peclet number	:	Forced convection for small Prandtl number
Schmidt number	:	Mass transfer

4.39 (b)

$$\text{Nusselt number} = \frac{hL}{k}$$
$$100 = \frac{h \times 0.5}{0.025}$$
or $$h = \mathbf{5\ W/m^2K}$$

4.41 (c)

The value of temperature gradient dT/dy decreases with the distance from the plate and at the top of boundary layer if becomes zero as the temperature equals to free stream temperature of the fluid.

4.43 (a)

$$Q = hA\Delta T = 75 \times 2 \times (150 - 50) = \mathbf{15\ kW}$$

4.44 (b)

Prandtl number: $Pr = \frac{\mu c_p}{K} = \frac{\nu}{\alpha}$

$$Pr = \frac{\text{Molecular momentum diffusivity}}{\text{Thermal diffusivity}}$$

Molecular momentum diffusivity is the reason for developing velocity flow field and thermal diffusivity is for temperature field.

4.45 (b)

The characteristic length for computing Grashof number in the case of horizontal cylinder is diameter 'D'.

4.46 (a)

Buoyancy forces which arises due to density changes of the fluid always act in the upper direction.

4.47 (c)

Prandtl number > 1

$$Pr = \left(\frac{\delta}{\delta_t}\right)^3$$
$$\frac{\delta}{\delta_t} > 1\ ;\ \delta > \delta_t$$

4.48 (b)

Average Nusselt number for natural convection on a vertical plate of length L,

$$Nu_L = 0.13\,(Gr_L\,Pr)^{1/3}$$
$$Gr = \frac{g\beta\Delta T L^3}{\nu^2}$$
$$Nu = \frac{hL}{k}$$

where Gr is Grashof number.

4.49 (a)

In a pipe, laminar flow in fully developed region with constant heat flux bulk mean temperature and pipe wall temperature increases in flow direction.

$$\therefore \quad \frac{dT_m}{dx} = \text{constant} = (\text{positive})$$

$\Rightarrow$ along x, T_m increases.

$$\because \quad T_S = T_m + \frac{\xi_s}{h}$$

$\because \frac{\xi_s}{h}$ is constant.

$\because T_m$ increases.

T_S also increases linearly.

4.50 (b)

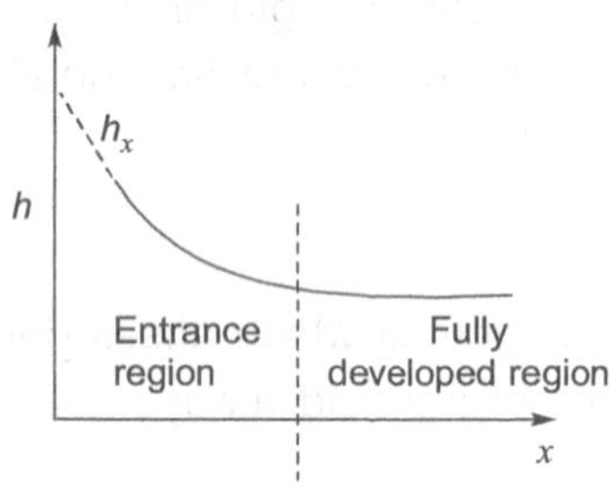

Heat transfer coefficient are highest at the tube inlet in laminar developing flow.

4.52 (a)

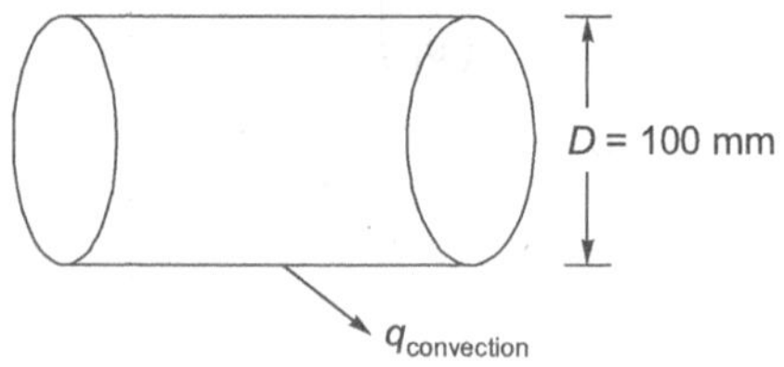

$$Nu = \frac{hD}{k_{air}} \Rightarrow 25 = \frac{h \times 0.1}{0.03}$$

$$\Rightarrow \quad h = 7.5 \text{ W/m}^2\text{K}$$

4.53 (d)

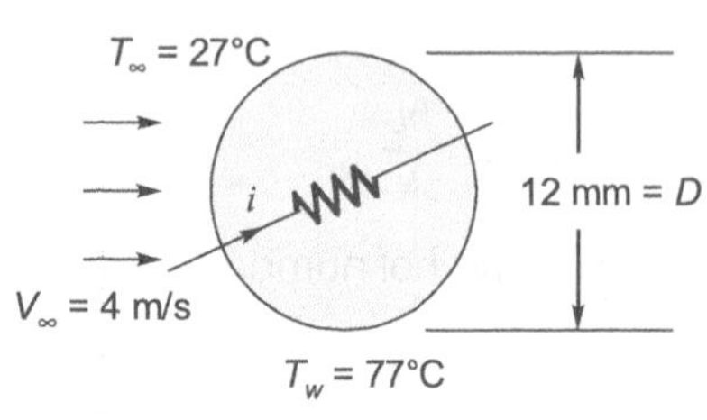

$$Nu = \frac{hD}{k_{air}} = 31.4$$

$$\Rightarrow \quad h = \frac{31.4 \times 0.026}{\left(\frac{12}{1000}\right)}$$

$$= 68.033 \text{ W/m}^2\text{K}$$

Heat lost by sphere = $hA\Delta T$

$$= h \times (4\pi R^2)\,(T_w - T_\infty)$$

$$= 68.033 \times 4 \times \pi \times \left(\frac{6}{1000}\right)^2 (77 - 27) \text{ watt}$$

$$= 1.539 \text{ watt}$$

4.54 (c)

$$\text{Nu} = f(\text{Re, Pr})$$

$$\frac{hD}{k} = f\left(\frac{VD\rho}{\mu}, \frac{\mu c_p}{k}\right)$$

4.55 (b)

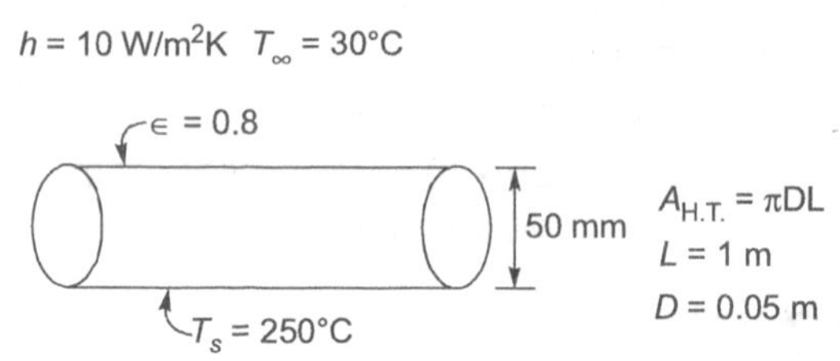

Total H.T. rate per unit length = H.T. rate by convection to air per unit length

+ H.T. rate by radiation to ambient per unit length

$$= \frac{hA(T_s - T_\infty)}{L} + \frac{\sigma \in A}{L}(T_s^4 - T_\infty^4)$$

$$= \frac{h\pi DL(T_s - T_\infty)}{L} + \frac{\sigma \in (\pi DL)(T_s^4 - T_\infty^4)}{L}$$

$= 10 \times \pi \times 0.5\,(250 - 30) + 5.67 \times 10^{-8} \times 0.8 \times \pi \times 0.05 \times (523^4 - 303^4)$

$= 818.8$ W/m

4.56 (c)

For liquid metals (like mercury), Prandtl number is very low since they have high thermal conductivity.

By definition

$$P_r = \frac{\nu}{\alpha} \quad \text{i.e. } \delta << \delta_t$$

δ = H.B.L. thickness;

δ_t = T.B.L. thickness

4.57 (b)

Momentum and energy transfers by advection in the condensate film are assumed to be negligible.

4.58 (d)

$$h = f(v, D, \rho, \mu, C_p, k)$$

4.59 (c)

The value of boundary layer thickness and local friction coefficient for turbulent flow is given by

$$\therefore \quad \delta = \frac{0.382x}{\text{Re}_x^{1/5}}; C_f = \frac{0.0592}{\text{Re}_x^{1/5}}$$

4.60 (a)

The Dittus-Boelter equation (for turbulent flow) is an explicit function for calculating the Nusselt number and is given by

$$Nu_d = 0.023\left(\text{Re}_d^{0.8}\right)(\text{Pr})^n$$

4.61 (a)

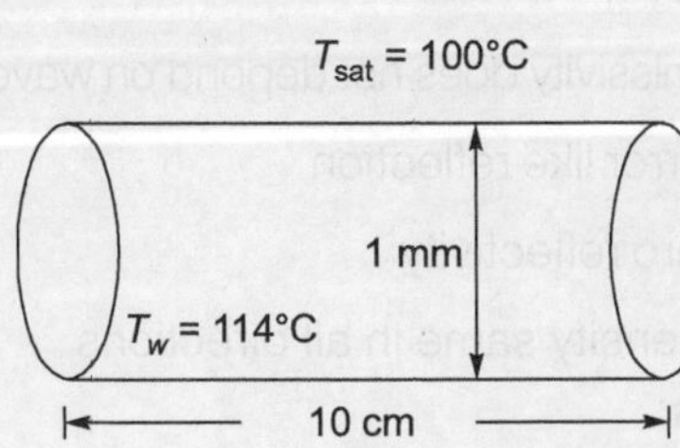

$$q_{\text{convection}} = h \times \pi\, DL \times (T_w - T_{\text{sat}})$$

$$\text{Power} = 5000 \times \pi \times \frac{1}{1000} \times \frac{10}{100} \times (114 - 100)$$

$$= 21.99 \text{ W}$$

4.62 (b)

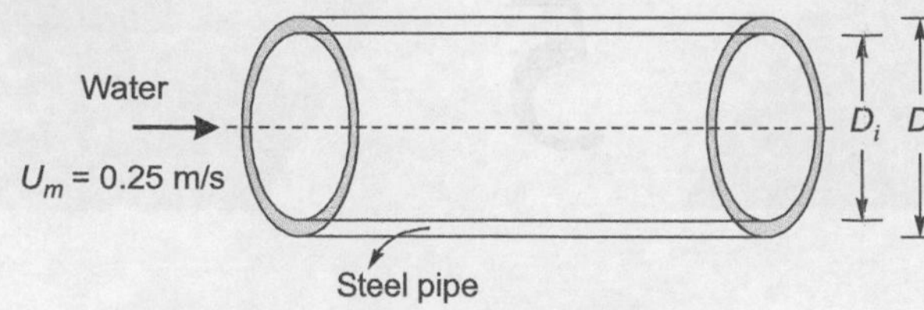

Inner diameter, D_i = 0.0525 m

Outer diameter, D_o= 0.06033 m

$$\rho = 960 \text{ kg/m}^3$$

$$\mu = 2.82 \times 10^{-4} \text{ kg/m-s,}$$

$$K = 0.68 \text{ W/m-°C}$$

$$\text{Pr} = 1.76, \text{ Nu} = 151.4$$

$$\text{Reynolds number, Re} = \frac{\rho U_m D_h}{\mu};$$

D_h = Hydraullic mean diameter

[Here : $D_h = D_i$]

$$\text{Re} = \frac{960 \times 0.25 \times 0.0525}{2.82 \times 10^{-4}}$$

$$\text{Re} = 44680.85$$

4.63 (a)

Nusselt number, Nu = 151.4

$$\frac{hD_i}{k} = 151.4$$

$$h = \frac{151.4 \times 0.68}{0.0525}$$

$$h = 1961 \text{ W/m}^2\text{°C}$$

4.65 (c)

- If $\frac{Gr}{\text{Re}^2} > 10$, then free convection is of primary importance.
- If $\frac{Gr}{\text{Re}^2} < 10$, then forced convection is of primary importance.

■■■■

5 Radiation

5.1 Heat transfer by radiation between two grey bodies of emissivity ε is proportional to (notations have their usual meanings)

(a) $\frac{(E_b - J)}{(1-\varepsilon)}$ (b) $\frac{(E_b - J)}{(1-\varepsilon)/\varepsilon}$

(c) $\frac{(E_b - J)}{(1-\varepsilon)^2}$ (d) $\frac{(E_b - J)}{(1-\varepsilon^2)}$

[ESE : 2000]

5.2 Solar radiation of 1200 W/m² falls perpendicularly on a gray opaque surface of emissivity 0.5. If the surface temperature is 50°C and surface emissive power is 600 W/m², the radiosity of that surface will be

(a) 600 W/m² (b) 1000 W/m²
(c) 1200 W/m² (d) 1800 W/m²

[ESE : 2000]

5.3 The shape factor of a hemispherical body placed on a flat surface with respect to itself is

(a) zero (b) 0.25
(c) 0.5 (d) 1.0

[ESE : 2001]

5.4 An enclosure consists of four surfaces 1, 2, 3, and 4. The view factors for radiation heat transfer (where the subscripts 1, 2, 3, 4 refer to the respective surfaces) are F_{11} = 0.1, F_{12} = 0.4 and F_{13}= 0.25. The surface areas A_1 and A_4 are 4 m² and 2 m² respectively. The view factor F_{41} is

(a) 0.75 (b) 0.50
(c) 0.25 (d) 0.10

[ESE : 2001]

5.5 Two large parallel grey plates with a small gap, exchange radiation at the rate of 1000 W/m² when their emissivities are 0.5 each. By coating one plate, its emissivity is reduced to 0.25. Temperatures remain unchanged. The new rate of heat exchange shall become

(a) 500 W/m² (b) 600 W/m²
(c) 700 W/m² (d) 800 W/m²

[ESE : 2002]

5.6 Two long parallel plates of same emissivity 0.5 are maintained at different temperatures and have radiation heat exchange between them. The radiation shield of emissivity 0.25 placed in the middle will reduce radiation heat exchange to

(a) 1/2 (b 1/4
(c) 3/10 (d) 3/5 [ESE : 2002]

5.7 Match **List-I** (Type of radiation) with **List-II** (Characteristic) and select the correct answer using the codes given below the lists:

List-I

A. Black body
B. Grey body
C. Specular
D. Diffuse

List-II

1. Emissivity does not depend on wavelength
2. Mirror like reflection
3. Zero reflectivity
4. Intensity same in all directions

Codes:

	A	B	C	D
(a)	2	1	3	4
(b)	3	4	2	1
(c)	2	4	3	1
(d)	3	1	2	4

[ESE : 2002]

5.8 **Assertion (A):** Heat transfer at high temperature is dominated by radiation rather than convection.
Reason (R): Radiation depends on fourth power of temperature while convection depends on unit power relationship.

(a) Both A and R are individually true and R is the correct explanation of A
(b) Both A and R are individually true but R is not the correct explanation of A
(c) A is true but R is false
(d) A is false but R is true

[ESE : 2002]

5.9 A body at 500 K cools by radiating heat to ambient atmosphere maintained at 300 K. When the body has cooled to 400 K, the cooling rate as a percentage of original cooling rate is about
(a) 31.1 (b) 41.5
(c) 50.3 (d) 80.4

[ESE : 2003]

5.10 Fraction of radiative energy leaving one surface that strikes the other surface is called
(a) radiative flux
(b) emissive power of the first surface
(c) view factor
(d) reradiation flux [ESE : 2003]

5.11 Consider the following statements in respect of automobile engine with thermosyphon cooling:
1. Heat transfer from gases to cylinder walls takes place by convection and radiation.
2. Most of the heat transfer from radiator to atmosphere takes place by radiation.
3. Most amount of heat transfer from radiator to atmosphere takes place by convection.
4. Heat transfer from cylinder walls takes place by conduction and convection.

Which of these statement are correct?
(a) 1, 2 and 4 (b) 1, 3 and 4
(c) 2, 3 and 4 (d) 1 and 2

[ESE : 2003]

5.12 Match **List-I** (Surface with Orientations) with **List-II** (Equivalent Emissivity) and select the correct answer using the codes given below the lists:

List-I
A. Infinite parallel planes
B. Body 1 completely enclosed by body 2 but body 1 is very small
C. Radiation exchange between two small gray bodies
D. Two concentric cylinders with large lengths

List-II
1. ε_1
2. $\dfrac{1}{\dfrac{1}{\varepsilon_1}+\dfrac{1}{\varepsilon_2}-1}$
3. $\dfrac{1}{\dfrac{1}{\varepsilon_1}+\left(\dfrac{A_1}{A_2}\right)\left(\dfrac{1}{\varepsilon_2}-1\right)}$
4. $\varepsilon_1\varepsilon_2$

Codes:

	A	B	C	D
(a)	3	1	4	2
(b)	2	4	1	3
(c)	2	1	4	3
(d)	3	4	1	2

[ESE : 2004]

5.13 Two spheres *A* and *B* of same material have radii 1 m and 4 m and temperature 4000 K and 2000 K respectively. Which one of the following statements is correct?
The energy radiated by sphere *A* is
(a) greater than that of sphere *B*
(b) less than that of sphere *B*
(c) equal to that of sphere *B*
(d) equal to double that of sphere *B*

[ESE : 2004]

5.14 For an opaque plane surface the irradiation, radiosity and emissive power are respectively 20, 12 and 10 W/m^2. What is the emissivity of the surface?
(a) 0.2 (b) 0.4
(c) 0.9 (d) 1.0 [ESE : 2004]

5.15 Two long parallel surfaces, each of emissivity 0.7 are maintained at different temperatures and accordingly have radiation exchange between them. It is desired to reduce 75% of this radiant heat transfer by inserting thin parallel shields of equal emissivity 0.7 on both sides. What would be the number of shields?
(a) 1 (b) 2
(c) 3 (d) 4 [ESE : 2004]

5.16 The earth receives at its surface radiation from the sun at the rate of 1400 W/m^2. The distance of centre of sun from the surface of earth is 1.5×10^{11} m and the radius of sun is 7.0×10^8 m.

What is approximately the surface temperature of the sun treating the sun as a black body?

(a) 3650 K (b) 4500 K
(c) 5800 K (d) 6150 K

[ESE : 2004]

5.17 What is the value of the shape factor F_{12} in a cylindrical cavity of diameter d and height h between bottom face known as surface 1 and top flat surface known as surface 2?

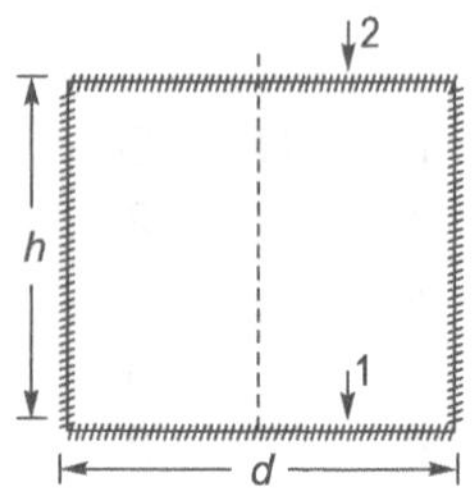

(a) $\frac{2h}{2h+d}$ (b) $\frac{d}{d+4h}$
(c) $\frac{4d}{4d+h}$ (d) $\frac{2d}{2d+h}$

[ESE : 2004]

5.18 A hemispherical surface 1 lies over a horizontal plane surface 2 such that convex portion of the hemisphere is facing sky. What is the value of the geometrical shape factor F_{12}?

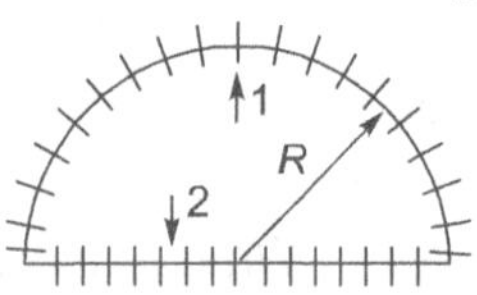

(a) 1/4 (b) 1/2
(c) 3/4 (d) 1/8

[ESE : 2004]

5.19 A small sphere of outer area 0.6 m^2 is totally enclosed by a large cubical hall. The shape factor of hall with respect to sphere is 0.004. What is the measure of the internal side of the cubical hall?

(a) 4 m (b) 5 m
(c) 6 m (d) 10 m

[ESE : 2004]

5.20 Match **List-I** with **List-II** and select the correct answer using the code given below the lists:

List-I

A. Radiation heat transfer
B. Conduction heat transfer
C. Forced convection
D. Transient heat flow

List-II

1. Fourier number
2. Wien's displacement law
3. Fourier law
4. Stanton number

Codes:

	A	B	C	D
(a)	2	1	4	3
(b)	4	3	2	1
(c)	2	3	4	1
(d)	4	1	2	3

[ESE : 2005]

5.21 Intensity of radiation at a surface in perpendicular direction is equal to:

(a) Product of emissivity of surface and $1/\pi$
(b) Product of emissivity of surface and π
(c) Product of emissive power of surface and $1/\pi$
(d) Product of emissive power of surface and π

[ESE : 2005]

5.22 What is the shape factor of a hemispherical body placed on a flat surface with respect to itself?

(a) Zero (b) 0.25
(c) 0.5 (d) 1.0

[ESE : 2005]

5.23 What is the value of the shape factor for two infinite parallel surface separated by a distance d?

(a) 0 (b) ∞
(c) 1 (d) d

[ESE : 2006]

5.24 If the temperature of a solid surface changes from 27°C to 627°C, then its emissive power changes which ratio?

(a) 6 : 1 (b) 9 : 1
(c) 27 : 1 (d) 81 : 1

[ESE : 2006]

5.25 Match **List-I** (Phenomenon) with **List-II** (Number/Law/Factor) and select the correct answer using the code given below the lists:

List-I

A. Radiation heat transfer
B. Conduction heat transfer

C. Forced convection
D. Transient heat flow

List-II
1. Biots number
2. View factor
3. Fourier's law
4. Stanton number

Codes:

	A	B	C	D
(a)	4	3	2	1
(b)	2	1	4	3
(c)	4	1	2	3
(d)	2	3	4	1

[ESE : 2006]

5.26 Consider the diagram given below. Which one of the following is correct?

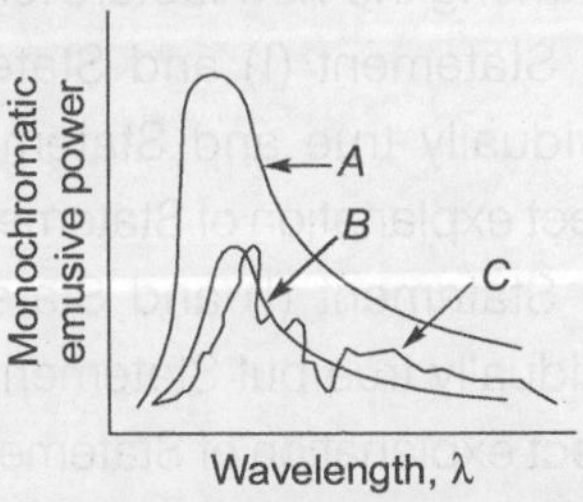

(a) Curve *A* is for gray body, Curve *B* is for black body, and Curve *C* is for selective emitter
(b) Curve *A* is for selective emitter, Curve *B* is for black body, and Curve *C* is for gray body
(c) Curve *A* is for selective emitter, Curve *B* is for gray body, and Curve *C* is for selective emitter
(d) Curve *A* is for black body, Curve *B* is for gray body, and Curve *C* is for selective emitter

[ESE : 2007]

5.27 What is the basic equation of thermal radiation from which all other equations of radiation can be derived?
(a) Stefan - Boltzmann equation
(b) Planck's equation
(c) Wien's equation
(d) Rayleigh - Jeans formula [ESE : 2007]

5.28 For a hemisphere, the solid angle is measured
(a) in radian and its maximum value is π.
(b) in degree and its maximum value is 180°.
(c) in steradian and its maximum value is 2π.
(d) in steradian and its maximum value is π.

[ESE : 2007]

5.29 For the radiation between two infinite parallel planes of emissivity ε_1 and ε_2 respectively, which one of the following is the expression for emissivity factor?

(a) $\varepsilon_1 \varepsilon_2$ (b) $\frac{1}{\varepsilon_1}+\frac{1}{\varepsilon_2}$

(c) $\frac{1}{\frac{1}{\varepsilon_1}+\frac{1}{\varepsilon_2}}$ (d) $\frac{1}{\frac{1}{\varepsilon_1}+\frac{1}{\varepsilon_2}-1}$

[ESE : 2007]

5.30 What is the radiation intensity in a particular direction?
(a) Radiant energy per unit time per unit area of the radiating surface
(b) Radiant energy per unit time per unit solid angle per unit area of the radiating surface
(c) Radiant energy per unit time per unit solid angle per unit projected area of the radiating surface in the given direction
(d) Radiant energy per unit time per unit projected area of the radiating surface in the given direction

[ESE : 2007]

5.31 What is the equivalent emissivity for radiant heat exchange between a small body (emissivity = 0.4) in a very large enclosure (emissivity = 0.5)?
(a) 0.5 (b) 0.4
(c) 0.2 (d) 0.1

[ESE : 2008]

5.32 The heat exchange between a small body having emissivity ε_1 and area A_1 and a large enclosure having emissivity ε_2 and area A_2 is given by:

$$q_{1-2} = A_1\,\varepsilon_1\,\sigma\,(T_1^4 - T_2^4)$$

What is the assumption for this equation?
(a) $\varepsilon_2 = 1$
(b) $\varepsilon_2 = 0$
(c) A_1 is very small as compared to A_2
(d) Small body is at centre of enclosure

[ESE : 2008]

5.33 What will be the view factor F_{21} for the geometry as shown in the figure below (sphere within a cube)?

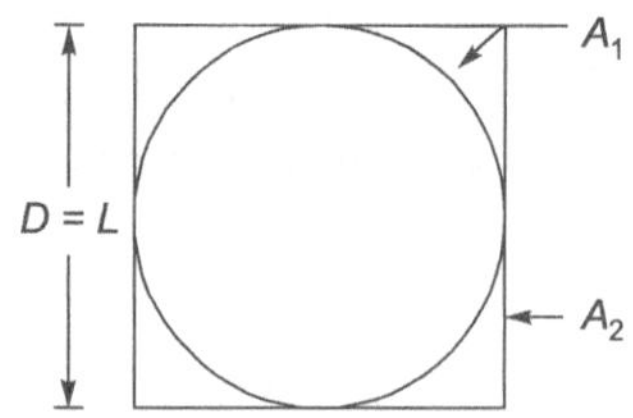

(a) $\frac{\pi}{2}$ (b) $\frac{\pi}{4}$

(c) $\frac{\pi}{3}$ (d) $\frac{\pi}{6}$

[ESE : 2009]

5.34 With reference to the below figure, the shape factor between 1 and 2 is

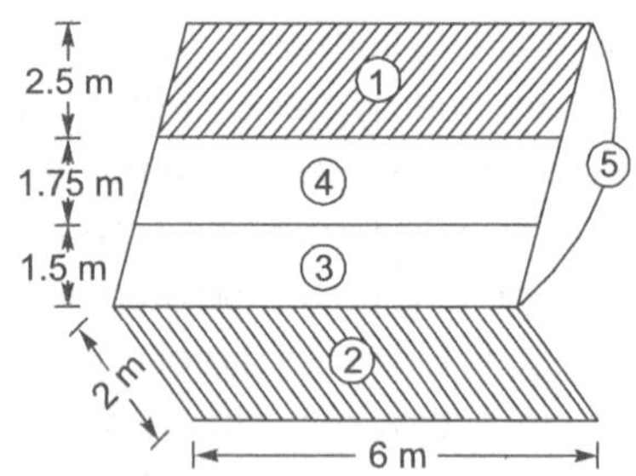

(a) 0.272 (b) 0.34
(c) 0.66 (d) data insufficient

[ESE : 2010]

5.35 Two radiating surfaces $A_1 = 6$ m^2 and $A_2 = 4$ m^2 have the shape factor $F_{1-2} = 0.1$; the shape factor F_{2-1} will be

(a) 0.18 (b) 0.15
(c) 0.12 (d) 0.10 [ESE : 2010]

5.36 The radiative heat transfer rate per unit area (W/m^2) between two plane parallel grey surfaces whose emissivity is 0.9 and maintained at 400 K and 300 K is

(a) 992 (b) 812
(c) 567 (d) 464

[ESE : 2010]

5.37 Ice is very close to a

(a) Gray body (b) Black body
(c) White body (d) Specular body

[ESE : 2011]

5.38 A satellite floats in deep space with very high velocity. It will continuously lose heat by

(a) convection
(b) conduction and convection
(c) radiation
(d) radiation and convection

[ESE : 2012]

5.39 An oxygen and nitrogen mixture has transmissivity for heat radiations that is practically equal to

(a) 1 (b) 0.5
(c) 2 (d) 0

[ESE : 2012]

5.40 **Statement (I):** The view factors from the surface, say A_i, of the enclosure to all surfaces of the enclosure, including to itself, when summed, should be equal to unity by definition of the view factor.

Statement (II): This is called the summation relation among the view factors for enclosure.

(a) Both Statement (I) and Statement (II) are individually true and Statement (II) is the correct explanation of Statement (I)
(b) Both Statement (I) and Statement (II) are individually true but Statement (II) is not the correct explanation of Statement (I)
(c) Statement (I) is true but Statement (II) is false
(d) Statement (I) is false but Statement (II) is true.

[ESE : 2012]

5.41 Which of the following statements are correct for a blackbody?

1. A blackbody continues to emit radiation even when it is in thermal equilibrium with its surroundings.
2. A black body is a perfect emitter.
3. Absorptivity, $\alpha = 1$ represents a black surface

(a) 1 and 2 only (b) 1 and 3 only
(c) 1, 2 and 3 (d) 2 and 3 only

[ESE : 2013]

5.42 What is net radiant interchange per square meter for two very large plates at temperature 800 K and 500 K respectively? Emissivities of hot and cold plates are 0.8 and 0.6 and Stefan-Boltzmann constant is 5.67×10^{-8} W/m^{2}K^4

(a) 1026 kW/m^2 (b) 10.26 kW/m^2
(c) 102.6 kW/m^2 (d) 1.026 kW/m^2

[ESE : 2013]

5.43 If a body is at 2000 K, the wavelength at which the body emits maximum amount of radiation is
(a) 1.45 μm (b) 1.45 cm
(c) 0.345 cm (d) 0.344 cm
[ESE : 2014]

5.44 An isothermal cubical (10 m × 10 m × 10 m) blackbody at 200° C is suspended in air. The total radiation emitted by this body to its surroundings will be
(a) 1702.9 kW (b) 1800.7 kW
(c) 54.4 kW (d) 28.38.1 kW
[ESE : 2014]

5.45 A 1 m diameter spherical cavity is maintained at a uniform temperature of 500 K. The emissivity of the material of the sphere is 0.5; one 10 mm diameter hole is drilled. The maximum rate of radiant energy streaming through the hole will be
(a) 2782 W (b) 0.139 W
(c) 1392 W (d) 0.278 W **[ESE : 2014]**

5.46 For a hemispherical furnace with a flat circular base of diameter *D*, the view factor from the dome to its base is
(a) 0.5 (b) 1
(c) 0 (d) 0.32 **[ESE : 2014]**

5.47 For the same type of shapes, the value of Radiation Shape Factor will be higher when surfaces are
(a) more closer only
(b) moved further apart
(c) smaller and held closer
(d) larger and held closer
[ESE : 2015]

5.48 During very cold weather conditions, cricket players prefer to wear white woolen sweaters rather than coloured woolen sweaters. The reason is that white wool comparatively :
1. Absorbs less heat from body
2. Emits less heat to the atmosphere

Which of the above statements is/are correct ?
(a) 1 only (b) 2 only
(c) Both 1 and 2 (d) Neither 1 nor 2
[ESE : 2016]

5.49 A body 1 in the form of a sphere of 2 cm radius at temperature T_1 is located in body 2, which is a hollow cube of 5 cm side and is at temperature T_2 ($T_2 < T_1$). The shape factor F_{21} for radiation heat transfer becomes :
(a) 0.34 (b) 0.43
(c) 0.57 (d) 0.63
[ESE : 2016]

5.50 Solar radiation of 1000 W/m^2 is incident on a grey opaque surface with emissivity of 0.4 and emissive power of 400 W/m^2. The radiosity of the surface will be :
(a) 940 W/m^2 (b) 850 W/m^2
(c) 760 W/m^2 (d) 670 W/m^2
[ESE : 2016]

5.51 Consider the following statements:
1. Absorptivity depends on wave length of incident radiation waves
2. Emissivity is dependent on wave length of incident radiation waves

Which of the above statements is/are correct ?
(a) 1 only (b) 2 only
(c) Both 1 and 2 (d) Neither 1 nor 2
[ESE : 2016]

5.52 The view factors F_{12} and F_{21}, for the sphere of diameter *d* and a cubical box of length *l* = *d* as shown in the figure, respectively, are

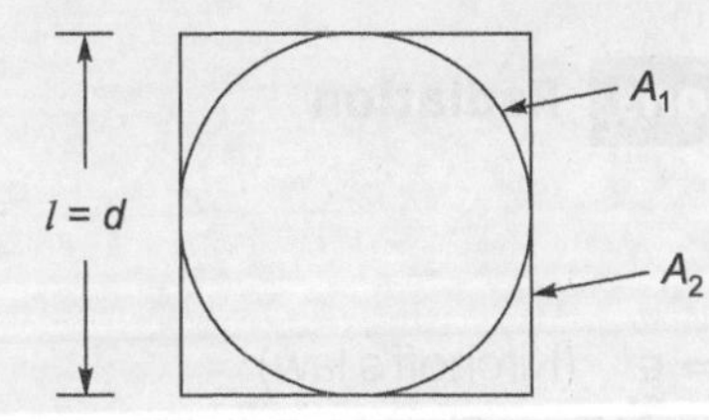

(a) 1 and $\frac{\pi}{3}$ (b) $\frac{\pi}{3}$ and 1
(c) 1 and $\frac{\pi}{6}$ (d) $\frac{\pi}{6}$ and 1
[ESE : 2017]

5.53 The temperature of a body of area 0.1 m^2 is 900 K. The wavelength for maximum monochromatic emissive power will be nearly
(a) 2.3 μm (b) 3.2 μm
(c) 4.1 μm (d) 5.0 μm **[ESE : 2019]**

5.54 A hemispherical furnace of radius 1.0 m has a roof temperature of T_1 = 800 K and emissivity ε_1 = 0.8. The flat floor of the furnace has a temperature T_2 = 600 K and emissivity ε_2 = 0.5. The view factor F_{12} from surface 1 to 2 will be:

(a) 0.3 (b) 0.4
(c) 0.5 (d) 0.6 [ESE : 2019]

5.55 The overall heat transfer coefficient due to convection and radiation for a steam maintained at 200°C running in a large room at 30°C is 17.95 W/m²K. If the emissivity of the pipe surface is 0.8; the value of $\sigma = 5.67 \times 10^{-8}$ W/m²K⁴; the heat transfer coefficient due to radiation will be nearly

(a) 17 W/m²K (b) 14 W/m²K
(c) 11 W/m²K (d) 8 W/m²K
[ESE : 2020]

5.56 If the monochromatic emissivity of a body is independent of wavelength, it is called

(a) transparent body (b) reflective body
(c) black body (d) gray body
[ESE : 2024]

5.57 Which one of the following equations is called Kirchhoff's identity?

(a) $\varepsilon = \dfrac{E}{E_b}$ (b) $\varepsilon = \dfrac{E_b}{E}$
(c) $\varepsilon = \infty$ (d) $\varepsilon = 1$
[ESE : 2024]

■■■■

Answers Radiation

5.1 (b)	5.2 (c)	5.3 (c)	5.4 (b)	5.5 (b)	5.6 (c)	5.7 (d)	5.8 (a)	5.9 (a)
5.10 (c)	5.11 (b)	5.12 (c)	5.13 (c)	5.14 (c)	5.15 (c)	5.16 (c)	5.17 (b)	5.18 (b)
5.19 (b)	5.20 (c)	5.21 (c)	5.22 (c)	5.23 (c)	5.24 (d)	5.25 (d)	5.26 (d)	5.27 (b)
5.28 (c)	5.29 (d)	5.30 (c)	5.31 (b)	5.32 (c)	5.33 (d)	5.34 (d)	5.35 (b)	5.36 (b)
5.37 (b)	5.38 (c)	5.39 (a)	5.40 (a)	5.41 (c)	5.42 (b)	5.43 (a)	5.44 (a)	5.45 (d)
5.46 (a)	5.47 (d)	5.48 (c)	5.49 (a)	5.50 (*)	5.51 (a)	5.52 (c)	5.53 (b)	5.54 (c)
5.55 (c)	5.56 (d)	5.57 (*)						

Explanations Radiation

5.1 (b)

$\alpha = \epsilon$ (Kirchoff's law)

$q = A(E - \alpha G)$ $\rho = 1 - \alpha = 1 - \epsilon$

Radiosity

$J = \epsilon E_b + (1 - \epsilon)G$

$$q = A\left(J - \frac{J - \epsilon E_b}{1 - \epsilon}\right) = \frac{E_b - J}{\left(\dfrac{1-\epsilon}{\epsilon A}\right)}$$

5.2 (c)

$J = \varepsilon E_b + \rho G = 600 + (0.5)(1200)$

$= \mathbf{1200\ W/m^2}$

5.3 (c)

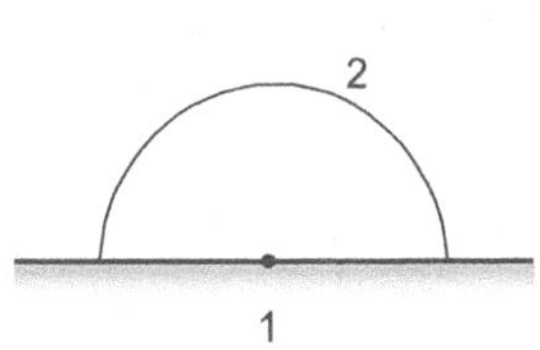

$F_{11} + F_{12} = 1$

$F_{11} = 0$

$\therefore F_{12} = 1$

$A_1F_{12} = A_2F_{21}$ $(F_{12} = 1)$

$\pi R^2 = F_{21} 2\pi R^2$

$F_{21} = \mathbf{0.5}$

$F_{22} = 1 - F_{21} = 0.5$

5.4 (b)

Surface enclosure theorem

$F_{11} + F_{12} + F_{13} + F_{14} = 0$

$F_{14} = 1 - (0.10 + 0.40 + 0.25)$

$F_{14} = 0.25$

$A_1F_{14} = A_4F_{41}$

$F_{41} = 0.25 \times 4/2 = \mathbf{0.50}$

5.5 (b)

$$\frac{Q_{new}}{Q_{old}} = \frac{(\text{Resistance})_{old}}{(\text{Resistance})_{new}}$$

$$= \frac{\left(\frac{2}{\varepsilon}-1\right)}{\left(\frac{1}{\varepsilon_1}+\frac{1}{\varepsilon_2}-1\right)} = \frac{\left(\frac{2}{0.5}-1\right)}{\left(\frac{1}{0.5}+\frac{1}{0.25}-1\right)}$$

$$Q_{new} = \frac{3}{5}\times 1000 = 600 \text{ W/m}^2$$

5.6 (c)

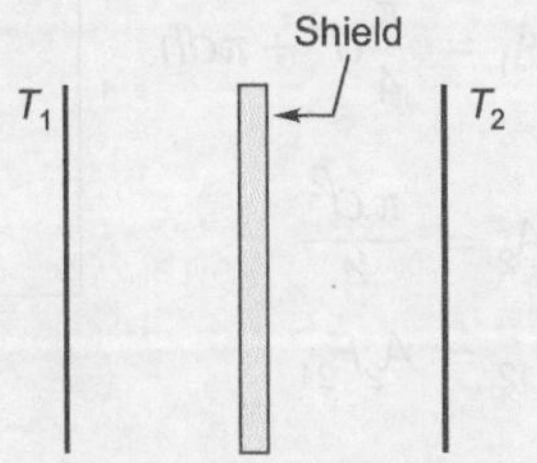

Without shield:

$$Q_1 = \frac{\sigma\left(T_1^4 - T_2^4\right)}{\frac{1}{\varepsilon_1}+\frac{1}{\varepsilon_2}-1} = \frac{\sigma\left(T_1^4 - T_2^4\right)}{\frac{1}{0.5}+\frac{1}{0.5}-1}$$

$$= \frac{\sigma\left(T_1^4 - T_2^4\right)}{3}$$

With shield:

$$Q_2 = \frac{\sigma\left(T_1^4 - T_2^4\right)}{\frac{1}{\varepsilon_1}+\frac{1}{\varepsilon_2}+\frac{2}{\varepsilon_S}-2}$$

$$= \frac{\sigma\left(T_1^4 - T_2^4\right)}{\frac{1}{0.5}+\frac{1}{0.5}+\frac{2}{0.25}-2} = \frac{\sigma\left(T_1^4 - T_2^4\right)}{10}$$

or $\frac{Q_2}{Q_1} = \frac{3}{10}$

$\therefore \quad Q_2 = \frac{3}{10}Q_1$

5.7 (d)

Black body : Zero reflectivity

Grey body : Emissivity does not depend on wavelength.

Specular surface : Mirror like behavior

Diffuse surface : Intensity same in all direction

5.8 (a)

Heat transfer at high temperature is dominated by radiation rather than convection because. $q_{rad} = \in A\,\sigma T^4$ heat transfer in radiation varies at fourth power of absolute temperature. $q_{conv} = hA\,\Delta T$.

5.9 (a)

$$\frac{\theta_1}{\theta_0}\times 100 = \frac{T_1^4 - T_\infty^4}{T_0^4 - T_\infty^4}\times 100 = \frac{(400)^4-(300)^4}{(500)^4-(300)^4}\times 100$$

$$= \mathbf{32.15\%}.$$

5.10 (c)

The view factor is the fraction of radiative energy that diffused from one surface element and strikes the other surface directly.

5.11 (b)

Temperature difference is lower so density difference creates convection.

5.12 (c)

Equivalent thermal resistance between two gray bodies exchainging heat

$$R_{th} = \frac{1-\varepsilon_1}{\varepsilon_1 A_1}+\frac{1}{A_1F_{12}}+\frac{1-\varepsilon_2}{\varepsilon_2 A_2}$$

(i) For infinite parallel plates

$F_{12} = 1; \quad A_1 = A_2 = A$

$$R_{th} = \frac{1}{A}\left[\frac{1}{\varepsilon_1}-1+1+\frac{1}{\varepsilon_2}-1\right] = \frac{1}{A}\left[\frac{1}{\varepsilon_1}+\frac{1}{\varepsilon_2}-1\right]$$

(ii) Concentric cylinders

$F_{12} = 1$

$$R_{th} = \frac{1}{A_1}\left[\frac{1}{\varepsilon_1}-1+1+\frac{A_1}{A_2}\left(\frac{1}{\varepsilon_2}-1\right)\right]$$

$$= \frac{1}{A_1}\left[\frac{1}{\varepsilon_1}+\frac{A_1}{A_2}\left(\frac{1}{\varepsilon_2}-1\right)\right]$$

(iii) A small gray body (i) in a large gray enclosure (ii)

$$R_{th} = \frac{1-\varepsilon_1}{\varepsilon_1 A_1}+\frac{1}{A_1F_{12}}+\frac{1-\varepsilon_2}{\varepsilon_2 A_2}$$

$A_1 << A_2$

$\frac{A_1}{A_2} = 0$ and $F_{12} = 1$

$$R_{th} = \frac{1}{A_1}\left[\frac{1}{\varepsilon_1}-1+1\right] = \frac{1}{\varepsilon_1 A_1}$$

5.13 (c)

$$q_A = \sigma.4\pi r_A^2.T_A^4 = 4\sigma\pi r_A^2 T_A^4$$
$$q_B = \sigma.4\pi r_B^2.T_B^4 = 4\sigma\pi r_B^2 T_B^4$$
$$\frac{q_A}{q_B} = \frac{r_A^2 T_A^4}{r_B^2 T_B^4}$$
$$= \frac{1^2 \times (4000)^4}{4^2 \times (2000)^4} = \frac{1 \times 2^4 \times 2000^4}{4^2 \times 2000^4} = 1$$

Hence, $q_A = q_B$

5.14 (c)

Radiosity: $J = \varepsilon E_b + (1-\varepsilon)G$
$$12 = (10) + (1-\varepsilon)20$$
$$12 = 10 + 20 - 20\varepsilon$$
$$9 = 10\varepsilon$$
$$\varepsilon = \frac{9}{10} = 0.9$$

5.15 (c)

$$q_{\text{with shield}} = \frac{1}{n+1} q_{\text{without shield}}$$
$$\left(1 - \frac{q_{\text{with shield}}}{q_{\text{without shield}}}\right) = 1 - \frac{1}{n+1}$$
$$\left(1 - \frac{3}{4}\right) = \frac{1}{n+1}$$
$$\frac{1}{4} = \frac{1}{n+1}$$
$\Rightarrow$ $n = $ **3**

5.16 (c)

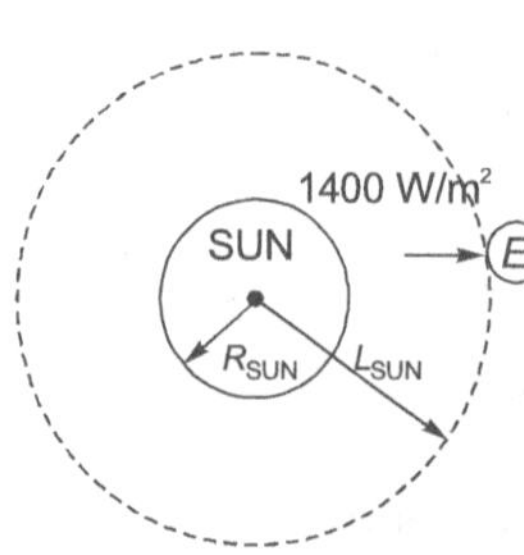

$R_{sun} = 7 \times 10^8$ m

$L_{sun} = 1.5 \times 10^{11}$ m

From the figure it is clear that energy radiated from sun is leaving a virtual sphere of radius L_{SUN}.

$\Rightarrow$ Energy radiated by Sun

= energy leaving from sphere of radius L_{sun}

$\Rightarrow \sigma T_{sun}^4 \times 4\pi R_{sun}^2$ = energy leaving per unit area of square × area of square

$$\Rightarrow \sigma T_{sun}^4 \times 4\pi R_{sun}^2 = 1400 \text{ W/m}^2 \times 4\pi L_{sun}^2$$

On solving ,

$$T_{sun} = 5802.63 \text{ K}$$

5.17 (b)

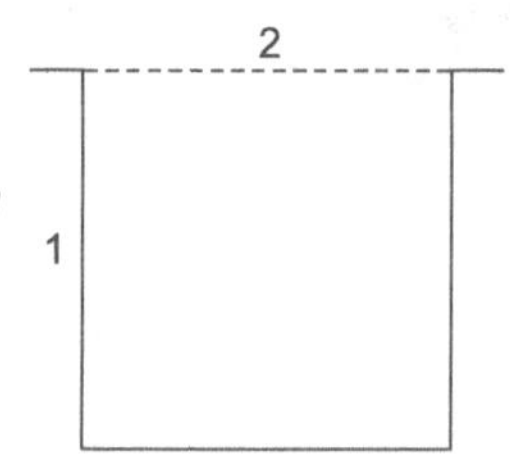

$$A_1 = \frac{\pi}{4}d^2 + \pi dh$$
$$A_2 = \frac{\pi.d^2}{4}$$
$$A_1F_{12} = A_2F_{21} \qquad (F_{21} = 1)$$
$$F_{12} = \frac{A_2}{A_1} \times 1 = \frac{\frac{\pi}{4}d^2}{\frac{\pi}{4}d^2 + \pi dh}$$
$$F_{12} = \frac{d}{d+4h}$$

5.18 (b)

$$F_{22} = 0; \quad F_{21} = 1$$
$$F_{12} = \left(\frac{A_2}{A_1}\right)F_{21} = \frac{1}{2}$$
$$F_{12} = \frac{1}{2}$$

5.19 (b)

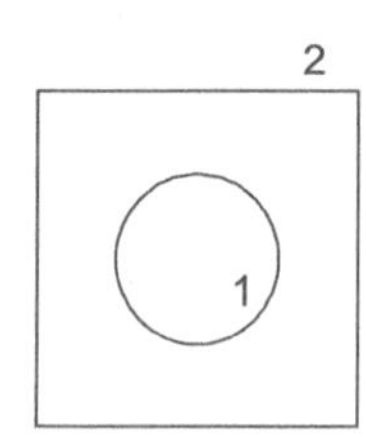

$$A_1 = 0.6 \text{ m}^2$$
$$F_{21} = 0.004$$
$$A_2 = 6a^2$$
(a is edge length)
$$F_{21} = 0.004$$
$$F_{11} + F_{12} = 1$$
$$F_{11} = 0$$
$$\therefore \quad F_{12} = 1$$
$$\therefore \quad A_1F_{12} = A_2F_{21}$$
$$0.6 \times 1 = 6a^2 \times 0.004$$
a = **5 m**

5.20 (c)

Fourier law : For conduction heat transfer

Fourier number $\left(F_0 = \dfrac{\alpha\tau}{L_c^2}\right)$: Useful in unsteady state heat transfer

Wiens displacement law : Gives theory of radiation

Stanton number : Useful in forced convection heat transfer.

5.21 (c)

Intensity of radiation, $I_n = \dfrac{E}{\pi}$

where E = Emissive power

5.22 (c)

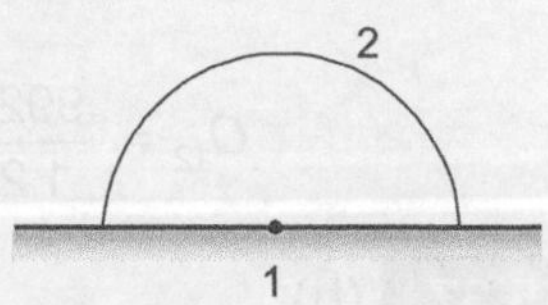

$$F_{11} + F_{12} = 1$$
$$F_{11} = 0$$
$$\therefore \quad F_{12} = 1$$
$$A_1F_{12} = F_{21}A_2$$
$$(\pi R^2) = F_{21}(2\pi R^2)$$
$$F_{21} = 0.5$$
$$F_{21} + F_{22} = 1$$
$$F_{22} = 1 - 0.5 = \mathbf{0.5}$$

5.23 (c)

For two infinite plane surface; energy emitted by the one surface is totally intercepted by the other surface.

$$\therefore \quad F_{12} = 1$$

5.24 (d)

Emissive power, $E_b = \sigma T^4$

$$\frac{E_{b_2}}{E_{b_1}} = \left(\frac{T_2}{T_1}\right)^4 = \left(\frac{627+273}{27+273}\right)^4 = \mathbf{81 : 1}$$

5.25 (d)

(a) Radiation heat transfer : View factor
(b) Conduction heat transfer : Fourier's law
(c) Forced convection : Stanton number
(d) Transient heat flow : Biot number

5.26 (d)

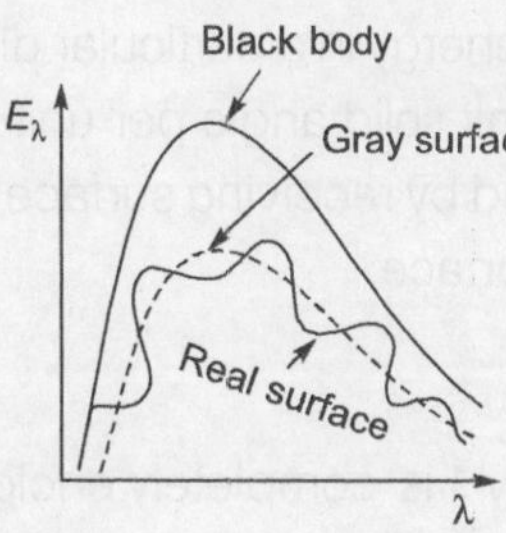

Monochromatic emission power of black surface is always more than that of any real surface at same temperature.

The monochromatic emissivity of real surface is different at different wavelength and because of irregular nature of the curve obtained for real surfaces, the heat flow estimation becomes very difficult. Hence, each real surface is approximated to an imaginary surface known as gray surface.

The surface, having same value of monochromatic emissivity at all the wavelength, which is equal to the average emissivity of real surface is know as gray surface.

5.27 (b)

Planck's distribution law is the mother equation for other radiation laws, since other can be derived from it.

5.28 (c)

Solid angle is defined as a portion of the space inside a sphere enclosed by a conical surface with the vertex of the cone at the centre of the sphere for a hemisphere, the solid angle is measured in a steradian and its maximum value is 2π.

5.29 (d)

Equivalent thermal resistance between two gray bodies exchainging heat

$$R_{th} = \frac{1-\varepsilon_1}{\varepsilon_1 A_1} + \frac{1}{A_1F_{12}} + \frac{1-\varepsilon_2}{\varepsilon_2 A_2}$$

(i) For infinite parallel plates

$$F_{12} = 1$$
$$A_1 = A_2 = A$$

$$R_{th} = \frac{1}{A}\left[\frac{1}{\varepsilon_1} - 1 + 1 + \frac{1}{\varepsilon_2} - 1\right] = \frac{1}{A}\left[\frac{1}{\varepsilon_1} + \frac{1}{\varepsilon_2} - 1\right]$$

5.30 (c)

Radiation intensity is defined as rate of emission of radiant energy in a particular direction per unit time per unit solid angle per unit projected area substanded by receiving surface with respect to emitting surface.

5.31 (b)

When body 1 is completely enclosed by body 2 and body 1 is small

Then equivalent emissivity is given by ε_1

So emissivity = 0.4

5.32 (c)

When body 1 is completely enclosed by body 2, body 1 is large.

$$\therefore \ \varepsilon \text{ is given by } \frac{1}{\frac{1}{\varepsilon_1}+\frac{A_1}{A_2}\left(\frac{1}{\varepsilon_2}-1\right)}$$

Since, $\frac{A_1}{A_2}$ is very less hence, whole term can be neglected.

we get,

$$Q_{1\text{-}2} = A_1\varepsilon_1\sigma(T_1^4 - T_2^4)$$

5.33 (d)

$$F_{11} + F_{12} = 1$$

$$\because \quad F_{11} = 0$$

$$0 + F_{12} = 1$$

$$\Rightarrow \quad F_{12} = 1$$

$$A_1 F_{12} = A_2 F_{21}$$

$$\Rightarrow \quad F_{21} = \frac{A_1}{A_2} = \frac{4\pi\left(\frac{D}{2}\right)^2}{6D^2} = \frac{\pi}{6}$$

5.34 (d)

This question is solved by superposition rule and it requires view factor value. Since, graph is not given, it can't be solved.

5.35 (b)

From the reciprocity theorem,

$$A_1F_{12} = A_2F_{21}$$

$$A_1 = 6\text{m}^2 \qquad (F_{12} = 0.1)$$

$$A_2 = 4\text{ m}^2$$

$$F_{2\text{-}1} = \frac{A_1}{A_2}F_{12}$$

$$F_{2\text{-}1} = \frac{6}{4}\times 0.1$$

$$F_{2\text{-}1} = \mathbf{0.15}$$

5.36 (b)

Radiative heat transfer per unit area between two grey parallel plate of emissivity 0.9,

$$Q_{12} = \frac{\sigma\left(T_1^4 - T_2^4\right)}{\frac{1}{\varepsilon_1}+\frac{1}{\varepsilon_2}-1}$$

Since $\varepsilon_1 = \varepsilon_2 = 0.9$

and $\sigma = 5.67\times10^{-8}$

$T_1 = 400$ K; $T_2 = 300$ K

$$Q_{12} = \frac{5.67\times10^{-8}\left(400^4 - 300^4\right)}{\frac{1}{0.9}+\frac{1}{0.9}-1}$$

$$Q_{12} = \frac{992.25}{1.222} = \mathbf{811.84\ W/m^2}$$

5.37 (b)

Ice is very close to black body (absorptivity)$_{ice}$; 0.985.

5.38 (c)

As there will be no mass surrounding satellite it will lose heat by radiation only.

5.41 (c)

(i) A black body is an idealized physical body that absorbs all incident electromagnetic radiation, regardless of frequency or angle of incidence.

(ii) A black body in thermal equilibrium emits electromagnetic radiation called black body.

(iii) It is an ideal emitter. It emits as much or more energy at every frequency than any other body at the thermal equilibrium.

(iv) It is isotropic diffuse emitter.

(v) It is perfect absorber with $\alpha = 1$.

5.42 (b)

$$\frac{Q}{A} = \frac{\sigma\left(T_1^4 - T_2^4\right)}{\frac{1}{\epsilon_1}+\frac{1}{\epsilon_2}-1} = \frac{5.67\times10^{-8}\left(800^4 - 500^4\right)}{\left(\frac{1}{0.6}+\frac{1}{0.8}-1\right)}$$

$$\therefore \quad \frac{Q}{A} = 10.26 \times 10^3 \text{ W/m}^2$$

$$\text{or} \quad = \mathbf{10.26 \text{ kW/m}^2}$$

5.43 (a)

As per Wien's Displacement law,

$\lambda_{max} T = 2898\ \mu$mK

$$\lambda_{max} = \frac{2898}{2000} = \mathbf{1.449\ \mu m}$$

5.44 (a)

$$Q_R = \sigma A T^4$$
$$A = 6 \times 10 \times 10 = 600 \text{ m}^2$$
$$\sigma = 5.67 \times 10^{-8} \text{ W/m}^2\text{K}^4$$
$$T = 200°/C = 200 + 273 = 473 \text{ K}$$
$$\therefore \quad Q_R = 5.67 \times 10^{-8} \times 600 \times (473)^4$$
$$= \mathbf{1702.86 \text{ kW}}$$

5.45 (d)

$$A = \frac{\pi d^2}{4} = \frac{3.14}{4} \times 0.01^2$$
$$= 0.785 \times 10^{-4} \text{ m}^2$$
$$\epsilon = 0.5;\ T = 500 \text{ K}$$
$$Q_R = \sigma A T^4 \text{ watt}$$
$$= 5.67 \times 10^{-8} \times 0.785 \times 10^{-4} \times 500^4$$
$$= \mathbf{0.278 \text{ W}}$$

5.46 (a)

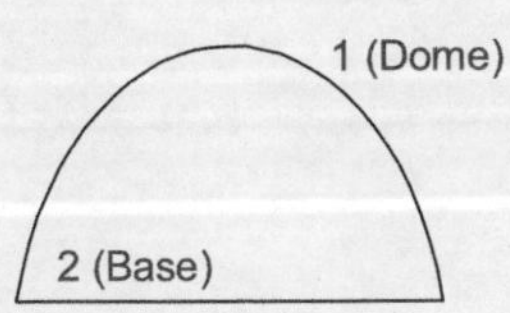

$$A_1 F_{12} = A_2 F_{21}$$
$$F_{21} + F_{22} = 1$$
$$F_{21} = 1 \qquad (\because F_{22} = 0)$$
$$\therefore \quad F_{12} = \frac{A_2}{A_1}$$

$$F_{21} = \frac{A_2}{A_1} = \frac{\pi r^2}{2\pi r^2} = \frac{1}{2} = \mathbf{0.5}$$

5.47 (d)

The value of radiation shape factor is high if surfaces are large and closer to each other.

5.48 (c)

While wool being good reflector hence bad emitter thus emits less heat to atmosphere.

5.49 (a)

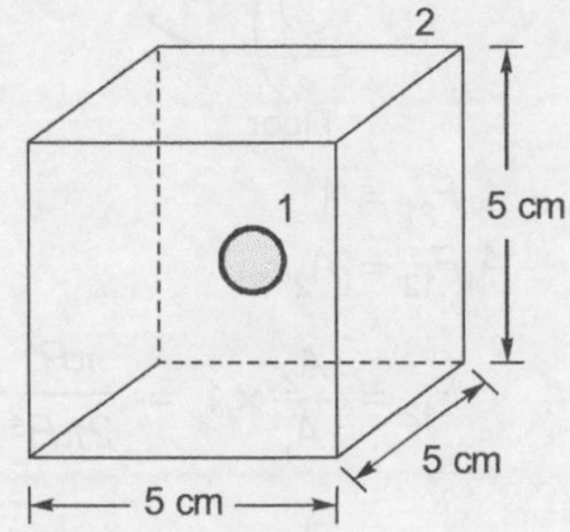

$$r = 2 \text{ cm}$$
$$F_{11} = 0$$
$$F_{12} + F_{11} = 1$$
$$F_{12} = 1$$
$$F_{21} = \frac{A_1 F_{12}}{A_2} = \frac{4\pi(0.02)^2}{6 \times (0.05)^2} = \mathbf{0.34}$$

5.52 (c)

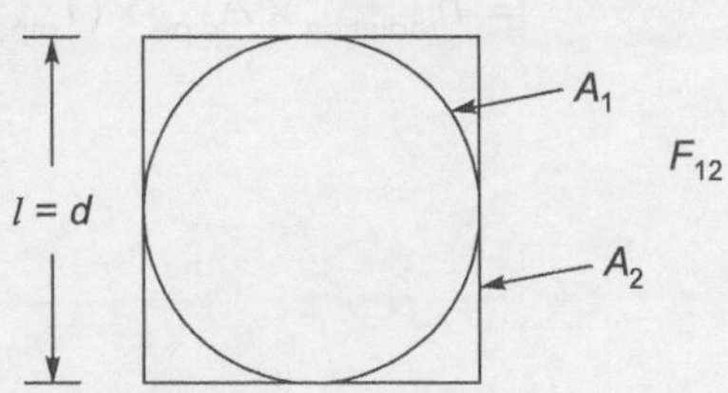

Sphere in a cube

From reciprocity relation,

$$A_1 F_{12} = A_2 F_{21}$$

$$F_{21} = \frac{A_1}{A_2} = \frac{4\pi\left(\frac{d}{2}\right)^2}{6d^2} = \frac{\pi}{6}$$

5.53 (b)

According to Wein's Displacement law,

$$\lambda_m T = \text{constant} = 2898\ \mu\text{m-K}$$
$$\therefore \quad \lambda_m \times 900 = 2898\ \mu\text{m-K}$$
$$\Rightarrow \quad \lambda_m = \frac{2898}{900} = 3.22\ \mu\text{m}$$

where, λ_m = Wavelength at which E_{b_λ} is maximum

5.54 (c)

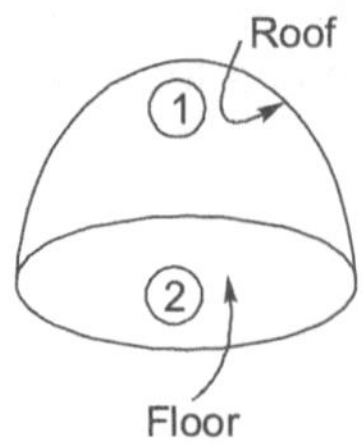

$$F_{21} = 1$$
$$A_1F_{12} = A_2F_{21}$$
$$F_{12} = \frac{A_2}{A_1} \times 1 = \frac{\pi R^2}{2\pi R^2} = \frac{1}{2}$$

5.55 (c)

$$T_{room} = 200°C + 273 = 473 \text{ K}$$
$$T_{pipe} = 30 + 273 = 303 \text{ K}$$
$$\epsilon_{pipe} = 0.8$$

Heat lost by pipe due to radiation to the room air

$$= \sigma A_{pipe}\, \epsilon_{pipe} \left(T_{pipe}^4 - T_{room}^4\right)$$
$$= h_{radiation} \times A_{pipe} \times (T_{pipe} - T_{room})$$

∴ Heat transfer coefficient due to radiation

$$h_{radiation} = \frac{\sigma \times \epsilon_{pipe} \left(T_{pipe}^4 - T_{room}^4\right)}{T_{pipe} - T_{room}}$$

$$= \frac{5.67 \times 10^{-8} \times 0.8\left[(200+273)^4 - (30+273)^4\right]}{(200-30)}$$

$$= 11.1067 \text{ W/m}^2\text{K}$$

5.56 (d)

For gray body, monochromatic emissivity is independent of wavelength, i.e.

$$\epsilon_\lambda = \text{Constant}$$

5.57 (*)

According to Kirchoff's identity, $\varepsilon = \alpha$

where, ε = Total emissivity

α = absorptivity

Note: There was a typing error in question paper given by UPSE. Option (c) should have been $\varepsilon = \alpha$.

6 Heat Exchangers

6.1 The equation of effectiveness $\varepsilon = 1 - e^{-NTU}$ of a heat exchanger is valid (*NTU* is number of transfer unit) in the case of
(a) boiler and condenser for parallel flow
(b) boiler and condenser for counter flow
(c) boiler and condenser for both parallel flow and counter flow
(d) gas turbine for both parallel flow and counter flow

[ESE : 2000]

6.2 Which one of the following heat exchangers gives parallel straight line pattern of temperature distribution for both cold and hot fluid?
(a) Parallel-flow with unequal heat capacities
(b) Counter-flow with equal heat capacities
(c) Parallel-flow with equal heat capacities
(d) Counter-flow with unequal heat capacities

[ESE : 2001]

6.3 In a counter-flow heat exchanger, the hot fluid is cooled from 110°C to 80°C by a cold fluid which gets heated from 30°C to 60°C *LMTD* for the heat exchanger is
(a) 20°C (b) 30°C
(c) 50°C (d) 80°C

[ESE : 2001]

6.4 In a counter flow heat exchanger, the product of specific heat and mass flow rate is same for the hot and cold fluids. If *NTU* is equal to 0.5 then the effectiveness of the heat exchanger is
(a) 1.0 (b) 0.5
(c) 0.33 (d) 0.2

[ESE : 2001]

Direction: Each of the next questions consists of two statements, one labelled as the **'Assertion (A)'** and the other as **'Reason (R)'**. You are to examine these two statements carefully and select the answers to these items using the codes given below:

Codes:
(a) Both **A** and **R** are individually true and **R** is the correct explanation of **A**
(b) Both **A** and **R** are individually true but **R** is not the correct explanation of **A**
(c) **A** is true but **R** is false
(d) **A** is false but **R** is true

6.5 **Assertion (A):** It is not possible to determine *LMTD* in a counter flow heat exchanger with equal heat capacity rates of hot and cold fluids.
Reason (R): Because the temperature difference is invariant along the length of the heat exchanger.

[ESE : 2002]

6.6 **Assertion (A):** A counter flow heat exchanger is themodynamically more efficient than the parallel flow type.
Reason (R): A counter flow heat exchanger has a lower *LMTD* for the same temperature conditions.

[ESE : 2003]

6.7 Match **List-I** (Heat Exchanger Process) with **List-II** (Temperature Area Diagram) and select the correct answer using the codes given below the lists:

List-I
A. Counter flow sensible heating
B. Parallel flow sensible heating
C. Evaporating
D. Condensing

List-II

1. 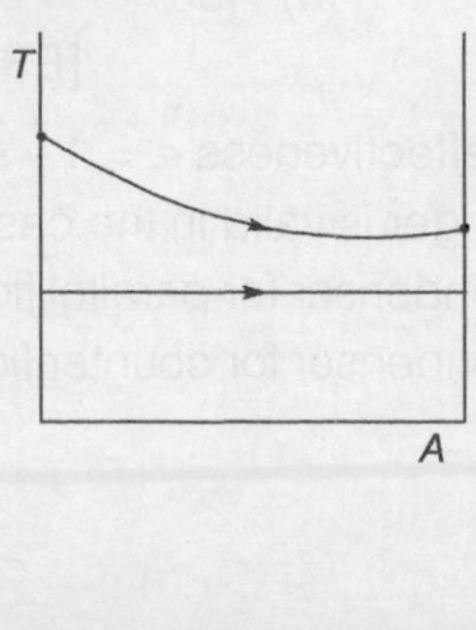

2. 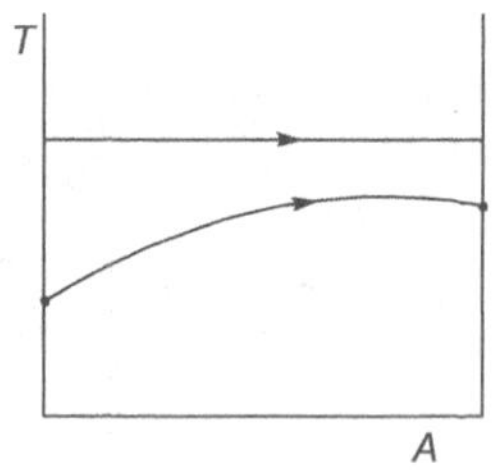

3. 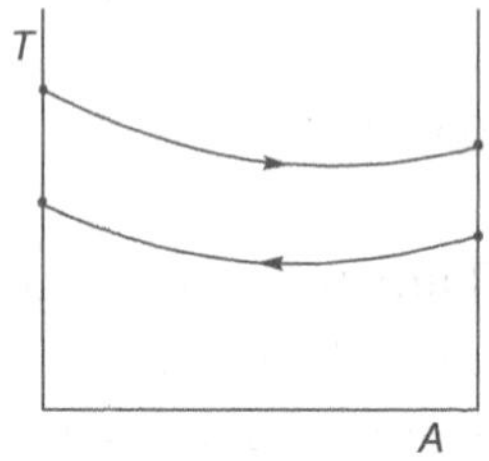

4. 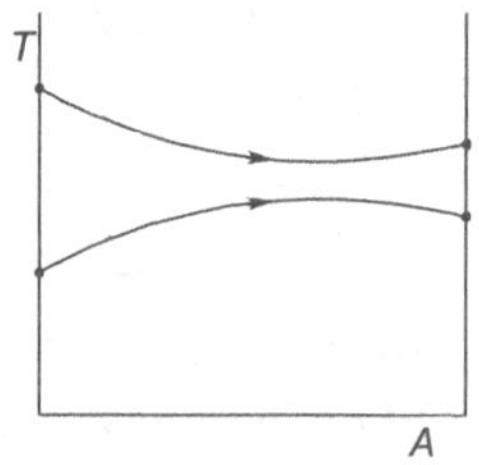

5. 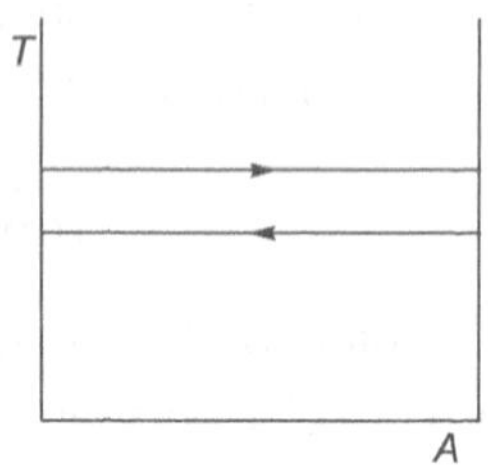

Codes:

	A	B	C	D
(a)	3	4	1	2
(b)	3	2	5	1
(c)	4	3	2	5
(d)	4	2	1	5

[ESE : 2004]

6.8 After expansion from a gas turbine, the hot exhaust gases are used to heat the compressed air from a compressor with the help of a cross flow compact heat exchanger of 0.8 effectiveness. What is the number of transfer units of the heat exchanger?

(a) 2 (b) 4
(c) 8 (d) 16

[ESE : 2005]

6.9 The equation of effectiveness $\epsilon = 1 - e^{-NTU}$ for a heat exchanger is valid in the case of

(a) boiler and condenser for parallel flow
(b) boiler and condenser for counter flow
(c) boiler and condenser for both parallel flow and counter flow
(d) gas turbine for both parallel flow and counter flow

[ESE : 2006]

6.10 Air can be best heated by steam in a heat exchanger of

(a) plate type
(b) double pipe type with fins on steam side
(c) double pipe type with fins on air side
(d) shell and tube type

[ESE : 2006]

6.11 Match **List-I** with **List-II** and select the correct answer using the code given below the lists:

List-I

A. Heat exchangers
B. Turbulent flow
C. Free convection
D. Radiation heat transfer

List-II

1. View factor **2.** Effectiveness
3. Nusselt number **4.** Eddy diffusivity

Codes:

	A	B	C	D
(a)	3	1	2	4
(b)	2	4	3	1
(c)	3	4	2	1
(d)	2	1	3	4

[ESE : 2007]

6.12 The temperature distribution curve for a heat exchanger as shown in the figure below (with usual notations) refers to which one of the following?

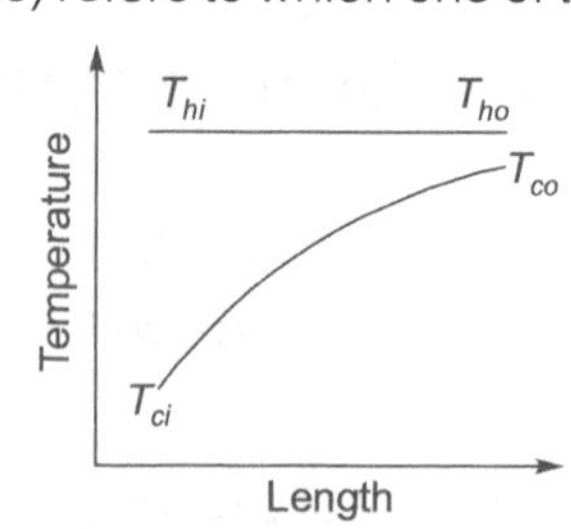

(a) Tubular parallel flow heat exchanger
(b) Tube in tube counter flow heat exchanger
(c) Boiler
(d) Condenser

[ESE : 2008]

6.13 The '*NTU*' (Number of Transfer Units) in a heat exchanger is given by which one of the following?

(a) $\frac{UA}{C_{min}}$ (b) $\frac{UA}{C_{max}}$

(c) $\frac{UA}{E}$ (d) $\frac{C_{max}}{C_{min}}$

U = Overall heat transfer coefficient

C = Heat capacity

E = Effectiveness

A = Heat exchange area **[ESE : 2008]**

6.14 Match **List-I** (Application) with **List-II** (Type of Heat Exchanger) and select the correct answer using the code given below the lists:

List-I	List-II
A. Gas to liquid	1. Compact
B. Space vehicle	2. Shell and tube
C. Condenser	3. Finned tube
D. Air pre-heater	4. Regenerative

Codes:

	A	B	C	D
(a)	2	4	3	1
(b)	3	1	2	4
(c)	2	1	3	4
(d)	3	4	2	1

[ESE : 2008]

6.15 In a balanced counter flow heat exchanger with $\dot{m}_h c_h = \dot{m}_c c_c$, the NTU is equal to 1.0. What is the effectiveness of the heat exchanger?

(a) 0.5 (b) 1.5

(c) 0.33 (d) 0.2 **[ESE : 2009]**

6.16 Consider the following statements:

1. Mass transfer refers to mass in transit due to a species concentration gradient in a mixture.
2. Must have a mixture of two or more species for mass transfer to occur.
3. The species concentration gradient is the driving potential for mass transfer.
4. Mass transfer by diffusion is analogous to heat transfer by conduction.

Which of thes statement are correct?

(a) 1, 2 and 3 only (b) 1, 2 and 4 only

(c) 2, 3 and 4 only (d) 1, 2, 3 and 4

[ESE : 2010]

6.17 In a heat exchanger, the hot liquid enters with a temperature of 180°C and leaves at 160°C. The cooling fluid enters at 30°C and leaves at 110°C. The capacity ratio of the heat exchanger is

(a) 0.25 (b) 0.40

(c) 0.50 (d) 0.55 **[ESE : 2010]**

6.18 For a balanced counter-flow heat exchanger, the temperature profiles of the two fluids are

(a) Parallel and non-linear

(b) Parallel and linear

(c) Linear but non-parallel

(d) Divergent from one another

[ESE : 2010]

6.19 Match **List-I** with **List-II** and select the correct answer using the code given below the lists:

List-I

A. Number of Transfer Units

B. Periodic flow heat exchangers

C. Phase change

D. Deposition on heat exchanger surface

List-II

1. Regenerators
2. Fouling factor
3. A measure of heat exchanger size
4. Condensers

Codes:

	A	B	C	D
(a)	3	4	1	2
(b)	2	4	1	3
(c)	3	1	4	2
(d)	2	1	4	3

[ESE : 2011]

6.20 Floating heads are provided in heat exchangers to

(a) Accommodate vapours released

(b) Decrease pressure drop

(c) Regulate the flow

(d) Avoid deformation of tubes due to thermal expansion **[ESE : 2011]**

6.21 The fouling factor in heat exchanger is defined as

(a) $R_f = U_{\text{dirty}} - U_{\text{clean}}$

(b) $R_f = \dfrac{1}{U_{\text{dirty}}} - \dfrac{1}{U_{\text{clean}}}$

(c) $\dfrac{1}{R_f} = \dfrac{1}{U_{\text{dirty}}} - \dfrac{1}{U_{\text{clean}}}$

(d) $\dfrac{1}{R_f} = U_{\text{dirty}} - U_{\text{clean}}$ **[ESE : 2012]**

6.22 In a heat exchanger, the hot gases enter with a temperature of 150°C and leave at 75°C.

The cold fluid enters at 25°C and leaves at 125°C. The capacity ratio of the exchanger is

(a) 0.65 (b) 0.75
(c) 0.85 (d) 0.95 [ESE : 2012]

6.23 What does *NTU* indicate?

(a) Effectiveness of heat exchanger
(b) Efficiency of heat exchanger
(c) Size of heat exchanger
(d) Temperature drop in heat exchanger

[ESE : 2012]

6.24 The correct sequence for different types of heat exchangers in the decreasing order of effectiveness is

(a) parallel-flow, counterflow, shell and tube and cross-flow
(b) cross-flow, counterflow, shell and tube and parallel-flow
(c) counterflow, shell and tube, cross-flow and parallel-flow
(d) counterflow, cross-flow, shell and tube and parallel-flow [ESE : 2012]

6.25 In a double-pipe counterflow heat exchanger, 1000 kg/h of oil having a specific heat 20 J/kg-K is cooled from 150°C to 125°C by 1250 kg/h of fluid of specific heat 16 J/kg-K. The fluid leaves the heat exchanger at 75°C. In such case, the temperature at which the cooling fluid is entering the heat exchanger and LMTD will be

(a) 25°C and 50°C (b) 50°C and 50°C
(c) 50°C and 75°C (d) 75°C and 50°

[ESE : 2012]

6.26 In a concentric double-pipe heat exchanger where one of the fluids undergoes phase change

(a) The two fluids should flow opposite to each other
(b) The two fluids should flow parallel to each other
(c) The two fluids should flow normal to each other
(d) The directions of flow of the two fluids are of no consequence

[ESE : 2014]

6.27 A counter flow shell and tube heat exchanger is used to heat water with hot exhaust gases. The water (c = 4180J/ kgK) flows at the rate of 2 kg/s and the exhaust gases (c =1000 J/kgK) flow at the rate of 5 kg/s. If the heat transfer surface area is 32 m^2 and the overall heat transfer coefficient is 200 W/m^{2}K, the the *NTU* of the heat exchanger is

(a) 4.5 (b) 2.4
(c) 8.6 (d) 1.28 [ESE : 2014]

6.28 In a two-fluid heat exchanger, the inlet and outlet temperatures of the hot fluid are 65°C and 40°C respectively. For the cold fluid, these are 15°C and 43°C. The heat exchanger is a

(a) Parallel flow heat exchanger
(b) Counter flow heat exchanger
(c) Mixed flow heat exchanger
(d) Phase-change heat exchanger

[ESE : 2014]

6.29 In a double-pipe heat exchanger, the cold fluid is water with inlet temperature 20°C and mass flow rate 20 kg/s and the hot fluid water inlet temperature 80°C and mass flow rate 10 kg/s. Assume that for water c_p = 4.2 kJ/kg°C, independent of temperature to which the cold fluid can be heated in a parallel flow and in a counter flow heat exchanger?

(a) 80°C in both parallel flow and counter flow
(b) 50°C in both parallel flow and counter flow
(c) 40°C in parallel flow and 50°C in counter flow
(d) 40°C in parallel flow and 80°C in counter flow

[ESE : 2014]

6.30 A counter flow shell and tube exchanger having an area of 32.5 m^2, is used to heat water with hot exhaust gases. The water (c_p = 4.16 kJ/kgK) flows at a rate of 2 kg/s while the exhaust gases (c_p = 1.03 kJ/kgK) flow at a rate of 5.15 kg/s. If the overall heat transfer surface coefficient is 200 W/m^{2}K, the NTU for the heat exchanger is

(a) 1.2 (b) 2.4
(c) 3.6 (d) 4.8

[ESE : 2015]

Direction: The following items consists of two statements, one labelled as **Assertion 'A'** and the other labelled as **Reason 'R'**. You are to examine these two statements carefully and select your answers to these items using the codes given below:

Codes:

(a) Both A and R are true and R is the correct explanation of A
(b) Both A and R are true but R is not a correct explanation of A
(c) A is true but R is false
(d) A is false but R is true

6.31 **Assertion (A):** A counter flow heat exchanger is more effective than a parallel flow heat exchanger.
Reason (R): For same temperature limits of hot and cold fluids, the overall heat transfer coefficient of counter flow heat exchanger is more than parallel flow heat exchanger.
[ESE : 2015]

6.32 **Assertion (A):** With heat exchanger gas turbine cycle, the cycle efficiency reduces as the pressure ratio increases.
Reason (R): As the pressure ratio increases, the delivery temperature from the compressor increase and ultimately will exceed that of the exhaust gas from the turbine.
[ESE : 2015]

6.33 Hot gases enter a heat exchanger at 200°C and leave at 150°C. The cold air enters at 40°C and leaves at 140°C. The capacity ratio of the heat exchanger will be :
(a) 0.40 (b) 0.45
(c) 0.50 (d) 0.52 **[ESE : 2016]**

6.34 A cross-flow type air heater has an area of 50 m^2. The overall transfer coefficient is 100 W/m^2K; and heat capacity of the stream, be it hot or cold, is 1000 W/K. What is the NTU ?
(a) 500 (b) 50
(c) 5 (d) 0.5 **[ESE : 2016]**

6.35 The effectiveness of a counter-flow heat exchanger has been estimated as 0.25. Hot gases enter at 200°C and leave at 75°C. Cooling air enters at 40°C. The temperature of the air leaving the unit will be :
(a) 60°C (b) 70°C
(c) 80°C (d) 90°C
[ESE : 2016]

6.36 In a counterflow heat exchanger, hot gases enter at 250°C and leave at 100°C. Atmospheric air enters at 50°C and leaves at 80°C. The effectiveness of the heat exchanger will be
(a) 0.20 (b) 0.25
(c) 0.30 (d) 0.75
[ESE : 2017]

6.37 Consider the following statements :
1. The efficiency of heat transfer in a condenser will improve by increase of the overall heat transfer coefficient.
2. The efficiency of heat transfer in a condenser will improve by increase of the velocity of flow of water in the tube.
3. The difference between the temperature of steam entering the condenser and the inlet water temperature should be maximum for maximum efficiency.

Which of the above statements are correct?
(a) 1 and 2 only (b) 1 and 3 only
(c) 2 and 3 Only (d) 1, 2 and 3
[ESE : 2017]

6.38 In a counter flow heat exchanger, hot gases enter the system at 200°C and leave at 80°C. The temperature of the outside air entering the unit is 35°C. Its temperature at the exit is 90°C. The heat exchanger has an effectiveness of
(a) 0.35 (b) 0.34
(c) 0.33 (d) 0.32
[ESE : 2018]

6.39 In a heat exchanger, 50 kg of water is heated per minute from 50°C to 110°C by hot gases which enter the heat exchanger at 250°C. The value of C_p for water is 4.186 kJ/kgK and for air is 1 kJ/kgK. If the flow rate of gases is 100 kg/min, the net change of enthalpy of air will be nearly
(a) 17.6 MJ/min (b) 15.0 MJ/min
(c) 12.6 MJ/min (d) 10.0 MJ/min
[ESE : 2020]

6.40 Which one of the following is having the highest value of fouling factor?

(a) Sea water (b) Refrigerating liquid
(c) Fuel oil (d) Industrial air

[ESE : 2022]

6.41 Water at the rate of 68 kg/min is heated from 35°C to 75°C by oil having a specific heat of 1.9 kJ/kg°C. The fluids are used in a counter flow double-pipe heat exchanger, and the oil enters the exchanger at 110°C and leaves at 75°C. The overall heat-transfer coefficient is 320 W/m^2°C. What is the total heat transfer? [Take specific heat of water as 4180 J/kgK]

(a) 78.3 kW (b) 189.5 kW
(c) 241.3 kW (d) 280.6 kW

[ESE : 2023]

6.42 Air circulates at the rate of 68 kg/s and is to be heated from 16°C to 34°C. What is the water mass flow for an air heater coil having hot water entering at 85°C and leaving at 74°C? (Take specific heat of air at constant pressure as 1.02 kJ/kgK and specific heat of water as 4.187 kJ/kgK)

(a) 19 kg/s (b) 17 kg/s
(c) 27 kg/s (d) 14 kg/s

[ESE : 2024]

■■■■

Answers Heat Exchangers

6.1 (c)	6.2 (b)	6.3 (c)	6.4 (c)	6.5 (d)	6.6 (c)	6.7 (a)	6.8 (b)	6.9 (c)
6.10 (c)	6.11 (b)	6.12 (d)	6.13 (a)	6.14 (b)	6.15 (a)	6.16 (d)	6.17 (a)	6.18 (b)
6.19 (c)	6.20 (d)	6.21 (b)	6.22 (b)	6.23 (c)	6.24 (c)	6.25 (c)	6.26 (d)	6.27 (d)
6.28 (b)	6.29 (c)	6.30 (a)	6.31 (c)	6.32 (a)	6.33 (c)	6.34 (c)	6.35 (c)	6.36 (d)
6.37 (a)	6.38 (*)	6.39 (c)	6.40 (c)	6.41 (b)	6.42 (c)			

Explanations Heat Exchangers

6.1 (c)

Heat capacity ratio C will be zero for both evaporator and condenser

$C = 0;\ \varepsilon = 1 - e^{-NTU}$

This is valid for both evaporator and condenser

6.2 (b)

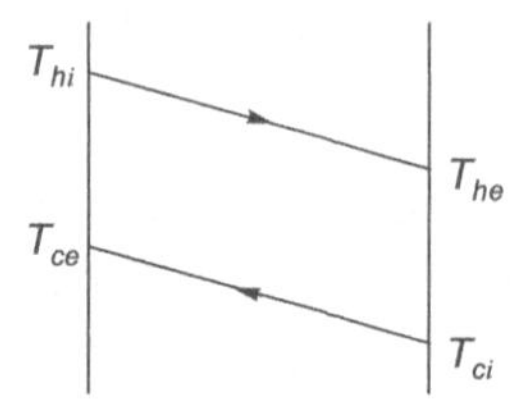

Counter flow with equal heat capacities.

6.3 (c)

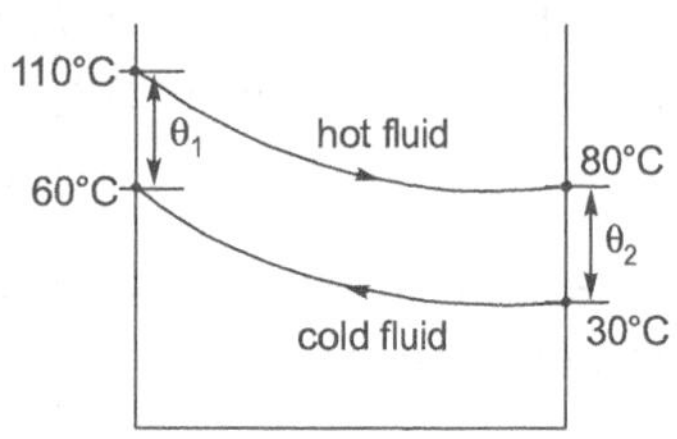

$$\theta_1 = \theta_2 = 50°C$$

Hence

$$LMTD = 50°C$$

6.4 (c)

$$C = 1$$

$$\therefore \quad \varepsilon = \frac{NTU}{1+NTU}$$

$$\varepsilon = \frac{0.5}{1.5} = \frac{1}{3} = 0.33$$

6.5 (d)

Counter flow heat exchanger with equal heat capacity rates of hot and cold fluids.

$$LMTD = \frac{\Delta T_i - \Delta T_e}{\ln\left(\frac{\Delta T_i}{\Delta T_e}\right)}$$

In this case, $\Delta T_i = \Delta T_e$

$$LMTD = 0/0$$

This value is indeterminate. The value of $LMTD$ can be found by applying L'Hospital rule.

$\therefore \quad LMTD =$ Either ΔT_i or ΔT_e

6.6 (c)

Counter flow has higher *LMTD* than parallel flow heat exchanger so they are more effective heat exchanger.

6.7 (a)

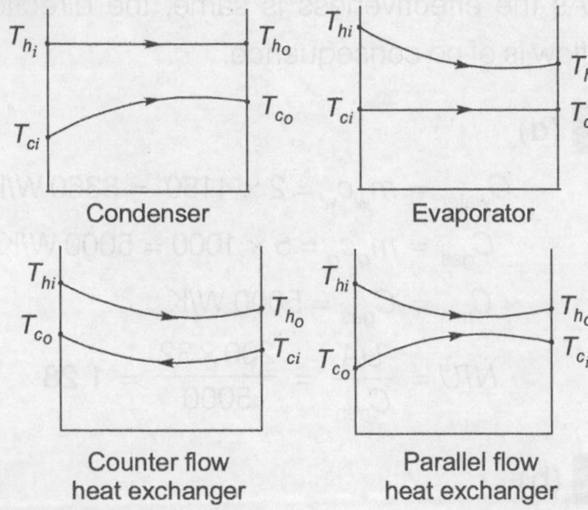

6.8 (b)

Since $C = 1$

$$\varepsilon = \frac{NTU}{1+NTU}$$

$$\therefore \quad 0.8 = \frac{NTU}{1+NTU}$$

$$NTU = 4$$

6.9 (c)

In condenser, the vapour gets condensed, hence the heat capacity of hot fluid $(mc_p)_h$ can be assumed to be infinity.

In evaporator, the liquid gets vaporized by taking the required energy. Hence the heat capacity of cold fluid $(mc_p)_c$ can be assumed to be infinity.

We get, $\varepsilon = 1 - e^{-NTU}$

6.10 (c)

If fins are to be used on a surface separating a gas and liquid, they are generally placed on gas side which is the side of lower heat transfer coefficient.

6.11 (b)

(a) Heat Exchangers : Effectiveness
(b) Turbulent flow : Eddy diffusivity
(c) Free convection : Nusselt number
(d) Radiation heat transfer : View factor

6.12 (d)

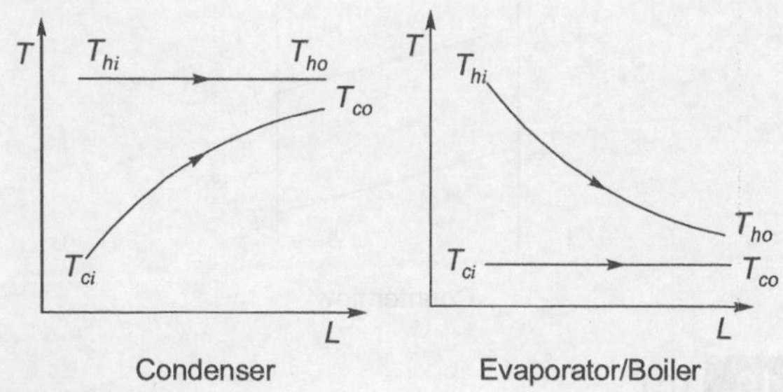

6.13 (a)

$$NTU = \frac{UA}{C_{min}}$$

6.14 (b)

Application	Type of HE
Gas to liquid	Finned tube
Space vehicle	Compact
Condenser	Shell and tube
Air pre-heater	Regenerative

6.15 (a)

Effectiveness of heat exchanger

$$\varepsilon = \frac{NTU}{NTU+1} = \frac{1}{1+1} = 0.5$$

6.16 (d)

The process of mass transfer takes place due to species concentration difference in a system or mixture and mass transfer by diffusion is analogous to heat transfer by conduction and it can be understood by the ficks law of diffusion.

6.17 (a)

$$t_{h_1} = 180°C \qquad t_{h_2} = 160°$$

$$t_{c_1} = 30°C \qquad t_{c_2} = 110°C$$

From the energy balance equation:

$$\dot{m}_h \times c_{ph}(t_{h_1} - t_{h_2}) = \dot{m}_c \times C_{pc}(t_{c_2} - t_{c_1})$$

Capacity ratio = Product of mass flow rate and specific heat

$$C_h(180-160) = C_c(110-30)$$

$$\frac{C_c}{C_h} = \mathbf{0.25}$$

6.18 (b)

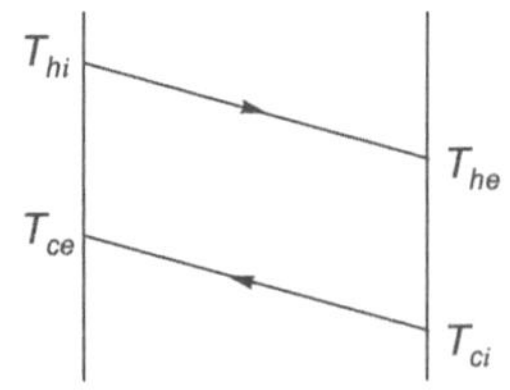

Counter flow

6.19 (c)

Number of transfer units is used as a measure of heat exchanger size. *NTU* is, a measure of effectiveness of heat exchanger.

$$NTU = \frac{UA}{C_{min}}$$

6.21 (b)

The resistance due to fouling = R_f

$$\therefore \quad \frac{1}{U_{dirty}} = \frac{1}{U_{clean}} + \frac{1}{1/R_f}$$

$$\therefore \quad R_f = \frac{1}{U_{dirty}} - \frac{1}{U_{clean}}$$

6.22 (b)

$$\text{Capacity ratio} = \frac{\text{Minimum heat capacity}}{\text{Maximum heat capacity}}$$

Since $(mc_p dT)_{hot} = (mc_p dT)_{cold}$

$$\therefore \quad \frac{(mc_p)_{cold}}{(mc_p)_{hot}} = \frac{(dT)_{hot}}{(dT)_{cold}}$$

$$= \frac{150-75}{125-25} = \mathbf{0.75}$$

6.25 (c)

$(mc_p dT)_{hot} = (mc_p dT)_{cold}$

$1000 \times 20 \times (150 - 125) = 1250 \times 16 \times (75 - T_i)$

$T_i = 50°C$

In this case since $(mc_p)_{hot} = (mc_p)_{cold}$

$$\therefore \quad LMTD = T_{h_e} - T_{c_i} = T_{h_i} - T_{c_e}$$

$$LMTD = 150 - 75 = 125 - 50 = \mathbf{75°C}$$

6.26 (d)

The capacity ratio becomes zero when one of the fluids undergoes phase change as in the case of condenser or evaporator. So, effectiveness is same for parallel and counter flow heat exchanger. As the effectiveness is same, the direction of flow is of no consequence.

6.27 (d)

$$C_{water} = m_w c_w = 2 \times 4180 = 8360 \text{ W/K}$$

$$C_{gas} = m_g c_g = 5 \times 1000 = 5000 \text{ W/K}$$

$$C_{min} = C_{gas} = 5000 \text{ W/K}$$

$$NTU = \frac{UA}{C_{min}} = \frac{200 \times 32}{5000} = \mathbf{1.28}$$

6.28 (b)

Since $T_{Ce} > T_{he}$

Possible only in counter flow heat exchanger.

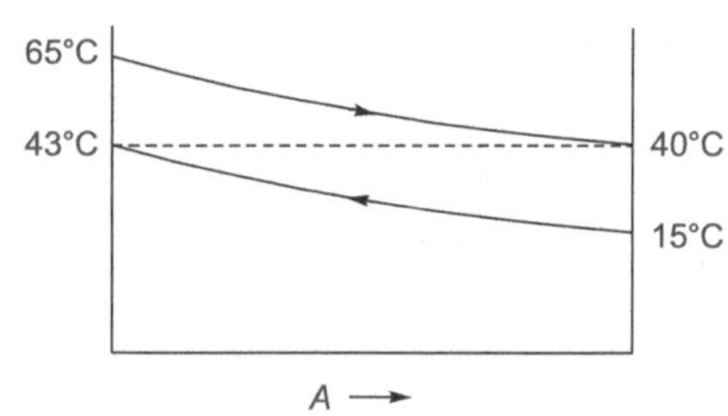

6.29 (c)

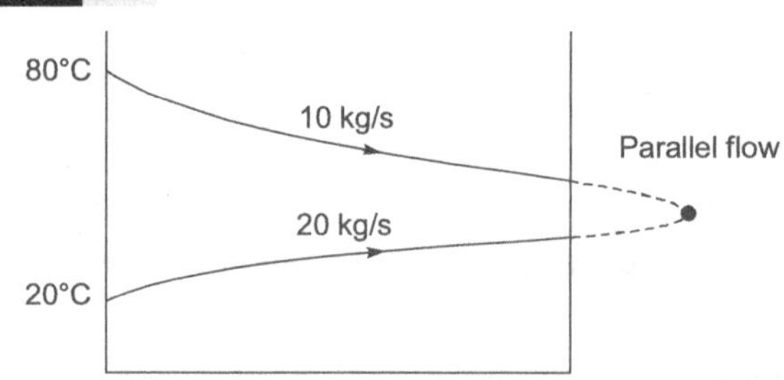

Maximum temperature of the cold fluid will be reached when both hot and cold attains the same temperature.

$\therefore \; 10 \times 4.2 \times [80 - T] = 20 \times 4.2 \times [T - 20]$

$$80 - T = 2T - 40$$

$$120 = 3T$$

$$T = 40°C$$

For counter flow heat exchanger.

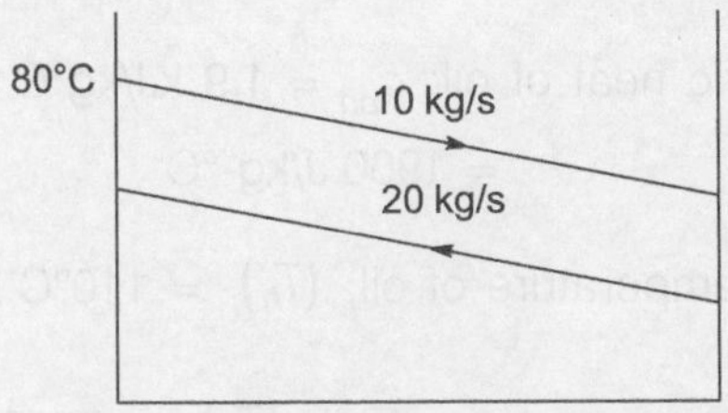

Maximum effectiveness will be

$$1 = \frac{20 \times 4.2 \times [T-20]}{10 \times 4.2 \times [80-20]}$$

$$60 = 2T - 40$$

$$2T = 100$$

$$T = \mathbf{50°C}$$

6.30 (a)

$$NTU = \frac{UA}{C_{min}}$$

$$= \frac{200 \times 32.5}{5.15 \times 1.03 \times 10^3}$$

$$= 1.225$$

6.31 (c)

Reason is wrong because the overall heat transfer coefficient of a heat exchanger mainly depends upon convective heat transfer coefficient on both hot and cold side which may be same in both the type of heat exchanger.

6.32 (a)

As the pressure ratio in Brayton cycle increases, the temperature difference between compressor exit air temperature and turbine exhaust gas temperature keeps on decreasing hence in the regenerator provided because of lesser temperature difference between hot and cold fluid good heat transfer can not be expected. Thus the recovery of heat from the exhaust gets reduced and thermal efficiency of the cycle decreases.

6.33 (c)

$$\Delta T_h = 200 - 150 = 50°$$

$$\Delta T_c = 140 - 40 = 100°$$

Heat loss by hot gases = Heat gained by cold gases

$$C_h \Delta T_h = C_c \Delta T_c$$

$$\frac{C_c}{C_h} = \frac{\Delta T_h}{\Delta T_c} = 0.5$$

6.34 (c)

$$NTU = \frac{UA}{C_{min}} = \frac{100 \times 50}{1000} = 5$$

6.35 (c)

$$\varepsilon = 0.25$$

$$0.25 = \frac{T_{c_e} - 40}{200 - 40}$$

$$T_{c_e} = 40 + 0.25 \times 160$$

$$= \mathbf{80°C}$$

6.36 (d)

$$T_{hi} = 250°C \quad T_{he} = 100°C$$

$$T_{ci} = 50°C \quad T_{ce} = 80°C$$

From energy balance equation:

$$\dot{m}_h c_{ph}(T_{hi} - T_{he}) = \dot{m}_c c_{pc}(T_{ce} - T_{ci})$$

$$\dot{m}_h c_{ph}(250 - 100) = \dot{m}_c c_{pc}(80 - 50)$$

$$\Rightarrow \quad \dot{m}_h c_{ph} < \dot{m}_c c_{pc}$$

Hence $\epsilon_{HE} = \frac{T_{hi} - T_{he}}{T_{hi} - T_{ci}} = \frac{250 - 100}{250 - 50}$

$$= \frac{150}{200} = 0.75$$

6.37 (a)

$$Q = UA\Delta T_m$$

As velocity of flow of water increases the convective H.T. coefficient on water side increases. As a result U value Increases

$$\left(\because \frac{1}{U} = \frac{1}{h_{steam}} + \frac{1}{h_{water\ side}}\right)$$

6.38 (*)

Counter flow heat exchanger

$$T_{h,i} = 200°C$$

$T_{c_1,i} = 35°C; \; T_{h_1,o} = 80°C; \; T_{c_1,o} = 90°C$

From energy balance

$$C_h\left(T_{h_1,i} - T_{h_1,o}\right) = C_c\left(T_{c_1,o} - T_{c_1,i}\right)$$

$$C_h\,(200 - 80) = C_c\,(90 - 35)$$

$$\frac{C_h}{C_c} = \frac{55}{120}$$

$$\Rightarrow \quad C_h = C_{min}$$

$$C_c = C_{max}$$

$$\in = \frac{C_h\left(T_{c,i} - T_{h_1,o}\right)}{C_{min}\left(T_{hi} - T_{ci}\right)} = \frac{200-80}{200-35}$$

$$\in = 0.7272$$

6.39 (c)

From energy balance equation for heat exchanger.

The rate of enthalpy decrease of water = The rate of enthalpy increase of air

$$= \dot{m}_{water} \times c_{p_{water}} \times \left(T_{water\ at\ exit} - T_{water\ at\ inlet}\right)$$

= 50 kg/min × 4.186 kJ/kgK × (100 – 50) k

= 12558 kJ/min = 12.6 MJ/min

6.40 (c)

Value of fouling factor is the highest for fuel oil and is around 0.0009 (m^2 k/W).

6.41 (b)

Given :

Mass of water $(\dot{m}_c) = 68$ kg/min $= \frac{68}{60}$ kg/s

Inlet temperature of water; $(T_c)_i = 35°C$

Outlet temperature of water; $(T_c)_o = 75°C$

Specific heat of oil; $c_{ph} = 1.9$ kJ/kg°C

= 1900 J/kg-°C

Inlet temperature of oil; $(T_h)_i = 110°C$

Outlet temperature of oil; $(T_h)_o = 75°C$

Overall heat-transfer coefficient;

$U = 320$ W/m^2-°C

Specific heat of water; $c_{pc} = 4180$ J/kg-K

Heat transfer rate:

$$q = \dot{m}_h c_{ph}\left(T_{h_i} - T_{h_o}\right) = \dot{m}_c c_{pc}\left(T_{c_o} - T_{c_i}\right)$$

$$q = \dot{m}_c c_{pc}\left(T_{c_o} - T_{c_i}\right)$$

$$q = \frac{68}{60} \times 4180 \times (75 - 35)$$

$$q = 189.493 \times 10^3 \text{ W}$$

$$q = 189.5 \text{ kW}$$

6.42 (c)

$\dot{m}_c = 68$ kg/s, $T_{c,i} = 16°C$, $T_{c,o} = 34°C$; $C_{pc} = 1020$ J/kgK

$T_{h,i} = 85°C$, $T_{h,o} = 74°C$; $C_{ph} = 4187$ J/kgK

Heat balance equation,

$$\dot{m}_c C_{ph}\left(T_{c,o} - T_{c,i}\right) = \dot{m}_h C_{ph}\left(T_{h,i} - T_{h,o}\right)$$

$$68 \times 1020(34 - 16) = \dot{m}_h \times 4187(85 - 74)$$

$$\dot{m}_h = 27.1072 \text{ kg/s}$$

■■■■

7 Boiling and Condensation

7.1 **Assertion (A):** If the heat flux in pool boiling over a horizontal surface is increased above the critical heat flux, the temperature difference between the surface and liquid decreases sharply.

Reason (R): With increasing heat flux beyond the value corresponding to the critical heat flux, a stage is reached when the rate of formation of bubbles is so high that they start to coalesce and blanket the surface with a vapour film.

(a) Both **A** and **R** are individually true and **R** is the correct explanation of **A**
(b) Both **A** and **R** are individually true but **R** is not the correct explanation of **A**
(c) **A** is true but **R** is false
(d) **A** is false but **R** is true

[ESE : 2003]

7.2 When a liquid flows through a tube with sub-cooled or saturated boiling, what is the process known?

(a) Pool boiling
(b) Bulk boiling
(c) Convection boiling
(d) Forced convection boiling

[ESE : 2009]

7.3 **Assertion (A):** Dropwise condensation is associated with higher heat transfer rate as compared to the heat transfer rate in film condensation.

Reason (R): In drop condensation there is free surface through which direct heat transfer takes place.

(a) Both **A** and **R** are individually true and **R** is the correct explanation of **A**
(b) Both **A** and **R** are individually true but **R** is not the correct explanation of **A**
(c) **A** is true but **R** is false
(d) **A** is false but **R** is true

[ESE : 2009]

7.4 For film-wise condensation on a vertical plane, the film thickness δ and heat transfer coefficient h vary with distance x from the leading edge as

(a) δ decreases, h increases
(b) Both δ and h increase
(c) δ increases, h decreases
(d) Both δ and h decrease

[ESE : 2010]

7.5 In the film established along a vertical plate during condensation of any vapour over the plates, the temperature distribution curve is

(a) Concave upwards
(b) Concave downwards
(c) Parabolic
(d) Straight line **[ESE : 2011]**

7.6 In spite of large heat transfer coefficients in boiling liquids, cavities are used advantageously when the entire surface is exposed to

(a) Nucleate boiling (b) Film boiling
(c) Transition boiling (d) Pool boiling

[ESE : 2011]

7.7 A pipe of 10 cm diameter and 10 m length is used for condensing steam on its outer surface. The average heat transfer coefficient h_h (when the pipe is horizontal) is n times the average heat transfer coefficient h_v (when the pipe is vertical). The value of n is:

(a) 2.44 (b) 3.34
(c) 4.43 (d) 5.34 **[ESE : 2016]**

7.8 Which one of the following regimes of boiling curve can be considered as reverse of condensation?

(a) Free convection boiling regime
(b) Nucleate boiling regime

(c) Transition boiling regime
(d) Film boiling regime [ESE : 2018]

7.9 In transition boiling, heat flux decreases due to which of the following?

1. Low value of film heat transfer coefficient at the surface during 100°C to 120°C surface temperature.
2. Major portion of heater surface is covered by vapour film which has smaller thermal conductivity as compared to liquid
3. Nucleate boiling occurs very fast.

Select the correct answer using the code given below

(a) 1 only (b) 2 only
(c) 3 only (d) 1, 2 and 3

[ESE : 2019]

7.10 Large heat transfer coefficients for vapour condensation can be achieved by promoting

(a) Film condensation
(b) Dropwise condensation
(c) Cloud condensation
(d) Dew condensation [ESE : 2020]

7.11 Consider the following statements for Nucleate boiling:

1. For water, the critical heat flux does not exceed 1 MW/m^2.
2. Nucleate boiling is the most desirable boiling regime in practice because of high heat transfer rates.
3. Heat flux increases at a higher rate with increase in temperature.

Which of the above statements is(are) correct?

(a) 1 only (b) 2 only
(c) 1 and 3 only (d) 2 and 3 only

[ESE : 2021]

7.12 In drop-wise condensation, the heat transfer rate is

(a) 5 times less than that in film-wise condensation.
(b) 15 times less than that in film-wise condensation.
(c) 25 times more than in film-wise condensation.
(d) 10 times more than that in film-wise condensation.

[ESE : 2021]

Answers Boiling and Condensation

7.1 (a) 7.2 (b) 7.3 (a) 7.4 (c) 7.5 (b) 7.6 (a) 7.7 (a) 7.8 (d) 7.9 (b)
7.10 (b) 7.11 (b) 7.12 (d)

Explanations Boiling and Condensation

7.1 (a)

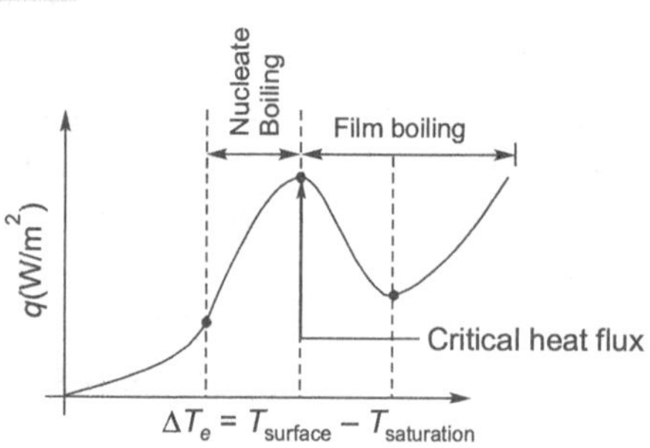

The trend of increase of heat flux with increase in excess temperature is observed upto critical heat flux point only. This is due to the fact that the bubble formation is very rapid and the bubbles blanket the heating surface and prevent the incoming fresh liquid from taking their place.

7.2 (b)

Pool Boiling: In this case the liquid above the hot surface is essentially stagnant and its motion near the surface is due to free convection.

Forced Convection Boiling: This refers to a situation where the fluid motion is induced by external means. This type of boiling occurs in water tube boilers involving forced convection.

Sub-Cooled or Local Boiling: In this case the liquid temperature is below the saturation temperature and bubbles are formed in the vicinity of heat surface.

Saturated Boiling: Here, the liquid temperature exceeds the saturation temperature.

7.4 (c)

For filmwise condensetion on a vertical plane,
Film thickness (δ) is given by

$$\delta = \left[\frac{4k\mu(t_{sat} - t_s)x}{\rho_l(\rho_l - \rho_v)gh_{fg}}\right]^{1/4}$$

Hence $\delta \propto (x)^{1/4}$
and local heat transfer coefficient is given by

$$h_x = \left[\frac{\rho_l(\rho_l - \rho_v)k^3 gh_{fg}}{4\mu x(t_{sat} - t_s)}\right]^{1/4}$$

Hence $h_x \propto \frac{1}{(x)^{1/4}}$

i.e. with increase in distance from the leading edge, film thickness (δ) increases and heat transfer coefficient decreases.

7.5 (b)

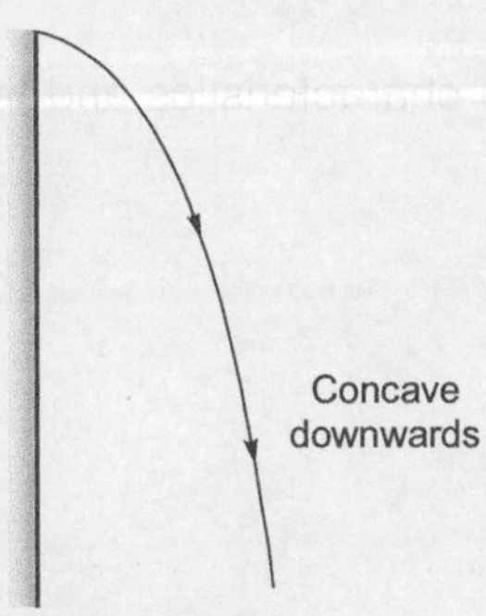

7.7 (a)

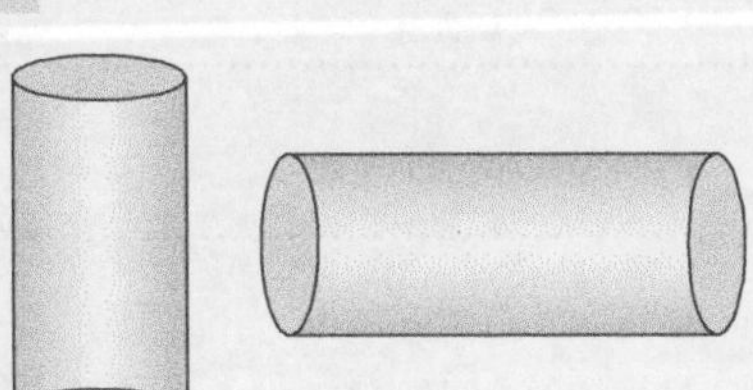

$$h_H = 0.725\left[\frac{\rho_l(\rho_l - \rho_v)k^3 gh_{fg}}{\mu_l(T_{sat} - T_s)D}\right]^{1/4}$$

$$h_H = 0.725\left(\frac{C}{D}\right)^{1/4}$$

$$h_v = 0.943\frac{C^{1/4}}{L^{1/4}}$$

$$L_H = 0.725\frac{C^{1/4}}{D^{1/4}}$$

$$h_H = nh_v$$

$$0.943\frac{C^{1/4}}{L^{1/4}} = n \times 0.725\frac{C^{1/4}}{D^{1/4}}$$

$$n = \mathbf{2.434}$$

7.10 (b)

Dropwise condensation offers lesser thermal resistance during phase change thus can enhance heat transfer rate between vapour and solid surface.

7.11 (b)

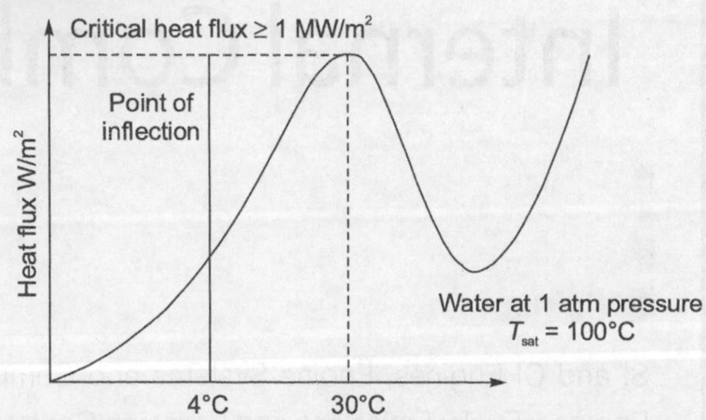

In the Nucleate boiling, beyond point of inflection, 'h' value decreases due to interference of bubbles formed on the surface of wire which will prevent the contact of liquid with wire. As such the rate of increase of heat flux with ΔT decrease, in this region there are vapour bubbles rise as jets (OR) columns ($\Delta T \simeq$ 10° to 30°C) and subsequently merge into slugs of vapour.

7.12 (d)

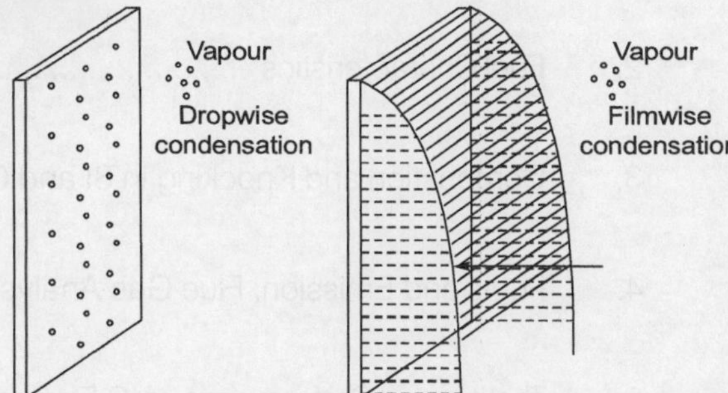

In filmwise condensation, the sliding film will act as barrier to the H.T. offering Thermal resistance between vapour and wall thereby reducing the HT rate as compared to that in dropwise condensation.

■■■■

UNIT

IV

Internal Combustion Engines

Syllabus

SI and CI Engines, Engine Systems and Components, Performance characteristics and testing of IC Engines; Fuels; Emissions and Emission Control.

Contents

Sl.	Topic	Page No.
1.	Air Standard Cycle	275
2.	Fuel Characteristics	285
3.	Combustion and Knocking in SI and CI Engine	289
4.	Fuels and Emission, Flue Gas Analysis	300
5.	Testing and Performance of IC Engine	309

OOOO

1 Air Standard Cycle

Direction: Each of the next questions consists of two statements, one labelled as the **'Assertion (A)'** and the other as **'Reason (R)'**. You are to examine these two statements carefully and select the answers to these items using the codes given below:

Codes:
(a) Both **A** and **R** are individually true and **R** is the correct explanation of **A**
(b) Both **A** and **R** are individually true but **R** is not the correct explanation of **A**
(c) **A** is true but **R** is false
(d) **A** is false but **R** is true

1.1 **Assertion (A):** The air standard efficiency of the diesel cycle decreases as the load is increased.
Reason (R): With increases of load, cut-off ratio increases.
[ESE : 2001]

1.2 **Assertion (A):** The CI engine is found to be more efficient than an SI engine.
Reason (R): Modern CI engines operate on a dual-cycle, which has an efficiency greater than the Otto cycle. [ESE : 2001]

1.3 Match **List-I** (Cycle operating between fixed temperature limits) with **List-II** (Characteristics of cycle efficiency η temperature) and select the correct answer using the codes given below the lists:

List-I
A. Otto cycle
B. Diesel cycle
C. Carnot cycle
D. Brayton cycle

List-II
1. η depends only upon temperature limits
2. η depends only on pressure limits
3. η depends on volume compression ratio
4. η depends on cut-off ratio and volume compression ratio

Codes:

	A	B	C	D
(a)	3	4	1	2
(b)	1	4	3	2
(c)	3	2	1	4
(d)	1	2	3	4

[ESE : 2001]

1.4 In an air-standard Diesel cycle; r is the compression ratio, ρ is the fuel cut-off ratio and γ is the adiabatic index $\left(\frac{c_p}{c_v}\right)$ Its air standard efficiency is given by

(a) $\eta = 1 - \left\{\frac{1}{\gamma \cdot r^{\gamma-1}} \cdot \frac{(\rho^{\gamma}-1)}{(\rho-1)}\right\}$

(b) $\eta = 1 - \left\{\frac{1}{\gamma \cdot r^{\gamma-1}} \cdot \frac{(\rho^{\gamma-1}-1)}{(\rho-1)}\right\}$

(c) $\eta = 1 - \left\{\frac{1}{\gamma \cdot r^{\gamma-1}} \cdot \frac{(\rho^{\gamma-1})}{(\rho-1)}\right\}$

(d) $\eta = 1 - \left\{\frac{1}{\gamma \cdot r^{\gamma}} \cdot \frac{(\rho^{\gamma-1}-1)}{(\rho-1)}\right\}$

[ESE : 2002]

1.5 (1)

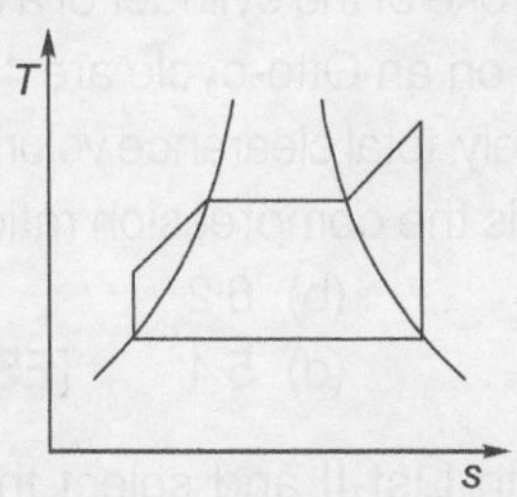

(2)

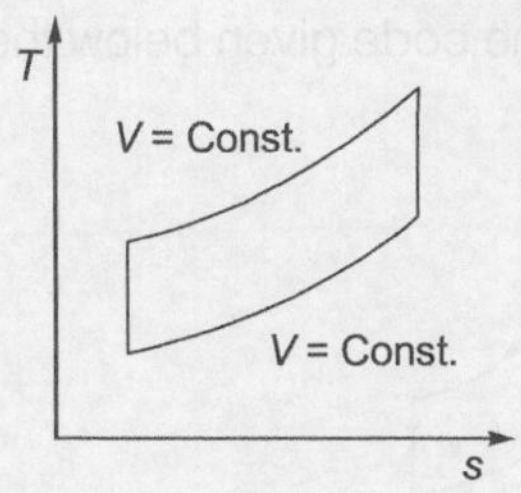

(3) 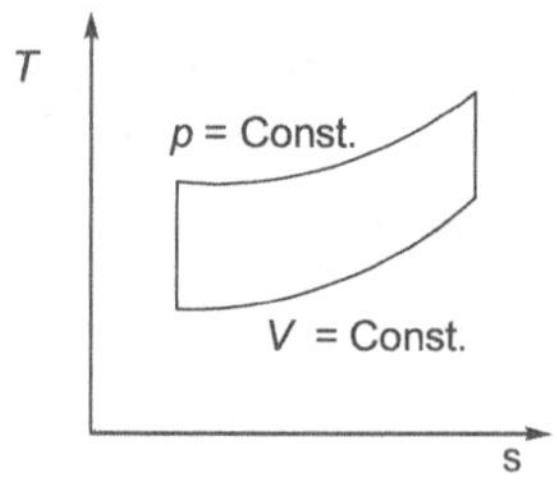

(4) 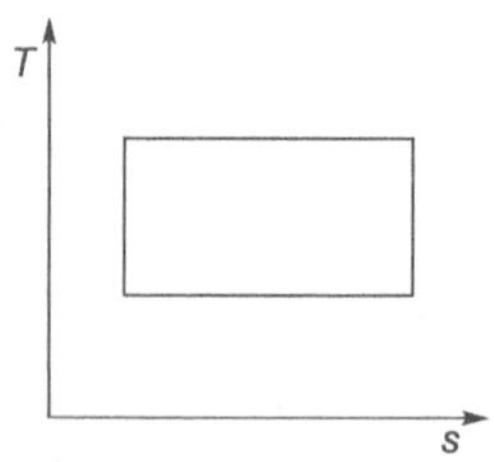

The correct sequence of the given four cycles on T-s plane in Figure (1), (2), (3) (4) is
(a) Rankine, Otto, Carnot and Diesel
(b) Rankine, Otto, Diesel and Carnot
(c) Otto, Rankine, Diesel and Carnot
(d) Otto, Rankine, Carnot and Diesel

[ESE : 2002]

1.6 In a standard dual air cycle, for a fixed amount of heat supplied and a fixed value of compression ratio, the mean effective pressure
(a) shall increases with increases in r_p (pressure ratio for constant volume heating) and decreases in r_c (constant pressure cut-off ratio)
(b) shall increases with decreases in r_p and increases in r_c
(c) shall remain independent of r_p
(d) shall remain independent of r_c

[ESE : 2003]

1.7 The bore and stroke of the cylinder of a 6-cylinder engine working on an Otto-cycle are 17 cm and 30 cm respectively, total clearance volume is 9225 cm^3, then what is the compression ratio ?
(a) 7·8 (b) 6·2
(c) 15·8 (d) 5·4 [ESE : 2009]

1.8 Match **List-I** with **List-II** and select the correct answer using the code given below the lists:

List-I

A. 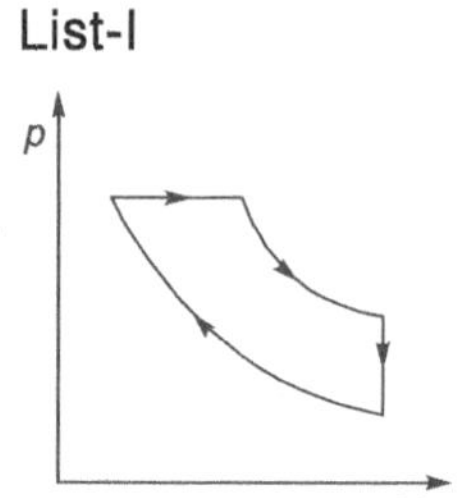

B. 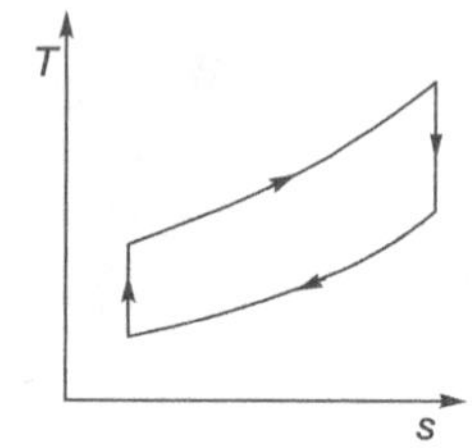

C. 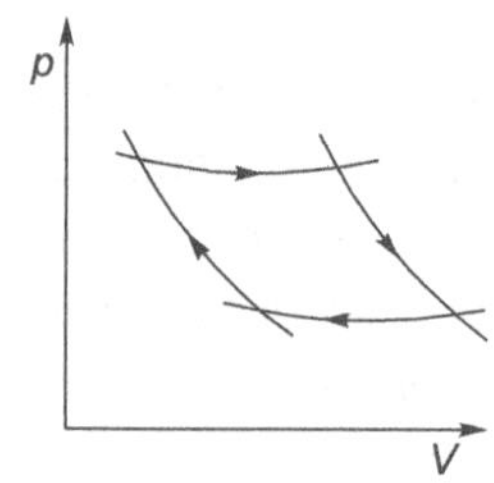

D. 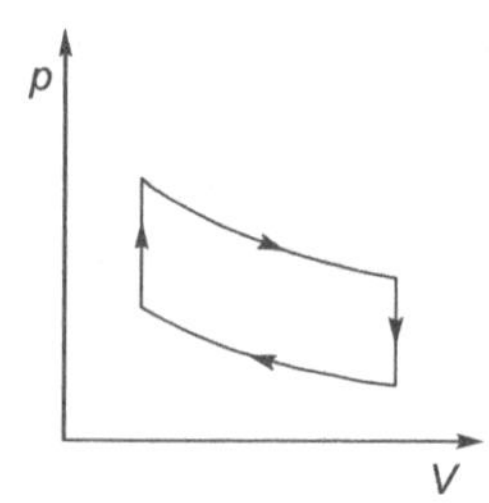

List-II
1. Diesel cycle
2. Carnot cycle
3. Brayton cycle
4. Otto cycle

Codes:

	A	B	C	D
(a)	4	2	3	1
(b)	1	2	3	4
(c)	4	3	2	1
(d)	1	3	2	4

[ESE : 2010]

1.9 An IC engine has a bore and stroke each equal to 2 units. The total area to calculate heat loss from the engine can be taken as
(a) 4π (b) 5π
(c) 6π (d) 8π [ESE : 2012]

1.10 For a heat engine working on an air-standard Diesel cycle the efficiency will
(a) increase as the fuel cut-off ratio is increased
(b) not change as the fuel cut-ratio is increased
(c) decrease as the fuel cut-off ratio is increased
(d) change as the fuel cut-off ratio is increased

[ESE : 2013]

1.11 For a given compression ratio, which of the following ideal cycles have equal thermal efficiency
(a) Diesel and Stirling (b) Stirling and Otto
(c) Otto and Joule (d) Joule and Diesel
[ESE : 2013]

1.12 An air standard Otto cycle has volumetric compression ratio of 6, the lowest cycle pressure of 0.1 MPa and operates between temperature limits of 27°C and 1569°C. What is the temperature after isentropic expansion if ratio of specific heats is 1.4?
(a) 600 K (b) 700 K
(c) 900 K (d) 1000 K [ESE : 2013]

1.13 In SI engines for higher thermal efficiency
(a) compression ratio should be high within the available limits
(b) heat liberation during combustion should be maximum
(c) surface to volume ratio should be high
(d) long flame travel distance
[ESE : 2013]

Directions: Each of the next questions consists of two statements, one labelled as the 'Statement (I)' and the other as 'Statement (II)'. You are to examine these two statements carefully and select the answers to these items using the codes given below:

Codes:
(a) Both Statement (I) and Statement (II) are individually true and Statement (II) is the correct explanation of Statement (I).
(b) Both Statement (I) and Statement (II) are individually true but Statement (II) is **NOT** the correct explanation of Statement (I).
(c) Statement (I) is true but Statement (II) is false
(d) Statement (I) is false but Statement (II) is true.

1.14 **Statement (I):** Though the composition of the working fluids change from air and fuel to combustion products in all internal combustion engines, the working cycle is still named as air standard cycle.
Statement (II): The composition of nitrogen in air and fuel mixture is high and does not undergo any chemical reaction in the combustion chamber.
[ESE : 2013]

1.15 **Statement (I):** Specific output of a SI engine is higher than the CI engine.
Statement (II): Its higher engine speed and ability to burn richer mixture results in increased energy conversion rate. [ESE : 2013]

1.16 For the same compression ratio, the efficiency of an air standard Otto cycle is
(a) More than the efficiency of an air standard Diesel cycle
(b) Less than the efficiency of an air standard Diesel cycle
(c) Equal to the efficiency of an air standard Diesel cycle
(d) None of the above [ESE : 2014]

1.17 An Otto cycle has a compression ratio of 8. If 250 kJ of work is extracted from the cycle, the heat rejected by the cycle is
(a) 500 kJ (b) 442.69 kJ
(c) 331.4 kJ (d) 192.69 kJ
[ESE : 2014]

1.18 An ideal spark ignition engine has a compression ratio of 9. What is its Air standard efficiency if ratio of specific heats is 1.5?
(a) 63% (b) 67%
(c) 70% (d) 72% [ESE : 2014]

1.19 A spark ignition engine has a compression ratio of 8 and the volume before compression is 0.9 m^3/kg. Net heat interaction per cycle is 1575 kJ/kg. What is the mean effective pressure?
(a) 20 kPa (b) 20 bar
(c) 2000 Pa (d) 2 bar
[ESE : 2015]

1.20 **Assertion (A):** The specific fuel consumption of a CI engine is lower than that of an SI engine.
Reason (R): For the same power, a CI engine is bigger is size than an SI engine.
(a) Both **A** and **R** are true and **R** is the correct explanation of **A**
(b) Both **A** and **R** are true but **R** is not a correct explanation of **A**
(c) **A** is true but **R** is false
(d) **A** is false but **R** is true

[ESE : 2015]

1.21 An ideal Otto-cycle works between minimum and maximum temperatures of 300 K and 1800 K. What is the compression ratio of the cycle for maximum work output when $\gamma = 1.5$ for this ideal gas ?

(a) 5 (b) 6
(c) 7 (d) 8 [ESE : 2016]

1.22 Consider the following statements :

1. The air standard efficiency of an Otto cycle is a function of the properties of the working substance (gas)
2. For the same compression ratio and same input, the thermal efficiency of an Otto cycle is more than that of a Diesel cycle
3. The thermal efficiency of a Diesel cycle increases with decrease of cutoff ratio

Which of the above statements are correct?

(a) 1 and 2 only (b) 1 and 3 only
(c) 2 and 3 only (d) 1, 2 and 3

[ESE : 2016]

1.23 Consider the following statements :

1. Both Otto and Diesel cycles are special cases of dual combustion cycle
2. Combustion process in IC engines is neither fully constant volume nor fully constant pressure process
3. Combustion process in ideal cycle is replaced by heat addition from internal source in closed cycle
4. Exhaust process is replaced by heat rejection in ideal cycle

Which of the above statements are correct ?

(a) 1, 2 and 3 only (b) 3 and 4 only
(c) 1, 2 and 4 only (d) 1, 2, 3 and 4

[ESE : 2016]

1.24 In an Otto cycle, air is compressed from 2.2 m^3 to 0.26 m^3 from an initial pressure of 1.2 kg/cm^2. The net output/cycle is 440 kJ.

What is the mean effective pressure of the cycle?

(a) 227 kPa (b) 207 kPa
(c) 192 kPa (d) 185 kPa

[ESE : 2016]

1.25 A 4-cylinder diesel engine running at 1200 r.p.m. developed 18.6 kW. The average torque when one cylinder was cut out was 105 N m. If the calorific value of the fuel was 42000 kJ/kg and the engine used 0.34 kg of diesel/kW hr, the indicated thermal efficiency was nearly

(a) 29% (b) 26%
(c) 31% (d) 23%

[ESE : 2017]

1.26 Consider the following statements for the air-standard efficiency of Diesel cycle:

1. For the same compression ratio, the efficiency decreases with increasing cutoff ratios.
2. For the same compression ratio and same heat input, Diesel cycle is more efficient than Otto cycle.
3. For constant maximum pressure and constant heat input, Diesel cycle is more efficient than Otto cycle.

Which of the above statements are correct?

(a) 1, 2 and 3 (b) 1 and 2 only
(c) 1 and 3 only (d) 2 and 3 only

[ESE : 2018]

1.27 If the intake air temperature of an IC engine increases, its efficiency will

(a) remain same (b) decrease
(c) increase (d) remain unpredictable

[ESE : 2018]

1.28 A diesel engine has a compression ratio of 14 and cutoff takes place at 6% of the stroke. The air standard efficiency will be

(a) 74.5 % (b) 60.5%
(c) 52.5% (d) 44.5% [ESE : 2019]

1.29 A fictitious pressure that, if it acted on the piston during the entire power stroke, would produce the same amount of net work as that produced during the actual cycle is called

(a) Quasi equivalent pressure
(b) Mean equivalent pressure
(c) Mean effective pressure
(d) Quasi static pressure

[ESE : 2020]

1.30 An ideal cycle based on the concept of combination of two heat transfer processes, one at constant volume and the other at constant pressure, is called
(a) Otto cycle (b) Dual cycle
(c) Diesel cycle (d) Carnot cycle
[ESE : 2020]

1.31 A 4-cylinder, 4-stroke single acting petrol engine consumes 6 kg of fuel per minute at 800 rpm when the air-fuel ratio of the mixture supplied is 9 : 1. The temperature is 650 K and pressure is 12.5 bar at the end of compression stroke. Take R = 300 Nm/kg.K, diameter of cylinder as 8 cm, stroke of cylinder as 10 cm. The compression ratio will be nearly
(a) 6.2 (b) 5.7
(c) 5.2 (d) 4.6
[ESE : 2020]

1.32 For the same compression ratio, the Brayton cycle efficiency is
(a) Same as the Diesel cycle efficiency
(b) Equal to the Otto cycle efficiency
(c) More than the Diesel cycle efficiency
(d) Less than the Otto cycle efficiency
[ESE : 2020]

1.33 A diesel engine has a compression ratio of 20 and cut-off takes place at 5% of the stroke. What is the cut-off ratio?
(a) 1.21 (b) 1.47
(c) 1.73 (d) 1.95
[ESE : 2021]

1.34 The cubic capacity of a four-stroke over-square spark-ignition engine is 275 cc. The clearance volume is 25 cc. What is the compression ratio of the engine?
(a) 8 (b) 10
(c) 12 (d) 14
[ESE : 2021]

1.35 The efficiency of any cycle increases with
(a) The decrease of maximum pressure and the constant of exhaust pressure.
(b) The decrease of maximum pressure and the decrease of exhaust pressure.
(c) The increase of maximum pressure and the decrease of exhaust pressure.
(d) The increase of maximum pressure and the constant of exhaust pressure.
[ESE : 2021]

Directions for the following **two (02)** items:

Read the following information and answer the **two** items that follows:

The bore and stoke of a water-cooled, vertical, single-cylinder, and four-stroke diesel engine are 80 mm and 110 mm respectively and the torque is 23.5 Nm.

1.36 What is the brake mean effective pressure of the engine?
(a) 1.01 bar (b) 2.01 bar
(c) 3.26 bar (d) 5.34 bar
[ESE : 2023]

1.37 What is the mean effective pressure if its rating is 4 kW at 1500 rpm?
(a) 3.26 bar (b) 4.26 bar
(c) 5.78 bar (d) 6.46 bar
[ESE : 2023]

1.38 **Statement (I)**: The simple air standard cycle analysis cannot predict the variation of thermal efficiency with mixture strength since air is assumed to be the working medium.
Statement (II): Fuel air cycle analysis suggests that the thermal efficiency will deteriorate as the mixture supplied to an engine is enriched.
(a) Both statement (I) and statement (II) are individually true statement (II) is the corect explanation of statement (I).
(b) Both statement (I) and statement (II) are individually true, but statement (II) is not the corect explanation of statement (I).
(c) Statement (I) is true, but Statement (II) is false.
(d) Statement (I) is false, but Statement (II) is true.
[ESE : 2023]

1.39 What is the compression ratio of the Otto cycle for a petrol engine with a cylinder bore of 50 mm, a stroke of 75 mm, and clearance volume of 21.3 cm^3?
(a) 4.9 (b) 5.9
(c) 6.9 (d) 7.9
[ESE : 2023]

1.40 Which type of gas power cycle has limitation of isochoric heat addition and rejection in piston cylinder arrangement?
(a) Carnot cycle (b) Otto cycle
(c) Diesel cycle (d) Dual cycle
[ESE : 2023]

1.41 The cubic capacity of a four-stroke over-square spark-ignition engine is 245 cc. The over-square ratio is 1.1. The clearance volume is 27.2 cc. What is the stroke length of the engine?
(a) 4.36 cm (b) 5.36 cm
(c) 6.36 cm (d) 7.36 cm
[ESE : 2023]

■■■■

Answers Air Standard Cycle

1.1 (a)	1.2 (b)	1.3 (a)	1.4 (a)	1.5 (b)	1.6 (a)	1.7 (d)	1.8 (d)	1.9 (c)
1.10 (c)	1.11 (b)	1.12 (c)	1.13 (a)	1.14 (b)	1.15 (a)	1.16 (a)	1.17 (d)	1.18 (b)
1.19 (b)	1.20 (b)	1.21 (b)	1.22 (d)	1.23 (c)	1.24 (a)	1.25 (a)	1.26 (c)	1.27 (b)
1.28 (b)	1.29 (c)	1.30 (b)	1.31 (a)	1.32 (b)	1.33 (d)	1.34 (c)	1.35 (c)	1.36 (d)
1.37 (c)	1.38 (b)	1.39 (d)	1.40 (b)	1.41 (c)				

Explanations Air Standard Cycle

1.1 (a)

$$\eta_{diesel} = 1-\frac{1}{r^{\gamma-1}}\left[\frac{\rho^{\gamma}-1}{\gamma(\rho-1)}\right]$$

$$r_c = \rho = \frac{T_3}{T_2}$$

Load $\uparrow T_3 \uparrow \rho \uparrow \eta \downarrow$

η of diesel engine increases as the load on the engine decreases and at limiting condition becomes equal to otto. With increase in load, cut off ratio increases.

1.2 (b)

CI engine has higher thermal efficiency due to higher compression ratio.

1.3 (a)

Otto cycle

$$\eta_{Otto} = 1-\frac{1}{r^{\gamma-1}}$$

$$r = \text{compression ratio} = \frac{V_1}{V_2}$$

$$\gamma = \text{specific heat ratio} = \frac{c_p}{c_v}$$

Diesel Cycle

$$\eta_{Disesel} = 1-\frac{1}{r^{\gamma-1}}\left(\frac{r_c^{\gamma}-1}{\gamma(r_c-1)}\right)$$

where

$$r = \text{compression ratio} = \frac{V_1}{V_2}$$

$$r_c = \text{cut off ratio} = \frac{V_3}{V_2}$$

Brayton cycle

$$\eta_{Brayton} = 1-\frac{1}{r_p^{(\gamma-1)/\gamma}}$$

where

$$r_p = \text{pressure ratio} = \frac{p_2}{p_1}$$

Carnot Cycle

$$\eta_{Carnot} = 1-\frac{T_2}{T_1}$$

T_2 = higher temperature
T_1 = lower temperature

1.5 (b)

Rankine cycle	:	Two adiabatic process and two constant pressure process
Otto cycle	:	Two adiabatic process and two constant volume process

Diesel cycle : Two adiabatic process, one constant volume and one constant pressure process

Carnot cycle : Two adiabatic process and two isothermal process

1.6 (a)

Mean effective pressure of dual cycle.

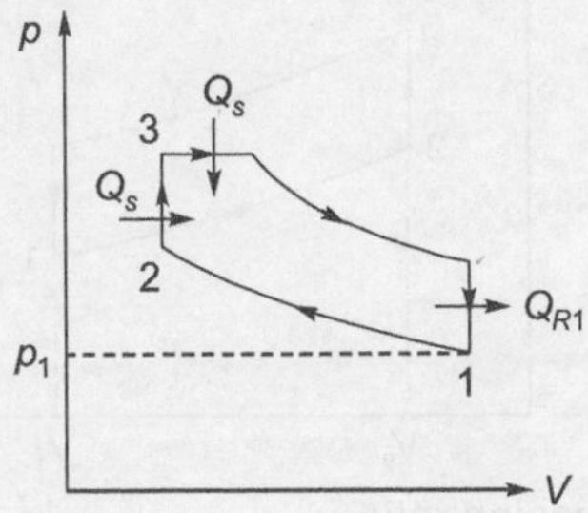

$$p_m = \frac{p_1\left[\gamma \cdot r_p \cdot r^{\gamma}(r_c - 1) + r^{\gamma}(r_p - 1) - r(r_p \cdot r_c^{\gamma} - 1)\right]}{(\gamma - 1)(r - 1)}$$

1.7 (d)

Clearance volume of a single cylinder

$$V_c = \frac{9225}{6} = 1537.5 \text{ cm}^3$$

Swept volume: $V_S = \frac{\pi}{4} d^2 L$

$$= \frac{\pi}{4} \times (17^2) \times 30 = 6805.95 \text{ cm}^3$$

Compression ratio,

$$r = 1 + \frac{V_S}{V_C} = 1 + \frac{6805.95}{1537.50}$$

$$= 5.42$$

1.8 (d)

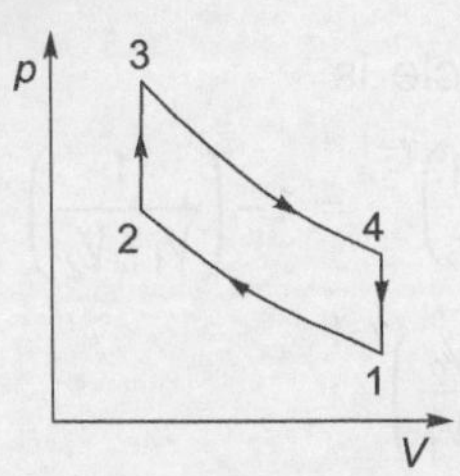

Otto cycle

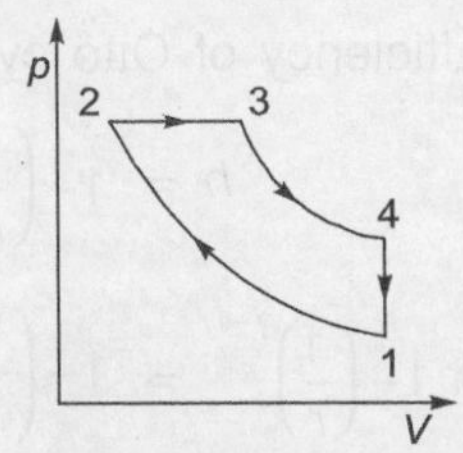

Diesel cycle

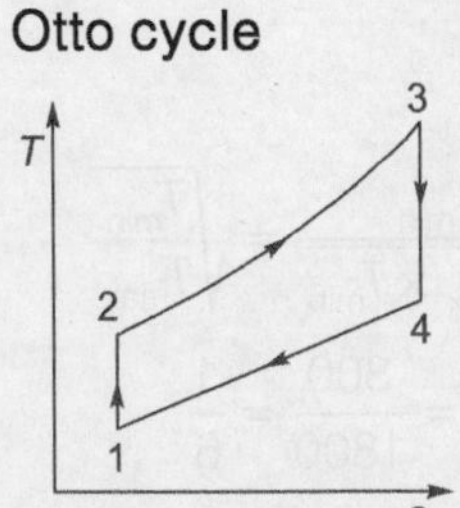

Brayton cycle

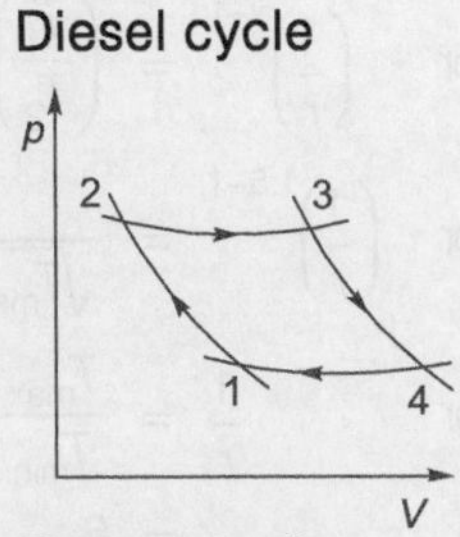

Carnot cycle

1.9 (c)

Area of wall of the engine is taken for heat transfer

$\therefore$ Total area $= \pi D \times L$

$= \pi \times 2 \times 2 +$ Area from top of cylinder and piston top

$$= 4\pi + \frac{\pi \times 2^2}{4} + \frac{\pi \times 2^2}{4}$$

$$= 6\pi$$

1.10 (c)

$$\eta_{diesel} = 1 - \frac{1}{r_k^{\gamma-1}}\left(\frac{r_c^{\gamma} - 1}{r_c - 1}\right)$$

for $\quad r_c = 1.65, \; \frac{r_c^{\gamma} - 1}{r_c - 1} = 1.563$

for $\quad r_c = 1.75, \; \frac{r_c^{\gamma} - 1}{r_c - 1} = 1.5853$

So, $\left(\frac{r_c^{\gamma} - 1}{r_c - 1}\right)$ increases with increase in cut-off ratio.

$\therefore \eta_{diesel}$ decreases

1.11 (b)

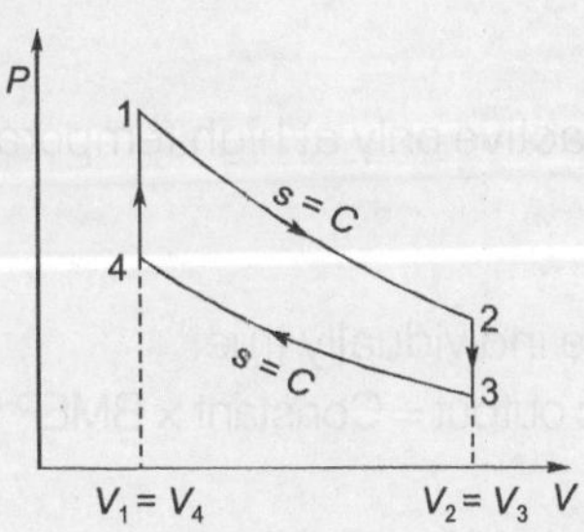

Otto Cycle

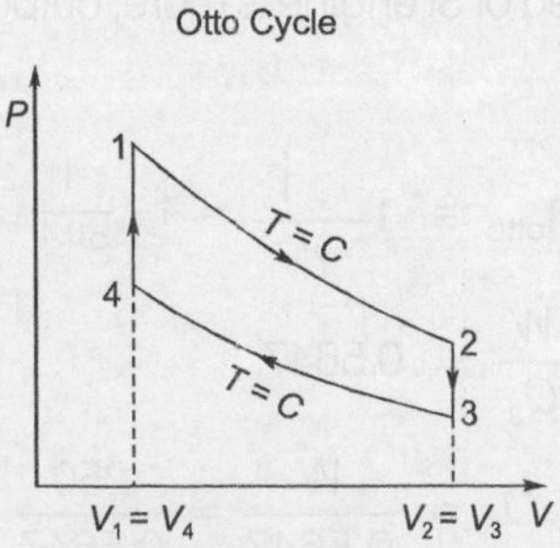

Stirling Cycle

Thus for a given compression ratio, $\frac{V_3}{V_4} = r$, both these cycles have equal thermal efficiency.

1.12 (c)

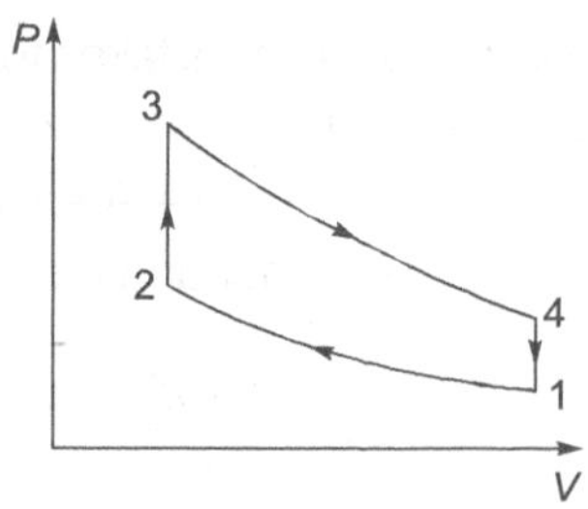

$$\frac{V_1}{V_2} = 6 = \frac{V_4}{V_3}$$

$$T_3V_3^{\gamma-1} = T_4V_4^{\gamma-1}$$

$$\therefore \quad T_4 = \frac{T_3}{\left(\frac{V_4}{V_3}\right)^{\gamma-1}} = \frac{T_3}{r^{\gamma-1}}$$

Given, $T_1 = 273 + 27 = 300$ K

$T_3 = 273 + 1569 = 1842$ K

So, $T_4 = \frac{1842}{6^{0.4}} = 899.55$ K

or $T_4 = 900$ K

1.13 (a)

$$\eta_{Otto} = 1 - \frac{1}{r^{\gamma-1}}$$

As $r\uparrow, \left(\frac{1}{r^{\gamma-1}}\right)\downarrow$ and hence $\eta_{Otto}\uparrow$.

Option (b), (c) and (d) are the knock parameters.

1.14 (b)

N_2 is reactive only at high temperature.

1.15 (a)

Both are individually true.

Specific output = Constant × BMEP × Mean Piston speed

As speed of SI engine is more, output is also more.

1.17 (d)

$$\eta_{otto} = 1 - \frac{1}{r^{r-1}} = 1 - \frac{1}{8^{0.4}} = 0.5647$$

$$\frac{W}{Q_s} = 0.5647$$

$$\Rightarrow \quad Q_s = \frac{W}{0.5647} = \frac{250}{0.5647} = 442.713 \text{ kJ}$$

$$Q_s - Q_R = W$$

or $Q_R = Q_c - W = 442.713 - 250$

$= \mathbf{192.71\ kJ}$

1.18 (b)

$$\eta = 1 - \frac{1}{r^{r-1}} = 1 - \frac{1}{9^{0.5}} = 1 - \frac{1}{3}$$

$= 0.6667 = \mathbf{67\%}$

1.19 (b)

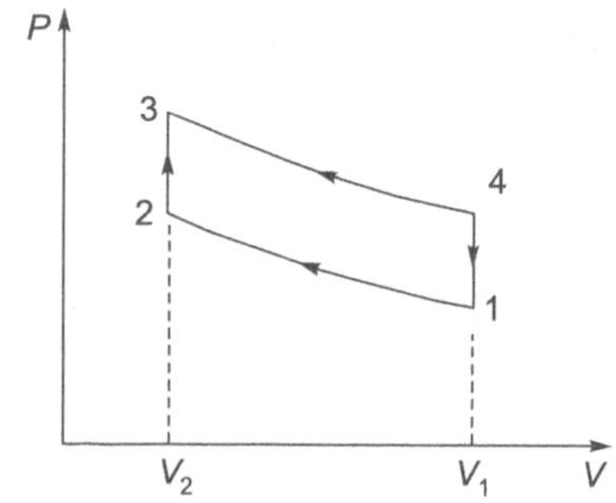

Compression ratio,

$r = 8$

$v_1 = 0.9$ m³/kg

$$r = \frac{v_1}{v_2}$$

$$\Rightarrow \quad v_2 = \frac{0.9}{8} = 0.1125 \text{ m}^3/\text{kg}$$

Work done = Mean effective pressure × swept volume $(v_1 - v_2)$

$1575 \times 10^3 = p_m \times (0.9 - 0.1125)$

$\Rightarrow \quad p_m = \mathbf{20\ bar}$

1.20 (b)

Both the statement are correct, but II statement is not correct explanation of first statement.

1.21 (b)

Efficiency of Otto cycle is

$$h = 1 - \left(\frac{1}{r}\right)^{\gamma-1} = 1 - \left(\frac{1}{V_1/V_2}\right)$$

$$\text{or } 1 - \left(\frac{1}{r}\right)^{\gamma-1} = 1 - \left(\frac{V_2}{V_1}\right)^{\gamma-1}$$

$$\text{or } \left(\frac{1}{r}\right)^{\gamma-1} = \left(\frac{T_1}{T_2}\right)$$

$$\text{or } \left(\frac{1}{r}\right)^{1.5-1} = \frac{T_{min}}{\sqrt{T_{max} \times T_{min}}} = \sqrt{\frac{T_{min}}{T_{max}}}$$

$$\text{or } \frac{1}{r} = \frac{T_{max}}{T_{min}} = \frac{300}{1800} = \frac{1}{6}$$

$\therefore \quad r = \mathbf{6}$

1.24 (a)

$$W = P_m \times V_s$$
$$V_s = (2.2 - 0.26)\ m^3 = 1.94\ m^3$$
$$(V_1 - V_2) = 1.94\ m^3$$
$$P_m = \frac{W}{V_s} = \frac{(440 \times 1000)\text{ Joules}}{1.94\,m^3}$$
$$= \frac{440 \times 10^3}{1.94}\ N/m^3 = \mathbf{227\ kPa}$$

1.25 (a)

$k = 4$; $N = 1200$ rpm; B.P. = 18.6 kW

$$T_{\text{per cyl.}} = 105\ \text{N-m}$$
$$(CV)_f = 42{,}000\ \text{kJ/kg}$$
$$bsfc = 0.34\ \text{kg/kW-hr} = \dot{m}_f/\text{B.P.}$$
$$\text{I.P} = 21.621\ \text{kW}$$
$$\eta_{ith} = \frac{IP}{HA/\sec} = 29.3\%$$

Use the equation bsfc $= \dfrac{m_f / hr}{\text{B. Power in kW}}$

1.26 (c)

Statement 1 ⟶ ✓

Statement 2 ⟶ ✗

Statement 3 ⟶ ✓

$$\eta_d = 1 - \left(\frac{1}{r}\right)^{\gamma-1} \times \frac{(\rho^\gamma - 1)}{\gamma(\rho - 1)}$$

1.27 (b)

As per UPSC it is (b) but actually it depends on type of engine.

1.28 (b)

$$r = 14, \quad \%p = 6\%, \ \gamma = 1.4$$
$$\Rightarrow \quad \rho = \%p(r-1) + 1 = 1.78$$
$$\eta_d = 1 - \left(\frac{1}{r}\right)^{\gamma-1} \times \frac{(\rho^\gamma - 1)}{\gamma(r-1)} = 60.4\%$$

1.31 (a)

If we take $\dot{m}_f$ = 6 kg/hr instead of per minute then only answer will match the options.

Given, $k = 4$, $\dot{m}_f$ = 6 kg/hr, N = 800 rpm, A/F = 9 : 1, T_2 = 650 K, P_2 = 12.5 bar, R = 300 J/kgK, $D = 8 \times 10^{-2}$ m, $L = 10 \times 10^{-2}$ m,

$$r = \frac{V_1}{V_2} = \frac{V_2 + V_3}{V_2} = 1 + \frac{V_3}{V_2} \quad \text{...(A)}$$

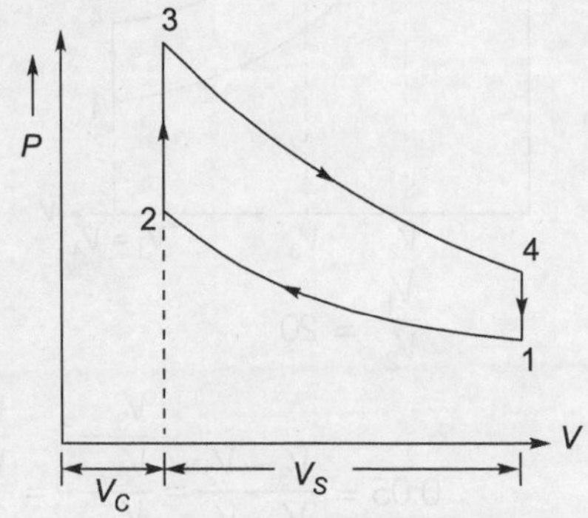

Mass of charge per hour $= \dot{m}_f + \dot{m}_a$

$$= \dot{m}_f + \dot{m}_f \times \frac{A}{F} = 6 + 6 \times 9$$
$$= 60\ \text{kg/hr} = 1\ \text{kg/min}$$

Number of suctions per minute =

$$\frac{N}{2} = \frac{800}{2} = 400$$

$$\therefore \text{ Mass/suction} = \frac{1\ \text{kg/min}^2}{400\ \text{suction/min}} = \frac{1}{400}$$
$$= 0.0025\ \text{kg/suction}$$

Mass/suction/cylinder = 0.000625

Now, $P_2V_2 = mrT_2$

$$V_2 = \frac{mrT_2}{P_2} = \frac{0.000625 \times 0.3 \times 650}{12.5 \times 100}$$
$$= 0.0000975\ m^3$$
$$V_s = \frac{\pi}{4} D^2 L$$
$$= \frac{\pi}{4}(8 \times 10^{-2})^2 \times 10 \times 10^{-2}$$
$$= 0.0005026\ m^3$$

From (A), $r = 1 + \dfrac{V_s}{V_2} = 1 + \dfrac{0.0005026}{0.0000975}$

$$= 6.154$$

1.32 (b)

Efficiency of Brayton cycle $= 1 - \dfrac{1}{r^{\gamma-1}}$

where, r = pressure ratio, R = compression ratio

Efficiency of Otto cycle $= 1 - \dfrac{1}{R^{\gamma-1}}$

Both are same if r and R are pressure ratios.

1.33 (d)

$$r = 20,\ c = 0.05$$

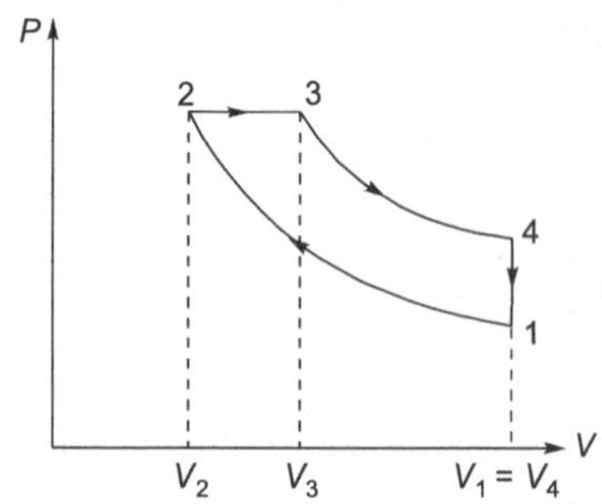

$$\frac{V_1}{V_2} = 20$$

$$0.05 = \frac{V_3 - V_2}{V_1 - V_2} = \frac{\frac{V_3}{V_2} - 1}{\frac{V_1}{V_2} - 1} = \frac{\frac{V_3}{V_2} - 1}{20 - 1}$$

$$\frac{V_3}{V_2} = 1.95$$

1.34 (c)

$$V_S = 275 \text{ cm}^3;\ V_C = 25 \text{ cm}^3$$

$$r = \frac{V_1}{V_2} = \frac{V_C + V_S}{V_C} = \frac{25 + 275}{25} = 12$$

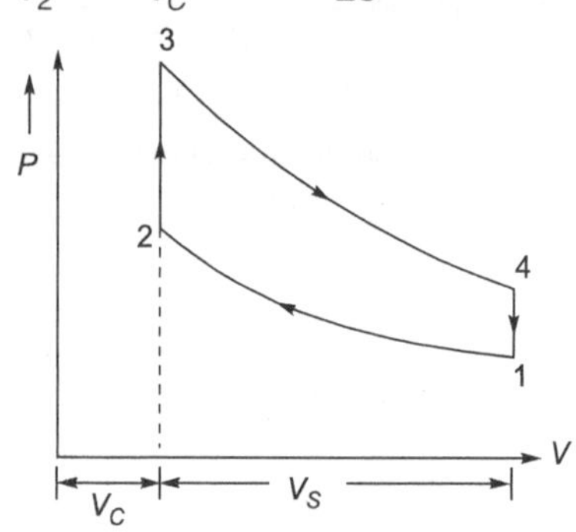

1.36 (d)

Given : Diameter of bore, d = 80 mm

Stroke length, l = 110 mm; Torque, T= 23.5 Nm

$$T\omega = P_m \times \frac{\pi}{4} d^2 \times l \times \frac{N}{2}$$

$$23.5 \times 2\pi \times N = P_m \times \frac{\pi}{4} \times 0.08^2 \times 0.11 \times \frac{N}{2}$$

$$P_m = 5.34 \times 10^5 \text{ Pa} = 5.34 \text{ bar}$$

1.37 (c)

$$4 \times 10^3 = p_m \times \frac{\pi}{4} \times 0.08^2 \times 0.11 \times \frac{1500}{2 \times 60}$$

$$p_m = 5.78 \times 10^5 \text{ Pa}$$

1.38 (b)

The air standard cycle analysis shows the general effect of only compression ratio on engine efficiency whereas the fuel-air cycle analysis gives the effect of variation of fuel air ratio, inlet pressure and temperature on the engine performance.

1.39 (d)

Given: Bore diameter, d = 50 mm

Stroke length, l = 75 mm

Clearance volume, V_c = 21.3 cm^3

$$\text{Swept volume, } V_s = \frac{\pi}{4} \times 5^2 \times 7.5 = 147.26 \text{ cm}^3$$

$$r = 1 + \frac{147.26}{21.3} = 7.91$$

1.40 (b)

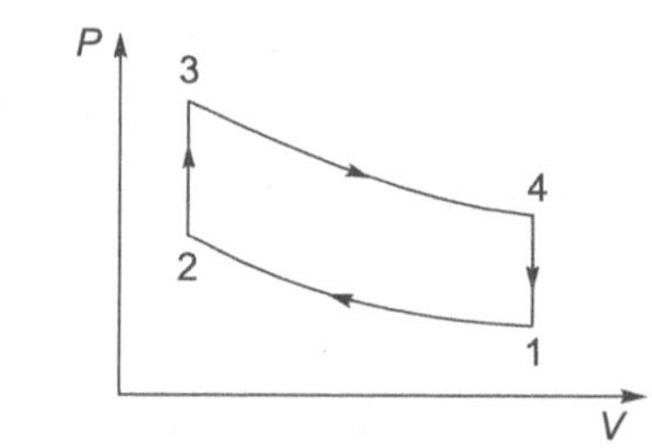

In case of Otto cycle, both heat addition and heat rejection take place at constant volume.

1.41 (c)

Given:

Cubic capacity, V_T= 245 cm^3

$$\text{Over square ratio, } \frac{d}{l} = 1.1$$

Clearance volume, V_C = 27.2 cm^3

Swept volume, V_s = 245 − 27.2 = 217.8 cm^3

$$\frac{\pi}{4} d^2 l = 217.8$$

$$\frac{\pi}{4} \times 1.1^2 \times l^3 = 217.8$$

$$l = 6.12 \text{ cm}$$

■■■■

2 Fuel Characteristics

2.1 Consider the following statements regarding the advantages of fuel injection over carburetion in SI engines:

1. Higher power output and increased volumetric efficiency.
2. Simple and inexpensive injection equipment.
3. Longer life of injection equipment.
4. Less knocking and reduced tendency for back-fire.

Select the correct answer using the codes given below:

(a) 1, 2 and 3 (b) 1, 2 and 4
(c) 2 and 3 (d) 1 and 4

[ESE : 2001]

2.2 **Assertion (A):** A pintle nozzle is employed to mix the fuel properly even with the slow air movement available with many open combustion chambers in CI engines.

Reason (R): The mixing of fuel and air is greatly affected by the nature of the air movement in the combustion chambers in CI engines.

(a) Both **A** and **R** are individually true and **R** is the correct explanation of **A**
(b) Both **A** and **R** are true but **R** is not the correct explanation of **A**
(c) **A** is true but **R** is false
(d) **A** is false but **R** is true

[ESE : 2002]

2.3 The stoichiometric air-fuel ratio for petrol is 15 : 1. What is the air-fuel ratio regarded for maximum power?

(a) 16 : 1 – 18 : 1 (b) 15 : 1 – 17 : 1
(c) 12 : 1 – 14 : 1 (d) 9 : 1 – 11 : 1

[ESE : 2004]

2.4 A 4-stroke diesel engine, when running at 2000 rpm has an injection duration of 1.5 ms. What is the corresponding duration of the crank angle in degrees?

(a) 18° (b) 9°
(c) 38° (d) 15°

[ESE : 2005]

2.5 For which of the following reasons, do the indirect injection diesel engines have higher specific output compared to direct injection diesel engine?

1. They have lower surface to volume ratio
2. They run at higher speeds.
3. They have higher air utilization factor.
4. They have lower relative heat loss.

Select the correct answer using the code given below:

(a) 1 and 2 (b) 2 only
(c) 2 and 3 (d) 3 and 4

[ESE : 2005]

2.6 Consider the following statements:

1. For a diesel cycle, the thermal efficiency decreases as the cut off ratio increases.
2. In a petrol engine the high voltage for spark is in the order of 1000 V.
3. The material for centre electrode in spark plug is carbon.

Which of these statements is/are correct?

(a) Only 1 (b) Only 1 and 2
(c) Only 2 and 3 (d) 1, 2 and 3

[ESE : 2006]

2.7 Where does mixing of fuel and air take place in case of diesel engine?

(a) Injection pump (b) Injector
(c) Engine cylinder (d) Inlet manifold

[ESE : 2006]

2.8 Piston compression rings are made of which one of the following?

(a) Cast iron (b) Bronze
(c) Aluminium (d) White metal

[ESE : 2007]

2.9 For a conventional SI engine, the value of fuel-air ratio in the normal operating range is

(a) 0.056 - 0.083 (b) 0.083 - 0.56
(c) 0.0056 - 0.83 (d) 0.056 - 0.83

[ESE : 2007]

2.10 The Stoichiometric air-fuel ratio by volume for combustion of carbon monoxide in air is

(a) 1.19 (b) 2.38
(c) 2.45 (d) 4.76 [ESE : 2010]

2.11 Pistons of diesel engines are usually cooled by

(a) Air (b) Water
(c) Lubricating oil (d) Fuel oil

[ESE : 2011]

2.12 Air injection in IC engine refers to injection of

(a) Air only
(b) Liquid fuel only
(c) Liquid fuel and air
(d) Supercharging air

[ESE : 2011]

Directions: The following items consists of two statements; one labelled as **'Statement (I)'** and the other as **'Statement (II)'**. You are to examine these two statements carefully and select the answers to these items using the codes given below:

Codes:

(a) Both Statement (I) and Statement (II) are individually true and Statement (II) is the correct explanation of Statement (I).
(b) Both Statement (I) and Statement (II) are individually true but Statement (II) is **NOT** the correct explanation of Statement (I).
(c) Statement (I) is true but Statement (II) is false
(d) Statement (I) is false but Statement (II) is true.

2.13 Statement (I): CI diesel engines use multi hole nozzle for fuel injection.

Statement (II): Single hole nozzles cause wall impingement and carbon deposits.

[ESE : 2013]

2.14 Statement (I): In common rail system, the nozzle construction must be closely matched to ensure equality of fuel discharge from cylinder to cylinder.

Statement (II): The discharge from the nozzles is regulated by the size of orifice and pressure drop.

[ESE : 2014]

2.15 In Diesel engines, the control of black smoke in exhaust can be achieved by:

1. Running the engine at lower load.
2. Maintaining the injection system perfect.
3. Using Diesel fuel of higher Cetane number.

Which of the above statements are correct?

(a) 1, 2 and 3 (b) 1 and 2 only
(c) 1 and 3 only (d) 2 and 3 only

[ESE : 2014]

2.16 Assertion (A): The cut off ratio of a Diesel engine cycle should be greater than one, but should be as low as possible.

Reason (R): Lower cut off ratio does improve the thermal efficiency but lowers the specific work output. Hence, the value of cut off ratio must be optimized.

(a) Both **A** and **R** are true and **R** is the correct explanation of **A**
(b) Both **A** and **R** are true but **R** is not a correct explanation of **A**
(c) **A** is true but **R** is false
(d) **A** is false but **R** is true

ESE : 2015]

Directions: The following items consists of two statements; one labelled as **'Statement (I)'** and the other as **'Statement (II)'**. You are to examine these two statements carefully and select the answers to these items using the codes given below:

Codes:

(a) Both Statement (I) and Statement (II) are individually true and Statement (II) is the correct explanation of Statement (I).
(b) Both Statement (I) and Statement (II) are individually true but Statement (II) is **NOT** the correct explanation of Statement (I).
(c) Statement (I) is true but Statement (II) is false
(d) Statement (I) is false but Statement (II) is true.

2.17 Statement (I) : To prevent knocking in SI engines the end gas should have a low density.
Statement (II) : Pre-ignition is caused due to detonation. **[ESE : 2016]**

2.18 Statement (I) : Air-blast injection in diesel engines could reduce engine efficiency.

Statement (II) : Air-blast injection in diesel engines is not instantaneous but happens when the piston moves outward with the injection valve remaining open for whatever reason.

[ESE : 2017]

2.19 Considering the following statements:
1. Recycling exhaust gases by partial mixing with the intake gases increases the emission of oxides of nitrogen from the engine.
2. The effect of increase in altitude of operation on the carburetor is to enrich the entire port-throttle operation.
3. When the carburetor throttle is suddenly opened, the air-fuel mixture may lean out temporarily resulting in engine stall.
4. Use of multi-venturi system makes airstream when the fuel is introduced at the main venturi throat.

Which of the above statements are correct?
(a) 1 and 3 (b) 1 and 4
(c) 2 and 3 (d) 2 and 4

[ESE : 2018]

2.20 Statement (I) : In air-blast injection, a separate compressor is used to create an air blast at a pressure of 6 MPa.
Statement (II) : The solid injection system is heavier as it needs increasing the fuel pressure to 30 MPa.
(a) Both Statement (I) and Statement (II) are individually true and Statement (II) is the correct explanation of Statement (I).
(b) Both Statement (I) and Statement (II) are individually true but Statement (II) is not the correct explanation of Statement (I).
(c) Statement (I) is true but Statement (II) is false.
(d) Statement (I) is false but Statement (II) is true.

[ESE : 2018]

2.21 A 1 g sample of fuel is burned in a bomb calorimeter containing 1.2 kg of water at an initial temperature of 25°C. After the reaction, the final temperature of the water is 33.2°C. The heat capacity of the calorimeter is 837 J/°C. The specific heat of water is 4.18 J/°C. The heat released by the fuel will be nearly
(a) 36 kJ/g (b) 42 kJ/g
(c) 48 kJ/g (d) 54 kJ/g

[ESE : 2019]

■■■■

Answers Carburation and Diesel Injection

2.1 (d)	2.2 (b)	2.3 (c)	2.4 (a)	2.5 (d)	2.6 (a)	2.7 (c)	2.8 (a)	2.9 (a)
2.10 (b)	2.11 (c)	2.12 (c)	2.13 (a)	2.14 (a)	2.15 (b)	2.16 (a)	2.17 (c)	2.18 (a)
2.19 (c)	2.20 (c)	2.21 (c)						

Explanations Carburation and Diesel Injection

2.1 (d)

(i) Increased volumetric efficiency and hence increase power and torque

(ii) Better distribution of mixture to each cylinder and hence lower specific fuel consumption

Cost of injection system in high and life is less.

2.2 (b)

Pintle nozzle has projection which protrudes through the mouth of nozzle body (hollow conical spray).

It results in good atomization and penetration. Swirl is given in combustion chambers for proper combustion

2.3 (c)

Slightly rich mixture (About equivalence ratio of 0.8) is used for maximum power.

2.4 (a)

$$\text{Time for injection} = \frac{\theta}{360} \times \frac{60}{N}$$

(N will be same for 2 or 4 stroke)

$$0.0015 = \frac{\theta}{360} \times \frac{60}{2000}$$

$$\theta = \frac{0.0015 \times 360 \times 2000}{60} = 18°$$

2.6 (a)

Voltage across the spark plug electrode is around 25000 V.

The best material for electrodes is platinum or platinum tungsten alloy but it is less common due to its high cost.

Mostly Nickel alloy is used for electrode.

2.7 (c)

Diesel is directly injected in cylinder where it gets mixed with air.

2.8 (a)

Since grey cast iron has properties of self lubrication and damping of small vibrations they are widely used for machine base, engine frames, drainage pipes, elevator and industrial furnace counter weights, pump housings cylinder and piston rings of IC engines, flywheel etc.

2.9 (a)

For SI engines air fuel ratio lies between 12 to 16.

2.11 (c)

Piston of diesel engine are usually cooled by lubricating oil. In addition, the lubricant on the cylinder walls must function as a seal to minimize the amount of gases of combustion that pass the piston rings and enter the crankcase.

2.13 (a)

For uniform impingement, CI diesel engine use multihole nozzle.

2.21 (c)

$$\begin{aligned} m_f(\text{C.V.}) &= m_w\, c_p\, \Delta T + W_{eq} \times \Delta T \\ &= 1.2 \times 4.18 \times 10^3 \times 8.2 \\ &\quad + 837 \times 8.2 \\ &= 41131.2 + 6863.4 \\ &= 47994.6 \text{ J} \approx 48000 \text{ J} \\ (\text{C.V.}) &= 48 \text{ kJ/g} \end{aligned}$$

■■■■

3 Combustion and Knocking in SI and CI Engine

3.1 Match **List-I** (Air-fuel ratio by mass) with **List-II** (Engine operation mode) and select the correct answer using the codes given below the lists:

	List-I		**List-II**
A.	10 : 1	1.	CI engine part load
B.	16 : 1	2.	SI engine part load
C.	35 : 1	3.	SI engine idling
D.	12.5 : 1	4.	CI engine full load
		5.	SI engine full load

Codes:

	A	B	C	D
(a)	3	2	1	5
(b)	4	2	1	5
(c)	3	1	2	4
(d)	4	1	2	3

[ESE : 2000]

Direction: Each of the next questions consists of two statements, one labelled as the **'Assertion (A)'** and the other as **'Reason (R)'**. You are to examine these two statements carefully and select the answers to these items using the codes given below:

Codes:

(a) Both **A** and **R** are individually true and **R** is the correct explanation of **A**

(b) Both **A** and **R** are individually true but **R** is not the correct explanation of **A**

(c) **A** is true but **R** is false

(d) **A** is false but **R** is true

3.2 **Assertion (A):** The CI engine is basically more suitable for supercharging than the SI engine.

Reason (R): In the CI engine super-charging tends to prevent diesel knocking.

[ESE : 2000]

3.3 **Assertion (A):** Knocking in SI engines is due to auto-ignition of the end charge while knocking in CI engines is due to auto-ignition of the first charge.

Reason (R): Spark ignition engines employ lower compression ratio than diesel engines and the fuel used has a calorific value lower than that of diesel oil.

[ESE : 2001]

3.4 Consider the following statements relevant to the ignition system of SI engine:

1. Too small a dwell angle will lead to the burning of condenser and contact points.
2. Too small a dwell angle will result in misfiring.
3. Too large a dwell angle will result in burning of condenser and contact points.
4. Too large a dwell will result in misfiring.

Which of these statements are correct?

(a) 1 and 2 (b) 2 and 3
(c) 3 and 4 (d) 4 and 1

[ESE : 2002]

3.5 Stoichometric air-fuel ratio by volume for combustion of methane in air is :

(a) 15 : 1 (b) 17.16 : 1
(c) 9.52 : 1 (d) 10.58 : 1

[ESE : 2002]

3.6 Auto-ignition time for petrol-air mixture is minimum when the ratio of actual fuel-air ratio and chemically correct fuel-air ratio is

(a) 0.8 (b) 1.0
(c) 1.2 (d) 1.5

[ESE : 2002]

3.7 In spark ignition engines knocking can be reduced by :

(a) increasing the compression ratio
(b) increasing the cooling water temperature
(c) retarding the spark advance
(d) increasing the inlet air temperature

[ESE : 2002]

3.8 The tendency of knocking in CI engine is reduced by

(a) high self-ignition temperature of fuel

(b) decrease in jacket water temperature
(c) injection of fuel just before TDC
(d) decrease in injection pressure

[ESE : 2002]

3.9 **Assertion (A):** Power by a four-stroke engine working on Otto cycle is higher than the power generated by a two-stroke engine for the same swept volume speed, temperature and pressure conditions.

Reason (R): In a four-stroke engine one cycle is completed in two revolutions.

(a) Both **A** and **R** are individually true and **R** is the correct explanation of **A**
(b) Both **A** and **R** are individually true but **R** is not the correct explanation of **A**
(c) **A** is true but **R** is false
(d) **A** is false but **R** is true

[ESE : 2003]

3.10 Which one of the following action(s) increase(s) the knocking tendency in the SI engine?

(a) Increasing mixture strength beyond equivalence ratio (ϕ) = 1.4
(b) Retarding the spark and increasing the compression ratio
(c) Increasing the compression ratio and reducing engine speed
(d) Increasing both mixture strength beyond equivalence (ϕ) = 1.4 and the compression ratio

[ESE : 2004]

3.11 Which of the following factor(s) increase(s) the tendency for knocking in CI engine?

(a) Increasing both the compression ratio and the coolant temperature
(b) Increasing both the speed and the injection advance
(c) Increasing the speed, injection advance and coolant temperature
(d) Increasing the compression ratio

[ESE : 2004]

3.12 Which of the following feature(s) is/are used in the combustion chamber design to reduce SI engine knock?

(a) Spark plug located away from exhaust value, wedge shaped combustion chamber and short flame travel distance
(b) Wedge shaped combustion chamber
(c) Wedge shaped combustion chamber and short flame travel distance
(d) Spark plug located away from exhaust valve, short flame travel distance and side valve design

[ESE : 2004]

3.13 Match **List-I** (SI Engine Operational Mode) with **List-II** (Air Fuel Ratio by Mass) and select the correct answer using the codes given below the lists:

	List-I		List-II
A.	Idling	**1.**	4 : 1
B.	Cruising	**2.**	10 : 1
C.	Maximum power	**3.**	12.5 : 1
D.	Cold starting	**4.**	15 : 1
		5.	14.8 : 1

Codes:

	A	B	C	D
(a)	2	4	3	1
(b)	5	4	1	3
(c)	2	3	5	1
(d)	5	3	1	4

[ESE : 2004]

3.14 Match **List-I** (SI Engine Operating Mode) with **List-II** (Approximate A/F Ratio) and select the correct answer using the code given below the lists:

	List-I		List-II
A.	Cold Start	**1.**	10 : 1
B.	Idling	**2.**	16 : 1
C.	Cruising	**3.**	13 : 1
D.	Maximum Power	**4.**	4 : 1
		5.	20 : 1

Codes:

	A	B	C	D
(a)	4	3	2	1
(b)	2	1	5	3
(c)	4	1	2	3
(d)	2	3	5	1

[ESE : 2005]

3.15 The knocking tendency in compression ignition engines increases with
(a) Increase of coolant water temperature
(b) Increase of temperature of inlet air
(c) Decrease of compression ratio
(d) Increase of compression ratio
[ESE : 2005]

3.16 Which of the following cannot be caused by a hot spark plug?
1. Pre-ignition 2. Post-ignition
3. Detonation 4. Run-on-ignition
Select the correct answer using the code given below:
(a) 1 and 4 (b) 2 only
(c) 2 and 3 (d) 3 only
[ESE : 2005]

3.17 Consider the following statements:
1. Supercharging increases the power output and increases the volumetric efficiency.
2. Supercharging is more suitable for SI engines than CI engines.
3. The limits of supercharging for an SI engine is set by knock while that for a CI engine is set by thermal loading.
Which of these statements are correct?
(a) 2 and 3
(b) 1, 2 and 3
(c) 1 and 3
(d) 1 and 2
[ESE : 2005]

3.18 Which of the following combustion chamber design features reduce(s) knocking in SI engines?
1. Spark plug located near the inlet valve
2. T-head.
3. Wedge shaped combustion chamber.
4. Short flame travel distance.
Select the correct answer using the code given below:
(a) 1 and 3 (b) 3 only
(c) 3 and 4 (d) 1 and 2
[ESE : 2005]

3.19 Consider the following statements:
In order to prevent detonation in a spark ignition engine, the charge away from the spark plug should have
1. Low temperature
2. Low density
3. Long ignition delay
Which of these statements is/are correct?
(a) Only 1 (b) Only 2
(c) Only 3 (d) 1, 2 and 3
[ESE : 2006]

3.20 The tendency of petrol to detonate in terms of octane number is determined by comparison of fuel with which of the following?
(a) Iso-octane
(b) Mixture of normal heptane and iso-octane
(c) Alpha methyl naphthalene
(d) Mixture of methane and ethane
[ESE : 2006]

3.21 Consider the following statements:
1. In the SI engines detonation occurs near the end of combustion whereas in CI engines knocking occurs near the beginning of combustion.
2. In SI engines no problems are encountered on accounts of pre-ignition.
3. Low inlet pressure and temperature reduce knocking tendency in SI engines but increases the knocking tendency in CI engines.
Which of these statements are correct?
(a) 1, 2 and 3 (b) Only 1 and 2
(c) Only 2 and 3 (d) Only 1 and 3
[ESE : 2006]

3.22 The purpose of employing supercharging for an engine is
(a) to provide forced cooling air
(b) to raise exhaust pressure
(c) to inject excess fuel for coping with higher load
(d) to supply an intake of air at a density greater than the density of the surrounding atmosphere [ESE : 2006]

3.23 The delay period in CI engine depends upon which of the following?
(a) Temperature and pressure in the cylinder at the time of injection
(b) Nature of the fuel mixture strength
(c) Relative velocity between the fuel injection and air turbulence pressure of residual gases
(d) All of the above [ESE : 2007]

3.24 Match **List-I** (SI engine Operational Mode) with **List-II** (*A-F* Ratio Supplied by the Carburetor) and select the correct answer using the code given below the lists:

	List-I		List-II
A.	Idling	1.	3
B.	Cruising	2.	10
C.	Maximum Power	3.	13
D.	Cold starting	4.	16
		5.	20

Codes:

	A	B	C	D
(a)	2	4	5	1
(b)	4	5	3	2
(c)	2	4	3	1
(d)	4	5	3	1

[ESE : 2007]

3.25 What is the main objective of supercharging of the engine?
(a) To reduce the mass of the engine per brake power
(b) To reduce space occupied by engine
(c) To increase the power output of engine
(d) All of the above
[ESE : 2007]

3.26 Consider the following statements:
Diesel knock can be reduced by
1. increasing the compression ratio.
2. increasing the engine speed.
3. increasing the injection retard.
4. decreasing the inlet air temperature.
Which of these statements are correct?
(a) 2 and 4 only (b) 1, 2 and 3 only
(c) 1 and 3 only (d) 1, 2, 3 and 4
[ESE : 2007]

3.27 For minimizing knocking tendency in SI engine, where should the spark plug be located?
(a) Near inlet valve
(b) Away from both the valves
(c) Near exhaust valve
(d) Midway between inlet and exhaust valves
[ESE : 2008]

3.28 In a SI engine, combustion stage-I takes 1 ms and combustion stage-II takes 1.5 ms when the engine runs at 1000 rpm. If stage-I time duration is independent of engine speed what will be the additional spark advance necessary when the engine speed is doubled?
(a) 0° (b) 6°
(c) 12° (d) 24°
[ESE : 2008]

3.29 Consider the following:
Carbon
Carbon-monoxide
Hydrogen, and
Sulphur
What is the amount of oxygen (in kg) required for complete combustion of each one of the above respectively?
(a) 1, 8, 8/3, 4/7 (b) 4/7, 1, 8, 8/3
(c) 8/3, 4/7, 8, 1 (d) 8, 1, 4/7, 8/3
[ESE : 2008]

3.30 Which of the following symptoms shows that the combustion is necessarily complete?
(a) Presence of free carbon in exhaust
(b) Presence of CO in exhaust
(c) Presence of oxygen in exhaust
(d) Presence of nitrogen in exhaust
[ESE : 2009]

3.31 What is the purpose of supercharging an engine?
(a) To increase the power output of engine
(b) To reduce specific fuel consumption
(c) To reduce the noise of the engine
(d) To improve cooling of cylinders
[ESE : 2009]

3.32 Detonation in the SI engine becomes severe during
(a) Low load and high speed
(b) High load and high speed
(c) Low load and low speed
(d) High load and low speed
[ESE : 2010]

3.33 Consider the following:
1. Increased cetane number.
2. Increased compression ratio.
3. Increased injection advance.
4. Increased air turbulence.
Which of the above factors reduce physical delay in the diesel engine?
(a) 1, 2 and 3 only (b) 2 and 4 only
(c) 2, 3 and 4 only (d) 1, 2, 3 and 4
[ESE : 2010]

3.34 Consider the following factors:
Diesel engine knock can be reduced by increasing
1. Engine speed.
2. Compression ratio.
3. Degree of supercharge.
4. Injection advance.

Which of these statements is/are correct?
(a) 1, 2, 3 and 4 (b) 2 and 3 only
(c) 1, 3 and 4 only (d) 2 only **[ESE : 2010]**

Direction: Each of the next questions consists of two statements, one labelled as the **'Assertion (A)'** and the other as **'Reason (R)'**. You are to examine these two statements carefully and select the answers to these items using the codes given below:

Codes:
(a) Both **A** and **R** are individually true and **R** is the correct explanation of **A**
(b) Both **A** and **R** are individually true but **R** is not the correct explanation of **A**
(c) **A** is true but **R** is false
(d) **A** is false but **R** is true

3.35 Assertion (A): In a diesel engine knocking decreases with reduced engine speed.
Reason (R): With increased engine speed, quantity of fuel injected during the delay period increases. **[ESE : 2010]**

3.36 Assertion (A) : In SI engines, higher compression ratio induces knocking.
Reason (R) : Higher compression ratio increases the temperature of the end mixture. **[ESE : 2011]**

3.37 By supercharging the diesel engine, the possibility of knocking
(a) Decreases (b) Increases
(c) Remains constant (d) None of the above **[ESE : 2011]**

3.38 Supercharging is the process of
(a) Supplying the intake of an engine with air at a density grater than the density of the surrounding atmosphere
(b) Providing forced cooling air
(c) Injection of compressed air to remove combustion products fully
(d) Raising exhaust pressure **[ESE : 2011]**

3.39 In a petrol engine, the tendency for detonation increases with
(a) retarded spark timing
(b) running the engine at high speed
(c) supercharging
(d) increasing the cooling rate **[ESE : 2012]**

3.40 Consider the following statements:
1. Diesel fuels are compared using an ignition delay criterion.
2. Cetane number is equal to the percentage of cetane plus 15% of the percentage of heptamethyl nonane in the fuel.
3. Cetane number of alphamethyl naphthalene is assigned the value of 15.

Which of these statements are correct?
(a) 1, 2 and 3 (b) 1 and 2 only
(c) 2 and 3 only (d) 1 and 3 only **[ESE : 2012]**

3.41 In an SI engine the first stage of combustion takes 1 ms and the second stage of combustion takes 2 ms at 1000 rpm of the engine. The spark timing is set at 15°C btdc. What will be the new spark timing required if the engine speed is doubled?
(a) 15°C btdc (b) 21°C btdc
(c) 27°C btdc (d) 39°C btdc **[ESE : 2013]**

3.42 Increase in the compression ratio in Otto cycle may cause
(a) misfiring (b) detonation
(c) knocking (d) longer ignition delay

3.43 Which of the following increase the diesel knocking tendency?
1. Decreased compression ratio
2. Decreased engine speed
3. Retarded injection timing
4. Supercharging

(a) 1 only (b) 2 only
(c) 2 and 3 (d) 3 and 4 **[ESE : 2013]**

3.44 Preignition in SI engines results in
1. Increase in the work of compression
2. Increase in the network of the process cycles
3. Decrease in fuel efficiencies
4. Decrease in fuel loss from the engine

(a) 1, 2 and 3 (b) 1, 3 and 4
(c) 1 and 3 only (d) 2 and 4 only

ESE : 2013]

3.45 Which of the following statements are correct for 'Diesel knock'?

1. A long ignition delay period increases tendency of Diesel knock.
2. The Diesel knock occurs near the end of combustion.
3. The rate of pressure rise in Diesel knock is lower than in detonation in spark ignition engines.

(a) 1, 2 and 3 (b) 1 and 2 only
(c) 2 and 3 only (d) 1 and 3 only

[ESE : 2013]

Directions: Each of the next consists of two statements, one labelled as the 'Statement (I)' and the other as 'Statement (II)'. You are to examine these two statements carefully and select the answers to these items using the codes given below:

Codes:

(a) Both Statement (I) and Statement (II) are individually true and Statement (II) is the correct explanation of Statement (I).
(b) Both Statement (I) and Statement (II) are individually true but Statement (II) is **NOT** the correct explanation of Statement (I).
(c) Statement (I) is true but Statement (II) is false
(d) Statement (I) is false but Statement (II) is true.

3.46 **Statement (I):** A good CI engine fuel is a bad SI engine fuel and a good SI engine fuel is a bad CI engines fuel.

Statement (II): A CI engine fuel requires high self-ignition temperature and a long delay period whereas a SI engine fuel requires low self-ignition temperature and short delay period. [ESE : 2013]

3.47 **Statement (I):** While designing the SI engine the designer tries to keep the compression ratio as high as possible.

Statement (II): Whereas while designing the CI engine the designer tries to keep the compression ratio as low as possible. [ESE : 2013]

3.48 **Statement (I):** An SI engine requires greater spark advance at lower loads.

Statement (II): Increased dilution by residual gases at lower loads reduces the combustion rate. [ESE : 2014]

3.49 **Statement (I):** In CI engines increase of load decreases the knocking tendency.

Statement (II): Increase of load increases the temperature of mixture and thereby decrease in delay angle. [ESE : 2014]

3.50 **Statement (I):** The four stroke cycle internal combustion reciprocating engines run at higher speeds than the two stroke cycle engines.

Statement (II): The separate exhaust and intake strokes of the four stroke engines provide greater opportunity for the dissipation of heat from critical parts such as piston.

[ESE : 2014]

3.51 The tendency of detonation is high in engines of larger cylinder diameter because of

(a) Higher intake pressure in larger cylinder
(b) Higher fuel/air ratio in larger cylinder
(c) Flame having to travel longer distance in larger cylinder
(d) Sparks are advanced more in larger cylinder

[ESE : 2015]

3.52 Consider the following statements:

1. Fuels of higher octane number can be employed at higher compression ratio
2. In CI engines, brake specific fuel consumption decreases with increasing load

Which of the above statements is/are correct?

(a) Both 1 and 2 (b) 1 only
(c) 2 only (d) Neither 1 nor 2

[ESE : 2015]

3.53 Which of the following factors can control detonation in spark ignition engines?

1. Increasing engine rpm
2. Advancing spark timing
3. Making fuel-air ratio very rich

(a) 1, 2 and 3 (b) 1 and 2 only
(c) 2 and 3 only (d) 1 and 3 only

[ESE : 2015]

3.54 Consider the following statements with regard to IC engines:

1. For best fuel economy of spark ignition engines, the fuel-air mixture should be lean
2. With supercharging, the specific consumption in compression ignition engines increases
3. With increase of load, knocking tendency in compression ignition engines decreases

Which of the above statements are correct?
(a) 1 and 2 only (b) 2 and 3 only
(c) 1 and 3 only (d) 1, 2 and 3

[ESE : 2015]

Direction: The following items consists of two statements, one labelled as **Assertion 'A'** and the other labelled as **Reason 'R'**. You are to examine these two statements carefully and select your answers to these items using the codes given below:

Codes:
(a) Both **A** and **R** are true and **R** is the correct explanation of **A**
(b) Both **A** and **R** are true but **R** is not a correct explanation of **A**
(c) **A** is true but **R** is false
(d) **A** is false but **R** is true

3.55 Assertion (A): In CI engine, increase of load decreases the knocking tendency.
Reason (R): Increase of load increases the temperature of mixture and thereby increase in delay angle. [ESE : 2015]

3.56 Assertion (A): Liquid-cooled engines are able to vary the size of their passage ways through the engine block, so that coolant flow may be tailored to the needs of each area. Locations with either high peak temperatures (narrow islands around the combustion chamber) or high heat flow (around exhaust ports) may require generous cooling. This reduces the occurrence of hot spots.
Reason (R): Air-cooled engines may also vary their cooling capacity by using more closely spaced cooling fins in that area, but this can make their manufacture difficult and expensive.

[ESE : 2015]

Directions: The following items consists of two statements; one labelled as **'Statement (I)'** and the other as **'Statement (II)'**. You are to examine these two statements carefully and select the answers to these items using the codes given below:

Codes:
(a) Both Statement (I) and Statement (II) are individually true and Statement (II) is the correct explanation of Statement (I).
(b) Both Statement (I) and Statement (II) are individually true but Statement (II) is **NOT** the correct explanation of Statement (I).
(c) Statement (I) is true but Statement (II) is false
(d) Statement (I) is false but Statement (II) is true.

3.57 Statement (I) : Knocking in Petrol engine is the auto-ignition of the rich mixture entering the combustion chamber.
Statement (II) : Knocking is due to high compression ratio.

[ESE : 2016]

3.58 Statement (I) : In Automotive Petrol engines during idling operation a rich mixture is required (*F/A* : 0.08)
Statement (II) : Rich mixture is required because mixture is diluted by products of combustion.

[ESE : 2016]

3.59 Statement (I) : Piston temperature profiles near full load are flattened in case of liquid cooled engines whereas for air cooled engines temperature profiles are steeper.
Statement (II) : The piston temperature profiles are different in nature for liquid cooled and air cooled engines because of the different values of heat capacities. [ESE : 2016]

3.60 Knocking in a spark ignition engine can be reduced by
1. retarding the spark 2. supercharging
3. increasing the engine speed
4. using a fuel of long straight chain structure

Select the correct answer using the code given below:
(a) 1 and 4 (b) 1 and 3
(c) 2 and 3 (d) 2 and 4

[ESE : 2017]

3.61 Consider the following statements :
Combustion chamber is
1. the volume between TDC and BDC during the combustion process
2. the space enclosed between the upper part of the cylinder and the top of the piston during the combustion process
3. the space enclosed between TDC and the top of the piston during the combustion process

Which of the above statements is /are correct?
(a) 1 only (b) 2 only
(c) 3 only (d) 1, 2 and 3

[ESE : 2019]

3.62 Which one of the following valves is provided for starting the engine manually, during cold weather conditions?
(a) Starting jet valve
(b) Compensating jet valve
(c) Choke valve
(d) Auxiliary air valve

[ESE : 2020]

3.63 In order to burn a fuel completely, which of the following basic conditions must be fulfilled?
1. Supply enough air for complete combustion of fuel.
2. Secure low turbulence for thorough mixing of fuel and air.
3. Maintain a furnace temperature high enough to ignite the incoming fuel air mixture.
4. Provide a furnace volume large enough to allow time for combustion to be completed.

Select the correct answer using the code given below:
(a) 1, 2 and 3 only (b) 1, 2 and 4 only
(c) 1, 3 and 4 only (d) 2, 3 and 4 only

[ESE : 2021]

Answers Combustion and Knocking in SI and CI Engine

3.1 (a)	3.2 (a)	3.3 (c)	3.4 (b)	3.5 (c)	3.6 (c)	3.7 (c)	3.8 (c)	3.9 (d)
3.10 (c)	3.11 (b)	3.12 (c)	3.13 (a)	3.14 (c)	3.15 (c)	3.16 (b)	3.17 (c)	3.18 (c)
3.19 (d)	3.20 (b)	3.21 (d)	3.22 (d)	3.23 (d)	3.24 (c)	3.25 (d)	3.26 (c)	3.27 (c)
3.28 (b)	3.29 (c)	3.30 (c)	3.31 (a)	3.32 (d)	3.33 (b)	3.34 (b)	3.35 (a)	3.36 (a)
3.37 (a)	3.38 (a)	3.39 (c)	3.40 (b)	3.41 (b)	3.42 (b)	3.43 (a)	3.44 (c)	3.45 (d)
3.46 (c)	3.47 (b)	3.48 (a)	3.49 (a)	3.50 (a)	3.51 (c)	3.52 (a)	3.53 (d)	3.54 (c)
3.55 (c)	3.56 (c)	3.57 (b)	3.58 (a)	3.59 (a)	3.60 (b)	3.61 (b)	3.62 (c)	3.63 (c)

Explanations Combustion and Knocking in SI and CI Engine

3.1 (a)

Under idling condition the pressure at intake manifold is less due to restriction to air-flow. When intake valve opens, the pressure differential between combustion chamber and intake manifold results in initial backward flow of exhaust gases as a result of this, the mixture of fuel-air ratio in combustion chamber is diluted. This results in poor combustion and hence rich mixture is required. The general air-fuel ratio is 9-10.

3.2 (a)

CI engine are most suitable for supercharging as it would increase the air density as well as temperature of intake air which reduces the delay period and permits more fuel to be injected.

3.3 (c)

In practical purpose $(C.V.)_{petrol} > (C.V.)_{diesel}$

3.4 (b)

Dwell angle – period measured in degree of cam rotation during which the contact point remains closed.

The dwell angle must be sufficiently large to allow magnetic saturation of primary coil.

Too **small dwell angle** will result in lower secondary voltage and hence poor spark or misfiring

Too **large dwell angle** will lead to burning of condenser and the contact points due to over saturation of winding

3.6 (c)

When the mixture is nearly 10% richer than stoichiometric, ignition lag is minimum.

3.7 (c)

Retarding spark will reduce the tendency of knocking because burning gases are not compressed. So less rise in temperature of end charge.

3.8 (c)

Injection of fuel just before TDC will have high temperature surrounding so less delay period. Other factors enlisted increase the delay period of fuel.

3.9 (d)

Power developed by two stroke engine will always be higher.

3.10 (c)

Increase in compression ratio results in increase in temperature and due to increase in temperature, knocking tendency increases. Decrease is engine speed results in decrease in turbulence and hence knocking tendency increases.

3.12 (c)

The spark plug should be located close to exhaust valve, so that it is not in the end region otherwise detonation will occur.

3.13 (a)

Idling: Under idling condition the pressure in intake manifold is less due to restriction to air-flow. When intake valve opens the pressure differential between combustion chamber and intake manifold results in initial backward flow of exhaust gases as a result of this, the mixture of fuel-air ratio in combustion chamber is diluted. This results in poor combustion and hence rich mixture is required. The general air-fuel ratio is 9-10.

Maximum power range: Engine requires richer mixture during peak power operation. To provide better power and to prevent over heating of exhaust valve the general air-fuel ratio are 12-13.5 ($\phi > 1$).

3.15 (c)

Decrease of compression ratio will reduce temperature and increase delay period. This increases the knocking tendency in CI engine.

3.16 (b)

Spark ignition caused by spark plug electrode can only be pre-ignition and not post-ignition.

3.17 (c)

Supercharging results in better scavenging and reduced exhaust gas temperature in engine. It improves volumetric efficiency. Supercharging is not suitable for SI engine because it leads to knocking.

3.19 (d)

Following factors reduce the knocking tendency in SI engine.

1. The temperature of end charge should be low.
2. The ignition delay of the end charge should be long in order to avoid pre-ignition of end charge.
3. The density of end charge should be low.

3.20 (b)

Two reference fuels used for octane number rating of fuel are iso-octane and normal heptane.

3.21 (d)

Pre-ignition increases the tendency of knocking in SI engine as burning charge will be compressed and lead to more increase in temperature.

3.22 (d)

Due to supercharging, greater mass flow-rate is provided. Thus more oxygen to support combustion is available in combustion chamber. This allows more fuel to be burned and more work to be done per cycle, increasing the power output of the engine.

3.23 (d)

Delay period is affected by the following factors.

(i) Fuel
(ii) Injection pressure or size of droplet
(iii) Injection advance angle
(iv) Compression ratio.
(v) Intake temperature
(vi) Jacket water temperature
(vii) Fuel temperature
(viii) Intake pressure or super charging
(ix) Speed
(x) Air-fuel ratio
(xi) Engine size
(xii) Type of combustion chamber

3.24 (c)

(a) Idling : 10
(b) Cruising : 16
(c) Maximum power : 13
(d) Cold starting : 3

3.25 (d)

Objective of supercharging:

(i) To increase the power output for a given weight and bulk of the engine. This is important for aircraft, marine and automative engine where weight and space are important.

(ii) To compensate for the loss of power due to altitude. This mainly relates to aircraft engines which lose power at an approximate rate of one percent 100 metres altitudes. This is also relevant for other engines which are used at high altitudes.

(iii) To obtain more power from an existing engine.

3.26 (c)

Factors tending to reduce knocking in CI engine.

S. No.	Factors	
1.	Self-ignition temperature of fuel	Low
2.	Time lag or delay period for fuel	High
3.	Compression ratio	High
4.	Inlet temperature and pressure	High
5.	Combustion chamber wall temp.	High
6.	Speed	Low
7.	Cylinder size	Large

3.27 (c)

The exhaust valve should be located close to the spark plug so that it is not in the end gas region, otherwise there will be a tendency to knock.

3.28 (b)

Necessary spark advance time

$= 1.5 - 1 = 0.5$ ms

New Engine speed = 1000 × 2 = 2000 rpm

∴ Necessary spark advance angle

$$= \frac{360° \times 2000}{60} \times 0.5 \times 10^{-3}$$

$= 6°$

3.29 (c)

$$C + O_2 \rightarrow CO_2 \qquad CO + \frac{1}{2}O_2 \rightarrow CO_2$$

$$H_2 + \frac{1}{2}O_2 \rightarrow H_2O \qquad S + O_2 \rightarrow SO_2$$

Amount of oxygen required for carbon $= \frac{32}{12} = \frac{8}{3}$

Amount of oxygen required for

$CO = \frac{16}{28} = \frac{4}{7}$

Amount of oxygen required for $H_2 = \frac{16}{2} = 8$

Amount of oxygen required for $S = \frac{32}{32} = 1$

3.30 (c)

$2C + O_2 \rightarrow 2CO$ (Incomplete combustion)
(Free carbon)

$$CO + \frac{1}{2}O_2 \rightarrow CO_2$$

Presence of nitrogen does not necessarily tell about complete combustion because NO_x formation takes place at very high temperature.

3.31 (a)

Supercharging is a process of increasing power output by supplying charge at higher pressure and result in higher temperature.

3.33 (b)

Factors reduce the delay period in Diesel engine.

1. Higher cetane number
2. Higher compression ratio
3. Low injection advance
4. Higher jacket water temperature

3.34 (b)

Diesel engine knock can be reduced by

1. Increasing compressing ratio
2. Decreasing injection advance
3. Increasing degree of supercharge
4. By decreasing engine speed

3.36 (a)

In SI engine, higher compression ratio induces knocking because it increases the temperature of end mixture and due to this the end mixture will ignite on its own.

3.37 (a)

Supercharging is the process of supplying air or fuel air mixture higher than the pressure at which engine naturally aspirates. By super charging, the possibility of knocking in diesel engine decreases.

3.39 (c)

Supercharging of a petrol engine increases the temperature of the end charge and hence chances of detonation increase.

3.40 (b)

Cetane number of alpha methyl napthalene is assigned the value of 'zero'.

3.43 (a)

In CI engine, to reduce the knocking tendency delay period should be shorter.

Increase in variable	Effect on delay period
Compression ratio	Reduces
Speed	Increases in terms of crank angle. Reduces in terms of millisecond.
Injection time advance	Reduces
Supercharging	Reduces

The correct option is (a)

3.44 (c)

Pre-ignition is generally caused by an overheated spot which may occur at the spark plug, combustion chamber deposits or exhaust valves. Consequences of preignition are:

1. Increases the time loss and hence reduces net work done.
2. Vicious cycle of preignition and detonation is formed leading to power loss.
3. Decrease in fuel efficiency.

3.45 (d)

1. Ignition delay should be shorter to avoid knocking.
2. Knocking in SI engine starts at the end of combustion while knocking in CI engine starts at the beginning of combustion.
3. $(P_{max})_{CI} = 1.3 \text{ to } 1.4 \times (P_{max})_{SI}$

3.46 (c)

CI engine fuel require shorter delay period to reduce knocking tendency. It should have lower SIT.

SI engine fuel require high SIT and long delay period.

3.47 (b)

Designer tries to keep compression ratio for SI engine as high as possible but it is limited by knocking and vice versa for CI engine.

3.51 (c)

The tendency of detonation is high in petrol engines of larger cylinder diameter because the time taken by the flame front to reach the end charge is more.

Option (c) is correct.

3.53 (d)

Detonation in spark ignition engines can be reduced by:

(i) Increasing engine rpm

(ii) Advancing spark timing

(iii) Making fuel-air ratio very rich.

3.58 (a)

(I) For idling conditions slow speeds more energy is needed to run the engine. Hence, rich mixture. Thus statement II is also correct.

3.59 (a)

(I) Higher the load, higher the temperature. Thus slope is less.

(II) More heat transfer takes places when cooling takes place by liquid. More change in temperature.

3.60 (b)

Statement 1 → correct

Statement 2 → wrong

Statement 3 → correct

Statement 4 → wrong

3.63 (c)

The objective of good combustion is to release all of the heat in the fuel. This is accomplished by controlling the "three T's" of combustion which are:

1. Temperature high enough to ignite and maintain ignition of the fuel.
2. Turbulence or intimate mixing of the fuel and oxygen, and
3. Time sufficient for complete combustion.

■■■■

4 Fuels and Emission, Flue Gas Analysis

4.1 Consider the following statements:

1. Octane rating of gasoline is based on iso-octane and n-heptane fuels which are paraffins.
2. Tetraethyl lead is added to gasoline to increases octane number.
3. Ethylene dibromide is added as scavenging agent to remove lead deposits on spark plugs.
4. Surface ignition need not necessarily cause knocking.

Which of these statements are correct?

(a) 1, 2, 3 and 4 (b) 2, 3 and 4
(c) 1 and 4 (d) 1, 2 and 3

[ESE : 2000]

4.2 Consider the following statements :

1. Recycling exhaust gases with intake increases emission of oxides of nitrogen from the engine.
2. When the carburettor throttle is suddenly opened , the fuel air mixture leans out temporarily causing engine stall.
3. The effect of increase in altitude on carburettor is to enrich the entire part-throttle operation.
4. Use of multiple venturi system makes it possible to obtain a high velocity air stream when the fuel is introduced at the main venturi throat.

Which of these statements are correct?

(a) 1 and 3 (b) 1 and 2
(c) 2 and 3 (d) 2 and 4

[ESE : 2000]

4.3 **Assertion (A):** Catalytic converters for reduction of oxides of nitrogen in engine exhaust cannot be used with leaded fuels.

Reason (R): Catalyst will be removed due to chemical corrosion by lead salts.

(a) Both **A** and **R** are individually true and **R** is the correct explanation of **A**
(b) Both **A** and **R** are individually true but **R** is not the correct explanation of **A**
(c) **A** is true but **R** is false
(d) **A** is false but **R** is true

[ESE : 2000]

4.4 Consider the following statements:

1. Motor gasoline is a mixture of various hydrocarbons with a major proportional being aromatic hydrocarbons.
2. Compressed natural gas is a mainly composed of methane.
3. Producer gas has a Predominant component of hydrogen with lesser proportion of carbon monoxide.
4. Cetane number of fuel used in diesel engines in India is in the range of 80 to 90.

Which of these statements are correct?

(a) 1 and 2 (b) 1 and 3
(c) 2, 3 and 4 (d) 1, 2, 3 and 4

[ESE : 2001]

4.5 With respect to IC Engine emissions, consider the following statements:

1. Evaporative emissions have no carbon monoxide and oxides of nitrogen.
2. Blow-by emissions are essentially carbon monoxide and suspended particulate matter.
3. Exhaust emissions contain 100% of carbon monoxide, 100% of oxide of nitrogen and around 50-55% of hydrocarbons emitted by the engine.
4. There are no suspended particulates in the exhaust.

Which of these statements are correct?

(a) 1 and 4 (b) 1 and 3
(c) 2 and 3 (d) 1, 2, 3 and 4

[ESE : 2001]

4.6 Consider the following statements regarding knock rating of SI engine fuels:

1. Iso-octane is assigned a rating of zero octane number.

2. Normal heptane is assigned a rating of hundred octane number.
3. Iso-octane is assigned a rating of hundred octane number.
4. Normal heptane is assigned a rating of zero octane number.

Which of these statements are correct?

(a) 1 and 2 (b) 2 and 3
(c) 3 and 4 (d) 4 and 1

[ESE : 2002]

4.7 Exhaust emission Vs Air fuel ratio curves for a petrol engine are shown in the given figure. The curve *C* represents

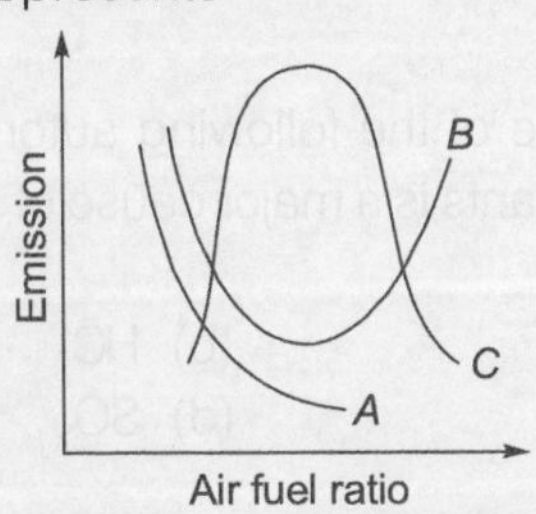

(a) Hydro carbon (b) Carbon dioxide
(c) Carbon monoxide (d) Oxides of nitrogen

[ESE : 2003]

4.8 **Assertion (A):** Octane number is used for rating of fuels in spark ignition engine.

Reason (R): Octane number of a fuel is defined as percentage by volume of iso-octane in a mixture of iso-octane and α-methyl naphthalene.

(a) Both **A** and **R** are individually true and **R** is the correct explanation of **A**
(b) Both **A** and **R** are individually true but **R** is not the correct explanation of **A**
(c) **A** is true but **R** is false
(d) **A** is false but **R** is true

[ESE : 2003]

4.9 Bomb calorimeter is used to determine the calorific value of

(a) Solid fuel only
(b) Gaseous fuels only
(c) Solid as well as gaseous fuels
(d) Solids as well as liquid fuels

[ESE : 2003]

4.10 Orsat apparatus is used to determine products of

(a) all constituents of fuel combustion by mass
(b) all constituents of fuel combustion by volume
(c) only dry constituents of combustion by mass
(d) only dry constituents of combustion by volume

[ESE : 2003]

4.11 Match **List-I** (Measuring Appliances) with **List-II** (Properties/Composition of Fuel) and select the correct answer using the codes given below the lists:

List-I

A. Hydrometer
B. Bomb calorimeter
C. Reid bomb
D. Orsat apparatus

List-II

1. Vapour pressure
2. Composition of products of combustion
3. Specific gravity
4. Heating value

Codes:

	A	B	C	D
(a)	2	1	3	4
(b)	3	4	1	2
(c)	2	4	3	1
(d)	3	1	2	4

[ESE : 2003]

4.12 Consider the following statements for NO_x emission form IC engines:

1. Formation of NO_x depends upon combustion temperature.
2. Formation of NO_x depends upon type of coolant used.
3. Exhaust gas recirculation is an effective means for control of NO_x.
4. Activated Platinum is used for reduction of NO_x.

Which of these statements are correct?

(a) 1 and 2 (b) 1, 2 and 3
(c) 2 and 4 (d) 1 and 3 [ESE : 2004]

4.13 Consider the following statements:

Exhaust emission of carbon monoxide from spark ignition engine is

1. Mainly fuel-air mixture strength dependent.
2. In the range of 0 to 0.05%.
3. Measured with the help of an instrument working on the principle of non-dispersive infrared analysis.
4. Controlled by the use of a two way catalytic convertor.

Which of these statements are correct?

(a) 1 and 4 (b) 2 and 3
(c) 1 and 3 (d) 1, 2, 3 and 4

[ESE : 2004]

4.14 An engine using octane-air mixture has N_2, O_2, CO_2, CO and H_2O as constituents in the exhaust gas. Which one of the following can be conceded?

(a) Supply mixture is stoichiometric
(b) Supply mixture has incomplete combustion
(c) Supply mixture is rich
(d) Supply mixture is lean [ESE : 2004]

4.15 Which one of the following orders is the correct order of passing the flue gases through the different absorbent (in the flasks) during analysis in Orsat apparatus?

(a) Potassium hydroxide solution, alkaline solution of pyrogallic and cuprous chloride solution
(b) Potassium hydroxide solution, cuprous chloride solution and alkaline solution of pyrogallic acid
(c) Alkaline solution of pyrogallic acid, cuprous chloride solution and potassium hydroxide solution
(d) Cuprous chloride solution, potassium hydroxide solution and alkaline solution of pyrogallic acid [ESE : 2004]

4.16 Which one of the following cannot be controlled by a three-way catalytic converter?

(a) HC emission (b) CO emission
(c) NO_x emission (d) SPM emission

[ESE : 2005]

4.17 The discharge of hydrocarbons from petrol automobile exhaust is minimum when the vehicle is

(a) Idling (b) Cruising
(c) Acceleration (d) Decelerating

[ESE : 2005]

4.18 Which one of the following fuels can be obtained by fermentation of vegetable matter?

(a) Benzene (b) Diesel
(c) Gasoline (d) Alcohol

[ESE : 2005]

4.19 Consider the following statements:

1. The gases measured directly by Orsat apparatus from a flue gas sample are CO_2, O_2 and N_2
2. Bomb calorimeter measures higher calorific value of fuel at constant pressure
3. For burning 1 kg of fuel (carbon) to carbon monoxide, the stoichiometric quantity of air required is 8/3 kg.

Which of these statements is/are correct?

(a) Only 1 (b) Only 2
(c) Only 3 (d) 1, 2 and 3

[ESE : 2006]

4.20 The Cetane number of automotive diesel fuel used in India is in which one of the following ranges?

(a) 30 - 40 (b) 41 - 50
(c) 51 - 60 (d) 61 - 70

[ESE : 2007]

4.21 Which one of the following automobile exhaust gas pollutants is a major cause of photochemical smog?

(a) CO (b) HC
(c) NO_X (d) SO_X

[ESE : 2007]

4.22 Which one of the following is represented by the molecular structure of the paraffin family of hydrocarbon fuel given below, for an IC Engine?

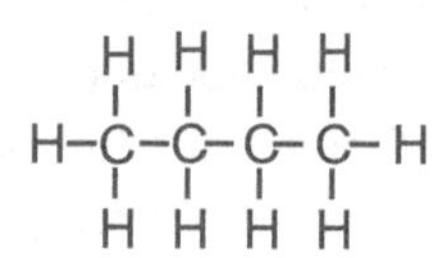

(a) Ethane (b) Propane
(c) Butane (d) Hexane

[ESE : 2007]

4.23 The Orsat apparatus, which gives volumetric percentage of four constituents of dry flue gas, is arranged for absorption of three gases and the fourth content being obtained by difference. Match **List-I** (Gases) with **List-II** (Solution for absorption/by difference) and select the correct answer using the code given below the lists:

List-I	**List-II**
A. Carbon dioxide	**1.** By difference
B. Carbon monoxide	**2.** Caustic soda
C. Oxygen	**3.** Pyrogallic acid
D. Nitrogen	**4.** Cuprous chloride

Codes:

	A	B	C	D
(a)	2	4	3	1
(b)	4	3	2	1

(c) 2 3 4 1
(d) 4 1 2 3 [ESE : 2007]

4.24 The three way catalytic converter cannot control which one of the following?
(a) HC emission (b) CO emission
(c) NO_x emission (d) PM emission
[ESE : 2008]

4.25 Oxides of nitrogen in the engine exhaust can be reduced by which of the following methods?
1. Decrease in compression ratio.
2. Exhaust gas recirculation.
3. Use of 5% lean mixture.
4. Use of oxidation catalysts in exhaust manifold.

Select the correct answer using the code given below:
(a) 1, and 2 only (b) 1 and 3 only
(c) 1, 2 and 4 (d) 2 and 4 only
[ESE : 2008]

4.26 Which one of the following set of materials is most commonly used in catalytic converters for CI engines?
(a) Platinum, palladium and rhodium
(b) Palladium, rhodium and ruthenium
(c) Rhodium, ruthenium and platinum
(d) Ruthenium, platinum and palladium
[ESE : 2008]

4.27 Consider the following:
1. EGR
2. Increased compression ratio.
3. Thermal reactor in the exhaust manifold.
4. Use of 5% lean mixture.

Which of these are used to control NO_x emission from engine exhaust?
(a) 1 only (b) 1, 2 and 4 only
(c) 2, 3 and 4 only (d) 1, 2, 3 and 4
[ESE : 2010]

4.28 Consider the following:
1. Catalytic converter.
2. Use of high compression ratio.
3. Oxidation catalyst in the exhaust manifold.
4. Use of high degree of supercharger.

By which of these techniques, can the oxides of nitrogen in the engine exhaust be controlled?
(a) 1 only (b) 2 and 4 only
(c) 2, 3 and 4 only (d) 1, 2 and 3 only
[ESE : 2010]

4.29 The higher calorific value of fuel occurs when water vapour
(a) Present in products of combustion is condensed to the initial temperature of the fuel
(b) Present in products of combustion is not condensed to the initial temperature of the fuel
(c) It not formed during combustion of a fuel
(d) Is formed during combustion of a fuel
[ESE : 2010]

4.30 When a hydrocarbon fuel burns is excess air Orsat analysis of products of combustion will show
(a) CO, CO_2 and N_2
(b) CO_2, CO, N_2 and H_2O
(c) CO_2, N_2, O_2 and H_2O
(d) CO_2, N_2 and O_2 [ESE : 2011]

4.31 What will be the products of combustion shown by an Orsat analyzer when natural gas burns in deficit air?
(a) CO_2, O_2 and N_2
(b) CO_2, CO and N_2
(c) CO_2, CO, N_2 and H_2O
(d) CO_2, CO, N_2 and O_2
[ESE : 2011]

4.32. Match **List-I** with **List-II** and select the correct answer using the code given below the lists:

List-I
A. The rich mixture which provides maximum power gives large amounts of
B. The condition like wall quenching and wall deposits results in
C. Addition of tetraethyl lead is being eliminated because of
D. The normal compression ratio is dropped from 10.5 : 1 to 8 : 1 to reduce

List-II
1. Hydrocarbon emission
2. NO_x in exhaust gases
3. Co and hydrocarbons in exhaust

4. Adverse effects on exhaust emission

Codes:

	A	B	C	D
(a)	2	4	1	3
(b)	3	4	1	2
(c)	2	1	4	3
(d)	3	1	4	2

[ESE : 2011]

4.33 In the SI engine, highest UBHC concentration is observed during

(a) Maximum load (b) Acceleration

(c) Deceleration (d) Idling

[ESE : 2011]

4.34 When a hydrocarbon fuel burns in insufficient air, the Orsat apparatus for flue gas analysis will show which of the following sets of gases?

(a) CO_2, N_2, O_2 and H_2O

(b) CO, CO_2 and N_2

(c) CO, CO_2 N_2 and H_2O

(d) CO, CO_2, N_2 and O_2

[ESE : 2012]

4.35 When water emulsion is used as the fuel in SI engine the emission of nitrogen oxides reduces significantly, this is due to

(a) water reacts with nitrogen oxides

(b) water does not contain nitrogen

(c) the combustion temperature decreased

(d) the combustion temperature increased

[ESE : 2013]

4.36 In spark ignition petrol engines maximum emission of NO_x takes place when the air fuel ratio (A/F) is

(a) very weak mixture and A/F ratio about 19

(b) weak mixture and A/F ratio about 17

(c) chemically correct mixture and A/F ratio about 15

(d) rich mixture and A/F ratio about 13

[ESE : 2013]

4.37 In an experiment to determine the higher calorific value of a fuel with bomb calorimeter, a few drops of water is dropped into the bomb before sealing. This is to

(a) help absorption of carbon dioxide gas

(b) absorb latent heat of steam formed by combustion of hydrogen

(c) saturate the air inside with water vapour initially

(d) avoid overheating of the contents by burnt gases

[ESE : 2013]

4.38 Consider the following statements for a combustion process:

1. The total mass of each chemical element in the reactants is preserved in the products
2. The presence of carbon monoxide in the products of combustion implies incomplete combustion

Which of the above statement is/are correct?

(a) 1 only (b) 2 only

(c) Neither 1 nor 2 (d) Both 1 and 2

[ESE : 2015]

4.39 Oxides of nitrogen in Petrol engine exhaust can be reduced by the following methods:

1. Use of 5% lean mixture
2. Advancing the spark timing
3. Reciculating a fraction of exhaust gas
4. Using an oxidation catalyst in the exhaust manifold

Which of the above statements is/are correct?

(a) 1 and 2 (b) 2 only

(c) 3 and 4 (d) 3 only

[ESE : 2015]

4.40 **Assertion (A):** A good CI engine fuel, like diesel oil, is a bad SI engine fuel and a good SI engine fuel, like petrol, is a bad CI engine fuel.

Reason (R): A good CI engine fuel requires high self-ignition temperature and a good SI engine fuel requires low self-ignition temperature.

(a) Both **A** and **R** are true and **R** is the correct explanation of **A**

(b) Both **A** and **R** are true but **R** is not a correct explanation of **A**

(c) **A** is true but **R** is false

(d) **A** is false but **R** is true

[ESE : 2015]

4.41 The Orsat apparatus gives

1. Volumetric analysis of dry products of combustion
2. Gravimetric analysis of dry products of combustion

Which of the above statements is/are correct ?

(a) 1 only
(b) 2 only
(c) Both 1 and 2
(d) Neither 1 nor 2

[ESE : 2016]

Directions: The following items consists of two statements; one labelled as **'Statement (I)'** and the other as **'Statement (II)'**. You are to examine these two statements carefully and select the answers to these items using the codes given below:

Codes:

(a) Both Statement (I) and Statement (II) are individually true and Statement (II) is the correct explanation of Statement (I).
(b) Both Statement (I) and Statement (II) are individually true but Statement (II) is **NOT** the correct explanation of Statement (I).
(c) Statement (I) is true but Statement (II) is false
(d) Statement (I) is false but Statement (II) is true.

4.42 **Statement (I) :** Automotive Petrol engines require Petrol of Octane number between 85-95.

Statement (II) : Automotive Diesel engines require Diesel oil of Cetane number between 85-95. [ESE : 2016]

4.43 **Statement (I) :** Coolant and antifreeze refer to the same product.

Statement (II) : Gas engines do not require cooling. [ESE : 2016]

4.44 Which of the following actions will help to reduce the black smoke emission of a diesel engine?

1. Run at lower load, i.e., derating
2. Have regular maintenance of the diesel engine, particularly of injection system
3. Use diesel oil of higher cetane number

Select the correct answer using the code given below.

(a) 1 and 2 only (b) 1 and 3 only
(c) 2 and 3 only (d) 1, 2 and 3

[ESE : 2018]

4.45 In Orsat apparatus, when the percentage of carbon dioxide, oxygen and carbon monoxide are known, the remaining gas is assumed to be

(a) Hydrogen (b) Sulphur dioxide
(c) Nitrogen (d) Air

[ESE : 2020]

4.46 Which one of the following statements is NOT correct?

(a) A simple and convenient apparatus used for the volumetric analysis of dry flue gases is known as orsat apparatus.
(b) The calorific values of liquid fuels are generally determined by using an orsat apparatus.
(c) The calorific values of solid and liquid fuels are generally determined by using a bomb calorimeter.
(d) In orsat apparatus, potassium hydroxide used as absorbent to absorb carbon dioxide.

[ESE : 2022]

4.47 What is the amount of air required to burn one kg of fuel and product of combustion for a fuel the percentage composition of which is given as C = 70%, H_2 = 30%?

(a) 16.54 kg (b) 17.54 kg
(c) 18.54 kg (d) 19.54 kg

[ESE : 2022]

■■■■

Answers Fuels and Emission, Flue Gas Analysis

4.1 (a)	4.2 (c)	4.3 (a)	4.4 (a)	4.5 (b)	4.6 (c)	4.7 (d)	4.8 (c)	4.9 (d)
4.10 (d)	4.11 (b)	4.12 (b)	4.13 (d)	4.14 (b)	4.15 (a)	4.16 (d)	4.17 (b)	4.18 (d)
4.19 (c)	4.20 (b)	4.21 (c)	4.22 (c)	4.23 (a)	4.24 (d)	4.25 (a)	4.26 (a)	4.27 (a)
4.28 (a)	4.29 (a)	4.30 (d)	4.31 (b)	4.32 (d)	4.33 (c)	4.34 (b)	4.35 (c)	4.36 (c)
4.37 (b)	4.38 (d)	4.39 (d)	4.40 (c)	4.41 (a)	4.42 (c)	4.43 (c)	4.44 (a)	4.45 (c)
4.46 (b)	4.47 (c)							

Explanations Fuels and Emission, Flue Gas Analysis

4.1 (a)

Ethylene dibromide added with tetraethyl lead to form volatile lead bromide compound after combustion has taken place and avoid lead deposit so they are also called Scavengers.

The surface ignition may or may not cause knocking.

4.2 (c)

Exhaust gas recirculation reduces the peak temperature so it reduces the NO_x formation.

At higher altitudes the density of air decreases and hence rich mixture is supplied at higher altitudes.

4.3 (a)

Due to leaded fuel it erodes the catalyst rhodium Rh of catalytic converter.

4.4 (a)

Composition of producer gas

CO	—	15-20%
H_2	—	15-20%
CH_4	—	2-6%
CO_2	—	7-10%
N_2	—	40-50%

4.6 (c)

Isooctane fuel has least knocking tendency thus assigned octane number 100 and normal heptane detonates easily thus assigned zero octane number.

4.8 (c)

Octane number of a fuel is defined as percentage by volume of iso-octane in a mixture of iso-octane and n-heptane.

4.12 (b)

Platinum/Rhodium metal matrix in honeycomb like structure used. If the coolant used is not effective then temperature will be high.

4.14 (b)

If O_2 and CO both are present in exhaust then, it shows incomplete combustion in I.C. engine.

4.16 (d)

3-way catalytic convertor

Palladium for CO, Rhodium for NO_x

Platinum for HC

It does not control the emission of particulate matter.

4.18 (d)

Fermentation is used for making alcohol.

4.19 (c)

The gases measured directly by orsat apparatus from a flue gas sample are CO_2, O_2 and CO.

Bomb calorimeter measures higher calorific value of fuel at constant volume.

4.20 (b)

In India, cetane number lies between 40 to 60.

4.21 (c)

NO_X is one of the primary cause of photochemical smog. Smog is formed by the photo chemical reaction of automobile exhaust and atmospheric air in presence of sunlight, NO_2 decomposes into NO and mono atomic oxygen.

NO_2 + energy from sunlight $\rightarrow$ NO + O + smog

$O + O_2 \rightarrow O_3$

4.22 (c)

Butane is a gas with formula C_4H_{10} i.e. alkane with four carbon atom. Butane are highly flammable, colorless easily liquified gas.

4.24 (d)

Inside catalytic converter, a catalyst stimulates a chemical reaction in which toxious by product of combustion carbon monoxide, unburned hydrocarbon and oxides of nitrogen are converted to less toxic or inert substance.

4.25 (a)

For the emission of oxides of nitrogen two factors are responsible which are high temperature and availability of oxygen.

By decreasing the compression ratio, peak temperature decreases, hence emission of NO_x decreases.

By recirculating a portion of the exhaust gas, the peak combustion temperature decreases since the inert gas serves as a heat sink and this also reduces the quantity of oxygen available for combustion.

And by using reduction catalysts in the exhaust manifold the NO_x emission can be reduced.

4.26 (a)

Platinum : for hydro-carbon (HC)

Palladium : for Carbon-mono-oxide.

Rhodium : for oxide of nitrogen (NO_x)

4.28 (a)

Formation of NO_x prevails when there is high temperature and availability of oxygen high compression ratio and use of high degree of supercharge increases the local temperature and hence NO_x formation tendency will increase but by using catalytic converter it can be controlled.

4.29 (a)

The calorific value or heating value of the fuel is the amout of energy liberated by the complete oxidation of a unit mass or volume of a fuel.

If a fuel contains hydrogen, water will be formed as one of the products of combustion. If this water is condensed, a large amount of heat will be released this is known as higher calorific value of fuel and if combustion product is not condensed and present in completely vapour phase it is called lower calorific value of fuel.

4.30 (d)

When a hydrocarbon fuel burns, in excess air of it means the content of air is more of in the exhaust for there will be formation CO_2, N_2 and O_2.

4.31 (b)

When a natural gas burns in a deficit air, the formation of carbon monoxide is must in the exhaust.

4.32 (d)

By the addition of tetraethyl lead, there is adverse effect on exhaust emission. It deteriorates the performance of catalytic converter.

In case of rich mixture by which we get maximum power, there are more chances of formation of carbon-monoxide in the exhaust.

4.33 (c)

Highest unburnt hydrocarbon concentration is observed during decleration.

4.34 (b)

When Orsat apparatus is used for flue gas analysis of a hydrocarbon fuel that burns is insufficient air, then there will be no oxygen and also it does not show H_2O. Hence it will show CO, CO_2 and N_2.

4.35 (c)

(i) NO_x formation is high at higher combustion temperature.

(ii) Use of water emulsion decreases the combustion temperature and hence NO_x.

4.36 (c)

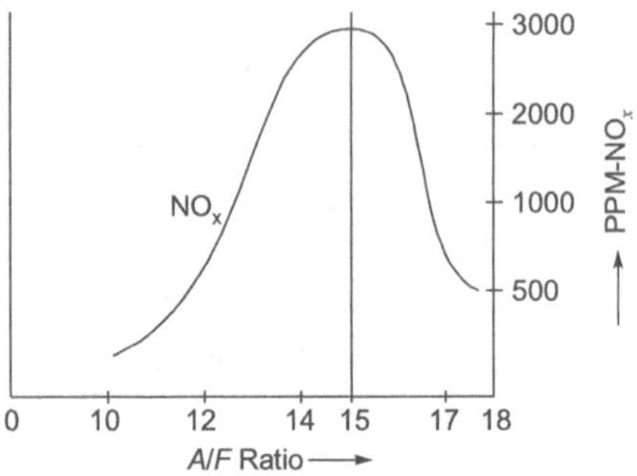

Maximum NO_x is formed at A/F ratio of 15. At lean and rich mixtures, the NO_x concentration is comparatively low.

4.37 (b)

Bomb calorimeter is used to determine the (High Calorific Value) HCV of solid as well as liquid fuels. Water is dropped to absorb latent heat of steam formed by combustion of hydrogen. When the water vapour present in the products of combustion gets condensed, the amount of heat produced per unit mass is known as higher heating value.

4.38 (d)

Mass and energy always remains constant. Carbon monoxide formed due to incomplete combustion otherwise carbondioxide will be formed.

4.40 (c)

In CI engine self ignitiion temperature of fuel should be low to avoid knocking while for SI engine same should be high. So statement II is wrong.

4.42 (c)

(I) Higher octane no; better is the running of petrol engines (loss knocking)

(II) Max cetane no, for diesel engines is around 55.

4.43 (c)

(I) If coolant and anti freeze is added that is why it is consider as same product.

(II) Gas engine is basically heavy power IC engine and is required cooling system for the dessipation of heat generation.

4.47 (c)

For complete combustion, oxygen required

$$\underbrace{C}_{12} + \underbrace{O_2}_{32} \rightarrow \underbrace{CO_2}_{44}$$

For burning 1 kg of carbon

$$= \frac{32}{12} = 2.666 \text{ kg of } O_2 \text{ is required}$$

For 70% carbon, O_2 required

$$= \frac{32}{12} \times 0.7 = 1.866 \text{ kg}$$

Similarly, $\underbrace{H_2}_{2} + \underbrace{\frac{1}{2}O_2}_{16} \rightarrow \underbrace{H_2O}_{18}$

For burning 2 kg of H_2 = 16 kg of O_2 is required

O_2 required for 30% H_2 = $\frac{16}{2} \times 0.3 = 2.4$ kg

Total O_2 required = 2.4 + 1.866 = 4.266

∴ Nitrogen required

$$= \frac{77}{23} \times 4.266 = 14.28 \text{ kg}$$

∴ Air mass required

$$= M_{O_2} + M_{N_2}$$
$$= 14.28 + 4.266 = 18.54 \text{ kg}$$

5 Testing and Performance of IC Engine

5.1 For the same maximum pressure and heat input, the most efficient cycle is

(a) Otto cycle (b) Diesel cycle

(c) Brayton cycle (d) Dual combustion cycle

[ESE : 2000]

5.2 Match **List-I** (Performance Parameter Y) with **List-II** (Curves labelled 1, 2, 3, 4 and 5) regarding a CI engine run at constant speed and select the correct answer using the codes given below the lists:

List-I

A. Total fuel consumption rate

B. Mechanical efficiency

C. Indicated power

D. Brake specific fuel consumption

List-II

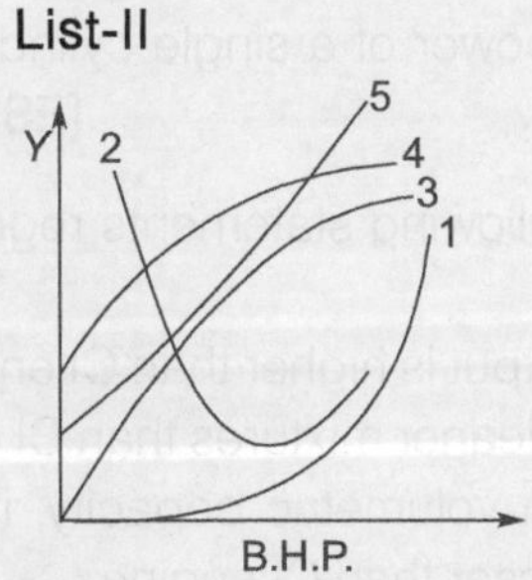

Codes:

	A	B	C	D
(a)	5	3	4	2
(b)	1	3	4	2
(c)	5	4	2	3
(d)	1	4	2	3

[ESE : 2001]

5.3 Match **List-I** with **List-II** and select the correct answer using the codes given below the lists:

List-I

A. Supercharging

B. Morse test

C. Heterogeneous combustion

D. Ignition quality of petrol

List-II

1. Multicylinder engine
2. CI engine
3. Calorific value
4. Aircraft engine
5. Octane number
6. Single cylinder SI engine

Codes:

	A	B	C	D
(a)	4	1	2	5
(b)	6	3	2	5
(c)	6	1	5	2
(d)	4	3	5	2

[ESE : 2001]

5.4 Match **List-I** with **List-II** and select the correct answer using the codes given below the lists:

List-I

A. Air standard efficiency of Otto cycle

B. Morse test

C. Constant volume cycle

D. Constant pressure heat addition

List-II

1. Mechanical efficiency
2. Diesel cycle
3. Brake thermal efficiency
4. Otto cycle
5. $1-\frac{1}{r^{\gamma-1}}$

Codes:

	A	B	C	D
(a)	5	1	4	2
(b)	3	5	2	4
(c)	3	5	4	2
(d)	5	1	2	4

[ESE : 2001]

5.5 The order of values of thermal efficiency of Otto, Diesel and Dual cycle, when they have equal compression ratio and heat rejection, is given by
(a) $\eta_{otto} > \eta_{diesel} > \eta_{dual}$
(b) $\eta_{diesel} > \eta_{dual} > \eta_{otto}$
(c) $\eta_{dual} > \eta_{diesel} > \eta_{otto}$
(d) $\eta_{otto} > \eta_{dual} > \eta_{diesel}$ [ESE : 2002]

5.6 The volumetric efficiency of a well designed SI engine is in the range of
(a) 40% – 50% (b) 50% – 60%
(c) 60% – 70% (d) 70% – 90% [ESE : 2002]

5.7 Variation of specific fuel consumption with fuel-air ratio for spark ignition engine is represented by which of the curves shown below?

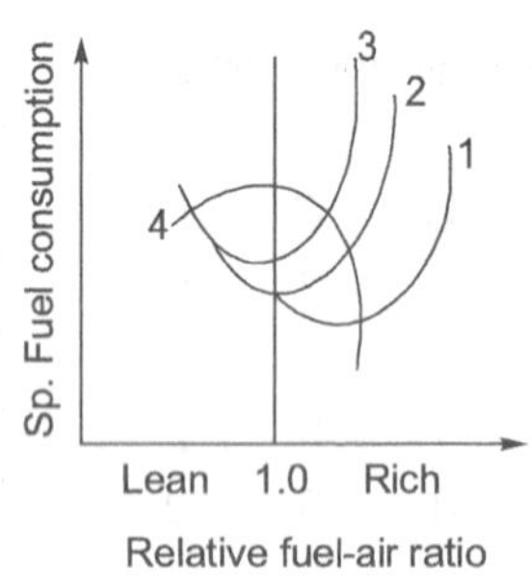

(a) Curve 1 (b) Curve 2
(c) Curve 3 (d) Curve 4 [ESE : 2002]

5.8 If the performance of diesel engines of different sizes, cylinder dimensions and power rating are to be compared, which of the following parameters can be used for such comparison?
(a) Swept volume
(b) Air fuel ratio
(c) Specific brake fuel consumption
(d) Volumetric efficiency [ESE : 2003]

5.9 An engine produces 10 kW brake power while working with a brake thermal efficiency of 30%. If the calorific value of the fuel used is 40,000 kJ/kg, then what is the fuel consumption?
(a) 1.5 kg/hour (b) 3.0 kg/hour
(c) 0.3 kg/hour (d) 1.0 kg/hour [ESE : 2005]

5.10 A 40 kW engine has a mechanical efficiency of 80%. If the frictional power is assumed to be constant with load, what is the approximate value of the mechanical efficiency at 50% of the rated load?
(a) 45% (b) 55%
(c) 65% (d) 75% [ESE : 2005]

5.11 For the same indicated work per cycle, mean speed and permissible fluctuation of speed, what is the size of flywheel required for a multi-cylinder engine in comparison to a single-cylinder engine?
(a) Bigger
(b) Smaller
(c) Same
(d) Depends on thermal efficiency of the engine [ESE : 2006]

5.12 Consider the following statements:
1. Supercharging increases the power output of an engine.
2. Supercharging increases the brake thermal efficiency considerably.
3. Supercharging helps scavenging of cylinders.

Which of these statements are correct?
(a) Only 1 and 2 (b) Only 2 and 3
(c) Only 1 and 3 (d) 1, 2 and 3 [ESE : 2006]

5.13 The main object of Morse test is to find out
(a) Performance of a petrol engine
(b) Performance of a diesel engine
(c) Frictional power of a petrol engine
(d) Indicated power of a single cylinder diesel engine [ESE : 2010]

5.14 Consider the following statements regarding SI engine:
1. Specific output is higher than CI engine.
2. It can burn leaner mixtures than CI engine.
3. For a given volumetric capacity, its power output is lesser than CI engine.
4. Its exhaust pollution is higher relative to CI engine.

Which of these statements is/are correct?
(a) 1, 2, 3 and 4 (b) 2 only
(c) 2, 3 and 4 only (d) 1 and 4 only [ESE : 2010]

5.15 Consider the following statements regarding a CI engine:
1. CI engine knock can be reduced by increasing compression ratio.
2. Thermal efficiency of a CI engine is lower relative to a SI engine.
3. CI engine has a higher specific output relative to SI engine

4. CI engines use leaner mixtures relative to SI engines.

Which of these statements are correct?

(a) 1 and 3 only (b) 2 and 4 only
(c) 1 and 4 only (d) 1, 2, 3 and 4

[ESE : 2010]

5.16 Which phenomena have the most adverse effect on volumetric efficiency when engine works at high speeds?

(a) Flow friction and chocking
(b) Ram effect and choking
(c) Flow friction and charge heating
(d) Charge heating and back flow [ESE : 2010]

5.17 Which of the following remarks pertaining to supercharging of IC engines are correct?

1. Ideal supercharging increases the output, but decreases the thermal efficiency slightly.
2. The indicated mean effective pressure and mechanical efficiency increase with supercharging.
3. The permissible amount of supercharging depends on the compression ratio, air-fuel mixture and design of engine.
4. Engine friction losses increases with supercharging.

(a) 1, 2, 3 and 4 (b) 1, 2 and 3 only
(c) 1, 2 and 4 only (d) 2, 3 and 4 only

[ESE : 2013]

5.18 The purpose of a thermostat in an engine cooling system is to

(a) prevent the coolant from boiling
(b) allows the engine to warm up quickly
(c) pressurize the system to raise the boiling point
(d) indicate to the driver, the coolant temperature

[ESE : 2013]

5.19 The mechanical efficiency of a single cylinder four stroke engine is 80%. If the frictional power is estimated to be 25 kW, the indicated power will be

(a) 100 kW (b) 125 kW
(c) 150 kW (d) 175 kW [ESE : 2014]

5.20 A single cylinder four stroke engine operating at 80% of mechanical efficiency develops a brake power of 60 kW. The indicated power and the power lost due to friction respectively are:

(a) 40 kW and 20 kW (b) 75 kW and 20 kW
(c) 40 kW and 20 kW (d) 75 kW and 15 kW

[ESE : 2014]

5.21 The following reasons are mentioned while recommending supercharging for the engines used in aeroplanes and submarines:

1. More volumetric efficiency, better combustion and increased power output.
2. Higher peak pressure, increased temperature and smaller size.

Which of the above reasons is/are correct?

(a) 1 only (b) 2 only
(c) Both 1 and 2 (d) Neither 1 nor 2

[ESE : 2014]

5.22 A single cylinder, 4-stroke cycle engine is fitted with a rope brake. The diameter of the brake wheel is 600 mm and the rope diameter is 26 mm. The dead weight on the brake is 200 N and the spring balance reads 30 N. If the engine runs at 450 rpm, what will be the brake power of the engine?

(a) 1.5 kW (b) 2.5 kW
(c) 3.5 kW (d) 4.5 kW [ESE : 2015]

5.23 Consider the following statements regarding supercharging of CI engines:

1. Supercharging results in quieter and smoother operation of a CI engine
2. Supercharging of a CI engine requires increase in valve overlap
3. The limit of supercharging for a CI engine is reached by thermal and mechanical loading

Which of the above statements are correct?

(a) 1, 2 and 3 (b) 1 and 2 only
(c) 1 and 3 only (d) 2 and 3 only

[ESE : 2015]

5.24 Consider the following statements pertaining to supercharging of engines:

1. The power output for a given engine increases
2. The loss of power due to altitude is compensated
3. The increase in supercharging pressure decreases the tendency to detonate in spark ignition engines
4. The mechanical efficiency of supercharged engines is quite high compared to naturally aspirated engines

Which of the above statements are correct?
(a) 1 and 4 (b) 3 and 4
(c) 2 and 3 (d) 1 and 2 [ESE : 2015]

5.25 A four-cylinder four-stroke SI engine develops an output of 44 kW. If the pumping work is 5% of the indicated work and mechanical loss is an additional 7%, then the power consumed in pumping work is :
(a) 50 kW (b) 25 kW
(c) 5.0 kW (d) 2.5 kW [ESE : 2016]

5.26 A single cylinder, four-stroke cycle oil engine is fitted with a rope brake. The diameter of the brake wheel is 600 mm and the rope diameter is 26 mm. The dead load on the brake is 200 N and the spring balance reads 30 N. If the engine runs at 600 rpm, what will be the nearest magnitude of the brake power of the engine ?
(a) 3.3 kW (b) 5.2 kW
(c) 7.3 kW (d) 9.2 kW
[ESE : 2016]

5.27 Consider the following statements regarding C.I. engine :
1. C.I. engines are more bulky than S.I. engines
2. C.I. engines are more efficient than S.I. engines
3. Lighter flywheels are required in C.I. engines

Which of the above statements are correct ?
(a) 1 and 3 only (b) 2 and 3 only
(c) 1 and 2 only (d) 1, 2 and 3
[ESE : 2016]

5.28 The objective of supercharging an engine is :
1. To reduce space occupied by the engine
2. To increase the power output of an engine when greater power is required

Which of the above statements are correct ?
(a) 1 only (b) 2 only
(c) Both 1 and 2 (d) Neither 1 nor 2
[ESE : 2016]

5.29 In a Morse test on a 2-cylinder, 2-stroke SI engine, the brake power is 9 kW and the BHP of individual cylinders with spark cutoff are 4.25 kW and 3.75 kW, respectively. The mechanical efficiency of the engine is
(a) 90% (b) 80%
(c) 52.5% (d) 45.5%
[ESE : 2017]

5.30 If the maximum pressure in both air-standard Otto and Diesel cycles is the same, then the relations for compression ratio r and the efficiency η between the two cycles are
(a) $r_{Diesel} > r_{Otto}$ and $\eta_{Diesel} > \eta_{Otto}$
(b) $r_{Otto} > r_{Diesel}$ and $\eta_{Diesel} > \eta_{Otto}$
(c) $r_{Diesel} > r_{Otto}$ and $\eta_{Otto} > \eta_{Diesel}$
(d) $r_{Otto} > r_{Diesel}$ and $\eta_{Otto} > \eta_{Diesel}$
[ESE : 2017]

5.31 If one cylinder of a diesel engine receives more fuel than the others, it is a serious condition for that cylinder and can be checked by
1. judging the seizure of the piston
2. checking incomplete combustion in that cylinder
3. checking cylinder exhaust temperature with a pyrometer

Which of the above is/are correct?
(a) 1 only (b) 2 only
(c) 3 only (d) 1, 2 and 3
[ESE : 2018]

5.32 Consider the following statements with reference to combustion and performance in a four-stroke petrol engine:
1. The auto-ignition temperature of petrol as a fuel is higher than that of diesel oil as a fuel.
2. The highest compression ratio of petrol engines is constrained by the possibility of detonation.
3. A petrol engine is basically less suitable for supercharging than a diesel engine.

Which of the above statements are correct?
(a) 1 and 2 only (b) 1 and 3 only
(c) 2 and 3 only (d) 1, 2 and 3
[ESE : 2018]

5.33 A four-stroke single-cylinder SI engine of 6 cm diameter and 10 cm stroke running at 4000 rpm develops power at a mean effective pressure of 10 bar. The power developed by the engine is

(a) 9.42 kW (b) 5.54 kW
(c) 4.92 kW (d) 2.94 kW

[ESE : 2018]

5.34 A four-stroke engine having a brake power of 105 kW is supplied with fuel at the rate of 4.4 kg/min for 10 minutes. The brake specific fuel consumption of the engine is

(a) 0.18 kg/kW-hr
(b) 0.25 kg/kW-hr
(c) 0.36 kg/kW-hr
(d) 0.42 kg/kW-hr

[ESE : 2018]

5.35 In IC engine

1. the ideal air capacity of a two-stroke engine is the mass of air required to concurrently fill the total cylinder volume at inlet temperature and exhaust pressure.
2. with increase in air-fuel ratio beyond the value for maximum power, there is a fall in power developed and this fall is more with higher values of air-fuel ratio
3. the volumetric efficiency of the engine depends on the design of intake and exhaust manifold

Which of the above are correct?

(a) 1 and 2 only (b) 1 and 3 only
(c) 2 and 3 only (d) 1, 2 and 3

[ESE : 2018]

5.36 The following are the results of a Morse test conducted on a four-cylinder, four-stroke petrol engine at a common constant speed in all cases: The brake power of the engine when all the cylinders are firing is 80 kW. The brake power of the engine when each cylinder is cut-off in turn is 55 kW, 55.5 kW, 54.5 kW and 55 kW, respectively. The mechanical efficiency of the engine when all the cylinders are firing will be

(a) 90% (b) 85%
(c) 80% (d) 75%

[ESE : 2018]

5.37 A 4-stroke diesel engine has length of 20 cm and diameter of 16 cm. The engine is producing power of 25 kW when it is running at 2500 r.p.m. The mean effective pressure of the engine will be nearly

(a) 5.32 bar (b) 4.54 bar
(c) 3.76 bar (d) 2.98 bar

[ESE : 2019]

5.38 A 4-stroke, 6-cylinder gas engine with a stroke volume of 1.75 litres develops 26.25 kW at 506 r.p.m. and the *MEP* is 600 kN/m^2. The number of misfires per minute per cylinder will be

(a) 3 (b) 4
(c) 5 (d) 6

[ESE : 2019]

5.39 A 2-stroke oil engine has bore of 20 cm, stroke 30 cm, speed 350 r.p.m., i.m.e.p. 275 kN/m^2, net brake load 610 N, diameter of brake drum 1 m, oil consumption 4.25 kg/hr, calorific value of fuel 44×10^3 kJ/kg. The indicated thermal efficiency will be

(a) 29.1% (b) 31.3%
(c) 33.5% (d) 35.7%

[ESE : 2019]

5.40 The mechanical efficiency of a single-cylinder four-stroke engine is 60%. The frictional power is estimated to be 30 kW. What is the indicated power?

(a) 120 kW (b) 75 kW
(c) 150 kW (d) 130 kW

[ESE : 2021]

5.41 A four-stroke petrol engine at full load delivers 100 kW. It requires 10 kW to rotate it without load at same speed. What is the mechanical efficiency at half load?

(a) 67.82% (b) 70.24%
(c) 77.32% (d) 83.33%

[ESE : 2021]

5.42 A 42.5 kW engine has a mechanical efficiency of 85%. If the frictional power is assumed to be constant with load, what is the mechanical efficiency at 60% of the load?

(a) 44.2% (b) 40.0%
(c) 66.3% (d) 77.3%

[ESE : 2023]

5.43 A single-cylinder engine running at 1800 rpm develops a torque of 8 N-m. The indicated power of the engine is 1.8 kW. What is the loss due to friction power as the percentage of brake power?

(a) 9.36% (b) 19.36%
(c) 29.36% (d) 39.36%

[ESE : 2023]

5.44 A gasoline engine working on Otto cycle consumes 8 litres of gasoline per hour and develops 25 kW. The specific gravity of gasoline is 0.75 and its calorific value is 44000 kJ/kg. What is the indicated thermal efficiency of the engine?

(a) 14.1% (b) 24.1%
(c) 34.1% (d) 44.1% [ESE : 2023]

Directions for the following **two (02)** items:

Read the following information and answer the **two** items that follows:

A four-cylinder diesel engine of 4-stroke type has stroke to bore-stroke ratio as 1.2 and the cylinder diameter is 12 cm. Based on the indicator card, the area of 30 cm^2 and length as half of stroke is given. The indicator spring constant is given as 20×10^3 kN/m^2 and engine is running at 2000 rpm.

5.45 What is mean effective pressure of the engine?

(a) 2.33×10^5 N/m^2 (b) 4.33×10^5 N/m^2
(c) 6.33×10^5 N/m^2 (d) 8.33×10^5 N/m^2

[ESE : 2023]

5.46 What is the indicated power for one cylinder?

(a) 22.6 kW (b) 38.4 kW
(c) 41.2 kW (d) 54.5 kW

[ESE : 2023]

5.47 A six-cylinder four-stroke diesel engine develops 125 kW at 3000 rpm. Its brake-specific consumption is 200 gm/kWh. What is the quantity of fuel to be injected per cycle per cylinder? (The specific gravity of the fuel is 0.85)

(a) 0.0726 cc/cycle (b) 0.0379 cc/cycle
(c) 0.0981 cc/cycle (d) 0.0545 cc/cycle

[ESE : 2024]

5.48 The Rolls-Royce CV12 turbocharged four-stroke direct injection diesel engine has a displacement of 26.1 litres. The engine has a maximum output of 900 kW at 2300 r.p.m. What is the brake mean effective pressure?

(a) 18 bar (b) 15 bar
(c) 10.15 bar (d) 21 bar [ESE : 2024]

5.49 A high-performance four-stroke SI engine has a compression ratio 10: 1. The ideal air standard Otto cycle efficiency is 0.602. The indicated efficiency is 55 percent of the corresponding ideal air standard Otto cycle efficiency. The engine mechanical efficiency is 85 percent, and the volumetric efficiency is 90 percent. The calorific value of the fuel is 44 MJ/kg. What is the engine arbitrary overall efficiency?

(a) 26.1% (b) 28.1%
(c) 29.5% (d) 30.5% [ESE : 2024]

Answers Testing and Performance of IC Engine

5.1 (b)	5.2 (a)	5.3 (a)	5.4 (a)	5.5 (d)	5.6 (d)	5.7 (c)	5.8 (c)	5.9 (b)
5.10 (c)	5.11 (b)	5.12 (d)	5.13 (c)	5.14 (d)	5.15 (c)	5.16 (c)	5.17 (d)	5.18 (b)
5.19 (b)	5.20 (d)	5.21 (a)	5.22 (b)	5.23 (a)	5.24 (d)	5.25 (d)	5.26 (a)	5.27 (c)
5.28 (c)	5.29 (a)	5.30 (a)	5.31 (c)	5.32 (d)	5.33 (a)	5.34 (b)	5.35 (d)	5.36 (c)
5.37 (d)	5.38 (a)	5.39 (a)	5.40 (b)	5.41 (d)	5.42 (d)	5.43 (b)	5.44 (c)	5.45 (d)
5.46 (a)	5.47 (d)	5.48 (a)	5.49 (b)					

Explanations Testing and Performance of IC Engine

5.3 (a)

Supercharging : Aircraft engine
Morse test : Multicylinder engine
Heterogenous combustion : CI engine
Ignition quality : Octane number

5.4 (a)

Air standard efficiency of Otto $= 1-\frac{1}{r^{\gamma-1}}$

Morse test is used to find *IP* by which we can find mechanical efficiency.

Otto cycle : Constant volume heat addition and heat rejection. Reversible adiabatic compression and expansion.

Diesel Cycle : Constant pressure heat addition.

5.5 (d)

For same compression ratio and heat rejection

$$\eta_{Otto} > \eta_{dual} > \eta_{disel}$$

5.6 (d)

The normal range of volumetric efficiency at full throttle for SI engines is between 80 to 85% where as for CI engines it is between 85 to 90%.

5.9 (b)

$BP = 10$ kW; $\eta_{bth} = 30\%$

$CV = 40000$ kJ/kg

$$\eta_{bth} = \frac{BP}{\dot{m}_f \times CV}$$

$$0.30 = \frac{10}{\dot{m}_f \times 40000}$$

$$\dot{m}_f = \frac{10}{0.3 \times 40000} \times 3600$$

$$= \frac{36000}{12000} = \textbf{3 kg/hr}$$

5.10 (c)

$BP = 40$ kW

$$\eta_{mech} = 0.8 = \frac{BP}{IP}$$

$$0.8 = \frac{40}{IP}$$

$IP = 50$ kW

Frictional power,

$FP = IP - BP$

$= 50 - 40 = 10$ kW

At 50% of the rated load

$BP = 20$ kW

$IP = BP + FP$

$= 20 + 10 = 30$ kW

$$\eta_{mech} = \frac{20}{30} \times 100 = \textbf{66.6\%}$$

5.11 (b)

Turning moment will be more uniform for multicylinder engine as compared to single cylinder, hence smaller size of flywheel is required for multicylinder engine as compared to single cylinder.

5.12 (d)

Effect of supercharging on performance of the engine:

1. The power output of a supercharged engine is higher than its naturally aspirated engine.
2. Mechanical efficiency is slightly improved.
3. Supercharging results in better scavenging and reduced exhaust gas temperature in the engine.

5.14 (d)

C.I. engine burn at leaner mixture than S.I. engine and for a given volumetric capacity, power output of S.I. engine will be more than C.I. engine.

5.17 (d)

(i) Supercharging increases power output and thermal efficiency. It does not increase the fuel consumption.

(ii) The increased pressure and temperature as a result of super-charging may lead to detonation, so fuel must have better antiknock property.

(iii) The permissible extent of supercharging depends upon the ability of the engine to withstand the increased gas loading and thermal stresses.
(iv) Durability, reliability and fuel economy are the main considerations that limit the degree of supercharging.
(v) Engine friction losses increases with supercharging due to inertia of moving parts.

5.18 (b)

Whenever the engine is started from cold, the coolant temperature has to be brought to the desired level in order to minimize the warm-up time. This purpose is achieved by a thermostat fitted in a system which initially prevents the circulation of water below a certain temperature through the radiator so that the water gets heated up quickly. When the preset temperature is reached, thermostat allows the water to flow through the radiator.

5.19 (b)

$$\eta_m = \frac{bp}{ip}$$

$$\Rightarrow \quad 0.8 = \frac{ip-fp}{ip} = \frac{ip-25}{ip}$$

$$\Rightarrow \quad 0.8\,ip - ip = -25$$

$$\Rightarrow \quad ip[1-0.8] = 25$$

$$\Rightarrow \quad ip = \frac{25}{0.2} = \mathbf{125\ kW}$$

5.20 (d)

$$\eta_m = 0.8,$$
$$bp = 60\text{ kW}$$
$$\frac{bp}{ip} = 0.8$$
$$\Rightarrow \quad ip = \frac{bp}{0.8} = \frac{60}{0.8} = 75\text{ kW}$$
$$fp = ip - bp = 75 - 60 = \mathbf{15\ kW}$$

5.21 (a)

Due to supercharging, thermal load on the various parts of the engine increases. Increased gas loading caused by supercharging necessitate the use of larger bearing areas and heavier engine components.

So, statement (2) is not correct.

5.22 (b)

$$\text{Break power: } BP = \frac{\pi DN(W-S)}{60}\text{ watt}$$

where, D = Effective break drum diameter
N = rpm; W = Dead weight
S = Spring balance reading
here, $D = 600 + 26 = 626$ mm
$N = 450$ rpm; $W = 200$ N; $S = 30$ N

$$\therefore \quad bp = \frac{3.14\times0.626\times450\times(200-30)}{60,000}$$
$$= \mathbf{2.5062\ kW}$$

5.23 (a)

In CI engines, any increase in pressure and temperature decreases the ignition delay and hence engine runs smoother.

So, option (a) is correct.

5.25 (d)

As per the given data,
Net power output,

$$W = (\text{Total output} - P_W) - \text{Mech. losses}$$

$$\text{or} \quad W_{net} = W - W\times\frac{5}{100} - W\times\frac{7}{100}$$
$$44 = 0.88\text{ W}$$
$$\text{or} \quad W = 50\text{ kW}$$
$$W_p = \frac{5}{100}W = \frac{5}{100}\times 50 = \mathbf{2.5\ kW}$$

5.26 (a)

For calculating the B. Power mechanically the engine speed must always be taken as 'N' only.

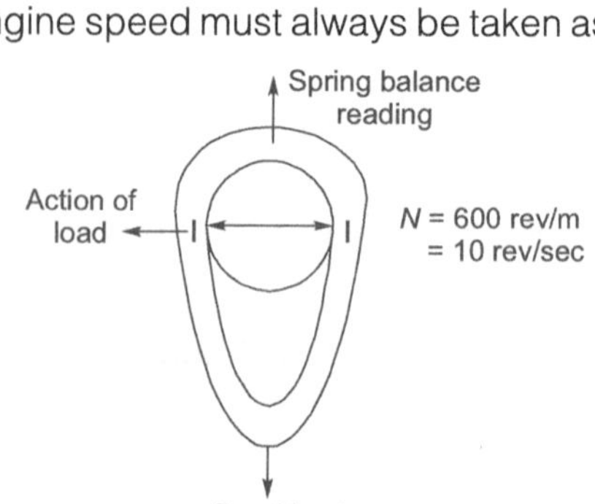

The load will act at the centre of the rope.
∴ Net dia. of rope
= Dia. of wheel + 2 × radius of rope = dia

$$\therefore \quad D_{brake} = (600 + 26) \text{ mm} = 626 \text{ mm} = 0.626 \text{ m}$$

$\therefore$ Brake power is,

$$bp = \frac{\text{Net Force} \times \pi D_{brake} \times N}{1000} \text{ kW} = \frac{(200-30)\times\pi\times 0.626\times 10}{1000} \text{ kW} = \mathbf{3.34 \text{ kW}}$$

5.27 (c)

Theoretical compect of Petrol and Diesel engines. Higher the comp. ratio, higher than η. (Diesel engines). Thus, bigger size and more bulky and higher η.

5.29 (a)

$$\left.\begin{aligned} IP_1 &= 4.75 \text{ kW} \\ IP_2 &= 5.25 \text{ kW} \end{aligned}\right\} I.P. = 10 \text{ kW}$$

$$\eta_{mech} = \frac{BP}{IP} = 90\%$$

5.30 (a)

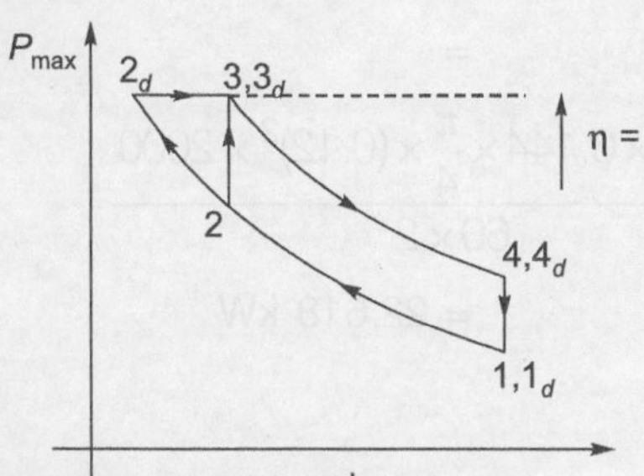

$r_d > r_o$ and $\eta_d > \eta_o$

5.33 (a)

$$\dot{V}_s = \frac{\pi}{4} d^2 \times L \times k \times \frac{N}{60\times 2}$$

$$\text{B.P.} = P_{bm} \times \dot{V}_s = 10 \times \dot{V}_s = 9.424 \text{ kW}$$

5.36 (c)

$$\text{B.P.} = 80 \text{ kW}$$
$$IP_1 = 25 \text{ kW}$$
$$IP_2 = 24.5 \text{ kW}$$
$$IP_3 = 25.5 \text{ kW}$$
$$IP_4 = 25 \text{ kW}$$
$$\text{I.P.} = IP_1 + IP_2 + IP_3 + IP_4 = 100 \text{ kW}$$
$$\eta_m = \frac{\text{B.P.}}{\text{I.P.}} = 80\%$$

5.37 (d)

$L = 20$ cm, $D = 16$ cm, B.P. = 25 kW, $N = 2500$ rpm

$$P_m = \frac{B.P.}{\dot{V}_s} = \frac{B.P.}{\frac{\pi}{4}\times D^2 \times L \times k \times \frac{N}{60\times 2}} = 2.98 \text{ bar}$$

5.38 (a)

B.P. = 26.25, N = 500 rpm,

P_{bm} = 600 kN/m^2, V_s = 1.75 litre

If, N = 506 rpm then B.P. = 26.25 kN

$\therefore$ It is 4-stroke gas engine, so $N_{eff} = N/2$

So, '3' misfiring occurs per minute per cylinder

5.39 (a)

2-stroke, D = 20 cm, L = 30 cm, N = 350 rpm, P_{im} = 275 kN/m^2

$$\text{IP} = P_{im} \times \dot{V}_s = 275 \times \left(\frac{\pi}{4} \times D^2 \times L \times k \times \frac{N}{60}\right)$$

$$\eta_{ith} = \frac{IP}{\dot{m}_f \times (\text{C.V.})_f} = 29.1\%$$

5.40 (b)

$$\eta_m = \frac{BP}{BP+FP} = \frac{IP-FP}{IP}$$

$$0.6 = 1 - \frac{FP}{IP}$$

$$\frac{FP}{IP} = 0.4$$

$$IP = 75 \text{ kW}$$

5.41 (d)

$$BP = 100 \text{ kW}$$
$$FP = 10 \text{ kW}$$

$$\eta_m = \frac{BP/2}{BP/2 + FP} = \frac{50}{50+10} \times 100 = 83.33\%$$

5.42 (d)

Given: $\eta_m = 0.85$

Let $\quad BP = x$

$\quad FP = y$

@ 100% load

$$\eta_m = \frac{BP}{BP+FP}$$

$$0.85 = \frac{x}{x+y},$$

$$x = 5.67\ y$$

$$x = 42.5\ \text{kW}$$

$$y = 7.495\ \text{kW}$$

$$\text{So, } \eta_m = \frac{0.6x}{0.+6x+y} = \frac{0.6\times 42.5}{0.6\times 42.5+4.795} = 0.7728$$

5.43 (b)

Brake power, $P = T\omega$

$$= \frac{8}{1000}\times 2\pi\times\frac{1800}{60} = 1.51\ \text{kW}$$

Friction power = 1.8 − 1.51 = 0.29 kW

$$\therefore \frac{\text{Friction power}}{\text{Brake power}} = \frac{0.29}{1.51} = 0.1921$$

5.44 (c)

$$\eta_{ith} = \frac{IP}{\dot{m}_f\times CV} \quad \text{...(i)}$$

$$\dot{m}_f = \dot{V}_f\times\rho_f = \frac{8}{3600}\times\times 10^{-3}\times 0.75\times 1000$$

From (i)

$$\eta_{ith} = \frac{25\times 3600}{8\times 0.75\times 44000} = 0.3409$$

5.45 (d)

Given: $\frac{L}{D} = 1.2$, D = 12 cm, Area of indicator diagram = 30 cm^2

$$\frac{L}{D} = 1.2$$

Length of indicator diagram (l) = $\frac{L}{2} = 7.2$ cm

$\Rightarrow \quad L = 14.4$ cm

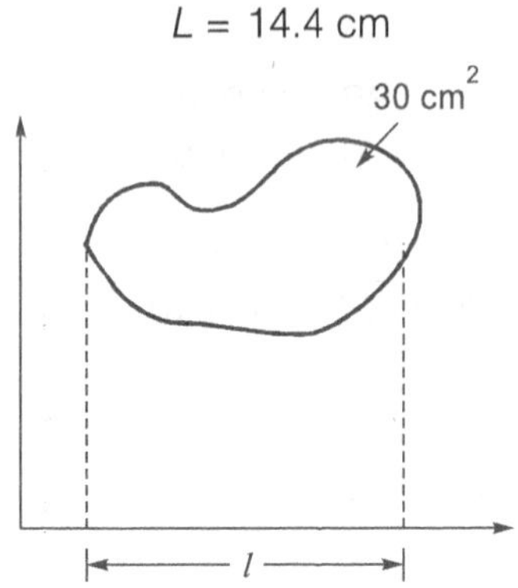

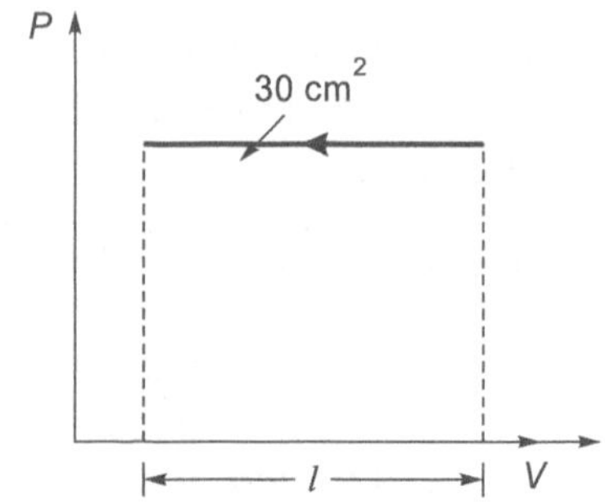

$$h\times l = 30\ \text{cm}^2$$

$$h = 4.16\ \text{cm}$$

$$\text{M.e.p} = h\times k$$

$$= 4.16\times 10^{-2}\times 20\times 10^3$$

$$= 833.33\ \text{kN/m}^2$$

$$= 8.33\times 10^5\ \text{N/m}^2$$

5.46 (a)

$$\text{I.P.} = \frac{PlanK}{60n}$$

$$\frac{IP}{k} = \frac{Plan}{60n}$$

$$= \frac{833.33\times 0.144\times\frac{\pi}{4}\times(0.12)^2\times 2000}{60\times 2}$$

$$= 22.618\ \text{kW}$$

5.47 (d)

Given : k = 6; $N_i = \frac{N}{2}$; BP = 125 kW; N = 3000 rpm; bsfc = 0.2 kg/kWh; S = 0.85

$$\text{bsfc} = \frac{\dot{m}_f(\text{kg/hr})}{BP(\text{kW})}$$

$$\Rightarrow \quad \dot{m}_f = 0.2\times 125 = 25\ \text{kg/hr}$$

$$= 0.4167\ \text{kg/min}$$

For 2 revolution → 1 cycle

3000 rev/min → 1500 cycles/min

$$\dot{m}_f = \frac{0.4167(\text{kg/min})}{1500\ \text{cycles/min}} = 0.0002778\ \text{kg/cycle}$$

Specific gravity, $S = \frac{\rho_{fuel}}{\rho_{water}}$

$$\Rightarrow \quad \rho_{fuel} = 0.85 \times 1000 = 850 \text{ kg/m}^3$$

$$\Rightarrow \quad \rho_{fuel} = \frac{m_f}{V_f}$$

$$V_f = \frac{m_f}{\rho_{fuel}} = \frac{0.0002778}{850} = 3.2682 \times 10^{-7} \text{ m}^3\text{/cycle}$$

$$V_f = 0.32682 \text{ cm}^3\text{/cycle}$$

$$V_f = \frac{0.32682}{6} = 0.0545 \text{ cm}^3\text{/cycle/cylinder}$$

5.48 (a)

Given: V_s = 26.1 litres; BP = 900 kW;
N = 2300 rpm; k (Number of cylinders) = 1

$$\text{BP} = \frac{P_{mep} V_s N.K}{60n}$$

$$\Rightarrow 900 \times 10^3 = \frac{P_{mep} \times 26.1 \times 10^{-3} \times 2300 \times 1}{60 \times 2}$$

$$\Rightarrow \quad P_{mep} = 17.99 \times 10^5 \text{ Pa}$$

$$= 17.99 \text{ bar} \simeq 18 \text{ bar}$$

5.49 (b)

Given: r = 10; $\eta_{air\ std.}$ = 0.602;
η_{ith} = 0.55 × $\eta_{air\ std.}$ η_m = 0.85; η_{vol} = 0.90;
CV_{fuel} = 44 MJ/kg

$$\eta_{overall} = \eta_{ith} \times \eta_m$$

$$= 0.55 \times \eta_{air\ std.} \times \eta_m$$

$$= 0.55 \times 0.602 \times 0.85$$

$$= 0.2814 \text{ or } 28.14\%$$

■■■■

UNIT

V

Refrigeration and Air-conditioning

Syllabus

Vapour compression refrigeration, Refrigerants and Working cycles, Condensers, Evaporators and Expansion devices, Other types of refrigeration systems like Vapour absorption, Vapour jet, Thermo electric and Vortex tube refrigeration. Psychometric properties and processes, Comfort chart, Comfort and industrial air conditioning, Load calculations and Heat pumps.

Contents

Sl.	Topic	Page No.
1.	Heat Pump and Refrigeration Systems	321
2.	Ideal, Vapour, Air Refrigeration Cycle and Refrigerants	331
3.	Refrigeration Equipments	355
4.	Psychrometry and Psychrometric Processes	360
5.	Cooling Load Calculations and Duct Design	379

OOOO

1 Heat Pump and Refrigeration Systems

1.1 The COP of a heat pump β_{hp} and the COP of a refrigerator β_{Ref} are related as

(a) $\beta_{hp} + \beta_{Ref} = 1$ (b) $\beta_{hp} - \beta_{Ref} = 1$
(c) $\beta_{Ref} - \beta_{hp} = 1$ (d) $\beta_{hp} - \beta_{Ref} = 0$

[ESE : 2000]

1.2 The working temperatures in the evaporator and condenser coils of a refrigerator are –30°C and 32°C respectively. If the actual refrigerator has a COP of 0.75 of the maximum, the required power input for a refrigerating effect of 5 kW is nearly,

(a) 1.7 kW (b) 2.94 kW
(c) 3.92 kW (d) 4.0 kW

[ESE : 2001]

1.3 **Assertion (A):** COP of heat pump is more than the COP of its refrigerator version.
Reason (R): Pumping of heat requires less work relative to extraction of heat from the evaporator.

(a) Both **A** and **R** are individually true and **R** is the correct explanation of **A**
(b) Both **A** and **R** are individually true but **R** is not the correct explanation of **A**
(c) **A** is true but **R** is false
(d) **A** is false but **R** is true

[ESE : 2002]

1.4 A heat pump for domestic heating operates between a cold system at 0°C and the hot system at 60°C. What is the minimum electric power consumption if the heat rejected is 80000 kJ/hr?

(a) 2 kW (b) 3 kW
(c) 4 kW (d) 5 kW

[ESE : 2003]

1.5 In a one ton capacity water cooler, water enters at 30°C at the rate of 200 litres per hour. The outlet temperature of water will be (specific heat of water = 4.18 kJ/kgK)

(a) 3.5°C (b) 6.3°C
(c) 23.7°C (d) 15°C

[ESE : 2003]

1.6 A refrigerator working on a reversed Carnot cycle has a COP of 4. If it works as a heat pump and consumes 1 kW, the heating effect will be

(a) 1 kW (b) 4 kW
(c) 5 kW (d) 6 kW

[ESE : 2003]

1.7 A heat engine with 30% efficiency drives a refrigerator of COP 5. What would be the net heat input to the engine for each MW of heat removed in the refrigerator?

(a) 66.67 kJ/s (b) 600 kJ/s
(c) 666.67 kJ/s (d) 6600 kJ/s

[ESE : 2004]

1.8 In a 0.5 TR capacity water cooler, water enters at 30°C and leaves at 15°C. What is the actual water flow rate?

(a) 50 litres/hour (b) 75 litres/hour
(c) 100 litres/hour (d) 125 litres/hour

[ESE : 2005]

1.9 In an Electrolux refrigerator

(a) Ammonia is absorbed in water
(b) Ammonia is absorbed in hydrogen
(c) Hydrogen is evaporated in ammonia
(d) Ammonia is evaporated in hydrogen

[ESE : 2005]

1.10 A heat pump is used to heat a house in the winter and then reversed to cool the house in the summer. The inside temperature of the house is to be maintained at 20°C. The heat transfer through the house wall is 7.9 kJ/s and the outside temperature in winter is 5°C. What is the minimum power (Approximate) required to drive the heat pump?

(a) 40.5 W (b) 405 W
(c) 42.5 W (d) 425 W

[ESE : 2006]

1.11 A refrigerating machine having coefficient of performance equal to 2 is used to remove heat at the rate of 1200 kJ/min. What is the power required for this machine?

(a) 80 kW (b) 60 kW
(c) 20 kW (d) 10 kW

[ESE : 2007]

1.12 The refrigerating efficiency, that is the ratio of actual COP to reversible COP of a refrigeration cycle is 0.8, the condenser and evaporator temperatures are 51°C and –30°C respectively. If cooling capacity of the plant is 2.4 kW then what is the work requirement?

(a) 1.00 kW (b) 1.33 kW
(c) 1.25 kW (d) 2.008 kW

[ESE : 2009]

1.13 **Assertion (A) :** An air-conditioner operating as a heat pump is superior to an electric resistance heater for winter heating.

Reason (R) : A heat pump rejects more heat than the heat equivalent of the heat absorbed.

(a) Both **A** and **R** are individually true and **R** is the correct explanation of **A**
(b) Both **A** and **R** are individually true but **R** is not the correct explanation of **A**
(c) **A** is true but **R** is false
(d) **A** is false but **R** is true [ESE : 2009]

1.14 Operating temperature of a cold storage is –2°C. From the surrounding at ambient temperature of 40°C heat leaked into the cold storage is 30 kW. If the actual COP of the plant is 1/10 of the maximum possible COP, then what will be the power required to pump out the heat to maintain the cold storage temperature at –2°C?

(a) 1.90 kW (b) 3.70 kW
(c) 20.28 kW (d) 46.50 kW

[ESE : 2009]

1.15 A refrigerator operates between the temperatures of –23°C and 27°C. If one *TR* = 3.5 kW, the minimum power required per *TR* to operate the refrigerator is

(a) 0.5 kW (b) 0.7 kW
(c) 0.9 kW (d) 1.0 kW [ESE : 2010]

1.16 An ideal refrigerator is operating between a condenser temperature of 37°C and an evaporator temperature of –3°C. If the machine is functioning as a heat pump, its coefficient of performance will be

(a) 6.00 (b) 6.75
(c) 7.00 (d) 7.75 [ESE : 2010]

1.17 A Carnot heat pump works between temperature limits of 277°C and 27°C. Its COP is

(a) 1.108 (b) 1.2
(c) 2.2 (d) 9.26 [ESE : 2010]

1.18 In an aircraft refrigeration system, the pressure at the cooling turbine outlet is equal to

(a) Ambient pressure
(b) Cabin pressure
(c) Compressor inlet pressure
(d) Evaporator pressure [ESE : 2011]

1.19 In gas cycle refrigeration, an engine is used instead of a throttle valve for pressure drop of the refrigerant, because

(a) enough cooling or temperature drop is not obtained by throttling
(b) there can be even heating of the gas if the temperature before throttling is not below the maximum inversion temperature
(c) there is inadequate content of temperature
(d) there can be leakage of gas [ESE : 2013]

1.20 In a 3 ton capacity water cooler, water enters at 30°C and leaves at 15°C steadily. What is the water flow rate per hour?

(a) 60 kg (b) 100 kg
(c) 602 kg (d) 2520 kg

[ESE : 2013]

1.21 A refrigerator that operates on a Carnot cycle is required to transfer 2000 kJ/min to the atmosphere at 27°C, where the low temperature reservoir is at 0°C. What is the power required?

(a) 200 W (b) 32.93 kW
(c) 200 kW (d) 3.33 kW

[ESE : 2015]

1.22 An ideal refrigerating machine works between the temperature limits of 45°C and –8°C. The power required per ton of refrigeration is
(a) 1.0 kW (b) 1.2 kW
(c) 0.8 kW (d) 0.7 kW
[ESE : 2015]

1.23 **Assertion (A):** The COP of an air conditioning plant is higher than the COP of a household refrigerator.
Reason (R): For the same condenser temperature, the suction pressure of the evaporator is higher in air conditioning plant than in household refrigerator.
(a) Both **A** and **R** are true and **R** is the correct explanation of **A**
(b) Both **A** and **R** are true but **R** is not a correct explanation of **A**
(c) **A** is true but **R** is false
(d) **A** is false but **R** is true
[ESE : 2015]

1.24 Two reversible refrigerators are arranged in series and their COPs are 5 and 6 respectively. The COP of composite refrigeration system would be:
(a) 1.5 (b) 2.5
(c) 3.5 (d) 4.5
[ESE : 2016]

1.25 In a refrigerator, the evaporator and condenser coil temperatures are –33°C and 27°C, respectively. Assuming that the maximum COP is realized, the required power input for a refrigerating effect of 4 kW is
(a) 8 kW (b) 4 kW
(c) 2 kW (d) 1 kW
[ESE : 2017]

1.26 The refrigeration system of an ice plant working between temperatures of –5°C and 25°C produces 20 kg of ice per minute from water at 20°C. The specific heat of water is 4.2 kJ/kg and latent heat of ice is 335 kJ/kg. The refrigeration capacity of the refrigeration plant is
(a) 9040 kJ/min (b) 8750 kJ/min
(c) 8380 kJ/min (d) 8010 kJ/min
[ESE : 2017]

1.27 The coefficient of performance of a heat pump working on reversed Carnot cycle is 6. If this machine works as a refrigerator with work input of 10 kW, the refrigerating effect will be
(a) 35 kW (b) 40 kW
(c) 45 kW (d) 50 kW
[ESE : 2018]

1.28 Consider the following statements:
1. Heat pumps and air conditioners have the same mechanical components.
2. The same system can be used as heat pump in winter and as air conditioner in summer.
3. The capacity and efficiency of a heat pump fall significantly at high temperature.

Which of the above statements are correct?
(a) 1 and 2 only (b) 1 and 3 only
(c) 2 and 3 only (d) 1, 2 and 3
[ESE : 2018]

1.29 An ideal refrigerator working on a reversed Rankine cycle has a capacity of 3 tons. The COP of the unit is found to be 6. The capacity of the motor required to run the unit is (take 1 T = 210 kJ/min)
(a) 1.85 kW (b) 1.75 kW
(c) 1.65 kW (d) 1.5 kW
[ESE : 2018]

1.30 A cold storage has capacity for food preservation at a temperature of – 3°C when the outside temperature is 27°C. The minimum power required to operate with a cooling load of 90 kW is
(a) 20 kW (b) 15 kW
(c) 10 kW (d) 5 kW
[ESE : 2018]

1.31 A cold storage is to be maintained at –5°C while the surroundings are at 35°C. The heat leakage from the surroundings into the cold storage is estimated to be 29 kW. The actual COP of the refrigeration plant used is $\frac{1}{3}^{rd}$ that of an ideal plant working between the same temperatures. The power required to drive the plant will be:

(a) 13 kW (b) 14 kW
(c) 15 kW (d) 16 kW

[ESE : 2019]

1.32 A reversed Carnot engine is used for heating a building. It supplies 210×10^3 kJ/hr of heat to the building at 20 °C. The outside air is at –5°C. The heat taken from the outside will be nearly

(a) 192×10^3 kJ/hr (b) 188×10^3 kJ/hr
(c) 184×10^3 kJ/hr d) 180×10^3 kJ/hr

[ESE : 2019]

1.33 A domestic food freezer maintains a temperature of –15°C. The ambient air temperature is 30°C. If heat leaks into the freezer at the continuous rate of 1.75 kJ/s, the least power necessary to pump this heat out continuously will be nearly

(a) 0.1 kW (b) 0.2 kW
(c) 0.3 kW (d) 0.4 kW

[ESE : 2020]

1.34 Ice is formed at 0°C from water at 20°C. The temperature of the brine is –8°C. The refrigeration cycle used is perfect reversed Carnot cycle. Latent heat of ice = 335 kJ/kg, and c_{pw} = 4.18. The ice formed per kWh will

(a) 81.4 kg (b) 76.4 kg
(c) 71.8 kg (d) 68.8 kg

[ESE : 2020]

1.35 A cold storage is to be maintained at –5°C while the surroundings are at 35°C. The heat leakage from the surroundings into the cold storage is estimated to be 29 kW. The actual COP of the refrigeration plant used is one-third of an ideal plant working between the same temperatures. The power required to drive the plant will be

(a) 10 kW (b) 11 kW
(c) 12 kW (d) 13 kW

[ESE : 2020]

1.36 The ideal gas-refrigeration cycle is the same as the

(a) Brayton cycle
(b) Reversed Brayton cycle
(c) Vapour compression refrigeration cycle
(d) Vapour absorption refrigeration cycle

[ESE : 2020]

1.37 A refrigerator operating on reversed Carnot cycle extracts 500 kJ/min heat from a refrigerated space being maintained at –16°C and rejects heat to the atmosphere at 27°C. What is the work input required to run the refrigerator?

(a) 42.46 kJ/min (b) 55.24 kJ/min
(c) 66.36 kJ/min (d) 83.66 kJ/min

[ESE : 2023]

1.38 A cold storage plant is required to store fish. The fish is supplied at a temperature of 30 °C. The fish is stored in cold storage which is maintained at –8 °C. If the plant requires 75 kW to drive it, what is the capacity of the plant? (Assume actual COP of the plant as 0.3 of the Carnot COP)

(a) 40.8 TR (b) 35.9 TR
(c) 44.8 TR (d) 54.6 TR

[ESE : 2024]

■■■■

Answers Heat Pump and Refrigeration Systems

1.1 (b) 1.2 (a) 1.3 (c) 1.4 (c) 1.5 (d) 1.6 (c) 1.7 (c) 1.8 (c) 1.9 (a)

1.10 (b) 1.11 (d) 1.12 (a) 1.13 (a) 1.14 (d) 1.15 (b) 1.16 (d) 1.17 (c) 1.18 (b)

1.19 (a) 1.20 (c) 1.21 (d) 1.22 (d) 1.23 (a) 1.24 (b) 1.25 (d) 1.26 (c) 1.27 (d)

1.28 (d) 1.29 (b) 1.30 (c) 1.31 (a) 1.32 (a) 1.33 (c) 1.34 (a) 1.35 (d) 1.36 (b)

1.37 (d) 1.38 (c)

Explanations Heat Pump and Refrigeration Systems

1.1 (b)

Let Q_2 = heat extracting from cold body

Q_1 = heat delivering to hot body

$T_1 > T_2$

$$(COP)_R = \frac{Q_2}{Q_1 - Q_2}$$

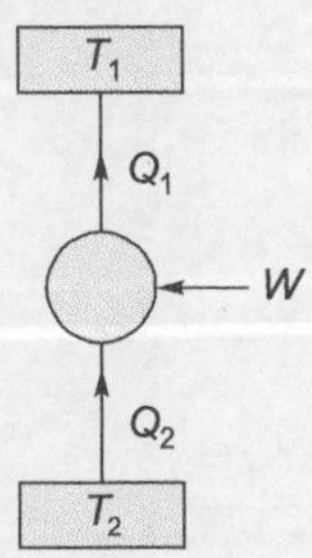

$$(COP)_{HP} = \frac{Q_1}{Q_1 - Q_2}$$

$\therefore$ **$(COP)_{HP} - (COP)_R = 1$**

1.2 (a)

$$(COP)_{max} = \frac{T_1}{T_2 - T_1} = \frac{243}{305 - 243} = 3.919$$

$$(COP)_{actual} = 0.75 \times (COP)_{max}$$

$$= 0.75 \times 3.919 = 2.939$$

Now :

$$(COP)_{actual} = \frac{\text{Actual refrigerating effect}}{\text{Power input}}$$

$$\text{Power input} = \frac{5}{2.939} = \mathbf{1.7\ kW}$$

1.3 (c)

$$(COP)_{HP} = (COP)_R + 1$$

The refrigerator can work as heat pump if the desired effect is altered. The work gets changed if the temperature difference gets changed, not the device.

1.4 (c)

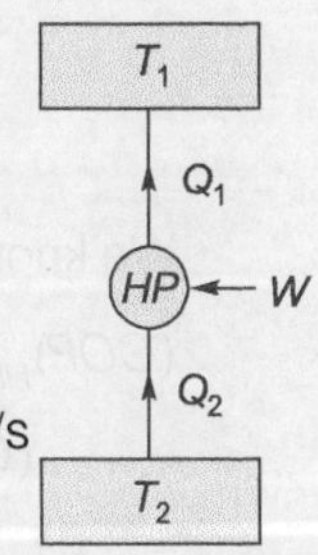

Given data:

$T_2 = 0°C = 273\ K$

$T_1 = 60°C = 333\ K$

$$Q_1 = 80000\ \text{kJ/hr} = \frac{80000}{3600}\ \text{kJ/s}$$

$$= 22.22\ kW$$

$$(COP)_{HP} = \frac{T_1}{T_1 - T_2} = \frac{333}{333 - 273} = 5.55$$

also $(COP)_{HP} = \dfrac{Q_1}{W}$

$$\therefore \quad 5.55 = \frac{22.22}{W} \text{ or } W = \frac{22.22}{5.55} = 4\ kW$$

1.5 (d)

Given data :

Capacity of water cooling

Q_2 = 1 ton = 1 × 3.5 kW = 3.5 kW

Water inlet temperature,

$T_{w1} = 30°C$

Volume flow rate,

V = 200 litre/h

$$= \frac{200}{1000 \times 3600} m^3/s$$

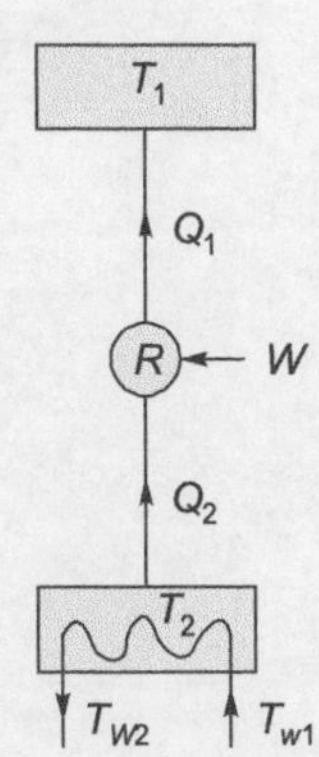

c_p = 4.18 kJ/kgK

Capacity of cooling water,

$$Q_2 = \dot{m}\, c_p (T_{w1} - T_{w2})$$

$$Q_2 = \rho\, \dot{V} c_p (T_{w1} - T_{w2})$$

$$3.5 = 1000 \times \frac{200}{1000 \times 3600} \times 4.18(30 - T_{w2})$$

$$15.07 = 30 - T_{w2}$$

or $T_{w2} = 30 - 15.07 = 14.93°C \simeq$ **15°C**

1.6 (c)

Given data :

$$(COP)_R = 4$$
$$W = 1 \text{ kW}$$

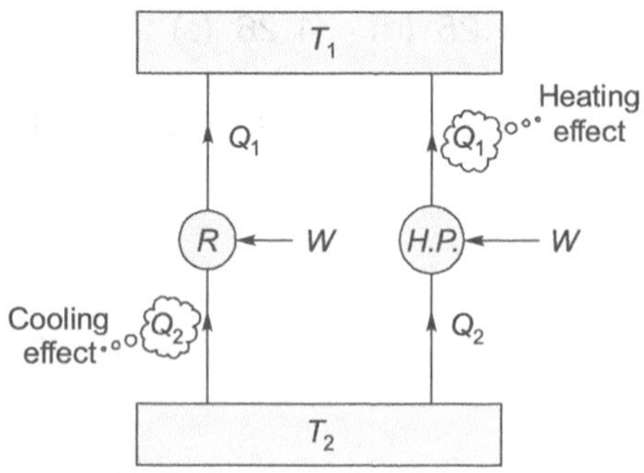

$$W = 1 \text{ kW}$$

We know that,

$$(COP)_{HP} - (COP)_R = 1$$
$$(COP)_{HP} - 4 = 1$$
$$(COP)_{HP} = 5$$

also $$(COP)_{HP} = \frac{\text{Heating effect } (Q_1)}{\text{Work input } (W)}$$

$\therefore$ $$5 = \frac{Q_1}{1}$$

or $Q_1 = $ **5 kW**

Alternatively

$$(COP)_R = 4$$
$$W = 1 \text{ kW}$$

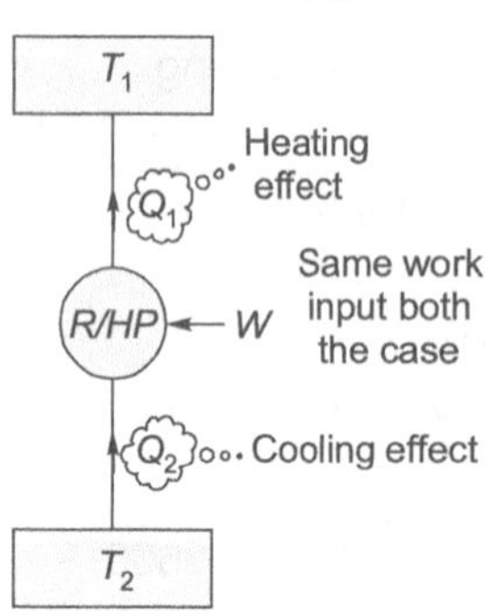

$$(COP)_R = \frac{\text{Cooling effect } (Q_2)}{\text{Work input } (W)}$$

$\therefore$ $$4 = \frac{Q_2}{1}$$

or $Q_2 = 4$ kW

Heating effect: $Q_1 = W + Q_2 = 1 + 4 = $ **5 kW**

1.7 (c)

Given data :

$$\eta = 30\% = 0.30$$
$$(COP)_R = 5$$

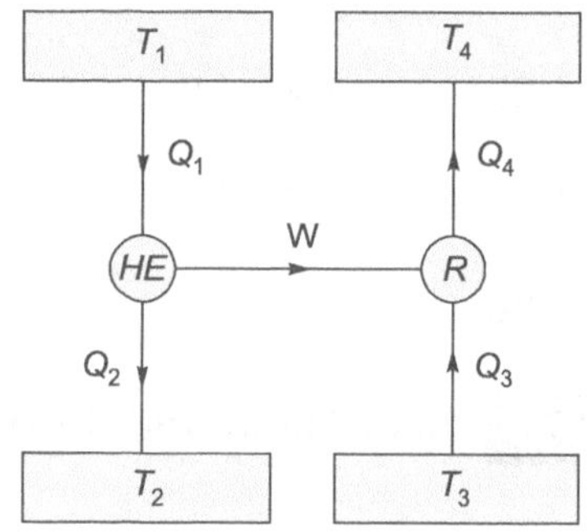

$$Q_3 = 1 \text{ MW} = 1000 \text{ kW}$$

$$(COP)_R = \frac{Q_3}{W}$$

$$5 = \frac{1000}{W} \text{ or } W = \frac{1000}{5} = 200 \text{ kJ/s}$$

$$\eta = \frac{W}{Q_1}$$

$$0.30 = \frac{200}{Q_1}$$

or $Q_1 = \frac{200}{0.30} = $ **666.67 kJ/s**

1.8 (c)

Capacity of water cooler,

$$Q = 0.5 \text{ TR} = 0.5 \times 3.5 \text{ kW}$$
$$= 1.75 \text{ kW} \quad \{\because 1 \text{ TR} = 3.5 \text{ kW}\}$$

Water inlet temperature,

$$T_{wi} = 30°C$$

Water outlet temperature,

$$T_{wo} = 15°C$$

Capacity of water cooler,

$$Q = \dot{m} c_p (T_{wi} - T_{wo})$$
$$Q = \rho \dot{V} c_p (T_{wi} - T_{wo})$$

where $\rho = 1000 \text{ kg/m}^3$

V is volume flow rate in m^3/s

$$c_p = 4.18 \text{ kJ/kgK}$$

$\therefore$ $1.75 = 1000 \times V \times 4.18(30 - 15)$

or $V = 2.791 \times 10^{-5} \text{ m}^3/\text{s}$

$= 2.791 \times 10^{-5} \times 1000 \times 3600$ litre/hr

$= 100.47$ litre/hr

$\simeq$ **100 litre/hour**

1.9 (a)

The water is used as a absorbent because it has ability to absorb ammonia refrigerant.

1.10 (b)

Given data

$$T_1 = 20°C = (20 + 273)\text{ K} = 293\text{ K}$$
$$Q_1 = 7.9\text{ kJ/s} = 7.9\text{ kW}$$
$$T_2 = 5°C = (5 + 273)\text{ K} = 278\text{ K}$$

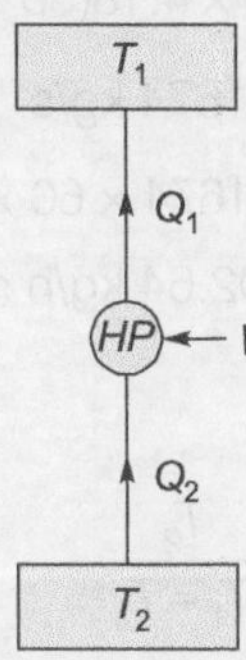

$$(COP)_{HP} = \frac{T_1}{T_1 - T_2} = \frac{293}{293 - 278}$$

also $$(COP)_{HP} = \frac{\text{Heating effect : } Q_1}{\text{Minimum work input : } W_{min}}$$

$$\therefore \quad 19.53 = \frac{7.9}{W_{min}}$$

or $$W_{min} = \frac{7.9}{19.53} = 0.4045\text{ kW}$$
$$= 404.5\text{ W} \simeq \mathbf{405\text{ W}}$$

1.11 (d)

$(COP)_R = 2$

Refrigeration capacity :

$$Q = 1200\text{ kJ/min}$$
$$= \frac{1200}{60}\text{ kJ/s} = 20\text{ kW}$$

$$(COP)_R = \frac{\text{Refrigeration capacity}(Q)}{\text{Power required }(P)}$$

$$2 = \frac{20}{P}$$

or $$P = \frac{20}{2} = \mathbf{10\text{ kW}}$$

1.12 (a)

Given data for refrigeration system,

$$\frac{(COP)_{act}}{(COP)_{rev}} = 0.8$$

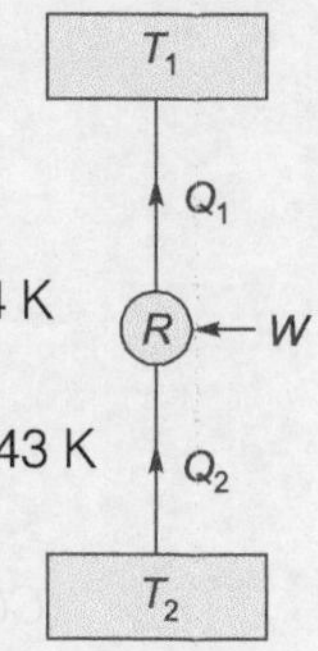

$$T_1 = 51°C$$
$$= (51 + 273)\text{ K} = 324\text{ K}$$
$$T_2 = -30°C$$
$$= (-30 + 273)\text{ K} = 243\text{ K}$$

Cooling capacity :

$$Q_2 = 2.4\text{ kW}$$

$$(COP)_{rev} = \frac{T_2}{T_1 - T_2} = \frac{243}{324 - 243} = 3$$

$$\therefore \quad (COP)_{act} = 0.8 \times (COP)_{rev} = 0.8 \times 3 = 2.4$$

also $$(COP)_{act} = \frac{\text{Cooling capacity }(Q_2)}{\text{Work-requirement }(W)}$$

$$2.4 = \frac{2.4}{W}$$

or $$W = \mathbf{1\text{ kW}}$$

1.14 (d)

Given data :

$$T_2 = -2°C = (-2 + 273)\text{ K}$$
$$= 271\text{ K}$$
$$T_1 = 40°C = (40 + 273)\text{ K}$$
$$= 313\text{ K}$$
$$Q_2 = 30\text{ kW}$$

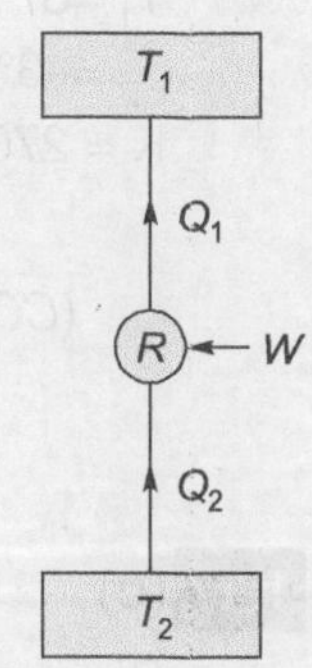

$$(COP)_{act} = \frac{(COP)_{max}}{10}$$

$$(COP)_{max} = \frac{T_2}{T_1 - T_2} = \frac{271}{313 - 271} = 6.45$$

$$\therefore \quad (COP)_{act} = \frac{6.45}{10} = 0.645$$

also $$(COP)_{act} = \frac{Q_2}{W}$$

$$\therefore \quad 0.645 = \frac{30}{W}$$

or $$W = \mathbf{46.51\text{ kW}}$$

1.15 (b)

$$T_2 = -23°C = (-23 + 273)\text{ K} = 250\text{ K}$$
$$T_1 = 27°C = (27 + 273)\text{ K} = 300\text{ K}$$
$$Q_2 = 3.5\text{ kW}$$

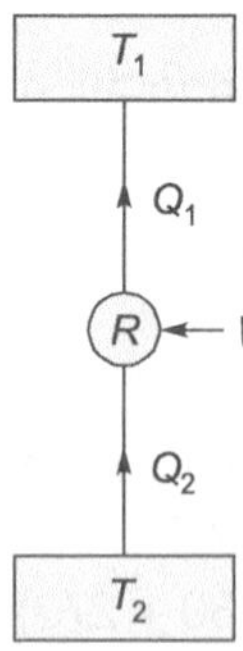

$$(COP)_R = \frac{T_2}{T_1 - T_2} = \frac{250}{300-250} = 5$$

$$\text{also } (COP)_R = \frac{\text{Refrigeration capacity}(Q_2)}{\text{Minimum power required } (W_{min})}$$

$$\therefore \quad 5 = \frac{3.5}{W_{min}}$$

$$\text{or} \quad W_{min} = \frac{3.5}{5} = \mathbf{0.7\ kW}$$

1.16 (d)

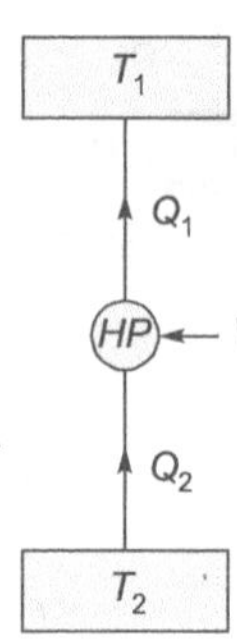

T_1 =37°C = (37 + 273) K = 310 K
T_2 =–3°C = (–3 + 273)
K = 270 K

$$(COP)_{HP} = \frac{T_1}{T_1 - T_2} = \frac{310}{310-270}$$

$$= \mathbf{7.75}$$

1.17 (c)

Given data :

$$T_1 = 277°C = (277 + 273)\ K = 550\ K$$
$$T_2 = 27°C = (27 + 273)\ K = 300\ K$$

$$(COP)_{HP} = \frac{T_1}{T_1 - T_2} = \frac{550}{550-300} = \mathbf{2.2}$$

1.18 (b)

In an aircraft refrigeration system, the pressure at the cooling turbine outlet is equal to the cabin pressure.

1.19 (a)

Since air is an ideal gas, hence Joule-Thomson coefficient ; $\mu = 0$
Therefore, no temperature drop is obtained by throttling.

1.20 (c)

Cooling capacity, Q = 3 ton = 3 × 3.5 kW

$\because$ 1 ton = 3.5 kW

Inlet temperature of water, T_{w1} = 30°C

Outlet temperature of water, T_{w2} = 15°C

Cooling capacity,

$$Q = \dot{m}\, c_p(T_{w1} - T_{w2})$$
$$10.5 = \dot{m} \times 4.18(30-15)$$
$$\text{or} \quad \dot{m} = 0.1674 \text{ kg/s}$$
$$= 0.1674 \times 60 \times 60 \text{ kg/h}$$
$$= 602.64 \text{ kg/h} \simeq \mathbf{602\ kg/h}$$

1.21 (d)

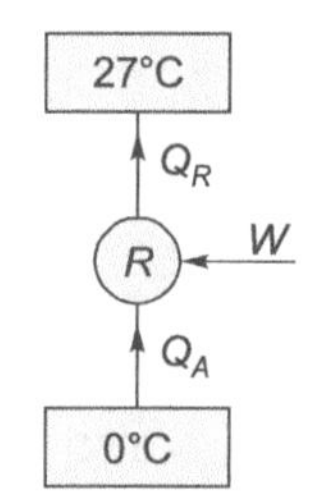

$$(COP)_R = \frac{T_2}{T_1 - T_2}$$
$$= \frac{273}{27} = 10.11$$
$$COP = \frac{Q_A}{W} = \frac{Q_A}{Q_R - Q_A}$$
$$\frac{Q_R}{Q_A} = \frac{1+COP}{COP} = \frac{11.11}{10.11} = 1.0989$$
$$\text{or} \quad Q_A = \frac{Q_R}{1.0989} = 30.33 \text{ kW}$$
$$\therefore \quad W = Q_R - Q_A$$
$$= 33.33 - 30.33 = \mathbf{3\ kW}$$

Option (d) is closest.

1.22 (d)

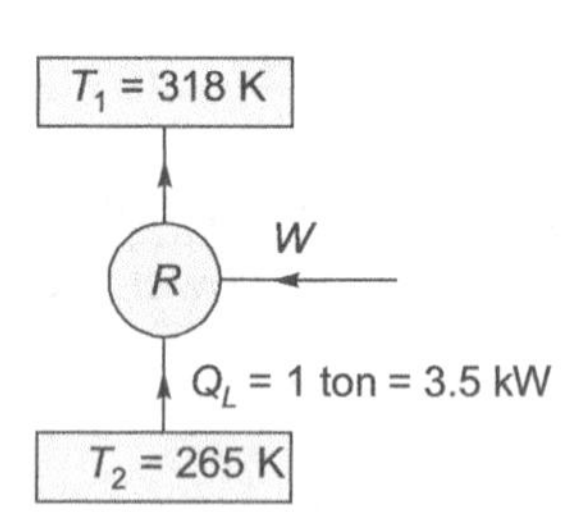

$$COP = \frac{T_2}{T_1 - T_2} = \frac{Q_L}{W}$$
$$W = \frac{3.5 \times (318-265)}{265} = \mathbf{0.7\ kW}$$

1.23 (a)

The evaporator temperature maintained in refrigerator is lower than that the evaporator temperature in air-conditioning. This is obtained by keeping lower evaporator pressure in refrigerator.

1.24 (b)

$$(COP)_{C\cdot C} = \frac{COP_1 \times COP_2}{1 + COP_1 + COP_2} = \frac{5 \times 6}{1 + 5 + 6}$$

$$= \mathbf{2.5}$$

1.27 (d)

$$COP_{HP} = 1 + COP_R$$

$$\Rightarrow \quad COP_R = 5 = \frac{RE}{10}$$

$$RE = 50 \text{ kW}$$

1.29 (b)

$$COP = \frac{DE}{W_{in}}$$

$$\Rightarrow \quad 6 = \frac{10.5}{W_{in}}$$

$$W_{in} = 1.75 \text{ kW}$$

1.30 (c)

$$COP = \frac{T_L}{T_H - T_L} = \frac{DE}{W_{in}} = \frac{90}{W_{in}}$$

$$W_{in} = 10 \text{ kW}$$

1.31 (a)

$T_L = -5°C = 268$ K, $T_H = 35°C = 308$ K

$$COP_{actual} = \frac{1}{3}(COP)_{ideal}$$

$$= \frac{1}{3}\left\{\frac{T_L}{T_H - T_L}\right\} = \frac{1}{3}\left\{\frac{268}{308 - 268}\right\} = 2.233$$

$$\text{Also,} COP_{actual} = \frac{R.C.}{\dot{W}_{input}} = \frac{29}{\dot{W}_{input}}$$

$$\dot{W}_{input} = 12.987 \approx 13 \text{ kW}$$

1.32 (a)

A heat supplied to room = 210 × 10³ kJ/Hr

$T_H = 20°C = 293$ K, $T_L = -5°C = 268$ K

Heat taken from ouside = $\dot{Q}_1$ = ?

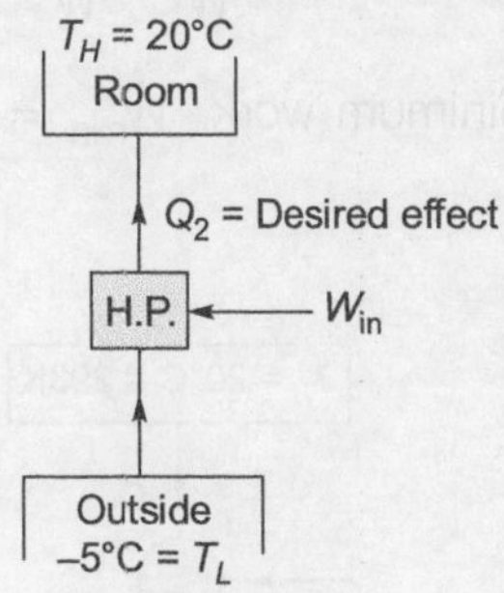

$$(COP)_{HP} = \frac{T_H}{T_H - T_L} = \frac{\dot{Q}_2}{W_{in}} = \frac{\dot{Q}_2}{\dot{Q}_2 - \dot{Q}_1}$$

$$= \frac{293}{293 - 268} = \frac{210 \times 10^3}{210 \times 10^3 - \dot{Q}_1}$$

$$\dot{Q}_1 = 192 \times 10^3 \text{ kJ/Hr}$$

Alternate solution:

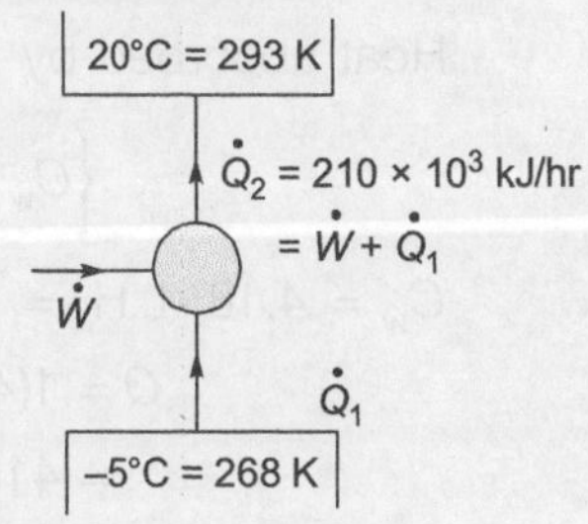

$$\oint_{rev} \frac{\delta Q}{T} = 0$$

$$\frac{\dot{Q}_1}{T_1} + \frac{\dot{Q}_2}{T_2} = 0$$

$$\frac{\dot{Q}_1}{268} + \frac{(-210 \times 10^3)}{293} = 0$$

$$\dot{Q}_1 = 192.081 \times 10^3 \text{ kJ/hr}$$

1.33 (c)

$$T_L = -15 + 273 = 258 \text{ K}$$

$$T_H = 30 + 273 = 303 \text{ K}$$

∵ To maintain continuous temperature,

Heat absorbed should be equal to rate of heat leaks into freezer so,

$$\dot{Q} = 1.75 \text{ kJ/s}$$

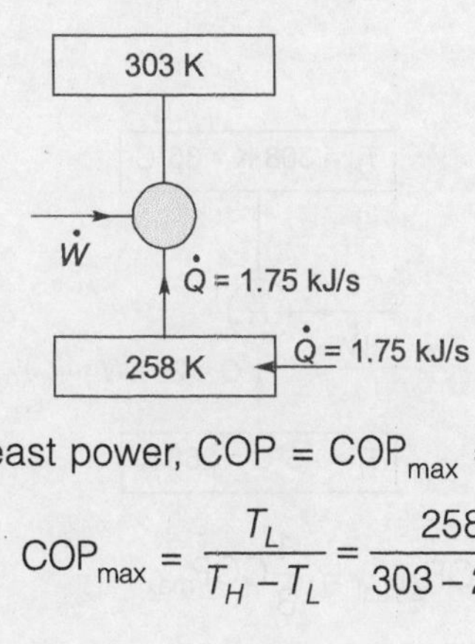

For least power, $COP = COP_{max} = COP_{Carnot}$

$$COP_{max} = \frac{T_L}{T_H - T_L} = \frac{258}{303 - 258} = 5.733$$

Also, $COP_{max} = \frac{\dot{Q}}{\dot{W}} = \frac{1.75}{\dot{W}}$

So minimum work, $\dot{W}_{min} = 0.305$ kW

1.34 (a)

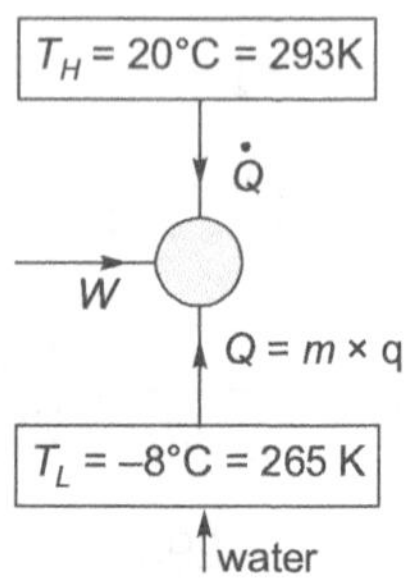

Let m kg ice is force per kWhr

$$\underset{water}{20°C} \rightarrow \underset{W}{0°C} \rightarrow \underset{ice}{0°C}$$

Heat absorbed by brine/kg of ice

$$\left\{C_w \times 20 + L.H.\right\}$$

$C_w = 4.18$, L.H. = 335 kJ/kg

$$Q = 1(4.18 \times 20 + 335)$$
$$= 418.6 \text{ kJ/kg}$$

[1 kWh = 3600 kJ]

Let heat rejected by Brine is to surroundings at 20°C

$$T_H = 20 + 273 = 293 \text{ K}$$

$$COP_{max} = \frac{Q}{W}$$

but $COP_{max} = \frac{T_L}{T_H - T_L} = \frac{265}{293 - 265} = 9.464$

$\therefore \quad 9.464 = \frac{Q}{W} = \frac{m \times q}{3600}$

$\therefore \quad m = 81.39$ kg

1.35 (d)

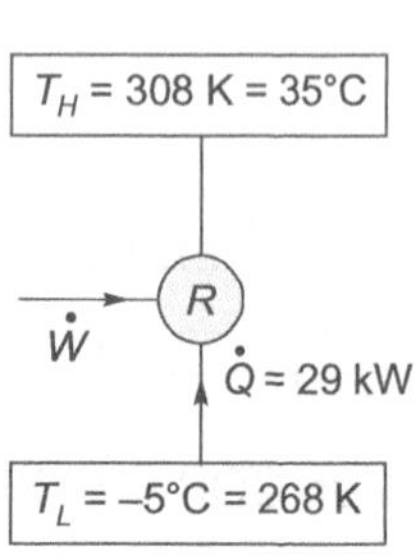

$$COP_{actual} = \frac{1}{3} COP_{max}$$

$$= \frac{1}{3}\left[\frac{268}{308 - 268}\right] = 2.233$$

But, $COP_{actual} = \frac{\dot{Q}}{\dot{W}}$

So, power required, $\dot{W} = 12.98$ kW

1.37 (d)

Given :

Refrigeration effect, R.E. = 500 kJ/min

Temperature of refrigerated space,

$$T_L = 273 - 16 = 257 \text{ K}$$

Temperature of atmosphere,

$$T_H = 273 + 27 = 300 \text{ K}$$

$$COP = \frac{R_E}{W} = \frac{T_L}{T_H - T_L}$$

$$\frac{500}{W} = \frac{257}{300 - 257}$$

$$W = 83.66 \text{ kJ/min}$$

1.38 (c)

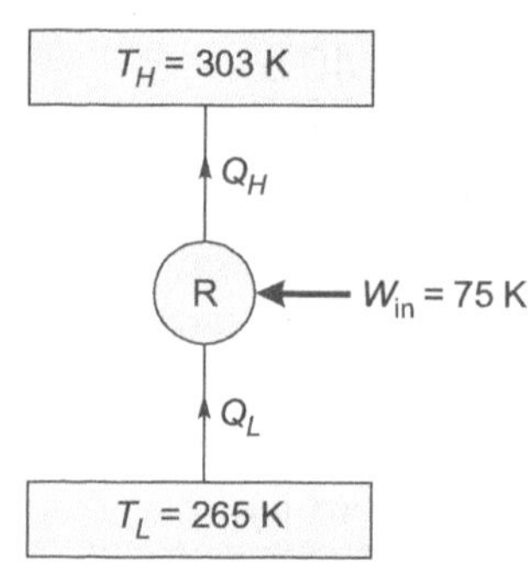

Given : $T_H = 30°C = 30 + 273 = 303$ K

$$T_L = -8°C = -8 + 273 = 265 \text{ K}$$

$$COP_{act} = 0.3 \times COP_{carnot}$$

$$= 0.3 \times \frac{265}{303 - 265} = 2.092$$

$\Rightarrow \quad COP_{act} = 2.092 = \frac{Q_L}{W_{in}}$

$\Rightarrow \quad Q_L = W_{in} \times 2.092$

$$= 75 \times 2.092 = 156.91 \text{ kW}$$

$\Rightarrow \quad Q_L = \frac{156.91}{3.5} = 44.83$ TR

$$\simeq 44.8 \text{ TR}$$

■■■■

2 Ideal, Vapour, Air Refrigeration Cycle and Refrigerants

2.1 Consider the following statements:

1. Azeotrops are the mixtures of refrigerants and behave like pure substances.
2. Isomers refrigerants are compounds with the same chemical formula but have different molecular structures.
3. The formula $n + p + q = 2$ m is used for unsaturated Chloro-fluoro-carbon compounds (m, n, p and q are the number of atoms of carbon hydrogen, fluorine and chlorine respectively).

Which of these statements are correct?

(a) 1 and 3 (b) 2 and 3
(c) 1 and 2 (d) 1, 2 and 3

[ESE : 2000]

2.2 The enthalpies at the beginning of compression, at the end of compression and at the end of condensation are respectively 185 kJ/kg, 210 kJ/kg and 85 kJ/kg. The *COP* of the vapour compression refrigeration system is

(a) 0.25 (b) 5.4
(c) 4 (d) 1.35

[ESE : 2000]

2.3 The discharge pressure of the compressor in the refrigeration system goes up due to the

(a) lower volumetric efficiency of the compressor
(b) formation of scale in the condenser
(c) large size of the condenser
(d) undercharge of the refrigerant

[ESE : 2000]

2.4 The leakage in a Freon based refrigeration system can be detected by using a/an

(a) oxyacetylene torch
(b) halide torch
(c) sulphur torch
(d) blue litmus paper

[ESE : 2000]

2.5 Consider the following statements:
Subcooling in the condenser of a refrigeration system is advisable when

1. expansion valve is at a higher elevation than condenser.
2. there is a large pressure drop in the line connecting condenser to the expansion valve.
3. the refrigeration effect is to be increased.
4. the compressor work is to be reduced.

Which of these statements are correct?

(a) 1 and 2 (b) 1, 3 and 4
(c) 2, 3 and 4 (d) 1, 2 and 3

[ESE : 2001]

2.6 Match **List-I** (Refrigerant) with **List-II** (Chemical constituent) and select the correct answer using the codes given below the lists:

List-I
A. R - 12
B. R - 22
C. R - 717
D. R - 113

List-II
1. Trichloro-trifluoro-ethane ($CCl_2F\text{-}CClF_2$)
2. Difluoro-monocholoro-methane (CHF_2Cl)
3. Ammonia (NH_3)
4. Difluoro-dichloro-methane (CCl_2F_2)

Codes:

	A	B	C	D
(a)	3	2	4	1
(b)	4	2	3	1
(c)	3	1	4	2
(d)	4	1	3	2

[ESE : 2001]

2.7 For a heat pump working on vapour compression cycle, enthalpy values of the working fluid at the end of heat addition process, at the end of compression process, at the end of heat rejection process, and at the end of iso-enthalpic expansion process are 195 kJ/kg, 210 kJ/kg, 90 kJ/kg and 90 kJ/kg respectively. The mass flow rate is 0.5 kg/s. Then the heating capacity of heat pump is, nearly

(a) 7.5 kW (b) 45 kW
(c) 52.5 kW (d) 60 kW

[ESE : 2001]

2.8 A solar-absorption refrigeration system has generator temperature of 87°C, evaporator temperature of –3°C, condenser and absorber temperatures of 27°C each, then its maximum possible COP is

(a) 10.0 (b) 9.0
(c) 1.80 (d) 1.50

[ESE : 2001]

2.9 The working temperatures in evaporator and condenser coils of a refrigerator are –23°C and 27°C respectively. The COP of the refrigerator is 0.8 of the maximum COP. For a power input of 1 kW, the refrigeration effect produced will be

(a) 4 kW (b) 5 kW
(c) 8 kW (d) 2.5 kW

[ESE : 2001]

2.10 A one ton capacity water cooler cools water steadily from 35°C to 20°C. The specific heat of water is 4.18 kJ/kgK. The water flow rate will be, nearly,

(a) 13.33 *l*/hr (b) 33.3 *l*/hr
(c) 200 *l*/hr (d) 250 *l*/hr

[ESE : 2001]

2.11 Experimental measurements on a refrigeration system indicate that rate of heat extraction by the evaporator, rate of heat rejection in the condenser and rate of heat rejection by the compressor body to environment are 70 kW, 90 kW and 5 kW respectively. The power input (in kW) required to operate the system is

(a) 15 (b) 20
(c) 25 (d) 75

[ESE : 2002]

2.12 Ozone depletion by CFCs occurs by breakdown of

(a) chlorine atoms from refrigerant by UV radiation and reaction with ozone in troposphere
(b) fluorine atoms from refrigerant by UV radiation and reaction with ozone in troposphere
(c) chlorine atoms from refrigerant by UV radiation and reaction with ozone in stratosphere
(d) fluorine atoms from refrigerant by UV radiation and reaction with ozone in stratosphere

[ESE : 2002]

2.13 A standard vapour compression refrigeration cycle consists of the following four thermodynamic processes in sequence

(a) isothermal expansion, isentropic compression, isothermal compression and isentropic expansion
(b) constant pressure heat addition, isentropic compression, constant pressure heat rejection and isenthapic expansion
(c) constant pressure heat addition, isentropic compression, constant temperature heat rejection and isentropic expansion
(d) isothermal expansion, constant pressure heat addition, isothermal compression and constant pressure heat rejection

[ESE : 2002]

2.14 Maximum possible COP of a solar absorption refrigeration system with generator temperature of 360 K, absorber temperature of 300 K, condenser temperature of 300 K and evaporator temperature of 270 K is

(a) 9 (b) 6
(c) 3 (d) 1.5

[ESE : 2002]

2.15 Subcooling heat exchanger is used in a refrigeration cycle. The enthalpies at condenser outlet and evaporator outlet are 78 and 182 kJ/kg respectively. The enthalpy at outlet of isentropic compressor is 230 kJ/kg and enthalpy of subcooled liquid is 68 kJ/kg. The COP of the cycle is

(a) 3.25 (b) 2.16
(c) 3.0 (d) 3.5 [ESE : 2002]

2.16 Assertion (A): The mass flow rate through a compressor for various refrigerants at same temperature and pressure, is proportional to their molecular weights.

Reason (R): According to Avogadro's law all gases have same number of moles in a given volume at same pressure and temperature.

(a) Both **A** and **R** are individually true and **R** is the correct explanation of **A**
(b) Both **A** and **R** are individually true but **R** is not the correct explanation of **A**
(c) **A** is true but **R** is false
(d) **A** is false but **R** is true

[ESE : 2002]

2.17 Which one of the following is the *p*-*V* diagram for air refrigeration cycle?

(a)
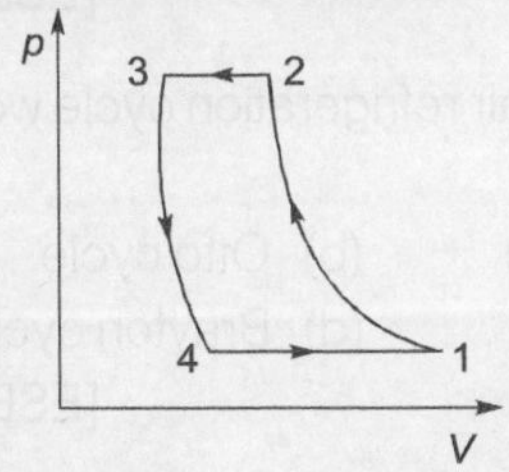

(b)
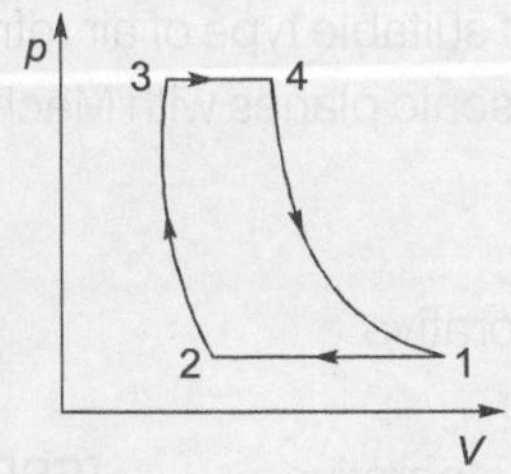

(c)
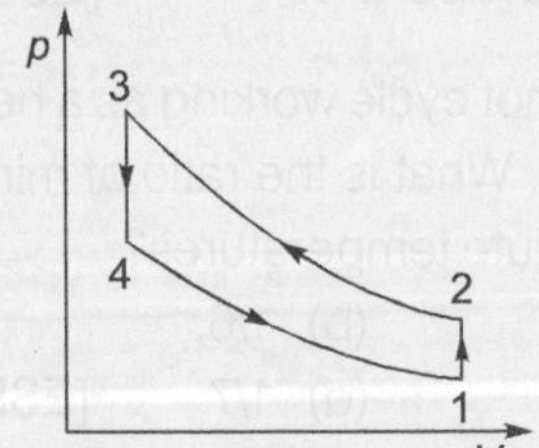

(d)
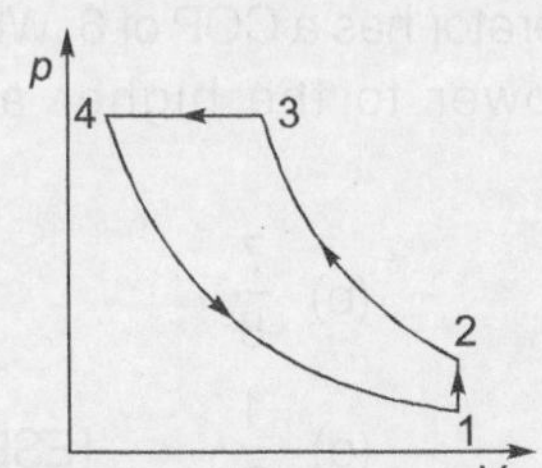

[ESE : 2003]

2.18 A Carnot refrigerator requires 1.5 kW/ton of refrigeration to maintain a region at a temperature of –30°C. The COP of the Carnot refrigerator is

(a) 1.42 (b) 2.33
(c) 2.87 (d) 3.26 [ESE : 2003]

2.19 Match **List-I** (Process) with **List-II** (Type) for Bell Coleman or Joule or Reverse Brayton cycle for gas cycle refrigeration and select the correct answer using the codes given below the lists:

List-I	List-II
A. Compression	1. Isobaric
B. Heat rejection	2. Isothermal
C. Expansion	3. Isentropic
D. Heat absorption	4. Isenthalpic

Codes:

	A	B	C	D
(a)	3	1	4	2
(b)	3	1	3	1
(c)	3	2	3	2
(d)	3	1	2	2

[ESE : 2003]

2.20 Theoretical maximum COP of a vapour absorption system (where, T_G = generator temperature, T_E = evaporator temperature, T_0 = environmental temperature) is

(a) $\frac{T_E}{T_G}\left(\frac{T_G - T_0}{T_0 - T_E}\right)$ (b) $\frac{T_E}{T_G}\left(\frac{T_0 - T_E}{T_G - T_0}\right)$

(c) $\frac{T_G}{T_E}\left(\frac{T_G - T_0}{T_0 - T_E}\right)$ (d) $\frac{T_G}{T_E}\left(\frac{T_0 - T_E}{T_G - T_0}\right)$

[ESE : 2003]

2.21 Oil separator is not required in refrigeration system if

(a) refrigerant and oil are immiscible at all the pressures and temperatures
(b) refrigerant and oil are immiscible at condensation pressure and temperature
(c) refrigerant and oil are miscible at all pressures and temperatures
(d) refrigerant and oil are miscible at condensation pressures and temperatures

[ESE : 2003]

2.22 Consider the following statements in respect of absorption refrigeration and vapour compression refrigeration system:

1. The former runs on low grade energy.
2. The pumping work in the former is negligible since specific volume of strong liquid solution is small.
3. The latter uses an absorber while former uses a generator.
4. The liquid pump alone replaces compressor of the latter.

Which of these statements are correct?
(a) 1 and 2 (b) 1 and 3
(c) 1 and 4 (d) 2 and 4
[ESE : 2003]

2.23 In a vapour compressor refrigeration system, the refrigeration capacity is 2100 kJ/min and heat rejection factor is 1.2. What will respectively be the heat rejected from the condenser and COP?
(a) 5040 kJ/min and 5 (b) 2520 kJ/min and 5
(c) 2520 kJ/min and 4 (d) 5040 kJ/min and 4
[ESE : 2004]

2.24 Assertion (A): Subcooling of refrigerant liquid increases the coefficient of performance of a refrigeration cycle.
Reason (R): Subcooling reduces the work requirement of a refrigeration cycle.
(a) Both **A** and **R** are individually true and **R** is the correct explanation of **A**
(b) Both **A** and **R** are individually true but **R** is not the correct explanation of **A**
(c) **A** is true but **R** is false
(d) **A** is false but **R** is true [ESE : 2004]

2.25 The values of enthalpy at the beginning of compression, at the end of compression and at the end of condensation are 85 kJ/kg, 210 kJ/kg and 185 kJ/kg, respectively, What is the value of the COP of the vapour compression refrigeration system?
(a) 0.25 (b) 5.4
(c) 4 (d) 1.35
[ESE : 2005]

2.26 A refrigerator based on reversed Carnot cycle works between two such temperatures that the ratio between the low and high temperature is 0.8. If a heat pump is operated between same temperature range, then what would be its COP?
(a) 2 (b) 3
(c) 4 (d) 5 [ESE : 2005]

2.27 A refrigerator works on reversed Carnot cycle producing a temperature of –40°C. Work done per TR is 700 kJ per ten minutes. What is the value of its COP?
(a) 3 (b) 4.5
(c) 5.8 (d) 7.0
[ESE : 2005]

2.28 A refrigeration plant uses a condenser with heat rejection ratio of 1.2. If the capacity of the plant is 210 kJ/min, then the value of the COP of the refrigeration plant is
(a) 3 (b) 5
(c) 7 (d) 9 [ESE : 2005]

2.29 In a vapour compression refrigeration plant, the refrigerant leaves the evaporator at 195 kJ/kg and the condenser at 65 kJ/kg. For 1 kg/s of refrigerant, the refrigeration effect is
(a) 70 kW (b) 100 kW
(c) 130 kW (d) 160 kW
[ESE : 2005]

2.30 A Bell-Coleman air refrigeration cycle works as a reversed
(a) Stirling cycle (b) Otto cycle
(c) Diesel cycle (d) Brayton cycle
[ESE : 2005]

2.31 Which is the most suitable type of air refrigeration system for supersonic planes with Mach number 3 or above?
(a) Boot-strap
(b) Simple evaporative
(c) Regenerative
(d) Boot-strap evaporative [ESE : 2005]

2.32 A reversed Carnot cycle working as a heat pump has a COP of 7. What is the ratio of minimum to maximum absolute temperatures?
(a) 7/8 (b) 1/6
(c) 6/7 (d) 1/7 [ESE : 2005]

2.33 A Carnot refrigerator has a COP of 6. What is the ratio of the lower to the higher absolute temperatures?
(a) $\frac{1}{6}$ (b) $\frac{7}{8}$
(c) $\frac{6}{7}$ (d) $\frac{1}{7}$ [ESE : 2006]

2.34 The leaks in a refrigeration system using freon are detected by
(a) A halide torch, which on detection produces greenish flame lighting
(b) Sulphur sticks, which on detection given while white smoke
(c) Using reagents
(d) Sensing reduction in pressure [ESE : 2006]

2.35 In a vapour compression refrigeration cycle for making ice, the condensing temperature for higher COP
(a) Should be near the critical temperature of the refrigerant
(b) Should be above the critical temperature of the refrigerant
(c) Should be much below the critical temperature of the refrigerant
(d) Could be of any value as it does not affect the COP **[ESE : 2006]**

2.36 The figure given below depicts saturation dome for water on the temperature-entropy plane. What is the temperature difference ΔT shown on a typical isobar line known as?

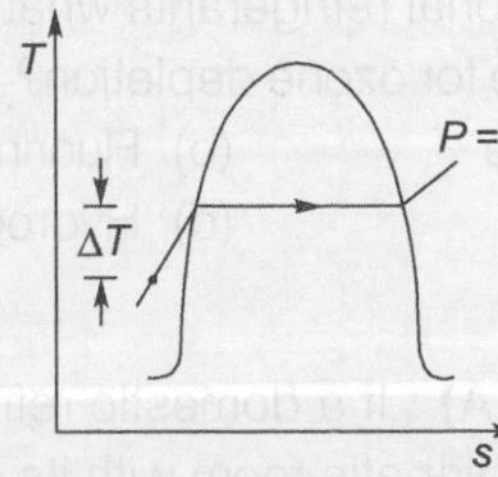

(a) Degree of wet bulb depression
(b) Degree of saturation
(c) Degree of subcooling
(d) Degree of reheat **[ESE : 2006]**

2.37 In a vapour absorption refrigeration, heat is rejected in
(a) Condenser only
(b) Generator only
(c) Absorber only
(d) Condenser and absorber **[ESE : 2006]**

2.38 In a vapour compression refrigeration plant, the enthalpy values at different points are :
(i) Enthalpy at exit of the evaporator = 350 kJ/kg
(ii) Enthalpy at exit of the compressor = 375 kJ/kg
(iii) Enthalpy at exit of the condenser = 225 kJ/kg
The refrigerating efficiency of the plant is 0.8. What is the power required per kW of cooling to be produced?
(a) 0.25 kW (b) 4.0 kW
(c) 12.5 kW (d) 11 kW
[ESE : 2006]

2.39 Where is an air refrigeration cycle generally employed?
(a) Domestic refrigerators
(b) Commercial refrigerators
(c) Air-Conditioning
(d) Gas liquefaction **[ESE : 2006]**

2.40 When the Brayton cycle working in the pressure limits of p_1 and p_2 is reversed and operated as a refrigerator, what is the ideal value of COP for such a cycle?

(a) $\left(\frac{p_2}{p_1}\right)^{\gamma-1} - 1$ (b) $\frac{1}{\left(\frac{p_2}{p_1}\right)^{\gamma-1} - 1}$

(c) $\frac{1}{\left(\frac{p_2}{p_1}\right)^{\frac{\gamma-1}{\gamma}} - 1}$ (d) None of these

[ESE : 2007]

2.41 A heat pump works on a reversed Carnot cycle. The temperature in the condenser coils is 27°C and that in the evaporator coils is –23°C. For a work input of 1 kW, how much is the heat pumped?
(a) 1 kW (b) 5 kW
(c) 6 kW (d) None of these
[ESE : 2007]

2.42 Selection of a refrigerant for a vapour compression system depends on which among the following?
(a) Toxicity
(b) Environmental effect
(c) Saturation pressure-temperature relationship
(d) All of the above **[ESE : 2007]**

2.43 For simple vapour compression cycle, enthalpy at suction = 1600 kJ/kg, enthalpy at discharge from the compressor = 1800 kJ/kg, enthalpy at exit from condenser = 600 kJ/Kg. What is the COP for this refrigeration cycle?
(a) 3.3 (b) 5.0
(c) 4.0 (d) 4.5 **[ESE : 2008]**

2.44 Match **List-I** (Cycle) with **List-II** (Equipment) and select the correct answer using the code given below the lists:

List-I
A. Air refrigeration
B. Vapour Compression refrigeration
C. Vapour absorption refrigeration
D. Steam jet refrigeration

List-II
1. Absorber
2. Flash chamber
3. Turbine
4. Compressor

Codes:

	A	B	C	D
(a)	3	2	1	4
(b)	3	4	3	2
(c)	3	4	1	2
(d)	1	2	3	4

[ESE : 2008]

2.45 What is an azeotrope?
(a) A non-halogenic refrigerant
(b) A refrigerant dissolved in alcohol
(c) A mixture of refrigerants without phase separation
(d) An eco-friendly refrigerant [ESE : 2008]

2.46 Environmental protection agencies advise against the use of Chloro-fluoro-carbon refrigerants because these react with
(a) Water vapour and cause acid rain
(b) Plants and cause green house effect
(c) Oxygen and cause its depletion
(d) Ozone layer and cause its depletion
[ESE : 2008]

2.47 Which one of the following expansion processes takes place in a vapour compression cycle?
(a) Polytropic process with change in temperature
(b) Adiabatic process with work transfer
(c) Isentropic process with change in enthalpy
(d) Adiabatic process with constant enthalpy
[ESE : 2009]

2.48 Which one of the following statements is correct with respect to the schematic diagram as shown below?

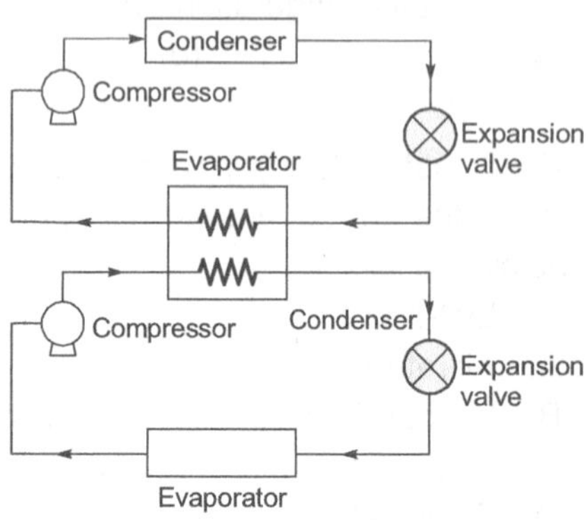

(a) Multi-evaporator vapour compression system of refrigeration
(b) Two stage compression, vapour compression refrigeration system
(c) Cascade system of vapour compression refrigeration system
(d) None of the above [ESE : 2009]

2.49 In vapour absorption refrigeration system heating in generator is done at 177°C, refrigeration in evaporator at –3°C and cooling in condenser at 27°C. Then what will be the maximum COP of the system?
(a) 1.5 (b) 2.0
(c) 3.0 (d) 4.0 [ESE : 2009]

2.50 In conventional refrigerants what is the element responsible for ozone depletion?
(a) Chlorine (b) Fluorine
(c) Carbon (d) Hydrogen
[ESE : 2009]

2.51 **Assertion (A) :** If a domestic refrigerator works inside an adiabatic room with its door open, the room temperature gradually decreases.
Reason (R) : Vapour compression refrigeration cycles have high COP compared to air refrigeration cycles.
(a) Both **A** and **R** are individually true and **R** is the correct explanation of **A**
(b) Both **A** and **R** are individually true but **R** is not the correct explanation of **A**
(c) **A** is true but **R** is false
(d) **A** is false but **R** is true [ESE : 2009]

2.52 In vapour compression refrigeration system, at entrance to which component the working fluid is superheated vapour?
(a) Evaporator (b) Condenser
(c) Compressor (d) Expansion valve
[ESE : 2009]

2.53 The effects of superheating of vapour in the evaporator and sub-cooling of condensate in the condenser, for the same compressor work
(a) Increase the COP
(b) Decrease the COP
(c) Superheating increase COP, but sub cooling decrease COP
(d) Superheating decrease COP, but subcooling increases COP
[ESE : 2010]

2.54 The vapour compression refrigeration cycle is an inherently irreversible cycle, because
(a) The compressor is non-ideal
(b) The evaporator is not frictionless
(c) The condensation process is not isothermal
(d) Of the use of expansion valve instead of an expansion engine [ESE : 2010]

2.55 **Assertion (A):** In a practical vapour compression refrigerator, the vapour should leave the evaporator with a definite amount of superheat.
Reason (R): It reduces the work done by the compressor.
(a) Both **A** and **R** are individually true and **R** is the correct explanation of **A**
(b) Both **A** and **R** are individually true but **R** is not the correct explanation of **A**
(c) **A** is true but **R** is false
(d) **A** is false but **R** is true [ESE : 2010]

2.56 A condenser of a refrigeration system rejects heat at a rate of 120 kW, while the compressor of the system consumes a power of 30 kW. The COP of the system will be
(a) 1/4 (b) 1/3
(c) 3 (d) 4 [ESE : 2011]

2.57 In a refrigeration plant, if the condenser temperature increases, the power input to the compressor will
(a) Increase (b) Decrease
(c) Remain the same (d) Be unpredictable
[ESE : 2011]

2.58 Lowest COP is of vapour
(a) Compression cycle with superheated vapour
(b) Compression cycle with dry compression
(c) Compression cycle with wet compression
(d) Absorption cycle
[ESE : 2011]

2.59 An ideal vapour compression refrigerator operates between a condenser pressure p_1 and an evaporator pressure p_2. Which of the following changes would increase its COP?
(a) Increasing p_1 by Δp and keeping p_2 constant
(b) Decreasing p_2 by Δp and keeping p_1 constant
(c) Adopting wet compression
(d) Subcooling the refrigerant
[ESE : 2012]

2.60 An absorption refrigeration system is supplied with energy Q_1 in the form of heat at temperature T_1, while it extracts energy Q_3 as heat from a cold space at temperature T_3. Then ambient temperature is T_2. Then the COP of the refrigerator is (ignoring the small amount of energy supplied in the pump)
(a) $\dfrac{T_1(T_2-T_1)}{T_3(T_1-T_2)}$ (b) $\dfrac{T_3(T_2-T_1)}{T_1(T_1-T_2)}$
(c) $\dfrac{T_3(T_1-T_2)}{T_1(T_2-T_3)}$ (d) $\dfrac{T_1(T_1-T_2)}{T_3(T_2-T_3)}$
[ESE : 2012]

2.61 Consider the following statements:
1. An ammonia absorption refrigerator has a COP more than 3 and is superior to vapour compression system.
2. Ammonia absorption machines are preferable where waste heat is available from an existing source.
3. Absorption refrigerator has no moving part and hence needs little maintenance.
4. The partial pressure of ammonia vapour varies, being high in the condenser and low in the evaporator.

Which of these statements are correct?
(a) 1, 2, 3 and 4 (b) 1, 2 and 3 only
(c) 2, 3 and 4 only (d) 1 and 4 only
[ESE : 2012]

2.62 A vapour compression refrigerator has a COP of 4 and extracts 10 kJ of heat from the cold reservoir. If this machine is worked as a heat pump , how much heat will it deliver to the environment?
(a) 2 kJ (b) 2.5 kJ
(c) 12.5 kJ (d) 25 kJ
[ESE : 2012]

2.63 A refrigerator based on reversed Carnot cycle works between refrigeration and condenser temperature of –23°C and 27°C. If a cooling rate of 5 kW is desired, what will be the required power in kW?

(a) 1 (b) 2.5
(c) 5 (d) 25

[ESE : 2012]

2.64 The following readings refer to a vapour compression refrigerator:
The enthalpy per kg of refrigerant flow, from *p-h* chart
at inlet to compressor — 1500 kJ/kg
at outlet to compressor — 1700 kJ/kg
at exit of condenser — 300 kJ/kg
The COP of the refrigerator is

(a) 3 (b) 4
(c) 5 (d) 6 [ESE : 2012]

2.65 Ammonia used as refrigerant is non-corrosive to

(a) iron and steel
(b) copper and copper alloys
(c) Both (a) and (b)
(d) Neither (a) nor (b) [ESE : 2012]

2.66 In a VCR plant, the refrigerant leaves the compressor and evaporator with enthalpy of 205 kJ/kg and 177 kJ/kg respectively. Enthalpy of the refrigerant leaving the condenser is 105 kJ/kg. If the mass flow rate of the refrigerant is 0.2 kg/s, the refrigeration effect will be

(a) 12.2 kW (b) 14.4 kW
(c) 16.4 kW (d) 20.2 kW

[ESE : 2012]

2.67 Statement (I): In a binary mixture charged refrigeration system, when a leakage occurs, first the high-pressure refrigerant leaks and then the low-pressure refrigerant leaks.

Statement (II): High-pressure refrigerant has the higher boiling point and low-pressure refrigerant has the lower boiling point.

(a) Both Statement (I) and Statement (II) are individually true and Statement (II) is the correct explanation of Statement (I)
(b) Both Statement (I) and Statement (II) are individually true but Statement (II) is not the correct explanation of Statement (I)
(c) Statement (I) is true but Statement (II) is false
(d) Statement (I) is false but Statement (II) is true.

[ESE : 2012]

2.68 The COP of a refrigerator working on a reversed Carnot cycle is 5. The ratio of the highest absolute temperature to the lowest absolute temperature would be

(a) 1.25 (b) 1.3
(c) 1.4 (d) 1.2 [ESE : 2012]

2.69 The most suitable pair of refrigerant and absorbent combination for solar refrigeration is

(a) ammonia and water
(b) ammonia and sodium thiocyanide
(c) water and lithium bromide
(d) R22 and dimethylformamide [ESE : 2012]

2.70 In an ideal vapour compression refrigeration cycle the refrigeration effect is 80 kJ/kg of refrigerant flow. The COP is 4. If the flow of the refrigerant is 2 kg/s, the heat rejected in the condenser is

(a) 50 kW (b) 100 kW
(c) 150 kW (d) 200 kW

[ESE : 2013]

2.71 A refrigerator operating on standard vapour compression cycle has COP = 5.5 and is driven by 40 kW compressor. Saturated liquid and saturated vapour enthalpies at condensing temperature of 30°C are 69.55 kJ/kg and 201.45 kJ/kg respectively. Saturated refrigerant vapour leaving the evaporator has enthalpy of 187.53 kJ/kg. Find refrigerant temperature at compressor discharge. (Take c_p of refrigerant vapour as 0.6155 kJ/kgK.)

(a) 40.2°C (b) 41.2°C
(c) 42.2°C (d) 43.2°C

[ESE : 2013]

2.72 In a vapour compression cycle, a good refrigerant should have a

(a) large latent heat of vaporization at condenser pressure
(b) large latent heat of evaporization at evaporator pressure
(c) condenser pressure close to critical pressure
(d) low critical pressure [ESE : 2013]

2.73 It is not a cause for discouragement of the use of Chlorofluorocarbon (CFC) refrigerants in domestic refrigerators and air-conditioners for

(a) their ozone depletion potential
(b) their global warming potential

(c) their tendency to migrate to the upper atmosphere by molecular diffusion
(d) their toxicity and non-availability
[ESE : 2013]

2.74 A Carnot heat pump works between 27°C and 327°C. What will be its COP?
(a) 0.09 (b) 1.00
(c) 1.09 (d) 2.0 [ESE : 2014]

2.75 In a vapour compression refrigeration system, the high pressure liquid from the condenser/receiver is cooled below its saturation temperature to
(a) Reduce the net work per cycle
(b) Reduce the net refrigerating effect
(c) Increase the net refrigerating effect
(d) Reduce the pressure on the high pressure side [ESE : 2014]

2.76 In an ideal Vapour Compression Refrigeration cycle the enthalpy values at salient points are as follows:
At inlet to compressor : 1500 kJ/kg
At outlet to compressor : 1800 kJ/kg
At inlet to evaporator : 300 kJ/kg
What is the COP of the cycle?
(a) 3 (b) 4
(c) 5 (d) 6 [ESE : 2014]

2.77 An air-conditioning system operating on the reversed Carnot cycle is required to remove heat from the room at a rate of 25 kW to maintain its temperature constant at 20°C. The temperature of the surroundings being 35°C, the power required to operate this air-conditioning system will be
(a) 1.28 kW (b) 4.02 kW
(c) 5.12 kW (d) 12.80 kW
[ESE : 2014]

2.78 Consider the following data referring to a refrigerator working on Vapour- compression refrigeration cycle:
1. Enthalpy at entry to compressor = 180kJ/kg
2. Enthalpy at exit from compressor = 210kJ/kg
3. Enthalpy at exit of condenser = 60 kJ/kg
What is the COP of the refrigerator?
(a) 2 (b) 3
(c) 4 (d) 5 [ESE : 2015]

2.79 In case of vapour compression refrigerator, if the condenser temperature of the refrigerant is closer to the critical temperature, then there will be
1. Excessive power consumption
2. High compression
3. Large volume flow
Which of the above statements are correct?
(a) 1 and 2 only (b) 1 and 3 only
(c) 2 and 3 only (d) 1, 2 and 3
[ESE : 2015]

2.80 Which of the following refrigeration systems is most suitable for solar cooling?
(a) Ejector refrigeration system
(b) Vapour absorption system
(c) Desiccant refrigeration system
(d) Vortex tube refrigeration system
[ESE : 2015]

2.81 **Assertion (A):** The main difference between vapour compression refrigeration cycle and Bell-Coleman gas refrigeration cycle is that in gas cycle, an expander is used in place of a throttle valve.
Reason (R): In throttling of a perfect gas, temperature remains constant.
(a) Both **A** and **R** are true and **R** is the correct explanation of **A**
(b) Both **A** and R are true but **R** is not a correct explanation of **A**
(c) **A** is true but **R** is false
(d) **A** is false but **R** is true [ESE : 2015]

2.82 Consider the following statements in respect of vapour compression refrigeration units:
1. In actual units the refrigerant leaving the evaporator is superheated
2. Superheating of refrigerant at exit of evaporator increases the refrigerating effect
3. The superheating of refrigerant increases the work of the compressor
Which of the above statements are correct ?
(a) 1 and 2 only (b) 1 and 3 only
(c) 2 and 3 only (d) 1, 2 and 3
[ESE : 2016]

2.83 In a vapour compression refrigerator, the heat rejected in condenser is 1500 kJ/kg of refrigerant flow and the work done by compressor is 250 kJ/kg. The COP of the refrigerator is:

(a) 5 (b) 6

(c) 7 (d) 8 **[ESE : 2016]**

2.84 A refrigeration plant is designed to work between –3°C and 27°C. The plant works on the Carnot cycle. If the same plant is used as a heat-pump system, then the COP of the heat pump becomes :

(a) 10 (b) 9

(c) 8 (d) 7 **[ESE : 2016]**

2.85 A refrigeration plant working on Carnot cycle is designed to take the load of 4 T of refrigeration. The cycle works between 2°C and 27°C. The power required to run the system is :

(a) 1.27 kW (b) 3.71 kW

(c) 5.71 kW (d) 7.27 kW

[ESE : 2016]

2.86 The COP of an ideal refrigerator of capacity 2.5 T is 5. The power of the motor required to run the plant is :

(a) 1.15 kW (b) 1.35 kW

(c) 1.55 kW (d) 1.75 kW **[ESE : 2016]**

2.87 The most suitable refrigeration system utilizing solar energy is :

(a) Ammonia-Water vapour absorption refrigeration system

(b) Lithium Bromide-Water vapour absorption refrigeration system

(c) Desiccant refrigeration system

(d) Thermoelectric refrigeration system

[ESE : 2016]

Direction: Each of the following items consists of two statements, one labelled as the **'Statement (I)'** and the other as **'Statement (II)'**. You are to examine these two statements carefully and select the answers to these items using the codes given below:

(a) Both Statement (I) and Statement (II) are individually true and Statement (II) is the correct explanation of Statement (I).

(b) Both Statement (I) and Statement (II) are individually true but Statement (II) is not the correct explanation of Statement (I).

(c) Statement (I) is true but Statement (II) is false.

(d) Statement (I) is false but Statement (II) is true.

2.88 **Statement (I) :** Use of non-azeotropic mixtures used as the refrigerant in a vapour-compression system improves the coefficient of performance.

Statement (II) : The increase in this coefficient is attributable to reduction in volume.

[ESE : 2017]

2.89 **Statement (I) :** Sub-cooling of a refrigerant liquid increases the coefficient of performance of a refrigeration cycle.

Statement (II) : Sub-cooling reduces the work requirement of the refrigeration cycle.

[ESE : 2017]

2.90 **Statement (I) :** In vapour-absorption system, larger the value of specific solution circulation, more the pump work.

Statement (II) : Higher solution circulation rates of poor as well as rich solutions need larger pressure drops in the system. **[ESE : 2017]**

2.91 The ordinates and abscissae of the diagram given for the vapour-compression refrigeration cycle represent

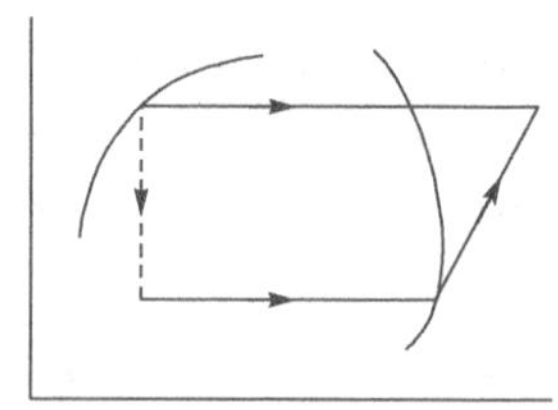

(a) pressure and volume

(b) temperature and entropy

(c) enthalpy and entropy

(d) pressure and enthalpy **[ESE : 2017]**

2.92 Consider the following statements for refrigeration and air-conditioning:

1. In a refrigerating machine, the heat exchanger that absorbs heat is connected to a conditioned space.

2. A refrigerating cycle operating reversibly between two heat reservoirs has the highest coefficient of performance.
3. The lower the refrigeration required and the higher the temperature of heat rejection to the surroundings, the larger the power consumption.

Which of the above statements are correct?
(a) 1 and 2 only (b) 1 and 3 only
(c) 2 and 3 only (d) 1, 2 and 3
[ESE : 2017]

2.93 Consider the following statements in respect of a vapour-absorption refrigeration cycle:
1. The absorption refrigeration cycle is generally used when waste heat is available from an existing source or when free energy like solar energy is to be used.
2. There are no moving parts in the absorption refrigeration plant except a small liquid pump.
3. The value of the coefficient of performance is nearly the same in both vapour-absorption and vapour-compression refrigeration plants.

Which of the above statements are correct?
(a) 1, 2 and 3 (b) 1 and 2 only
(c) 1 and 3 only (d) 2 and 3 only
[ESE : 2017]

2.94 In the working of a vapour-compression refrigeration plant, the following enthalpies are recorded at salient points in the cycle :
1. Enthalpy at inlet to compressor (saturated vapour), h_1 = 300kJ/kg
2. Enthalpy at outlet of compressor (after isentropic compression), h_2 = 330 kJ/kg
3. Enthalpy at exit of condenser (saturated liquid), h_3 = 150 kJ/kg

What is the COP of the plant?
(a) 3 (b) 4
(c) 5 (d) 6 [ESE : 2017]

2.95 Consider the following statements in respect of (*l*) the temperature of the medium, (*m*) the refrigerant and. (*n*) the condenser and absorption system-in a refrigeration unit:
1. Temperature of the medium being cooled must be below that of the evaporator.
2. Refrigerant leaves the condenser as liquid.
3. Any solar thermally operated absorption system is capable only of intermittent operation.

Which of the above statements are correct?
(a) 1 and 2 only (b) 1 and 3 only
(c) 2 and 3 only (d) 1, 2 and 3
[ESE : 2017]

2.96 In a vapour absorption refrigerator, the temperatures of evaporator and ambient air are 10°C and 30°C, respectively. For obtaining COP of 2 for this system, the temperature of the generator is to be nearly
(a) 90°C (b) 85°C
(c) 80°C (d) 75°C [ESE : 2018]

2.97 The following data refer to a vapour compression refrigerator:
Enthalpy at compressor inlet = 1200 kJ/kg
Enthalpy at compressor outlet = 1400 kJ/kg
Enthalpy at condenser outlet = 200 kJ/kg The COP of the refrigerator is
(a) 7 (b) 6
(c) 5 (d) 4 [ESE : 2018]

2.98 Consider the following statements regarding vapour absorption system in the field of refrigeration:
1. In ammonia-water absorption system, ammonia is the refrigerant.
2. In water-lithium bromide system, water is the refrigerant.
3. Ammonia-water absorption reaction is endothermic.
4. The amount of ammonia absorbed by water is inversely proportional to the temperature of ammonia.

Which of the above statements are correct?
(a) 1, 2 and 3 (b) 1, 3 and 4
(c) 1, 2 and 4 (d) 2, 3 and 4
[ESE : 2018]

2.99 **Statement (I)** : Referring to vapour compression refrigeration system, the coefficient of performance (COP) of a domestic refrigerator is less than that of a comfort air-conditioning plant.

Statement (II) : In domestic refrigerator, the work required for pumping the same amount of heat is more than that in an air-conditioning plant because of greater difference between condenser and evaporator temperatures.

(a) Both Statement (I) and Statement (II) are individually true and Statement (II) is the correct explanation of Statement (I).
(b) Both Statement (I) and Statement (II) are individually true but Statement (II) is not the correct explanation of Statement (I).
(c) Statement (I) is true but Statement (II) is false.
(d) Statement (I) is false but Statement (II) is true.

[ESE : 2018]

2.100 In an Electrolux refrigerator, a thermo-siphon bubble pump is used to lift the
(a) weak aqua solution from the generator to the separator
(b) weak aqua solution from the separator to the absorber
(c) strong aqua solution from the generator to the separator
(d) strong aqua solution from the generator to the evaporator

[ESE : 2019]

2.101 A Freon 12 simple saturation cycle operates at temperatures of 35°C and –15°C for the condenser and evaporator. If the refrigeration effect produced by the cycle is 111.5 kJ/kg and the work required by the compressor is 27.2 kJ/kg, the value of COP will be nearly
(a) 4.1 (b) 3.6
(c) 3.1 (d) 2.6

[ESE : 2020]

2.102 In an absorption type refrigeration system, heating in generator, refrigeration in evaporator and cooling by cooling water in condenser, take place at 95°C, –5°C and 30°C respectively. What is the maximum COP of the system?
(a) 1.17 (b) 1.35
(c) 1.52 (d) 1.78

[ESE : 2021]

2.103 A refrigerator operates on Reversed Carnot cycle. What is the power required to drive the refrigerator between temperatures of 42°C and 4°C, if heat at the rate of 2 kJ/s is extracted from the low temperature region?
(a) 0.174 kW (b) 0.374 kW
(c) 0.274 kW (d) 0.474 kW

[ESE : 2021]

2.104 Freon-12 is used in a simple saturation cycle, with suction saturation temperature of –10°C and condensing saturation temperature of 30°C. If the clearance volume is 6% of the stroke volume, what is the volumetric efficiency? (Consider specific volume at suction and discharge to be 0.07815 m^3/kg and 0.025 m^3/kg respectively)
(a) 87.24% (b) 71.31%
(c) 64.85% (d) 55.43%

[ESE : 2021]

2.105 Relative ozone destruction efficiency of R-12 is
(a) 0.29 (b) 0.86
(c) 0.05 (d) 0.57 [ESE : 2021]

2.106 **Statement I :** The vapour absorption system uses heat energy to change the condition of the refrigerant from the evaporator.

Statement II : The load variations do not affect the performance of a Vapour absorption system.

(a) Both Statement (I) and Statement (II) are individually true and Statement (II) is the correct explanation of Statement (I).
(b) Both Statement (I) and Statement (II) are individually true but Statement (II) is not the correct explanation of Statement (I).
(c) Statement (I) is true but Statement (II) is false.
(d) Statement (I) is false but Statement (II) is true. [ESE : 2022]

2.107 The performance test of an air conditioning unit rated as 140.7 kW (40 TR) seems to be indicating poor cooling. The test on heat rejection to atmosphere in its condenser shows the following.

Cooling water flow rate : 4 L/s

Water temperature : Inlet 30°C and outlet 40°C

Power input to motor : 48 kW (95% efficiency);

What is the actual refrigerating capacity of the unit?

[Take C_w = 4.1868]

(a) 34.7 TR (b) 45.6 TR
(c) 52.6 TR (d) 48.6 TR

[ESE : 2022]

2.108 Which one of the following is used in aircraft refrigeration?
(a) Vapour compression cycle refrigeration
(b) Gas cycle refrigeration
(c) Vapor absorption cycle refrigeration
(d) Steam ejector cycle refrigeration

[ESE : 2022]

2.109 An ammonia ice plant operates between a condenser temperature of 35°C and a evaporator temperature of –15°C. It produces 10 tons of ice per day from water at 30°C to ice at – 5°C. Assume simple saturation cycle, what is the capacity of the refrigeration plant? (Take specific heat of ice is 1.94 kJ/kgK. Specific heat of water is 4.1868 kJ/kgK; latent heat of fusion of ice at 0°C is 335 kJ/kg)
(a) 54.43 kW (b) 64.32 kW
(c) 74.52 kW (d) 84.23 kW

[ESE : 2022]

2.110 Consider the following statements in a vapor compression refrigeration system:
It is observed that an increase in condenser pressure, similarly results in
1. a decrease in the refrigerating capacity.
2. an increase in power consumption.
3. an increase in volumetric efficiency.

Which of the above statements are correct?
(a) 1 and 3 only (b) 2 and 3 only
(c) 1 and 2 only (d) 1, 2 and 3

[ESE : 2022]

2.111 Which one of the following is the designation for dichloro-tetrafluoro-ethane refrigerant used in refrigeration system?
(a) R114 (b) R116
(c) R113 (d) R118

[ESE : 2022]

2.112 Which of the following are highly explosive and flammable in a refrigeration system?
1. Methane
2. Butane
3. R134a

Select the correct answer using the code given below:
(a) 2 and 3 only (b) 1 and 2 only
(c) 1 and 3 only (d) 1, 2 and 3

[ESE : 2022]

2.113 A hydrocarbon refrigerator operating on equal proportions of propane and isobutane by mass works on simple saturation cycle. The condenser and evaporator pressure are 14 bar and 1.4 bar. What is the approximate theoretical COP of the system?
[Take properties: h_1 = 545 kJ/kg;
h_3 = 341.15 kJ/kg = h_4;
S_1 = 2.47 kJ/kgK = S_2;
h_2 = 661 kJ/kg and T_2 = 63°C]
(a) 1.8 (b) 5.4
(c) 3.6 (d) 2.7

[ESE : 2022]

2.114 The use of a single refrigerant in a simple vapor compression cycle for the production of low temperatures is limited by which of the following reasons?
1. Solidification temperature of the refrigerant.
2. Extremely low pressures in the evaporator and large suction volumes if a high-boiling refrigerant is selected.
3. Very high-pressure ratio, therefore a low coefficient of performance.

Select the correct answer using the code given below:
(a) 1 and 2 only (b) 2 and 3 only
(c) 1 and 3 only (d) 1, 2 and 3

[ESE : 2022]

■■■■

Answers Ideal, Vapour, Air Refrigeration Cycle and Refrigerants

2.1	(d)	2.2	(c)	2.3	(b)	2.4	(b)	2.5	(d)	2.6	(b)	2.7	(d)	2.8	(d)	2.9	(a)
2.10	(c)	2.11	(c)	2.12	(c)	2.13	(b)	2.14	(d)	2.15	(c)	2.16	(a)	2.17	(a)	2.18	(b)
2.19	(b)	2.20	(a)	2.21	(c)	2.22	(a)	2.23	(b)	2.24	(c)	2.25	(c)	2.26	(d)	2.27	(a)
2.28	(b)	2.29	(c)	2.30	(d)	2.31	(c)	2.32	(c)	2.33	(c)	2.34	(a)	2.35	(c)	2.36	(c)
2.37	(d)	2.38	(a)	2.39	(d)	2.40	(c)	2.41	(c)	2.42	(d)	2.43	(b)	2.44	(c)	2.45	(c)
2.46	(d)	2.47	(d)	2.48	(c)	2.49	(c)	2.50	(a)	2.51	(d)	2.52	(b)	2.53	(a)	2.54	(d)
2.55	(c)	2.56	(c)	2.57	(a)	2.58	(d)	2.59	(d)	2.60	(c)	2.61	(c)	2.62	(c)	2.63	(a)
2.64	(d)	2.65	(a)	2.66	(b)	2.67	(c)	2.68	(d)	2.69	(a)	2.70	(d)	2.71	(c)	2.72	(b)
2.73	(d)	2.74	(d)	2.75	(c)	2.76	(b)	2.77	(a)	2.78	(c)	2.79	(d)	2.80	(b)	2.81	(a)
2.82	(d)	2.83	(a)	2.84	(a)	2.85	(a)	2.86	(d)	2.87	(a)	2.88	(*)	2.89	(c)	2.90	(a)
2.91	(d)	2.92	(d)	2.93	(b)	2.94	(c)	2.95	(c)	2.96	(c)	2.97	(c)	2.98	(c)	2.99	(a)
2.100	(a)	2.101	(a)	2.102	(b)	2.103	(c)	2.104	(a)	2.105	(b)	2.106	(b)	2.107	(a)	2.108	(b)
2.109	(a)	2.110	(c)	2.111	(a)	2.112	(b)	2.113	(a)	2.114	(d)						

Explanations Ideal, Vapour, Air Refrigeration Cycle and Refrigerants

2.1 (d)

For unsaturated compound
$2m = p + q + n$

2.2 (c)

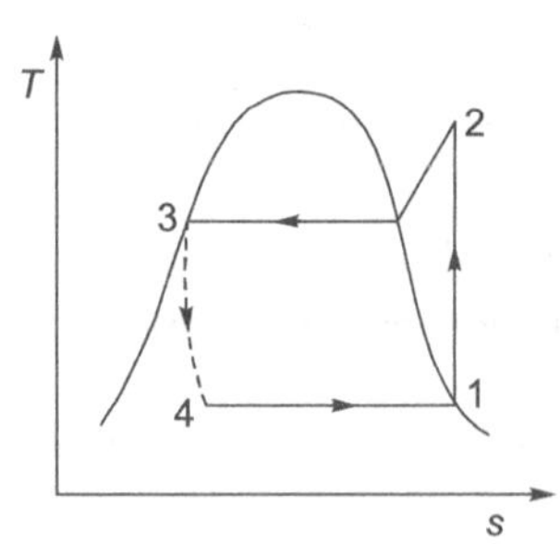

$$h_1 = 185 \text{ kJ/kg}$$
$$h_2 = 210 \text{ kJ/kg}$$
$$h_3 = h_4 = 85 \text{ kJ/kg}$$

$$\text{COP} = \frac{h_1 - h_4}{h_2 - h_1}$$
$$= \frac{185 - 85}{210 - 185} = \frac{100}{25} = 4$$

2.4 (b)

The leakage of Freon based refrigeration system can be detected by soap solution, a halide torch or an electronic detector.

2.5 (d)

Subcooling has no effect on compressor work but only increases refrigeration effect.

2.6 (b)

R-12 : CCl_2F_2
R - 22 : CHF_2Cl
R - 717 : NH_3
R - 113 : $CCl_2F{-}CClF_2$

2.7 (d)

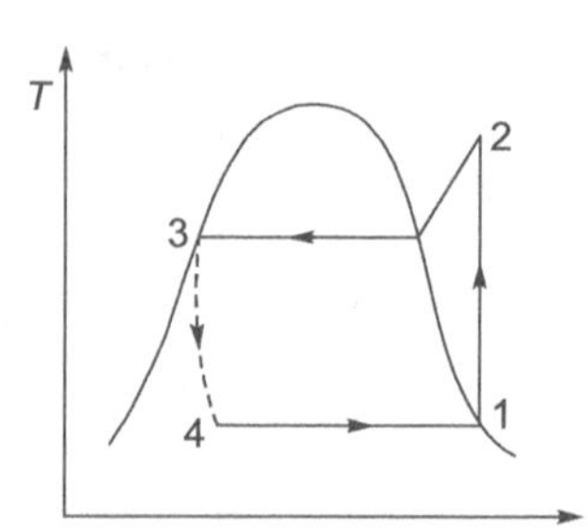

$h_1 = 195$ kJ/kg; $h_2 = 210$ kJ/kg

$h_3 = 90$ kJ/kg; $h_4 = 90$ kJ/kg

$$\text{Desired heating effect} = \dot{m}(h_2 - h_3) = 0.5 \times (210 - 90) = 120 \times 0.5 = \mathbf{60\ kW}$$

2.8 (d)

Given: $T_G = 360$ K; $T_E = 270$ K; $T_C = 300$ K

$$\text{COP} = \frac{T_E}{T_C - T_E} \times \frac{T_G - T_C}{T_G} = \frac{270}{30} \times \frac{60}{360} = \mathbf{1.5}$$

2.9 (a)

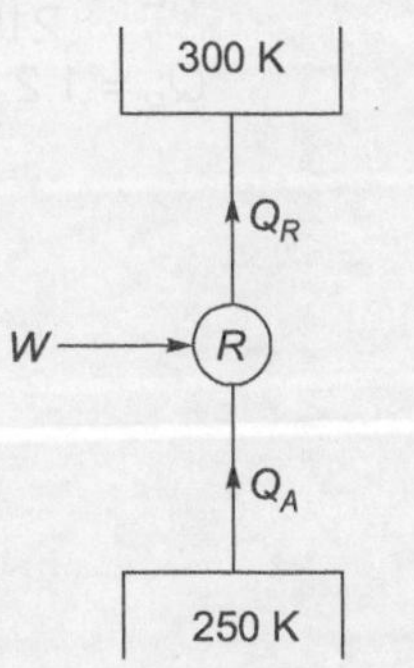

$$(\text{COP})_{max} = \frac{250}{300 - 250} = 5$$

$$(\text{COP})_{actual} = 0.8 \times 5 = 4$$

$$4 = \frac{\text{Re}}{W_{in}} = \frac{\text{Re}}{1}$$

$$Re = \mathbf{4\ kW}$$

2.10 (c)

1 TR = 3.5 kW

$$3.5 = \dot{m} \times 4.18 \times 15$$

$$\therefore \quad \dot{m} = 0.05582 \text{ kg/s}$$

$$\dot{V} = \frac{\dot{m}}{\rho} = \frac{0.05582}{1000} = 0.05582 \times 10^{-3} \text{ m}^3\text{/s} \approx \mathbf{200\ litre/hr}$$

2.11 (c)

For a cycle

$$\Sigma Q = \Sigma W$$

$$70 - 90 - 5 = W_{in}$$

$$W_{in} = \mathbf{-25\ kW}$$

2.12 (c)

Ozone depletion by CFCs occurs by breakdown of chlorine atoms from refrigerant by UV radiation and reaction with zone in stratosphere.

2.13 (b)

Process 1–2: Isentropic compression

Process 2–3: Constant pressure heat rejection

Process 3–4: Isenthalpic expansion

Process 4–1: Constant pressure heat addition

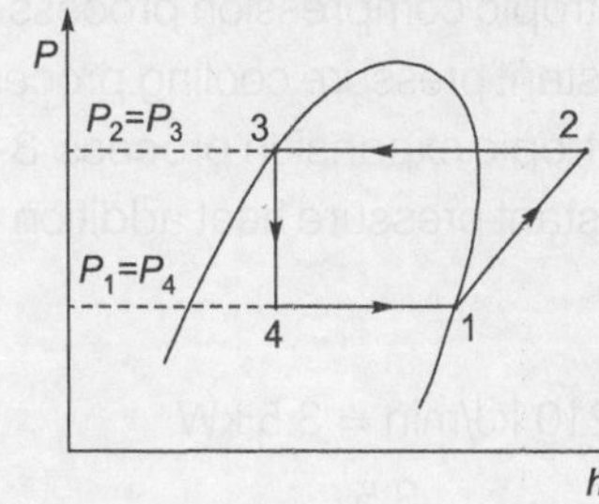

2.14 (d)

$$\text{COP} = \frac{T_E}{T_C - T_E} \times \frac{T_G - T_C}{T_G} = \frac{270}{300 - 270} \times \frac{360 - 300}{360} = 1.5$$

2.15 (c)

Given data

$h'_3 = 78$ kJ/kg; $h'_1 = 182$ kJ/kg

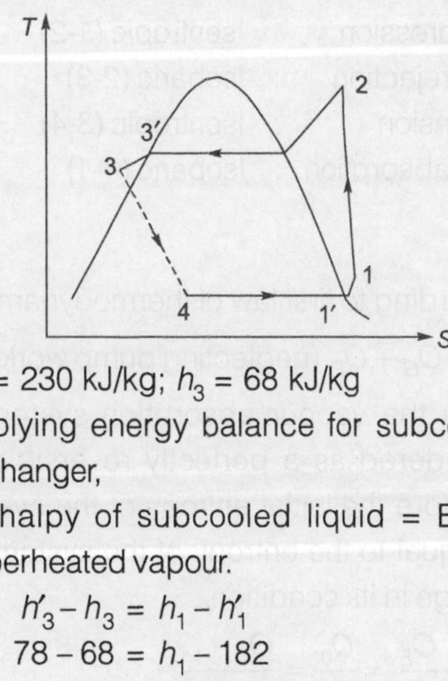

$h_2 = 230$ kJ/kg; $h_3 = 68$ kJ/kg

Applying energy balance for subcooling heat exchanger,

Enthalpy of subcooled liquid = Enthalpy of superheated vapour.

$$h'_3 - h_3 = h_1 - h'_1$$

$$78 - 68 = h_1 - 182$$

$$h_1 = 192 \text{ kJ/kg}$$

or $$COP = \frac{\text{Refrigerating effect } (RE)}{\text{Work input } (W)} = \frac{h'_1 - h_4}{h_2 - h_1} = \frac{182 - 68}{230 - 192} = \mathbf{3}$$

2.17 (a)

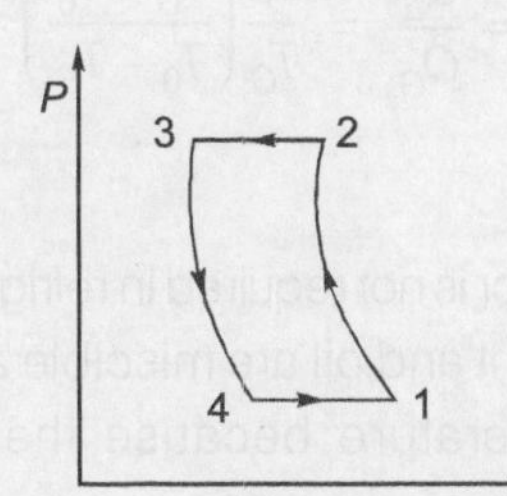

Air refrigeration cycle is Bell-Coleman cycle (Reversed Brayton or Joule cycle). It consists of
- Isentropic compression process 1-2
- Constant pressure cooling process 2-3
- Isentropic expansion process 3-4
- Constant pressure heat addition 4-1

2.18 (b)

1 TR = 210 kJ/min = 3.5 kW

$$\therefore\ (COP)_{Carnot} = \frac{3.5}{1.5} = 2.33$$

2.19 (b)

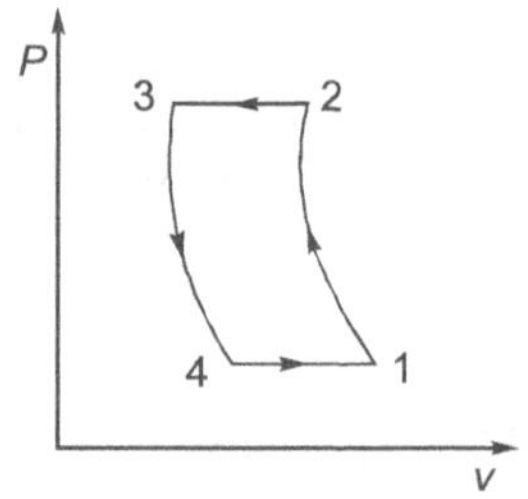

Compression	:	Isentropic (1-2)
Heat rejection	:	Isobaric (2-3)
Expansion	:	Isentropic (3-4)
Heat absorption	:	Isobaric (4-1)

2.20 (a)

According to first law of thermodynamics

$Q_0 = Q_G + Q_E$ (neglecting pump work W_p)

Since the vapour absorption system can be considered as a perfectly reversible system, therefore the initial entropy of the system must be equal to the entropy of the system after the change in its condition.

$$\frac{Q_G}{T_G} + \frac{Q_E}{T_E} = \frac{Q_0}{T_O} = \frac{Q_G + Q_E}{T_O}$$

$$\Rightarrow\quad Q_G\left(\frac{1}{T_G} - \frac{1}{T_0}\right) = Q_E\left(\frac{1}{T_0} - \frac{1}{T_E}\right)$$

$$\therefore\quad Q_G = Q_E\left(\frac{T_0 - T_E}{T_E}\right)\left(\frac{T_G}{T_G - T_0}\right)$$

$$(COP)_{max} = \frac{Q_E}{Q_G} = \frac{T_E}{T_G}\left(\frac{T_G - T_0}{T_0 - T_E}\right)$$

2.21 (c)

Oil separator is not required in refrigeration system if refrigerant and oil are miscible at all pressure and temperature because they do not get separated in evaporator and fouling doesn't take place.

2.22 (a)

Absorber and generator both are used in absorption refrigeration system.
Compressor is replaced by an absorber, a pump, generator and a pressure reducing valve.

2.23 (b)

Heat rejection factor =

$$\frac{\text{Heat rejected in condenser}}{\text{Refrigeration effect}}$$

$$1.2 = \frac{Q_R}{2100}$$

$Q_R = 1.2 \times 2100 = 2520$ kJ/min

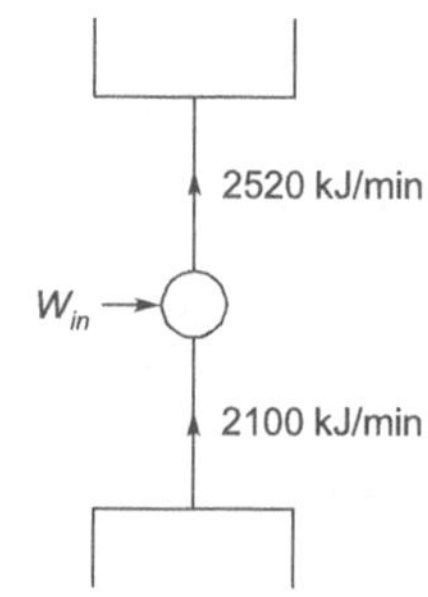

$W_{in} = 2520 - 2100 = 420$ kJ/min

$$COP = \frac{2100}{420} = \mathbf{5}$$

2.24 (c)

Subcooling increases the refrigeration effect but it has no effect on the work input.

2.25 (c)

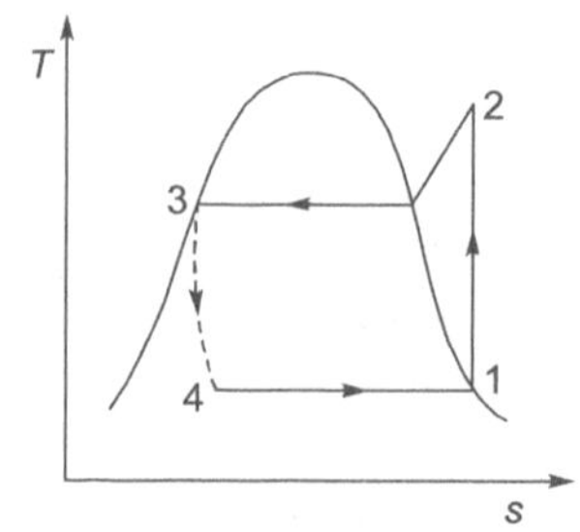

Given: $h_1 = 185$ kJ/kg; $h_2 = 210$ kJ/kg
$h_4 = 85$ kJ/kg

$$COP = \frac{h_1 - h_4}{h_2 - h_1}$$

$$COP = \frac{185 - 85}{210 - 185} = \frac{100}{25} = 4$$

2.26 (d)

$$\frac{T_L}{T_H} = 0.8$$

$$(COP)_{HP} = \frac{T_H}{T_H - T_L} = \frac{1}{1 - T_L/T_H}$$

$$= \frac{1}{1 - 0.8} = \frac{1}{0.2} = 5$$

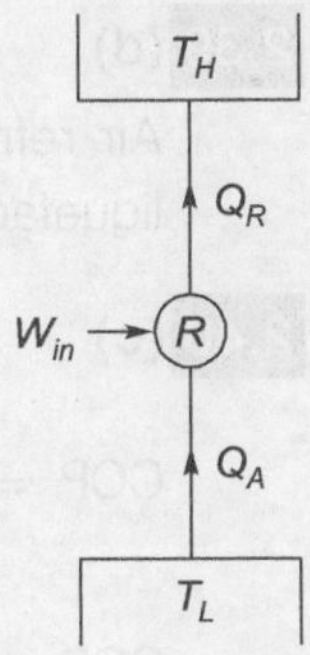

2.27 (a)

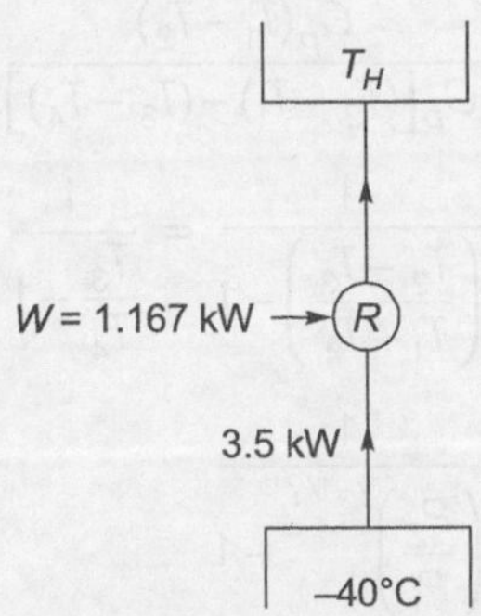

$$W = \frac{700}{10}\text{kJ/min} = \frac{700}{10 \times 60}\text{kJ/s}$$

$$\text{Now, } (COP)_R = \frac{RE}{W} = \frac{3.5 \times 60}{700/10}$$

$$= \frac{3.5 \times 10 \times 60}{700} = 3$$

2.28 (b)

$$HRR = \frac{Q_C}{R_E} = \frac{R_E + W}{R_E} = 1 + \frac{W}{R_E}$$

$$HRR = 1 + \frac{1}{COP}$$

$$\text{Then } 1.2 = 1 + \frac{1}{COP}$$

$$\therefore \quad COP = 5$$

2.29 (c)

Refrigeration effect = mass flow rate (Enthalpy at evaporator exit – Enthalpy at condenser exit)
= 1 × (195 – 65) = 130 kW

2.30 (d)

A Bell-coleman air refrigeration cycle works as reversed Brayton cycle or Joule cycle.

2.32 (c)

$$(COP)_{HP} = \frac{T_H}{T_H - T_L}$$

$$7 = \frac{1}{1 - T_L/T_H}$$

$$\frac{1}{7} = 1 - \frac{T_L}{T_H}$$

$$\frac{T_L}{T_H} = 1 - \frac{1}{7} = \frac{6}{7}$$

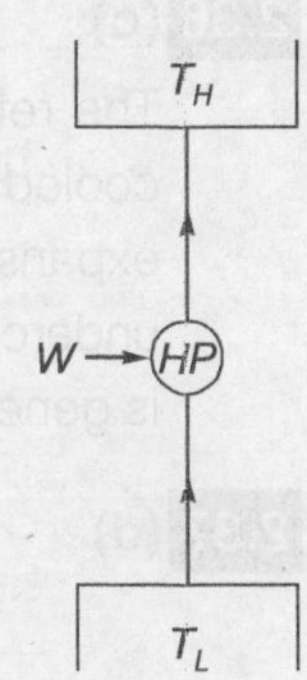

2.33 (c)

$$(COP)_R = \frac{T_L}{T_H - T_L}$$

$$6 = \frac{T_L}{T_H - T_L}$$

$$= \frac{1}{T_H/T_L - 1}$$

$$\frac{1}{6} = \frac{T_H}{T_L} - 1$$

$$\frac{T_H}{T_L} = 1 + \frac{1}{6} = \frac{7}{6}$$

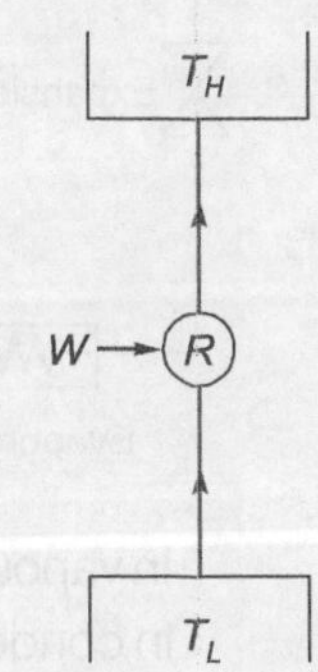

2.34 (a)

Halide torch method is used for detection of Freon leaks. It produces greenish flame lighting.

2.35 (c)

When condensor temperature is near or above the critical temperature, then work done by the compressor increases and COP decreases. When condensor temperature is below the critical temperature then compressor work will be less and refrigeration effect will increase i.e. COP is higher.

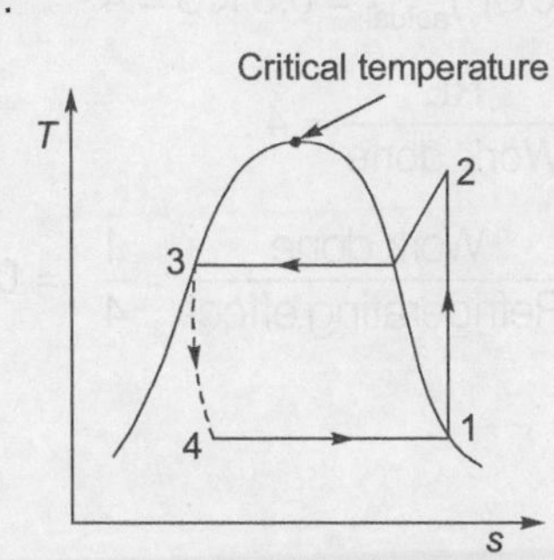

2.36 (c)

The refrigerant, after condensation process is cooled below the saturation temperature before expansion by throttling. Such a process is called undercooling or subcooling of the refrigerant and is generally done along the liquid line.

2.37 (d)

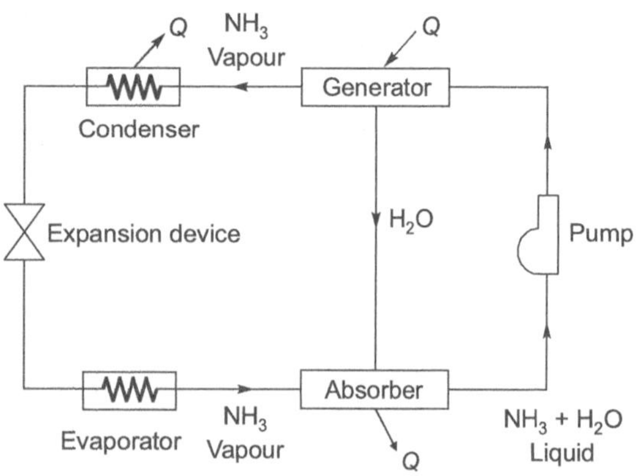

In vapour absorption refrigeration, heat is rejected in condensor and absorber.

2.38 (a)

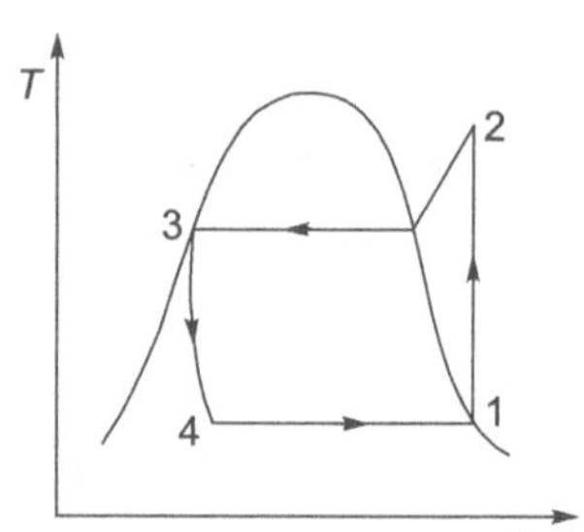

$h_1 = 350$ kJ/kg; $h_2 = 375$ kJ/kg

$h_3 = h_4 = 225$ kJ/kg

$$COP = \frac{h_1 - h_4}{h_2 - h_1} = \frac{350 - 225}{375 - 350} = \frac{125}{25} = 5$$

$$\text{Refrigerating efficiency} = \frac{(COP)_{actual}}{(COP)_{Theoretical}}$$

$$\Rightarrow \quad (COP)_{actual} = 0.8 \times 5 = 4$$

$$\Rightarrow \quad \frac{RE}{\text{Work done}} = 4$$

$$\Rightarrow \quad \frac{\text{Work done}}{\text{Refrigerating effect}} = \frac{1}{4} = \mathbf{0.25}$$

2.39 (d)

Air refrigeration cycle is used widely for gas liquefaction.

2.40 (c)

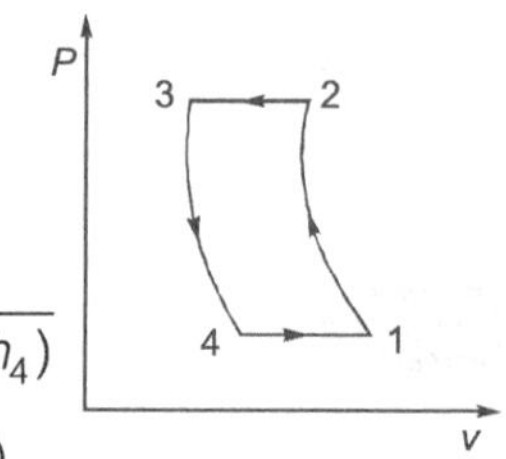

$$COP = \frac{RE}{Win}$$

$$COP = \frac{h_1 - h_4}{(h_2 - h_1) - (h_3 - h_4)}$$

$$= \frac{C_p(T_1 - T_2)}{C_p[(T_2 - T_1) - (T_3 - T_4)]}$$

$$COP = \frac{1}{\left(\frac{T_2 - T_3}{T_1 - T_4}\right) - 1} = \frac{1}{\frac{T_3}{T_4} - 1}$$

$$COP = \frac{1}{\left(\frac{P_2}{P_1}\right)^{\gamma - 1/\gamma} - 1}$$

2.41 (c)

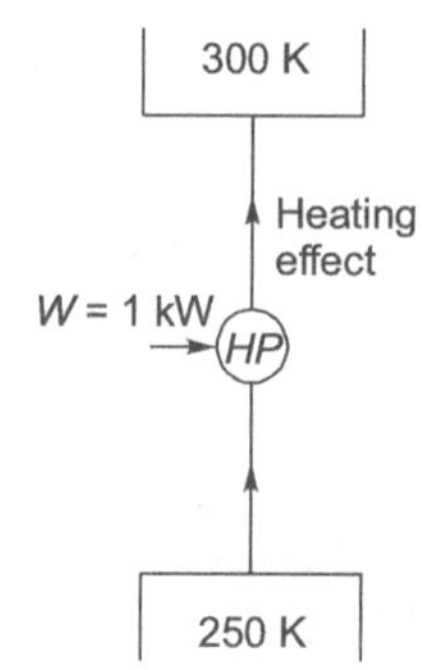

$$COP = \frac{300}{300 - 250} = 6$$

$$6 = \frac{\text{Heating effect}}{W}$$

$$6 = \frac{\text{Heating effect}}{1}$$

Hence heating effect = 6 kW

2.43 (b)

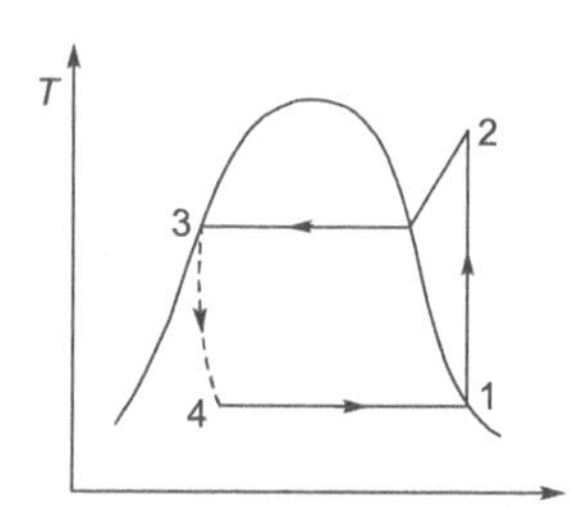

$h_1 = 1600$ kJ/kg

$h_2 = 1800$ kJ/kg

$h_3 = h_4 = 600$ kJ/kg

$$COP = \frac{RE}{W_{in}} = \frac{h_1 - h_4}{h_2 - h_1} = \frac{1600 - 600}{1800 - 1600} = \frac{1000}{200}$$

$$= \mathbf{5}$$

2.45 (c)

Azeotrope is a mixture of refrigerants without phase separation.

2.46 (d)

CFC refrigerants react with ozone layer and cause its depletion.

2.49 (c)

Given that

$$T_G = 177°C = 450 \text{ K}$$
$$T_C = 27°C = 300 \text{ K}$$
$$T_E = -3°C = 270 \text{ K}$$

COP of Refrigerator

$$= \eta_{H.E} \times COP_R$$

$$= \left(1 - \frac{T_C}{T_G}\right)\left(\frac{T_E}{T_C - T_E}\right) = \left(1 - \frac{300}{450}\right)\left(\frac{270}{300 - 270}\right)$$

$$= 3$$

2.50 (a)

In conventional refrigerants, chlorine element is responsible for ozone depletion.

2.56 (c)

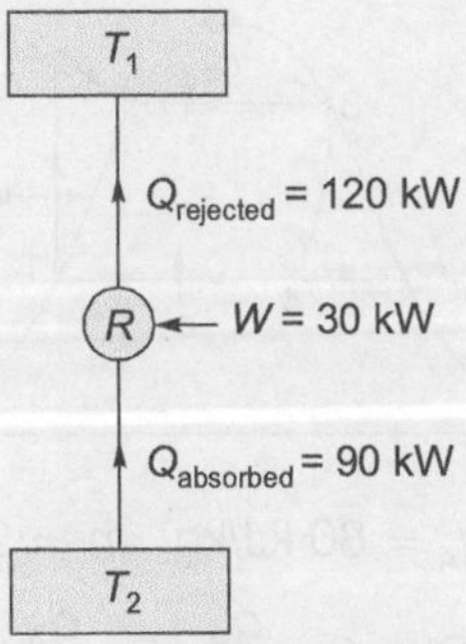

$$(COR)_R = \frac{\text{Desired effect}}{\text{Work required}}$$

$$Q_{abs} + W = 120 \text{ kW}$$
$$Q_{abs} = 120 - W$$
$$Q_{ab} = 120 - 30 = 90 \text{ kW}$$

$$\therefore \quad COP = \frac{90}{30} = 3$$

2.57 (a)

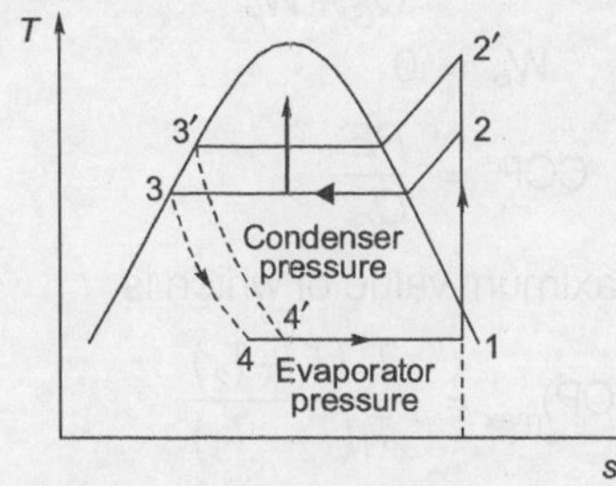

As condenser pressure increases, the work input to the compressor will increases.

2.58 (d)

Vapour absorption cycle operates on heat energy and has the least COP.

2.59 (d)

Only subcooling the refrigerant will increase the COP of the system.

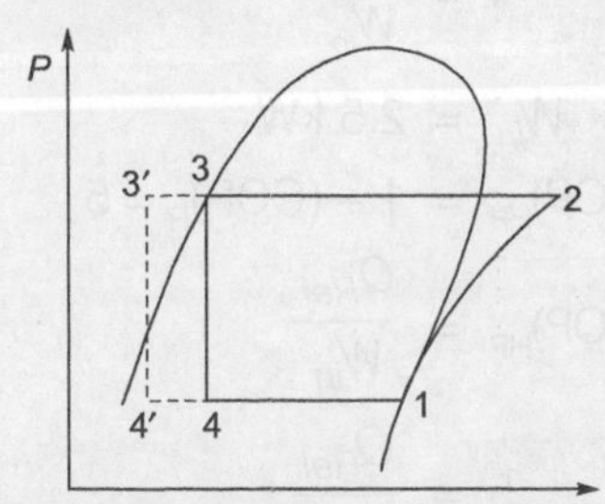

As shown in figure, on subcooling 3-3′ the RE will increase by 4′-4 without increase in work input to the compressor i.e., 1-2 hence the COP increases.

2.60 (c)

In absorption refrigeration

1 = Generator

C = Condenser

A = Absorber

3 = Storage space

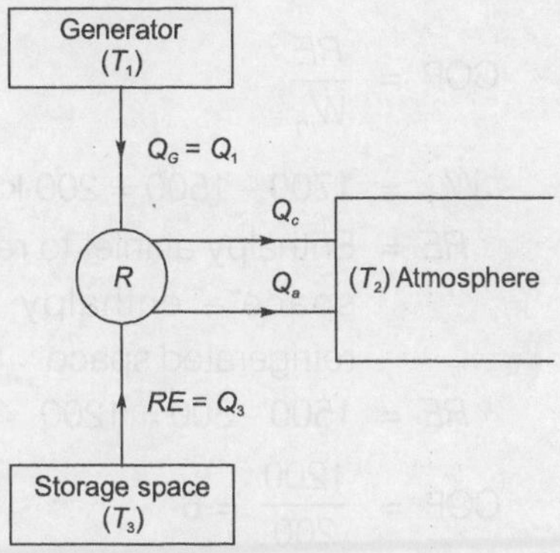

$$\text{COP} = \frac{RE}{Q_G + W_P}$$

$$W_P \approx 0$$

$$\therefore \quad \text{COP} = \frac{RE}{Q_G}$$

The maximum value of which is

$$(\text{COP})_{max} = \frac{T_3(T_1 - T_2)}{T_1(T_2 - T_3)}$$

2.61 (c)

The COP of an absorption system is less generally and hence it is inferior to vapour compression system interms of COP.

2.62 (c)

$$(\text{COP})_R = \frac{RE}{\dot{W}_{in}}$$

$$4 = \frac{10}{\dot{W}_{in}}$$

$$\dot{W}_{in} = 2.5 \text{ kW}$$

$$(\text{COP})_{HP} = 1 + (\text{COP})_R = 5$$

$$\therefore \quad (\text{COP})_{HP} = \frac{Q_{Rej}}{\dot{W}_{in}}$$

$$5 = \frac{Q_{Rej}}{2.5}$$

$$Q_{Rej} = \mathbf{12.5 \text{ kJ kW}}$$

2.63 (a)

$$\text{COP} = \frac{T_L}{T_H - T_L} = \frac{250}{300 - 250}$$

$$\text{COP} = 5$$

Also, $\text{COP} = \dfrac{RE}{\dot{W}_{in}}$

$$\therefore \quad \dot{W}_{in} = \frac{5}{5} = \mathbf{1 \text{ kW}}$$

2.64 (d)

$$\text{COP} = \frac{RE}{\dot{W}_{in}}$$

$$\dot{W}_{in} = 1700 - 1500 = 200 \text{ kJ/kg}$$

RE = Enthalpy at inlet to refrigerated space – enthalpy at exit to refrigerated space

$$RE = 1500 - 300 = 1200$$

$$\therefore \quad \text{COP} = \frac{1200}{200} = \mathbf{6}$$

2.65 (a)

Ammonia is corrosive to copper and copper alloys but is non corrosive to iron and steel.

2.66 (b)

RE = Enthalpy of refrigerant leaving evaporator – Enthalpy of refrigerant leaving the condenser

$$RE = 177 - 105 = 72 \text{ kJ/kg}$$

$$\dot{RE} = 72 \times 0.2 = \mathbf{14.4 \text{ kW}}$$

2.67 (c)

A refrigerant having high vapour pressure will have a lower boiling point and vice-versa.

2.68 (d)

$$\text{COP} = \frac{T_L}{T_H - T_L}$$

$$5 = \frac{T_L}{T_H - T_L}$$

$$0.2 = \frac{T_H}{T_L} - 1$$

$$\frac{T_H}{T_L} = \mathbf{1.2}$$

2.70 (d)

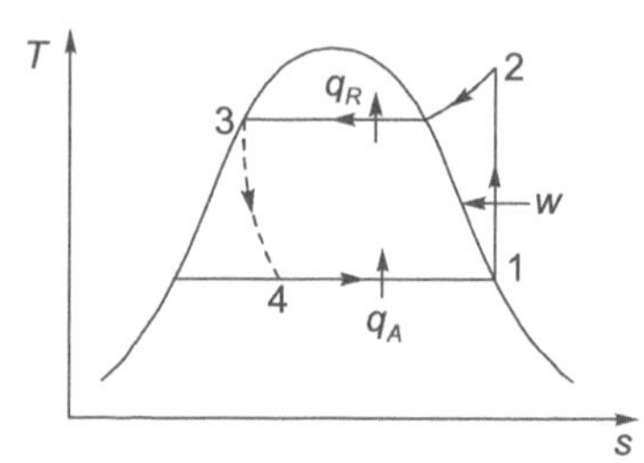

$$q_A = 80 \text{ kJ/kg}, \ \dot{m} = 2 \text{ kg/s}$$

$$\text{COP} = 4 = \frac{q_A}{w} = \frac{q_A}{q_R - q_A}$$

$$\therefore \quad \frac{q_R}{q_A} - 1 = \frac{1}{\text{COP}}$$

$$\therefore \quad q_R = 80\left(1 + \frac{1}{4}\right) = 80 \times \frac{5}{4} = 100 \text{ kJ/kg}$$

So, $q_R = 100 \times 2 = 200 \text{ kJ/s}$
$= \mathbf{200 \text{ kW}}$

2.71 (c)

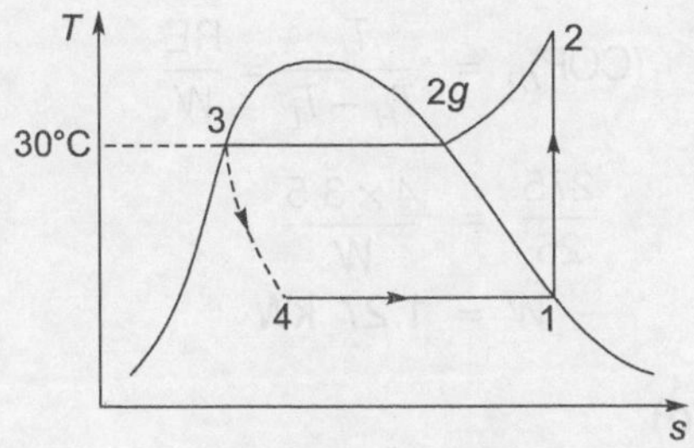

Given, $h_1 = 187.53$ kJ/kg

$h_{2g} = 201.45$ kJ/kg

$h_3 = 69.55$ kJ/kg $= h_4$

COP = **5.5**

$$\text{COP} = \frac{q_A}{W_{net}}$$

$$\therefore \quad \dot{m}_f = \frac{5.5 \times 40}{(h_1 - h_4)} = \frac{5.5 \times 40}{(187.53 - 69.55)}$$

$$= 1.8647 \text{ kg/s}$$

$$\dot{m}(h_2 - h_1) = 40 \text{ kW}$$

$$h_2 = \frac{40}{1.8647} + 187.53 = 208.98 \approx 209$$

$$h_2 - h_{2g} = c_p(T_2 - T_{2g})$$

$$\therefore \quad T_2 = \frac{h_2 - h_{2g}}{c_p} + T_{2g} = \frac{209 - 201.45}{0.6155} + 30$$

$$= \mathbf{42.26°C}$$

2.72 (b)

(i) Both the evaporator and condenser pressure should be above atmospheric pressure and as near to the atmospheric pressure as possible.

(ii) A refrigerant should have high latent heat of vaporization at the evaporator pressure. The high latent heat results in high refrigerating effect per kg of refrigerant circulated.

2.73 (d)

Except (d), all are responsible factor for global warming by one or other way.

2.74 (d)

$$(\text{COP})_{HP} = \frac{T_H}{T_H - T_L} = \frac{327 + 273}{300} = \frac{600}{300} = 2$$

2.75 (c)

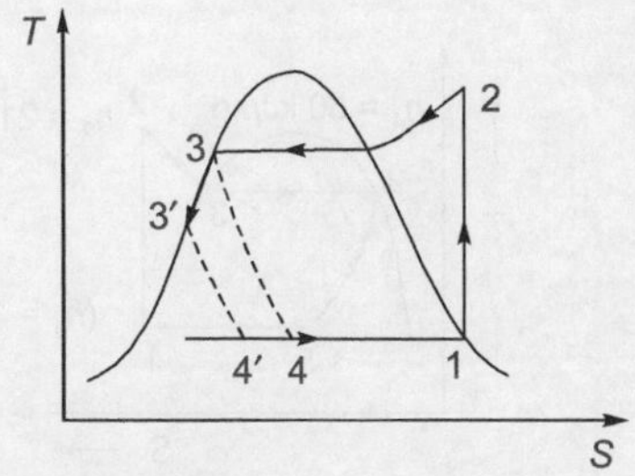

3 → saturated liquid

3′ → subcooled liquid

$(h_1 - h_4) < (h_1 - h_4')$

$\therefore$ q_A [i.e. heat extraction = $(h_1 - h_4')$] increases.

2.76 (b)

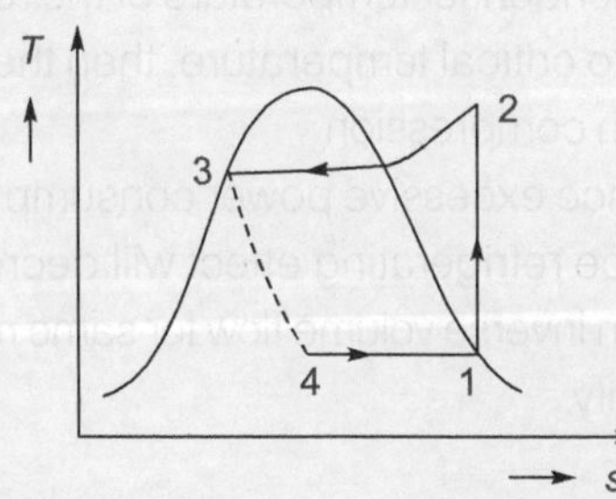

Given, $h_1 = 1500$ kJ/kg; $h_2 = 1800$ kJ/kg

$h_3 = h_4 = 300$ kJ/kg

$$\text{COP} = \frac{h_1 - h_4}{h_2 - h_1} = \frac{1500 - 300}{1800 - 1500}$$

$$= \frac{1200}{300} = 4$$

2.77 (a)

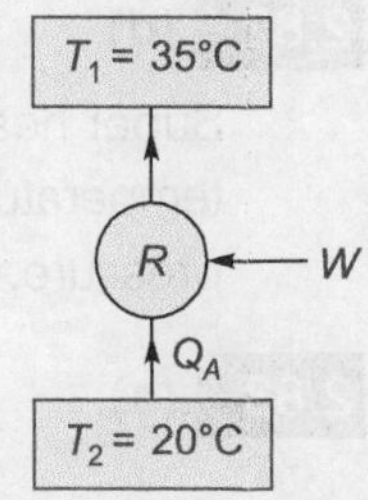

$$T_1 = 35° = 308 \text{ K}$$

$$T_2 = 20° = 293 \text{ K}$$

$$Q_A = 25 \text{ kW}$$

$$\text{COP} = \frac{293}{15} = 19.533$$

$$\text{COP} = \frac{Q_A}{W}$$

$$\therefore \quad W = \frac{Q_A}{\text{COP}} = \frac{25}{19.533} = \mathbf{1.2798 \text{ kW}}$$

2.78 (c)

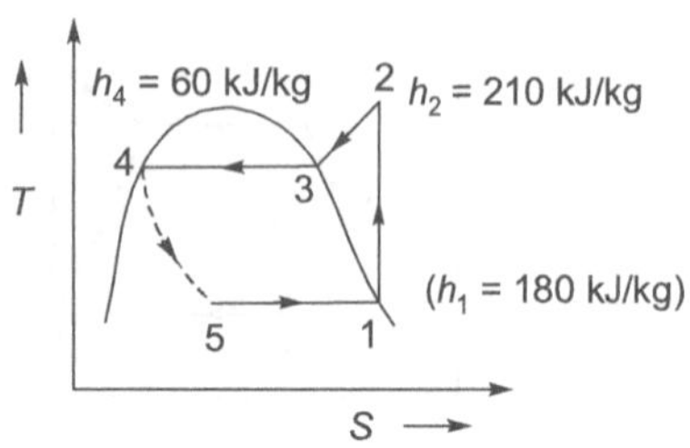

∴ Process 4-5 is an isenthalpic process

So, $h_5 = h_4 = 60$ kJ/kg

$$\therefore \quad \text{COP} = \frac{h_1 - h_5}{h_2 - h_1} = \frac{180-60}{210-180} = \frac{120}{30} = 4$$

2.79 (d)

If the condenser temperature of the refrigerant is close to critical temperature, then there will be

→ High compression

→ Hence excessive power consumption.

→ Since refrigerating effect will decrease. It will result in inverse volume flow for same refrigeration capacity.

2.81 (a)

In Bell-Coleman cycle, expander is used instead of throttle valve as here refrigerant can be treated as an ideal gas and we know enthalpy of an ideal gas is a function of temperature and if throttling is used, there will be no any temperature drop and here refrigerant wont pickup any heat from storage space.

2.82 (d)

Super heating is the process of increasing the temperature above saturated vapour at constant pressure.

2.83 (a)

$$\text{COP} = \frac{\text{RE}}{W} = \frac{Q_R - W}{W}$$

$$\Rightarrow \quad \text{COP} = \frac{1500-250}{250} = \mathbf{5}$$

2.84 (a)

$$(\text{COP})_{HP} = \frac{T_H}{T_H - T_L} = \frac{300}{30} = \mathbf{10}$$

2.85 (a)

$$(\text{COP})_R = \frac{T_L}{T_H - T_L} = \frac{\text{RE}}{W}$$

$$\frac{275}{25} = \frac{4 \times 3.5}{W}$$

$$W = \mathbf{1.27\ kW}$$

2.86 (d)

1 ton = 3.5 kW

$$(\text{COP})_R = \frac{\text{RE}}{W} \Rightarrow 5 = \frac{2.5 \times 3.5}{P}$$

$$P = \mathbf{1.75\ kW}$$

2.94 (c)

$$\text{COP} = \frac{h_1 - h_4}{h_2 - h_1} = \frac{150}{30} = 5$$

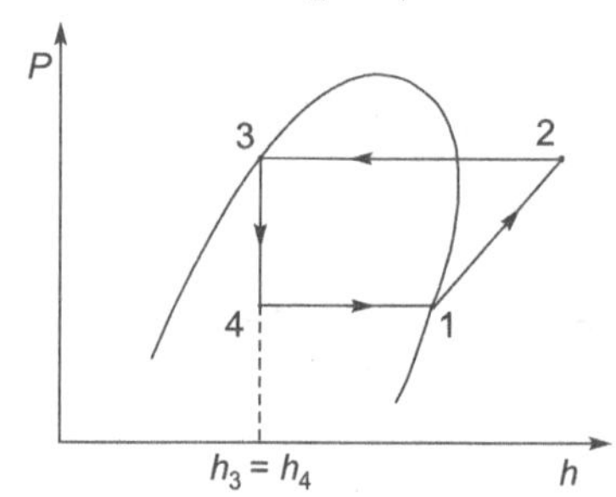

2.96 (c)

In VARS,

Given: T_E = 10° = 283 K

T_O = 30° = 303 K

T_G → T_G]E
30°C → 303 K
10°C → 283 K]R

$$\text{COP} = \left(1 - \frac{T_0}{T_G}\right)\left(\frac{T_E}{T_0 - T_E}\right)$$

$$\left(1 - \frac{303}{T_G}\right)\left(\frac{283}{303-283}\right) = 2$$

$$T_G = 79.8°\text{C} \simeq 80°\text{C}$$

2.97 (c)

$h_3 = h_4 = 200$ kJ/kg

$h_2 = 1400$ kJ/kg; $h_1 = 1200$ kJ/kg

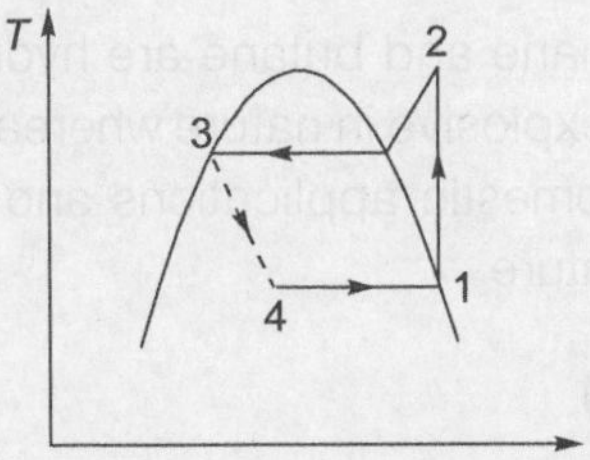

$$\text{COP} = \frac{h_1 - h_4}{h_2 - h_1} = \frac{1200 - 200}{1400 - 1200} = 5$$

2.101 (a)

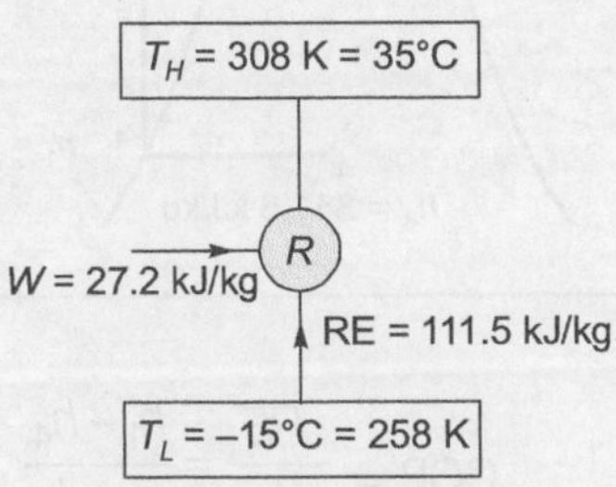

$$\text{COP} = \frac{RE}{W} = \frac{111.5}{27.2} = 4.099$$

2.102 (b)

$$T_G = 95 + 273 = 368 \text{ K}$$
$$T_E = 268 \text{ K}$$
$$T_O = 303 \text{ K}$$
$$(\text{COP})_{max} =$$
$$\frac{T_E(T_G - T_O)}{T_G(T_O - T_E)} = \frac{268(368 - 303)}{368(303 - 268)} = 1.352$$

2.103 (c)

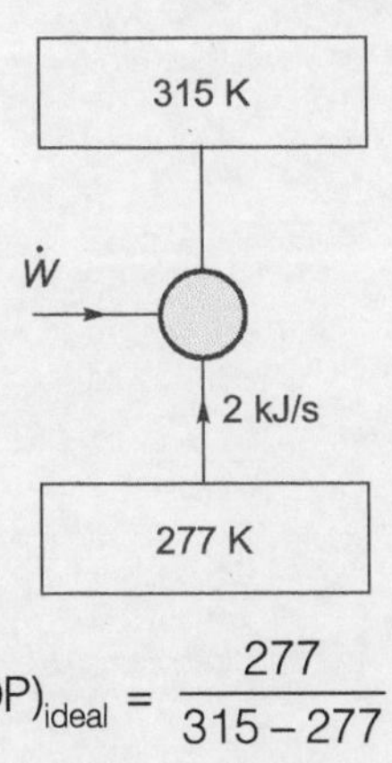

$$(\text{COP})_{ideal} = \frac{277}{315 - 277}$$
$$\frac{2}{\dot{W}} = \frac{277}{315 - 277}$$
$$\dot{W} = 0.274 \text{ kW}$$

2.104 (a)

$$T_E = -10°\text{C} = +263 \text{ K}$$
$$T_C = +30°\text{C} = 303 \text{ K}$$
$$V_C = 0.06\, V_S$$
$$P_1V_1^n = P_2V_2^n$$
$$\left(\frac{P_2}{P_1}\right)^{1/n} = \frac{V_1}{V_2}$$
$$\eta_V = 1 + 0.06 - 0.06\left(\frac{V_1}{V_2}\right)$$
$$= 1 + 0.06 - 0.06\left(\frac{0.07815}{0.025}\right)$$
$$\eta_V = 0.87244 = 87.24\%$$

2.105 (b)

Refrigerant	Lifetime (Yrs)	Relative Ozone Destruction Efficiency
R-11	45	~100%
R-12	100	~82%
R-113	85	~85%
R-114	190	~58%
R-115	1020	~50%

2.107 (a)

$$\text{RC} = 140.7 \text{ kW}$$
$$\rho_w = 1 \text{ kg/L} = 1000 \text{ kg/m}^3$$
$$\dot{Q}_C = \dot{Q}_{water} = \dot{m}_w C_w (40 - 30)$$
$$= 4 \times 4.18 \times 10$$
$$= 167.472 \text{ kJ/s}$$

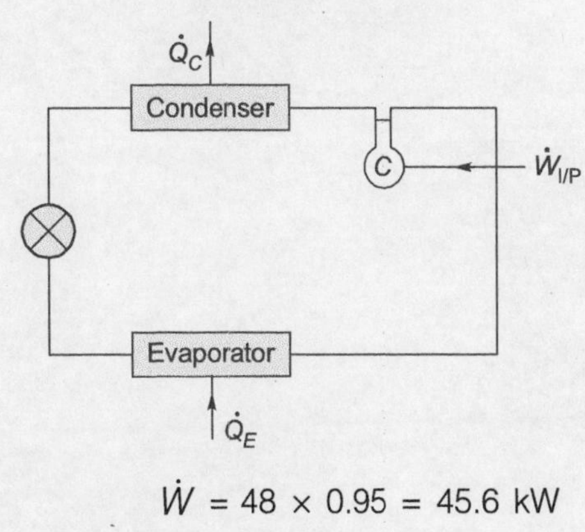

$$\dot{W} = 48 \times 0.95 = 45.6 \text{ kW}$$
$$\dot{Q}_E + \dot{W} = \dot{Q}_C$$
$$\dot{Q}_E = 121.872 \text{ kW} = 34.655 \text{ TR}$$

2.109 (a)

Given: m = 10 tons ice, C_{ice} = 1.94 kJ/kgK, Latent heat of water = 335 kJ/kg,
C_{water} = 4.1868 kJ/kgK, t = 24 hours

$$30°C \xrightarrow{\text{water}} 0°C \longrightarrow 0°C \xrightarrow{\text{ice}} -10°C$$

$$Q =$$

$$10 \times 10^3 (4.1868 \times 30 + 335 + 1.94 \times 10)$$

$$= 4800.039 \text{ MJ}$$

$$RC = \frac{Q}{t} = 55.55 \text{ kW}$$

2.111 (a)

$Cl_2F_4C_2$

$Cl_qF_PH_nC_m$

R - (m – 1) (n + 1) (p)

R - (2 – 1) (0 + 1) (4)

R114

Note: Tetra = 4

2.112 (b)

Methane and butane are hydrocarbons which are explosive in nature whereas R-134a is used in domestic applications and is not explosive in nature.

2.113 (a)

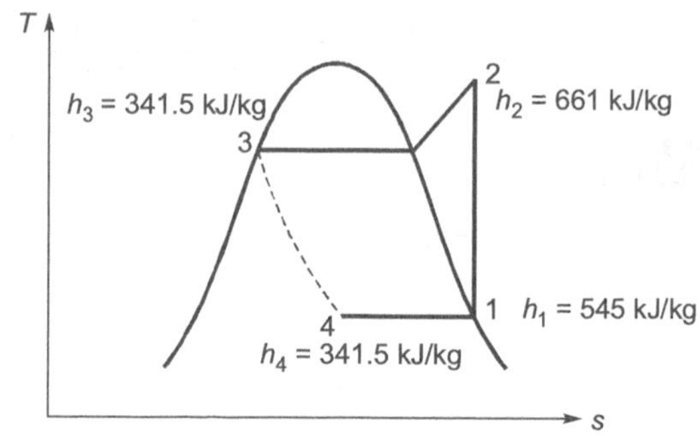

$$\text{COP} = \frac{RE}{W_{i/p}} = \frac{h_1 - h_4}{h_2 - h_1}$$

$$= \frac{545 - 341.15}{661 - 545} = 1.757$$

3 Refrigeration Equipments

3.1 Consider the following statements:
The pressure in a horizontal capillary tube of a refrigeration system decreases due to the
1. Frictional resistance offered by the tube wall.
2. Acceleration of refrigerant in the tube.
3. Heat transfer from the tube wall.
4. Decrease in the potential energy.

Which of these statements are correct?
(a) 1 and 4 (b) 2, 3 and 4
(c) 1, 2 and 3 (d) 1 and 2

[ESE : 2000]

3.2 Consider the following statements regarding refrigerants :
1. Refrigerant NH_3 is used in reciprocating compressors.
2. Refrigerant CO_2 is used in reciprocating compressors.
3. Refrigerant R-11 is used in reciprocating compressors.

Which of these statements are correct?
(a) 1 and 3 (b) 1 and 2
(c) 2 and 3 (d) 1, 2 and 3

[ESE : 2000]

3.3 Match **List-I** (Expansion device) with **List-II** (Operation) and select the correct answer using the codes given below the lists:

List-I
A. Float valve
B. Automatic expansion valve
C. Internally equalized thermostatic expansion valve
D. Externally equalized thermostatic expansion valve

List-II
1. Constant degree of superheat at evaporator exit pressure
2. Constant degree of superheat at evaporator inlet pressure
3. Constant level of refrigerant in the evaporator
4. Constant pressure in the evaporator

Codes:

	A	B	C	D
(a)	1	2	4	3
(b)	3	2	4	1
(c)	3	4	2	1
(d)	1	4	2	3

[ESE : 2001]

3.4 Match **List-I** (Refrigeration equipment) with **List-II** (Characteristic) and select the correct answer using the codes given below the lists:

List-I
A. Hermetically sealed compressor
B. Semi-hermetically sealed compressor
C. Open type compressor
D. Expansion device

List-II
1. Capillary tube
2. Both compressor and motor enclosed in a shell or casing
3. Both compressor and motor enclosed in a shell or casing with a removable cylinder cover
4. Driving motor not enclosed in a shell or casing and connected to the shaft driving the compressor

Codes:

	A	B	C	D
(a)	1	4	3	2
(b)	2	3	4	1
(c)	1	3	4	2
(d)	2	4	3	1

[ESE : 2002]

3.5 The sensing bulb of the thermostatic expansion valve is located at the
(a) exit of the evaporator
(b) inlet of the evaporator
(c) exit of the condenser
(d) inlet of the condenser [ESE : 2002]

3.6 When a refrigerator system is started from ambient conditions, the evaporator temperature decreases from ambient temperature to design value. This period is known as a pull-down period. The power requirement of compressor during pull-down
(a) decreases continuously
(b) increases continuously
(c) remains constant
(d) increases and then decreases

[ESE : 2003]

3.7 **Assertion (A):** Decrease of pressure and increase of temperature of the refrigerant in the suction pipeline connecting the evaporator to the reciprocating compressor reduces the refrigerating capacity of the system.
Reason (R): Decrease of pressure and increase of temperature of the refrigerant in the suction pipeline connecting the evaporator to the compressor reduces the volumetric efficiency of the reciprocating compressor.
(a) Both **A** and **R** are individually true and **R** is the correct explanation of **A**
(b) Both **A** and **R** are individually true but **R** is not the correct explanation of **A**
(c) **A** is true but **R** is false
(d) **A** is false but **R** is true

[ESE : 2003]

3.8 In a domestic refrigerator periodic defrosting is required because frosting
(a) causes corrosion of materials
(b) reduces heat extraction
(c) overcools food stuff
(d) partially blocks refrigerant flow

[ESE : 2004]

3.9 Which one of the following is responsible for the operation of a thermostatic expansion valve?
(a) Pressure changes in evaporator
(b) Temperature changes in evaporator
(c) Degree of superheat in evaporator
(d) Degree of subcooling in evaporator

[ESE : 2005]

3.10 For small installations of refrigeration systems (up to 35 kW), which type of condenser is used?
(a) Shell and tube type
(b) Shell and coil type
(c) Double tube type
(d) Air cooled type

[ESE : 2006]

3.11 Which type of valve is generally used in reciprocating refrigerant compressors?
(a) Mushroom valve (b) Puppet valve
(c) Plate valve (d) Throttle valve

[ESE : 2006]

3.12 What is hunting of thermostatic expansion valve?
(a) A variation of evaporator load with degree of superheat
(b) A variation in pressure of the evaporator with variation of load
(c) Alternate over-feeding and starving of refrigerant flow in the evaporator
(d) This is not used in connection with expansion valve

[ESE : 2008]

3.13 Match **List-I** (Devices) with **List-II** (Process undergone by air) and select the correct answer using the code given below the lists:

List-I
A. Cooling tower
B. Air coolers
C. Evaporator coil
D. Air cooled condenser

List-II
1. Heating
2. Cooling and dehumidification
3. Cooling and humidification
4. Heating and humidification

Codes:

	A	B	C	D
(a)	2	1	4	3
(b)	4	3	2	1
(c)	2	3	4	1
(d)	4	1	2	3

[ESE : 2008]

3.14 Which of the following is not an essential component of any refrigeration system, where refrigeration effect is produced by vaporization of refrigerant?

(a) Compressor (b) Condenser
(c) Evaporator (d) Expansion device

[ESE : 2011]

3.15 In gas cycle refrigeration system, an expander replaces the throttle valve of a vapour compression system, because

(a) The pressure drop obtained is not adequate
(b) The drop in temperature by throttling the gas is very small
(c) It reduces the irreversibility in the system
(d) Heat loss is significantly decreased with the expander

[ESE : 2011]

3.16 In the expansion valve of a refrigerator, the following property remains constant

(a) entropy (b) enthalpy
(c) internal energy (d) product pv

[ESE : 2013]

3.17 Consider the following functions:

1. Minimizing friction
2. Sealing the gas between suction and discharge ports
3. As a coolant to transfer heat from the crankcase to the compressor shell
4. To dampen the noise generated by moving parts

Which of the above functions do lubricants in refrigeration systems perform?

(a) 1, 2, 3 and 4 (b) 1 and 2 only
(c) 2 and 3 (d) 3 and 4 only

[ESE : 2014]

3.18 The choice of a refrigerant depends upon :

1. Refrigerating capacity
2. Type of compressor used (reciprocating, centrifugal or screw)
3. Service required (whether for air conditioning, cold storage or food freezing)

Which of the above statements is/are correct ?

(a) 1 and 3 only (b) 1 only
(c) 3 only (d) 1, 2 and 3

[ESE : 2016]

3.19 Consider the following statements :
The presence of air inside condensers

1. remains as a non-condensable gas
2. reduces the condensing coefficient
3. tends to cling to the surface
4. introduces large thermal resistance

Which of the above statements are correct?

(a) 1, 2, 3 and 4 (b) 1, 2 and 3 only
(c) 3 and 4 only (d) 1, 2 and 4 only

[ESE : 2017]

3.20 In a flooded evaporator refrigerator, an accumulator at the suction side of the compressor is provided to

(a) collect the vapours
(b) detect any liquids in the vapour
(c) retain the refrigeration effect as originally working
(d) collect the liquid refrigerant and preclude its reversion to the compressor

[ESE : 2018]

3.21 The compressor of an ammonia refrigerating machine has a volumetric efficiency of 85% and swept volume of 0.28 m^3/min. Ammonia having a dry specific volume of 0.25 m^3/kg enters the compressor with a dryness fraction of 0.7. The mass flow rate of ammonia through the machine is

(a) 1.28 kg/min (b) 1.36 kg/min
(c) 1.42 kg/min (d) 1.54 kg/min

[ESE : 2018]

3.22 Consider the following statements:

1. The operation of a refrigerator unit at more than one temperature can be accomplished by using different throttling valves and a separate compressor for each temperature range.
2. The refrigerated space must be maintained above the ice point to prevent freezing.
3. In domestic refrigerators, the refrigerant is throttled to a higher pressure in the freezer followed by full expansion in the refrigerated space.

Which of the above statements are correct?

(a) 1 and 2 only (b) 1 and 3 only
(c) 2 and 3 only (d) 1, 2 and 3

[ESE : 2018]

3.23 Which one of the following compressors will be used in vapour compression refrigerator for plants up to 100 tonnes capacity?
(a) Reciprocating compressor
(b) Rotary compressor
(c) Centrifugal compressor
(d) Double-acting compressor [ESE : 2019]

3.24 Consider the following statements :
An expansion device in a refrigeration system
1. reduces the pressure from the condenser to the evaporator
2. regulates the flow of the refrigerant to the evaporator depending on the load
3. is essentially a restriction offering resistance to flow

Which of the above statements are correct?
(a) 1 and 2 only (b) 1 and 3 only
(c) 2 and 3 only (d) 1, 2 and 3
[ESE : 2019]

3.25 An air cooled condenser has 6 m^2 of surface with a removal of 50 $kJ.hr^{-1}.m^{-2}.°C^{-1}$. What is the refrigerant temperature to dissipate 5235 kJ/hr, if the room temperature is 25°C?
(a) 24.31°C (b) 35.82°C
(c) 42.45°C (d) 56.94°C
[ESE : 2021]

3.26 A capillary tube in a one-ton R22 air conditioner has a bore of 2.3 mm. Saturated liquid from the condenser enters at a temperature of 48°C and flows adiabatically through the tube until its temperature is 5°C. The friction factor is given as $f = 0.32/Re^{0.25}$. What is the length of the capillary tube? (Take intermediate sections at 40°C, 30°C, 20°C and 10°C and the mass flow rate of the refrigerant as 0.02417 kg/s)
(a) 13.41 $kg.s^{-1}.m^{-1}$ (b) 16.42 $kg.s^{-1}.m^{-1}$
(c) 19.42 $kg.s^{-1}.m^{-1}$ (d) 22.46 $kg.s^{-1}.m^{-1}$
[ESE : 2022]

3.27 An R134a thermostatic expansion valve, not equipped with an external equalizer, has a superheat setting of 7 °C, while supplying the refrigerant to the evaporator at 0°C. The power fluid is same as refrigerant. What is the difference in pressure in opposite sides of the diaphragm required to open the valve? (Consider evaporator pressure at 0°C and 7°C as 2.928 bar and 3.748 bar respectively)
(a) 2.928 bar (b) 3.748 bar
(c) 0.82 bar (d) 6.676 bar
[ESE : 2024]

■■■■

Answers Refrigeration Equipments

3.1 (d)	3.2 (b)	3.3 (c)	3.4 (b)	3.5 (a)	3.6 (d)	3.7 (a)	3.8 (b)	3.9 (c)
3.10 (c)	3.11 (c)	3.12 (c)	3.13 (b)	3.14 (a)	3.15 (b)	3.16 (b)	3.17 (a)	3.18 (d)
3.19 (a)	3.20 (d)	3.21 (b)	3.22 (*)	3.23 (a)	3.24 (d)	3.25 (c)	3.26 (a)	3.27 (c)

Explanations Refrigeration Equipments

3.5 (a)

The thermostatic expansion valve consists of a needle valve and a seat, a metallic diaphragm, spring and an adjusting screw. In addition to this, it has a feeler or thermal bulb which is mounted on the suction line near the outlet of evaporator coil. The feeler bulb is partly filled with the same liquid refrigerant as used in the refrigeration system.

3.6 (d)

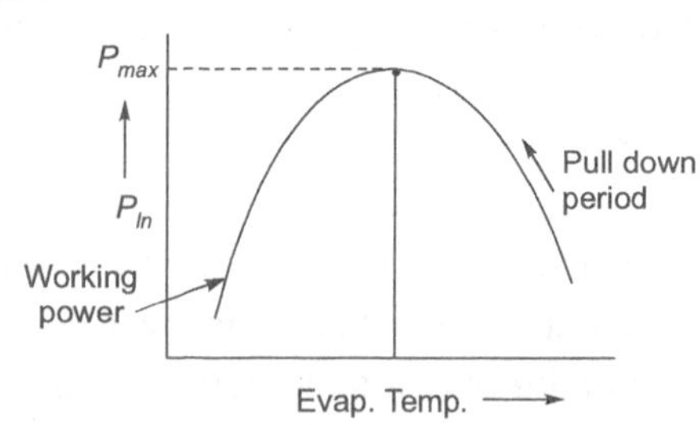

Hence, during pull down period power input increases then decreases.

3.8 (b)

When air is cooled below its 'dew point temperature', it gives up the moisture and deposits on nearest colder surface. Since the colder surface in a refrigerated area is evaporator, therefore the moisture deposits on the surface of the evaporator. When the temperature of the evaporator falls below 0°C, the moisture deposited on the surface of the evaporator freezes and forms a coating of frost. If this frost is not removed periodically, it acts as an insulator and retards the heat transfer rate.

3.9 (c)

Thermostatic expansion valve is also called a constant superheat value because it maintains a constant degree of superheat of the vapour refrigerant at the end of the evaporator coil by controlling the flow of liquid refrigerant through the evaporator.

3.10 (c)

Shell and tube type - 2 to 1000 tons
Shell and coil type - upto 50 tons
Double tube type - 10 kW
Air cooled type - 0.5 to 5 tons

3.11 (c)

Plate valve is used in the reciprocating refrigerant compressor due to less inertia effect.

3.12 (c)

Alternate overfeeding and starving of refrigerant flow in the evaporator call hunting of thermostatic expansion valve.

3.13 (b)

Cooling tower : Heating and humidification
Air coolers : Cooling and humidification
Evaporator coil : Cooling and dehumidi-fication
Air cooled condenser : Heating

3.15 (b)

In gas refrigeration system, an expander replaces. Throttling is an irreversible adiabatic process and constant enthalpy process so the drop in temperature by throttling the gas is very small.

3.16 (b)

Expansion valve of a refrigerator is of constant restriction type. The expansion is done at constant enthalpy with pressure drop from condenser to evaporator pressure.

3.17 (a)

In refrigeration systems, oil must perform certain functions other than minimizing friction, such as sealing the gas between suction and discharge ports, acting as a coolant to transfer heat from the crankcase to the compressor shell and damper the noise generated by moving parts.

3.20 (d)

To prevent the entry of liquid ref. into the compressor which may cause valve damage.

3.21 (b)

$$\eta_v = \frac{\dot{m}v_1}{\text{Swept volume}}$$

$$0.85 = \frac{\dot{m} \times v_1}{0.28}$$

$$v = v_f + xv_{fg} = v_f + x(v_g - v_f)$$

$$v_g >>>> .\ v_f$$

$$v = xv_g = 0.7 \times 0.25 = 0.175 \text{ m}^3\text{/kg}$$

$$\therefore \quad 0.85 = \frac{\dot{m} \times 0.175(\text{m}^3/\text{kg})}{0.28(\text{m}^3/\text{min})}$$

$$\dot{m} = 1.36 \text{ kg/min}$$

3.22 (*)

Only 1st statement is correct.

3.25 (c)

$$Q = hA\Delta T$$

$$= hA\,(T_{ref} - T_{room})$$

$$5235 = 50 \times 6 \times (T_{ref} - 25) \text{ kJ/hr}$$

$$T_{ref} = 42.45°\text{C}$$

3.27 (c)

At 7°C, $P_{evap,2} = 3.748$ bar

At 0°C, $P_{evap,1} = 2.928$ bar

∴ The difference in pressure in opposite sides of the diaphragm required to open the valve

$$P_{evap,2} - P_{evap1} = 3.748 - 2.928 = 0.82 \text{ bar}$$

■■■■

4 Psychrometry and Psychrometric Processes

4.1 When warm saturated air is cooled

(a) excess moisture condenses

(b) excess moisture condenses but relative humidity remains unchanged

(c) excess moisture condenses and specific humidity increases but relative humidity remains unchanged

(d) specific humidity increases and relative humidity decreases

[ESE : 2000]

4.2 When dry-bulb and wet-bulb temperatures are identical, it means that the

(a) air is fully saturated and dew-point temperature has reached

(b) air is fully saturated

(c) dew-point temperature has reached and humidity is 100%

(d) partial pressure of water vapour is equal to total pressure [ESE : 2000]

4.3 In a cooling tower, the minimum temperature to which water can be cooled is equal to the

(a) dew point temperature of the air at the inlet

(b) dry bulb temperature of the air at the inlet

(c) thermodynamic wet bulb temperature of the air at the inlet

(d) mean of the dew point and dry bulb temperature of the air at the inlet

[ESE : 2001]

4.4 If p_v is the partial pressure of vapour, p_s is the partial pressure of vapour for saturated air and p_b is the barometric pressure, the relationship between relative humidity 'ϕ' and degree of saturation 'μ' is given by

(a) $\mu = \phi\left(\frac{p_b - p_s}{p_b - p_v}\right)$ (b) $\mu = \phi\left(\frac{p_b - p_v}{p_b - p_s}\right)$

(c) $\mu = \phi\left(\frac{p_v}{p_b}\right)$ (d) $\mu = \phi\left(\frac{p_v}{p_s}\right)$

[ESE : 2001]

4.5 If p_a and p_v denote respectively the partial pressure of dry air and that of water vapour in moist air, the specific humidity of air is given by

(a) $\frac{p_v}{p_a + p_v}$ (b) $\frac{p_v}{p_a}$

(c) $\frac{0.622\, p_v}{p_a}$ (d) $\frac{0.622 p_v}{p_a + p_v}$

[ESE : 2001]

4.6 Consider the following statements :

When dry bulb and thermodynamic wet bulb temperatures are same

1. humidity ratio is 100%.
2. partial pressure of water vapour equals total pressure.
3. air is fully saturated.
4. dew point temperature is reached.

Select the correct statements(s) using the codes given below :

(a) 3 only (b) 1 and 2

(c) 3 and 4 (d) 1, 2, 3 and 4

[ESE : 2001]

4.7 **Assertion (A):** Effective temperature, an index of comfort, is defined as that temperature of saturated air at which one would experience the same feeling of comfort as experienced in the actual environment.

Reason (R): Comfort does not depend on humidity and air velocity.

(a) Both **A** and **R** are individually true and **R** is the correct explanation of **A**

(b) Both **A** and **R** are individually true but **R** is not the correct explanation of **A**

(c) **A** is true but **R** is false

(d) **A** is false but **R** is true

[ESE : 2001]

4.8 In case A, most air is adiabatically saturated and in case B, moist air is isobarically saturated. The saturation temperatures in cases A and B are respectively

(a) dry bulb temperature and wet bulb temperature
(b) dew point temperature and wet bulb temperature
(c) wet bulb temperature and dew point temperature
(d) wet bulb temperature and dry bulb temperature

[ESE : 2002]

4.9 For cooling and dehumidifying of unsaturated moist air, it must be passed over a coil at a temperature

(a) of adiabatic saturation of incoming stream
(b) which is lower than the dew point of incoming stream
(c) which lies between dry bulb and wet bulb temperature
(d) which lies between wet bulb and dew point temperature of incoming stream

[ESE : 2002]

4.10 If the volume of moist air with 50% relative humidity is isothermally reduced to half its original volume, then relative humidity of moist air becomes

(a) 25% (b) 60%
(c) 75% (d) 100%

[ESE : 2003]

4.11 Which of the following properties increase(s) during sensible heating of air-water vapour mixture?

1. Relative humidity
2. Humidity ratio
3. Wet bulb temperature
4. Specific enthalpy of air-vapour mixture

Select the correct answer from the codes given below:

(a) 1 and 2 (b) 3 only
(c) 2 and 3 (d) 3 and 4

[ESE : 2003]

4.12 When the wet and dry bulb temperatures are identical, which of the following statements is/are true?

1. Air is fully saturated.
2. Dew point temperature is reached.
3. Humidity ratio is unity.
4. A Partial pressure of vapour equals total pressure.

Select the correct answer from the code given below:

(a) 1 only (b) 1 and 2
(c) 3 and 4 (d) 1, 2, 3 and 4

[ESE : 2003]

4.13 Upon which of the following factors does the effective temperature for human comfort depend?

1. Dry bulb temperature
2. Humidity ratio
3. Air velocity
4. Mean radiation temperature

Select the correct answer from the codes given below:

(a) 1 and 2 (b) 1, 3 and 4
(c) 2, 3 and 4 (d) 1, 2, 3 and 4

[ESE : 2003]

4.14 Air at 35°C DBT and 25°C dew point temperature passes through the water shower whose temperature is maintained at 20°C. What is the process involved?

(a) Cooling and humidification
(b) Sensible cooling
(c) Cooling and dehumidification
(d) Heating and humidification [ESE : 2004]

4.15 Consider the following statements:

1. The specific humidity is the ratio of the mass of water vapour to the mass of dry air in a given volume of the mixture.
2. The relative humidity of the atmospheric air is the ratio of the actual mass of the water vapour in a given volume to that which it would have if it were saturated at the same temperature.
3. The degree of saturation is defined as the ratio of the specific humidity of a mixture to the specific humidity of the mixture when saturated at the same temperature.

Which of these statements are correct?

(a) 1 and 2 (b) 2 and 3
(c) 1 and 3 (d) 1, 2 and 3

[ESE : 2004]

4.16 Which one of the following is correct?

(a) Effective temperature is the index which correlates the combined effects of air dry bulb temperature, air humidity and air movement upon human comfort

(b) The value of effective temperature in winter and summer should be same for human comfort

(c) Effective temperature and wet bulb temperature are the same

(d) The value of the effective temperature should be higher in winter than in summer for comfort

[ESE : 2004]

4.17 Which one of the following are normally desired comfort conditions in an air-conditioning system?

(a) 25°C DBT and 50% RH

(b) 22°C DBT and 90% RH

(c) 15°C BDT and 75% RH

(d) 15°C DBT and 40% RH

[ESE : 2004]

4.18 Match **List-I** with **List-II** and select the correct answer using the code given below the lists:

List-I

A. Degree of saturation

B. Dry bulb temperature

C. Wet bulb temperature

D. Dew point temperature

List-II

1. Measure of latent enthalpy of moist air

2. Measure of total enthalpy of moist air

3. Measure of the capacity of air to absorb moisture

4. Measure of sensible enthalpy of moist air

Codes:

	A	B	C	D
(a)	2	1	3	4
(b)	3	4	2	1
(c)	2	4	3	1
(d)	3	1	2	4

[ESE : 2005]

4.19 When the wet bulb and dry bulb temperatures are equal, which of the following statements is/are correct?

1. Air is fully saturated.
2. Dew point temperature is reached.
3. Partial pressure of vapour equals to the total pressure.
4. Humidity ratio is 100%.

Select the correct answer using the code given below:

(a) 1 and 2 (b) 1 only

(c) 1, 2 and 4 (d) 2 and 3

[ESE : 2005]

4.20 Air at dry bulb temperature of 35°C and dew point temperature of 25°C passes through an air washer whose temperature is maintained at 20°C. What is the nature of the process involved?

(a) Cooling and humidification

(b) Sensible cooling

(c) Heating and dehumidification

(d) Cooling and dehumidification [ESE : 2005]

4.21 The optimum effective temperature for human comfort is

(a) higher in winter than that in summer

(b) lower in winter than that in summer

(c) same in winter and summer

(d) not dependent on season [ESE : 2005]

4.22 On which factor(s), does the heat lost by the human body in the process of radiation depend?

(a) Temperature only

(b) Temperature and air motion

(c) Temperature and relative humidity

(d) Relative humidity and air motion

[ESE : 2005]

4.23 An air-conditioned room of volume 10 m^3 has infiltration of air equivalent to 3 air changes per hour. Density of air is 1.2 kg/m^3, specific heat C_P is 1 kJ/kgK and temperature difference between room and ambient air is 20 K. What is the sensible heat load due to infiltrated air?

(a) 60 kJ/hr (b) 12 kJ/hr

(c) 0.45 kW (d) 0.2 kW [ESE : 2005]

4.24 What are the general comfort conditions in an air-conditioning system?

(a) 20°C DBT, 80% RH

(b) 24°C DBT, 60% RH

(c) 25°C DBT, 40% RH

(d) 25°C DBT, 100% RH [ESE : 2006]

4.25 The heat load from the occupants in air-conditioning load calculation is a source of

(a) Sensible heat only
(b) Latent heat only
(c) Both sensible and latent heat
(d) None of these [ESE : 2006]

4.26 Which one of the following is correct for the process 1-2 shown below?

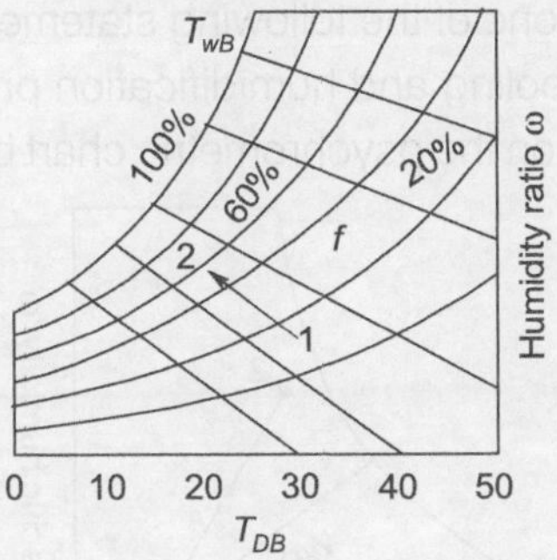

(a) The partial pressure of water vapour in air remains constant
(b) Specific humidity of air remains constant
(c) Enthalpy of air remains constant
(d) Dry bulb temperature of air remains constant [ESE : 2006]

4.27 Effective temperature depends on dry bulb temperature and
(a) Wet bulb temperature only
(b) Relative humidity
(c) Specific humidity
(d) Wet bulb temperature, and air motion [ESE : 2006]

4.28 Dry bulb temperature and wet bulb temperature is 25°C each, and velocity of air passing over human body is 6 m/min. If velocity increases to 20 m/min, then which one of the following is correct?
(a) The effective temperature decreases
(b) The effective temperature remains the same
(c) The effective temperature increases
(d) The change in effective temperature cannot be estimated with the given information [ESE : 2006]

4.29 The wet bulb depression is zero, when relative humidity is equal to
(a) 100 % (b) 60 %
(c) 40 % (d) Zero [ESE : 2006]

4.30 Consider the following statements:
Air washer can work as
1. Humidifier only
2. Dehumidifier only
3. Filter only
Which of these statements is/are correct?
(a) only 1 (b) only 2 and 3
(c) only 1 and 3 (d) 1, 2 and 3 [ESE : 2006]

4.31 A human body feels comfortable when the heat produced by the metabolism of human body is equal to
(a) Heat dissipated to the surroundings
(b) Heat stored in the human body
(c) Sum of (a) and (b)
(d) Difference of (a) and (b) [ESE : 2006]

4.32 Atmospheric air at 35°C and 60% RH can be brought to 20°C and 60% RH by
(a) Cooling and dehumidification process
(b) Cooling and humidification process
(c) Adiabatic saturation process
(d) Sensible cooling process [ESE : 2006]

4.33 On a psychrometric chart, what does a vertical downward line represent?
(a) Adiabatic saturation
(b) Sensible cooling
(c) Dehumidification
(d) Humidification [ESE : 2008]

4.34 Which of the following properties decreases(s) with sensible heating of air-water vapour mixture?
1. Relative humidity.
2. Humidity ratio.
3. Specific enthalpy of air-vapour mixture.
4. Wet bulb temperature.
Select the correct answer using the code given below:
(a) 1 only (b) 1 and 3
(c) 2 and 3 (d) 2 and 4 [ESE : 2008]

4.35 Consider the following statements regarding psychrometric processes:
1. Sensible heating is a process in which moisture content remains unchanged.
2. In the dehumidification process the dew point temperature remains same.

3. The process of adding moisture at constant dry bulb temperature is known as pure humidification process.

Which of these statements is/are correct?

(a) 1 and 2 (b) 1 and 3
(c) 2 and 3 (d) 1 only

[ESE : 2008]

4.36 Consider the following statements:
In a psychrometric chart
1. vertical lines indicate wet bulb temperature.
2. horizontal lines indicate relative humidity.
3. sensible heating or cooling is represented by an inclined line.

Which of these statements is/are correct?

(a) 1 only (b) 2 only
(c) 3 only (d) 1, 2 and 3

[ESE : 2008]

4.37 During sensible cooling of moist air, its relative humidity
(a) increases
(b) does not change
(c) decreases
(d) affects specific humidity [ESE : 2008]

4.38 In a sensible heating or cooling process
(a) dry bulb temperature remains constant
(b) wet bulb temperature remains constant
(c) the humidity ratio remains constant
(d) the relative humidity remains constant

[ESE : 2008]

4.39 On psychrometric chart, the constant wet bulb temperature lines coincide with
(a) constant relative humidity lines
(b) constant enthalpy lines
(c) constant dew point temperature lines
(d) constant volume lines [ESE : 2008]

4.40 In a psychrometric chart, what does a vertical downward line represent?
(a) Sensible cooling process
(b) Adiabatic saturation process
(c) Humidification process
(d) Dehumidification process [ESE : 2009]

4.41 In a sample of moist air at standard atmospheric pressure of 101.325 kPa and 26°C the partial pressure of water vapour is 1.344 kPa. If the saturation pressure of water vapour is 3.36 kPa at 26°C, then what are the humidity ratio and relative humidity of moist air sample?
(a) 0.00836 and 1.32%
(b) 0.00836 and 40%
(c) 0.01344 and 1.32%
(d) 0.01344 and 40% [ESE : 2009]

4.42 Which one of the following statements is correct for a cooling and humidification process 1-2 as shown on the psychrometric chart below?

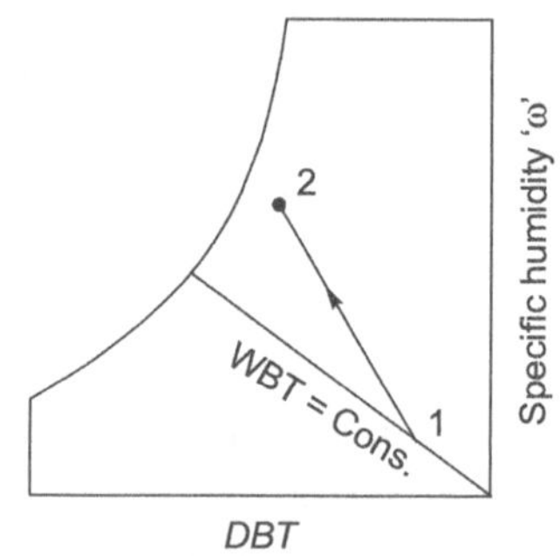

(a) WBT decreases in the process
(b) The total enthalpy increases in the process
(c) The total enthalpy remains constant in the process
(d) It is an adiabatic saturation process

[ESE : 2009]

4.43 What is the saturation temperature at the partial pressure of water vapour in the air-water vapour mixture called?
(a) Dry bulb temperature
(b) Wet bulb temperature
(c) Dew point temperature
(d) Saturation temperature [ESE : 2009]

4.44 The humidity ratio of atmospheric air at 28°C DBT and 760 mm of Hg is 0·016 kg of vapour per kg of dry air. What is the partial pressure of water vapour?
(a) 2.242 kN/m^2 (b) 2.535 kN/m^2
(c) 3.535 kN/m^2 (d) 4.242 kN/m^2

[ESE : 2009]

4.45 When moist air passes through a bed of silica gel, the
(a) Dry bulb temperature of air decreases
(b) Dry bulb temperature of air increases
(c) Specific humidity of air increases
(d) it undergoes adiabatic saturation

[ESE : 2010]

4.46 In air-water vapour mixture, the partial pressure of water vapour corresponds to the saturation temperature called
(a) Dew point temperature
(b) Wet bulb temperature
(c) Dry bulb temperature
(d) Adiabatic saturation temperature
[ESE : 2010]

4.47 As an index of comfort, the temperature of saturated air at which a person would experience the same feeling of comfort as experienced in the actual unsaturated environment is called the
(a) Comfort temperature
(b) Effective temperature
(c) Wet bulb temperature
(d) Soothing temperature [ESE : 2010]

4.48 A certain air has DBT 35°C and DPT of 20°C, the corresponding saturation pressure of water being 5.628 kPa and 2.33 kPa respectively. When the atmospheric pressure is assumed as 1.0132 bar, the specific humidity of air will be
(a) 2.5×10^{-3} (b) 7.8×10^{-3}
(c) 14.7×10^{-3} (d) 25×10^{-3}
[ESE : 2010]

4.49 In summer, air may be cooled and dehumidified by spraying chilled water to air in the form of mist. The minimum temperature to which air may be cooled is the
(a) Wet bulb temperature
(b) Adiabatic saturation temperature
(c) Apparatus dew point
(d) Dry bulb temperature [ESE : 2011]

4.50 For a given dry bulb temperature, as the relative humidity decreases, the wet bulb temperature will
(a) Increase
(b) Decrease
(c) Be the same
(d) Depend on other factors
[ESE : 2011]

4.51 In winter air-conditioning, the process is
(a) Heating, humidification and cooling
(b) Heating, humidification and heating
(c) Heating, dehumidification and heating
(d) Cooling, dehumidification and heating
[ESE : 2011]

4.52 If air flows over a cooling coil, dehumidification of air will take place if the coil surface temperature is below the following of the entering air
(a) Wet bulb temperature
(b) Dry bulb temperature
(c) Dew point temperature
(d) Adiabatic saturation temperature
[ESE : 2011]

4.53 In an adiabatic saturation process of air
(a) The enthalpy remains constant
(b) The temperature remains constant
(c) The absolute humidity remains constant
(d) The relative humidity remains constant
[ESE : 2011]

4.54 If the specific humidity of moist air remains the same but its dry bulb temperature increases, its dew point temperature
(a) Remains the same
(b) Increases
(c) Decreases
(d) May increase or decrease depending on its relative humidity
[ESE : 2011]

4.55 When air is adiabatically saturated, the temperature attained is the
(a) Dew point temperature
(b) Dry bulb temperature
(c) Wet bulb temperature
(d) Apparatus Dew point temperature
[ESE : 2011]

4.56 When unsaturated air undergoes adiabatic saturation, which of the following properties decrease(s)?
1. Relative humidity
2. Specific humidity
3. Dry-bulb temperature
4. Wet-bulb temperature
(a) 1, 2, 3 and 4 (b) 2 only
(c) 3 only (d) 4 only [ESE : 2012]

4.57 A stream of moist air at dry-bulb temperature of 40°C and dew-point temperature of 25°C passes through a water shower whose temperature is maintained at 20°C. The stream of air will undergo a process of
(a) sensible cooling

(b) evaporative cooling
(c) cooling and humidification
(d) cooling and dehumidification [ESE : 2012]

4.58 When only sensible heat is added to a sample of air, the process is represented on the psychrometric chart as a
(a) horizontal line moving towards left direction
(b) vertical line moving towards upward direction
(c) horizontal line moving towards right direction
(d) vertical line moving, towards downward direction [ESE : 2012]

4.59 During the sensible cooling process, specific humidity
(a) remains constant (b) increases
(c) decreases (d) Unpredicatable
[ESE : 2012]

4.60 The comfort conditions in air conditioning system are defined by
(a) 22°C dbt and 60% RH
(b) 25°C dbt and 100% RH
(c) 20°C dbt and 75% RH
(d) 27°C dbt and 75% RH [ESE : 2012]

4.61 The saturation temperature of water vapour at its partial pressure is equal to
(a) wet bulb temperature
(b) dew point temperature
(c) adiabatic saturation temperature
(d) dry bulb temperature
[ESE : 2013]

4.62 When unsaturated air is adiabatically saturated, which of the following properties decrease?
1. Dry bulb temperature
2. Wet bulb temperature
3. Relative humidity
4. Specific humidity
(a) 1 only
(b) 1 and 2 only
(c) 2 and 3 only
(d) 3 and 4 only [ESE : 2013]

4.63 It is desired to condition the outside air from 80% relative humidity and 45°C dry bulb temperature to 55% relative humidity and 25°C dry bulb temperature. The practical arrangement will be
(a) heating and humidification
(b) heating and dehumidification
(c) cooling and humidification
(d) cooling and dehumidification
[ESE : 2013]

4.64 Which parameters are the independent variables in the construction of psychrometric plot for a particular value of barometric pressure?
(a) Dry bulb temperature and specific humidity
(b) Wet bulb temperature and relative humidity
(c) Dry bulb temperature and wet bulb temperature
(d) Dry bulb temperature and enthalpy
[ESE : 2013]

4.65 Which one of the following statements is correct?
(a) During heating and humidification process, humidity ratio decreases
(b) During cooling and dehumidification process, humidity ratio increases
(c) During cooling and dehumidification process, dry bulb temperature increases.
(d) During heating and dehumidification process, dry bulb temperature increases.
[ESE : 2014]

4.66 Specific humidity is defined as Mass of
(a) Water vapour contained in air-vapour mixture per kg of dry air
(b) Water vapour contained per kg of air-vapour mixture
(c) Dry air contained per kg of air-vapour mixture
(d) None of the above
[ESE : 2014]

4.67 Consider the following statements for sensible heating. In this process:
1. Wet bulb temperature increases
2. Relative humidity decreases
3. Vapour pressure remains constant
Which of the above statements are correct?
(a) 1, 2 and 3 (b) 1 and 2 only
(c) 1 and 3 only (d) 2 and 3 only
[ESE : 2014]

4.68 In a Psychrometric chart, which of the following statement(s) is/are correct?
(a) It is used to determine properties of refrigerants
(b) It cannot determine WBT and DBT

(c) It is seldom used for air conditioning design
(d) It provides plots for moist air conditioning

[ESE : 2015]

4.69 In a Psychrometric chart, horizontal lines represent constant
(a) Humidity ratio and vertical lines represent constant dry bulb temperature
(b) Humidity ratio and vertical lines represent constant wet bulb temperature
(c) Dry bulb temperature and vertical lines represent constant absolute humidity ratio
(d) Wet bulb temperature and vertical lines represent constant humidity ratio

[ESE : 2015]

4.70 Air is passed through a cooling coil at a temperature of –5°C. If the temperature of air drops from 25°C to 10°C, the Bypass factor of the coil is
(a) 0.2 (b) 0.5
(c) 0.7 (d) 1.0

[ESE : 2015]

4.71 In sensible cooling of moist air, its physical properties vary as follows:
1. The wet bulb temperature decreases
2. The dew point temperature remains constant
3. The relative humidity increases

Which of the above statements are correct?
(a) 1, 2 and 3 (b) 1 and 2 only
(c) 1 and 3 only (d) 2 and 3 only

[ESE : 2015]

4.72 A desert cooler having a cooling efficiency of 70% reduces the temperature of atmospheric air from 37°C to 30°C. The wet bulb temperature of the air is
(a) 24°C (b) 25°C
(c) 26°C (d) 27°C

[ESE : 2015]

4.73 Consider the following statements with regard to air-conditioning systems:
1. In adiabatic saturation process, air-vapour mixture undergoes a process of constant relative humidity
2. Wet bulb temperature of air whose relative humidity is 100 is equal to the dew point temperature
3. In winter air conditioning, the process is heating and humidification
4. For designing air conditioning ducts, equal friction method automatically reduces the air velocity in the duct in the direction of flow

Which of the above statements are correct?
(a) 1 and 2 (b) 1 and 4
(c) 2 and 3 (d) 3 and 4

[ESE : 2015]

4.74 **Assertion (A):** Humidity ratio of moist air is the ratio of the mass of water vapour to the mass of moist air in a given value of air-water vapour mixture.
Reason (R): For any given barometric pressure, humidity ratio is a function of the dew point temperature alone.
(a) Both **A** and **R** are true and **R** is the correct explanation of **A**
(b) Both **A** and **R** are true but **R** is not a correct explanation of **A**
(c) **A** is true but **R** is false
(d) **A** is false but **R** is true

[ESE : 2015]

4.75 Consider the following statements :
1. Gases have a very low critical temperature
2. Gases can be liquefied by isothermal compression
3. In engineering problems, water vapour in atmosphere is treated as an ideal or perfect gas

Which of the above statements are correct ?
(a) 1 and 2 only (b) 2 and 3 only
(c) 1 and 3 only (d) 1, 2 and 3

[ESE : 2016]

4.76 In an air-conditioning plant, air enters the cooling coil at 27°C. The coil surface temperature is –5°C. If the bypass factor of the unit is 0.4, the air will leave the coil at:
(a) 5.6°C (b) 7.8°C
(c) 9.2°C (d) 11.2°C

[ESE : 2016]

4.77 The wet bulb and dry bulb temperatures of an air sample will be equal when :
1. Air is fully saturated
2. Dew point temperature is reached

3. Partial pressure of vapour equals the total pressure
4. Humidity ratio is 100%

Which of the above statements are correct ?

(a) 1 and 2 (b) 2 and 3
(c) 3 and 4 (d) 1 and 4

[ESE : 2016]

4.78 Air at 25°C DBT and 80% RH is passed over a cooling coil whose surface temperature is 10°C which is below DPT of the air. If the air temperature coming out of the cooling coil is 15°C, then the bypass factor of the cooling coil is :

(a) 0.56 (b) 0.67
(c) 0.76 (d) 0.87

[ESE : 2016]

4.79 Consider the following statements for the appropriate context:

1. The Relative Humidity of air remains constant during sensible heating or cooling
2. The Dew Point Temperature of air remains constant during sensible heating or cooling
3. The total enthalpy of air remains constant during adiabatic cooling
4. It is necessary to cool the air below its Dew Point Temperature for dehumidifying

Which of the above statements are correct ?

(a) 1, 2 and 3 (b) 1, 2 and 4
(c) 3 and 4 only (d) 2, 3 and 4

[ESE : 2016]

4.80 In a psychrometric chart, relative humidity lines are :

(a) Curved
(b) Inclined and straight but non-uniformly spaced
(c) Horizontal and non-uniformly spaced
(d) Horizontal and uniformly spaced

[ESE : 2016]

4.81 **Statement (I) :** Effective temperature is an index which correlates the combined effect of air temperature, air humidity and air movement upon human thermal comfort.

Statement (II) : Thermal comfort is not affected by mean radiant temperature.

(a) Both Statement (I) and Statement (II) are individually true and Statement (II) is the correct explanation of Statement (I).
(b) Both Statement (I) and Statement (II) are individually true but Statement (II) is **NOT** the correct explanation of Statement (I).
(c) Statement (I) is true but Statement (II) is false
(d) Statement (I) is false but Statement (II) is true.

[ESE : 2016]

4.82 Two air streams with mass flow rates of 36 kg/min and 14 kg/min with respective enthalpies of 36 kJ/kg da and 50 kJ/kg da are mixed. The enthalpy of the mixture is nearly

(a) 64 kJ/kg da (b) 55 kJ/kg da
(c) 46 kJ/kg da (d) 40 kJ/kg da

[ESE : 2017]

4.83 Air at 30°C and 1 bar has a specific humidity of 0.016 kg/kg of dry air. By considering the saturation pressure of water vapour at 30°C as 4.246 kPa, the relative humidity of the air will be

(a) 66.1% (b) 60.2%
(c) 58.8% (d) 56.8%

[ESE : 2017]

4.84 Consider the following statements in respect of an evaporative cooling process :

1. The wet-bulb temperature remains constant.
2. The dew-point temperature remains constant.
3. The enthalpy remains constant.

Which of the above statements are correct?

(a) 1, 2 and 3 (b) 1 and 2 only
(c) 2 and 3 only (d) 1 and 3 only

[ESE : 2017]

4.85 In an air-handling unit, air enters the cooling coil at a temperature of 30°C. The surface temperature of the coil is –10°C. If the bypass factor of the coil is 0.45, then the temperature of the air at the exit will be

(a) 6°C (b) 8°C
(c) 10°C (d) 12°C [ESE : 2018]

4.86 Consider the following statements:

1. The relative humidity of air does not change with temperature as long as specific humidity remains constant.
2. Dew-point temperature is the temperature at which condensation begins when air is cooled at constant volume.
3. Saturated air passing over a water surface does not cause change of air temperature.
4. For saturated air, dry-bulb, wet-bulb and dew-point temperatures are identical.

Which of the above statements are correct?

(a) 1 and 2 (b) 2 and 3
(c) 1 and 4 (d) 3 and 4

[ESE : 2018]

4.87 The enthalpy of moist air with normal notations is given by

(a) $h = (1.005 + 1.88\ W)\ t + 2500\ W$
(b) $h = 1.88\ Wt + 2500\ W$
(c) $h = 1.005\ W\ t$
(d) $h = (1.88 + 1.005\ W)\ t + 2500\ W$

[ESE : 2019]

4.88 If the relative humidity of atmospheric air is 100%, then the wet-bulb temperature will be

(a) more than dry-bulb temperature
(b) equal to dew-point temperature
(c) equal to dry-bulb temperature
(d) less than dry-bulb temperature

[ESE : 2019]

4.89 If the atmospheric conditions are 20°C, 1.013 bar and specific humidity of 0.0095 kg/kg of dry air, the partial pressure of vapour will be nearly

(a) 0.076 bar (b) 0.056 bar
(c) 0.036 bar (d) 0.016 bar

[ESE : 2020]

4.90 In air-conditioning systems, air may be cooled and dehumidified by

1. Spraying chilled water to air in the form of fine mist.
2. Circulating chilled water or brine in a tube placed across the air flow.
3. Placing the evaporator coil across the air flow.

Which of the above statements are correct ?

(a) 1 and 2 only (b) 1 and 3 only
(c) 2 and 3 only (d) 1, 2 and 3

[ESE : 2020]

4.91 A room having dimensions of 5 m x 5 m x 3 m contains air at 25°C and 100 kPa at a relative humidity of 75%. The corresponding value of p_s is 3.169 kPa. The partial pressure of dry air will be nearly

(a) 106 kPa (b) 98 kPa
(c) 86 kPa (d) 78 kPa

[ESE : 2020]

4.92 A measure of feeling warmth or coolness by the human body in response to the air temperature, moisture content and air motion is called

(a) Dry bulb temperature
(b) Effective temperature
(c) Wet bulb temperature
(d) Dew point temperature

[ESE : 2020]

4.93 The phenomenon that enables cooling towers to cool water to a temperature below the dry bulb temperature of air is termed as

(a) Chemical dehumidification
(b) Adiabatic evaporative cooling
(c) Cooling and dehumidification
(d) Sensible cooling

[ESE : 2020]

4.94 If the air is initially at dry bulb temperature 35°C and wet bulb temperature 26.1°C as it enters an air washer which has a humidifying efficiency of 85%, then what is the final dry bulb temperature of air washed with recirculated spray water?

(a) 26.81°C (b) 27.43°C
(c) 32.83°C (d) 30.49°C

[ESE : 2021]

4.95 Which of the following principles and processes involved in the production of low temperatures are correct?

1. Adiabatic demagnetization
2. Thermoelectric cooling
3. Reversible adiabatic expansion of a gas
4. Irreversible adiabatic expansion of a real gas

Select the correct answer using the code given below:

(a) 1 and 3 only (b) 1, 3 and 4 only
(c) 1, 2 and 3 only (d) 1, 2, 3 and 4

[ESE : 2022]

Directions for the following **two (02)** items:

Read the following information and answer the **two** items that follows:

For the atmospheric air at room temperature of 30°C and relative humidity of 60%, the saturation pressure at 30°C is given as 0.0425 bar.

4.96 What is the partial pressure of air?
(a) 0.4255 bar (b) 0.6755 bar
(c) 0.7895 bar (d) 0.9875 bar

[ESE : 2023]

4.97 What is the humidity ratio?
(a) 0.1606 kg/kg of dry air
(b) 0.01606 kg/kg of dry air
(c) 0.2606 kg/kg of dry air
(d) 0.02606 kg/kg of dry air [ESE : 2023]

4.98 If the partial pressure of water vapour for the mixture of dry air and water vapour is 12.79 mm Hg and its total pressure is 736 mm Hg, then what is the specific humidity?
(a) 0.068 kg w.v./kg d.a.
(b) 0.011 kg w.v./kg d.a.
(c) 0.023 kg w.v./kg d.a.
(d) 0.044 kg w.v./kg d.a. [ESE : 2024]

4.99 On a particular day, the atmospheric air records a partial pressure of water vapour 9.62 mm Hg and the saturation pressure of vapour is 31.8 mm Hg. The barometric pressure is observed to be 756 mm Hg. What are the relative humidity and the specific humidity respectively?
(a) 30.25%, 0.008 (b) 30.25%, 0.014
(c) 35.87%, 0.014 (d) 35.87%, 0.008

[ESE : 2024]

■■■■

Answers Psychrometry and Psychrometric Processes

4.1 (b) 4.2 (a) 4.3 (c) 4.4 (a) 4.5 (c) 4.6 (c) 4.7 (c) 4.8 (c) 4.9 (b)
4.10 (d) 4.11 (d) 4.12 (b) 4.13 (d) 4.14 (c) 4.15 (d) 4.16 (a) 4.17 (a) 4.18 (b)
4.19 (a) 4.20 (d) 4.21 (b) 4.22 (a) 4.23 (d) 4.24 (b) 4.25 (c) 4.26 (c) 4.27 (d)
4.28 (a) 4.29 (a) 4.30 (d) 4.31 (c) 4.32 (a) 4.33 (c) 4.34 (a) 4.35 (b) 4.36 (*)
4.37 (a) 4.38 (c) 4.39 (b) 4.40 (d) 4.41 (b) 4.42 (b) 4.43 (c) 4.44 (b) 4.45 (b)
4.46 (a) 4.47 (b) 4.48 (c) 4.49 (c) 4.50 (b) 4.51 (b) 4.52 (c) 4.53 (a) 4.54 (a)
4.55 (c) 4.56 (c) 4.57 (d) 4.58 (c) 4.59 (a) 4.60 (a) 4.61 (b) 4.62 (a) 4.63 (d)
4.64 (a) 4.65 (d) 4.66 (a) 4.67 (a) 4.68 (d) 4.69 (a) 4.70 (b) 4.71 (a) 4.72 (d)
4.73 (c) 4.74 (d) 4.75 (c) 4.76 (b) 4.77 (a) 4.78 (*) 4.79 (d) 4.80 (a) 4.81 (c)
4.82 (d) 4.83 (c) 4.84 (d) 4.85 (b) 4.86 (d) 4.87 (a) 4.88 (b) & (c) 4.89 (d)
4.90 (d) 4.91 (b) 4.92 (b) 4.93 (b) 4.94 (b) 4.95 (d) 4.96 (d) 4.97 (b) 4.98 (b)
4.99 (a)

Explanations Psychrometry and Psychrometric Processes

4.1 (b)

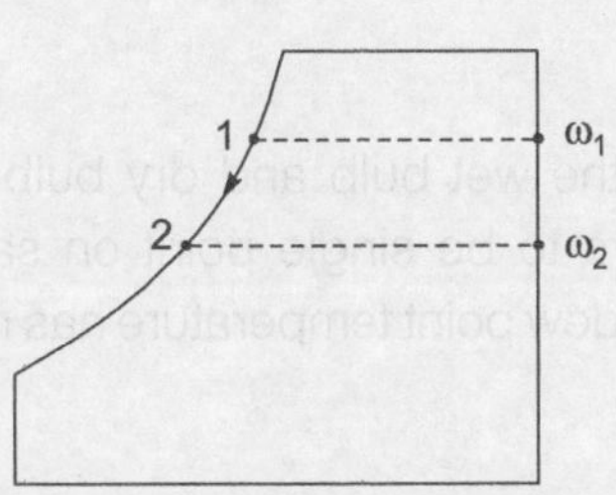

When warm saturated air is cooled, specific humidity decreases and relative humidity remains constant.

4.4 (a)

$$\mu = \frac{\omega}{\omega_s} = \frac{0.622\, p_v}{p - p_v}$$

$$= \frac{\dfrac{0.622\, p_v}{p - p_v}}{\dfrac{0.622\, p_{vs}}{p - p_{vs}}}$$

$$\mu = \frac{p_v}{p_{vs}}\left(\frac{p - p_{vs}}{p - p_v}\right) = \phi\left(\frac{p - p_{vs}}{p - p_v}\right)$$

4.5 (c)

$$\text{Humidity ratio} = \frac{m_v}{m_a} = \frac{R_a p_v}{R_v p_a}$$

$$= \frac{0.287\, p_v}{0.416\, p_a} = \frac{0.622\, p_v}{p_a} = \frac{0.622\, p_v}{p_s - p_v}$$

4.6 (c)

For saturated condition

DBT = WBT = DPT

And for saturated air relative humidity (ϕ) = 100%

4.7 (c)

Factors affecting human comfort:

(1) Effective temperature

(2) Heat production

(3) Heat and moisture losses from the human body

(4) Moisture contact of air

(5) Quality and quantity of air

(6) Air Motion

(7) Hot and cold surface

(8) Air stratification

4.8 (c)

When air is saturated adiabatically, we get wet bulb temperature and if isobarically, then we get dew point temperature.

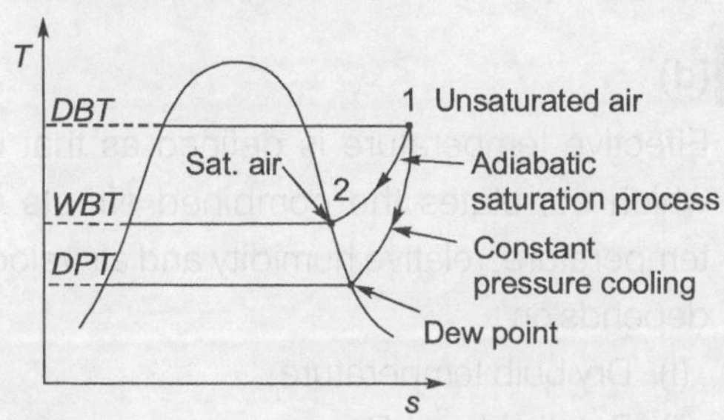

4.9 (b)

Cooling and dehumidification: The dehumidification of air is only possible when the effective surface temperature of the cooling coil is less than the dew point temperature of the air entering the coil.

4.10 (d)

$$\phi = \frac{p_v}{p_{vs}}$$

$$p_{v_1} v_1 = RT$$

$$p_{v_2}\left(\frac{v_1}{2}\right) = RT$$

$$\frac{\phi_1}{\phi_2} = \frac{RT}{v_1} \Big/ \frac{2RT}{v_1}$$

$$= \frac{RT}{v_1} \times \frac{v_1}{2RT} = \frac{1}{2}$$

$$\phi_2 = 2 \times 50 = \mathbf{100\%}$$

4.11 (d)

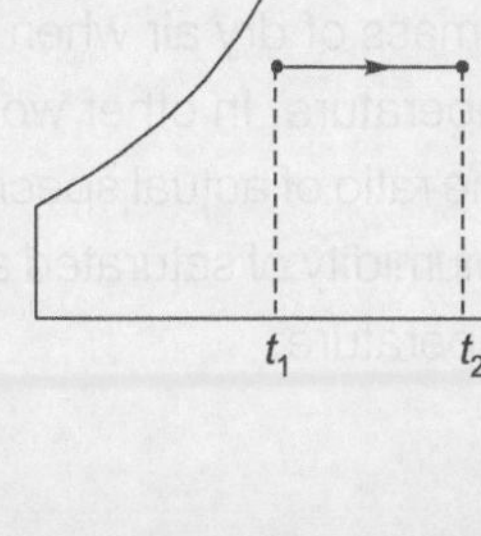

During sensible heating of air-water vapour mixture

1. Wet bulb temperature increases
2. Dry bulb temperature increases
3. Relative humidity decreases
4. Specific enthalpy increases
5. Humidity ratio remains same

4.12 (b)

Wet bulb temperature is equal to dry bulb temperature when air is fully saturated and dew point temperature is about to reached.

4.13 (d)

Effective temperature is defined as that index which correlates the combined effects of air temperature, relative humidity and air velocity. It depends on

(i) Dry bulb temperature
(ii) Relative humidity
(iii) Air Velocity
(iv) Mean radiation temperature.

4.14 (c)

As the temperature of water shower is less than dew point temperature hence, water vapour starts condensing. Thus, process involved is cooling and dehumidification.

4.15 (d)

Specific humidity or humidity ratio: It is the mass of water vapour present in1 kg of dry air (in the air-vapour mixture). It may also be defined as the ratio of mass of water vapour to the mass of dry air in a given volume of the air-vapour mixture.
Relative humidity: Relative humidity is the ratio of actual mass of water vapour in a given volume of moist air to the mass of water vapour in the same volume of saturated air at the same temperature and pressure.
Degree of saturation or percentage humidity: It is the ratio of actual mass of water vapour in a unit mass of dry air to the mass of water vapour in the same mass of dry air when it is saturated at same temperature. In other words, it may be defined as the ratio of actual specific humidity to the specific humidity of saturated air at the same dry bulb temperature.

4.17 (a)

Comfort conditions in air conditioning system are 25°C DBT & 50% RH

4.19 (a)

When the wet bulb and dry bulb are equal, it appears to be single point on saturation line, hence dew point temperature has reached.

4.20 (d)

Cooling will essentially take place because dry bulb temperature of air is more than water temperature. Since the dew point temperature of air is more than temperature of water so dehumidification of air takes place.

4.21 (b)

In winter the optimum effective temperature is 19°C where as in summer this temperature is 22°C.

4.22 (a)

The loss or gain by radiation (Q_R) from the body to the surrounding depends upon the mean radiant temperature. It is the average surface temperature of the surrounding objects when properly weighted and varies from place to place inside the room. When the mean radiant temperature is lower than the dry bulb temperature of the air in the room, Q_R is positive i.e., the body will undergo a radiant heat loss on the other hand, if the mean radiant temperature is higher than the dry bulb temperature of air in the room. Q_R is negative i.e., the body will undergo a radiant heat gain.

4.23 (d)

$$\text{cmm} = 10 \times \frac{3}{60} = 0.5 \text{ m}^3/\text{min}$$

$$\Delta T = 20 \text{ K}$$

$$\text{SHL} = 0.0204 \text{ cmm} (\Delta T)$$

$$= 0.0204 \times 0.5 \times 20 = \mathbf{0.204 \text{ kW}}$$

4.24 (b)

Comfort conditions in air-conditioning system are 24°C DBT and 60% RH.

4.26 (c)

Since enthalpy lines are parallel or coinciding with constant wet bulb temperature lines. The line shown on the figure is along constant enthalpy.

4.27 (d)

Effective temperature is governed by the comfort feeling of a human. It is controlled by heat content (dry bulb temperature) and moisture content (wet bulb temperature) the motion of air is necessary to carry heat dissipated from the body.

4.28 (a)

Higher air velocities require less difference between out-door effective temperature and inside effective temperature. Air velocity in the air conditioned system should be low enough to avoid the objectionable noise and drafty feeling and it should be high enough to carry out the heat from the human bodies.

4.29 (a)

Wet bulb depression = DBT – WBT

For saturated air, DBT = WBT hence, wet bulb depression is zero.

4.30 (d)

Air washer can work for humidifier, dehumidifier and filter also.

4.31 (c)

A human body feels comfortable when the heat produced by metabolism of human body is equal to the sum of the heat dissipated to the surrounding and the heat stored in human body by raising the temperature body.

Heat produced by the body = Heat dissipated to surroundings + Heat stored in body

4.32 (a)

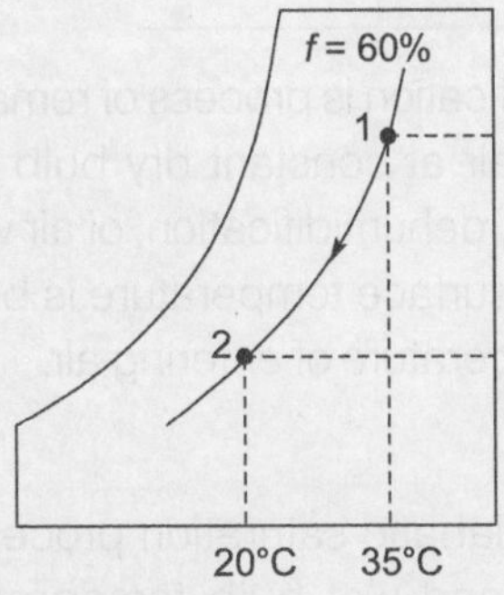

From the above figure, process 1-2 is cooling and dehumidification process.

4.33 (c)

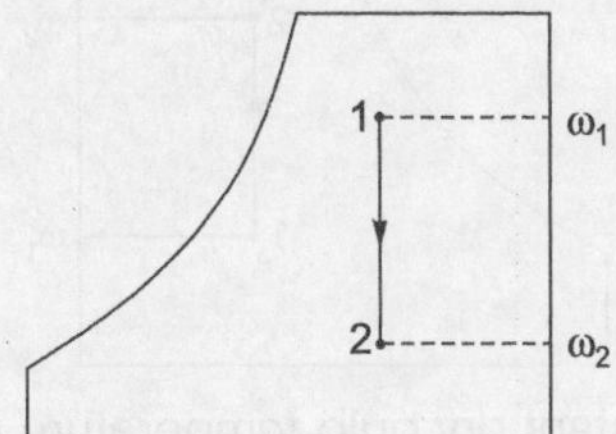

4.34 (a)

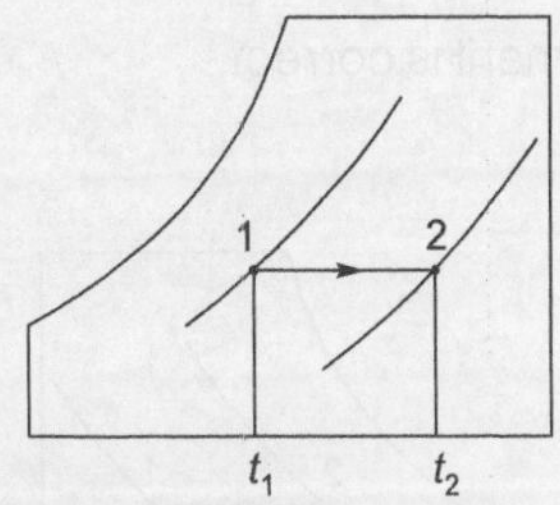

During sensible heating of air

1. Dry bulb temperature increases
2. Wet bulb temperature increases
3. Specific humidity remains constant
4. Relative humidity decreases
5. Enthalpy increases

4.35 (b)

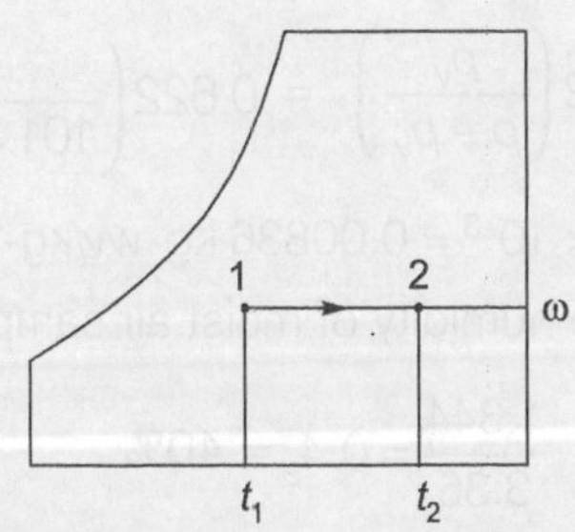

hence during sensible heating specific humidity remains constant.

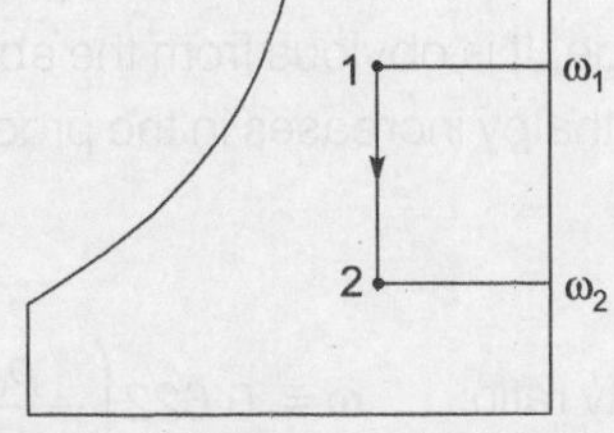

During dehumidification, dew point temperature decreases

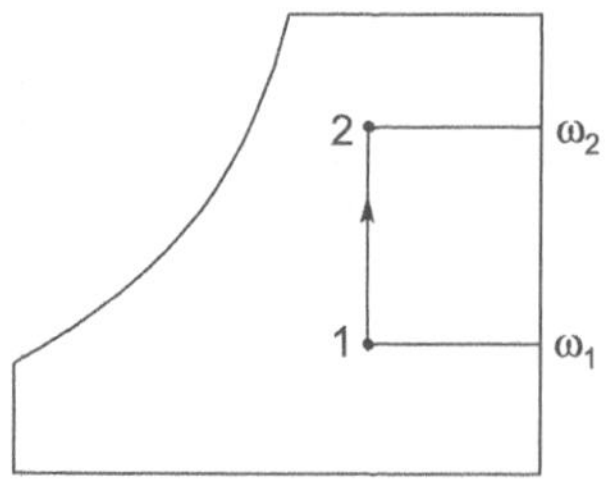

At constant dry bulb temperature if moisture in added, it is known as humidification.

4.36 (*)

No statement is correct.

4.37 (a)

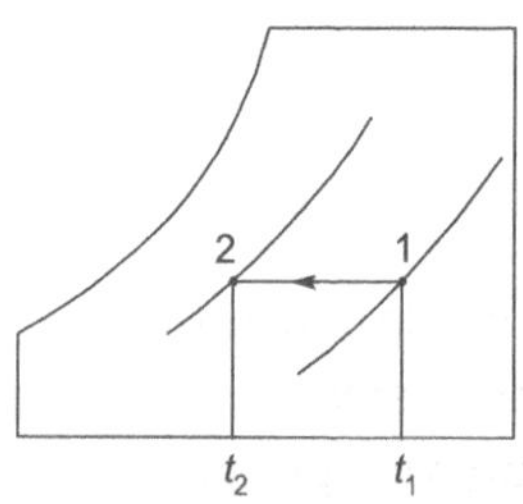

During sensible cooling relative humidity increases.

4.41 (b)

Humidity ratio of air

$$= 0.622\left(\frac{p_V}{p - p_V}\right) = 0.622\left(\frac{1.344}{101.325 - 1.344}\right)$$

$= 8.36 \times 10^{-3} = 0.00836$ kg-*wv*/kg-*da*

Relative humidity of moist air sample

$$= \frac{p_V}{p_{V_s}} = \frac{1.344}{3.36} = 0.4 = \mathbf{40\%}$$

4.42 (b)

In a psychrometric chart constant enthalpy lines and constant wet bulb temperature lines converge. It is obvious from the above diagram total enthalpy increases in the process.

4.44 (b)

Humidity ratio: $\omega = 0.622\left(\frac{p_V}{p - p_V}\right)$

$$0.016 = 0.622\left(\frac{p_V}{100 - p_V}\right)$$

or $p_v = \mathbf{2.5\ kN/m^2}$

4.45 (b)

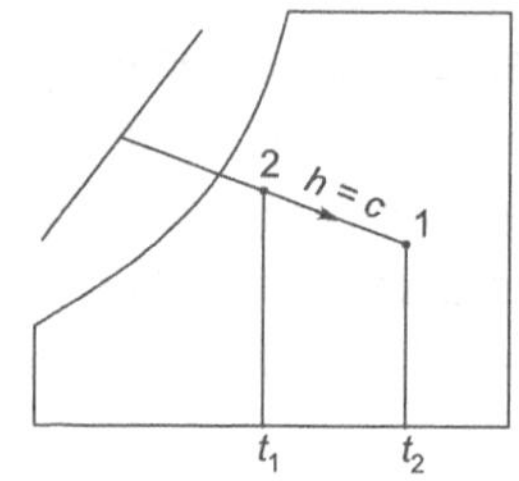

Silica gel is used in adiabatic chemical dehumidification. In this dry bulb temperature of air increases, humidity decreases and enthalpy remains constant.

4.47 (b)

Effective temperature is the temperature of saturated air at which a person would experience same feeling of comfort as in unsaturated environment.

4.48 (c)

DBT = 35°C; DPT = 20°C

$p_v = 2.33$ kPa; $p_{vs} = 5.628$ kPa

$p = 101.32$ kPa

Specific humidity,

$$\omega = 0.622\frac{p_v}{p - p_v} = 0.622\frac{2.33}{101.32 - 2.33}$$

$= \mathbf{14.64 \times 10^{-3}}$ **kg air/kg of dry air**

4.51 (b)

In winter air conditioning, for comfort, heating and humidification process is to be done.

4.52 (c)

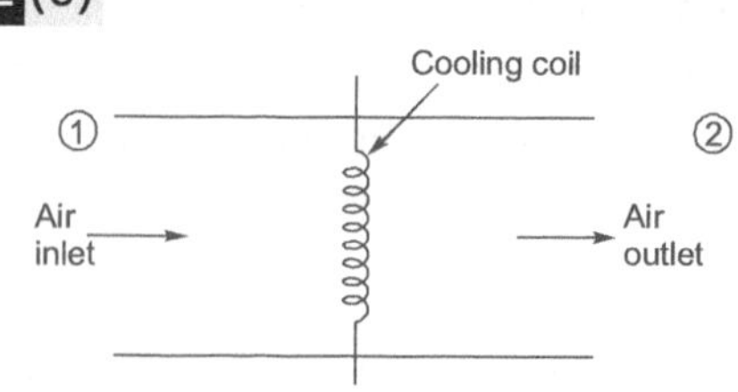

Dehumidification is process of remaining mixture from the air at constant dry bulb temperature. Therefore, dehumidification, of air will take place if the coil surface temperature is below the dew point temperature of entering air.

4.53 (a)

For an adiabatic saturation process of air. The enthalpy and wet bulb temeprature remains same.

4.54 (a)

From the psychrometric chart, we have seen that dew point temperature remains same.

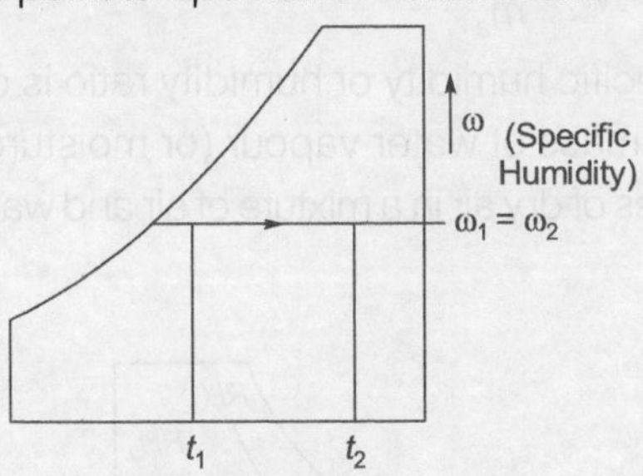

Dew point temperature line are horizontal and parallel to specific humidity line

4.55 (c)

When the air is adiabatically saturated the temperature attained is the wet bulb temperature. Wet bulb temperature is the lowest temperature that can be reached by evaporation of water only.

4.56 (c)

During 1-2 i.e., adiabatic saturation

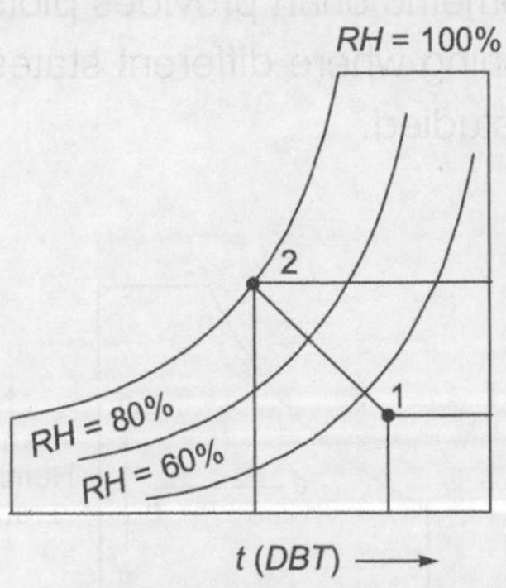

Relative Humidity (RH) increases

Specific Humidity (ω) increases

DBT(t) decreases

WBT remains constant as it follows constant enthalpy lines.

4.57 (d)

As the water temperature is below dew-point temperature of the moist air hence some water vapour will condense and hence the stream of air will undergo a process of cooling and dehumidification.

4.58 (c)

Process 1-2 represents sensible heating.

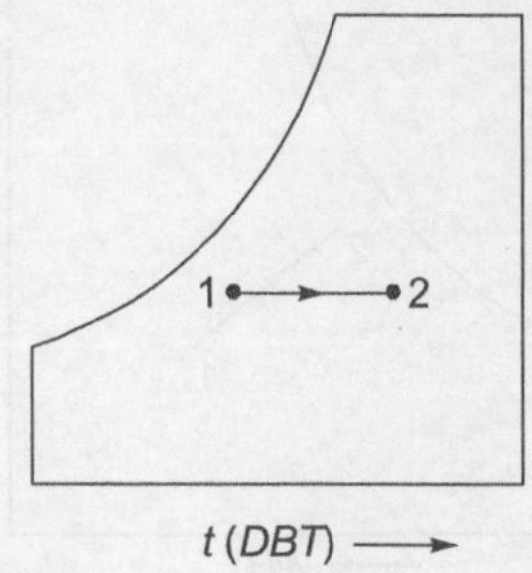

4.59 (a)

During sensible cooling i.e., process 1-2, the specific humidity remains constant.

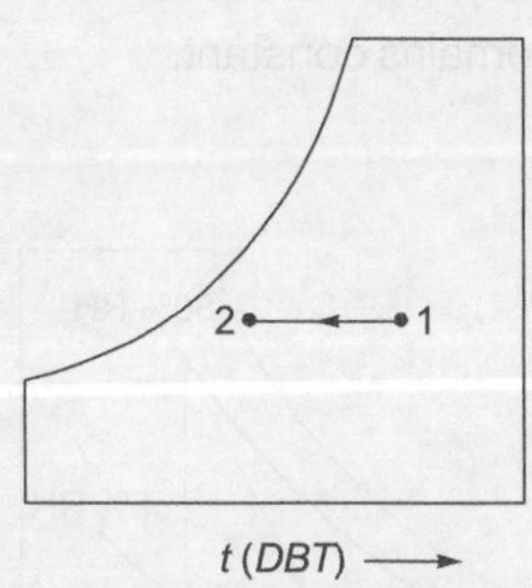

4.61 (b)

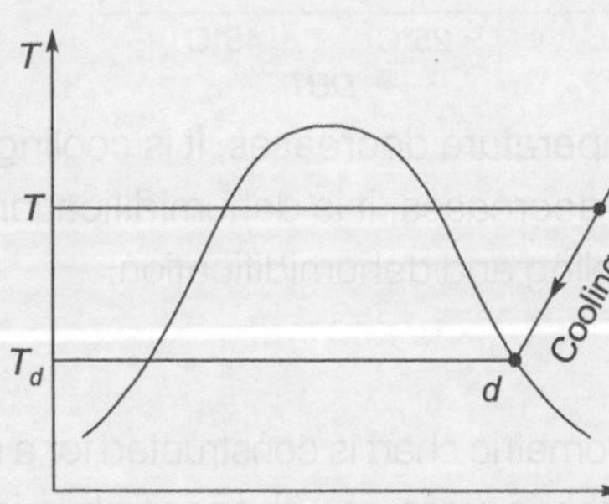

If sample of unsaturated moist air containing superheated water vapour is cooled (at constant pressure), the mixture will eventually reach the saturation temperature t_d of water vapour corresponding to its partial pressure, P_v at which first few drop of dew will be formed. This temperature t_d is called the DPT. DPT is the temperature to which moist air must be cooled at constant pressure before condensation of moisture takes place.

4.62 (a)

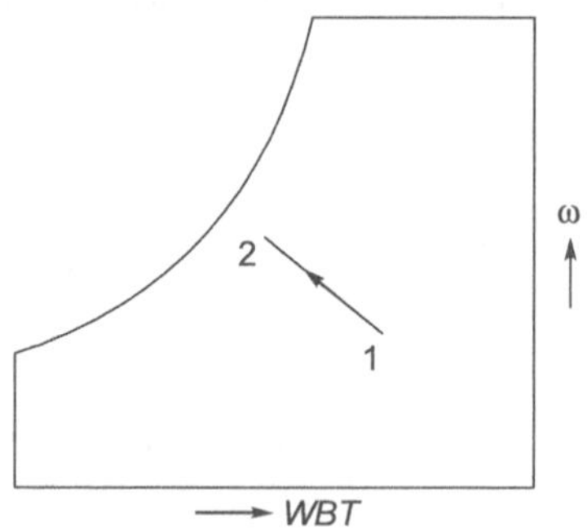

1→ 2 : Adiabatic saturation

RH increases

DBT decreases

Specific humidity increases

WBT remains constant.

4.63 (d)

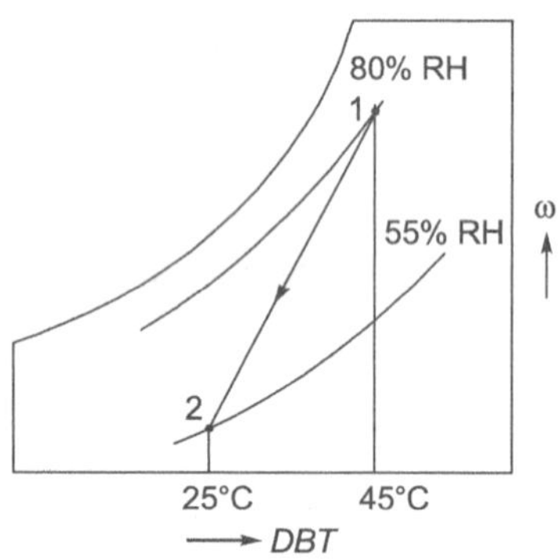

As temperature decreases, it is cooling.

As RH decreases, it is dehumidification.

So, cooling and dehumidification.

4.64 (a)

Psychrometric chart is constructed for a standard atmospheric pressure with specific humidity along the ordinate and the DBT along the abscissa. DBT and *w* are independent variables.

4.65 (d)

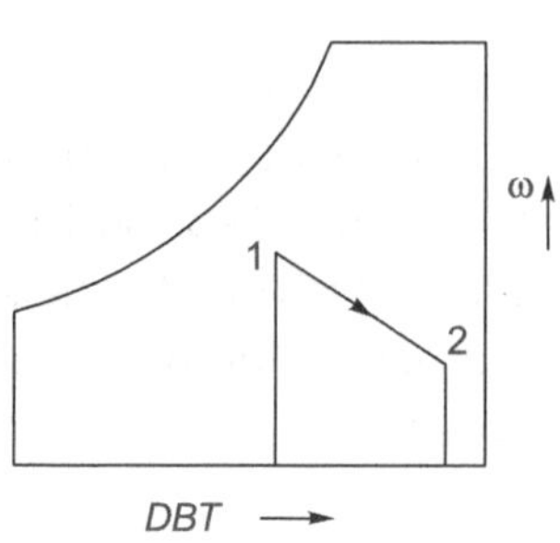

$DBT_2 > DBT_1$

4.66 (a)

$$\omega_v = \frac{m_v}{m_a}$$

Specific humidity or humidity ratio is defined as the mass of water vapour (or moisture) per unit mass of dry air in a mixture of air and water vapour.

4.67 (a)

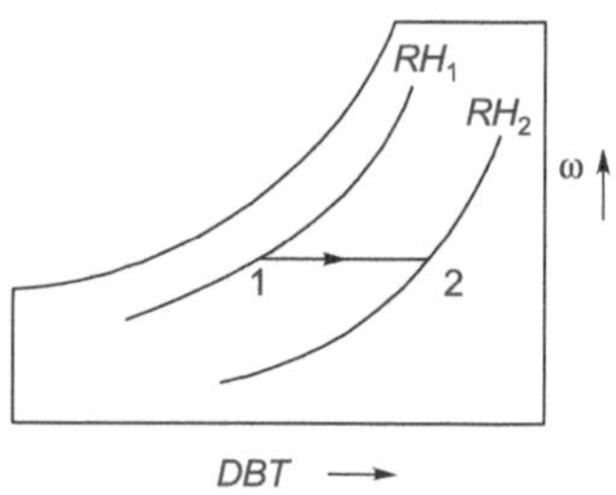

$RH_1 > RH_2$

$t_{db_1} < t_{db_2}$

$WBT_2 > WBT_1$

ω = constant = $f(p_v)$

$\therefore$ p_v = constant

4.68 (d)

Psychrometric chart provides plots for most air conditioning where different states of moist air can be studied.

4.69 (a)

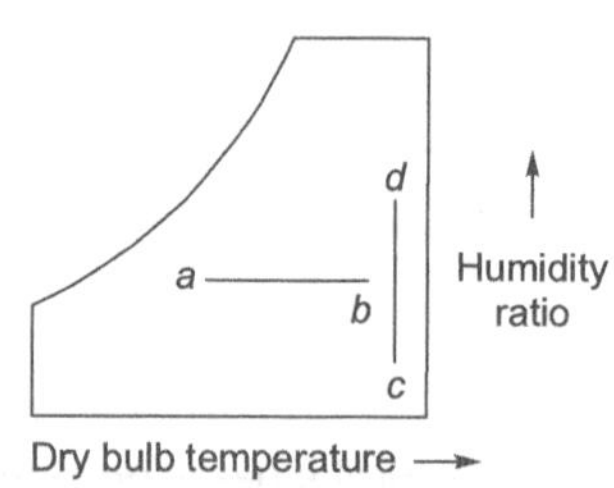

$a - b$ Represent humidity ratio and $c–d$ vertical lines represent dry bulb temperature.

4.70 (b)

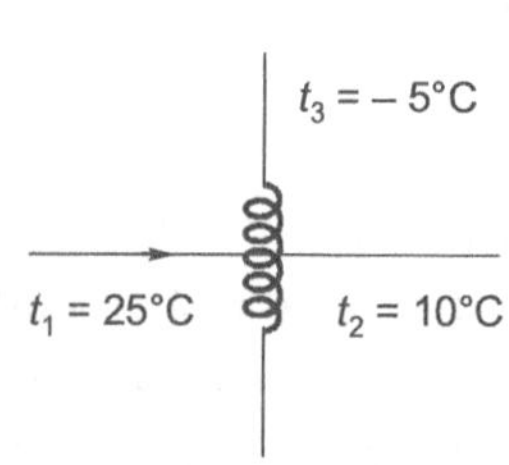

$$\text{By pass factor} = \frac{t_2 - t_3}{t_1 - t_3}$$

$$= \frac{10-(-5)}{25-(-5)} = \frac{15}{30} = \mathbf{0.5}$$

4.71 (a)

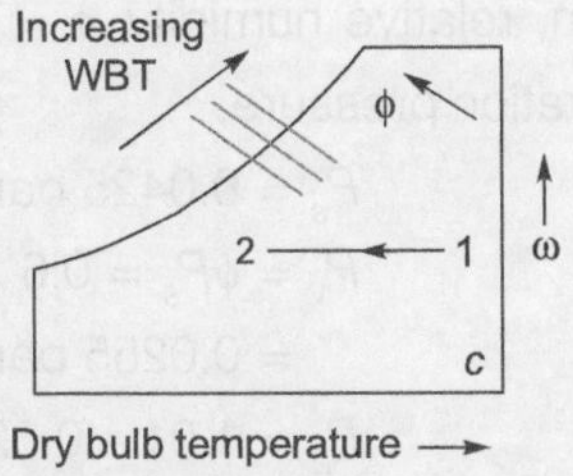

As one moves from 1 to 2.

→ WBT decreases

→ As dew point temperature are straight horizontal lines.

∴ DPT remains constant.

→ Relative humidity increases.

4.72 (d)

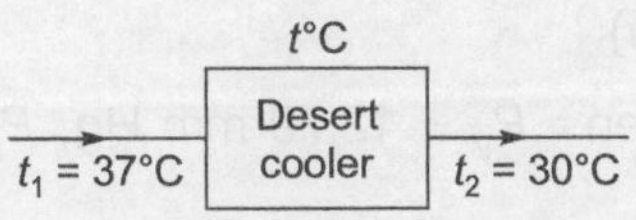

$$\therefore \quad \eta = 1 - \left(\frac{t_2 - t}{t_1 - t}\right)$$

$$0.7 = 1 - \left(\frac{30 - t}{37 - t}\right)$$

$$\therefore \quad t = \mathbf{27°C}$$

∴ (*d*) is the correct answer.

4.74 (d)

Specific humidity or Humidity ratio is ratio of mass of water vapour to mass of dry air, not a moist air.

∴ Statement 1 is false.

4.75 (c)

Gas can't be liquified by isothermal comp. since liquid can't exist above critical temperature.

4.76 (b)

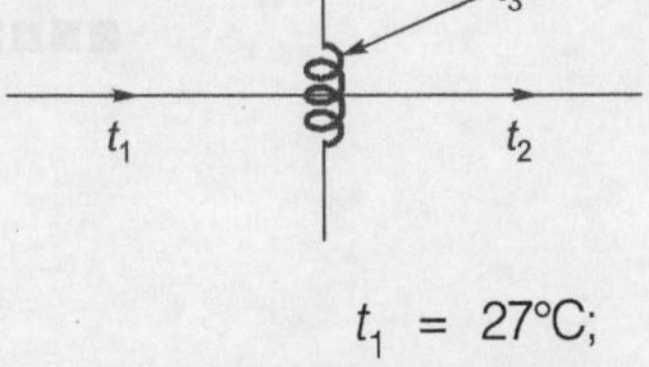

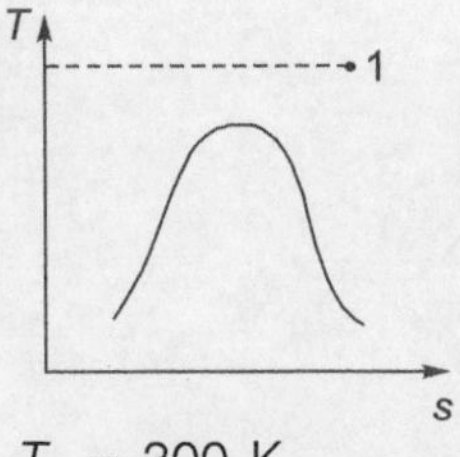

$$t_1 = 27°C; \qquad T_1 = 300 \text{ K}$$

$$t_3 = -5°C; \qquad T_3 = 268 \text{ K}$$

$$\text{BPF} = 0.4$$

$$\text{BPF} \Rightarrow 0.4 = \frac{t_2 - t_3}{t_1 - t_3} = \frac{t_2 - 268}{300 - 268}$$

$$t_2 = 280.8 \text{ K} = \mathbf{7.8°C}$$

4.78 (*)

Wrong option none of these

Answer is 0.33

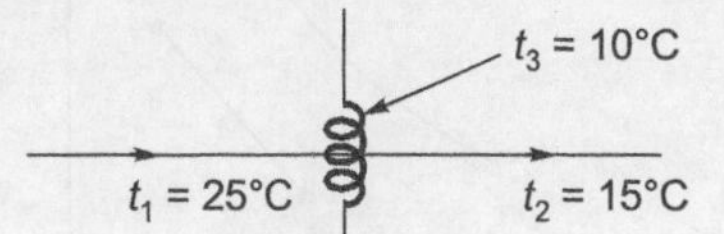

$$\text{BPF} = \frac{15 - 10}{25 - 10} = \frac{5}{15} = \mathbf{0.33}$$

4.79 (d)

Sensible heating → It is the process of increasing the temperature at constant specific humidity.

Sensible cooling → It is the process of decreasing the temperature at constant specific humidity.

4.80 (a)

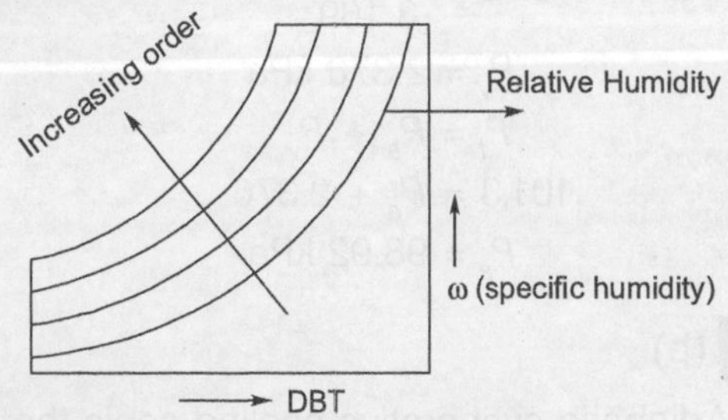

4.85 (b)

$$\text{BPF} = \frac{X - (-10)}{30 - (-10)} \Rightarrow 0.45 = \frac{X + 10}{40}$$

$$18 = X + 10$$

$$X = 8°C$$

4.89 (d)

$$\text{DBT} = 20°C = 293 \text{ K}$$

$$P_t = 1.013 \text{ bar}$$

Specific humidify,

$$w = 0.622 \times \frac{P_v}{P_t - P_v}$$

$$0.0095 = 0.622 \times \frac{P_v}{1.013 - P_v}$$

$$0.0152 = \frac{P_v}{1.013 - P_v}$$

$$0.0153 - 0.0152 P_v = P_v$$

$$0.0153 = 1.0152 P_v$$

Partial pressure of water vapour,

$$P_v = 0.016 \text{ bar}$$

4.90 (d)

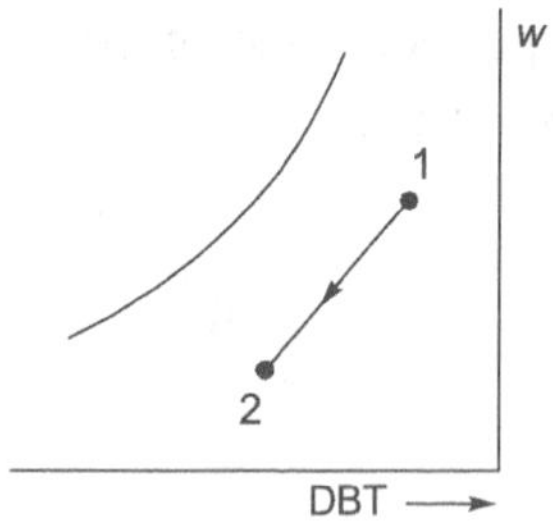

Spraying water less than DPT of air can also cool and dehumidify the air.

4.91 (b)

Relative humidity is given by,

$$\phi = \frac{P_v}{P_s}$$

where, P_s = 3.169 KPa, P_V = partial pressure of water vapour

$$0.75 = \frac{P_v}{3.169}$$

$$P_v = 2.376 \text{ kPa}$$

$$P_t = P_a + P_v$$

$$101.3 = P_a + 2.376$$

$$P_a = 98.92 \text{ kPa}$$

4.93 (b)

Adiabatic evaporative cooling cools the water below the dry bulb temperature.

4.94 (b)

$$\text{DBT} = 35°\text{C}$$

$$\text{WBT} = 26.1°\text{C}$$

$$0.85 = \frac{35-t}{35-26.1}$$

$$t = 27.43°\text{C}$$

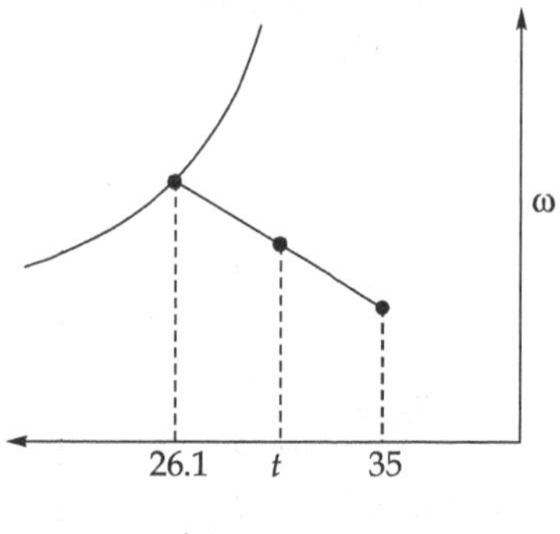

4.96 (d)

Given, relative humidity, ϕ = 60%

Saturation pressure,

$$P_s = 0.0425 \text{ bar}$$

$$P_v = \phi P_s = 0.6 \times 0.0425$$

$$= 0.0255 \text{ bar}$$

$$P_a = 1.01 - 0.0255 = 0.9845 \text{ bar}$$

4.97 (b)

$$\text{Humidity ratio, } \omega = 0.622 \times \frac{P_v}{P_b - P_v}$$

$$= 0.622 \times \frac{0.0255}{1.01 - 0.0255}$$

$$= 0.0161$$

4.98 (b)

Given : P_V = 12.79 mm Hg; P_t = 736 mm Hg

$$\therefore \text{Specific humidity, } \omega = 0.622 \times \frac{P_v}{P_t - P_v} 2$$

$$= 0.622 \times \frac{12.79}{736 - 12.79} = 0.011 \text{ kg w.v./kg d.a.}$$

4.99 (a)

Given : P_V = 9.62 mm Hg; P_{VS} = 31.8 mm Hg; P_t = 756 mm Hg

$$\therefore \text{Relative humidity, } \phi = \frac{P_V}{P_{VS}} = \frac{9.62}{31.8}$$

$$= 0.3025 = 30.25\%$$

$$\text{Specific humidity, } \omega = 0.622 \times \frac{P_V}{P_t - P_V}$$

$$= 0.622 \times \frac{9.62}{756 - 9.62}$$

$$= 0.008 \text{ kg w.v./kg d.a.}$$

5 Cooling Load Calculations and Duct Design

5.1 Consider the following parameters:
1. Dry-bulb temperature
2. Humidity ratio
3. Air velocity
4. Solar radiation intensity

Which of these parameter are taken into account for determining effective temperature for human comfort?
(a) 1 and 2 (b) 1 and 4
(c) 2, 3 and 4 (d) 1, 2 and 3

[ESE : 2000]

5.2 The desirable air velocity in the occupied zone for comfort for summer air-conditioners is in the range of
(a) 6 - 7 m/minute (b) 4 - 5 m/minute
(c) 2 - 3 m/minute (d) 0.5 - 1.5 m/minute

[ESE : 2000]

5.3 An air-conditioned room of volume 10 m^3 has infiltration of air equivalent to 3 air changes. Density of air is 1.2 kg/m^3, specific heat c_p is 1 kJ/kgK and temperature difference between room and ambient air is 20 K. The sensible heat load due to infiltrated air is
(a) 60 kJ/hr (b) 12 kJ/hr
(c) 6 kW (d) 0.2 kW

[ESE : 2000]

5.4 Consider the following statements:
1. The recommended outside air required per person for an auditorium is approximately 0.25 m^3/min
2. Outside air for ventilation purposes causes sensible heat load and also latent heat load
3. The sensible heat factor for an auditorium is generally kept as 0.7

Which of these statements are correct?
(a) 1 and 2 (b) 2 and 3
(c) 1 and 3 (d) 1, 2 and 3

[ESE : 2000]

5.5 Consider the following statements:
The typical air velocities in the ducts of air conditioning systems are
1. lower in residential buildings as compared to those of public buildings.
2. higher in residential buildings as compared to those of public buildings.
3. higher in industrial buildings as compared to those of public buildings.
4. equal in all types of buildings.

Which of these statements is/are correct?
(a) 1 only (b) 1 and 3
(c) 2 and 3 (d) 4 only

[ESE : 2000]

5.6 The by-pass factor of single cooling coil in an air-conditioner is 0.7. The by-pass factor, if three such cooling coils with the same apparatus due point are kept one behind the other, will be
(a) 0.210 (b) 0.292
(c) 0.343 (d) 0.412

[ESE : 2001]

5.7 For an air-conditioned space, RTH = 100 kW, RSHF = 0.75, volume flow rate is equal to 100 m^3/min and indoor design specific humidity is 0.01 kg per kg of dry air. The specific humidity of supply air is
(a) 0.010 (b) 0.0075
(c) 0.005 (d) 0.0025

[ESE : 2001]

5.8 Which one of the following statements is true for air conditioning duct design?
(a) Static regain method is used when the duct work is extensive, total pressure drop is low and flow is balanced
(b) Static regain method is used, when the duct work is extensive, total pressure drop is high and flow is unbalanced

(c) Equal friction method is used, when the duct work is extensive, total pressure drop is low and flow is balanced

(d) Equal friction method is used, when duct work is extensive, total pressure drop is low and flow is unbalanced

[ESE : 2001]

5.9 For air-conditioning system, the outdoor and indoor design dry bulb temperatures are 45°C and 25°C respectively. The space to be air-conditioned is 20 m x 30 m x 5 m and infiltration is estimated to be one change. If the density and specific heat of air are 1.2 kg of dry air per m^3 and 1.02 kJ/kg of dry air °C, then the sensible heat load due to infiltration is, nearly

(a) 122.4 kW (b) 61.2 kW
(c) 12.24 kW (d) 20.4 kW

[ESE : 2001]

5.10 The latent heat load in an auditorium is 25% of the sensible heat load. The value of Sensible Heat Factor (SHF) is equal to

(a) 0.25 (b) 0.5
(c) 0.8 (d) 1.0

[ESE : 2002]

5.11 For an office building the outdoor design conditions are 45°C DBT and humidity ratio of 0.015. The indoor design conditions are 25°C DBT and 0.01 humidity ratio. The supply air state is 15°C DBT and 0.007 humidity ratio. If the supply air flow rate is 1000 m^3/min and fresh air flow rate is 100 m^3/min, room sensible and room latent heat loads are respectively,

(a) 408 kW and 400 kW
(b) 408 kW and 150 kW
(c) 204 kW and 400 kW
(d) 204 kW and 150 kW [ESE : 2002]

5.12 In a system : Metabolic rate = M, work done by man = W, rate of convective, radiative and evaporative heat losses = Q and rate of heat storage = S. Then heat exchange between man and his environment is given by

(a) $M + W = Q + S$ (b) $M - W = Q - S$
(c) $M + W = Q - S$ (d) $M - W = Q + S$

[ESE : 2002]

5.13 Moist air enters the cooling coil with mass flow rate of 10 kg-da/s at dry bulb temperature of 30°C and humidity ratio of 0.017 kg v/kg da. It leaves the cooling coil at dry bulb temperature of 16°C and humidity ratio of 0.008 kg v/kg da. If specific heat of humid air is 1.02 kJ/kg da K and latent heat of water vapour is 2500 kJ/kg v. The sensible and latent heat transfer of cooling coil are, respectively

(a) 140 kW and 25000 kW
(b) 142.8 kW and 2.25 kW
(c) 142.8 kW and 225 kW
(d) 225 kW and 142.8 kW [ESE : 2003]

5.14 An air-conditioned room has length, width and height of 20 m, 30 m and 4 m respectively. The infiltration is assumed to be one air change. The outdoor and indoor dry bulb temperatures are 40°C and 25°C respectively. The sensible heat load due to infiltration is

(a) 734 kW (b) 12.24 kW
(c) 0.204 kW (d) 10 kW

[ESE : 2003]

5.15 Consider the following statements in respect of the contraction and expansion in air conditioning ducts:

1. Pressure drop is more in contraction than in expansion.
2. Pressure drop is more in expansion than in contraction.
3. Static pressure increases (regain) in expansion.
4. Static pressure increases (regain) in contraction.

Which of these statements are correct?

(a) 1 and 2 (b) 1, 2 and 3
(c) 1 and 3 (d) 2 and 4

[ESE : 2003]

5.16 Instantaneous cooling loads are not equal to instantaneous heat gains because

(a) Heat gains are offset by cooling provided by the AC system
(b) Indoor temperatures are lower
(c) Comfort conditions are maintained in the space
(d) Of the storage effect in the construction material of walls and roof [ESE : 2003]

5.17 Atmospheric air at dry bulb temperature of 15°C enters a heating coil whose surface temperature is maintained at 40°C. The air leaves the heating coil at 25°C. What will be the by-pass factor of the heating coil?
(a) 0.376 (b) 0.4
(c) 0.6 (d) 0.67 [ESE : 2004]

5.18 In an air-conditioning plant the refrigeration load on the coil is 100 TR. The mass and enthalpy of air leaving the coil are 420 kg/min and 40 kJ/kg respectively. The enthalpy of the air at the inlet to the coil under these conditions is
(a) 80 kJ/kg (b) 90 kJ/kg
(c) 100 kJ/kg (d) 102.5 kJ/kg
[ESE : 2004]

5.19 Velocity of air passing through a rectangular duct and a circular duct is same. Which one of the following is the correct expression for the equivalent diameter of the circular duct in respect of a rectangular duct for the same pressure loss per unit length? (*a* and *b* are the length and breadth of the rectangular duct cross-section)
(a) $\frac{a+b}{ab}$ (b) $\frac{2ab}{a+b}$
(c) $\frac{2a}{a-b}$ (d) $\frac{2b}{a+b}$ [ESE : 2004]

5.20 If coefficient of contraction at the vena contracta is equal to 0.62, then what will be the dynamic loss coefficient in sudden contraction in air-conditioning duct?
(a) 0.25 (b) 0.375
(c) 0.55 (d) 0.65 [ESE : 2004]

5.21 Air-conditioning has to be done for a hall whose RSH = 50 kW and RLH = 50 kW. There are no other sources of heat addition or leakages. What is the value of the RSHF?
(a) 0.25 (b) 0.5
(c) 0.75 (d) 1.00 [ESE : 2005]

5.22 In order to have a low by-pass factor of a cooling coil, the fin spacing and the number of tube rows should be
(a) Wide apart and high, respectively
(b) Wide apart and low, respectively
(c) Close and high, respectively
(d) Close and low, respectively [ESE : 2005]

5.23 For an air-conditioned space, RTH = 100 kW, RSHF = 0.75, volume flow rate = 100 m^3/min, and indoor design specific humidity is 0.01 kg/kg of dry air. What is the specific humidity of the supply air?
(a) 0.010 (b) 0.0075
(c) 0.005 (d) 0.0025
[ESE : 2005]

5.24 A passive method to keep the house comfortably warm by solar conditioning in cold climatic condition is to paint the
(a) Eastern wall of the house by black paint on its outer side
(b) Eastern wall of the house by black paint on its inner side
(c) Southern wall of the house by black paint on its outer side
(d) Southern wall of the house by black paint on its inner side
[ESE : 2005]

5.25 Consider the following statement pertaining to duct design:
1. Aspect ratio of ducts should be high.
2. In the equal friction method of design, use of dampers cannot be eliminated by any means.
3. The static regain method is not suitable for long ducts.
4. The velocity reduction method is employed only in simple systems.

Which of these statements are correct?
(a) 1 and 2
(b) 3 and 4
(c) 1 and 3
(d) 2 and 4
[ESE : 2006]

5.26 What is Sol-air temperature?
(a) It is equal to the sum of outdoor air temperature, and absorbed total radiation divided by outer surface convective heat transfer coefficient
(b) It is equal to the absorbed total radiation divided by convective heat transfer coefficient at outer surface
(c) It is equal to the total incident radiation divided by convective heat transfer coefficient at outer surface

(d) It is equal to the sum of indoor air temperature and absorbed total radiation divided by convective heat transfer coefficient at outer surface

[ESE : 2006]

5.27 What is the sensible heat factor during the heating and humidification process equal to?

(a) $\frac{H_1 + H_2}{H_3 - H_1}$ (b) $\frac{H_2 - H_1}{H_3 - H_1}$

(c) $\frac{H_1 + H_2}{H_1 - H_2}$ (d) $\frac{H_3 - H_1}{H_2 - H_1}$

[ESE : 2006]

5.28 Equal friction method of designing ducts is preferred

(a) when system is balanced
(b) when system is not balanced
(c) only for return ducts
(d) for any system

[ESE : 2008]

5.29 The atmospheric air at 760 mm of Hg, dry bulb temperature 15°C and wet bulb temperature 11°C enters a heating coil whose temperature is 41°C. If the by-pass factor of the heating coil is 0.5, then what will be the dry bulb temperature of the air leaving the coil ?

(a) 28°C (b) 29°C
(c) 30°C (d) 26°C [ESE : 2009]

5.30 A classroom is to be air-conditioned by obtaining the comfort conditions of 22°C DBT and 55% RH from outdoor conditions of 32°C DBT and 22°C WBT. The weight of outside air supplied is 30 kg/min. The comfort conditions required are achieved first by chemical dehumidification and then by cooling with a cooling coil as shown in the psychrometric chart below.

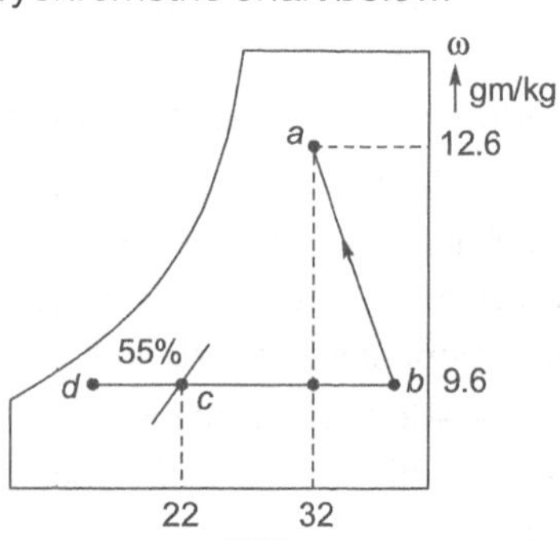

What is the capacity of the dehumidification in kg/hr?

(a) 3.2 (b) 5.4
(c) 6.8 (d) 9.5

[ESE : 2009]

5.31 Air enters a rectangular duct measuring 30 × 40 cm with a velocity of 8.5 m/s and a temperature of 40°C. Kinematic viscosity of the air is 16.95 × 10^{-6} m^2/s. What will be the Reynolds number ?

(a) 1.72×10^5 (b) 2.58×10^5
(c) 0.86×10^5 (d) 0.72×10^5

[ESE : 2009]

5.32 For designing air-conditioning ducts, equal friction method

(a) Ensures same velocity in the duct all through in all branches
(b) Ensures constant static pressure at all terminals in the duct
(c) Automatically reduces the air-velocity in the duct in the direction of flow
(d) Does all of the above

[ESE : 2010]

5.33 The room sensible heat loss is 30,000 kJ/h and the latent heat loss is 20,000 kJ/h. Then the sensible heat factor is

(a) –0.6 (b) 0.6
(c) 0.3 (d) 3.0 [ESE : 2013]

5.34 If air at 20°C is heated to 25°C using heater with a surface temperature of 30°C then, the by-pass factor is

(a) 2.0 (b) 0.5
(c) 1.0 (d) None of the above

[ESE : 2013]

5.35 The atmospheric air at dry bulb temperature 15°C enters a heating coil maintained at 40°C. If the air leaves the heating coil at 30°C, the coil efficiency is equal to

(a) 0.2 (b) 0.4
(c) 0.6 (d) 0.75

[ESE : 2013]

5.36 In the air-conditioned space, fan load is considered as

(a) latent heat gain

(b) sensible heat gain
(c) both sensible and latent heat gain
(d) Not considered as load

[ESE : 2013]

5.37 Equal friction method for designing air-conditioning duct
(a) ensures constant static pressure at all terminals in the duct
(b) ensures same velocity in the duct all through in all branches
(c) automatically reduces the air velocity in the duct in the direction of flow
(d) does not of the above

[ESE : 2013]

Directions: The following items consists of two statements; one labelled as **'Statement (I)'** and the other as **'Statement (II)'**. You are to examine these two statements carefully and select the answers to these items using the codes given below:

Codes:
(a) Both Statement (I) and Statement (II) are individually true and Statement (II) is the correct explanation of Statement (I).
(b) Both Statement (I) and Statement (II) are individually true but Statement (II) is **NOT** the correct explanation of Statement (I).
(c) Statement (I) is true but Statement (II) is false
(d) Statement (I) is false but Statement (II) is true.

5.38 Statement (I): Immersion chilling in tanks of slush ice is better than the slow cool air chilling for poultry products.
Statement (II): Cool air chilling of poultry products cause a moisture loss of 1 to 2 per cent while immersion chilling can cause a moisture absorption of 4 to 15 percent.

[ESE : 2013]

5.39 Statement (I) : In an air-conditioned room, the reflective coating should be on the inside of the window.
Statement (II): Window pane glass is transparent to solar radiation.

[ESE : 2014]

5.40 Assertion (A): The estimation of exact cooling load calculations of space to be conditional is important to know because it involves both the initial cost and operating cost.
Reason (R): The outside heat gain to conditioned space is called external load and the heat gain from inside the conditional space is called internal load.

[ESE : 2015]

5.41 The latent heat load in an auditorium is 25% of sensible heat load. The value of sensible heat factor is
(a) 0.3 (b) 0.5
(c) 0.8 (d) 1.0

[ESE : 2016]

5.42 Statement (I) : In air-conditioning, the atmospheric air (mixture of dry air and water vapour) can be considered as mixture of two ideal gases.
Statement (II) : In the temperature range used in air-conditioning, the partial pressure of the water vapour is very low and it follows the ideal gas relation with negligible error.

[ESE : 2018]

5.43 During an air-conditioning of a plant, the room sensible heat load is 40 kW and room latent heat load is 10 kW, ventilation air is 25% of supply air. At full load, the room sensible heat factor will be
(a) 0.9 (b) 0.8
(c) 0.7 (d) 0.6

[ESE : 2019]

5.44 Consider the following statements for sensible heat factor:
1. Sensible heat factor will be negative if sensible heat and latent heat are both negative.
2. Sensible heat factor will be negative if sensible heat is negative and latent heat is positive.
3. Sensible heat factor will be negative if sensible heat is positive and latent heat is negative.
4. Sensible heat factor will be negative if sensible heat and latent heat are both positive.

Which of the above statements are correct?
(a) 2 and 3 only (b) 1 and 2 only
(c) 1 and 3 only (d) 2 and 4 only

[ESE : 2021]

5.45 Thermal comfort depends on environmental and physiological factors. Which one of the following is representing the physiological factor?
(a) Air temperature (dry bulb) (b) Relative humidity
(c) Radiation (d) Amount of clothing (insulation)

[ESE : 2023]

Answers Cooling Load Calculations and Duct Design

5.1 (d)	5.2 (a)	5.3 (d)	5.4 (d)	5.5 (b)	5.6 (c)	5.7 (c)	5.8 (c)	5.9 (d)
5.10 (c)	5.11 (d)	5.12 (d)	5.13 (c)	5.14 (b)	5.15 (d)	5.16 (d)	5.17 (c)	5.18 (b)
5.19 (b)	5.20 (b)	5.21 (b)	5.22 (c)	5.23 (c)	5.24 (b)	5.25 (b)	5.26 (a)	5.27 (d)
5.28 (a)	5.29 (a)	5.30 (b)	5.31 (a)	5.32 (c)	5.33 (b)	5.34 (b)	5.35 (c)	5.36 (b)
5.37 (c)	5.38 (a)	5.39 (d)	5.40 (b)	5.41 (c)	5.42 (a)	5.43 (b)	5.44 (a)	5.45 (d)

Explanations Cooling Load Calculations and Duct Design

5.1 (d)

Factor affecting effective temperature for human comfort
(i) Dry bulb temperature
(ii) Relative humidity
(iii) Air velocity

5.2 (a)

Desirable air velocity for comfort is in the range of 5-8 m/min.

5.3 (d)

$$SHL = mc_p.\Delta T = (3 \times 10) \times 1.2 \times 1 \times 20$$

$$= 720 \text{ kJ/hr} = \frac{720}{3600} \text{ kJ/s} = \mathbf{0.2\ kW}$$

5.4 (d)

Outside air causes sensible heat gain and latent heat gain. Sensible heat factor for auditorium is 0.7.

5.5 (b)

Air velocity in the ducts of air-conditioning system depends on number of person.

5.6 (c)

Equivalent by pass factor = $n_1 \times n_2 \times n_3$

(here $n_1 = n_2 = n_3 = 0.7$)

$$= n^3 = (0.7)^3 = \mathbf{0.343}$$

5.7 (c)

$$0.75 = \frac{SH}{TH}$$

$$\Rightarrow SH = 0.75 \times TH$$

$$SH = 0.75 \times 100 = 75 \text{ kW}$$

$$LH = 100 - 75 = 25 \text{ kW}$$

$$25 = 50 \text{ (cmm)} \times \Delta\omega$$

$$\Delta\omega = \frac{25}{50 \times 100} = 0.005$$

$$0.01 - \omega_s = 0.005$$

$$\omega_s = 0.01 - 0.005$$

$$= \mathbf{0.005\ kg/kg\ of\ dry\ air}$$

5.8 (c)

Equal friction method is employed, when the duct work is extensive, total pressure drop is low and flow is balanced because this is most economical.

5.9 (d)

$$SHL = 0.0204 \times \text{(cmm)} \times \Delta T$$

$$= 0.0204 \times \left[1 \times \frac{(20 \times 30 \times 5)}{60}\right] \times (45 - 25)$$

$$= 0.0204 \times 50 \times 20 = \mathbf{20.4\ kW}$$

5.10 (c)

Given LH = 0.25 SH

$$\text{Now, SHF} = \frac{\text{SH}}{\text{SH}+\text{LH}} = \frac{1}{1.25} = \mathbf{0.8}$$

5.11 (d)

$$\text{SHL} = 0.0204 \times \text{cmm} \times \Delta T$$

$$= 0.0204 \times 1000 \times (25 - 15) = 204 \text{ kW}$$

$$\text{LHL} = 50 \text{ cmm} \times \Delta\omega$$

$$= 50 \times 1000 \times (0.01 - 0.007)$$

$$= 50 \times 1000 \times 0.003 = \mathbf{150\ kW}$$

5.12 (d)

A human body feels comfortable when the heat produced by metabolism of human body is equal to the sum of the heat dissipated to the surrounding and the heat stored in human body by raising the temperature of the body.

$$M = W + Q + S$$

$$M - W = Q + S$$

where,

M is metabolic rate

W is work done by man

Q is rate of convective, radiative and evaporative heat losses

S is rate of heat storage

5.13 (c)

$$Q_S = m_a c_{pa} \Delta T = 10 \times 1.02 \times 14 = 142.8 \text{ kW}$$

$$Q_L = m_a h_{fg} \Delta\omega = 10 \times 2500 \times (0.017 - 0.008)$$

$$= \mathbf{225\ kW}$$

5.14 (b)

$$\text{SHL} = 0.0204 \times \text{cmm} \times \Delta T$$

$$= 0.0204 \times \left[1 \times \frac{20 \times 30 \times 4}{60}\right] \times (40 - 25)$$

$$= \mathbf{12.24\ kW}$$

5.15 (d)

$$\text{The pressure loss in contractions} = C_2 C_r \left(\frac{V_2}{4.04}\right)^2$$

$$\text{The pressure loss in expansion} = C_2 C_r \left(\frac{V_2}{4.04}\right)^2$$

$$(V_2)_{\text{contraction}} > (V_2)_{\text{expansion}}$$

$$\text{but } (C_r)_{\text{contraction}} << (C_r)_{\text{expansion}}$$

$$\text{So, } (P_L)_{\text{expansion}} > (P_L)_{\text{contraction}}$$

when the enlargement is accompanied with pressure loss, the increase in static pressure or static regain is reduced by amount of the pressure loss. In such a case, static pressure regain.

$$S_{PR} = P_{S_2} - P_{S_1} = (P_{V_1} - P_{V_2}) - P_L = R(P_{V_1} - P_{V_2})$$

(where R is the static regain)

when the enlargement is not accompanied with pressure loss. Then there will be full conversion of velocity pressure into static pressure. In such a case, static pressure regain.

$$\text{SPR} = P_{s2} - P_{S1}, = (P_{V_1} - P_{V_2})$$

5.17 (c)

$$\text{By pass factor} = \frac{t_3 - t_2}{t_3 - t_1}$$

where,

t_3= Temperature of heating coil = 40°C

t_2= Exit temperature of air = 25°C

t_1= Temperature of air at inlet = 15°C

$$\text{BPF} = \frac{40 - 25}{40 - 15} = \mathbf{0.6}$$

5.18 (b)

$$100 \text{ TR} = 100 \times 210 = 21000 \text{ kJ/min}$$

$$\text{Now refrigeration capacity} = m_r (h_i - h_f)$$

$$21000 = 420 (h_i - 40)$$

$$h_i = \mathbf{90\ kJ/kg}$$

5.19 (b)

Equivalent diameter,

$$D = \frac{4 \times \text{Area}}{\text{Wetted perimeter}} = \frac{2ab}{a+b}$$

5.21 (b)

$$\text{RSHF} = \frac{\text{RSH}}{\text{RSH}+\text{RLH}} = \frac{50}{50+50} = \mathbf{0.5}$$

5.22 (c)

Low bypass factor signifies better cooling coil. If the coils are closely spaced and number of tube rows are high then maximum air will have contact with the coil for longer period. Hence efficiency improves.

5.23 (c)

$$\text{RSHF} = \frac{\text{RSH}}{\text{RTH}}$$

$$\therefore \quad \text{RSH} = 0.75 \times 100 = 75 \text{ kW}$$

$$\text{RLH} = 100 - 75 = 25 \text{ kW}$$

$$\text{RLH} = 50 \text{ cmm } (\omega_1 - \omega_2)$$

$$\omega_2 = \mathbf{0.005 \text{ kg/kg of dry air}}$$

5.25 (b)

The aspect ratio, for rectangular ducts, should not be greater than 8 in any case

In friction loss method no damper is required to balance them.

Static regain method is not possible to design economically very long branches and the branches very near to the fan for complete region.

The velocity reduction method of designing ducts is usually adopted for very simple system.

5.26 (a)

Sol air temperature, $t_e = t_0 + \frac{I\alpha}{f_0}$

where t_0 = Outside air temperature

$I\alpha$ = Total radiation absorbed

f_0 = Outside film coefficient

5.27 (d)

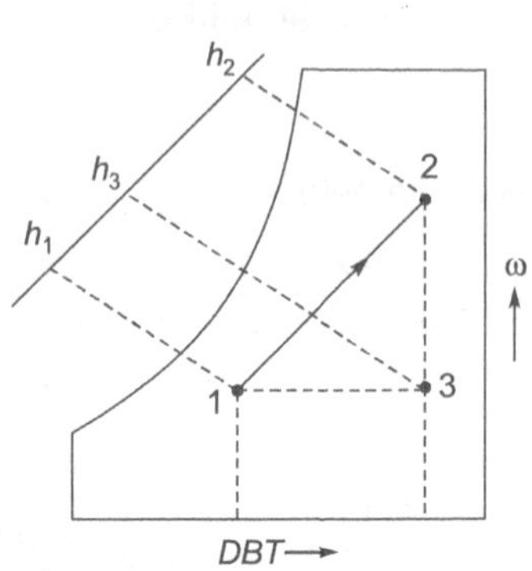

$$\text{SHF} = \frac{\text{SH}}{\text{SH} + \text{LH}}$$

Sensible heat (SH) = $H_3 - H_1$

Latent heat (LH) = $H_2 - H_3$

$$\text{SHF} = \frac{H_3 - H_1}{(H_3 - H_1) + (H_2 - H_3)}$$

$$\text{SHF} = \frac{H_3 - H_1}{H_2 - H_1}$$

5.29 (a)

Let the DBT of leaving coil = T°C

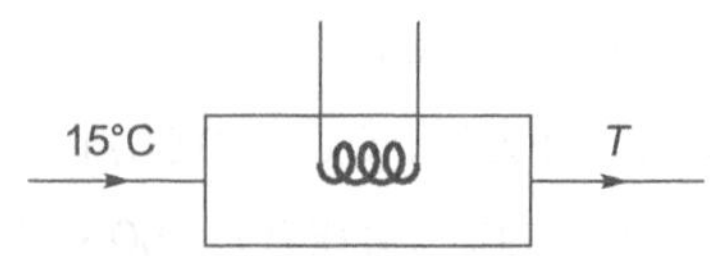

$$\frac{41 - T}{41 - 15} = 0.5$$

$$\Rightarrow \quad 41 - T = 13$$

$$\Rightarrow \quad T = \mathbf{28°C}$$

5.30 (b)

Capacity of dehumidification in kg per hour

$$= m \times (\omega_1 - \omega_2)$$

$$= (30 \times 60)(12.6 - 9.6) \times 10^{-3}$$

$$= \mathbf{5.4 \text{ kg/hour}}$$

5.31 (a)

Equivalent diameter of rectangular duct

$$= \frac{2ab}{a+b} = \frac{2 \times 30 \times 40}{30 + 40}$$

$$= 34.28 \text{ cm} = 0.3428 \text{ m}$$

Reynolds number,

$$\text{Re} = \frac{\text{UD}}{\nu} = \frac{8.5 \times 0.3428}{16.95 \times 10^{-6}}$$

$$= \mathbf{1.72 \times 10^5}$$

5.33 (b)

$$\text{SHF} = \frac{\text{RSHL}}{\text{RSHL} + \text{LHL}}$$

$$= \frac{30000}{30000 + 20000}$$

$$= \frac{3}{5} = \mathbf{0.6}$$

5.34 (b)

$$BPF = \frac{\text{Surface temp.} - \text{Air temp.}}{\text{Surface temp.} - \text{Inlet air temp}}$$

$$= \frac{30 - 25}{30 - 20} = \frac{5}{10} = \mathbf{0.5}$$

5.35 (c)

$$BPF = \frac{40 - 30}{40 - 15} = \frac{10}{25} = 0.4$$

Coil efficiency = 1 – BPF = **0.6**

5.36 (b)

Sensible heat is due to temperature difference. Due to fan, heat is removed from air, so it is sensible heat load.

5.37 (c)

Equal friction method of duct design:
In this method, the frictional pressure drop per unit length of the duct is maintained constant throughout the duct system. This method of sizing ducts, automatically, reduces the air velocity in the direction of flow. This method is recommended because of its simplicity.

5.38 (a)

Immersion chilling is more effective but it requires large quantity of water to flash the heat. Slush ice is more effective for moisture absorption.

5.41 (c)

Given data :

$$LH = 0.25\, SH$$

$$\therefore \quad SHF = \frac{SH}{SH + 0.25\, SH} = \mathbf{0.8}$$

5.43 (b)

RSH = 40 kW

Total heat (TH) = RSH + RLH = 50 kW

RLH = 10 kW

$$RSHF = \frac{RSH}{TH} = \frac{40}{50} = 0.8$$

5.45 (d)

Thermal comfort depends on environmental and physiological factors.

- **Environmental:** Air temperature (dry bulb), relative humidity, air speed and radiation (mean radiant temperature)
- **Physiological:** Metabolic rate and amount of clothing (insulation).

■■■■

UNIT

VI

Turbo Machinery

Syllabus

Reciprocating and Rotary pumps, Pelton wheel, Kaplan and Francis Turbines, velocity diagrams, Impulse and Reaction principles, Steam and Gas Turbines, Theory of Jet Propulsion – Pulse jet and Ram Jet Engines, Reciprocating and Rotary Compressors – Theory and Applications.

Contents

Sl.	Topic	Page No.
1.	Impact of Jet and Hydraulic Turbines	389
2.	Hydraulic Pumps	408
3.	Hydraulic Machines	427
4.	Air Compressors (Reciprocating and Rotary)	432
5.	Steam Turbines	454
6.	Gas Turbines	477
7.	Jet Propulsion and Compressible Flow	485

OOOO

1 Impact of Jet and Hydraulic Turbines

1.1 Consider the following turbines/wheels:
1. Francis turbine
2. Pelton wheel with two or more jets
3. Pelton wheel with a single jet
4. Kaplan turbine

The correct sequence of these turbines/wheels in increasing order of their specific speeds is
(a) 2, 3, 1, 4 (b) 3, 2, 1, 4
(c) 2, 3, 4, 1 (d) 3, 2, 4, 1
[ESE : 2000]

1.2 The gross head available to a hydraulic power plant is 100 m. The utilised head in the runner of the hydraulic turbine is 72 m. If the hydraulic efficiency of the turbine is 90%, the pipe friction head is estimated to be
(a) 20 m (b) 18 m
(c) 16.2 m (d) 1.8 m **[ESE : 2000]**

1.3 The cavitation number of any fluid machinery is defined as $\sigma = \frac{p-p'}{\rho V^2/2}$ (p is absolute pressure, ρ is density and V is free stream velocity)
The symbol p' denotes
(a) static pressure of fluid
(b) dynamic pressure of fluid
(c) vapour pressure of fluid
(d) shear stress of fluid **[ESE : 2000]**

1.4 Consider the following statements :
A water turbine governor
1. helps in starting and shutting down the turbo unit
2. controls the speed of turbine set to match it with the hydroelectric system
3. sets the amount of load which a turbine unit has to carry

Which of these statements are correct?
(a) 1, 2 and 3 (b) 1 and 2
(c) 2 and 3 (d) 1 and 3 **[ESE : 2000]**

1.5 The maximum number of jets generally employed in an impulse turbine without jet interference
(a) 4 (b) 6
(c) 8 (d) 12 **[ESE : 2001]**

1.6 For a water turbine, running at constant head and speed, the operating characteristic curves in the given figure show that upto a certain discharge 'q' both output power and efficiency remain zero.

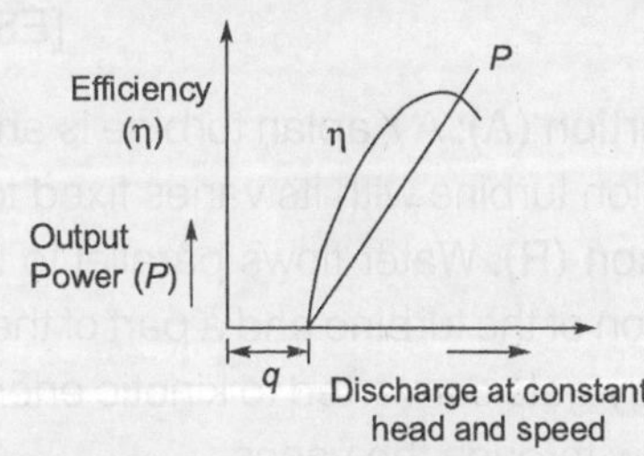

The discharge q is required to
(a) overcome initial inertia
(b) overcome initial friction
(c) keep the hydraulic circuit full
(d) keep the turbine running at no load
[ESE : 2001]

1.7 Consider the specific speed ranges of the following types of turbines:
1. Francis 2. Kaplan
3. Pelton

The sequence of their specific speed in increasing order is
(a) 1, 2, 3 (b) 3, 1, 2
(c) 3, 2, 1 (d) 2, 3, 1 **[ESE : 2001]**

1.8 The level of runner exit is 5 m above the tail race, and atmospheric pressure is 10.3 m. The pressure at the exit of the runner for a divergent draft tube can be
(a) 5 m (b) 5.3 m
(c) 10 m (d) 10.3 m
[ESE : 2001]

1.9 Consider the following statements:
A surge tank provided on the penstock connected to a water turbine
1. helps in reducing the water hammer.

2. stores extra water when not needed.
3. provides increased demand of water.

Which of these statements are correct?

(a) 1 and 3 (b) 2 and 3
(c) 1 and 2 (d) 1, 2 and 3

[ESE : 2001]

1.10 The gross head on a turbine is 300 m. The length of penstock supplying water from reservoir to the turbine is 400 m. The diameter of the penstock is 1 m and velocity of water through penstock is 5 m/s. If coefficient of friction is 0.0098, the net head on the turbine would be, nearly

(a) 310 m (b) 295 m
(c) 200 m (d) 150 m

[ESE : 2001]

1.11 **Assertion (A):** A Kaplan turbine is an axial flow reaction turbine with its vanes fixed to the hub.

Reason (R): Water flows parallel to the axis of rotation of the turbine and a part of the pressure energy gets converted to kinetic energy during its flow through the vanes.

(a) Both **A** and **R** are individually true and **R** is the correct explanation of **A**
(b) Both **A** and **R** are individually true but **R** is not the correct explanation of **A**
(c) **A** is true but **R** is false
(d) **A** is false but **R** is true [ESE : 2001]

1.12 A symmetrical stationary vane experiences a force F of 100 N as shown in the figure below, when the mass flow rate of water over the vane is 5 kg/s with a velocity v of 20 m/s without friction.

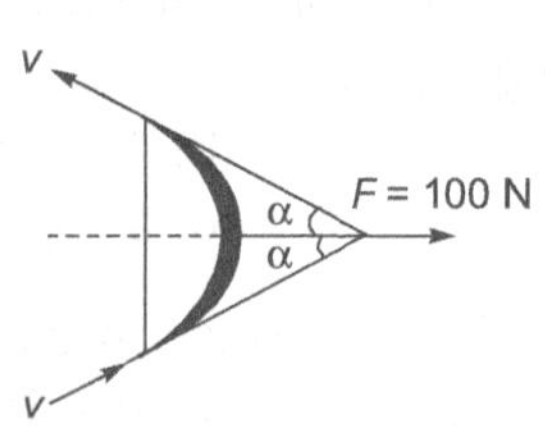

The angle α of the vane is

(a) zero (b) 30°
(c) 45° (d) 60°

[ESE : 2001]

1.13 The force of impingement of a jet on a vane increases if the

(a) vane angle is increased
(b) vane angle is decreased
(c) pressure is reduced
(d) vane is moved against the jet

[ESE : 2002]

1.14 A draft tube is used in a reaction turbine

(a) to guide water downstream without splashing
(b) to convert residual pressure energy into kinetic energy
(c) to convert residual kinetic energy into pressure energy
(d) to streamline the flow in the tail race

[ESE : 2002]

1.15 Match **List-I** (Hydraulic turbine) with **List-II** (Application area) and select the correct answer using the codes given below the lists:

List-I

A. Pelton turbine
B. Francis turbine
C. Kaplan turbine

List-II

1. Low head, large discharge
2. Medium head, medium discharge
3. High head, low discharge

Codes:

	A	B	C
(a)	2	3	1
(b)	2	1	3
(c)	3	1	2
(d)	3	2	1

[ESE : 2002]

1.16 Consider the following statements:

1. Pelton wheel is a tangential flow impulse turbine.
2. Francis turbine is an axial flow reaction turbine.
3. Kaplan turbine is a radial flow reaction turbine.

Which of these statements is/are correct?

(a) 1 and 3 (b) 1 only
(c) 2 only (d) 3 only

[ESE : 2002]

1.17 Efficiency of Pelton wheel shall be maximum if the ratio of jet velocity to tangential velocity of the wheel is

(a) 1/2 (b) 1
(c) 2 (d) 4

[ESE : 2002]

1.18 If H is the head available for a hydraulic turbine the power, speed and discharge, respectively are proportional to

(a) $H^{1/2}, H^{1/2}, H^{3/2}$ (b) $H^{3/2}, H^{1/2}, H^{1/2}$
(c) $H^{1/2}, H^{3/2}, H^{1/2}$ (d) $H^{3/2}, H^{1/2}, H$

[ESE : 2002]

1.19 The maximum efficiency in the case of Pelton wheel is (angle of deflection of the jet = 180° – β)

(a) $\frac{1-\cos\beta}{2}$ (b) $\frac{1+\cos\beta}{2}$
(c) $\frac{\cos\beta}{2}$ (d) $\frac{1+\cos\beta}{4}$

[ESE : 2002]

1.20 **Assertion (A):** The specific speed of a Pelton turbine is low.
Reason (R): Pelton turbine works under a high head and handles low discharge.

(a) Both A and R are individually true and R is the correct explanation of A
(b) Both A and R are individually true but R is not the correct explanation of A
(c) A is true but R is false
(d) A is false but R is true [ESE : 2002]

1.21 Which one of the following graphs correctly represents the relations between Head and Specific speed for Kaplan and Francis turbine?

(a) 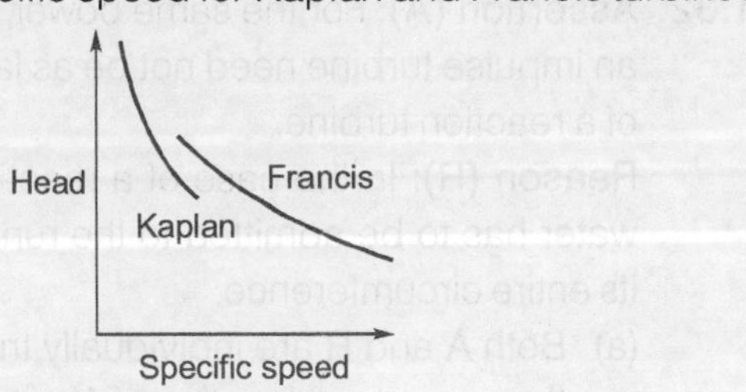

(b) 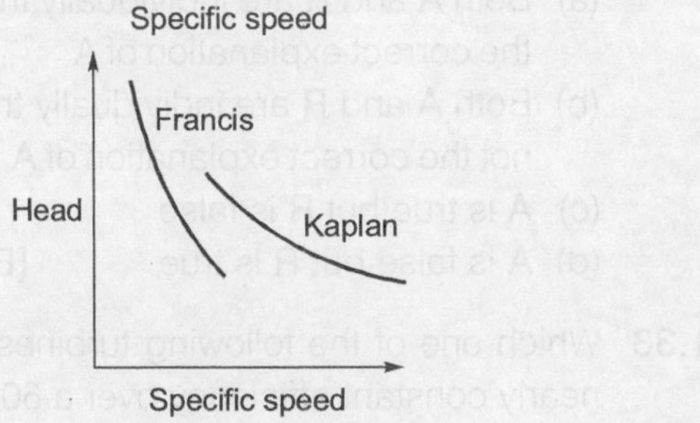

(c) 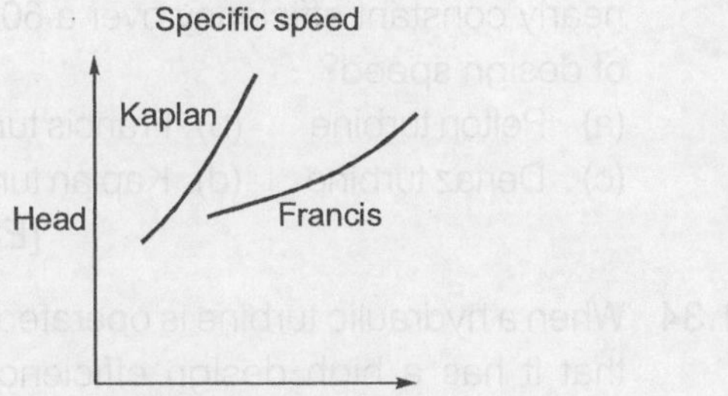

(d) 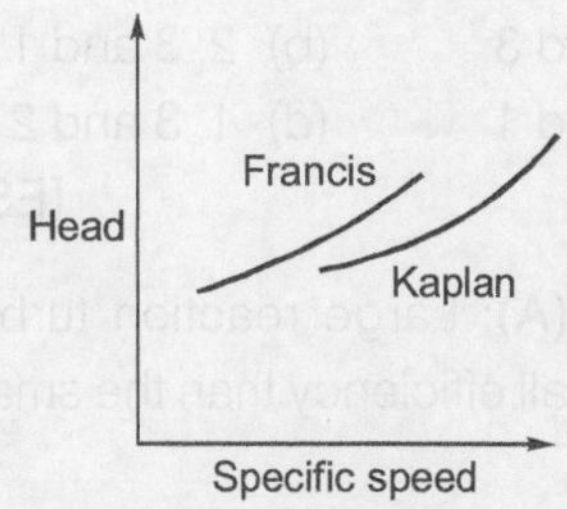

[ESE : 2003]

1.22 Euler equation of turbine giving energy transfer per unit mass E_0 (where U, V_w, V_r and V represents the peripheral, whirl, relative and absolute velocities respectively. Suffix 1 and 2 refer to the turbine inlet and outlet respectively) is given by

(a) $E_0 = U_1V_{w1} - U_2V_{w2}$
(b) $E_0 = U_1V_{r1} - U_2V_{r2}$
(c) $E_0 = U_1V_1 - U_2V_2$
(d) $E_0 = V_1V_{w1} - V_2V_{w2}$

[ESE : 2003]

1.23 Consider the following types of water turbines:

1. Bulb turbine
2. Francis turbine
3. Kaplan turbine
4. Pelton turbine

The correct sequence of order in which the operating head decreases while developing the same power is

(a) 4 , 2, 3, 1 (b) 3, 4, 1, 2
(c) 2, 1, 4, 3 (d) 1, 3, 2, 4

[ESE : 2003]

1.24 Two Pelton wheels A and B have the same specific speed and are working under the same head. Wheel A produces 400 kW at 1000 rpm. If B produces 100 kW, then its rpm is

(a) 4000 (b) 2000
(c) 1500 (d) 1250

[ESE : 2003]

1.25 Consider the following energies associated with a Pelton turbine:

1. Mechanical energy
2. Kinetic energy
3. Potential energy

The correct sequence of energy conversion starting from the entry of fluid is

(a) 1, 2 and 3 (b) 2, 3 and 1
(c) 3, 2 and 1 (d) 1, 3 and 2

[ESE : 2003]

1.26 **Assertion (A):** Large reaction turbines have higher overall efficiency than the small reaction turbines.

Reason (R): The mechanical efficiency of small reaction turbines is higher than that of larger ones.

(a) Both **A** and **R** are individually true and **R** is the correct explanation of **A**
(b) Both **A** and **R** are individually true but **R** is not the correct explanation of **A**
(c) **A** is true but **R** is false
(d) **A** is false but **R** is true

[ESE : 2003]

1.27 Match **List-I** (Specific speed) with **List-II** (Expression/Magnitude) and select the correct answer using the codes given below the lists:

List-I
A. Specific speed of turbine
B. Specific speed of pump
C. Specific speed of Pelton wheel
D. Specific speed of Francis turbine

List-II
1. $\frac{N\sqrt{Q}}{H^{3/4}}$ 2. $\frac{N\sqrt{P}}{H^{5/4}}$
3. 50 – 250 4. 10 – 50

Codes:

	A	B	C	D
(a)	3	4	1	2
(b)	3	2	1	4
(c)	2	1	4	3
(d)	1	2	4	3

[ESE : 2004]

1.28 An impulse turbine operating with a single nozzle has a specific speed of 5. What will be the approximate specific speed of the turbine if the turbine is operated with one more additional nozzle of the same size?

(a) 4 (b) 6
(c) 7 (d) 10 [ESE : 2004]

1.29 What is the range of the speed ratio for Kaplan turbine for its most efficient operation?

(a) 0.10 to 0.30 (b) 0.43 to 0.65
(c) 0.85 to 1.20 (d) 1.40 to 2.00

[ESE : 2004]

1.30 Match **List-I** (Flow parameter) with **List-II** (Type of turbine) and select the correct answer using the codes given below the lists:

List-I
A. High head
B. Axial flow
C. Mixed flow
D. High specific speed

List-II
1. Francis turbine
2. Pelton wheel
3. Kaplan turbine

Codes:

	A	B	C	D
(a)	1	2	3	1
(b)	1	2	3	2
(c)	2	3	1	3
(d)	2	2	1	3

[ESE : 2004]

1.31 A Francis turbine is coupled to an alternator to generate electricity with a frequency of 50 Hz. If the alternator has 12 poles, then the turbine should be regulated to run at which one of the following constant speeds?

(a) 250 rpm (b) 500 rpm
(c) 600 rpm (d) 1000 rpm

[ESE : 2004]

1.32 **Assertion (A):** For the same power, the rotor of an impulse turbine need not be as large as that of a reaction turbine.

Reason (R): In the case of a reaction turbine, water has to be admitted to the runner around its entire circumference.

(a) Both **A** and **R** are individually true and **R** is the correct explanation of **A**
(b) Both **A** and **R** are individually true but **R** is not the correct explanation of **A**
(c) **A** is true but **R** is false
(d) **A** is false but **R** is true [ESE : 2004]

1.33 Which one of the following turbines exhibits a nearly constant efficiency over a 60% to 140% of design speed?

(a) Pelton turbine (b) Francis turbine
(c) Deriaz turbine (d) Kaplan turbine

[ESE : 2005]

1.34 When a hydraulic turbine is operated, it is found that it has a high design efficiency and this efficiency remains constant over wide range of regulation from the design condition. What is the type of this turbine?

(a) Pelton (b) Francis
(c) Kaplan (d) Propeller

[ESE : 2005]

1.35 If α is the blade angle at the outlet, then the maximum hydraulic efficiency of an ideal impulse turbine is

(a) $\frac{1+\cos\alpha}{2}$ (b) $\frac{1-\cos\alpha}{2}$
(c) $\frac{1-\sin\alpha}{2}$ (d) $\frac{1+\sin\alpha}{2}$

[ESE : 2005]

1.36 Which of the following types of turbine is/are suitable for tidal power plants?

1. Tubular turbine 2. Kaplan turbine
3. Bulb turbine 4. Francis turbine

Select the correct answer using the codes given below:

(a) 1 only (b) 1 and 3
(c) 2 and 4 (d) 4 only

[ESE : 2005]

1.37 What should be the ratio of blade speed to jet speed for the maximum efficiency of a Pelton wheel?

(a) 1/4 (b) 1/2
(c) 3/4 (d) 1 [ESE : 2005]

1.38 A pelton wheel with single jet rotates at 600 rpm. The velocity of the jet from the nozzle is 100 m/s. If the ratio of the vane velocity to jet velocity is 0.44, what is the diameter of the Pelton wheel?

(a) 0.7 m (b) 1.4 m
(c) 2.1 m (d) 2.8 m [ESE : 2005]

1.39 Which of the following water turbines maintain a high efficiency over a long range of the part load?

1. Francis turbine 2. Kaplan turbine
3. Pelton turbine 4. Propeller turbine

Select the correct answer using the codes given below:

(a) 1 and 4 (b) 2 and 3 only
(c) 1, 2 and 3 (d) 2, 3 and 4

[ESE : 2006]

1.40 Which of the following advantages is/are possessed by a Kaplan turbine over a Francis turbine?

1. Low frictional losses.
2. Part load efficiency is considerably high.
3. More compact and smaller in size.

Select the correct answer using the codes given below:

(a) 1 only (b) 1 and 2 only
(c) 2 and 3 only (d) 1, 2 and 3

[ESE : 2006]

1.41 In an axial flow impulse turbine, energy transfer takes place due to

(a) change in relative kinetic energy
(b) change in absolute kinetic energy
(c) change in pressure energy
(d) change in energy because of centrifugal force

[ESE : 2006]

1.42 Which of the following water turbines does not require a draft tube?

(a) Propeller turbine (b) Pelton turbine
(c) Kaplan turbine (d) Francis turbine

[ESE : 2006]

1.43 A circular jet of water impinges on a vertical flat plate and bifurcates into two circular jets of half the diameter of the original as shown in the figure below.

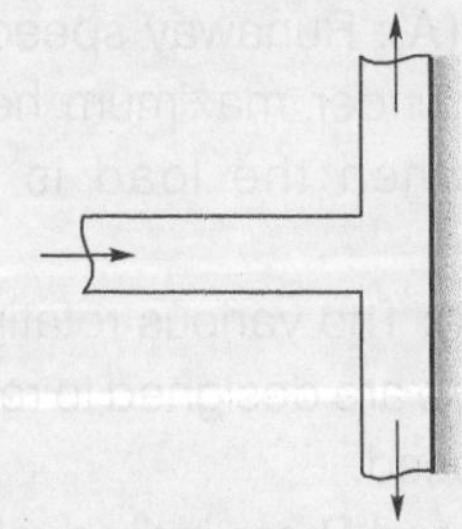

After hitting the plate,

(a) the jets move at equal velocity which is twice of the original velocity
(b) the jets move at equal velocity which is 3 times of original velocity
(c) data given is insufficient to calculate velocities of the two outgoing jets
(d) the jets move at equal velocity which is equal to original velocity [ESE : 2006]

1.44 Specific speed of a fluid machine

(a) refers to the speed of a machine of unit dimensions
(b) is a type-number representative of its performance

(c) is specific to the particular machine
(d) depends only upon the head under which the machine operates [ESE : 2007]

1.45 The overall efficiency of a Pelton turbine is 70%. If the mechanical efficiency is 85%, what is its hydraulic efficiency?
(a) 82.4% (b) 59.5%
(c) 72.3% (d) 81.5% [ESE : 2007]

1.46 The use of a draft tube in a reaction type water turbine helps to
(a) prevent air from entering
(b) increase the flow rate
(c) convert the kinetic energy to pressure energy
(d) eliminate eddies in the downstream
[ESE : 2007]

1.47 The specific speed (N_s) of a water turbine is expressed by which one of the following equations?
(a) $N_s = \frac{N\sqrt{P}}{H^{5/4}}$ (b) $N_s = \frac{N\sqrt{P}}{H^{3/4}}$
(c) $N_s = \frac{N\sqrt{Q}}{H^{5/4}}$ (d) $N_s = \frac{N\sqrt{Q}}{H^{3/4}}$
[ESE : 2007]

1.48 **Assertion (A):** Runaway speed of a turbine is the speed under maximum head at full gate opening when the load is disconnected suddenly.
Reason (R): The various rotating components of the turbine are designed to remain safe at the runaway speed.
(a) Both **A** and **R** are individually true and **R** is the correct explanation of **A**
(b) Both **A** and **R** are individually true but **R** is not the correct explanation of **A**
(c) **A** is true but **R** is false
(d) **A** is false but **R** is true [ESE : 2007]

1.49 Consider the following statements in respect of Kaplan turbine :
1. It is a reaction turbine.
2. It is a mixed flow turbine.
3. It has adjustable blades.
Which of these statements are correct?
(a) 1, 2 and 3 (b) 2 and 3 only
(c) 1 and 3 only (d) 1 and 2 only
[ESE : 2008]

1.50 If the number of jets in Pelton turbine is n, then the specific speed is
(a) $\propto n^2$ (b) $\propto n$
(c) $\propto n^{1/2}$ (d) Independent of n
[ESE : 2008]

1.51 What does Euler's equation of turbo machines relate to?
(a) Discharge and power
(b) Discharge and velocity
(c) Head and power
(d) Head and velocity [ESE : 2008]

1.52 Which one of the following statements is **not** correct in respect of hydraulic turbines?
(a) (Speed) is proportional to (1/Diameter)
(b) (Power) is proportional to $(\text{Speed})^3$
(c) (Power) is proportional to $(\text{Head})^{3/2}$
(d) (Speed) is proportional to $(\text{Head})^{1/2}$
[ESE : 2008]

1.53 Consider the following statements pertaining to specific speed of turbo machines:
1. Specific speed varies with shape of the runner and other parts of the machine.
2. Machines with higher specific speeds are limited to low heads.
3. Specific speed is dimensionless and is independent of variation of type of fluid used.
Which of these statements are correct?
(a) 1, 2 and 3 (b) 1 and 2 only
(c) 2 and 3 only (d) 1 and 3 only
[ESE : 2008]

1.54 A hydraulic reaction turbine working under a head of 16 m develops 640 kW of power. What is the unit power of the turbine?
(a) 10 kW (b) 40 kW
(c) 60 kW (d) 160 kW
[ESE : 2009]

1.55 A Francis turbine working at 400 rpm has a unit speed of 50 rpm and develops 500 kW of power. What is the effective head under which this turbine operates?
(a) 62.5 m (b) 64.0 m
(c) 40.0 m (d) 100 m [ESE : 2009]

1.56 The speed ratio of a Pelton wheel operating under a head of 900 m is 0.45. What is the peripheral velocity of the turbine wheel?

(a) 28 m/s (b) 96 m/s
(c) 42 m/s (d) 60 m/s [ESE : 2009]

1.57 Match **List-I** with **List-II** and select the correct answer using the codes given below the lists :

List-I

A. Pelton turbine
B. Francis turbine
C. Propeller turbine
D. Kaplan turbine

List-II

1. Specific speed from 300 to 1000 + axial flow with fixed runner vanes.
2. Specific speed from 10 to 50 + tangential flow.
3. Specific speed from 60 to 300 + mixed flow.
4. Specific speed from 300 to 1000 + axial flow with adjustable runner vanes.

Codes :

	A	B	C	D
(a)	2	1	3	4
(b)	4	1	3	2
(c)	2	3	1	4
(d)	4	3	1	2

[ESE : 2009]

1.58 Specific speed of a turbomachine
(a) is the speed of a machine having unit dimensions
(b) relates the shape rather than size of machine
(c) remains the same under different conditions of operations
(d) depends only upon the head under which the machine operates [ESE : 2010]

1.59 Consider the following statements regarding a draft tube used in water turbines:
1. It reduces the discharge velocity of water to minimize the loss of kinetic energy at the outlet.
2. It permits the turbine to be set above the tail race without any appreciable drop in available head.
3. It is used in both impulse and reaction type of water turbines.

Which of these statements is/are correct?
(a) 1, 2 and 3 (b) 1 and 2 only
(c) 2 and 3 only (d) 1 only
[ESE : 2010]

1.60 A jet of oil with relative density 0.7 strikes normally a plate with a velocity of 10 m/s. The jet has an area of 0.03 m^2. The force exerted on the plate by the jet is
(a) 210 N (b) 2.1 kN
(c) 20.6 kN (d) 206 N [ESE : 2010]

1.61 Which of the following hydraulic turbines are reaction turbines?
1. Francis
2. Kaplan
3. Propeller

(a) 1,2 and 3 (b) 1 and 2 only
(c) 2 and 3 only (d) 1 and 3 only
[ESE : 2011]

1.62 A Kaplan turbine is a
(a) outward flow reaction turbine
(b) inward flow impulse turbine
(c) low head axial flow turbine
(d) high head mixed flow turbine
[ESE : 2011]

1.63 In a reaction turbine
(a) It is possible to regulate the flow without loss
(b) It must be placed at the foot of the fall and above the tail race
(c) Work done is purely by the change in the kinetic energy of the jet
(d) Only part of the head is converted into velocity before the water enters the wheel
[ESE : 2011]

1.64 Assertion (A) : A draft tube is used to reduce the pressure at the runner exit in order to get the increased value of working head.

Reason (R) : A portion of the exit kinetic energy is recovered which otherwise goes waste to the tail race.
(a) Both **A** and **R** are individually true and **R** is the correct explanation of **A**
(b) Both **A** and **R** are individually true but **R** is not the correct explanation of **A**
(c) **A** is true but **R** is false
(d) **A** is false but **R** is true [ESE : 2011]

1.65 The specific speed of a hydraulic turbine depends upon
(a) speed and power developed
(b) discharge and power developed

(c) speed and head of water
(d) speed, power developed and head of water
[ESE : 2012]

1.66 When a hydraulic turbine is operated, it is found that it has high design efficiency and this efficiency remains constant over a wide range of regulation from the design condition. The turbine is

(a) Francis turbine (b) Propeller turbine
(c) Pelton turbine (d) Kaplan turbine
[ESE : 2012]

1.67 An adjustable blade propeller turbine is called as

(a) Banki turbine
(b) Pelton turbine
(c) Kaplan turbine
(d) Francis-Pelton turbine [ESE : 2012]

1.68 A turbine is working under a head of 200 m. The power developed by the turbine is 100 kW and discharge through the turbine is 0.125 m^3/s. In such case, the ratio of unit power to unit discharge for the turbine will be

(a) 4000 (b) 16000
(c) 160×10^3 (d) None of the above
[ESE : 2013]

1.69 **Statement (I):** Cavitation occurs at the entrance of the turbine while it occurs at the exit of the pump.
Statement (II): In both the cases local pressure at these points falls below the vapour pressure of the flowing liquid.

(a) Both Statement (I) and Statement (II) are individually true and Statement (II) is the correct explanation of Statement (I).
(b) Both Statement (I) and Statement (II) are individually true but Statement (II) is **NOT** the correct explanation of Statement (I).
(c) Statement (I) is true but Statement (II) is false
(d) Statement (I) is false but Statement (II) is true. [ESE : 2013]

1.70 Consider the following statements:

1. The wheel can be operated freely in air
2. Pressure at the exit of the nozzle is atmospheric
3. Pressure does not vary along the moving vanes
4. Change in direction of momentum imparts thrust over moving vanes

Which of the above statements are applied to impulse turbine?

(a) 1, 2 and 3 only (b) 1, 2 and 4 only
(c) 3 and 4 only (d) 1, 2, 3 and 4 only
[ESE : 2014]

1.71 A water jet 0.0015 m^2 in area issues from a nozzle with 15 m/s velocity. It is made to impinge perpendicular on to a plate that moves away from the jet with a velocity of 5 m/s. The force on the plate due to this impact is

(a) 150 N (b) 1470 N
(c) 340 N (d) 900 N [ESE : 2014]

1.72 Consider the following statements with regard to hydraulic turbines:

1. Kaplan turbines are most efficient at part load operations.
2. If n is the number of jets in a Pelton turbine, then the specific speed is proportional to n^2.
3. The flow ratio of Francis turbines are in the range of 0.1 – 0.3.

Which of the above statements is/are correct?

(a) 1, 2 and 3 (b) 1 and 2 only
(c) 1 and 3 only (d) 2 and 3 only
[ESE : 2014]

Directions: Each of the following items consists of two statements, one labelled as 'Statement (I)' and the other as 'Statement (II)'. You are to examine these two statements and select the answers to these items using the code given below.

Codes:

(a) Both Statement (I) and Statement (II) are individually true and Statement (II) is the correct explanation of Statement (I).
(b) Both Statement (I) and Statement (II) are individually true but Statement (II) is not the correct explanation of Statement (I).
(c) Statement (I) is true but Statement (II) is false.
(d) Statement (I) is false but Statement (II) is true

1.73 **Statement (I)** : In Impulse turbines pressure change occurs only in the nozzles of the machine. The pressure of liquid does not change while flowing through the rotor of the machine.
Statement (II): The pressure of liquid changes while it flows through the rotor of the machine in reaction turbine.

[ESE : 2014]

1.74 **Statement (I)** : An impulse turbine can run without change in its hydraulic efficiency even if its casing is damaged.

Statement (II): An impulse turbine will not have draft tube.

[ESE : 2014]

1.75 Formation and collapse of vapour bubbles are believed to be the root cause for cavitations in hydraulic turbines. Most favourable condition for the formation of bubbles is set in the turbines at

(a) penstock/Nozzle
(b) guide vanes/Inlet of the runner
(c) vanes receiving impact of jet
(d) outlet of the runner/Entrance of the draft tube

[ESE : 2014]

1.76 A jet of water issues from a nozzle with a velocity of 20 m/s and it impinges normally on a flat plate moving away from it at 10 m/s. If the cross sectional area of the jet is 0.01 m^2 and the density of water is taken as 1000 kg/m^3, then the force developed on the plate will be

(a) 100 N (b) 200 N
(c) 1000 N (d) 2000 N

[ESE : 2015]

1.77 Mainly hydraulic turbine are used to drive the electrical alternators which require maintaining the peripheral speed constant even at part load conditions to avoid the change in frequency of electric power. The governing of the hydraulic turbine is done by

(a) Controlling the flow area
(b) Controlling the velocity
(c) Using the fly wheel
(d) Combined control of flow area and velocity

[ESE : 2015]

Directions: Each of the following items consists of two statements, one labelled as 'Statement (I)' and the other as 'Statement (II)'. You are to examine these two statements and select the answers to these items using the code given below.

Codes:

(a) Both Statement (I) and Statement (II) are individually true and Statement (II) is the correct explanation of Statement (I).
(b) Both Statement (I) and Statement (II) are individually true but Statement (II) is not the correct explanation of Statement (I).
(c) Statement (I) is true but Statement (II) is false.
(d) Statement (I) is false but Statement (II) is true

1.78 **Statement (I)** : Outward radial flow turbines do race inherently.

Statement (II) : In outward radial flow turbines, the centrifugal head impressed upon the exiting water leads to flow increase.

[ESE : 2017]

1.79 **Statement (I)** : The speed of a governed water turbine will remain constant irrespective of load.

Statement (II) : In governing, the water supply is regulated to maintain the speed constant.

(a) Both Statement (I) and Statement (II) are individually true and Statement (II) is the correct explanation of Statement (I).
(b) Both Statement (I) and Statement (II) are individually true but Statement (II) is not the correct explanation of Statement (I).
(c) Statement (I) is true but Statement (II) is false.
(d) Statement (I) is false but Statement (II) is true

[ESE : 2017]

1.80 Which of the following statements are correct?

1. The specific speed of a turbine is the speed at which a homologous turbine develops 1 mhp under unit head at its maximum efficiency.
2. The specific speed is a dimensionless parameter used for the selection of turbines.
3. The function of guide vanes in reaction turbines is to minimize shock at entry of the fluid onto the runner blades.

Select the correct answer using the code given below.

(a) 1, 2 and 3 (b) 2 and 3 only
(c) 1 and 2 only (d) 1 and 3 only

[ESE : 2017]

1.81 The hydraulic efficiency of a turbine is the ratio of

(a) mechanical energy in the output shaft at coupling and hydrodynamic energy available from the fluid

(b) mechanical energy supplied by the rotor and hydrodynamic energy available from the fluid
(c) useful hydrodynamic energy in the fluid at final discharge and mechanical energy supplied to the rotor
(d) useful hydrodynamic energy in the fluid at final discharge and mechanical energy supplied to the shaft and coupling

[ESE : 2019]

1.82 In a lawn sprinkler, water leaves the jet with an absolute velocity of 2 m/s and the sprinkler arms are 0.1 m in length. The sprinkler rotates at a speed of 120 r.p.m. The utilization factor of this sprinkler will be nearly

(a) 0.72 (b) 0.64
(c) 0.56 (d) 0.49

[ESE : 2019]

1.83 While designing a Pelton wheel, the velocity of wheel 'u' is

(a) $K_u\sqrt{gH}$ (b) $2K_u\sqrt{gH}$
(c) $K_u\sqrt{2gH}$ (d) $2K_u\sqrt{2gH}$

where, K_u = speed ratio;
H = net head on turbine; g = gravity

[ESE : 2020]

1.84 The turbines of the same shape will have the same

(a) Thomas number (b) Reynolds number
(c) Specific speed (d) Rotational speed

[ESE : 2020]

1.85 An inward flow reaction turbine has an external diameter of 1 m and its breadth at inlet is 250 mm. If the velocity of flow at inlet is 2 m/s and 10% of the area of flow is blocked by blade thickness, the weight of water passing through the turbine will be nearly

(a) 10 kN/s (b) 14 kN/s
(c) 18 kN/s (d) 22 kN/s

[ESE : 2020]

1.86 When blade speed ratio is zero, no work is done because the distance travelled by the blade is zero even if the torque on the blade

(a) is minimum (b) is zero
(c) is maximum (d) remains the same

[ESE : 2020]

1.87 A 120 mm diameter jet of water is discharging from a nozzle into the air at a velocity of 40 m/s. The power in the jet with respect to a datum at the jet will be

(a) 380 kW (b) 360 kW
(c) 340 kW (d) 320 kW

[ESE : 2020]

1.88 A turbine develops 8000 kW when running at 1000 rpm. The head on the turbine is 30 m. If the head is reduced to 18 m. What is the speed developed by the turbine?

(a) 67.46 rpm (b) 95.24 rpm
(c) 54.67 rpm (d) 77.46 rpm

[ESE : 2021]

1.89 In hydraulic turbines, if the energy available at inlet is only kinetic energy, then that type of turbine is

(a) Reaction turbine (b) Impulse turbine
(c) Francis turbine (d) Kaplan turbine

[ESE : 2021]

1.90 In Francis turbine, as the water discharge is radial at the outlet, the velocity whirl at the outlet becomes

(a) 1 (b) 0
(c) ∞ (d) 0.5

[ESE : 2021]

1.91 Consider the following statements regarding Euler turbine equation:

1. The axial component produces a bending of the shaft which is taken by the journal bearings.
2. The axial force only can cause the rotation of the runner and produce work.
3. The tangential force only can cause the rotation of the runner and produce work.
4. The radial component produces a bending of the shat which is taken by the journal bearings.

Which of the above statements are correct?

(a) 1 and 2 only (b) 1 and 3 only
(c) 3 and 4 only (d) 2 and 4 only

[ESE : 2022]

1.92 Which one of the following turbines is the most popularly used one in the medium head range of 60 m–300 m?
(a) Kaplan turbine (b) Francis turbine
(c) Pelton turbine (d) Deriaz turbine
[ESE : 2022]

1.93 Match the following:

List-I (Turbine types)	List-II (Flow directions)
A. Deriaz turbine	1. Mixed flow
B. Propeller turbine	2. Tangential flow
C. Francis turbine	3. Axial flow
D. Pelton turbine	4. Diagonal flow

Select the correct matching using the code given below:

Codes:

	A	B	C	D
(a)	4	3	1	2
(b)	3	4	2	1
(c)	2	1	4	3
(d)	1	2	3	4

[ESE : 2022]

1.94 A Kaplan turbine plant develops 2000 kW under a head of 8 m while running at 100 rpm. The discharge is 250 m^3/s. The tip diameter of the runner is 6.5 m and the hub to tip ratio is 0.43. What is the flow ratio based on tip speed? (Take acceleration due to gravity as 9.81 m/s^2)
(a) 0.43 (b) 0.58
(c) 0.62 (d) 0.73
[ESE : 2022]

Directions for the following **two (02)** items:
Read the following information and answer the **two** items that follows:
A Pelton turbine is driven by two jets, generating 4.0 MW at 375 rev/min. The effective head at the nozzles is 200 m of water and the nozzle velocity coefficient, K_N = 0.98. The axes of the jets are tangent to a circle 1.5 m in diameter. The relative velocity of the flow across the buckets is decreased by 15 percent and the water is deflected through an angle of 150° neglecting bearing and windage losses.

1.95 What is the Jet speed ratio?
(a) 0.2798 (b) 0.4798
(c) 0.6798 (d) 0.8798
[ESE : 2023]

1.96 What is the runner efficiency of the Pelton turbine?
(a) 62.55% (b) 76.46%
(c) 86.66% (d) 92.46%
[ESE : 2023]

1.97 A turbine develops 9000 kW when running at a speed of 150 r.p.m. and under a head of 30 m. What is the specific speed of the turbine?
(a) 202 rpm (b) 152 rpm
(c) 189 rpm (d) 217 rpm
[ESE : 2024]

1.98 Consider the following regarding unit quantities:
1. Unit specific speed is defined as the speed of a turbine working under unit head.
2. Unit discharge is defined as the discharge passing through a turbine, which is working under a unit head.
3. Unit power is defined as the power developed by a turbine working under unit speed.

Which of the above statements is/are correct?
(a) 1 and 3 only (b) 2 and 3 only
(c) 2 only (d) 3 only
[ESE : 2024]

1.99 Consider regarding the following characteristic curves of hydraulic turbines:
1. Main characteristic curves are obtained by maintaining a constant speed and a constant gate opening on a turbine.
2. Operating characteristic curves are obtained by maintaining a constant head on a turbine.
3. Constant efficiency curves are known as Muschel curves.

Which of the above statements is/are correct?
(a) 1 and 2 only (b) 2 and 3 only
(c) 2 only (d) 3 only
[ESE : 2024]

■■■■

Answers Impact of Jet and Hydraulic Turbines

1.1	(b)	1.2	(a)	1.3	(c)	1.4	(a)	1.5	(b)	1.6	(b)	1.7	(b)	1.8	(b)	1.9	(d)
1.10	(b)	1.11	(d)	1.12	(d)	1.13	(b)	1.14	(c)	1.15	(d)	1.16	(b)	1.17	(c)	1.18	(b)
1.19	(b)	1.20	(a)	1.21	(b)	1.22	(a)	1.23	(a)	1.24	(b)	1.25	(c)	1.26	(c)	1.27	(c)
1.28	(c)	1.29	(d)	1.30	(c)	1.31	(b)	1.32	(d)	1.33	(d)	1.34	(c)	1.35	(a)	1.36	(b)
1.37	(b)	1.38	(b)	1.39	(b)	1.40	(d)	1.41	(b)	1.42	(b)	1.43	(a)	1.44	(b)	1.45	(a)
1.46	(c)	1.47	(a)	1.48	(b)	1.49	(c)	1.50	(c)	1.51	(c)	1.52	(a)	1.53	(b)	1.54	(a)
1.55	(b)	1.56	(d)	1.57	(c)	1.58	(b)	1.59	(b)	1.60	(b)	1.61	(a)	1.62	(c)	1.63	(d)
1.64	(a)	1.65	(d)	1.66	(d)	1.67	(c)	1.68	(a)	1.69	(d)	1.70	(d)	1.71	(a)	1.72	(c)
1.73	(b)	1.74	(b)	1.75	(d)	1.76	(c)	1.77	(a)	1.78	(a)	1.79	(a)	1.80	(d)	1.81	(b)
1.82	(c)	1.83	(c)	1.84	(c)	1.85	(b)	1.86	(c)	1.87	(b)	1.88	(*)	1.89	(b)	1.90	(b)
1.91	(c)	1.92	(b)	1.93	(a)	1.94	(d)	1.95	(b)	1.96	(c)	1.97	(a)	1.98	(c)	1.99	(d)

Explanations Impact of Jet and Hydraulic Turbines

1.1 (b)

Pelton wheel with single jet have less specific speed than Pelton wheel with two or more jets.

1.2 (a)

$$H_G = 100 \text{ m}$$

Euler's head, $H_e = 72$ m

$$\eta_h = 90\% = 0.9 = \frac{H_e}{H}$$

$$0.9 = \frac{72}{H}$$

or $$H = 80 \text{ m}$$

$$H_G = H + h_f$$

$$100 = 80 + h_f$$

or $$h_f = 100 - 80 = \mathbf{20\ m}$$

1.3 (c)

Cavitation number expresses the relationship between the difference of a local absolute pressure from the vapour pressure and the kinetic energy per unit volume, and is used to characterize the potential of flow to cavitate.

1.5 (b)

The maximum number of nozzle used is six while four jet Pelton wheel widely preferred.

1.8 (b)

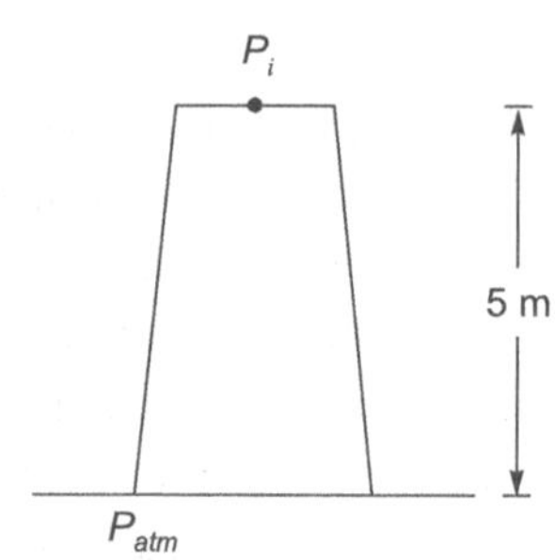

$$\frac{P_1}{\rho g} + \frac{V_1^2}{2g} + z = \frac{P_{atm}}{\rho g} + \frac{V^2}{2g}$$

$$\frac{P_1}{\rho g} + \frac{V_1^2}{2g} + 5 = 10.3$$

$$\frac{P_1}{\rho g} = 5.3 - \frac{V_1^2}{2g}$$

If $$V_1 = 0, \ \frac{P_1}{\rho g} = \mathbf{5.3\ m}$$

1.9 (d)

When the load on the generator decreases the governor reduces the rate of flow of water striking the runner in order to maintain the constant speed for the runner. But the sudden reduction of the rate of flow in the penstock may lead to setting up of water hammer in the pipe, which may cause excessive inertia in pipeline due to which the pipe may burst. Two devices viz, deflector and relief valve are thus provided to avoid the sudden reduction of the rate of flow in the penstock. But neither of these devices are of any assistance when the load on the generator increases and the turbine is in need of move water. Thus in order to fulfil both the above noted requirements, in addition to the above noted devices certain other devices such as surge tank and forebay are usually employed. Surge tanks are employed in the care of high and medium head hydro-power plants where the penstock is very long, and forebays are suitable for medium and low head-power plants where the length of the penstock is short.

1.10 (b)

Given data: $H_G = 300$ m; $l = 400$ m; $d = 1$ m
$V = 5$ m/s; $f = 0.0098$

Darcy's formula,

$$h_f = \frac{4flV^2}{2gd} = \frac{4 \times 0.0098 \times 400 \times (5)^2}{2 \times 9.81 \times 1}$$

$$= 19.98 \text{ m} \approx 20 \text{ m}$$

Net head: $H = H_G - h_f = 300 - 20 = 280$ m

No option is matching

If 0.0098 is a friction factor,

$$\text{then} \quad h_f = \frac{flV^2}{2gd} = \frac{0.0098 \times 400 \times (5)^2}{2 \times 9.81 \times 1}$$

$$= 4.99 \text{ m} \approx 5 \text{ m}$$

Net head: $H = H_G - h_f = 300 - 5 = \mathbf{295\ m}$

1.12 (d)

$$F = 2mV\cos\alpha$$

$$100 = 2 \times 5 \times 20 \cos\alpha$$

or $\cos\alpha = 1/2$

or $\alpha = \mathbf{60°}$

1.13 (b)

$$\text{Force} = \rho a (V - u)^2 (1 + \cos\theta)$$

where ρ = Density; a = Area of jet

V = Velocity of jet; u = Velocity of vane

θ = Vane of angle

1.14 (c)

The draft tube has two purposes as follows if

(i) It permits a negative or suction head to be established at the runner exit, thus making it possible to install the turbine above the tail race level without loss head.

(ii) It converts a large proportion of velocity energy rejected from the runner into useful pressure energy.

1.15 (d)

(a) Pelton wheel : High head, low discharge

(b) Francis turbine : Medium head, Medium discharge

(c) Kaplan turbine : Low head, large discharge

1.17 (c)

For maximum efficiency

$$\text{Tangential velocity of wheel} = \frac{\text{Jet velocity}}{2}$$

1.18 (b)

$$\text{Speed: } V = \sqrt{2gH}$$

$$\therefore \quad V \propto H^{1/2}$$

$$\text{Discharge: } Q = AV$$

$$\therefore \quad Q \propto D^2\sqrt{H}$$

$$\therefore \quad Q \propto H^{1/2}$$

Now,

$$\text{Power: } P = \rho QgH$$

$$P \propto \sqrt{H} \times H$$

$$P \propto H^{3/2}$$

1.19 (b)

$$\eta = \frac{2u}{V_1}(V_1 - u)(1 + \cos\beta)$$

For maximum efficiency

$$u = \frac{V_1}{2}$$

$$\therefore \quad \eta_{max} = \frac{1 + \cos\beta}{2}$$

1.20 (a)

$$\text{Specific speed: } N_s = \frac{N\sqrt{P}}{H^{5/4}}$$

Here, N = Speed in rpm

P = Power in kW

H = Head available in m

1.21 (b)

Francis turbine is medium head and medium specific speed turbine.

Kaplan turbine is low head and high specific speed turbine.

1.22 (a)

$$E_0 = u_1 V_{w1} - u_2 V_{w2}$$

1.23 (a)

Type of turbine	Operating head (m)
Pelton	300 m and above
Francis	60 m to 300 m
Kaplan	20 m to 60 m
Bulb	2 m to 20 m

1.24 (b)

$$N_S = \frac{N\sqrt{P}}{H^{5/4}}$$

For same specific speed and working under the same head i.e. $N_{S1} = N_{S2}$ & $H_1 = H_2$

$$N\sqrt{P} = \text{constant}$$

$$\therefore \quad N_1\sqrt{P_1} = N_2\sqrt{P_2}$$

$$N_2 = 1000\sqrt{\frac{400}{100}} = \mathbf{2000\ rpm}$$

1.25 (c)

As water flows down the height its potential energy gets converted into kinetic energy, when water hits the runner, the kinetic energy gets converted into rotational mechanical energy of runner.

1.26 (c)

A is true but **R** is false.

Mechanical efficiency of small reaction turbine is lower than that of larger ones and also large turbines are more efficient than small turbine. Mechanical loss are caused due to frictional losses in the bearings of the turbine shaft and of the accessaries coupled to the turbine. There loses reduce the output and a accounted by the term mechanical efficiency. The loss is less than 1 percent. The percentage decreases with the increase in the size of the plant and lubrication.

1.27 (c)

Type of turbine	Specific speed
Francis turbine	40-300
Pelton turbine	9-40

1.28 (c)

$$N_s = \frac{N\sqrt{P}}{H^{5/4}} \quad \text{for single jet}$$

$$= 5$$

$$N_{s,\,multi} = \frac{N\sqrt{nP}}{H^{5/4}} \quad \text{for multi jet}$$

$$= N_s\sqrt{n} = 5 \times \sqrt{2}$$

$$= 7.07 \approx 7$$

1.30 (c)

Type of flow	Turbine
Axial flow	Kaplan/Propeller
Tangential flow	Pelton
Mixed flow	Francis

1.31 (d)

Number of poles,

$$p = 12$$

$$N = \frac{120f}{p}$$

where $f = 50$ Hz

$$\therefore \quad N = \frac{120 \times 50}{12} = 500 \text{ rpm}$$

1.33 (d)

Kaplan turbine has adjustable blades, which can be automatically adjust according to the flow to achieve high efficiency over a wide range of flow.

1.34 (c)

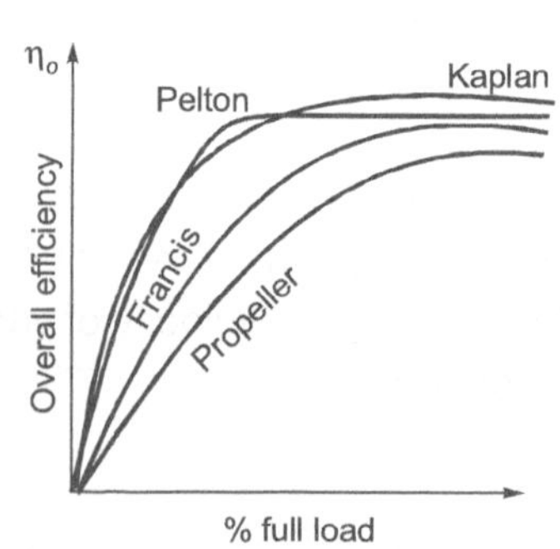

1.35 (a)

Hydraulic efficiency of Pelton wheel

$$= \frac{2u(v-u)(1+k\cos\alpha)}{v^2}$$

Here, k = Friction factor

u = Blade speed/bucket speed

v = Jet speed

α = Blade angle

For maximum efficiency the bucket speed should be halt of jet speed *i.e.* $u = \frac{v}{2}$

Thus, respective maximum efficiency is

$$\eta_{max.} = \left(\frac{1+\cos\alpha}{2}\right)$$

And, If bucket is frictionless, $k = 1$

1.36 (b)

The tubular turbine and bulb turbine are quite suitable for tidal plants and small canal plants.

1.37 (b)

$$\eta_{hydraulic} = \frac{2u(v-u)(1+k\cos\alpha)}{v^2}$$

For maximum efficiency, differentiate with respect to u,

$$\Rightarrow \quad \frac{d\eta_h}{du} = \frac{2(1+k\cos\alpha)}{v^2}(v-2u) = 0$$

$$\Rightarrow \quad u = \frac{v}{2}$$

Here, u = Blade speed; v = Jet speed

α = Blade angle; k = Friction factor

1.39 (b)

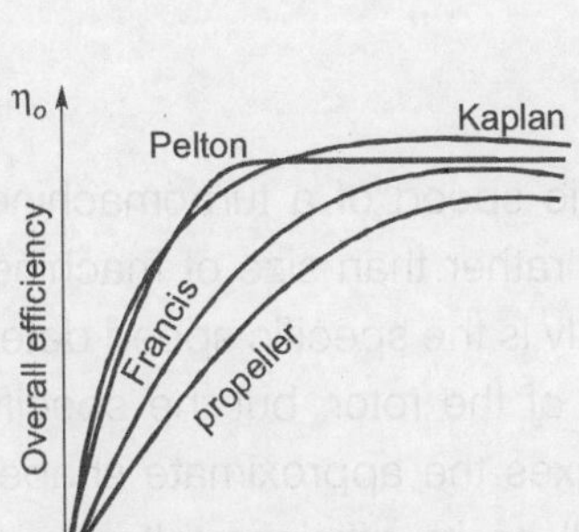

Kaplan and pelton wheels give almost maximum efficiency for wide range of load, while at full load all turbines have almost same efficiency.

1.41 (b)

$$\text{Work done by jet} = \frac{1}{2}m(v_1^2 - v_2^2)$$

where, v_1 = Absolute velocity before impact

v_2 = Absolute velocity after impact

m = Mass flow rate

1.42 (b)

Pelton turbine does not require draft tube because it utilizes the kinetic energy of water.

1.43 (a)

It is in accordance with the law of conservation of momentum i.e. initial momentum is equal to final momentum.

1.44 (b)

Specific speed is of a fluid machine is a type number representative of its performance.

In order to compare the performance of different pumps, it is necessary to have some term which will be common to all the centrifugal pump the term ordinarily used for this purpose is the specific speed. This is a type characteristic and can be used to predict the behaviour of one pump based on tests of similar, but different sized pumps. The specific speed is a characteristic of pumps which can be used as a basis for comparison of the performance of different pump.

1.45 (a)

$$\eta_o = \eta_h \eta_m$$

or

$$\eta_h = \frac{\eta_o}{\eta_m}$$

$$\therefore \text{Hydraulic: } \eta_h = \frac{0.70}{0.85} \times 100 = \mathbf{82.4\%}$$

1.46 (c)

The draft tube has two purpose as follows:

(i) It permits a negative or suction head to be established at the runner exit, thus making it possible to install the turbine above the tail race level without loss of head.

(ii) It converts a large proportion of velocity energy rejected from the runner into useful pressure energy.

1.47 (a)

The specific speed of water turbine is the speed of geometrical similar turbine which produced unit power under the unit head.

Specific speed, $N_s = \frac{N\sqrt{P}}{H^{5/4}}$

1.48 (b)

Run away speed is defined as the maximum speed the turbine may attain when it gets suddenly unloaded. It is necessary to know the runway speed so that the rotating parts of the turbine are designed for enough strength to withstand the corresponding high tensile stresses and vibrations.

1.49 (c)

Kaplan turbine is axial flow reaction turbine. It has got adjustable blades. Whereas propeller turbines have fixed blades.

1.50 (c)

Specific speed,

$$N_s = \frac{N\sqrt{nP}}{H^{5/4}}$$

where n = Number of jets

P = Power produced per jet

$\therefore \quad N_s \propto n^{1/2}$

1.51 (c)

Euler's equation of turbo machines relate to head and power.

1.52 (a)

$(DN) \propto H^{1/2}$

Power $\propto D^5 N^3$

Power $\propto D^2 H^{3/2}$

$N \propto H^{1/2}$

1.53 (b)

1. Specific speed varies with shapes of the runner and other parts of the machine.
2. Machine with higher specific speeds are limited to low heads.
3. Specific speed is not dimensionless quantity.

1.54 (a)

Unit power of the turbine,

$$P_u = \frac{P}{H^{3/2}} = \frac{640}{(16)^{3/2}} = \frac{640}{64}$$

$$= \mathbf{10\ kW}$$

1.55 (b)

Unit speed of the turbine is given by

$$\frac{N}{\sqrt{H}} = 50$$

$$\Rightarrow \frac{400}{\sqrt{H}} = 50$$

$$\Rightarrow H = \mathbf{64\ m}$$

1.56 (d)

$$\text{Speed ratio} = \frac{u}{\sqrt{2gH}}$$

$$\Rightarrow 0.45 = \frac{u}{\sqrt{2\times9.81\times900}}$$

$$\Rightarrow u = 0.45\sqrt{2\times9.81\times900}$$

$$= 59.79 \approx \mathbf{60\ m/s}$$

1.57 (c)

Type of turbine	Specific speed	Type of flow
Pelton	$N_s < 40$	Tangential flow
Modern Francis	$40 < N_s < 300$	Mixed flow
Propeller	$N_s > 300$	Axial flow (Fixed vane)
Kaplan	$N_s > 300$	Axial flow

1.58 (b)

Specific speed of a turbomachine relates the shape rather than size of machine.

Not only is the specific speed determine by the shape of the rotor, but the specific speed by itself fixes the approximate shape of the rotor as well as its efficiency. It may therefore be assumed for all practical purposes that the specific speed is a function only of the rotor shape, at least in hydraulic machine.

1.59 (b)

Draft tube must be airtight and under all condition of operation its lower end must be submerged below the level of water however it facilitates the installation of the turbine runner at any location above or below the tail race level without loss of head.

1.60 (b)

Force exerted by the jet on the plate

$= \rho AV(V_1 - V_2) \qquad \because V_2 = 0$

$= \rho AV^2 = 0.7 \times 10^3 \times 0.03 \times (10)^2$

$= 2100 \text{ N} = \mathbf{2.1 \text{ kN}}$

1.61 (a)

All Francis, Kaplan and Propeller are of reaction turbine type.

In a reaction turbine both pressure energy and kinetic energy are available at inlet.

1.62 (c)

Kaplan turbine is a low head axial flow turbine with adjustable vane.

1.63 (d)

In reaction turbine both pressure energy and kinetic energy are available at inlet.

1.65 (d)

$$\text{Specific speed, } N_s = \frac{N\sqrt{P}}{H^{5/4}}$$

1.67 (c)

Kaplan turbine has blades mounted on servo motor which can be adjusted in accordance to the flow to maintain high efficiency.

1.68 (a)

$$\text{Unit power, } P_u = \frac{P}{H^{3/2}}$$

$$\text{Unit discharge, } Q_u = \frac{Q}{\sqrt{H}}$$

$$\frac{P_u}{Q_u} = \frac{P}{Q}\frac{H^{1/2}}{H^{3/2}} = \frac{100}{0.125}\frac{(200)^{1/2}}{(200)^{3/2}} \times 10^3$$

$$= \mathbf{4000}$$

1.70 (d)

In an impulse turbine:

(i) All the available energy of water is converted into kinetic energy or velocity head by passing it through a contracting nozzle provided at the end of the penstock.

(ii) The work coming out of nozzle is formed into a free jet which impinges on a series of buckets of the runner thus causing it to revolve.

(iii) The runner revolves freely in air.

(iv) The water in contact remains at atmospheric pressure.

(v) Change in momentum is responsible for work done.

So, all statements are correct.

1.71 (a)

$$F_x = \rho Q (v - u)^2$$
$$= 1000 \times 0.0015 \times (15 - 5)^2$$
$$= 1000 \times 0.0015 \times 100$$
$$= \mathbf{150 \text{ N}}$$

1.72 (c)

If n is the number of jets in a Pelton turbine, then the specific speed is proportional to $\sqrt{n}$ (not n^2). So, statement 2 is wrong.

1.75 (d)

Negative or suction head is most likely to be established at the runner exit due to decreasing pressure of water through runner passage.

1.76 (c)

Given data:

$V = 20$ m/s; $u = 10$ m/s; $a = 0.01$ m^2

$\rho = 1000$ kg/m^3

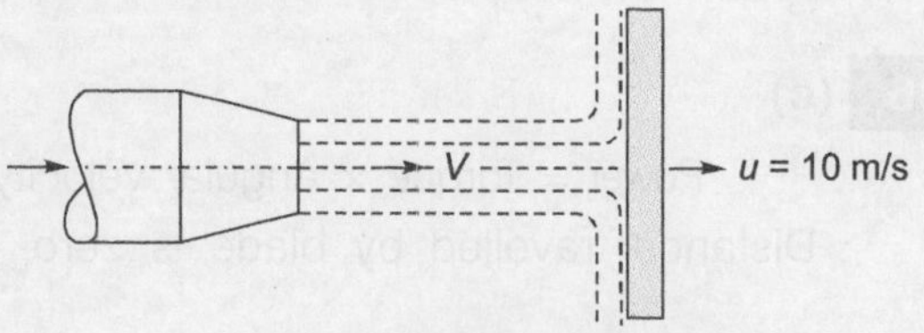

$$F = \rho a (V - u)^2$$
$$= 1000 \times 0.01 (20 - 10)^2$$
$$= \mathbf{1000 \text{ N}}$$

1.81 (b)

$$\eta_H = \frac{\text{Runner power}}{\text{Water power}}$$

1.82 (c)

Velocity of water,

$$V = 2 \text{ m/s}$$

Length of sprinkler arms = 0.1 m

$$N = 120 \text{ rpm}$$

$$u = \frac{\pi DN}{60} = \frac{\pi \times 0.2 \times 120}{60}$$

$$= 1.25 \text{ m/s}$$

$\therefore$ $D = 2 \times$ length of sprinkler arms

$$\text{Utilization factor} = \frac{\text{Ideal work output}}{\text{Energy available with water}}$$

$$= \frac{\frac{V_w u}{g}}{\frac{V_w u}{g} + \frac{V^2}{2g}} = \frac{2 \times 1.25}{2 \times 1.25 + \frac{2^2}{2}}$$

$$= 0.56$$

1.83 (c)

Speed ratio, $K_u = \dfrac{U}{\sqrt{2gH}}$

1.84 (c)

The specific speed remains same for all the similar turbine.

1.85 (b)

Given, $d_1 = 1$ m; $b_1 = 0.25$ m

$$V_{F_1} = 2 \text{ m/s}$$

$$k = 0.9$$

= coefficient of blade thickness

$$Q = k \cdot \pi d_1 b_1 V_{F_1}$$

$$Q = 0.9 \times \pi \times 1 \times 0.25 \times 2$$

$$= 1.413 \text{ m}^3\text{/sec}$$

Weight of water passing through turbine = $\rho g Q$ = 1000 × 9.81 × 1.413 = 14 kN/s

1.86 (c)

Power = Torque × Angular velocity

$\because$ Distance travelled by blade is zero, this implies $\dfrac{U}{V} = 0 \Rightarrow \omega = 0$

Hence, $P = 0$ even if T = maximum.

1.87 (b)

$$d = 120 \text{ mm} = 0.12 \text{ m}$$

$$V = 40 \text{ m/s}$$

$$\text{Power} = \frac{1}{2}\dot{m}V_1^2 = \frac{1}{2}\rho a V_1^3$$

$$= \frac{1}{2} \times 1000 \times \left(\frac{\pi}{4} \times 0.12^2\right) \times 40^3$$

Power = 360 kW

1.88 (*)

$$P_1 = 8000 \text{ kW}$$

$$N_1 = 1000 \text{ rpm}$$

$$H_1 = 30 \text{ m}, H_2 = 18 \text{ m}$$

$$N_2 = ?$$

$$\frac{N_1}{\sqrt{H_1}} = \frac{N_2}{\sqrt{H_2}}$$

$$\frac{1000}{\sqrt{30}} = \frac{N_2}{\sqrt{18}}$$

$$N_2 = 774.6 \text{ rpm}$$

Note : The answer will match with options only when N = 100 rpm is considered.

1.90 (b)

For Francis turbine,

$V_{w2} = 0$

1.94 (d)

Given : $P = 2000$ kW; $H = 8$ m, N = 1000 rpm, $Q = 250$ m^3/sec, $D_0 = 6.5$ m, $D_b = 0.43 \times D_0 = 2.8$ m

$$K_{f_2} = ? = \frac{V_{f_1}}{\sqrt{2gH}}$$

$$Q = \frac{\pi}{4}\left(D_0^2 - D_b^2\right) \times V_{f_1}$$

$$V_{f_1} = 9.25 \text{ m/s}$$

$$K_f = \frac{V_{f_1}}{\sqrt{2gH}}$$

$$= \frac{9.25}{\sqrt{2 \times 9.81 \times 8}} = 0.73$$

1.95 (b)

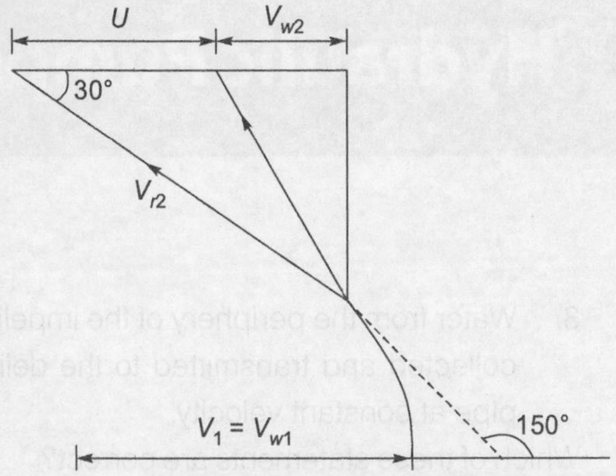

Given: Power = 4.40 MW, N = 375 rpm

Runner diameter (D) = 1.5 m

$$k_N = 0.98$$

$$\frac{V_{r2}}{V_{r1}} = 0.85, \; H = 200 \text{ m}$$

$$u = \frac{\pi DN}{60} = \frac{\pi \times 1.5 \times 375}{60} = 29.452 \text{ m/s}$$

$$V_1 = k_N\sqrt{2gH}$$

$$= 0.98\sqrt{2 \times 9.81 \times 200} = 61.357 \text{ m/s}$$

$$\text{Jet speed ratio} = \frac{u}{V_1} = \frac{29.452}{61.357} = 0.48$$

1.96 (c)

Runner efficiency (η)

$$= \frac{(V_1 - u)(1 + k\cos\theta) \times 2 \times u}{V_1^2}$$

$$= \frac{(61.357 - 29.452)(1 + 0.85\cos(30°)) \times 2 \times 29.452}{61.357^2}$$

$$b = 0.8666 = 86.66\%$$

1.97 (a)

Given: P = 9000 kW, N = 150 rpm, H = 30 m

Specific speed of turbine is given as

$$N_s = \frac{N\sqrt{P}}{H^{5/4}} = \frac{150 \times \sqrt{9000}}{(30)^{5/4}} = 202.68 \text{ rpm}$$

$$\simeq 202 \text{ rpm}$$

1.98 (c)

- Unit discharge is defined as the discharge passing through a turbine, which is working under a unit head.
- Unit power is defined as the power developed by a turbine working under unit head and gives maximum efficiency.

1.99 (d)

- Main characteristic curves or constant head curves are obtained by maintaining a constant head and a constant gate opening on a turbine.
- Operating characteristic curves or constant speed curves are obtained by maintaining a constant speed by changing the load on the turbine.

■■■■

2 Hydraulic Pumps

2.1 Which one of the following pairs of formulae represents the specific speeds of turbine and pump respectively? (Notations have their usual meaning)

(a) $\frac{NQ^{1/2}}{H^{3/4}}$ and $\frac{NP^{1/2}}{H^{5/4}}$

(b) $\frac{NQ^{1/2}}{H^{3/4}}$ and $\frac{NP^{1/2}}{H^{3/4}}$

(c) $\frac{NP^{1/2}}{H^{3/4}}$ and $\frac{NQ^{1/2}}{H^{5/4}}$

(d) $\frac{NP^{1/2}}{H^{5/4}}$ and $\frac{NQ^{1/2}}{H^{3/4}}$

[ESE : 2000]

2.2 Match **List-I** (Outlet vane angle β_2) with **List-II** (Curves labelled 1, 2 and 3 in the given figure) for a pump and select the correct answer using the code given below the lists:

List-I

A. $\beta_2 < 90°$

B. $\beta_2 = 90°$

C. $\beta_2 > 90°$

List-II

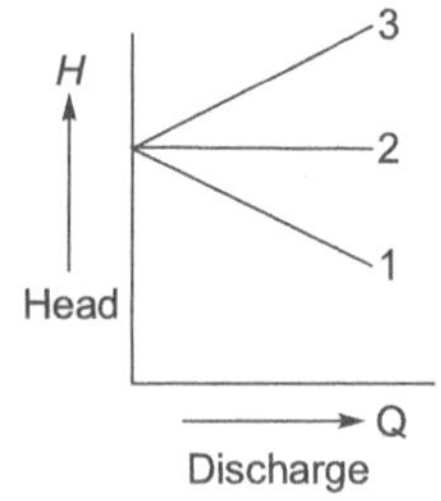

Codes:

	A	B	C
(a)	1	2	3
(b)	1	3	2
(c)	2	1	3
(d)	3	2	1

[ESE : 2000]

2.3 Consider the following statements regarding the volute casing of a centrifugal pump:

1. Loss of head due to change in velocity is eliminated.
2. Efficiency of the pump is increased.
3. Water from the periphery of the impeller is collected and transmitted to the delivery pipe at constant velocity.

Which of these statements are correct?

(a) 1, 2 and 3 (b) 1 and 2
(c) 2 and 3 (d) 1 and 3

[ESE : 2000]

2.4 **Assertion (A):** The efficiency of a pump is generally less than that of a turbine.
Reason (R): Although the losses in the two types of machines are of the same kind, the losses in pumps are more due to eddies and turbulence.

(a) Both **A** and **R** are individually true and **R** is the correct explanation of **A**
(b) Both **A** and **R** are individually true but **R** is not the correct explanation of **A**
(c) **A** is true but **R** is false
(d) **A** is false but **R** is true

[ESE : 2000]

2.5 Consider the following statements pertaining to a centrifugal pump:

1. The manometric head is the head developed by pump.
2. The suction pipe has, generally, a larger diameter as compared to the discharge pipe.
3. The suction pipe is provided with a foot valve and a strainer.
4. The delivery pipe is provided with a foot valve and a strainer.

Which of these statements are correct?

(a) 1, 2, 3, and 4 (b) 1 and 2
(c) 2 and 3 (d) 1 and 3

[ESE : 2001]

2.6 If a reciprocating pump having a mechanical efficiency of 80% delivers water at the rate of 80 kg/s with a head of 30 m, the brake power of the pump is

(a) 29.4 kW (b) 20.8 kW
(c) 15.4 kW (d) 10.8 kW

[ESE : 2001]

2.7 In fluid machinery, the relationship between saturation temperature and pressure decides the process of

(a) flow separation (b) turbulent mixing
(c) cavitation (d) water hammer

[ESE : 2001]

2.8 Volumetric flow rate Q, acceleration due to gravity g and head H form a dimensionless group, which is given by

(a) $\sqrt{\frac{gH^5}{Q}}$ (b) $\frac{Q}{\sqrt{gH^5}}$

(c) $\frac{Q}{\sqrt{g^3H}}$ (d) $\frac{Q}{\sqrt{g^2H}}$

[ESE : 2002]

2.9 A pump running at 1000 rpm consumes 1 kW and generates head of 10 m of water. When it is operated at 2000 rpm, its power consumption and head generated would be

(a) 4 kW, 50 m of water
(b) 6 kW, 20 m of water
(c) 3 kW, 30 m of water
(d) 8 kW, 40 m of water

[ESE : 2002]

2.10 A centrifugal pump gives maximum efficiency when its blades are

(a) bent forward (b) bent backward
(c) straight (d) wave shaped

[ESE : 2002]

2.11 In utilizing scaled models in the designing of turbo-machines, which of the following relationship must be satisfied?

(a) $\frac{H}{ND^3}$ = constant; $\frac{Q}{N^2D^2}$ = constant

(b) $\frac{Q}{D^2\sqrt{H}}$ = constant; $\frac{H}{N^3D}$ = constant

(c) $\frac{P}{QH}$ = constant; $\frac{H}{N^2D^2}$ = constant

(d) $\frac{NQ^{1/2}}{H^{3/2}}$ = constant; $\frac{NP^{1/2}}{H^{3/4}}$ = constant

[ESE : 2002]

2.12 The correct sequence of the centrifugal pump components through which the fluid flows is

(a) impeller, suction pipe, foot valve and strainer, delivery pipe
(b) foot valve and strainer, suction pipe, impeller, delivery pipe
(c) impeller, suction pipe, delivery pipe, foot valve and strainer
(d) suction pipe, delivery pipe, impeller, foot valve and strainer

[ESE : 2002]

2.13 A centrifugal pump driven by a directly coupled 3 kW motor of 1450 rpm speed, is proposed to be connected to another motor of 2900 rpm speed. The power of the motor should be

(a) 6 kW (b) 12 kW
(c) 18 kW (d) 24 kW

[ESE : 2002]

2.14 Match **List-I** (Type of pumps) with **List-II** (Associated features) and select the correct answer using the codes given below the lists:

List-I

A. Centrifugal pump
B. Gear pump
C. Reciprocating pump
D. Turbine pump

List-II

1. Air vessel
2. Draft tube
3. Guide vanes
4. Rotary pump
5. Rotor having blades

Codes:

	A	B	C	D
(a)	4	2	5	3
(b)	5	4	1	2

(c) 4 2 3 1
(d) 5 4 1 3 [ESE : 2003]

2.15 Match **List-I** (Industrial needs) with **List-II** (Type of pump) and select the correct answer using the codes given below the lists:

List-I

A. Combustible fluid to be pumped
B. High head but small discharge needed
C. Low head but large discharge needed
D. High head and high discharge needed

List-II

1. Single stage centrifugal pump
2. Multistage centrifugal pump
3. Positive displacement pump
4. Jet pump

Codes:

	A	B	C	D
(a)	3	2	1	4
(b)	4	3	1	2
(c)	3	1	4	2
(d)	4	3	2	1

[ESE : 2003]

2.16 Priming is necessary in
(a) centrifugal pumps to lift water from a greater depth
(b) centrifugal pumps to remove air in the suction pipe and casing
(c) hydraulic turbine to remove air in the turbine casing
(d) hydraulic turbine to increase the speed of turbine and to generate more power

[ESE : 2003]

2.17 Consider the following statements regarding air vessels provided in reciprocating pump installations:
1. The air vessels are fitted both on suction and delivery sides.
2. The air vessels are fitted far from the pump cylinder.
3. The air vessels save energy by reducing the friction loss.

Which of these statements are correct?
(a) 1, 2 and 3 (b) 1 and 2
(c) 2 and 3 (d) 1 and 3

[ESE : 2003]

2.18 A pump is installed at a height of 5 m above the water level in the sump. Frictional loss on the suction side is 0.6 m. If the atmospheric pressure is 10.3 m of water and vapour pressure head is 0.4 m (abs), the NPSH (Net Positive Suction Head) will be
(a) 3.7 m (b) 4 m
(c) 4.3 m (d) 4.6 m

[ESE : 2003]

2.19 On the assumption that a double suction impeller is the equivalent of two single suction impellers placed back to back, it is customary to base the specific speed of the double suction pump on
(a) one half of the total capacity
(b) three fourth of the total capacity
(c) full total capacity
(d) Double the total capacity [ESE : 2003]

2.20 **Assertion (A):** Multistage centrifugal pumps are only of the radial flow type.
Reason (R): In a multistage centrifugal pump, two or more impellers are keyed to a single shaft and enclosed in the same casing, the radial inlet to successive impellers being made through guide vanes.
(a) Both **A** and **R** are individually true and **R** is the correct explanation of **A**
(b) Both **A** and **R** are individually true but **R** is not the correct explanation of **A**
(c) **A** is true but **R** is false
(d) **A** is false but **R** is true [ESE : 2003]

2.21 A centrifugal pump needs 1000 W of power when operating at 1500 rpm. What is the power requirement if the speed of the pump is increased to 3000 rpm?
(a) 2000 W (b) 4000 W
(c) 6500 W (d) 8000 W [ESE : 2004]

2.22 Two centrifugal pumps have impellers with outer dimensions of each equal to twice the inner dimensions. The inner diameter of the second impeller is three times the inner diameter of the first one. What is the speed ratio N_2/N_1 of pumps, if the pumps are required to develop the same manometric head to start delivery of water?

(a) 9 (b) 4
(c) 1/2 (d) 1/3 [ESE : 2004]

2.23 Water is required to be lifted by a 10 kW pump from a depth of 100 m. If the pump is unable to lift the water, then which one of the following is correct?
(a) A greater capacity pump has to be used.
(b) A larger diameter delivery pipe has to be used.
(c) A larger diameter suction pipe has to be used.
(d) A multistage pump has to be used.
[ESE : 2004]

2.24 A centrifugal pump operating at 1000 rpm develops a head of 30 m. If the speed is increased to 2000 rpm and the pump operates with the same efficiency, what is the head developed by the pump?
(a) 60 m (b) 90 m
(c) 120 m (d) 150 m [ESE : 2004]

2.25 Two centrifugal pumps *A* and *B* operate at their maximum efficiencies at 1000 rpm and 500 rpm respectively. Against the same delivery head, pump *A* discharges 1 m^3/s and pump *B* discharges 4 m^3/s respectively. What is the ratio of specific speeds $(N_S)_A : (N_S)_B$?
(a) 1 : 2 (b) 1 : 1
(c) 1 : 4 (d) 4 : 1 [ESE : 2004]

2.26 When a fluid passes from the inlet to exit of the rotor in a centrifugal pump, tangential momentum
(a) increases and energy increases
(b) decreases and energy increases
(c) remains unchanged and energy decreases
(d) increases and energy remains unchanged
[ESE : 2004]

2.27 Which one of the following statements is relevant to the specific speed of a centrifugal pump?
(a) Head developed is unity and discharge is unity
(b) Head developed is unity and power absorbed is unity
(c) Discharge is unity and power absorbed is unity
(d) Each of head developed, power absorbed and discharge is equal to unity
[ESE : 2004]

2.28 In positive displacement pumps, the slip can sometimes be negative when the actual discharge is greater than the theoretical discharge. This happens in
(a) small suction pipes coupled with a low delivery head
(b) small suction pipes coupled with a medium delivery head
(c) long suction pipes coupled with a low delivery head
(d) long suction pipes with medium delivery head
[ESE : 2004]

2.29 Match **List-I** (Type of pump) with **List-II** (Liquid handled) and select the correct answer using the codes given below the lists:

List-I
A. Closed impeller pump
B. Semi-open impeller pump
C. Open impeller pump

List-II
1. Sandy water
2. Acids
3. Sewage water

Codes:

	A	B	C
(a)	1	3	2
(b)	3	1	2
(c)	2	3	1
(d)	1	2	3

[ESE : 2004]

Direction: Each of the next questions consists of two statements, one labelled as the **'Assertion (A)'** and the other as **'Reason (R)'**. You are to examine these two statements carefully and select the answers to these items using the codes given below:

Codes:
(a) Both **A** and **R** are individually true and **R** is the correct explanation of **A**

(b) Both **A** and **R** are individually true but **R** is not the correct explanation of **A**
(c) **A** is true but **R** is false
(d) **A** is false but **R** is true

2.30 **Assertion (A):** The volute casing of a centrifugal pump helps in creating the high velocity head necessary for enabling water flow upwards to a higher level.
Reason (R): The water flows through a diverging passage in the volute chamber.

[ESE : 2004]

2.31 **Assertion (A):** For higher specific speeds, radial flow pumps have the greatest efficiency.
Reason (R): Pumps having larger discharge under smaller heads have higher specific speeds.

[ESE : 2004]

2.32 Why is a minimum of Net Positive Suction Head required for a hydraulic pump?
(a) To prevent cavitation
(b) To increase discharge
(c) To increase suction head
(d) To increase efficiency [ESE : 2005]

2.33 For a given centrifugal pump,
(a) the discharge varies directly as the speed
(b) the head varies inversely as the speed
(c) the power varies as the square of the speed
(d) the discharge varies as the square of the speed

[ESE : 2005]

2.34 In *M-L-T* system, what is the dimension of specific speed for a rotodynamic pump?

(a) $L^{-\frac{3}{4}}T^{\frac{3}{2}}$ (b) $M^{\frac{1}{2}}L^{\frac{1}{4}}T^{-\frac{5}{2}}$

(c) $L^{\frac{3}{4}}T^{-\frac{3}{2}}$ (d) $L^{\frac{3}{4}}T^{\frac{3}{2}}$

[ESE : 2006]

2.35 If N_{s1} is the specific speed of a pump handling larger discharges at low heads and N_{s2} is that of a pump handling low discharges at high heads, then which one of the following is correct?
(a) $N_{s1} < N_{s2}$
(b) $N_{s1} > N_{s2}$
(c) $N_{s1} = N_{s2}$
(d) All the above three are possible under different situations.

[ESE : 2006]

2.36 If in a pump, the discharge is halved, then assuming that the speed remains unchanged, what would be the ratio of the heads H_2/H_1?

(a) $\sqrt{1/3}$ (b) $\sqrt{2/3}$

(c) $\sqrt[3]{0.25}$ (d) $\sqrt[3]{0.5}$

[ESE : 2007]

2.37 The power absorbed by a hydraulic pump is directly proportional to which one of the following? (*N* is the rotational speed of the pump)
(a) N (b) N^2
(c) N^3 (d) N^4 [ESE : 2007]

2.38 The vanes of a centrifugal pump are generally
(a) radial (b) curved backward
(c) curved forward (d) twisted

[ESE : 2007]

2.39 Which one of the following pumps is not a positive displacement pump?
(a) Reciprocating pump
(b) Centrifugal pump
(c) Vane pump
(d) Lobe pump

[ESE : 2007]

2.40 Why is an air vessel used in a reciprocating pump?
(a) To obtain a continuous supply of water at uniform rate
(b) To reduce suction head
(c) To increase the delivery head
(d) To reduce cavitation

[ESE : 2008]

2.41 Centrifugal pumps have which of the following advantages?
1. Low initial cost
2. Compact, occupying less floor space
3. Easy handling of highly viscous fluids

Select the correct answer using the codes given below:

(a) 1, 2 and 3 (b) 1 and 2 only
(c) 1 and 3 only (d) 2 and 3 only

[ESE : 2008]

2.42 Two pumps can operate independently at heads H_1, H_2 and discharge Q_1, Q_2 respectively. If the pumps are connected in parallel, then what are the resulting discharge (Q) and head (H)?

(a) $Q = Q_1 + Q_2, H = H_1 + H_2$
(b) $Q = Q_1 - Q_2, H = H_1 - H_2$
(c) $Q = Q_1 = Q_2, H = H_1 = H_2$
(d) $Q = Q_1 + Q_2, H = H_1 = H_2$

[ESE : 2008]

2.43 Why is multi-staging in centrifugal pumps used?

(a) For high flow rate
(b) For high head
(c) For high speed
(d) For high efficiency

[ESE : 2008]

2.44 The volute casing of a centrifugal pump has which of the following functions?

1. Eliminating loss of head due to change in velocity after exit from impeller.
2. Directs the flow towards the delivery pipe.
3. Converts a part of the velocity head to pressure head.
4. Gives a constant velocity of flow.

Select the correct answer using the codes given below:

(a) 1, 2 and 4 (b) 2 and 3 only
(c) 1 and 4 only (d) 2 and 4 only

[ESE : 2008]

2.45 A centrifugal pump with radial vane tips at the outlet has an impeller of 100 mm outer diameter. If the rotative speed is 3000 rpm and manometric efficiency 0.8 then what is the net head developed?

(a) 10 m (b) 20 m
(c) 30 m (d) 40 m

[ESE : 2009]

2.46 A mixed flow pump is driven by a 8 kW motor running at 1000 rpm. It delivers water at the rate of 1000 liters/min against a total head of 25 m. What is the specific speed of the pump in meter-minutes?

(a) 90 (b) 50
(c) 45 (d) 75 [ESE : 2009]

Direction: Each of the next consists of two statements, one labelled as the **'Assertion (A)'** and the other as **'Reason (R)'**. You are to examine these two statements carefully and select the answers to these items using the codes given below:

Codes:

(a) Both **A** and **R** are individually true and **R** is the correct explanation of **A**
(b) Both **A** and **R** are individually true but **R** is not the correct explanation of **A**
(c) **A** is true but **R** is false
(d) **A** is false but **R** is true

2.47 **Assertion (A) :** Increase in static suction lift of centrifugal pump may cause cavitation.
Reason (R) : Available Net Positive Suction Head increases with increase in static suction lift.

[ESE : 2009]

2.48 **Assertion (A):** Before starting a centrifugal pump it is to be properly primed.
Reason (R): Priming provides lubrication to the propeller shaft.

[ESE : 2010]

2.49 A pump, having an efficiency of 90%, lifts water to a height of 155 m at the rate of 7.5 m^3/s. The friction head loss in the pipe is 13 m. The required pump power, in kW, will be

(a) 13,730 (b) 1,373
(c) 137.3 (d) 13.73

[ESE : 2011]

2.50 A reciprocating water pump delivers 100 litres of water per second against a suction head of 5 m and a delivery head of 15 m. The power required to drive the pump is near about

(a) 10 kW (b) 15 kW
(c) 20 kW (d) 25 kW

[ESE : 2011]

2.51 Air vessels are used in reciprocating pumps in order to

(a) increase the delivery head
(b) reduce suction head
(c) minimize delivery head fluctuation
(d) reduce accelerating head

[ESE : 2011]

2.52 **Statement (I):** Radial-flow machines are used mainly for pumps where multi-staging will be useful to increase the delivery head.

Statement (II): Axial-flow pumps are useful where the head is low and discharge is large.

(a) Both Statement (I) and Statement (II) are individually true and Statement (II) is the correct explanation of Statement (I).
(b) Both Statement (I) and Statement (II) are individually true but Statement (II) is **NOT** the correct explanation of Statement (I).
(c) Statement (I) is true but Statement (II) is false
(d) Statement (I) is false but Statement (II) is true.

[ESE : 2012]

2.53 The preferred type of pump for small discharge and high head is

(a) centrifugal type (b) reciprocating type
(c) axial-flow type (d) radial-flow type

[ESE : 2012]

2.54 The specific speed $N_S = \frac{N\sqrt{Q}}{H^{3/4}}$ for a double-suction pump is to be evaluated. The discharge would be taken as

(a) half the actual discharge
(b) actual discharge
(c) double the actual discharge
(d) square the actual discharge

[ESE : 2012]

2.55 In a reciprocating pump, air vessels are used to

(a) smoothen the flow
(b) reduce suction head
(c) increase delivery head
(d) reduce acceleration head [ESE : 2013]

2.56 Consider the following statements pertaining to centrifugal pumps:

1. The inlet to the impeller of a centrifugal pump is always axial, while the outlet may be radial or inclined.
2. The impeller may be shrouded on both sides with an eye at the centre and vanes curved backwards.
3. Impeller of double entry type has a balanced radial thrust.
4. Un-shrouded and part shrouded impeller are used only where high efficiency is unimportant.

Which of these statement are correct?

(a) 1, 2, 3 and 4 (b) 1, 2 and 3 only
(c) 1, 2 and 4 only (d) 2, 3 and 4 only

[ESE : 2013]

2.57 A centrifugal pump is fully primed, but on starting it fails to deliver fluid. The probable reasons are listed below:

1. Leaky foot valve or suction line
2. Suction head is very low
3. Insufficient motor speed
4. Damaged or closed delivery valve

Which of these reasons are correct?

(a) 1, 2, 3 and 4 (b) 1, 2 and 3 only
(c) 2, 3 and 4 only (d) 1, 3 and 4 only

[ESE : 2013]

2.58 For a given centrifugal pump

(a) head varies inversely as square of speed
(b) discharge varies directly as speed
(c) discharge varies directly as square of speed
(d) power varies directly as fifth power of speed

[ESE : 2013]

2.59 A centrifugal pump is to discharge 0.118 m^3/s at speed of 1450 rpm against a head of 25 m. The impeller diameter is 25 cm, its width at outlet is 5 cm and manometric efficiency is 75%. The vane angle at outer periphery of impeller is

(a) 56.77° (b) 59.77°
(c) 61.77° (d) 48.77° [ESE : 2013]

2.60 Two geometrically similar pumps are running at 1000 rpm speed (both). If one pump has impeller diameter of 0.3 m and discharges 20 lps against 20 m head, and the other pump gives half of this discharges rate; calculate head and diameter of second pump

(a) 12.5 m and 0.12 m
(b) 10.5 m and 0.12 m
(c) 10.5 m and 0.23 m
(d) 12.5 m and 0.23 m [ESE : 2013]

2.61 The ratio of power outlet of the pump to the power input to the pump is known as

(a) mechanical efficiency
(b) static efficiency
(c) overall efficiency
(d) manometric efficiency [ESE : 2014]

2.62 A pump is defined as a device which converts
(a) hydraulic energy into mechanical energy
(b) mechanical energy into hydraulic energy
(c) kinetic energy into mechanical energy
(d) none of the above [ESE : 2014]

2.63 The specific speed of a pump is defined as the speed of the unit of such a size that it
(a) delivers unit discharge at unit head
(b) requires unit power to develop unit head
(c) delivers unit discharge at unit power
(d) produces unit power with unit head available [ESE : 2014]

2.64 Negative slip occurs in reciprocating pumps, when delivery pipe is
(a) long and suction pipe is short and pump is running at low speed
(b) long and suction pipe is short and pump is running at high speed
(c) short and suction pipe is long and pump is running at low speed
(d) short and suction pipe is long and pump is running at high speed [ESE : 2014]

2.65 **Statement (I) :** Non-dimensional performance curves are applicable to any pump in the homologous series.
Statement (II) : Viscosity of water varies with temperature causing cavitations on suction side.
Codes:
(a) Both Statement (I) and Statement (II) are individually true and Statement (II) is the correct explanation of Statement (I)
(b) Both Statement (I) and Statement (II) are individually true but Statement (II) is ***NOT*** the correct explanation of Statement (I)
(c) Statement (I) is true but Statement (II) is false
(d) Statement (I) is false but Statement (II) is true [ESE : 2016]

2.66 A centrifugal pump lifts 0.0125 m^3/s of water from a well with a static lift of 30 m. If the brake power of the driving electric motor is 5 kW, what is the overall efficiency of the pump-set?
(a) 57.6% (b) 63.9%
(c) 65.3% (d) 73.6% [ESE : 2017]

2.67 The mechanical efficiency of a centrifugal pump is the ratio of
(a) manometric head to the energy supplied by the impeller per kN of water
(b) energy supplied to the pump to the energy available at the impeller
(c) actual work done by the pump to the energy supplied to the pump by the prime mover
(d) energy available at the impeller to the energy supplied to the pump by the prime mover [ESE : 2018]

2.68 Consider the following advantages of rotary pumps compared to reciprocating pumps:
1. Steady discharge which increases with decrease in head
2. Suitable for handling fluids with suspended solid particles
3. Less bulky than displacement pumps positive
4. Can be started with open delivery with least load

Which of the above advantages are correct?
(a) 1, 2 and 3 (b) 1, 2 and 4
(c) 1, 3 and 4 (d) 2, 3 and 4 [ESE : 2018]

2.69 A centrifugal pump is required to lift 0.0125 m^3/s of water from a well with depth 30 m. If rating of the pump motor is 5 kW, and the density of water is 1000 kg/m^3, the efficiency of the pump will be nearly
(a) 82% (b) 74%
(c) 66% (d) 58% [ESE : 2020]

2.70 Which of the following statements regarding hydraulic pumps are correct ?
1. The gear pump consists of two close-meshing gear wheels which rotate in opposite directions.
2. In vane pump, as the rotor rotates, the vanes follow the contours of the casing.
3. The leakage is more in vane pump compared to gear pump.

(a) 1, 2 and 3 (b) 1 and 2 only
(c) 1 and 3 only (d) 2 and 3 only [ESE : 2020]

2.71 A centrifugal pump has an impeller of 30 cm outer diameter. The vane tips are radial at the outlet. For a rotative speed of 1450 rpm what is the manometric head developed?
[Assume a manometric efficiency of 82% and take g = 9.81 m/s^2)

(a) 37.24 m (b) 43.38 m
(c) 29.46 m (d) 32.88 m

[ESE : 2021]

2.72 A single-acting reciprocating pump, running at 50 rpm delivers 0.00736 m^3/s of water. The diameter of the piston is 200 mm and stroke length is 300 mm. What is the percentage slip of the pump?

(a) 5.29% (b) 6.29%
(c) 7.29% (d) 8.29%

[ESE : 2021]

2.73 A pump discharges a liquid into a tank at the rate of 0.032 m^3/s. The tank, 1.5 m in diameter and 4.20 m in height, can hold 3500 kg of liquid. The density of the liquid and mass flow rate of the liquid handled by the pump are respectively.

(a) 471.57 kg/m^3 and 16 kg/s
(b) 471.57 kg/m^3 and 15 kg/s
(c) 481.57 kg/m^3 and 16 kg/s
(d) 481.57 kg/m^3 and 15 kg/s

[ESE : 2021]

2.74 A pump impeller is 375 mm in diameter and it discharges water with velocity components of 2 m/s and 12 m/s in the radial and tangential directions respectively. The impeller is surrounded by a concentrical cylindrical chamber with parallel sides, the outer diameter being 450 mm. If the flow in the chamber is a free spiral vortex, what are the tangential velocity and radial velocity at the outlet of the chamber respectively?

(a) 12 m/s and 1.67 m/s
(b) 10 m/s and 1.67 m/s
(c) 12 m/s and 1.76 m/s
(d) 10 m/s and 1.76 m/s

[ESE : 2021]

2.75 Which one of the following types of impellers is used to handle highly solid-laden liquids like concrete and slurry?

(a) Fully Open Impeller
(b) Semi-Enclosed Impeller
(c) Fully-Enclosed Impeller
(d) Quarter Open Impeller

[ESE : 2021]

2.76 In a single reciprocating pump without air vessel, the ratio of the average frictional head to the maximum frictional head in the delivery pipe is

(a) 1/2 (b) 1/3
(c) 2/3 (d) 3/4

[ESE : 2021]

2.77 Consider the following statements regarding torque converter:

1. A torque converter is a modified form of fluid coupling.
2. A reciprocating pump is used in a torque converter.

Which of the above statements is/are correct?

(a) 1 only (b) 2 only
(c) Both 1 and 2 (d) Neither 1 nor 2

[ESE : 2022]

2.78 In a double acting pump, the work saved by fitting air vessels is about

(a) 19.2% (b) 29.2%
(c) 39.2% (d) 50.2%

[ESE : 2022]

2.79 A single acting reciprocating pump has a bore of 200 mm and a stroke of 350 mm and runs at 45 rpm. The suction head is 8 m and the delivery head is 20 m. What is the theoretical discharge of water if the slip is 12%? (Take acceleration due to gravity as 9.81 m/s^2)

(a) 7.260 l/s (b) 6.956 l/s
(c) 3.456 l/s (d) 4.586 l/s

[ESE : 2022]

2.80 A double-acting reciprocating pump has indicator diagram with area 40 cm^2 and length 8 cm. The bore diameter and stroke of the pump are 15 cm and 20 cm respectively. The pump motor runs at 100 rpm and the indicator spring constant is given as 1.5 × 10^8 Pa/m. What is the power required to drive a double acting reciprocating pump?

(a) 33.45 kW (b) 44.36 kW
(c) 66.34 kW (d) 88.36 kW

[ESE : 2023]

2.81 Consider the following statements regarding centrifugal pumps :

1. The rotating part of a centrifugal pump is called runner.
2. Volute casing is spiral type in which area of flow increases gradually.
3. In vortex casing, a vortex chamber is introducing between the casing and impeller.
4. Foot valve is fitted at the lower end of suction pipe.

Which of the above statements are correct?

(a) 1, 2 and 3 only (b) 2, 3 and 4 only
(c) 1, 3 and 4 only (d) 1, 2, 3 and 4

[ESE : 2024]

2.82 Consider the following statements :

1. The horizontal distance between the centre line of the pump and the water surface in the tank to which water is delivered is known as delivery head
2. The sum of suction head and delivery head is known as static head.
3. The manometric head is defined as the head against which centrifugal pump has to work.

Which of the above statements are correct?

(a) 1 and 3 only (b) 2 and 3 only
(c) 1 only (d) 2 only

[ESE : 2024]

2.83 Consider the following statements regarding efficiencies of a centrifugal pump:

1. The ratio of the manometric head to the head imparted by the impeller to the water is known as manometric efficiency.
2. The ratio of the power available at the impeller to the power at the shaft of the centrifugal pump is known as overall efficiency.
3. The ratio of the power output of the pump to the power input to the pump is known as mechanical efficiency.

Which of the above statements is/are correct?

(a) 1 and 3 only (b) 2 and 3 only
(c) 1 only (d) 3 only

[ESE : 2024]

2.84 The outlet velocity of flow and outlet velocity of whirl of an impeller of a centrifugal pump are 5 m/s and 12 m/s respectively. What is the resultant velocity at outlet?

(a) 17 m/s (b) 7 m/s
(c) 11 m/s (d) 13 m/s

[ESE : 2024]

■■■■

Answers Hydraulic Pumps

2.1	(d)	2.2	(a)	2.3	(c)	2.4	(a)	2.5	(c)	2.6	(a)	2.7	(c)	2.8	(b)	2.9	(d)
2.10	(b)	2.11	(c)	2.12	(b)	2.13	(d)	2.14	(d)	2.15	(b)	2.16	(b)	2.17	(d)	2.18	(c)
2.19	(a)	2.20	(b)	2.21	(d)	2.22	(d)	2.23	(d)	2.24	(c)	2.25	(b)	2.26	(a)	2.27	(a)
2.28	(c)	2.29	(c)	2.30	(d)	2.31	(d)	2.32	(a)	2.33	(a)	2.34	(c)	2.35	(b)	2.36	(c)
2.37	(c)	2.38	(b)	2.39	(b)	2.40	(a)	2.41	(b)	2.42	(d)	2.43	(b)	2.44	(b)	2.45	(b)
2.46	(a)	2.47	(c)	2.48	(c)	2.49	(a)	2.50	(c)	2.51	(d)	2.52	(b)	2.53	(b)	2.54	(a)
2.55	(d)	2.56	(c)	2.57	(d)	2.58	(b)	2.59	(b)	2.60	(d)	2.61	(c)	2.62	(b)	2.63	(a)
2.64	(d)	2.65	(b)	2.66	(d)	2.67	(d)	2.68	(c)	2.69	(b)	2.70	(b)	2.71	(b)	2.72	(b)
2.73	(b)	2.74	(b)	2.75	(a)	2.76	(c)	2.77	(a)	2.78	(c)	2.79	(a)	2.80	(d)	2.81	(b)
2.82	(b)	2.83	(c)	2.84	(d)												

Explanations Hydraulic Pumps

2.1 (d)

$$\text{Specific speed of turbine} = \frac{N\sqrt{P}}{H^{5/4}}$$

$$\text{Specific speed of pump} = \frac{N\sqrt{Q}}{H^{3/4}}$$

2.2 (a)

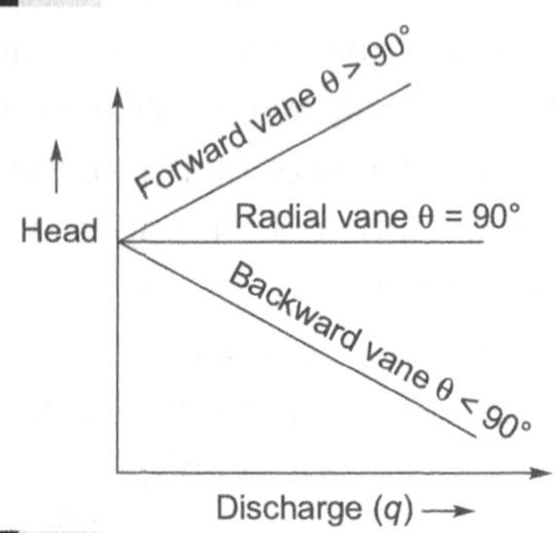

2.3 (a)

In a volute pump the impeller is surrounded by a spiral shaped casing which is known as volute chamber. The shape of the casing is such that the sectional area of flow around the periphery of the impeller gradually increases towards the delivery pipe. This increase in the cross-sectional area results in developing a uniform velocity throughout the casing.

In volute type of casing there is only a slight increase in the efficiency of the pump, because a considerable loss of energy takes place in eddies developed in the casing.

2.4 (a)

Apart from common losses in turbine and pump, the pump has additional friction and eddy losses in diffusers.

2.6 (a)

$$B.P. = \frac{\dot{m}gh}{\eta_m} = \frac{80\times 9.81\times 30}{0.8} = \mathbf{29.4\ kW}$$

2.7 (c)

Hydrodynamic cavitation describes the process of vaporization bubble generation and bubble implosion which occur in a flowing liquid as a result of decrease and subsequent increase in pressure. The vaporization will occur only if the pressure drops below the vapour pressure at the corresponding saturation temperature.

2.8 (b)

$$\pi_1 = H^a g^b Q$$

$$\pi = L^{a+b+3}T^{-2b-1}$$

$$\therefore\ b = -\frac{1}{2},\quad a = -\frac{5}{2}$$

$$\pi_1 = \frac{Q}{\sqrt{gH^5}}$$

2.9 (d)

Given data

Condition - 1

$N_1 = 1000$ rpm; $P_1 = 1$ kW; $H_1 = 10$ m

Condition - 2

$N_2 = 2000$ rpm; $P_2 = ?$

$H_2 = ?$

We know that

Head coefficient,

$$C_H = \frac{gH}{N^2D^2}$$

$$(C_H)_1 = (C_H)_2$$

$$\frac{gH_1}{N_1^2D_1^2} = \frac{gH_2}{N_2^2D_2^2}$$

$$\frac{H_1}{N_1^2} = \frac{H_2}{N_2^2}\quad \because D_1 = D_2 \text{ same given pump}$$

$$\frac{10}{(1000)^2} = \frac{H_2}{(2000)^2}$$

or $H_2 = \mathbf{40\ m}$

we know that

Power coefficient,

$$C_p = \frac{P}{\rho N^3D^5}$$

$$(C_p)_1 = (C_p)_2$$

$$\frac{P_1}{\rho N_1^3D_1^5} = \frac{P_1}{\rho N_2^3D_2^5}$$

or $$\frac{P_1}{N_1^3} = \frac{P_2}{N_2^3}$$

$$\frac{1}{(1000)^3} = \frac{P_2}{(2000)^3}$$

or $P = \mathbf{8\ kW}$

Alternatively

Unit speed,

$$N_u = \frac{N}{\sqrt{H}}$$

$$(N_u)_1 = (N_u)_2$$

$$\frac{N_1}{\sqrt{H_1}} = \frac{N_2}{\sqrt{H_2}}$$

$$\frac{1000}{\sqrt{10}} = \frac{2000}{\sqrt{H_2}}$$

or $H_2 = \mathbf{40\ m}$

Unit power,

$$P_u = \frac{P}{H^{3/2}}$$

$$(P_u)_1 = (P_u)_2$$

$$\frac{P_1}{H_1^{3/2}} = \frac{P_2}{H_2^{3/2}}$$

$$\frac{1}{(10)^{3/2}} = \frac{P_2}{(40)^{3/2}}$$

or $P_2 = (4)^{3/2} = 4^{1.5} = \mathbf{8\ kW}$

Note : Unit quantities :

1. Unit discharge: $Q_u = \dfrac{Q}{\sqrt{H}}$

2. Unit speed: $N_u = \dfrac{N}{\sqrt{H}}$

3. Unit power: $P_u = \dfrac{P}{H^{3/2}}$

Unit quantities are equally used for both the given turbines and pumps.

2.10 (b)

Backward curved vanes are commonly used (through the *H-Q* curve is declining) because these pump operate at high efficiency than in case of forward curved vanes.

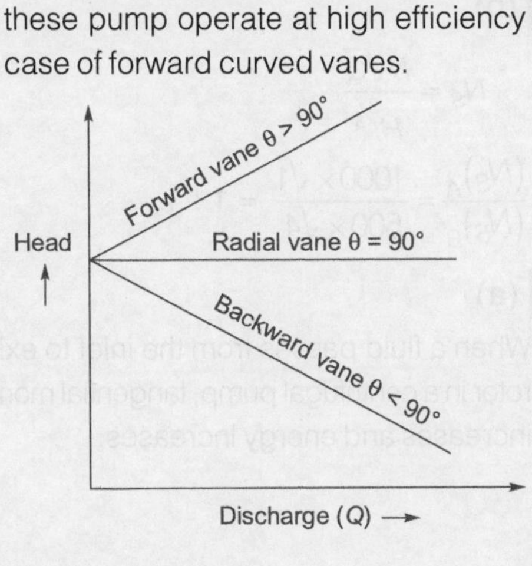

2.11 (c)

$$P = \rho QgH$$

$$\therefore \quad \frac{P}{QH} = \text{Constant}$$

and $\quad u = \dfrac{\pi DN}{60} \propto \sqrt{2gH}$

$$\frac{H}{N^2D^2} = \text{constant}$$

2.12 (b)

The correct sequence of the centrifugal pump component through whicih the fluid flows is

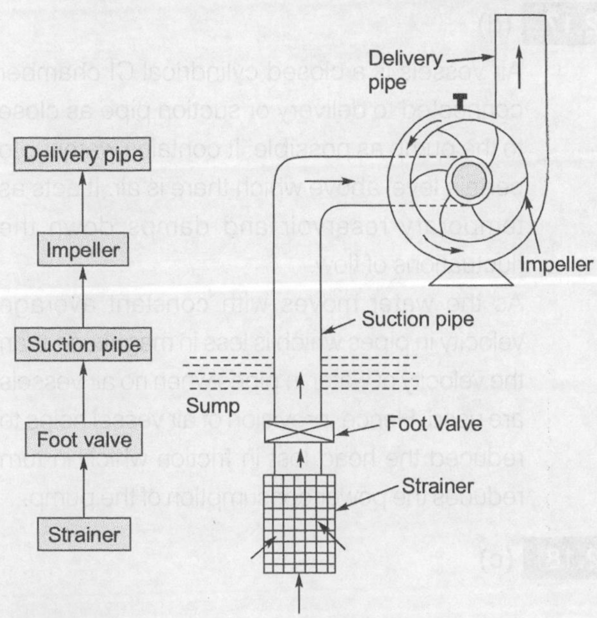

2.13 (d)

$P \propto N^3$ (for constant diameter)

$$\frac{P_2}{P_1} = \left(\frac{N_2}{N_1}\right)^3 = \left(\frac{2900}{1450}\right)^3$$

$$P_2 = 3 \times 8 = \mathbf{24\ kW}$$

2.14 (d)

(a) Centrifugal pump : Rotor having blades

(b) Gear pump : Rotary pump

(c) Reciprocating pump : Air vessel

(d) Turbine pump : Guide vanes

2.15 (b)

(a) Combustible fluid to be pumped : Jet pump.

(b) High head but small discharge needed : Positive displacement pump.

(c) Low head but large discharge needed : Single stage centrifugal.

(d) High head and high discharge needed : Multistage centrifugal.

2.16 (b)

Filling the pump casing and the suction pipe with the liquid before it is started is known as priming. Unless the centrifugal pump is primed, it will not start as there is no centrifugal head impressed on the liquid. Positive displacement type reciprocating or rotary pump do not need priming because the air, if any, in the cylinder or casing is driven out by the piston or vanes.

2.17 (d)

Air vessels is a closed cylindrical CI chamber connected to delivery or suction pipe as close to the pump as possible. It contains water upto certain level above which there is air. It acts as temporary reservoir and damps down the fluctuations of flow.

As the water moves with constant average velocity in pipes which is less in magnitude than the velocity of water in pipes when no air vessels are used. Hence, provision of air vessel helps to reduced the head lost in friction which in turn reduces the power consumption of the pump.

2.18 (c)

$$NPSH = \frac{p_0}{\rho.g} - \text{(Static head + frictional}$$

head in suction + vapour pressures)

= 10.3 – (5 + 0.6 +0.4) = **4.3 m**

2.19 (a)

Specific speed of pump: $N_s = \dfrac{N\sqrt{Q}}{H^{3/4}}$

when N = pump speed in rpm

Q = pump flow rate at best efficiency (for double suction impeller divide total pump flow by two)

H = manometric head at best efficiency point.

It is customary to base the specific speed of the double suction pump on one half of the total capacity.

2.20 (b)

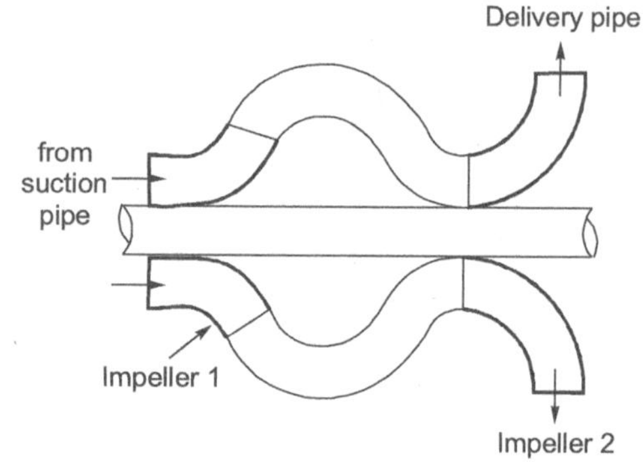

2.21 (d)

$$P \propto N^3$$

$$\therefore \quad P_2 = 1000 \times \left(\frac{3000}{1500}\right)^3 = \mathbf{8000\ W}$$

2.22 (d)

Given $D_0 = 2D_i$, $d_0 = 2d_i$

and $d_i = 3D_i$

$\therefore$ $d_0 = 6\,D_i$

$$N \propto \frac{1}{D}$$

$$\therefore \quad \frac{N_2}{N_1} = \frac{D_0}{d_0} = \frac{2D_i}{6D_i} = \frac{1}{3}$$

2.23 (d)

Multistage centrifugal pump is used for high head and high discharge.

2.24 (c)

$$H \propto N^2$$

$$\therefore \quad H_2 = 30 \times \left(\frac{2000}{1000}\right)^2 = \mathbf{120\ m}$$

2.25 (b)

$$N_S = \frac{N\sqrt{Q}}{H^{3/4}}$$

$$\frac{(N_S)_A}{(N_S)_B} = \frac{1000 \times \sqrt{1}}{500 \times \sqrt{4}} = \mathbf{1:1}$$

2.26 (a)

When a fluid passes from the inlet to exit of the rotor in a centrifugal pump, tangential momentum increases and energy increases.

Considering the interaction of fluid and rotor it will be seen that axial and radial component do not contribute to the rotation. The axial component C_a, produces thrust and the radial component C_r, produces the radial force, the tangential component C_t, produces rotational effects.

2.27 (a)

The specific speed of the pump is defined as the speed of geometrical similar pump which deliver unit discharge under the unit head.

$$N_s = \frac{N\sqrt{Q}}{H^{3/4}}$$

2.28 (c)

For most of positive displacement pump, the actual discharge is less than the theoretical discharge in this case coefficient of discharge (C_d) of the pump is less than one and the slip of the pump is positive. However, in some cases the actual discharge of the pump may be more than theoretical discharge, in which case C_d will be more than one and the slip be negative , which is known as negative slip.

This may happen in the case of pump having long suction pipe and low delivery head, especially when these are running at high speed. This is to because for such pumps the interia pressure in the suction pipe will be large in comparison to the pressure on the outside of the delivery valve, which may cause, the delivery valve to open before the suction stroke is completed. Some liquid is thus pushed directly into this delivery even before the delivery stroke is commenced, which results in making the actual discharge more than the theoretical discharge.

2.29 (c)

Open impeller: In this impeller, vanes are attached to the hub i.e. each vane is supported only by the hub. There is no shroud plate or side wall. Such impeller are used for centrifugal pumps handling contaminated heterogeneous, highly viscous fluids like water with sand, mud, paper pump, acid slurry, sewage etc. The open construction of impellers minimizes. The possibility of clogging impeller and resultant frequent overhaul. Such pumps are often referred as non-clogging pumps. This impeller dealing with dense fluids is robust in construction and is made of forged steel or cast bronze.

Semi-open impeller: This impeller is provided with a shroud plate or wall on one side only. The impeller vanes are thus supported from the hub side as well as from the side of the shroud plate. This gives more rigidity to vanes against bending. Such impellers are used for centrifugal pump dealing with thick, heterogeneous fluids. Due to only one shroud plate, cleaning of impeller vanes is easy.

Closed impeller: This impeller has shroud plates on both sides of vanes, enclosing the fluids passage. The vanes get support from hub as well as from two side plates. However, vanes are no more accessible for cleaning purpose. Such impeller are suitable for only thinner fluids with low viscosity, clear homogeneous liquid etc.

2.30 (d)

The shape of the volute casing is nearly involute or spiral around the impeller. Such a cross section gradually increases from "neck" or "tongue" towards the delivery outlet in the forward direction of rotation. The gradually increasing area of cross-section performs diffuser action converting most of the kinetic energy of the fluid discharged by the impeller into pressure head.

2.31 (d)

A is false but **R** is true.

For higher specific speed axial flow pumps are used as they give higher efficiency when flow is axial and specific speed is high.

2.32 (a)

The Net Positive Suction Head (*NPSH*) is defined as the absolute pressure head at the inlet to the pump, minus the vapour pressure head (in absolute units) corresponding to the temperature of the liquid pumped, plus the velocity head at this point. The *NPSH* should always be positive to keep away the cavitation of $NPSH > 0$.

2.33 (a)

The dimensionless number for centrifugal pump is $\frac{Q}{ND^3}$.

Hence discharge is directly proportional to speed.

2.34 (c)

The specific speed of the pump,

$$N_s = \frac{N\sqrt{Q}}{H^{3/4}} = \frac{T^{-1}(L^3T^{-1})^{1/2}}{L^{3/4}} = \frac{L^{3/2}T^{-3/2}}{L^{3/4}} = L^{3/4}T^{-3/2}$$

2.35 (b)

$$N_S = \frac{N\sqrt{Q}}{H^{3/4}}$$

2.36 (c)

If speed remain same then

$$Q \propto H^{3/2}.$$

$$\therefore \quad \frac{H_1}{H_2} = \left(\frac{Q_1}{Q_2}\right)^{2/3} = (2)^{2/3}$$

$$\frac{H_2}{H_1} = (0.5)^{2/3}$$

$$\frac{H_2}{H_1} = (0.25)^{1/3} = \sqrt[3]{0.25}$$

2.37 (c)

$$U_1 = \frac{\pi DN}{60}$$

For a specific pump, we can write

$$U_1 \propto N$$

$$V_{f_1} \propto N$$

But $\quad Q = \pi D_1 \times b_1 \times V_{f_1}$

$$Q \propto V_{f_1}$$

$$Q \propto N$$

Now $\quad V_{f_1} \propto \sqrt{H}$

$$N \propto \sqrt{H}$$

$$\therefore \quad H \propto N^2$$

Power: $\quad P = \rho gQH$

$$P \propto N \times N^2$$

$$P \propto N^3$$

2.38 (b)

In practice, the power required to drive the pump is proportional to $Q \times H$. If $H \propto Q$; the load on prime mover fluctuates considerably. While when $H \propto \frac{1}{Q}$; The load on the prime mover almost remains constant, hence it is preferred to have backward curved vanes.

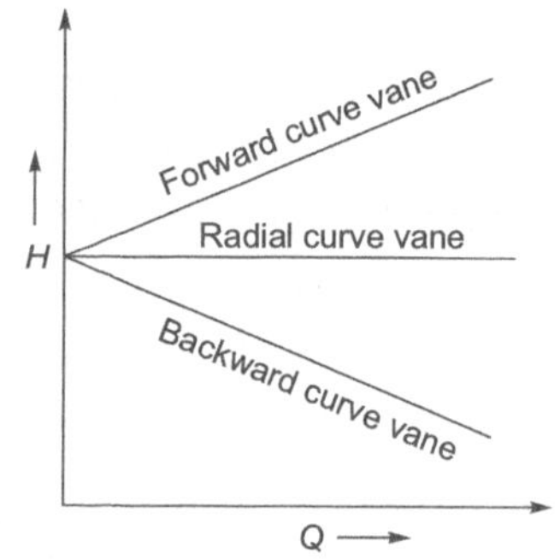

2.39 (b)

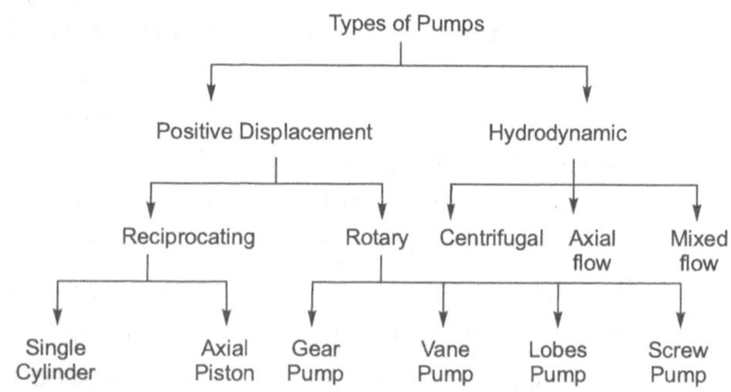

2.40 (a)

Air vessel are fitted to delivery side of reciprocating pump due to following reason

(a) Constant rate of discharge can be ensured

(b) Power consumption can be reduced because it reduces friction losses

2.41 (b)

Centrifugal pumps involves

(i) Low initial cost

(ii) Compact, occupying less floor space

Reciprocating pumps are used for high viscous fluids.

2.42 (d)

If the pumps are connected in parallel $Q = Q_1 + Q_2$ and if the pumps are connected in series $H = H_1 + H_2$

2.43 (b)

Multi staging in centrifugal pumps are used to obtain high head.

2.44 (b)

The volute casing of a centrifugal pump has following function

(i) Directs the flow towards the delivery pipe

(ii) Converts a part of the velocity head to pressure head

2.45 (b)

$$u_2 = \frac{\pi D_2 N}{60} = \frac{3.14 \times 0.1 \times 3000}{60}$$

$$= 15.7 \text{ m/s}$$

For Radial Vane Tips

$$V_{w2} = u_2 = 15.7 \text{ m/s}$$

$$\eta_H = \frac{gH}{V_{w2} u_2}$$

$$\frac{gH}{V_{w2} u_2} = 0.8$$

or $$H = \frac{0.8 \times u_2^2}{9.81} = \frac{0.8 \times (15.7)^2}{9.81} = \mathbf{20.10 \text{ m}}$$

2.46 (a)

Specific speed of pump,

$$N_s = \frac{N\sqrt{Q}}{H^{3/4}} = \frac{1000\sqrt{1.000}}{(25)^{3/4}}$$

$$= 89.44 \approx \mathbf{90 \text{ units}}$$

2.47 (c)

A is true but R is false.

NPSH decreases with increase in h_s.

2.48 (c)

Priming is done in the centrifugal pump to generate sufficient head and not for lubrication.

2.51 (d)

Refer to explanation of Question 2.1

2.52 (b)

Axial flow pump have a large discharge but a relatively low head per stage.

2.53 (b)

Reciprocating pump gives high pressure rise but low discharge.

2.55 (d)

Purpose of an air vessel:

- To obtain a continuous supply of liquid at a uniform rate.
- To save a considerable amount of work in overcoming the frictional resistance in the suction and delivery pipes.
- To run the pump a at high speed without separation.
- To act like an intermediate reservoir.

2.56 (c)

Impeller of double entry type has a balanced axial thrust not radial thrust. Other statements at a valid pertaining to centrifugal pumps.

2.57 (d)

If suction head is too low, then pump delivers too much liquid and take too much power. Other statements are valid.

2.58 (b)

For centrifugal pump:

$$\frac{H_m}{N^2 D^2} = \text{Constant}$$

∴ Head varies directly as square of speed.

$$\frac{Q}{ND^3} = \text{Constant}$$

∴ Discharge varies directly as speed.

$$\frac{P}{N^3 D^5} = \text{Constant}$$

∴ Power varies directly as cubic of speed.

2.59 (b)

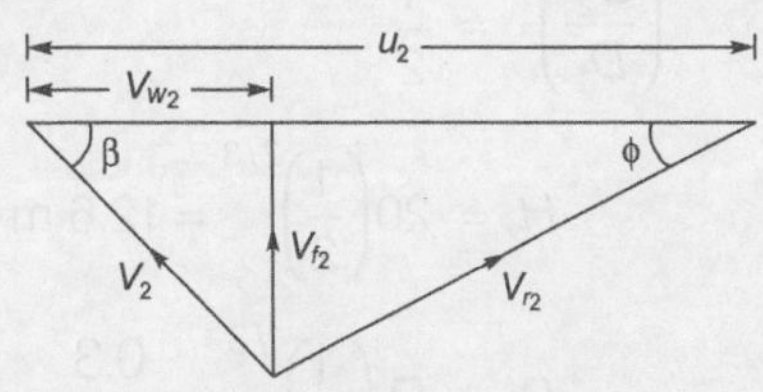

Outlet Velocity Triangle

Given, $Q = 0.118 \text{ m}^3/\text{s}$; $N = 1450$ rpm

$H_m = 25$ m; $D_2 = 0.25$ m; $b_2 = 0.05$ m

$$\eta_{mano} = 0.75$$

$$V_{f_2} = \frac{Q}{\pi D_2 b_2}$$

$$= \frac{0.118}{3.14 \times 0.25 \times 0.05} = 3 \text{ m/s}$$

$$\tan\phi = \frac{V_{f_2}}{u_2 - V_{w_2}}$$

$$\eta_{mano} = \frac{gH_m}{V_{w_2}u_2}$$

We have

$$u_2 = \frac{\pi D_2 N}{60}$$

$$= \frac{3.14 \times 0.25 \times 1450}{60} = 19 \text{ m/s}$$

$$\therefore \quad V_{w_2} = \frac{9.81 \times 25}{0.75 \times 19} = 17.237 \text{ m/s}$$

$$\therefore \quad \tan\phi = \frac{3}{19 - 17.237} = \frac{3}{1.763}$$

$$\therefore \quad \phi = \mathbf{59.55°}$$

2.60 (d)

$$\frac{H}{N^2D^2} = \text{Constant}$$

$$\frac{H_1}{D_1^2} = \frac{H_2}{D_2^2}$$

$$\therefore \quad H_2 = H_1\left(\frac{D_2}{D_1}\right)^2$$

$$\frac{Q}{ND^3} = \text{Constant}$$

$$\therefore \quad \frac{Q_2}{D_2^3} = \frac{Q_1}{D_1^3}$$

$$\therefore \quad Q_2 = Q_1\left(\frac{D_2}{D_1}\right)^3$$

$$\therefore \quad \left(\frac{D_2}{D_1}\right)^3 = \frac{1}{2}$$

$$\therefore \quad H_2 = 20\left(\frac{1}{2}\right)^{2/3} = 12.6 \text{ m}$$

$$D_2 = D_1\left(\frac{1}{2}\right)^{1/3} = \frac{0.3}{2^{1/3}} = \mathbf{0.238 \text{ m}}$$

2.61 (c)

The overall efficiency of the pump is defined as the ratio of the power output from the pump to the power input from the prime mover driving the pump.

2.63 (a)

The specific speed of a centrifugal pump may be defined as the speed in revolutions per minute of a geometrically similar pump of such a size that under corresponding condition it would deliver 1 litre of liquid per second against a head of 1 meter.

$$N_s = \frac{N\sqrt{Q}}{H_m^{3/4}}$$

2.64 (d)

Negative slip may happen in the case of pumps having long suction pipe and low delivery head, especially when these are running at high speed. This is so because for such pumps the inertia pressure in the suction pipe will be large in comparison to the pressure on the outside of the delivery valve, which may cause the delivery valve to open before the suction stroke is completed. Some liquid is thus pushed directly into the delivery pipe even before the delivery stroke is commenced, which results in making the actual discharge more than the theoretical discharge.

2.66 (d)

Given data: $Q = 0.0125 \text{ m}^3\text{/s}$; $H = 30$ m
$P = 5 \text{ kW} = 5000 \text{ W}$

$$\eta_o = \frac{\rho QgH}{P}$$

Assume efficiency of electric motor = 100%

$$= \frac{1000 \times 0.0125 \times 9.81 \times 30}{5000}$$

$$= 0.7357 = 73.57\%$$

2.67 (d)

$$\eta_{mech} = \frac{\text{Impeller power}}{\text{Shaft power}}$$

2.68 (c)

Rotary pump gives steady discharge and discharge increase with decrease in head.
Reciprocating pump is suitable for fluid having solid particles.

2.69 (b)

Given, $Q = 0.0125\ m^3/sec$; $H_s = 30\ m$

Efficiency shaft of power = 5 kW

Pump is given by,

$$\eta_0 = \frac{\rho g Q H_m}{S_p}$$

$$= \frac{1000 \times 9.81 \times 0.0125 \times 30 \times 100}{5000}$$

$$= 74\%$$

2.70 (b)

The volumetric efficiency of vane pump (upto 95%) is higher compared to gear pump.

2.71 (b)

$$d_2 = 30\ cm = 0.3\ m$$

$$\phi = 90° \Rightarrow V_{w2} = u_2$$

$$N = 1450\ rpm$$

$$\eta_{mano} = 82\%$$

$$\eta_{mano} = \frac{gH_m}{V_{w2}u_2}$$

$$u_2 = \frac{\pi d_2 N}{60} = \frac{\pi \times 0.3 \times 1450}{60} = 22.77\ m/s$$

$$\eta_{mano} = 0.82 = \frac{9.81 \times H_m}{22.77^2}$$

$$H_m = 43.36\ m$$

2.72 (b)

$$N = 50\ rpm$$

$$Q_{act} = 0.00736\ m^3/sec$$

$$D = 0.2\ m$$

$$L = 0.3\ m$$

$$\%\ slip = ? = \frac{Q_{th} - Q_{act}}{Q_{th}}$$

$$Q_{th} = \frac{ALN}{60} = \frac{\left(\frac{\pi}{4}D^2\right) \times L \times N}{60}$$

$$= \frac{\left(\frac{\pi}{4} \times 0.2^2\right) \times 0.3 \times 50}{60}$$

$$= 0.00785\ m^3/sec$$

$$\%\ slip = \frac{0.00785 - 0.00736}{0.00785}$$

$$= 6.29\%$$

2.73 (b)

$$Q = 0.032\ m^3/sec$$

$$m = 3500\ kg$$

$$\text{Density} = \frac{\text{Mass}}{\text{Vol. of tank}} = \frac{3500}{\text{Area} \times \text{Height of tank}}$$

$$= \frac{350}{\left(\frac{\pi}{4} \times 1.5^2\right) \times 4.2} = 471.57\ kg/m^3$$

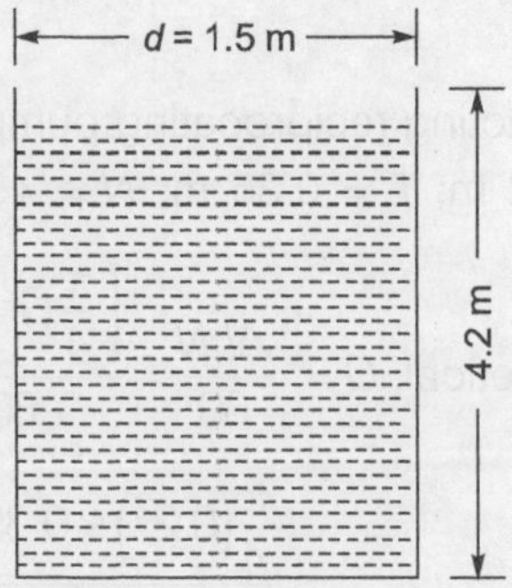

$$\text{Mass flow rate} = \text{Density} \times Q$$

$$= 471.57 \times 0.032$$

$$\text{Mass flor rate} = 15\ kg/sec$$

2.74 (b)

$$d_2 = 375\ mm, \quad d_3 = 450\ mm$$

$$V_{f2} = 2\ m/s, \ V_{w2} = 12\ m/s$$

For free vortex,

$$Vr = \text{Constant}$$

$$V_{w2}r_2 = V_{w3}r_3$$

$$12 \times \frac{375}{2} = V_{w2} \times \frac{450}{2}$$

$$V_{w3} = 10\ m/s$$

$$V_{f2}r_2 = V_{f3}r_3$$

$$2 \times \frac{375}{2} = V_{f3} \times \frac{450}{2}$$

$$V_{f3} = 1.67\ m/s$$

2.76 (c)

Friction head for delivery pipe,

$$h_{fd} = \frac{fLV_d^2}{2gd_d} = \frac{fL_d}{2gd_d}\left(\frac{A}{A_d} r\omega \sin\theta\right)^2$$

Maximum h_{fd} is at ($\theta = \pi/2$) and hence

$$h_{fdm} = \frac{fL_d}{2gd_s}\left(\frac{A}{A_d} r\omega\right)^2$$

$$h_{fda} = \text{Average } h_{fd} = (2/3)h_{fdm}$$

2.78 (c)

Work saved by air vessel in single acting pump is 84.8%

For double acting work saved is 39.2%.

2.79 (a)

Single acting reciprocating pump,

$D = 0.2$ m; $L = 0.35$ m; $N = 45$ rpm

$$\text{Theoretical, } Q = \frac{ALN}{60} = \frac{\left(\frac{\pi}{4}D^2\right)LN}{60}$$

$$= \frac{\frac{\pi}{4}(0.2^2)\times 0.35\times 45}{60}$$

Theoretical, $Q = 8.24$ l/s

If we calculate actual discharge which is

$$Q_{act} = 0.88 \times Q_{th}$$

$$Q_{act} = 7.26\ l/s$$

2.80 (d)

$$\text{Height } (h) = \frac{\text{Area of indicator diagram}}{\text{Length of indicator diagram}} = \frac{40}{8}$$

$$h = 5 \text{ cm} = 5 \times 10^{-2} \text{ m}$$

$$P_{mep} = h \times \text{Spring constant}$$

$$= 5 \times 10^{-2} \times 1.5 \times 10^8 \text{ Pa/m}$$

$$= 7500 \text{ kPa}$$

$$\text{Power} = \frac{P_{mep} \times L \times A \times N \times K}{60}$$

$$= \frac{7500 \times 0.2 \times \frac{\pi}{4} \times (0.15)^2 \times 100 \times 2}{60}$$

$$\simeq 88.357 \text{ kW}$$

2.81 (b)

- In case of centrifugal pumps, the rotating part is called impeller.
- In case of turbine, the rotating part is called runner.
- Volute casing is spiral type in which area of flow increases gradually.
- In vortex casing, a vortex chamber is introducing between the casing and impeller.
- Foot valve is fitted at the lower end of suction pipe.

2.82 (b)

- The vertical distance between the centreline of the pump and the water surface in the tank to which water is delivered is known as delivery head.
- The sum of suction head and delivery head is known as static head.
- The manometric head is defined as the head against which centrifugal pump has to work.

2.83 (c)

- The ratio of the manometric head to the head imparted by the impeller to the water is known as manometric efficiency.
- The ratio of the power available at the impeller to the power at the shaft of the centrifugal pump is known as mechanical efficiency.
- The ratio of the power output of the pump to the power input to the pump is known as overall efficiency.

2.84 (d)

Given: $V_{F2} = 5$ m/s, $V_{w2} = 12$ m/s

At the outlet of an impeller,

$$V_2 = \sqrt{V_{w2}^2 + V_{F2}^2} = \sqrt{12^2 + 5^2} = \sqrt{13^2} = 13 \text{ m/s}$$

■■■■

3 Miscellaneous Hydraulic Machines

3.1 Consider the following statements regarding a torque converter:

1. Its maximum efficiency is less than that of the fluid coupling.
2. It has two runners and a set of stationary vanes interposed between them.
3. It has two runners.
4. The ratio of secondary to primary torque is zero for the zero value of angular velocity of secondary.

Which of these statements are correct?

(a) 1 and 2 (b) 3 and 4
(c) 1 and 4 (d) 2 and 4

[ESE : 2000]

3.2 In a fluid coupling, the torque transmitted is 50 kNm, when the speed of the driving and driven shaft is 900 rpm and 720 rpm respectively. The efficiency of the fluid coupling will be

(a) 20% (b) 25%
(c) 80% (d) 90%

[ESE : 2001]

3.3 Consider the following statements regarding the fluid coupling :

1. Efficiency increases with increase in speed ratio.
2. Neglecting friction the output torque is equal to input torque.
3. At the same input speed, higher slip requires higher input torque.

Which of these statements are correct?

(a) 1, 2 and 3 (b) 1 and 2
(c) 2 and 3 (d) 1 and 3

[ESE : 2001]

3.4 A hydraulic coupling transmits 1 kW of power at an input speed of 200 rpm, with a slip of 2%. If the input speed is changed to 400 rpm, the power transmitted with the same slip is

(a) 2 kW (b) 1/2 kW
(c) 4 kW (d) 8 kW

[ESE : 2001]

3.5 Fluid flow machines are using the principle of either (i) supplying energy to the fluid, or (ii) extracting energy from the fluid. Some fluid flow machines are combination of both (i) and (ii) they are classified as

(a) compressors (b) hydraulic turbines
(c) torque converters (d) wind mills

[ESE : 2002]

3.6 A hydraulic press has a ram of 20 cm diameter and a plunger of 5 cm diameter. The force required at the plunger to lift a weight of 16×10^4 N shall be

(a) 256×10^4 N (b) 64×10^4 N
(c) 4×10^4 N (d) 1×10^4 N

[ESE : 2002]

3.7 **Assertion (A):** A hydraulic ram is a device used to lift water from deep wells.

Reason (R): Hydraulic ram works on the principle of water hammer.

(a) Both **A** and **R** are individually true and **R** is the correct explanation of **A**
(b) Both **A** and **R** are individually true but **R** is not the correct explanation of **A**
(c) **A** is true but **R** is false
(d) **A** is false but **R** is true

[ESE : 2002]

3.8 An accumulator is a device to store

(a) sufficient quantity of liquid to compensate the change in discharge
(b) sufficient energy to drive the machine when the normal energy source does not function
(c) sufficient energy in case of machines which work intermittently to supplement the discharge from the normal source

(d) liquid which otherwise would have gone to waste

[ESE : 2003]

3.9 Which one of the following pairs is not correctly matched?

(a) Centrifugal pump : Rotating blades in the motor create centrifugal head

(b) Reciprocating pump : Positive displacement pump

(c) Turbine pump : Centrifugal pump with guide vanes

(d) Gear pump : Gear teeth work like rotating blades to create centrifugal head

[ESE : 2004]

3.10 The function of which of the following hydraulic devices is analogous to that of the flywheel of a reciprocating engine and an electric storage battery?

(a) Hydraulic ram

(b) Hydraulic accumulator

(c) Hydraulic intensifier

(d) Hydraulic jack

[ESE : 2005]

3.11 In a jet pump

(a) kinetic energy of fluid is converted into potential energy

(b) energy of high velocity stream is converted into pressure energy

(c) energy of high pressure fluid is converted into energy of low pressure fluid

(d) potential energy of fluid is converted into kinetic energy

[ESE : 2006]

3.12 In a hydraulic coupling, what is the ratio of speed of the turbine runner to that of the pump impeller to maintain circulatory motion of oil?

(a) < 1 (b) = 1

(c) > 1 (d) Can be any value

[ESE : 2007]

3.13 A turbine pump is basically a centrifugal pump equipped additionally with

(a) backward curved blades

(b) vaned diffusion casing

(c) inlet guide blades

(d) adjustable blades

[ESE : 2008]

3.14 Which one of the following is correct for hydraulic coupling?

(a) It connects two shafts rotating at about the same speed

(b) It connects two shafts running at different speeds

(c) It is used to augment the torque to the driven shaft

(d) It is used to connect the centrifugal pump and its electrical motor for efficient operation

[ESE : 2008]

3.15 Which one of the following graphs represents the characteristics of a torque converter ? (Where suffix r stands for turbine runner and P stands for pump impeller)

(a)

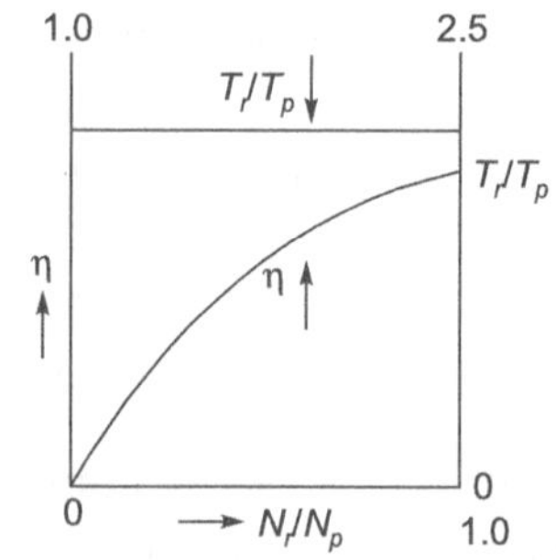

(b)

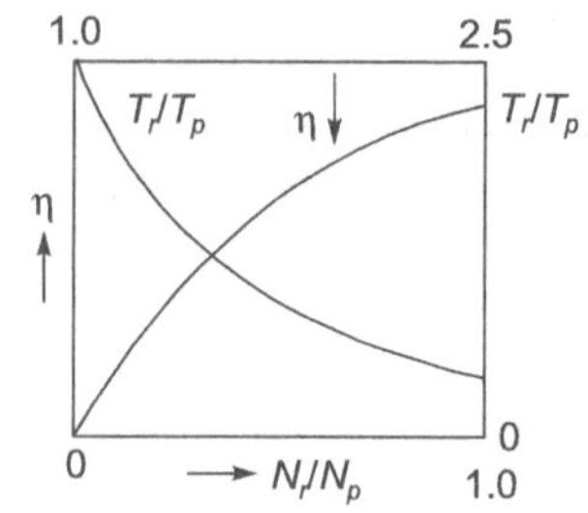

(c)

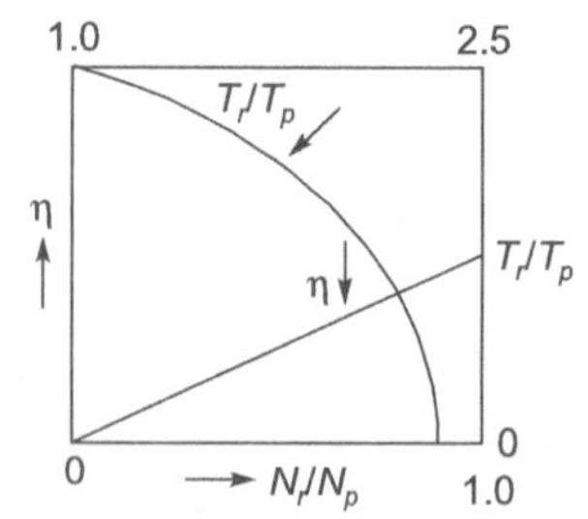

(d)

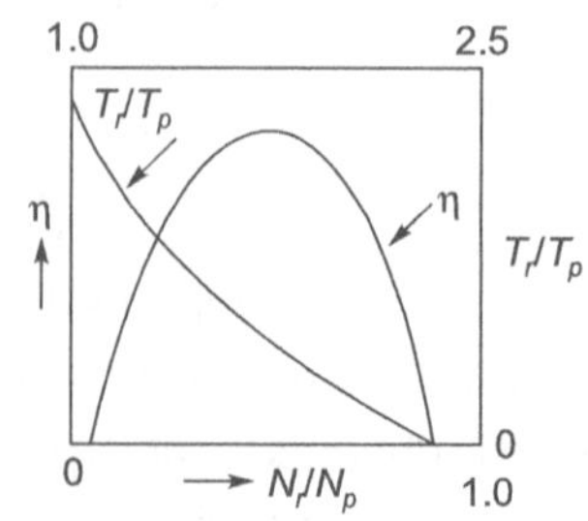

[ESE : 2009]

3.16 Which one of the following combination represents the power transmission systems?
(a) Pump, hydraulic accumulator, hydraulic intensifier and hydraulic coupling
(b) Pump, turbine, hydraulic accumulator and hydraulic coupling
(c) Turbine, accumulator, intensifier and hydraulic coupling
(d) Accumulator, intensifier, hydraulic coupling and torque converter

[ESE : 2009]

3.17 The component of torque converter that allows torque multiplication is
(a) turbine (b) impeller
(c) stator (d) freewheel

[ESE : 2013]

3.18 Jet pumps are often used in process industry for their
(a) larger capacity
(b) high efficiency
(c) capacity to transport gases, liquids and mixtures of both
(d) none of the above

[ESE : 2015]

3.19 In a hydraulic coupling
(a) the magnitudes of input and output torques are equal
(b) the magnitude of input torque is greater than output torque
(c) the magnitude of input torque is less than output torque
(d) the magnitude of input torque is negligible as compared to output torque

[ESE : 2015]

3.20 The condition for power transmission by flow through a pipeline to be maximum is that the loss of head of the flow due to friction throughout the pipeline length is :
(a) one-third of the total head at inlet end
(b) one-fourth of the total head at inlet end
(c) three-fourth of the total head at inlet end
(d) one-half of the total head at inlet end

[ESE : 2016]

3.21 Consider the following statements:
1. The distinguishing features of a radial flow reaction turbine are— (i) only a part of the total head of water is converted into velocity head before it reaches the runner and (ii) the flow-through water completely fills all the passages in the runner.
2. Kaplan turbine is essentially a propeller working in reverse, and its blades are so mounted that all the blade angles can be adjusted simultaneously by means of suitable gearing even as the machine is in operation.
3. A draft tube is a pipe of gradually increasing cross-sectional area which must be airtight, and under all conditions of operation, its lower end must be submerged below the level of the discharged water in the tailrace.

Which of the above statements are correct?
(a) 1 and 2 only (b) 1 and 3 only
(c) 2 and 3 only (d) 1, 2, and 3

[ESE : 2018]

3.22 The specific speed of a turbine is the speed of an imaginary turbine, identical with the given turbine, which
(a) delivers unit discharge under unit speed
(b) delivers unit discharge under unit head
(c) develops unit discharge under unit speed
(d) develops unit power under unit head

[ESE : 2018]

3.23 According to aerofoil theory, the guide angle of Kaplan turbine blades is defined as the angle between
(a) lift and resultant force
(b) drag and resultant force
(c) lift and tangential force
(d) lift and drag

[ESE : 2018]

3.24 The specific speed of a hydraulic turbine depends on
(a) speed and power developed
(b) speed and water head
(c) discharge and power developed
(d) speed, head and power developed

[ESE : 2018]

3.25 A one fourth scale model of a pump was tested in a laboratory at 1000 rpm. The head developed and power input at the best efficiency point were found to be 7 m and 25 kW respectively. What is the speed of the prototype if the prototype pump has to operate against a head of 36 m?

(a) 467 rpm (b) 567 rpm
(c) 667 rpm (d) 767 rpm

[ESE : 2022]

■■■■

Answers Miscellaneous Hydraulic Machines

3.1	(b)	3.2	(c)	3.3	(a)	3.4	(a)	3.5	(c)	3.6	(d)	3.7	(d)	3.8	(c)	3.9	(d)
3.10	(b)	3.11	(b)	3.12	(a)	3.13	(b)	3.14	(b)	3.15	(d)	3.16	(d)	3.17	(c)	3.18	(c)
3.19	(a)	3.20	(a)	3.21	(d)	3.22	(d)	3.23	(a)	3.24	(d)	3.25	(b)				

Explanations Miscellaneous Hydraulic Machines

3.2 (c)

$$\text{Efficiency}: \eta = \frac{N_2}{N_1} = \frac{720}{900} = \mathbf{0.80}$$

3.3 (a)

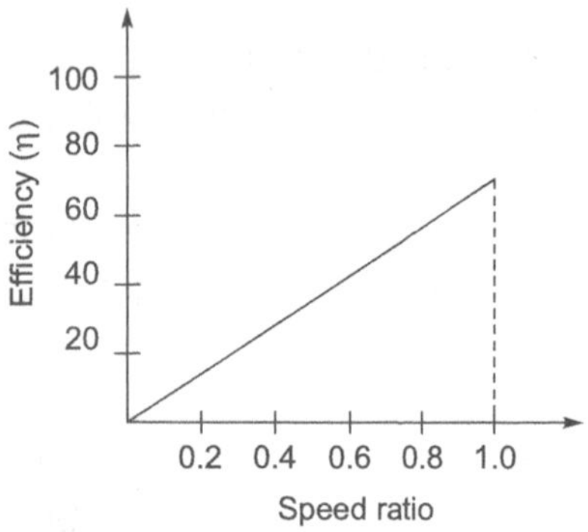

Efficiency-speed ratio curve for a fluid coupling. The efficiency of fluid coupling is large, usually in excess of 94 percent. A typical efficiency versus speed ratio curve for a fluid coupling is shown in figure. The efficiency of the fluid coupling starts at zero and increases uniformly with the speed ratio until $\eta = 95$ percent and then reduces to zero.

As the torque transmitted increases, the slip increases and the efficiency drops.

3.4 (a)

Same slip means same torque

$$\Rightarrow \quad \frac{P}{\omega} = \text{constant}$$

$$\therefore \quad P_2 = P_1\frac{\omega_2}{\omega_1} = 1\times\frac{400}{200}$$

$$= 2 \text{ kW}$$

3.5 (c)

Torque convertor acts as a liquid coupling. It transmit power/energy by means of fluid.

3.6 (d)

$$\frac{F_1}{A_1} = \frac{F_2}{A_2}$$

$$\Rightarrow \quad \frac{16\times10^4}{\pi\times(20)^2} = \frac{F_2}{\pi(5)^2}$$

$$\therefore \quad F_2 = \mathbf{1\times10^4\ N}$$

3.7 (d)

The hydraulic ram is a type of pump in which the energy of large quantity of water falling through small height is utilized to lift a small quantity of this water to greater height. No external power is, therefore, required to operate this pump.

The working of a hydraulic ram is based on the principle of water hammer or inertia pressure developed in the supply pipe.

3.8 (c)

An accumulator is a device to store sufficient energy in case of machines which work intermittently to supplement the discharge from the normal source.

3.9 (d)

In gear pump, as the gear rotate they separate on the intake side of pump, creating a void and suction which is filled by fluid. The fluid is carried by the gears to the discharge side of pump, where the meshing of gear displaces the fluid. It has the ability to pump highly viscous fluid.

3.10 (b)

Hydraulic accumulator is a device which stores the pump delivery when it is not required by the system and makes available to the system whenever required.

3.12 (a)

Circulatory motion of oil is required to transit power from pump impeller to turbine runner.

Efficiency of coupling

$$= \frac{\text{Output power of turbine}}{\text{Input power of pump}}$$

$$= \frac{T\omega_t}{T\omega_p} = \frac{\omega_t}{\omega_p}$$

[∵ Torque is constant in hydraulic coupling]

Because efficiency will be always less than 1, thus

$$\frac{\omega_t}{\omega_p} < 1$$

3.13 (b)

A turbine pump in basically a centrifugal pump equipped additionally with vaned diffusion casing.

3.14 (b)

A hydraulic coupling connects two shafts running at different speeds.

3.16 (d)

Turbine and pump convert one form of energy to another form, thus not a power transmission system.

3.17 (c)

A hydraulic torque converter consists of series of fixed guide values which are provided between turbine runner and pump impeller. The liquid exerts a torque on the driven unit through these stationary guide vanes as it flows from pump to turbine runner.

3.19 (a)

In hydraulic couplings, input and output torque are equal but speeds of output shaft is less than input shaft. It is torque converter where output torque can be less or more than input.

3.21 (d)

- In reaction turbine, head is partially converted into velocity.
- In Kaplan turbine, the orientation of moving blade angle can also be adjusted.
- The draft tube is always immersed in tail race.

3.24 (d)

$$N_S = \frac{N\sqrt{P}}{H^{5/4}}$$

3.25 (b)

$$\frac{D_m}{D_P} = \frac{1}{4}$$

Model	Prototype
$N = 1000$ rpm	$N = ?$
$H = 7$ m	$H = 36$
$P = 24$ kW	

$$\left.\frac{H}{D^2N^2}\right|_M = \left.\frac{H}{D^2N^2}\right|_P$$

$$N_P^2 = \frac{H_P}{H_m}\times\left(\frac{D_m}{D_P}\right)^2 \times N_m^2$$

$$= \frac{36}{7}\times\left(\frac{1}{4}\right)^2 \times 1000^2$$

$$N_P = \frac{6}{\sqrt{7}}\times\frac{1}{4}\times 1000$$

$$= \frac{1500}{\sqrt{7}} = \frac{1500}{2.64}$$

$$N_P = 567 \text{ rpm}$$

■■■■

4 Air Compressors (Reciprocating and Rotary)

4.1 The capacity of an air compressor is specified as 3 m^3/min. It means that the compressor is capable of

(a) supplying 3 m^3 of compressed air per minute
(b) compressing 3 m^3 of free air per minute
(c) supplying 3 m^3 of compressed air at NTP
(d) compressing 3 m^3 of standard air per minute

[ESE : 2000]

4.2 A two-stage compressor takes in air at 1.1 bars and discharges at 20 bars. For maximum efficiency, the intermediate pressure is

(a) 10.55 bars (b) 7.33 bars
(c) 5.5 bars (d) 4.7 bars

[ESE : 2000]

4.3 Which one of the following pairs of features and compressors type is not correctly matched?

(a) Vane compressor : Intake and delivery ports compression is attained by back flow and internal compression cylindrical rotor set to eccentric casing
(b) Reciprocating compressor : Intermittent discharge requires receiver, produces high pressure, slow speed and lubrication problems
(c) Centrifugal compressor : Continuous flow, radial flow, handles large volume, much higher speed and fitted into design of aeroengines
(d) Axial flow compressor : Successive pressure drops through contracting passages, blades are formed from a number of circular arcs, axial flow

[ESE : 2000]

4.4 Consider the following statements:
The volumetric efficiency of a reciprocating compressor can be enhanced by

1. heating the intake air.
2. decreasing the clearance volume.
3. cooling the intake air.

Which of these statements is/are correct?

(a) 1 alone (b) 1 and 2
(c) 2 and 3 (d) 3 alone

[ESE : 2000]

4.5 Reciprocating compressors are provided with

(a) simple disc/plate valve
(b) poppet valve
(c) spring-loaded disc valve
(d) solenoid valve

[ESE : 2000]

4.6 Consider the following statements:
In centrifugal compressors, there is a tendency of increasing surge when

1. the number of diffuser vanes is less than the number of impeller vanes.
2. the number of diffuser vanes is greater than the number of impeller vanes.
3. the number of diffuser vanes is equal to the number of impeller vanes.
4. mass flow is greatly in excess of that corresponding to the design mass flow.

Which of these statements is/are correct?

(a) 1 and 4 (b) 2 alone
(c) 3 and 4 (d) 2 and 4

[ESE : 2000]

4.7 In an axial flow compressor design, velocity diagrams are constructed from the experimental data of aerofoil cascades. Which one of the following statements in this regard is correct?

(a) Incidence angle of the approaching air is measured from the trailing edge of the blade
(b) δ is the deviation angle between the angle of incidence and tangent to the camber line
(c) The deflection ε of the gas stream while passing through the cascade is given by $\varepsilon = \alpha_1 - \alpha_2$
(d) ε is the sum of the angle of incidence and camber less any deviation angle, i.e., $\varepsilon = i + \theta - \delta$

[ESE : 2000]

4.8 If n is the polytropic index of compression and p_2/p_1 is the pressure ratio for a three-stage compressor with ideal intercooling, the expression for total work of three stage is

(a) $\frac{3n}{(n-1)} p_1 V_1 \left\{ \left(\frac{p_2}{p_1} \right)^{(n-1)/n} - 1 \right\}$

(b) $\frac{n}{(n-1)} p_1 V_1 \left\{ \left(\frac{p_2}{p_1} \right)^{(n-1)/3n} - 1 \right\}$

(c) $\frac{n}{(n-1)} p_1 V_1 \left\{ \left(\frac{p_2}{p_1} \right)^{(n-1)/n} - 1 \right\}$

(d) $\frac{3n}{(n-1)} p_1 V_1 \left\{ \left(\frac{p_2}{p_1} \right)^{(n-1)/3n} - 1 \right\}$

[ESE : 2001]

4.9 The flow in the vaneless space between the impeller exit and diffuser inlet of a centrifugal compressor can be assumed as

(a) free vortex
(b) forced vortex
(c) solid body rotation
(d) logarithmic spiral [ESE : 2001]

4.10 Which portion of the centrifugal compressor characteristics shown in the figure is difficult to obtain experimentally?

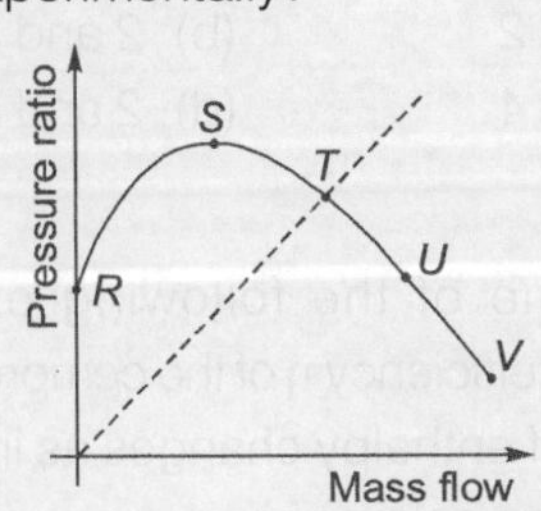

(a) RS (b) ST
(c) TU (d) UV

[ESE : 2001]

4.11 Consider the following statements regarding the axial flow in an air compressor:

1. Surging is a local phenomenon while stalling affects the entire compressor.
2. Stalling is a local phenomenon while surging affects the entire compressor.
3. The pressure ratio of an axial compressor stage is smaller than that of a centrifugal compressor stage.

Which of these statements are correct?

(a) 1, 2 and 3 (b) 1 and 2
(c) 2 and 3 (d) 1 and 3

[ESE : 2001]

4.12 The pressure rise in the impeller of centrifugal compressor is achieved by

(a) the decrease in volume and diffusion action
(b) the centrifugal action and decrease in volume
(c) the centrifugal and diffusion action
(d) the centrifugal and push-pull action

[ESE : 2002]

4.13 Compared to axial compressors centrifugal compressors are more suitable for

(a) high head, low flow rate
(b) low head, low flow rate
(c) low head, high flow rate
(d) high head, high flow rate

[ESE : 2002]

4.14 Stalling of blades in axial-flow compressor is the phenomenon of

(a) air steam blocking the passage
(b) motion of air at sonic velocity
(c) unsteady, periodic and reversed flow
(d) air stream not able to follow the blade contour

[ESE : 2002]

4.15 The inlet and exit velocity diagrams of a turbomachine rotor are shown.

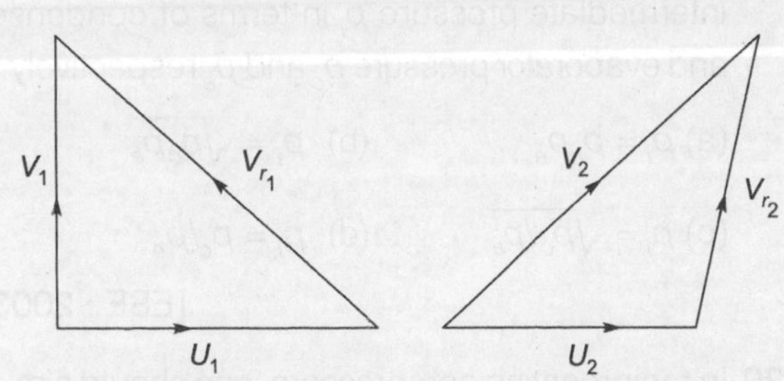

The turbomachine is:

(a) an axial compressor with backward curved blades
(b) a radial compressor with backward curved blades
(c) a radial compressor with forward curved blades
(d) an axial compressor with forward curved blades

[ESE : 2002]

4.16 In an axial flow compressor:
α_1 = exit angle from stator
β_1 = inlet angle to rotor
α_2 = inlet angle to stator
β_2 = outlet angle from rotor
The condition to have a 50% degree of reaction is
(a) $\alpha_1 = \beta_2$
(b) $\alpha_2 = \beta_1$
(c) $\alpha_1 = \beta_2$ & $\beta_1 = \alpha_2$
(d) $\alpha_1 = \alpha_2$ & $\beta_1 = \beta_2$
[ESE : 2002]

4.17 The ratio of the clearance volume to the displacement volume of a R12 reciprocating compressor is 0.05. Specific volume at inlet and outlet of compressor are 0.04 and 0.02 m^3/kg respectively. Volumetric efficiency of the compressor is
(a) 95.0% (b) 47.5%
(c) 38.0% (d) 19.0%
[ESE : 2002]

4.18 Roots blower is an example of
(a) Reciprocating (positive displacement) compressor
(b) Rotary (positive displacement) compressor
(c) Centrifugal compressor
(d) Axial compressor [ESE : 2003]

4.19 In a two-stage compressor with ideal intercooling, for the work requirement to be minimum, the intermediate pressure p_i in terms of condenser and evaporator pressure p_c and p_e respectively is
(a) $p_i = p_c p_e$ (b) $p_i = \sqrt{p_c p_e}$
(c) $p_i = \sqrt{p_c / p_e}$ (d) $p_i = p_c / p_e$
[ESE : 2003]

4.20 In reciprocating compressors, one should aim at compressing the air
(a) adiabatically (b) isentropically
(c) isothermally (d) polytropically
[ESE : 2004]

4.21 Which one of the following is the correct expression for the degree of reaction for an axial-flow air compressor?
(a) $\dfrac{\text{Work input to the rotor}}{\text{Work input to the stage}}$
(b) $\dfrac{\text{Change of enthalpy in the rotor}}{\text{Change of enthalpy in the stage}}$
(c) $\dfrac{\text{Pressure rise in the rotor}}{\text{Pressure rise in the stage}}$
(d) $\dfrac{\text{Isentropic work}}{\text{Actual work}}$ [ESE : 2004]

4.22 For a two-stage reciprocating air compressor, the suction pressure is 1.5 bar and the delivery pressure is 54 bar. The value of the ideal intercooler pressure is
(a) 6 bar (b) 9 bar
(c) 27.75 bar (d) $9/\sqrt{2}$ bar
[ESE : 2004]

4.23 Consider the following statements with reference to supercharging of I.C. Engine
1. Reciprocating compressor are invariably used for high degree of supercharging.
2. Rotary compressor like roots blower are quite suitable for low degree of supercharging.
3. Axial flow compressor are most commonly employed for supercharging diesel engine used in heavy duty transport vehicles.
4. Centrifugal compressor are used for turbo-charging.

Which of these statements are correct?
(a) 1 and 2 (b) 2 and 3
(c) 1 and 4 (d) 2 and 4
[ESE : 2004]

4.24 Which one of the following expresses the isentropic efficiency η of the compression process in terms of enthalpy changes as indicated in the figure given below?

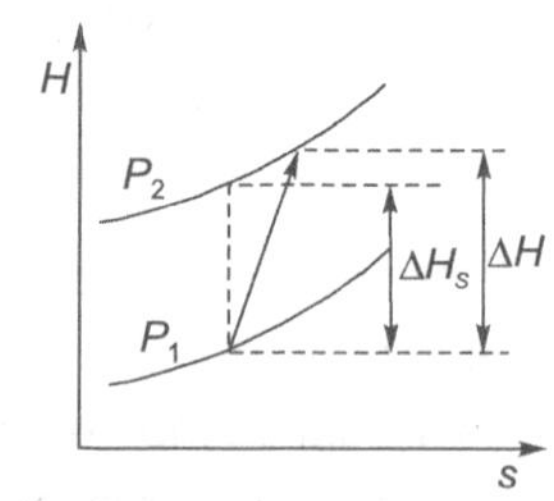

(a) $\eta = \Delta H_s / \Delta H$
(b) $\eta = \Delta H / \Delta H_s$
(c) $\eta = (\Delta H - \Delta H_s)/\Delta H$
(d) $\eta = (\Delta H - \Delta H_s)/\Delta H_s$ [ESE : 2005]

4.25 While flowing through the rotor blades in an axial flow air compressor, the relative velocity of air
(a) continuously decreases
(b) continuously increases
(c) first increases and then decreases
(d) first decreases and then increases
[ESE : 2005]

4.26 For centrifugal compressors, which one of the following is the correct relationship between pressure coefficient (ϕ_p), slip factor (ϕ_s), work input factor (ϕ_w) and isentropic efficiency (η_a)?
(a) $\phi_p = \dfrac{\phi_s \times \phi_w}{\eta_a}$ (b) $\phi_p = \dfrac{\phi_w}{\phi_s \times \eta_a}$
(c) $\phi_p = \phi_s \times \phi_w \times \eta_a$ (d) $\phi_p = \dfrac{\phi_s \times \eta_a}{\phi_w}$
[ESE : 2005]

Direction: Each of the next qustions consists of two statements, one labelled as the **'Assertion (A)'** and the other as **'Reason (R)'**. You are to examine these two statements carefully and select the answers to these items using the codes given below:

Codes:
(a) Both **A** and **R** are individually true and **R** is the correct explanation of **A**
(b) Both **A** and **R** are individually true but **R** is not the correct explanation of **A**
(c) **A** is true but **R** is false
(d) **A** is false but **R** is true

4.27 Assertion (A): A vane type rotary compressor is a rotodynamic machine.
Reason (R): A rotodynamic machine is one in which a fluid flows freely through the rotating part of the machine. **[ESE : 2005]**

4.28 Assertion (A): In multi-stage compressors, the polytropic efficiency is always greater than the isentropic efficiency.
Reason (R): Higher the pressure ratio, the greater is the polytropic efficiency. **[ESE : 2005]**

4.29 In a radial blade centrigugal compressor, the velocity of blade tip is 400 m/s and slip factor is 0.9. Assuming the absolute velocity at inlet to be axial, the work done per kg of flow is
(a) 36 kJ (b) 72 kJ
(c) 144 kJ (d) 360 kJ **[ESE : 2005]**

4.30 Which of the following statements are correct for multi-staging in a reciprocating air compressor?
1. It decreases the volumetric efficiency.
2. The work done can be reduced.
3. A small high-pressure cylinder is required.
4. The size of flywheel is reduced.

Select the correct answer using the code given below:
(a) 1, 2 and 3 (b) 2, 3 and 4
(c) 1, 3 and 4 (d) 1, 2 and 4
[ESE : 2006]

4.31 Consider the following statements:
For a large aviation gas turbine, an axial flow compressor is usually preferred over centrifugal compressor because
1. the maximum efficiency is higher.
2. the frontal area is lower.
3. the pressure rise per stage is more.
4. the cost is lower.

Which of these statements are correct?
(a) 1 and 4 (b) 1 and 2
(c) 1, 2 and 3 (d) 2, 3 and 4
[ESE : 2006]

4.32 In an axial flow compressor, when the degree of reaction is 50%, it implies that
(a) Work done in compression will be the least
(b) 50% stages of the compressor will be ineffective
(c) Pressure after compression will be optimum
(d) The compressor will have symmetrical blades
[ESE : 2006]

4.33 In a centrifugal compressor, how can the pressure ratio be increased?
(a) Only by increasing the tip speed
(b) Only by decreasing the inlet temperature
(c) By both (a) and (b)
(d) Only by increasing the inlet temperature
[ESE : 2006]

4.34 What is the preferred intercooler pressure for a two stage air compressor working between the suction pressure p_s and the delivery pressure p_d?
(a) $(p_s + p_d)/2$ (b) $(p_d - p_s)/2$
(c) $(p_s \times p_d)^{1/2}$ (d) $(p_s \times p_d)^{1/4}$
[ESE : 2006]

4.35 Consider the following statements:
Volumetric efficiency of a reciprocating air compressor increases with
1. increase in clearance ratio.
2. decrease in delivery pressure.
3. multi-staging.

Which of these statements is/are correct?
(a) Only 1 and 2 (b) Only 2 and 3
(c) Only 3 (d) 1, 2 and 3

[ESE : 2006]

4.36 What does application of centrifugal air compressors lead to?
(a) Large frontal area of aircraft
(b) Higher flow rate through the engine
(c) Higher aircraft speed
(d) Lower frontal area of the aircraft

[ESE : 2006]

4.37 The power required to drive a turbo-compressor for a given pressure ratio decreases when
(a) Air is heated at entry
(b) Air is cooled at entry
(c) Air is cooled at exit
(d) Air is heated at exit

[ESE : 2006]

4.38 Which of the following can be the cause/causes of an air-cooled compressor getting overheated during operation?
1. Insufficient lubricating oil.
2. Broken valve strip.
3. Clogged intake filter.

Select the correct answer using the code given below:
(a) Only 3 (b) Only 1 and 2
(c) Only 2 and 3 (d) 1, 2 and 3

[ESE : 2006]

4.39 Match **List-I** (Name of Equipment) with **List-II** (Pressure Ratio) and select the correct answer using the code given below the lists:

	List-I		List-II
A.	Fan	1.	1.1
B.	Blower	2.	2.5
C.	Centrifugal air compressor	3.	4
D.	Axial flow air compressor	4.	10

Codes:

	A	B	C	D
(a)	2	1	3	4
(b)	1	2	3	4
(c)	1	2	4	3
(d)	2	1	4	3

[ESE : 2007]

4.40 Which of the following techniques are employed for control of reciprocating compressors?
1. Throttle control
2. Clearance control
3. Blowing air to waste

Select the correct answer using the codes given below:
(a) 1, 2 and 3 (b) 1 and 2 only
(c) 2 and 3 only (d) 1 and 3 only

[ESE : 2007]

4.41 Consider the following statements in respect of axial flow air compressors:
1. An axial flow air compressor is often described as a reversed reaction turbine.
2. With 50% degree of reaction, the velocity diagrams are symmetrical.

Which of these statements is/are correct?
(a) 1 only (b) 2 only
(c) Both 1 and 2 (d) Neither 1 nor 2

[ESE : 2007]

4.42 Which one of the following statements is correct for reciprocating air compressor?
(a) Its volumetric efficiency increases with increasing clearance ratio
(b) Its volumetric efficiency increases with increasing pressure ratio
(c) Its volumetric efficiency does not vary with change in clearance ratio and pressure ratio
(d) Its volumetric efficiency decreases with increasing clearance ratio and pressure ratio, both **[ESE : 2007]**

4.43 Stalling phenomena in an axial flow compressor stage is caused due to which one of the following?
(a) Higher mass flow rate than the designed value
(b) Lower mass flow rate than the designed value
(c) Higher mass flow rate or non-uniformity in the blade profile
(d) Lower mass flow rate or non-uniformity in the blade profile **[ESE : 2007]**

4.44 In a multi-stage axial flow compressor with equal temperature rise in all stages, the pressure ratio in the subsequent stages
(a) Remains constant (b) Increases gradually
(c) Decreases (d) Increases rapidly
[ESE : 2007]

4.45 Which one of the following types of impeller vanes are most commonly used in centrifugal type compressors?
(a) Forward curved (b) Radial
(c) Backward curved (d) Tangential
[ESE : 2008]

4.46 A centrifugal compressor is suitable for which of the following?
(a) High pressure ratio, low mass flow
(b) Low pressure ratio, low mass flow
(c) High pressure ratio, high mass flow
(d) Low pressure ratio, high mass flow
[ESE : 2008]

4.47 The suction pressure is 1 bar and delivery pressure is 125 bar. What is the ideal intermediate pressure at the end of first stage for a 3-stage air compressor?
(a) 25 bar (b) 5 bar
(c) 10 bar (d) 20 bar
[ESE : 2008]

4.48 The ratio of the static enthalpy rise in the rotor to the static enthalpy rise in the stage of an axial flow compressor defined as
(a) Power input factor
(b) Flow coefficient
(c) Temperature coefficient
(d) Degree of reaction
[ESE : 2008]

4.49 Which one of the following statements is correct?
(a) Reciprocating compressors are used to supply large quantities of air at a lower pressure ratio.
(b) Centrifugal compressors are used to supply large quantities of air at a lower pressure ratio.
(c) Centrifugal compressors are used to supply small quantities of air at a lower pressure ratio.
(d) Centrifugal compressors cannot be run at high speed because of impeller, diffuser and casing.
[ESE : 2009]

4.50 Which one of the following graphs shows the correct representation of the processes for a two stage air compressor with perfect intercooling and no pressure drop in the intercooler ?

(a) 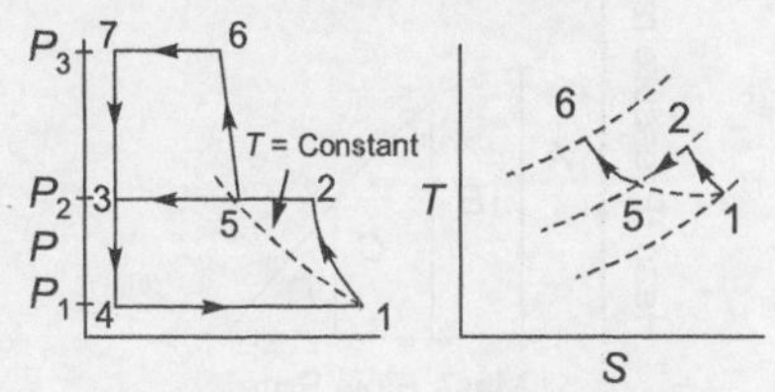

(b) 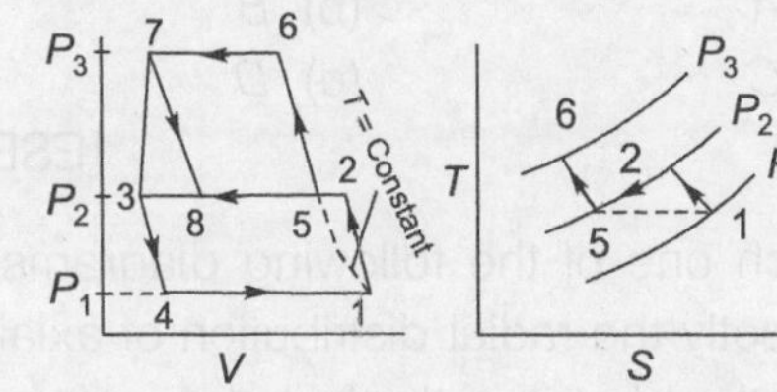

(c) 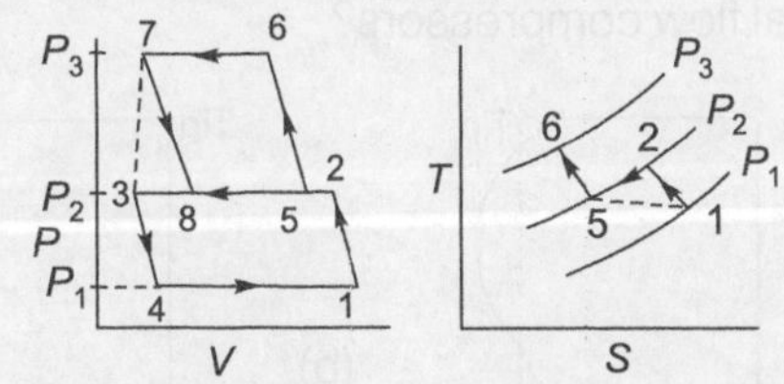

(d) 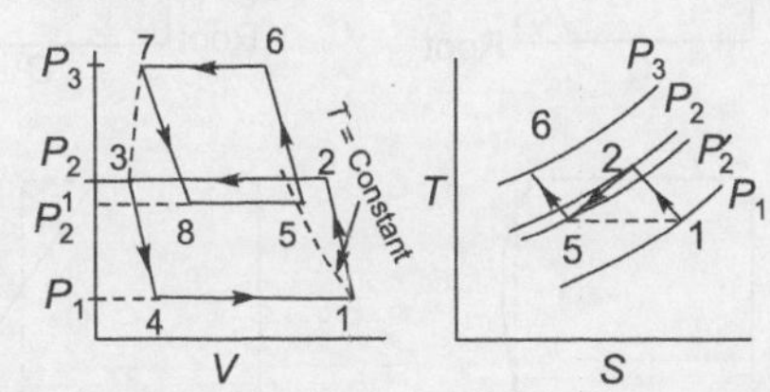

[ESE : 2009]

4.51 Consider the following statements pertaining to axial flow compressors:
1. Like centrifugal compressor, axial flow compressors are limited by surge at low mass flow rates.
2. Axial flow compressors experience choking at low flow rates.
3. The design point of axial flow compressors is close to the surge limit.
4. As mass flow diminishes the compressor blades stall causing flow separation.

Which of these statements is/are correct?
(a) 1 and 2 only (b) 1, 2 and 3
(c) 1, 3 and 4 (d) 3 and 4 only
[ESE : 2009]

4.52 In the graph as shown below, for an axial flow compressor, surging is likely to occur in which one of the following zones?

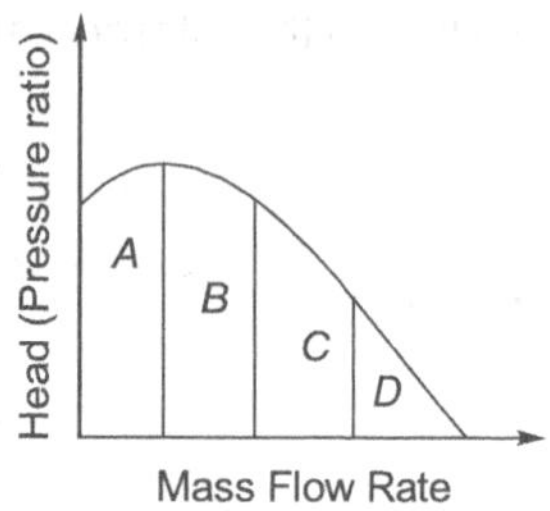

(a) *A* (b) *B*
(c) *C* (d) *D*

[ESE : 2009]

4.53 Which one of the following diagrams depicts correctly the radial distribution of axial velocity over the blades in the last stage of multi-stage axial flow compressors?

(a)
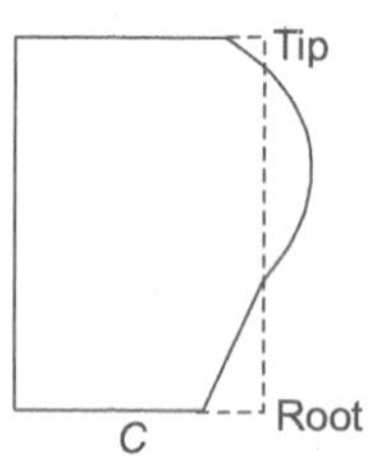

(b)
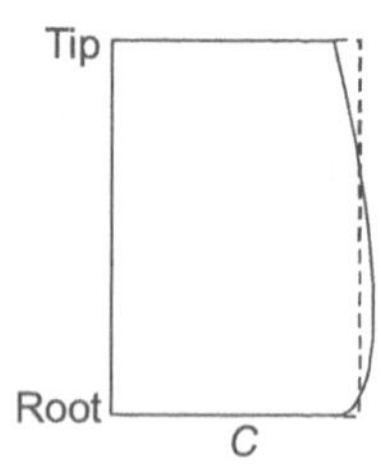

(c)
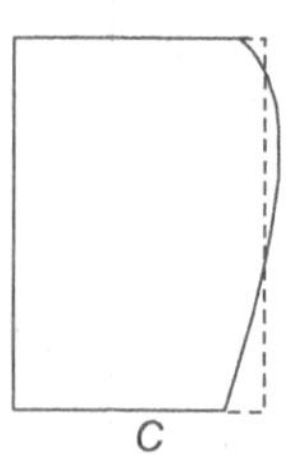

(d)
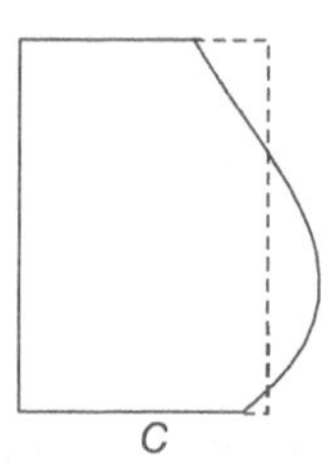

[ESE : 2009]

4.54 Assertion (A) : Multi-staging compression is done only in reciprocating compressors.

Reason (R) : Reciprocating compressors are used to compress large pressure ratio at low discharge.

(a) Both **A** and **R** are individually true and **R** is the correct explanation of **A**
(b) Both **A** and **R** are individually true but **R** is not the correct explanation of **A**
(c) **A** is true but **R** is false
(d) **A** is false but **R** is true

[ESE : 2009]

4.55 Consider the following statements:

1. $p_2 = \dfrac{p_1 + p_3}{2}$
2. $p_2 = \sqrt{p_1 p_3}$
3. Intercooling is perfect.
4. Work in the LP cylinder is twice the work in the HP cylinder.

Which of these statements are correct for maximum compression efficiency in a 2-stage compressor?

(a) 1 and 3 only (b) 2 and 3 only
(c) 2 and 4 only (d) 1, 3 and 4

[ESE : 2010]

4.56 The following statements refer to axial flow compressors as compared to centrifugal compressors:

1. Can be designed for higher pressure ratios than centrifugal compressors.
2. Maximum efficiency higher than centrifugal compressors but in a narrow speed range.
3. More suitable for aviation gas turbines due to lower frontal area.
4. Lighter in mass than centrifugal compressors.

Which of these statements are correct?

(a) 1 and 3 only (b) 2 and 3 only
(c) 1, 2 and 3 only (d) 1, 2, 3 and 4

[ESE : 2010]

4.57 Consider the following statements regarding a reciprocating air compressor:

1. The effect of clearance is to reduce the volumetric efficiency.
2. The clearance has no effect on work done per kg of air delivered.
3. The volumetric efficiency decreases with increasing pressure ratio.

Which of these statements is/are correct?

(a) 1, 2 and 3 (b) 1 and 2 only
(c) 2 and 3 only (d) 3 only

[ESE : 2010]

4.58 In a two-stage reciprocating air compressor, the suction and delivery pressures are 1 and 4 bar respectively. For maximum efficiency, the intercooler pressure is

(a) 1.5 bar (b) 2.0 bar
(c) 2.5 bar (d) 3.0 bar

[ESE : 2010]

4.59 **Assertion (A):** The performance parameter 'Polytropic efficiency' is used for axial flow gas turbines and air compressors.

Reason (R): Polytropic efficiency is dependent on the pressure ratio.

(a) Both A and R are individually true and R is the correct explanation of A
(b) Both A and R are individually true but R is not the correct explanation of A
(c) A is true but R is false
(d) A is false but R is true

[ESE : 2010]

4.60 Which of the following statements is not correct for the volumetric efficiency of a reciprocating air compressor?

(a) It decreases with increase in ambient temperature
(b) It increases with decrease in pressure ratio
(c) It increases with decrease in clearance ratio
(d) It decreases with increase in delivery pressure [ESE : 2011]

4.61 In a single stage reciprocating air compressor, the work done on air to compress it from suction pressure to delivery pressure will be minimum when the compression is

(a) Isothermal process
(b) Adiabatic process
(c) Polytropic process
(d) Constant pressure process

[ESE : 2011]

4.62 In an axial flow compressor, stalling is the phenomenon of

(a) Air motion at sonic velocity
(b) Air stream blocking the passage
(c) Reversed flow of air
(d) Air stream not following the blade contour

[ESE : 2011]

4.63 **Assertion (A)** : In centrifugal compressors, sometimes guide vanes are provided at inlet.

Reason (R) : The guide vanes provide prewhirl which helps in restricting the Mach number at inlet to an acceptable value, below supersonic.

(a) Both A and R are individually true and R is the correct explanation of A
(b) Both A and R are individually true but R is not the correct explanation of A
(c) A is true but R is false
(d) A is false but R is true

[ESE : 2011]

4.64 For minimum work input in a two-stage compression process the intermediate pressure is the

(a) Arithmetic mean of suction and discharge pressure
(b) Logarithmic mean of suction and discharge pressure
(c) Geometric mean of suction and discharge pressure
(d) hyperbolic mean of suction and discharge pressure [ESE : 2011]

4.65 The power consumed by a reciprocating air compressor will be minimum, if the compression follows the thermodynamic

(a) isentropic process
(b) adiabatic process
(c) isothermal process
(d) polytropic process [ESE : 2012]

4.66 Losses in a centrifugal compressor are due to

(a) impeller channel losses only
(b) inlet losses only
(c) Both (a) and (b)
(d) Neither (a) nor (b) [ESE : 2012]

4.67 Compressors used in gas turbines are of

(a) reciprocating type
(b) centrifugal type
(c) axial-flow type
(d) All of the above [ESE : 2012]

4.68 Stalling of blades in axial-flow compressors is the phenomenon of

(a) air stream blocking the passage
(b) motion of air at sonic velocity
(c) unsteady periodic and reverse flow
(d) air stream not able to flow over the blade contour [ESE : 2012]

4.69 The multistage compression of air as compared to single-stage compression

(a) improves volumetric efficiency for the given pressure ratio
(b) reduces work done per kg of air

(c) gives more uniform torque
(d) All of the above [ESE : 2012]

4.70 For a 2-stage air compressor for the maximum efficiency of compressor, consider the following statements:

1. $p_2 = \frac{1}{2}(p_1 + p_3)$
2. $p_2 = \sqrt{p_1.p_3}$
3. Intercooling between the stages is complete
4. Work is equally shared by the two stages

Which of these statements are correct?
(a) 1 and 4 only (b) 1, 3 and 4
(c) 2, 3 and 4 (d) 2 and 3 only
[ESE : 2012]

4.71 In the centrifugal compressor the work input is equal to sum of
(a) pressure head, relative head and dynamic head
(b) dynamic head, centrifugal head and relative head
(c) pressure head, centrifugal head and dynamic head
(d) pressure head, centrifugal head and relative head [ESE : 2013]

4.72 For a centrifugal compressor with radial vanes, slip factor is the ratio of
(a) isentropic work to Euler work
(b) whirl velocity to the blade velocity at the impeller exit
(c) stagnation pressure to static pressure
(d) isentropic temperature rise to actual temperature rise [ESE : 2013]

4.73 The specific speed of a centrifugal compressor is generally
(a) less than that of reciprocating compressor
(b) independent of compressor type, but depends only on size of compressor
(c) higher than that of axial compressor
(d) more than specific speed of reciprocating compressor but less than axial compressor
[ESE : 2013]

4.74 In a centrifugal compressor, the highest Mach number leading to shock wave in the fluid flow occurs at
(a) diffuser inlet radius
(b) diffuser outlet radius
(c) impeller inlet radius
(d) impeller outlet radius [ESE : 2013]

4.75 In a centrifugal compressor, an increase in speed at a given pressure ratio causes
(a) increase inflow and increase in efficiency
(b) increase in flow and decrease in efficiency
(c) decrease in flow and increase in efficiency
(d) decrease in flow and decrease in efficiency
[ESE : 2013]

4.76 In an axial flow compressor, the ratio of pressure rise in the rotor blades to the pressure rise in the compressor in one stage is known as
(a) pressure coefficient
(b) work factor
(c) degree of reaction
(d) slip factor [ESE : 2013]

Directions: The following items consist of two statements; one labelled as **'Statement (I)'** and the other as **'Statement (II)'**. You are to examine these two statements carefully and select the answers to these items using the codes given below:

Codes:
(a) Both Statement (I) and Statement (II) are individually true and Statement (II) is the correct explanation of Statement (I).
(b) Both Statement (I) and Statement (II) are individually true but Statement (II) is **NOT** the correct explanation of Statement (I).
(c) Statement (I) is true but Statement (II) is false
(d) Statement (I) is false but Statement (II) is true.

4.77 **Statement (I):** The index of compression and the index of expansion are different in reciprocating compressors.
Statement (II): Reciprocating compressors are used to compress a large pressure ratio and low discharge.
[ESE : 2013]

4.78 **Statement (I)** : The term surge indicates a phenomenon of instability which takes place at low flow values and which involves an entire system including not only the centrifugal compressor, but also the group of components traversed by the fluid upstream and downstream of it.
Statement (II): Choking is defined as separation of fluid from the rotor blades of centrifugal compressor. [ESE : 2014]

4.79 Following points express the effect of keeping high clearance volume for the cylinders in reciprocating compressor. Which one of the following points is disagreeable?

(a) By increasing clearance volume volumetric efficiency decreases
(b) By increasing clearance volume power consumption increases
(c) By increasing the clearance volume, chances of piston striking the cylinder head gets reduced
(d) By increasing clearance volume maximum compression pressure value decreases

[ESE : 2014]

4.80 Consider the following statements:

1. Stalling is the separation of flow from the blade surface.
2. Surging leads to physical damage due to impact loads and high frequency vibration.
3. Mass flow rate is minimum if choking occurs.

Which of the above statements are correct?

(a) 1, 2 and 3 (b) 1 and 3 only
(c) 1 and 2 only (d) 2 and 3 only

[ESE : 2014]

4.81 Which of the following statements are correct?

1. Velocity compounded impulse turbine gives less speed and less efficiency
2. For an ideal centrifugal compressor, the pressure produced depends on impeller velocity and diameter.
3. While flowing through the rotor blades in a gas turbine, the relative velocity of gas continuously decreases
4. While flowing through the rotor blades in an axial flow compressor, the relative velocity of air continuously decreases

Which of the above statements are correct?

(a) 1 and 3 (b) 2 and 3
(c) 1 and 4 (d) 2 and 4

[ESE : 2014]

4.82 What is the power required to drive a centrifugal air compressor, when impeller diameter is 0.45 m and N is 7200 rpm?

(a) 28.78 kW/kg/s (b) 30.78 kW/kg/s
(c) 27.78 kW/kg/s (d) 26.78 kW/kg/s

[ESE : 2014]

4.83 Cooling of reciprocating compressor cylinder:

1. Increases the volumetric efficiency
2. Increases the work input
3. Decreases the volumetric efficiency
4. Decreases the work input

Which of the above statements are correct?

(a) 1 and 2 (b) 2 and 3
(c) 3 and 4 (d) 1 and 4

[ESE : 2014]

4.84 For a multistage reciprocating compressor; which of the following statements are correct?

1. It decreases volumetric efficiency
2. The work of compression is reduced
3. The high pressure cylinder is smaller in size

(a) 1 and 2 only (b) 2 and 3 only
(c) 1 and 3 only (d) 1, 2 and 3

[ESE : 2015]

4.85 Consider the following statements about two-stage reciprocating compressors:

1. For the same pressure ratio the volumetric efficiency of a two-stage compressor is more than that of single stage compressor.
2. A two-stage compressor requires minimum work when inter-cooling is perfect.
3. The intercooler pressure p_2 is the average of inlet pressure p_1, and delivery pressure p_3.

Which of the above statements are correct?

(a) 1 and 3 only (b) 1 and 2 only
(c) 2 and 3 only (d) 1, 2 and 3

[ESE : 2015]

4.86 For minimum compression work in a 2-stage reciprocating air compressor, which of the following expressions gives the ratio of low pressure cylinder to high pressure cylinder diameters?

(a) $\left(\frac{p_3}{p_1}\right)^{\frac{1}{4}}$ (b) $\left(\frac{p_3}{p_1}\right)^{\frac{1}{3}}$

(c) $\left(\frac{p_3}{p_1}\right)^{\frac{1}{2}}$ (d) $\left(\frac{p_3}{p_2}\right)^{\frac{1}{4}}$

where p_1, p_2 and p_3 are suction, intermediate and delivery pressures respectively.

[ESE : 2015]

4.87 The multistage compression of air as compared to single stage compression
(a) Improves volumetric efficiency for the given pressure ratio
(b) Reduces work done per kg of air
(c) Gives more uniform torque
(d) All of the above

[ESE : 2015]

4.88 Surging is the phenomenon of
(a) Steady, periodic and reversed flow
(b) Unsteady, periodic and reversed flow
(c) Unsteady, periodic and uniform flow
(d) I-dimensional steady and uniform flow

[ESE : 2015]

4.89 In an axial flow compressor, the ratio of pressure in the rotor blades to the pressure rise in the compressor in one stage is known as
(a) Work factor
(b) Slip factor
(c) Degree of reaction
(d) Pressure coefficient

[ESE : 2015]

4.90 In the following diagram, for axial flow compressors, surging is likely to occur in

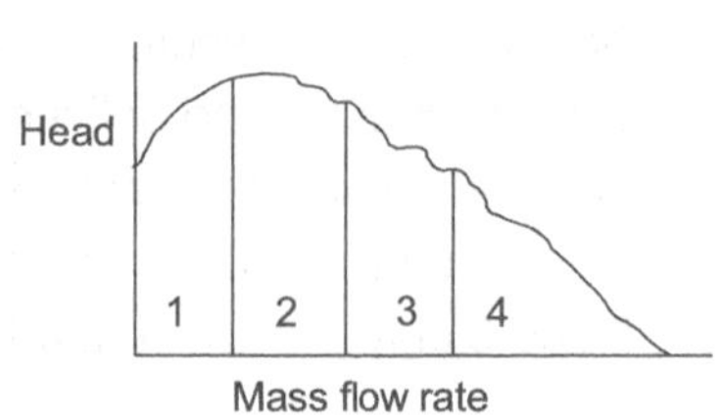

(a) 4th zone (b) 2nd zone
(c) 1st zone (d) 3rd zone

[ESE : 2015]

4.91 In a centrifugal compressor, an increase in speed at a given pressure ratio causes
(a) Increase in flow and increase in efficiency
(b) Increase in flow and decrease in efficiency
(c) Decrease in flow and decrease in efficiency
(d) Decrease in flow and increase in efficiency

[ESE : 2015]

4.92 What is the correct sequence in increasing order of air handling/compressing machines based on the pressure ratio?
(a) Air blower, axial flow fan, centrifugal compressor and reciprocating compressor
(b) Axial flow fan, centrifugal compressor, air blower and reciprocating compressor
(c) Air blower, centrifugal compressor, axial flow fan, and reciprocating compressor
(d) Axial flow fan, air blower, centrifugal compressor and reciprocating compressor

[ESE : 2015]

4.93 The head developed is maximum (keeping other parameters such as rotor diameter, speed, width, inlet angle, etc. constant) for a centrifugal compressor with
(a) Rotor with backward curved blades
(b) Rotor with forward curved blades
(c) Rotor with radial blades
(d) All of the above

[ESE : 2015]

4.94 **Assertion (A):** Axial flow air compressors need many stages to develop high pressure ratios.
Reason (R): The amount of turning of air flow in blade row is limited by the occurrence of separation, a phenomenon caused by adverse pressure gradient.
(a) Both **A** and **R** are individually true and **R** is the correct explanation of **A**
(b) Both **A** and **R** are individually true but **R** is not the correct explanation of **A**
(c) **A** is true but **R** is false
(d) **A** is false but **R** is true

[ESE : 2015]

4.95 For a two stage compressor, the ratio of diameters of L.P. cylinder to H.P. cylinder is equal to :
(a) Square of the ratio of final pressure to initial pressure
(b) The ratio of final pressure to initial pressure
(c) The square root of the ratio of final pressure to initial pressure
(d) Cube root of the ratio of final pressure to initial pressure

[ESE : 2016]

4.96 Which one of the following methods can be adopted to obtain isothermal compression in an air compressor ?
(a) Increasing the weight of the compressor
(b) Interstage heating
(c) Atmospheric cooling

(d) Providing appropriate dimensions to the cylinder [ESE : 2016]

4.97 Consider the following statements :
The compression process in a centrifugal compressor is comparable with :
1. Reversible and adiabatic
2. Irreversible and adiabatic
Which of the above statements is/are correct?
(a) Both 1 and 2 (b) Neither 1 nor 2
(c) 1 only (d) 2 only
[ESE : 2016]

4.98 A portable compressor is taken from a place where the barometric pressure is 750 mm Hg and the average intake temperature is 27°C to a mountainous region where the barometric pressure is 560 mm Hg and temperature is 7°C. The reduction in mass output of the machine is:
(a) 80% (b) 60%
(c) 40% (d) 20% [ESE : 2016]

4.99 The ratio of static enthalpy rise in the rotor to the static enthalpy rise in the stage of an axial flow compressor is defined as :
(a) Power input factor
(b) Flow coefficient
(c) Temperature coefficient
(d) Degree of reaction [ESE : 2016]

4.100 The performance of a single stage reciprocating air compressor is evaluated by its :
(a) Isentropic efficiency
(b) Isothermal efficiency
(c) Adiabatic efficiency
(d) Volumetric efficiency
[ESE : 2016]

4.101 In a two stage reciprocating air-compressor with a suction pressure of 2 bar and delivery pressure of 8 bar, the ideal intercooler pressure will be:
(a) 10 bar (b) 6 bar
(c) 4 bar (d) 3 bar
[ESE : 2016]

4.102 Statement (I) : The volume of air taken into the cylinder of a reciprocating air compressor is less than the stroke volume of the cylinder.
Statement (II) : Air that has been compressed to clearance volume expands to larger volumes during the suction stroke.
(a) Both Statement (I) and Statement (II) are individually true and Statement (II) is the correct explanation of Statement (I).
(b) Both Statement (I) and Statement (II) are individually true but Statement (II) is not the correct explanation of Statement (I).
(c) Statement (I) is true but Statement (II) is false.
(d) Statement (I) is false but Statement (II) is true. [ESE : 2017]

4.103 Consider the following statements:
The volumetric efficiency of a reciprocating compressor can be improved by
1. decreasing the clearance volume
2. cooling the intake air
3. heating the intake air
Which of the above statements is/are correct?
(a) 2 only (b) 3 only
(c) 1 and 2 (d) 1 and 3
[ESE : 2017]

4.104 Consider the following statements in respect of maximum efficiency of a two-stage reciprocating compressor :
1. The pressure ratios are same for each stage.
2. The work done is same in each stage.
3. The intercooling is perfect.
Which of the above statements are correct?
(a) 1 and 2 only (b) 1 and 3 only
(c) 2 and 3 only (d) 1, 2 and 3
[ESE : 2017]

4.105 Consider the following statements for single-stage reciprocating compressors:
1. Isothermal process is the most desirable process for compression.
2. The size of clearance volume provided in the compressor has no effect on work done per kg of air delivered.
3. The volumetric efficiency of the compressor decreases with increasing pressure.
Which of the above statements are correct?
(a) 1 and 2 only (b) 1 and 3 only
(c) 2 and 3 only (d) 1, 2 and 3
[ESE : 2017]

4.106 Air is drawn in a compressor at the rate of 0.8 kg/s at a pressure of 1 bar and temperature of 20°C, and is delivered at a pressure of 10 bar and temperature of 90°C. This air delivery is through and exit valve of area 2×10^{-3} m^2. If R is 287 kJ/kgK, the exit velocity of the air is

(a) 41.7 m/s (b) 35.8 m/s
(c) 29.7 m/s (d) 27.3 m/s

[ESE : 2018]

4.107 The clearance volume in reciprocating air compressor is provided

(a) to reduce the work done per kg of air delivered
(b) to increase the volumetric efficiency the compressor
(c) to accommodate the valves in the head of the compressor
(d) to create turbulence in the air to be delivered

[ESE : 2018]

4.108 In axial flow pumps and compressors, the combined velocity diagram with common base is used to determine change in

(a) absolute velocity $(V_2 - V_1)$
(b) relative velocity $(V_{r_2} - V_{r_1})$
(c) tangential velocity $(U_2 - U_1)$
(d) whirl velocity $(V_{u_2} - V_{u_1})$ **[ESE : 2019]**

4.109 An air compressor compresses atmospheric air at 0.1 MPa and 27°C by 10 times of air inlet pressure. During compression, the heat lost to the surrounding is estimated to be 5% of compression work. Air enters the compressor with a velocity of 40 m/s and leaves with 100 m/s. The inlet and exit cross-sectional areas are 100 cm^2 and 20 cm^2 respectively. The temperature of air at the exit from the compressor will be

(a) 1498 K (b) 1574 K
(c) 1654 K (d) 1726 K **[ESE : 2019]**

4.110 A compressor delivers 4 m^3 of air having a mass of 5 kg. The specific weight and specific volume of air being delivered will be nearly

(a) 12.3 N/m^3 and 0.8 m^3/kg
(b) 14.6 N/m^3 and 0.4 m^3/kg
(c) 12.3 N/m^3 and 0.4 m^3/kg
(d) 14.6 N/m^3 and 0.8 m^3/kg **[ESE : 2019]**

4.111 In centrifugal compressors, there exists a loss of energy due to the mismatch of direction of relative velocity of fluid at inlet with inlet blade angle. This loss is known as

(a) frictional loss (b) incidence loss
(c) clearance loss (d) leakage loss

[ESE : 2019]

4.112 A centrifugal compressor develops a pressure ratio of 5 and air consumption of 30 kg/s. The inlet temperature and pressure are 15 °C and 1 bar respectively. For an isentropic efficiency of 0.85, the power required by the compressor will be nearly

(a) 5964 kW (b) 5778 kW
(c) 5586 kW (d) 5397 kW

[ESE : 2019]

4.113 A single acting two-stage air compressor deals with 4 m^3/min of air at 1.013 bar and 15°C with a speed of 250 rpm. The delivery pressure is 80 bar. If the inter cooling is complete, the intermittent pressure after first stage will be

(a) 9 bar (b) 8 bar
(c) 7 bar (d) 6 bar **[ESE : 2020]**

4.114 The actual and theoretical COP of rolling piston compressor are 3.6 and 4.7 respectively. What is the relative COP?

(a) 8.3 (b) 16.92
(c) 1.3 (d) 0.76 **[ESE : 2021]**

4.115 Surge is a condition which occurs

(a) in forward movement of aircraft
(b) beyond the stable limit of operation
(c) in forward motion of aircraft with speed above sonic velocity
(d) due to the blast of air produced in a compressor

[ESE : 2022]

4.116 Consider the following statements regarding stalling in axial compressor:

1. The flow separation occurring on the suction side of the blade is called negative stalling.
2. The flow separation occurring on the pressure side of the blade is called positive stalling.
3. The rotating stall decreases the effectiveness of the compressor.
4. The delivery pressure significantly drops with large stalling leading to flow reversal.

Which of the above statements are correct?

(a) 1 and 2 only (b) 2 and 3 only
(c) 3 and 4 only (d) 1 and 4 only

[ESE : 2022]

4.117 During a trial on single acting single stage compression the following obervations are made:

Dimensions of cylinder : 10 cm bore and 8 cm stroke,

Speed of rotation : 500 rpm,

Barometer reading : 76 cm Hg

Atmospheric temperature : 27°C

Delivery air temperature = 130°C,

Free air delivery = 15 m^3/hr

Spring balance of dynamometer type (electric motor) reading : 10 kg

Radius of arm of spring balance : 30 cm

Take mechanical efficiency = 0.90.

What is the volumetric efficiency of the compressor?

(a) 38.49% (b) 52.56%
(c) 63.84% (d) 79.62% [ESE : 2023]

4.118 The following data refers to an axial flow compressor:

β_1 = 60°, turning angle = 30°, degree reaction 50%, speed 36000 rpm, mean diameter = 140 mm, inlet pressure = 2 bar and inlet temperature = 57°C. What is blade mean speed?

(a) 140.36 m/s (b) 263.89 m/s
(c) 313.85 m/s (d) 413.85 m/s

[ESE : 2023]

4.119 The indicated work of a single-stage reciprocating air compressor is 250 kJ/min. What is the indicated power?

(a) 250 kW (b) 0.24 kW
(c) 15 kW (d) 4.16 kW

[ESE : 2024]

4.120 Consider the following statements:

1. Thermal efficiency is the ratio of the net work output to the heat supplied.
2. Compressor isentropic efficiency is the ratio of the work output in isentropic compression to the ideal work required.
3. Turbine isentropic efficiency is the ratio of the actual work output to the isentropic work output.

Which of the above statements are correct?

(a) 1 and 2 only (b) 2 and 3 only
(c) 1 and 3 only (d) 1, 2 and 3

[ESE : 2024]

■■■■

Answers Air Compressors (Reciprocating & Rotary)

4.1 (b)	4.2 (d)	4.3 (d)	4.4 (c)	4.5 (a)	4.6 (b)	4.7 (c)	4.8 (d)	4.9 (d)
4.10 (a)	4.11 (c)	4.12 (c)	4.13 (c)	4.14 (d)	4.15 (c)	4.16 (c)	4.17 (a)	4.18 (b)
4.19 (b)	4.20 (c)	4.21 (b)	4.22 (b)	4.23 (d)	4.24 (a)	4.25 (a)	4.26 (c)	4.27 (d)
4.28 (c)	4.29 (c)	4.30 (b)	4.31 (b)	4.32 (d)	4.33 (c)	4.34 (c)	4.35 (b)	4.36 (a)
4.37 (b)	4.38 (d)	4.39 (b)	4.40 (d)	4.41 (c)	4.42 (d)	4.43 (d)	4.44 (c)	4.45 (b)
4.46 (d)	4.47 (b)	4.48 (d)	4.49 (b)	4.50 (b)	4.51 (c)	4.52 (a)	4.53 (d)	4.54 (d)
4.55 (b)	4.56 (c)	4.57 (a)	4.58 (b)	4.59 (c)	4.60 (a)	4.61 (a)	4.62 (d)	4.63 (a)
4.64 (c)	4.65 (c)	4.66 (c)	4.67 (c)	4.68 (d)	4.69 (d)	4.70 (c)	4.71 (b)	4.72 (b)
4.73 (d)	4.74 (c)	4.75 (b)	4.76 (c)	4.77 (b)	4.78 (c)	4.79 (b)	4.80 (c)	4.81 (c)
4.82 (a)	4.83 (c)	4.84 (b)	4.85 (b)	4.86 (a)	4.87 (d)	4.88 (b)	4.89 (c)	4.90 (c)
4.91 (b)	4.92 (d)	4.93 (b)	4.94 (a)	4.95 (*)	4.96 (c)	4.97 (d)	4.98 (d)	4.99 (d)
4.100 (b)	4.101 (c)	4.102 (a)	4.103 (c)	4.104 (d)	4.105 (d)	4.106 (a)	4.107 (c)	4.108 (d)
4.109 (a)	4.110 (a)	4.111 (b)	4.112 (a)	4.113 (a)	4.114 (d)	4.115 (b)	4.116 (c)	4.117 (d)
4.118 (b)	4.119 (d)	4.120 (d)						

Explanations Air Compressors (Reciprocating & Rotary)

4.1 (b)

Capacity of air compressor is expressed in terms of free air delivery.

4.2 (d)

$$P_2 = \sqrt{P_1 P_3} = \sqrt{20 \times 1.10} = \textbf{4.7 bar}$$

4.5 (a)

Two type of compressor valves are used :
(i) Plate type valve
(ii) Reed type valve

4.6 (b)

In centrifugal compressor, there is a tendency for increasing the surge when mass flow rate is less than the design mass flow rate and the number of diffuser vanes is greater than the number of impeller vanes.

4.8 (d)

Workdone for N stage

$$W = N\frac{n}{n-1}P_1V_1\left\{\left(\frac{P_2}{P_1}\right)^{\frac{n-1}{Nn}} - 1\right\}$$

When $N = 3$

$$\therefore \quad W = \frac{3n}{n-1}P_1V_1\left\{\left(\frac{P_2}{P_1}\right)^{\frac{n-1}{3n}} - 1\right\}$$

4.9 (d)

There is clearance between the impeller and vane leading edge, called vaneless space. In the vaneless space, the flow follows an approximately logarithmic spiral path to leading edge of the vanes after which it is constrained by the diffuser channel to follow the blade shape curvature.

4.11 (c)

The pressure ratio of an axial compressor per stage is 1.2 : 1 and for centrifugal compressor 4.5 : 1

Surging produces noise and vibrations in compressor hence it effect the entire compressor.

4.12 (c)

In centrifugal compressor, rotation of the impeller caused air to be compressed by centrifugal action into the reduced volume at the tip of the compressor.

4.13 (c)

Centrifugal compressor is more suitable for low head and high flow rate.

4.14 (d)

The phenomena in which air stream fails to follow the blade contour is called stalling.

4.15 (c)

For centrifugal compressor, forward blading velocity triangles

4.16 (c)

For axial flow compressor having degree of reaction 50%.

$\alpha_1 = \beta_2, \beta_1 = \alpha_2$

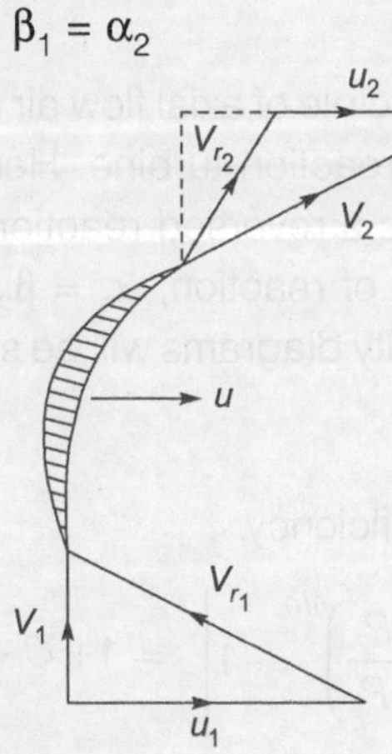

4.17 (a)

Volumetric efficiency of compressor is

$$\eta_v = 1 + C - C\left(\frac{V_1}{V_2}\right)$$

$$= 1 + 0.05 - 0.05 \times \left(\frac{0.04}{0.02}\right) = 0.95 = \mathbf{95\%}$$

4.18 (b)

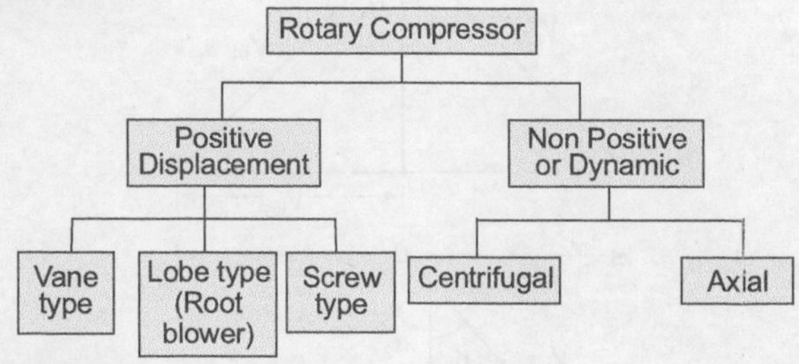

4.19 (b)

For minimum compressor work, the pressure ratios and works are equal for all stages in intercooling.

i.e., $P_i = \sqrt{P_c P_e}$

4.20 (c)

Due to minimum work input is required in isothermal compression. It is the ideal process for compression.

4.21 (b)

$$\text{Degree of reaction, } R = \frac{(\Delta h)_{\text{rotor}}}{(\Delta h)_{\text{rotor}} + (\Delta h)_{\text{stator}}}$$

4.22 (b)

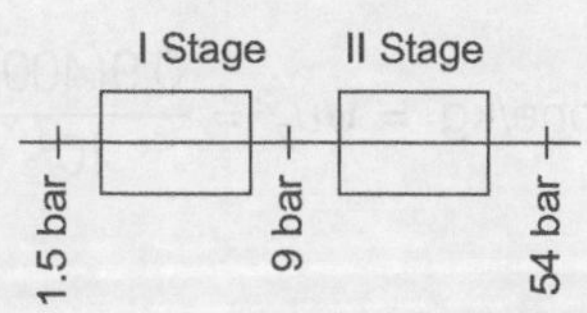

$$P_I = \sqrt{P_1 P_2} = \sqrt{1.5 \times 54} = \mathbf{9\ bar}$$

4.24 (a)

Isentropic efficiency

$$\eta_{\text{isen}} = \frac{(\Delta H)_s}{\Delta H}$$

4.25 (a)

After passing through the divergence passage formed between rotor blades, which increase its absolute velocity, the air will come out relative velocity V_{r_2} at an angle β_2 which is less than β_1. This turns the air towards axial direction. Some pressure rise will take place in rotor due to decreases in relative velocity at exit i.e. V_{r_2} will be less V_{r_1}.

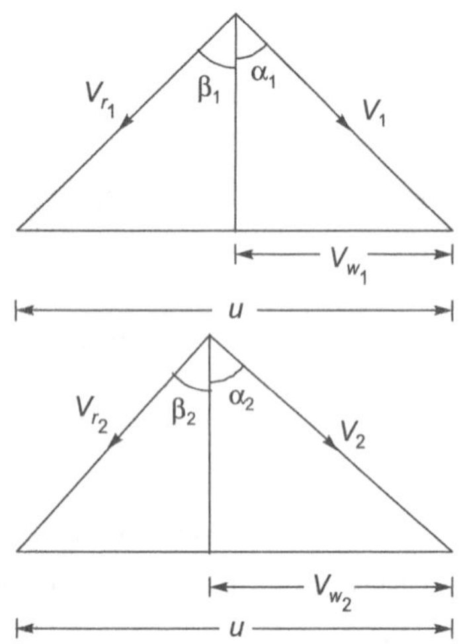

4.26 (c)

$$\eta_a = \frac{\phi_p}{\phi_s \times \phi_w}$$

$$\therefore \quad \phi_p = \phi_s \times \phi_w \times \eta_a$$

4.27 (d)

Vane type rotary compressor works on the principle of positive displacement i.e., pressure increase by decrease in volume.

4.29 (c)

$$\text{Work done/kg} = \psi u_2^2 = \frac{0.9(400)^2}{10^3} = 144 \text{ kJ}$$

4.30 (b)

Advantages of multi-stage compression are

(i) Improved overall volumetric efficiency.

(ii) A reduction in work required per stroke, and therefore the total driving power.

(iii) Size and strength of cylinder can be adjusted to suit volume and pressure of gas.

(iv) Multi-cylinder give more uniform torque and better mechanical balance thus needing smaller flywheel

4.31 (b)

Due to small frontal area, axial flow compressor is suitable for air craft, while the pressure ratio per stage in axial flow compressor is less than centrifugal compressor.

4.32 (d)

For 50% reaction turbine $\alpha_1 = \beta_2$ and $\beta_1 = \alpha_2$ Hence blading design on this basis is referred to as symmetrical blading.

4.35 (b)

Multi-stage compression is used to improved the overall volumetric efficiency.

$$\eta_v = 1 + C - C\left(\frac{P_2}{P_1}\right)^{1/n}$$

By decreasing the delivery pressure (P_2) the volumetric efficiency of a reciprocating air compressor can be increased.

4.36 (a)

Centrifugal air compressor increases the frontal area of air craft which increases drag. So axial flow compressors are used in aircraft.

4.37 (b)

Density of air at low temperature, will be high.

4.40 (d)

For balancing supply and demand, following device are used in the reciprocating compressor.
(i) Throttle control, (ii) Blowing air to water

4.41 (c)

Working principle of axial flow air compressor is opposite to reaction turbine. Hence it is often described as a reversed reaction turbine with 50% degree of reaction, $\alpha_1 = \beta_2$ and $\beta_1 = \alpha_2$. Hence velocity diagrams will be symmetrical.

4.42 (d)

Volumetric efficiency,

$$\eta = 1 - \frac{V_c}{V_s}\left[\left(\frac{P_2}{P_1}\right)^{1/n} - 1\right] = 1 + C - C\left(\frac{P_2}{P_1}\right)^{1/n}$$

Hence volumetric efficiency decreases with increasing clearance ratio and pressure ratio, both.

4.46 (d)

A centrifugal compressor is suitable for low pressure ratio and high mass flow.

4.47 (b)

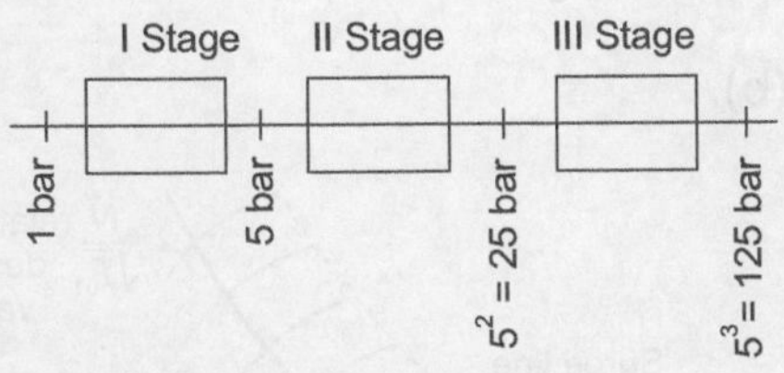

Hence the ideal intermediate pressure at the end of first stage is 5 bar.

4.48 (d)

Degree of reaction in axial flow compressor

$$= \frac{\text{Pressure rise in rotor}}{\text{Pressure rise in stage}}$$

$$= \frac{\text{Static enthalpy rise in rotor}}{\text{Static enthalpy rise in stage}}$$

4.52 (a)

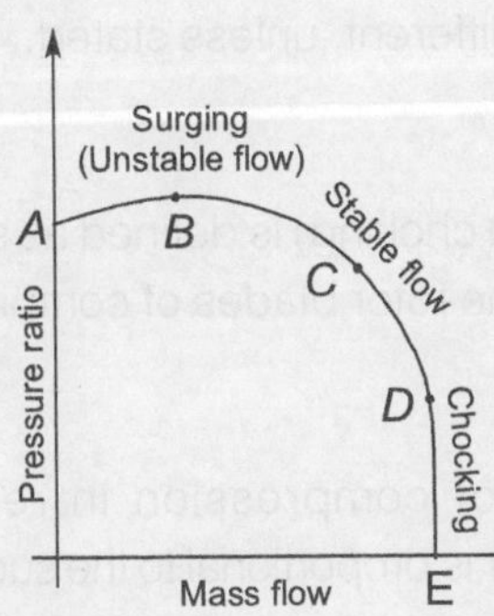

Hence *A-B* represents surging.

4.55 (b)

For perfect cooling between two stage of compressor

$$P_2 = \sqrt{P_1 P_3}$$

where, P_2 = Intermidiate pressure,

P_1 = Inlet pressure and P_3 = Delivery pressure

4.56 (c)

Axial flow compressor can be designed for higher pressure ratio because multistaging can be done. But multistaging is difficult in centrifugal compressor and due to large number of stages axial flow compressors are heavier than centrifugal compressor.

4.58 (b)

$$P_2 = \sqrt{P_1 P_3} = \sqrt{1 \times 4} = 2.0 \text{ bar}$$

4.60 (a)

Volumetric efficiency of a reciprocating compressor is

$$\eta_v = 1 + C - C\left(\frac{P_2}{P_1}\right)^{1/n}$$

$$C = \frac{V_c}{V_s}, \text{ clearance ratio}$$

P_2 = Delivery pressure

P_1 = Inlet pressure

4.61 (a)

By compressing the work using isothermal process, we have to provide minimum work.

4.62 (d)

Stalling is the phenomenon in which air stream fails to follow the blade contour.

4.63 (a)

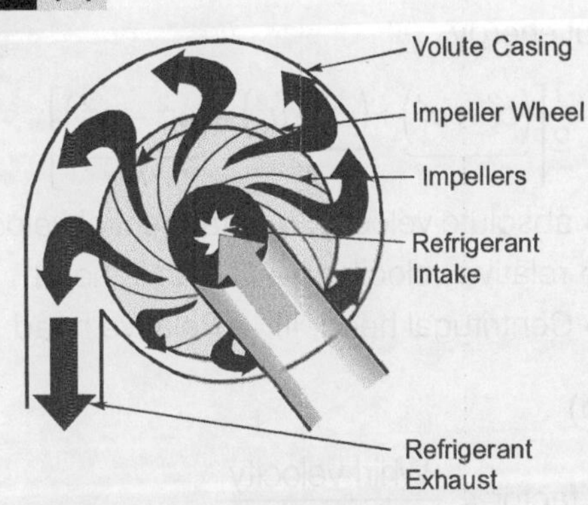

Centrifugal compressors, sometimes referred to as radial compressors, are a subclass of dynamic axisymmetric work-absorbing turbomachinery. Guide vanes provide prewhirl which helps in keeps the Mach number at inlet to an acceptable value below supersonic.

4.64 (c)

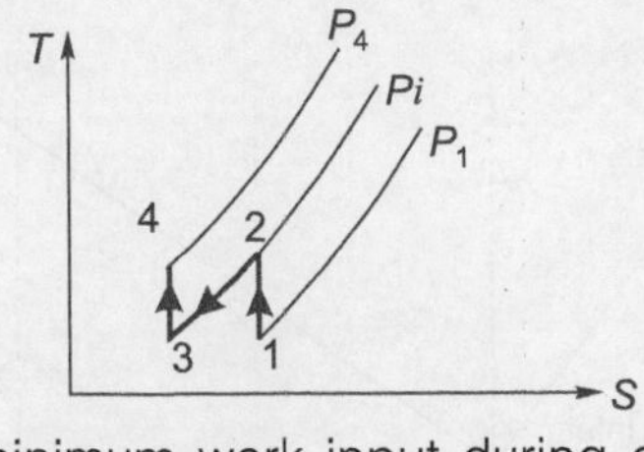

For minimum work input during compression process,

$$P_i = \sqrt{P_1 P_4}$$

4.65 (c)

Isothermal curve has minimum area under *P-V* curve and hence the work will be minimum.

4.68 (d)

When air stream does not follows the contour of the blade stalling occurs.

4.70 (c)

For 2-stage air compressor the pressure between the two stage should be geometric mean of the inlet and exit pressure of the system

i.e., $P_2 = \sqrt{P_1.P_3}$

Also, for maximum efficiency the intercooling between the two stages should be complete so the process is closer to the isothermal and the work input is minimum.

4.71 (b)

Input energy,

$$E = \frac{1}{2}\left[\underbrace{\left(c_2^2 - c_1^2\right)}_{I} + \underbrace{\left(u_2^2 - u_1^2\right)}_{II} + \underbrace{\left(w_1^2 - w_2^2\right)}_{III}\right]$$

$c \rightarrow$ absolute velocity; $u \rightarrow$ tangential velocity
$w \rightarrow$ relative velocity; I $\rightarrow$ Dynamic head
II $\rightarrow$ Centrifugal head; III $\rightarrow$ Relative head

4.72 (b)

$$\text{Slip factor} = \frac{\text{Whirl velocity}}{\text{Blade velocity}}$$

4.73 (d)

$$N_s|_{\text{Axial compressor}} > N_s|_{\text{Centrifugal}} > N_s|_{\text{Reciprocating}}$$

4.74 (c)

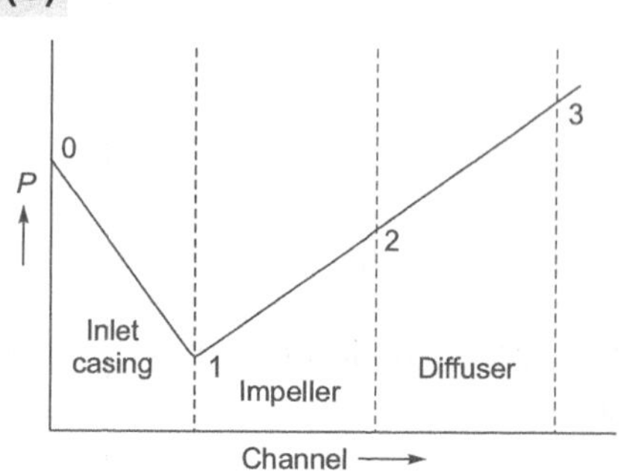

(pressure rise across centrifugal compressor)

Pressure at 1 is minimum and so, relative Mach number is highest at impeller inlet.

4.75 (b)

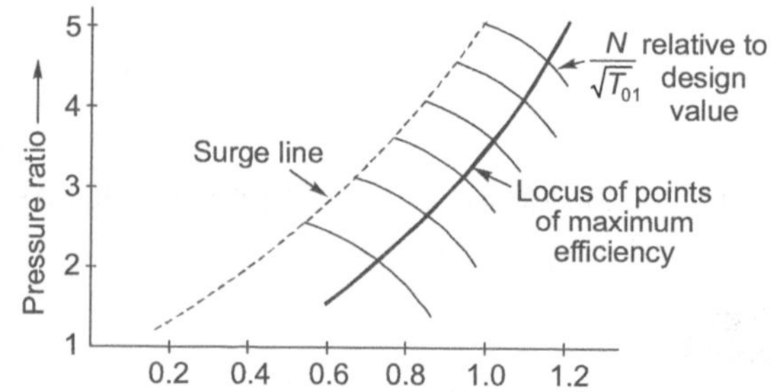

4.76 (c)

Degree of reaction (*R*)

$$= \frac{\text{Actual change of enthalpy in rotor}}{\text{Actual change of enthalpy in stage}}$$

4.77 (b)

Index are different, unless stated.

4.78 (c)

Stalling (not choking) is defined as separation of fluid from the rotor blades of compressors.

4.79 (b)

The work of compression in reciprocating compressor is proportional to the suction volume. The clearnace gas merely acts like a spring alternatively expanding and contracting. In practice, a large clearance volume results in a low volumetric efficiency and hence large cylinder dimensions, increased contact area between the piston and cylinder, and so, increased friction and work.

4.80 (c)

Mass flow rate is maximum if choking occurs. So, statement (3) is not correct.

4.82 (a)

$$u = \frac{\pi DN}{60} = \frac{3.14 \times 0.45 \times 7200}{60}$$

$= 169.56$ m/s

$W = \phi u^2 = u^2$ (Assuming no slip i.e. $\phi = 1$)

$= 28750.29$ J/kg

$= 28.75$ kJ/kg or kW/kg/s

4.84 (b)

(i) $\eta_v = 1 + C - C\left(\frac{P_2}{P_1}\right)^{1/n}$

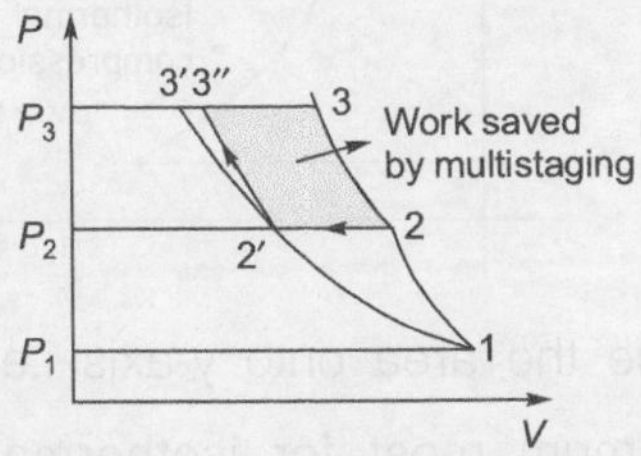

Low pressure ratio in a stage (p_2/p_1) will increase the volumetric efficiency.

(ii) By cooling the air between the stages of compression, the compression can be brought to isothermal and power input to the compressor can be reduced considerably.

(iii) The low pressure cylinder is made lighter and high pressure cylinders are made smaller for reduced pressure ratio in each stage.

4.85 (b)

Intercooler pressure p_2 is the geometric mean of inlet pressure p_1 and delivery pressure p_3.

i.e. $p_2 = \sqrt{p_1 p_3}$

4.86 (a)

$$P_1V_1 = P_2V_2$$

$$\Rightarrow \quad P_1 \times \frac{\pi}{4} d_1^2 \times l_1 = P_2 \times \frac{\pi}{4} d_2^2 \times l_2$$

$$\Rightarrow \quad \frac{d_1}{d_2} = \sqrt{\frac{P_2}{P_1}} \qquad (l_1 = l_2) \quad \text{(i)}$$

Also for minimum work

$$P_2 = \sqrt{P_1 P_3} \qquad \text{(ii)}$$

From equation (i) and (ii)

$$\frac{d_1}{d_2} = \left(\frac{P_3}{P_1}\right)^{\frac{1}{4}}$$

4.87 (d)

$$\eta_{volume} = 1 + C - C\left(\frac{p_2}{p_1}\right)^{\frac{1}{n}}$$

(i) Low pressure ratio in a cylinder, improves volumetric efficiency.

(ii) By cooling the air between the stages of compression, the compression work input can be reduced considerably.

(iii) Due to phasing of operation in stages, in a multistage compressor, the negative and positive forces are balanced to a larger extent. Thus, more uniform torque and better mechanical balance can be achieved. Hence option (d) is correct.

4.88 (b)

Surging is the phenomenon of unsteady, periodic and reversed flow which affects whole machine.

4.89 (c)

Degree of reaction

$$= \frac{\text{Pressure rise in rotor blades}}{\text{Pressure rise in the one stage of compressor}}$$

4.90 (c)

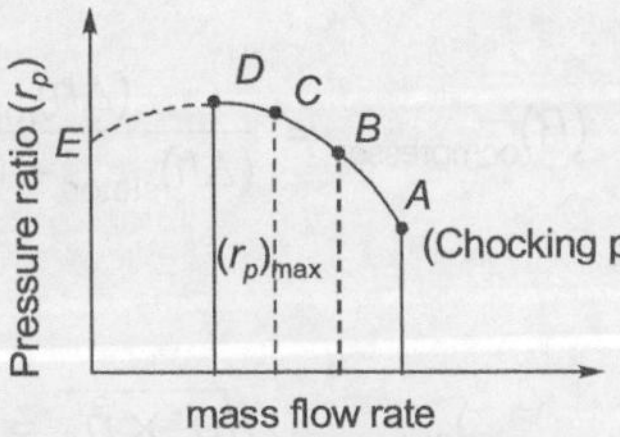

Point $A \rightarrow$ Chocking point

Zone DE $\rightarrow$ Surging zone:

4.91 (b)

By increase in speed at a given pressure ratio, mass flow rate increases but-efficiency decreases due to various frictional losses.

4.92 (d)

Since reciprocating compressor is given at the end of each option which can handle less amount of air, so in the question it should be asked decreasing order of air handling machines.

Axial flow fan > Air blower > Centrifugal compressor > reciprocating compressor.

4.93 (b)

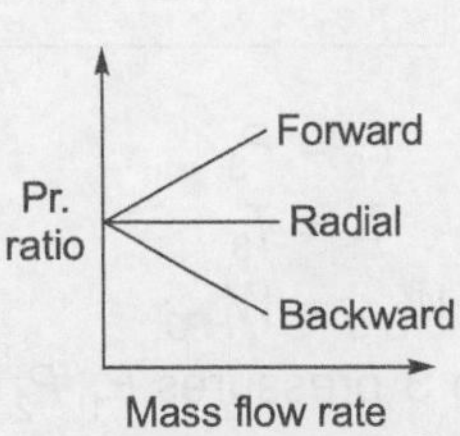

Pressure ratio $\propto$ Head developed

4.94 (a)

Several rows of airfoil blades are used to achieve large pressure rise in the compressor. The circular area for air flow is gradually reduced from the inlet to the outlet of the compressor.

4.95 (*)

Power will be 1/4.

4.96 (c)

In air compressor (reciprocating), cooling must be provided (by Fins) to approach isothermal compression to reduce work input.

4.99 (d)

$$(R)_{\text{compressor}} = \frac{(\Delta h)_{\text{rotor}}}{(\Delta h)_{\text{stator}} + (\Delta h)_{\text{rotor}}}$$

4.101 (c)

$$(P_{\text{intercooler}})_{2\text{ stage}} = \sqrt{p_s \times p_d} = \sqrt{2 \times 8} = 4$$

4.103 (c)

$$\eta_{\text{vol}} = 1 + C - C\left(\frac{p_d}{p_s}\right)^{\frac{1}{n}}$$

where $C = \frac{V_C}{V_S}$, clearance ratio

4.104 (d)

Perfect intercooling means:

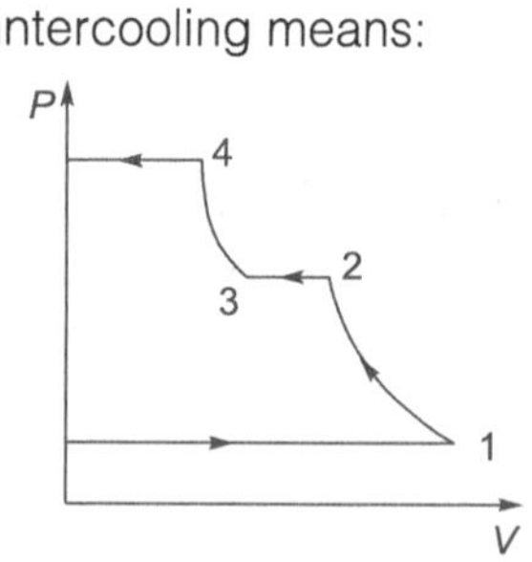

$$P_2 = P_3$$

and $T_1 = T_3$

$\Rightarrow$ $W_{LPC} = W_{HPC}$

and all the 3 pressures P_1, P_2 and P_4 are in G.P.

i.e., $\frac{P_2}{P_1} = \frac{P_4}{P_3}$

4.105 (d)

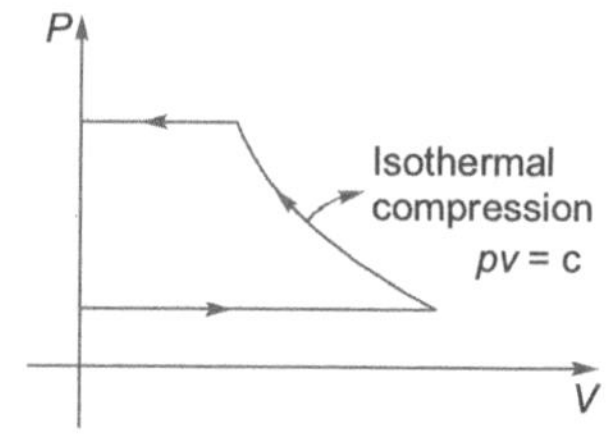

Since the area onto y-axis i.e. $\int v\,dp$ being minimum most for isothermal compression $\frac{W \cdot D}{\text{Cycle}}$ is least.

$$\eta_{\text{vol}} = 1 + C - C\left(\frac{P_d}{P_s}\right)^{\frac{1}{n}}$$

As p_d increases $\Rightarrow \eta_{\text{vol}}$ decreases.

4.106 (a)

Given: $\dot{m} = 0.8$ kg/s,

$P_1 = 1$ bar, $T_1 = 293$ K

$P_2 = 10$ bar

$T_2 = 363$ K

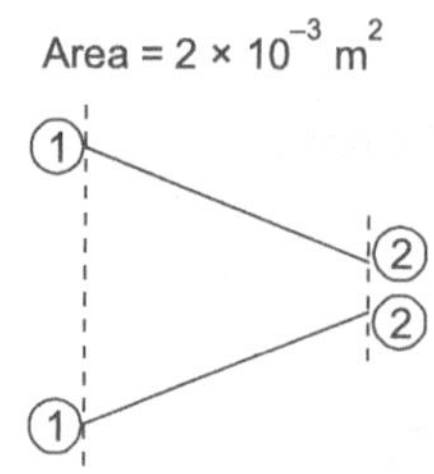

$$\dot{m} = \rho_2 A_2 v_2$$

$$0.8 = \frac{P_2}{RT_2} \times 2 \times 10^{-3} \times v_2$$

$$0.8 = \frac{10 \times 100}{0.287 \times 363} \times 2 \times 10^{-3} \times v_2$$

$$v_2 = 41.67 \text{ m/s}$$

Value of R is wrong actual value of R is 0.287 kJ/kgK

4.109 (a)

$P_1 = 0.1$ MPa, $T_1 = 300$ K, $P_2 = 1$ MPa, $T_2 = ?$
$A_1 = 100$ cm^2, $A_2 = 20$ cm^2, $V_1 = 40$ m/s, $V_2 = 100$ m/s

By continuity equation

$$\rho_1 A_1 V_1 = \rho_2 A_2 V_2$$

$$\frac{P_1}{RT_1} \cdot A_1 \cdot V_1 = \frac{P_2}{RT_2} \cdot A_2 \cdot V_2$$

$$\frac{0.1 \times 100 \times 40}{300} = \frac{1 \times 20 \times 100}{T_2}$$

$$T_2 = 1500 \text{ K}$$

4.110 (a)

(i) $$\rho \cdot g = \frac{m}{V} \cdot g = \frac{5}{4} \times 9.81$$

$$= 2.26 \approx 12.3 \text{ N/m}^3$$

(ii) $$\frac{v}{m} = \frac{4}{5} = 0.8 \text{ m}^3/\text{kg}$$

4.112 (a)

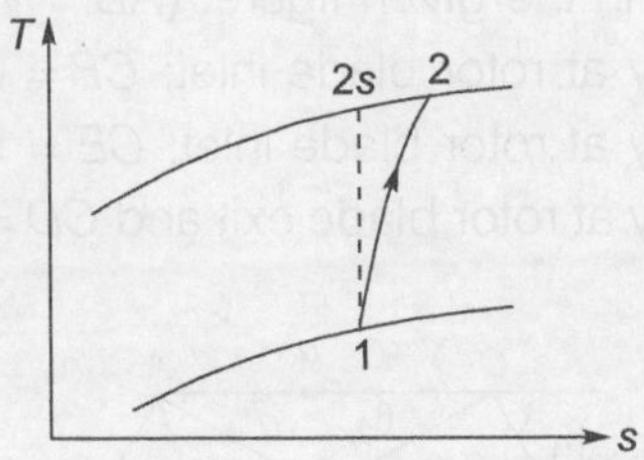

$$\frac{T_{2s}}{T_1} = \left(\frac{P_2}{P_1}\right)^{\frac{\gamma-1}{\gamma}} = 5^{\frac{0.4}{1.4}} = 1.583$$

$$T_{2s} = 288 \times 1.583 = 456.1 \text{ K}$$

$$\frac{T_{2s} - T_1}{T_2 - T_1} = \eta_c = 0.85$$

$$T_2 - T_1 = 197.81 \text{ K}$$

$$W = mc_p\Delta T = 30 \times 1.005 \times 197.81$$

$$W = 5964 \text{ kW}$$

4.113 (a)

For two stage air compressor with perfect intercooling, the intermittent.

Pressure (P_i) for optimum work input is,

$$P_i = \sqrt{P_1 P_2} = \sqrt{80} = 9 \text{ bar}$$

4.114 (d)

$$\text{Relative COP} = \frac{\text{Actual COP}}{\text{Theoretical COP}} = \frac{3.6}{4.7} = 0.76$$

4.117 (d)

Actual volume entered = Free air delivered

$$= \frac{15}{3600} \text{ m}^3/\text{s}$$

Theoretical swept volume,

$$V_s = \frac{\pi}{4} \times (0.1)^2 \times (0.08) \times \frac{500}{60}$$

$$= 5.235 \times 10^{-3} \text{ m}^3/\text{s}$$

Volumetric efficiency,

$$\eta = \frac{15/3600}{5.235 \times 10^{-3}} = 0.796$$

$$= 79.6\%$$

4.118 (b)

Given :Mean diameter;

$$D_m = 140 \text{ mm}$$

$$N = 36000 \text{ rpm}$$

$$u = \frac{\pi \times D_m \times N}{60}$$

$$= \frac{\pi \times 140 \times 36000}{1000 \times 60}$$

$$= 263.89 \text{ m/s}$$

4.119 (d)

$$\text{Indicated power} = \frac{250 \text{ kJ/min}}{60} = 4.1667 \text{ kW}$$

4.120 (d)

- Thermal efficiency $= \dfrac{\text{Net work output}}{\text{Heat supplied}}$
- Compressor isentropic efficiency

$$= \frac{\text{Work input required in isentropic compression}}{\text{Actual work required}}$$

- Turbine isentropic efficiency

$$= \frac{\text{Actual work output}}{\text{Isentropic work output}}$$

■■■

5 Steam Turbines

5.1 Partial admission steams turbine refers to the situation where the

(a) steam is admitted partially into the blades through nozzles
(b) nozzles occupy the complete circumference leading into the blade annulus
(c) nozzles do not occupy the complete circumference leading into the blade annulus
(d) steam is admitted partially into the blades directly

[ESE : 2000]

5.2 Consider the following statements regarding a 100% reaction turbine:

1. Change in absolute velocity of steam across the moving blades is zero.
2. Change in absolute velocity of steam across the moving blades is negative.
3. Enthalpy drop in fixed blades is zero.

Which of these statements is/are correct?

(a) 1 only (b) 2 only
(c) 2 and 3 (d) 1 and 3

[ESE : 2000]

5.3 Which one of the following pairs is not correctly matched?

(a) Internal efficiency of steam turbine : Product of stage efficiency and re- heat factor
(b) Stage efficiency of a turbine : Ratio of adiabatic heat drop to the isentropic heat drop per stage
(c) Dryness fraction of steam within a stage : Decreases due to reheating
(d) Steam condensation during expansion through the turbine : Enhances blade erosion

[ESE : 2000]

5.4 Velocity triangle for a reaction turbine stage is shown in the given figure. ($AB = v_1$ = absolute velocity at rotor blade inlet; $CB = w_1$ = relative velocity at rotor blade inlet; $CE = w_2$ = relative velocity at rotor blade exit and $CD = CB$)

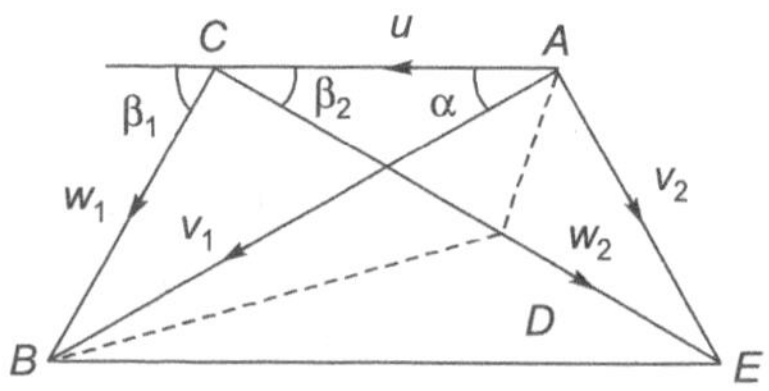

The ratio of reaction force to impulse force is

(a) CE/CB (b) CD/CE
(c) DE/BD (d) AE/AB

[ESE : 2000]

5.5 Consider the following statements:

1. Throttle governing improves quality of steam in the last few stages.
2. Internal efficiency of steam is not seriously affected by throttle governing.
3. Throttle governing is better than nozzle governing.

Which of these statements are correct?

(a) 1, 2 and 3 (b) 1 and 3
(c) 2 and 3 (d) 1 and 2

[ESE : 2000]

5.6 Which one of the following statements is correct?

(a) Reheat factor is zero if efficiency of the turbine is close to unity.
(b) Lower the efficiency, higher will be the reheat factor.
(c) Reheat factor is independent of steam conditions at turbine inlet.
(d) Availability of reheat is higher at low pressure end.

[ESE : 2000]

5.7 In a steam power plant, the ratio of the isentropic heat drop in the prime mover to the amount of heat supplied per unit mass of steam is known as
(a) stage efficiency
(b) degree of reaction
(c) Rankine efficiency
(d) relative efficiency
[ESE : 2000]

5.8 Consider the following statements:
The reheat cycle helps to reduce
1. fuel consumption
2. steam flow
3. the condenser size
Which of these statements are correct?
(a) 1 and 2 (b) 1 and 3
(c) 2 and 3 (d) 1, 2 and 3
[ESE : 2000]

5.9 **Assertion (A):** Modern turbines have velocity compounding at the initial stages and pressure compounding in subsequent stages.
Reason (R): Excessive tip leakage occurs in the high pressure region of reaction blading.
(a) Both **A** and **R** are individually true and **R** is the correct explanation of **A**
(b) Both **A** and **R** are individually true but **R** is not the correct explanation of **A**
(c) **A** is true but **R** is false
(d) **A** is false but **R** is true
[ESE : 2000]

5.10 The temperature-entropy diagram for a steam turbine power plant, operating on the Rankine cycle with reheat and regenerative feed heating is shown in the given figure. If m denotes the fraction of steam bled for feed heating, the work developed in the turbine per kg steam entering the turbine at state 5 is

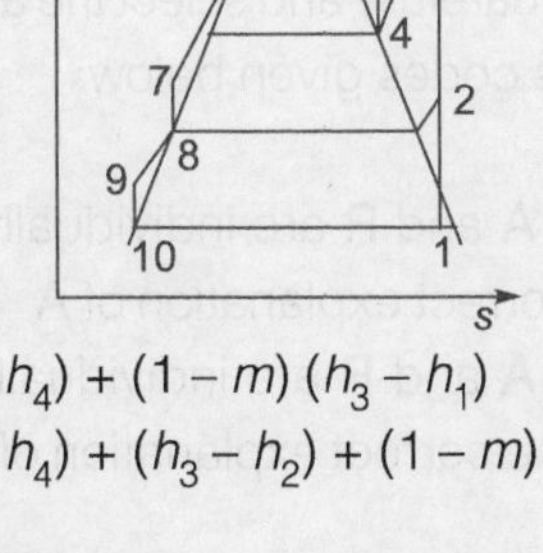

(a) $(h_5 - h_4) + (1 - m)(h_3 - h_1)$
(b) $(h_5 - h_4) + (h_3 - h_2) + (1 - m)(h_2 - h_1)$
(c) $2h_5 - h_4 - h_2 + (1 - m)(h_2 - h_1)$
(d) $(h_5 - h_4) + (1 - m)(h_3 - h_2)$
[ESE : 2001]

5.11 Match **List-I** (Machines) with **List-II** (Features) and select the correct answer using the codes given below the lists:

List-I
A. Steam engine
B. Impulse turbine
C. Reaction turbine
D. Centrifugal compressor

List-II
1. Velocity compounding
2. Diagram factor
3. Continuous pressure drop
4. Isentropic efficiency

Codes:

	A	B	C	D
(a)	3	4	2	1
(b)	2	1	3	4
(c)	2	4	3	1
(d)	3	1	2	4

[ESE : 2001]

5.12 For maximum blade efficiency of a single-stage impulse turbine, the blade speed ratio, (α is the angle made by absolute velocity at inlet) should be
(a) $\cos 2\alpha$ (b) $\frac{\cos 2\alpha}{2}$
(c) $\frac{\cos \alpha}{2}$ (d) $\frac{2}{\cos \alpha}$
[ESE : 2001]

5.13 The given figure shows the variation of certain steam parameter in case of a simple impulse turbine. The curve *A-B-C* represents the variation of

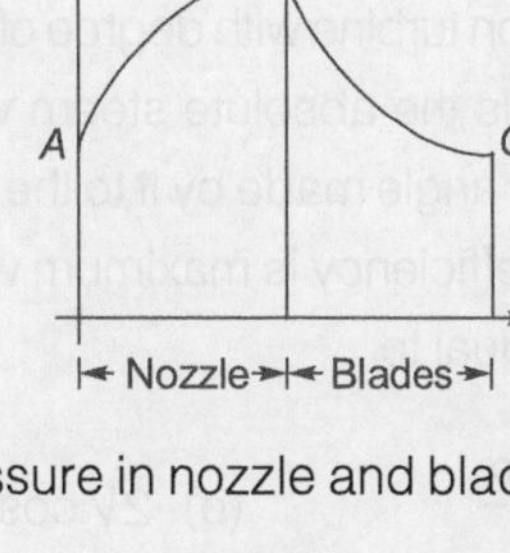

(a) pressure in nozzle and blades

(b) velocity in nozzle and blades
(c) temperature in nozzle and blades
(d) enthalpy in nozzle and blades

[ESE : 2001]

Direction: Each of the next questions consists of two statements, one labelled as the **'Assertion (A)'** and the other as **'Reason (R)'**. You are to examine these two statements carefully and select the answers to these items using the codes given below:

Codes:
(a) Both **A** and **R** are individually true and **R** is the correct explanation of **A**
(b) Both **A** and **R** are individually true but **R** is not the correct explanation of **A**
(c) **A** is true but **R** is false
(d) **A** is false but **R** is true

5.14 **Assertion (A):** Reaction blading is commonly used in intermediate and low pressure parts of steam turbines.
Reason (R): Reaction blading gives higher efficiency than impulse blading. [ESE : 2001]

5.15 **Assertion (A):** In conventional impulse steam turbine designs, only two rows of moving blades are used in a Curtis stage.
Reason (R): As the number of rows of moving blades in a Curtis stage increases, the effectiveness of the later rows decreases.

[ESE : 2001]

5.16 **Assertion (A):** With Throttle governing of a steam turbine, the turbine power is reduced by reduction in the available heat drop together with decrease in the rate of steam flow.
Reason (R): The pressure and the rate of steam flow are simultaneously decreased with the help of a throttle valve. [ESE : 2001]

5.17 For a reaction turbine with degree of reaction equal to 50%, (V is the absolute steam velocity at inlet and α is the angle made by it to the tangent on the wheel) the efficiency is maximum when the blade speed is equal to

(a) $\frac{V\cos\alpha}{2}$ (b) $2V\cos\alpha$
(c) $V\cos^2\alpha$ (d) $V\cos\alpha$

[ESE : 2002]

5.18 Match **List-I** (Equipment) with **List-II** (Application area) and select the correct answer using the codes given below the lists:

List-I
A. Anticipatory gear
B. Labyrinth
C. Inverted T-attachment
D. Deaerator

List-II
1. Sealing system
2. Steam power plant
3. Turbine governing system
4. Blades

Codes:

	A	B	C	D
(a)	4	2	3	1
(b)	3	1	4	2
(c)	4	1	3	2
(d)	3	2	4	1

[ESE : 2002]

5.19 In a reaction turbine the heat drop in fixed blade is 8 kJ/kg and the total heat drop per stage is 20 kJ/kg. The degree of reaction is
(a) 40% (b) 66.7%
(c) 60% (d) 25%

[ESE : 2002]

5.20 The main advantage of a reheat Rankine cycle is
(a) Reduced moisture content in L.P. side of turbine
(b) increased efficiency
(c) reduced load on condenser
(d) reduced load on pump

[ESE : 2002]

Direction: Each of the next questions consists of two statements, one labelled as the **'Assertion (A)'** and the other as **'Reason (R)'**. You are to examine these two statements carefully and select the answers to these items using the codes given below:

Codes:
(a) Both **A** and **R** are individually true and **R** is the correct explanation of **A**
(b) Both **A** and **R** are individually true but **R** is not the correct explanation of **A**

(c) A is true but R is false
(d) A is false but R is true

5.21 Assertion (A): The performance of a simple Rankine cycle is not sensitive to the efficiency of the feed pump.
Reason (R): The net work ratio is practically unity for a Rankine cycle.
[ESE : 2002]

5.22 Assertion (A): Rankine efficiency would approach Carnot cycle efficiency by providing a series of regenerative feed heating
Reason (R): With regenerative feed heating, expansion through the turbine approaches an isentropic process.
[ESE : 2002]

5.23 Employing superheated steam in turbines leads to
(a) Increase in erosion of blading
(b) Decrease in erosion of blading
(c) No erosion in blading
(d) No change in erosion of blading
[ESE : 2003]

5.24 Steam enters a De laval steam turbine with an inlet velocity of 30 m/s and leaves with an outlet velocity of 10 m/s. The work done by 1 kg of steam is
(a) 400 N-m (b) 600 N-m
(c) 800 N-m (d) 1200 N-m
[ESE : 2003]

5.25 In a 50% reaction stage, absolute velocity angle at inlet is 45°, mean peripheral speed is 75 m/s and the absolute velocity at the exit is axial. The stage specific work is
(a) 2500 m^2/s^2 (b) 3270 m^2/s^2
(c) 4375 m^2/s^2 (d) 5625 m^2/s^2
[ESE : 2003]

5.26 In a reaction turbine stage enthalpy drop in the stator blades is 4.62 kJ/kg and that in the rotor blades is 2.38 kJ/kg. The degree of reaction of the stage is
(a) 0.52 (b) 0.43
(c) 0.34 (d) 0.26 [ESE : 2003]

5.27 An emergency governor of a steam turbine trips the turbine when
1. Shaft exceeds 100% of its rated speed
2. Condenser becomes hot due to inadequate cooling water circulation
3. Lubrication system fails
4. Balancing of turbine is not proper

Select the correct answer from the codes given below:
(a) 1, 2 and 3 (b) 2, 3 and 4
(c) 3, 4 and 1 (d) 4, 1 and 2
[ESE : 2003]

5.28 In a regenerative feed heating cycle, the economic number of the stages of regeneration
(a) increases as the initial pressure and temperature increase
(b) decreases as the initial pressure and temperature increase
(c) is independent of the initial pressure and temperature
(d) depends only on the condenser pressure
[ESE : 2003]

5.29 Match **List-I** (Blades) with **List-II** (Features) and select the correct answer using the codes given below the lists:

List-I
A. Ceramic blades
B. Steam turbine blades
C. Alloy steel blades
D. Compressor blades

List-II
1. High creep strength
2. Forged and machined
3. Precision cast
4. Thick at mid chord
5. Thin trailing edge

Codes:

	A	B	C	D
(a)	2	1	5	4
(b)	3	4	5	1
(c)	2	4	3	5
(d)	3	2	1	5

[ESE : 2003]

5.30 Euler equation of turbine giving energy transfer per unit mass E_0 (where U, V_w, V_r and V represents the peripheral, whirl, relative and absolute velocities respectively. Suffix 1 and 2 refer to the turbine inlet and outlet respectively) is given by
(a) $E_0 = U_1 V_{w_1} - U_2 V_{w_2}$
(b) $E_0 = U_1 V_{r_1} - U_2 V_{r_2}$

(c) $E_0 = U_1 V_1 - U_2 V_2$

(d) $E_0 = V_1 V_{w_1} - V_2 V_{w_2}$

[ESE : 2003]

5.31 Consider the following statements pertaining to the features of a regenerative steam cycle plant as compared to a non-regenerative plant:

1. It increases the cycle efficiency.
2. It requires a bigger boiler.
3. It requires a smaller condenser.

Which of these statements are correct?

(a) 1, 2 and 3 (b) 1 and 2
(c) 2 and 3 (d) 1 and 3

[ESE : 2003]

Direction: Each of the next questions consists of two statements, one labelled as the **'Assertion (A)'** and the other as **'Reason (R)'**. You are to examine these two statements carefully and select the answers to these items using the codes given below:

Codes:

(a) Both **A** and **R** are individually true and **R** is the correct explanation of **A**
(b) Both **A** and **R** are individually true but **R** is not the correct explanation of **A**
(c) **A** is true but **R** is false
(d) **A** is false but **R** is true

5.32 **Assertion (A):** An ideal regenerative Rankine cycle power plant with saturated steam at the inlet to the turbine has same thermal efficiency as Carnot cycle working between the same temperature limits.

Reason (R): The change in entropy of steam during expansion in the turbine is equal to the change in entropy of the feed water during sensible heating at steam generator pressure.

[ESE : 2003]

5.33 **Assertion (A):** Impulse staging is commonly employed in high pressure part and reaction staging in intermediate low pressure parts of the steam turbine.

Reason (R): The tip leakage across moving blades is less in impulse staging as the pressure drop is small and there can be large pressure drop across fixed blades and nozzles.

[ESE : 2003]

5.34 Which one of the following is the feature of pressure-compounding (Rateau staging)?

(a) Low efficiency at low rotational speeds
(b) High efficiency with low fluid velocities
(c) High efficiency with high fluid velocities
(d) Low efficiency at high rotational speeds

[ESE : 2004]

5.35 In Parson's reaction turbines, the velocity diagram triangles at the inlet and outlet are which of the following?

(a) Asymmetrical (b) Isosceles
(c) Right-angled (d) Congruent

[ESE : 2004]

5.36 In Parson's turbine if α is nozzle angle, then the maximum efficiency of the turbine is

(a) $\dfrac{2\cos\alpha}{1+\cos\alpha}$ (b) $\dfrac{2\cos^2\alpha}{1+\cos^2\alpha}$

(c) $\dfrac{2\cos^2\alpha}{1-\cos\alpha}$ (d) $\dfrac{\cos^2\alpha}{1+2\cos^2\alpha}$

[ESE : 2004]

5.37 What is the value of the reheat factor in multi-stage turbine?

(a) 1.03 to 1.04 (b) 1.10 to 1.20
(c) 0.90 to 1.00 (d) 1.20 to 1.25

[ESE : 2004]

5.38 In which one of the following steam turbines, steam is taken from various points along the turbine, solely for feed-water heating?

(a) Extraction turbine
(b) Bleeder turbine
(c) Regenerative turbine
(d) Reheat turbine [ESE : 2004]

5.39 A 4-row velocity compounded steam turbine develops total 6400 kW. What is the power developed by the last row?

(a) 200 kW (b) 400 kW
(c) 800 kW (d) 1600 kW

[ESE : 2005]

5.40 A power plant, which uses a gas turbine followed by steam turbine for power generation, is called

(a) Topping cycle (b) Bottoming cycle
(c) Brayton cycle (d) Combined cycle

[ESE : 2005]

5.41 Which one of the following statements is not correct for a regenerative steam cycle?
(a) It increases the thermodynamic efficiency
(b) It reduces boiler capacity for a given output
(c) It reduces temperature stresses in the boiler due to hotter feed
(d) The efficiency increases with increases number of feed heaters [ESE : 2005]

5.42 **Assertion (A):** Throttle governing is used only in small steam turbines.
Reason (R): At part loads, the efficiency of steam turbine reduces considerably with throttle governing.
(a) Both **A** and **R** are individually true and **R** is the correct explanation of **A**
(b) Both **A** and **R** are individually true but **R** is not the correct explanation of **A**
(c) **A** is true but **R** is false
(d) **A** is false but **R** is true [ESE : 2005]

5.43 The degree of reaction of an impulse turbine
(a) is less than zero
(b) is greater than zero
(c) is equal to zero
(d) increases with steam velocity at the inlet

[ESE : 2005]

5.44 The compounding of steam turbine is done
(a) to improve efficiency
(b) to reduce the speed of rotor
(c) to reduce exit losses
(d) to increase the turbine output [ESE : 2005]

5.45 Which one of the following is used to bring down the speed of an impulse steam turbine to practical limits?
(a) A centrifugal governor
(b) Compounding of the turbine
(c) A large flywheel
(d) A gear box [ESE : 2006]

5.46 Consider the following for a steam turbine power plant:
1. Reduction in blade erosion.
2. Increase in turbine speed.
3. Increase in specific output.
4. Increase in cycle efficiency.

Which of the above occur/occurs due to reheating of steam?
(a) only 1 (b) 1 and 2
(c) 1, 3 and 4 (d) 2 and 3

[ESE : 2006]

5.47 What is the ratio of the isentropic work to Euler's work known as?
(a) Pressure coefficient
(b) Slip factor
(c) Work factor
(d) Degree of reaction

[ESE : 2006]

5.48 Which one of the following expresses the maximum blade efficiency of a Parson's turbine?

(a) $\dfrac{2\cos^2\alpha}{1+\cos^2\alpha}$ (b) $\dfrac{\cos^2\alpha}{1+2\cos^2\alpha}$

(c) $\dfrac{\cos\alpha}{1+\cos^2\alpha}$ (d) $\dfrac{\cos\alpha}{2}$

Where α is the jet angle at the entrance.

[ESE : 2006]

5.49 When is the greatest economy obtained in a regenerative feed heating cycle?
(a) Steam is extracted from only one suitable point of a steam turbine
(b) Steam is extracted only from the last stage of a steam turbine
(c) Steam is extracted only from the first stage of a steam turbine
(d) Steam is extracted from several places in different stages of steam turbines

[ESE : 2006]

5.50 In a reaction turbine the enthalpy drop in a stage is 60 units, the enthalpy drop in the moving blades is 32 units. The degree of reaction is
(a) 0.533 (b) 0.284
(c) 0.466 (d) 1.875

[ESE : 2006]

5.51 Consider the following statements in respect of impulse steam turbine
1. Blade passages are of constant cross-section.
2. Partial admission of steam is permissible.
3. Axial thrust is only due to change in flow velocity of steam at inlet and outlet of moving blade.

Which of these statement are correct?

(a) 1, 2 and 3 (b) Only 1 and 2
(c) Only 2 and 3 (d) Only 1 and 3

[ESE : 2006]

5.52 **Assertion (A):** Nozzle control governing cannot be used in reaction steam turbines.

Reason (R): In reaction steam turbines, full admission of steam must take place.

(a) Both **A** and **R** are individually true and **R** is the correct explanation of **A**
(b) Both **A** and **R** are individually true but **R** is not the correct explanation of **A**
(c) **A** is true but **R** is false
(d) **A** is false but **R** is true

[ESE : 2006]

5.53 Blade erosion in steam turbines takes place

(a) Due to high temperature steam
(b) Due to droplets in steam
(c) Due to high rotational speed
(d) Due to high flow rate

[ESE : 2007]

5.54 Throttle governing in steam turbines

(a) Leads to significant pressure loss
(b) Increases the efficiency
(c) Increases heat losses
(d) Decreases steam temperature

[ESE : 2007]

5.55 Which one of the following statements is correct?

(a) Efficiency of the Carnot cycle for thermal power plant is high and work ratio is also high in comparison to the Rankine cycle.
(b) Efficiency of the Carnot cycle is high and work ratio is low comparison to the Rankine cycle.
(c) Efficiency of the Carnot cycle is low and work ratio is also low in comparison to the Rankine cycle.
(d) Both the cycles have same efficiencies and work ratio.

[ESE : 2007]

5.56 For maximum blade efficiency (utilization factor), the work done (J/kg) in a single stage 50% reaction turbine is

(a) $2u^2$ (b) $1/u^2$
(c) u^3 (d) u^2

[ESE : 2007]

5.57 In ideal regenerative cycle the temperature of steam entering the turbine is same as that of

(a) water entering the turbine
(b) water leaving the boiler
(c) steam leaving the turbine
(d) water at any section of the turbine

[ESE : 2008]

5.58 In a simple impulse turbine the nozzle angle at the entrance is 30°. For maximum diagram efficiency. The blade-speed ratio is

(**Note**: sin 30° = 0.5, cos 30° = 0.866, sin 15° = 0.259, cos 15° = 0.966)

(a) 0·259 (b) 0·75
(c) 0·5 (d) 0·433

[ESE : 2009]

5.59 The pressure and velocity diagram as shown in the figure for a steam turbine refers to which one of the following:

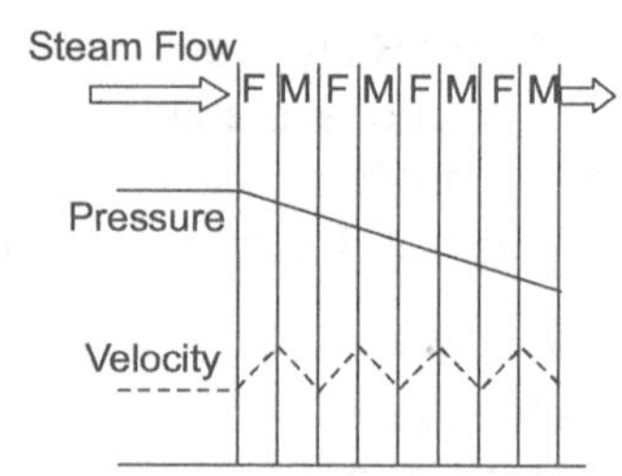

(Where : M–moving blade, F–fixed blade)

(a) Impulse turbine – Velocity compounded
(b) Impulse turbine – Pressure compounded
(c) Impulse turbine – Pressure and Velocity compounded
(d) Reaction turbine stages

[ESE : 2009]

5.60 If the enthalpy drop in the moving blades and fixed blades of a steam turbine is 10 kJ/kg and 15 kJ/kg respectively then the degree of reaction is

(a) 67% (b) 60%
(c) 40% (d) 33%

[ESE : 2009]

5.61 Consider the following statements:

For a velocity-compounded impulse steam turbine stage having two rows of moving blades

1. The work done in each row of moving blades is equal.
2. The efficiency of a velocity-compounded stage is less than that of a pressure-compounded stage.

3. The velocity-compounded stage is often used in the first stage of a multi-stage impulse turbine.

Which of these statements is/are correct?

(a) 1 and 3 only (b) 2 only
(c) 2 and 3 only (d) 1, 2 and 3

[ESE : 2010]

5.62 Assertion (A): Single stage impulse steam turbines are not used in practice.

Reason (R): Single stage impulse steam turbines have very low revolutions per minute.

(a) Both **A** and **R** are individually true and **R** is the correct explanation of **A**
(b) Both **A** and **R** are individually true but **R** is not the correct explanation of **A**
(c) **A** is true but **R** is false
(d) **A** is false but **R** is true

[ESE : 2010]

5.63 Willans line represents

(a) Total steam consumption vs. power output with throttle governing
(b) Total steam consumption vs. power output with cutoff governing
(c) Behaviour of supersaturated steam through nozzles
(d) Condensation of steam while flowing through the turbine

[ESE : 2011]

Direction: Each of the next questions consists of two statements, one labelled as the **'Assertion (A)'** and the other as **'Reason (R)'**. You are to examine these two statements carefully and select the answers to these items using the codes given below:

Codes:

(a) Both **A** and **R** are individually true and **R** is the correct explanation of **A**
(b) Both **A** and **R** are individually true but **R** is not the correct explanation of **A**
(c) **A** is true but **R** is false
(d) **A** is false but **R** is true

5.64 Assertion (A) : In steam turbines, supersaturated flow means that the vapour does not condense immediately as it crosses the dry saturated line.

Reason (R) : The mass flow with supersaturation flow is greater than the mass flow with isentropic flow. [ESE : 2011]

5.65 Assertion (A) : The pressure compounded impulse steam turbine is the most efficient type of impulse turbine.

Reason (R) : It is because the ratio of blade velocity to steam velocity remains constant.

[ESE : 2011]

5.66 In impulse steam turbine, the enthalpy drop of steam takes place in

(a) nozzles (b) rotor blades
(c) stator blades (d) exhaust pipe

[ESE : 2012]

5.67 Which of the following statements are correct with respect to regenerative feed heating in a steam cycle?

1. It increases cycle efficiency
2. It increases specific output
3. It reduces the condenser load
4. It improves the steam quality at the end of expansion

(a) 1 and 3 only (b) 2 and 4 only
(c) 2 and 3 only (d) 1, 2, 3 and 4

[ESE : 2013]

5.68 Without reducing the fluid flow rate, the speed of an impulse steam turbine can be brought down to practical limits by which of the following methods?

1. Large flywheel
2. Centrifugal governor
3. Velocity compounding
4. Steam bleeding

(a) 1, 2 and 3 (b) 2, 3 and 4
(c) 2 only (d) 3 only

[ESE : 2013]

5.69 To improve the quality of steam at turbine exit which of the following will be used?

1. Reheat cycle
2. Increase the maximum pressure when maximum and minimum temperatures are fixed
3. Use superheated steam, instead of saturated steam when the maximum and minimum pressures are fixed

(a) 1, 2 and 3 (b) 1 and 2 only
(c) 1 and 3 only (d) 2 and 3 only

[ESE : 2013]

5.70 In a half-degree reaction Parson's turbine, operating at design conditions, the enthalpy drop of steam in one stage of the turbine occurs
(a) Entirely in the fixed blades
(b) Entirely in the moving blades
(c) Half in the fixed blades and half in the moving blades
(d) None of the above [ESE : 2014]

5.71 Reheating of steam in a steam power plant:
1. Increases the cycle efficiency
2. Reduces the turbine speed
3. Reduces blade erosion
4. Increases specific output

Which of above statements are correct?
(a) 1, 2, 3 and 4 (b) 1, 2 and 4 only
(c) 2 and 4 only (d) 1, 3 and 4 only
[ESE : 2014]

5.72 **Statement (I) :** Modern turbines have velocity compounding at the initial stages and pressure compounding in subsequent stages.
Statement (II): Excessive tip leakage occurs in the high pressure region of reaction blading.
(a) Both Statement (I) and Statement (II) are individually true and Statement (II) is the correct explanation of Statement (I)
(b) Both Statement (I) and Statement (II) are individually true but Statement (II) is not the correct explanation of Statement (I)
(c) Statement (I) is true but Statement (II) is false
(d) Statement (I) is false but Statement (II) is true.
[ESE : 2014]

5.73 Consider the following statements:
1. Carnot, Ericsson and Stirling cycles are ideal power cycles that are completely reversible
2. Ericsson cycle is not a practical engine cycle
3. Stirling cycle is the only practical power cycle among the above
4. All these cycles have the same thermal efficiency

Which of the above statements are correct?
(a) 1, 2, 3 and 4 (b) 1, 2 and 4 only
(c) 2, 3 and 4 only (d) 1, 2 and 3 only
[ESE : 2015]

5.74 In order to determine the quality of wet steam by a separating and throttling calorimeter, the steam should be first separated and then throttled such that the final state is
(a) Saturated vapour only
(b) Superheated vapour only
(c) At a pressure higher than the original pressure
(d) A mixture of saturated liquid and vapour
[ESE : 2015]

5.75 The work done in a steady flow process is equal to $-\int vdp$. In the Rankine cycle, the turbine work is much greater than the pump work because
(a) The specific volume of water is much higher than that of steam
(b) The specific volume of steam is much higher than that of water
(c) The pressure drop in the turbine is much higher than that in the pump
(d) There is less irreversibility in the turbine than in the pump [ESE : 2015]

5.76 Consider the following statements:
1. The only practical way of improving efficiency of Otto cycle is to increase the compression ratio of an internal combustion engine
2. Ericsson cycle needs heat transfer in all the processes
3. Ericsson and Stirling cycles employ regenerative heat exchangers for reversible heat transfer
4. Atkinson cycle has a greater specific work than a comparable Otto cycle engine

Which of the above statements are correct?
(a) 1, 2, 3 and 4 (b) 1, 2 and 4 only
(c) 2, 3 and 4 only (d) 1, 2 and 3 only
[ESE : 2015]

5.77 Consider the following statements regarding cycles:
1. Stirling cycle consists of two isothermal and two adiabatic processes.
2. In vapour compression cycle, the refrigerant is in the form of dry saturated vapour before entering compressor.

3. Diesel cycle consists of one constant pressure; one constant volume and two isentropic processes

Which of the above statements are correct?

(a) 1, 2 and 3 (b) 1 and 2 only
(c) 1 and 3 only (d) 2 and 3 only

[ESE : 2015]

5.78 The efficiency of the vapour power Rankine cycle can be increased by

1. Increasing the temperature of the working fluid at which heat is added
2. Increasing the pressure of the working fluid at which heat is added
3. Decreasing the temperature of the working fluid at which heat is rejected

Which of the above statements are correct?

(a) 1 and 2 only (b) 1 and 3 only
(c) 2 and 3 only (d) 1, 2 and 3

[ESE : 2015]

5.79 Consider the following statements with regards to steam turbines:

1. A single stage impulse turbine has a nozzle angle α. The maximum blade efficiency of the turbine will be $\cos^2 \alpha$
2. For a reaction steam turbine with identical stator and rotor blades, the blade velocity for maximum blade efficiency is equal to inlet steam velocity
3. Velocity compounded impulse steam turbine gives less speed and less efficiency

Which of the above statements are correct?

(a) 1 and 2 only (b) 1 and 3 only
(c) 2 and 3 only (d) 1, 2 and 3

[ESE : 2015]

5.80 The optimum ratio of blade speed to tangential component of jet speed for the de Laval and Parsons turbine are

(a) 1 for both
(b) 1/2 for de Laval turbine and 1 for Parsons turbine
(c) 1 for de Laval turbine and 1/2 for Parsons turbine
(d) 1/2 for both

[ESE : 2015]

5.81 A steam turbine in which a part of the steam after expansion is used for process heating and the remaining steam is further expanded for power generation is/are :

1. Impulse turbine
2. Pass out turbine

(a) 1 only (b) 2 only
(c) Both 1 and 2 (d) Neither 1 nor 2

[ESE : 2016]

5.82 Consider the following statements :

1. Thermal efficiency of the simple Steam or Rankine cycle can be improved by increasing the maximum system pressure and temperature
2. Increasing the superheat of the steam improves the specific work and decreases the moisture content of exhaust steam
3. Increasing maximum system pressure always increases the moisture content at the turbine exhaust
4. Lowering the minimum system pressure increases the specific work of the cycle

Which of the above statements are correct ?

(a) 1, 2 and 3 (b) 1, 2 and 4
(c) 2, 3 and 4 (d) 1, 3 and 4

[ESE : 2016]

5.83 **Statement (I) :** Providing reheat in a Rankine cycle would increase the efficiency of the cycle.

Statement (II) : Reheat in Rankine cycle reduces specific steam consumption.

(a) Both Statement (I) and Statement (II) are individually true and Statement (II) is the correct explanation of Statement (I).
(b) Both Statement (I) and Statement (II) are individually true but Statement (II) is not the correct explanation of Statement (I).
(c) Statement (I) is true but Statement (II) is false.
(d) Statement (I) is false but Statement (II) is true.

[ESE : 2017]

5.84 Consider the following statements in respect of regenerative Rankine cycle:

1. Regeneration increases the efficiency of the cycle.
2. The boiler capacity is increased for a given output.
3. The capacity of the condenser is reduced.

Which of the above statements are correct?

(a) 1, 2 and 3 (b) 1 and 2 only
(c) 1 and 3 only (d) 2 and 3 only

[ESE : 2017]

5.85 The total power developed by a three-stage velocity compounded impulse steam turbine is 900 kW. The power magnitudes developed in the first and the second stages are, respectively

(a) 500 kW and 300 kW
(b) 100 kW and 300 kW
(c) 500 kW and 100 kW
(d) 100 kW and 100 kW

[ESE : 2017]

5.86 Which one of the following methods is more effective to improve the efficiency of the Rankine cycle used in thermal power plant?

(a) Increasing the condenser temperature
(b) Decreasing the condenser temperature
(c) Decreasing the boiler temperature
(d) Increasing the boiler temperature

[ESE : 2018]

5.87 Consider the following statements regarding Reheat Rankine Steam Cycle:

1. The main purpose of reheat in Rankine cycle is to increase the efficiency of the cycle.
2. In practice, the reheat is generally limited to one point of expansion.
3. Due to reheat, the steam rate (specific steam consumption) is reduced.

Which of the above statements are correct?

(a) 1 and 2 only (b) 2 and 3 only
(c) 1 and 3 only (d) 1, 2 and 3

[ESE : 2018]

Directions: The following items consist of two statements; one labelled as **'Statement (I)'** and the other as **'Statement (II)'**. You are to examine these two statements carefully and select the answers to these items using the codes given below:

Codes:

(a) Both Statement (I) and Statement (II) are individually true and Statement (II) is the correct explanation of Statement (I).
(b) Both Statement (I) and Statement (II) are individually true but Statement (II) is **NOT** the correct explanation of Statement (I).
(c) Statement (I) is true but Statement (II) is false
(d) Statement (I) is false but Statement (II) is true.

5.88 **Statement (I) :** the condenser in a steam power plant is always filled with a mixture of water, steam and air.

Statement (II) : Slightly wet steam enters the condenser wherein the pressure is below the atmospheric conditions, causing some leakage of air through the glands and also the release of some air dissolved in the boiler feedwater.

[ESE : 2018]

5.89 **Statement (I) :** Rankine efficiency of a steam power plant increases in winter compared to summer.

Statement (II) : The increase in Rankine efficiency is due to lower condenser temperature.

[ESE : 2018]

5.90 **Statement (I) :** Reheating between the high-pressure and low-pressure turbines increases the turbine work output.

Statement (II) : The constant pressure lines on T-s diagram diverge away from the origin.

[ESE : 2018]

5.91 Consider the following statements regarding compounding in steam turbines :

1. In impulse turbine, steam pressure remains constant between ends of the moving blades.
2. In reaction turbine, steam pressure drops from inlet to outlet of the blade.
3. In velocity compounding, partial expansion of steam takes place in the nozzle and further expansion takes place in the rotor blades.

Which of the above statements are correct?

(a) 1 and 2 only (b) 1 and 3 only
(c) 2 and 3 only (d) 1, 2 and 3

[ESE : 2019]

5.92 Which one of the following statements is correct with respect to axial flow 50% reaction turbine?

(a) The combined velocity diagram is symmetrical.
(b) The outlet absolute velocity should not be axial for maximum utilization.

(c) Angles of both stator and rotor are not identical.
(d) For maximum utilization, the speed $\frac{U}{v_1} = \sin^2 \alpha$. [ESE : 2019]

5.93 In a steam turbine, the nozzle angle at the inlet is 18°. The relative velocity is reduced to the extent of 6% when steam flows over the moving blades. The output of the turbine is 120 kJ/kg flow of steam. If the blades are equiangular, the speed ratio and the absolute velocity of steam at inlet for maximum utilization are nearly
(a) 0.42 and 230.2 m/s
(b) 0.48 and 230.2 m/s
(c) 0.42 and 515.1 m/s
(d) 0.48 and 515.1 m/s [ESE : 2019]

5.94 A turbine in which steam expands both in nozzle as well as in blades is called as
(a) impulse reaction turbine
(b) reciprocating steam turbine
(c) gas turbine
(d) Curtis turbine
[ESE : 2019]

5.95 Consider the following statements regarding reaction turbine:
1. Blade shape is aerofoil type and its manufacturing is difficult.
2. It is suitable for small power.
3. Leakage losses are less compared to friction losses.

Which of the above statements is/are correct?
(a) 1 only (b) 2 only
(c) 3 only (d) 1, 2 and 3
[ESE : 2019]

5.96 Which of the following are applied (used) ways of compounding steam turbines?
1. Pressure compounding
2. Temperature compounding
3. Velocity compounding

(a) 1, 2 and 3 (b) 1 and 2 only
(c) 2 and 3 only (d) 1 and 3 only
[ESE : 2020]

5.97 The process of abstracting steam at a certain section of the turbine and subsequently using it for heating feed water supplied to the boiler is called
(a) Reheating (b) Regeneration
(c) Bleeding (d) Binary vapour cycle
[ESE : 2020]

5.98 In an axial flow turbine, the utilization factor has an absolute maximum value of unity, for any degree of reaction if the value of nozzle angle α is
(a) 270° (b) 180°
(c) 90° (d) 0°
[ESE : 2020]

5.99 The steam turbine can be governed by the following methods except
(a) Reaction governing
(b) Throttle governing only
(c) Nozzle control governing only
(d) Combination of throttle and nozzle control governing [ESE : 2021]

5.100 Consider the following statements for steam turbines:
1. The ratio of actual enthalpy drop to isentropic enthalpy drop is known as mechanical efficiency.
2. The ratio of enthalpy drop in moving blades to enthalpy drop in the stage is known as degree of reaction.
3. Rateau turbine is the example of reaction turbine.
4. Curtis turbine is the example of impulse turbine.

Which of the above statements are correct?
(a) 2 and 4 only (b) 1 and 3 only
(c) 2 and 3 only (d) 1, 2, 3 and 4
[ESE : 2021]

5.101 In an impulse steam turbine, steam is accelerated through a nozzle from rest. It enters the nozzle at 9.8 bar dry and saturated. The height of the blade is 10 cm and the nozzle angle is 15°. Mean blade velocity is 144 m/s. The blade velocity ratio is 0.48 and blade velocity coefficient is 0.97. What is the isentropic heat drop?
(a) 28.4 kJ/kg (b) 38.4 kJ/kg
(c) 48.9 kJ/kg (d) 58.9 kJ/kg
[ESE : 2023]

5.102 A single stage impulse turbine has equal blade angles and nozzle angle of 30°. What is the maximum possible blade efficiency, if the blade velocity coefficient is 0.85?
(a) 33.46% (b) 45.63%
(c) 55.46% (d) 69.37%
[ESE : 2023]

5.103 Consider the following statements:
1. Impulse turbine stage has pressure drop occurring in moving blades only.
2. Reaction turbine stage has pressure drop occurring in both fixed as well as moving blades.
3. Reaction turbines have complete admission of steam or steam being admitted all around the rotor through fixed blade ring.

Which of the above statements are correct?
(a) 1 and 2 only (b) 2 and 3 only
(c) 1 and 3 only (d) 1, 2 and 3
[ESE : 2023]

5.104 Which type of steam turbine is also known as Curtis turbine?
(a) Pressure compounded impulse turbine.
(b) Velocity compounded impulse turbine.
(c) Pressure-velocity compounded impulse turbine.
(d) Cross-flow compound turbine.
[ESE : 2023]

5.105 A reaction turbine has mean blade speed of 180 m/s, blade speed to steam velocity ratio of 0.8, outlet angles of fixed and moving blades as 30° and 35°, specific volume at outlet of fixed blade as 0.5 m^3 and at moving blade outlet as 0.6 m^3. Areas at exit of fixed blade and moving blades are same. Consider the efficiency of blades as 90% when considered as nozzles and $K^2 = 0.88$, where K is blade velocity coefficient. What is the axial velocity component at the inlet?
(a) 92.5 m/s (b) 98.5 m/s
(c) 101.2 m/s (d) 112.5 m/s
[ESE : 2023]

5.106 Multi-cylinder turbines which have their rotors mounted on one and the same shaft and coupled to a single generator are known as
(a) single-shaft turbines
(b) topping turbines
(c) multi-axial turbines
(d) back pressure turbines
[ESE : 2024]

5.107 Consider the following regarding steam turbines:
1. In reaction turbine, the steam expands in the stationary nozzles and attains high velocity.
2. In impulse turbine, the steam enters the fast moving blades on the rotor from stationary nozzles; if the steam is superheated before allowing it to expand, the Rankine cycle efficiency may be increased
3. In impulse turbine, the resulting high velocity steam impinges against the blades which alter the direction of steam jet thus changing the momentum of jet and causing impulsive force on the blades.
4. In steam turbine, the steam consumption does not increase with increase in years of service.

Which of the above statements are correct?
(a) 1 and 2 only (b) 2 and 3 only
(c) 3 and 4 only (d) 1 and 4
[ESE : 2024]

5.108 Consider the following statements regarding steam power plants :
1. An increase in the initial pressure of steam raises the thermal efficiency.
2. The thermal efficiency will be dropped by raising the initial temperature of the steam without changing the pressure.
3. Intermediate reheating of steam improves the thermal efficiency of the plant.
4. An increase in the initial pressure of steam improves the thermal efficiency of steam power plant and also wetness fraction of such steam decreases at the end of expansion.

Which of the above statements are correct?
(a) 1 and 2 only (b) 2 and 3 only
(c) 3 and 4 only (d) 1 and 3
[ESE : 2024]

5.109 Consider the following statements regarding steam turbines:

1. Blade efficiency is the ratio of the work done on the blade per second to the energy entering the blade per second.
2. Internal efficiency is the ratio of the heat converted into useful work to the total adiabatic heat drop.
3. Net efficiency is the ratio of brake thermal efficiency to diagram thermal efficiency.

Which of the above statements are correct?

(a) 1 and 3 only (b) 2 and 3 only
(c) 1 and 2 only (d) 1, 2 and 3

[ESE : 2024]

5.110 Consider the following statements :

1. Bleeding is the process of draining steam from the turbine at certain point during its expansion.
2. There is a usual practice in bleeding installation to allow the bled steam to mix with the preheated air, after that the mixture of steam and air proceeds to the boiler.
3. The objective of governing is to keep the turbine speed fairly constant irrespective of load.

Which of the above statements are correct?

(a) 1 and 3 only (b) 2 and 3 only
(c) 1 and 2 only (d) 1, 2 and 3

[ESE : 2024]

5.111 Consider that the Rankine cycle efficiency is improved in the following ways:

1. By reducing heat addition in boiler
2. By increasing steam turbine expansion work.
3. By reducing feed pump work

Which of the above are correct?

(a) 1 and 2 only (b) 2 and 3 only
(c) 1 and 3 only (d) 1, 2 and 3

[ESE : 2024]

5.112 Isentropic expansion efficiency is expressed as

(a) Ideal work in expansion – Actual work in expansion

(b) Actual work in expansion + Ideal work in expansion

(c) $\dfrac{\text{Ideal work in expansion}}{\text{Actual work in expansion}}$

(d) $\dfrac{\text{Actual work in expansion}}{\text{Ideal work in expansion}}$

[ESE : 2024]

5.113 In the reheat cycle, the principal advantage of reheat is to increase the quality of steam at

(a) turbine exhaust (b) turbine inlet
(c) feed pump inlet (d) feed pump exhaust

[ESE : 2024]

■■■■

Answers Steam Turbine

5.1	(a)	5.2	(d)	5.3	(c)	5.4	(c)	5.5	(d)	5.6	(b)	5.7	(a)	5.8	(a)	5.9	(d)
5.10	(b)	5.11	(b)	5.12	(c)	5.13	(b)	5.14	(b)	5.15	(a)	5.16	(a)	5.17	(d)	5.18	(b)
5.19	(c)	5.20	(a)	5.21	(a)	5.22	(c)	5.23	(b)	5.24	(a)	5.25	(d)	5.26	(c)	5.27	(b)
5.28	(a)	5.29	(d)	5.30	(a)	5.31	(a)	5.32	(c)	5.33	(b)	5.34	(b)	5.35	(d)	5.36	(b)
5.37	(a)	5.38	(c)	5.39	(b)	5.40	(d)	5.41	(b)	5.42	(d)	5.43	(c)	5.44	(b)	5.45	(b)
5.46	(c)	5.47	(c)	5.48	(a)	5.49	(d)	5.50	(a)	5.51	(a)	5.52	(a)	5.53	(b)	5.54	(a)
5.55	(b)	5.56	(d)	5.57	(b)	5.58	(d)	5.59	(d)	5.60	(c)	5.61	(c)	5.62	(c)	5.63	(a)
5.64	(b)	5.65	(a)	5.66	(a)	5.67	(a)	5.68	(d)	5.69	(c)	5.70	(c)	5.71	(d)	5.72	(d)
5.73	(b)	5.74	(b)	5.75	(b)	5.76	(a)	5.77	(d)	5.78	(d)	5.79	(b)	5.80	(b)	5.81	(b)
5.82	(d)	5.83	(d)	5.84	(a)	5.85	(a)	5.86	(b)	5.87	(b)	5.88	(a)	5.89	(a)	5.90	(a)
5.91	(a)	5.92	(a)	5.93	(d)	5.94	(a)	5.95	(a)	5.96	(d)	5.97	(c)	5.98	(d)	5.99	(a)
5.100	(a)	5.101	(c)	5.102	(d)	5.103	(b)	5.104	(b)	5.105	(d)	5.106	(a)	5.107	(b)	5.108	(*)
5.109	(c)	5.110	(a)	5.111	(b)	5.112	(d)	5.113	(a)								

Explanations Steam Turbine

5.1 (a)

In partial admission, steam is admitted through some set of nozzles. The steam enters at few sector and not througout the full set of nozzles.

5.2 (d)

Degree of reaction

$$= \frac{\text{Enthalpy drop in moving blade}}{\text{total entalpy drop}}$$

for 100% degree of reaction

Total enthalpy drop = Enthalpy drop in moving blade.

Hence enthalpy drop in fixed blade is zero.

Now when change of absolute velocity of steam across the moving blades is zero. Then total work done by turbine is equal to change in relative K.E. in moving blade then degree of reaction will be 100%.

5.3 (c)

Reheating increase dryness fraction of steam within a stage.

5.5 (d)

Nozzle control governing is more efficient method of governing compared with throttle governing specially at part loads.

5.6 (b)

For multi-stage steam turbine:

$$\text{Reheat factor} = \frac{\text{Cummulative enthalpy drop}}{\text{Isentropic enthalpy drop}}$$

It depends on:

(i) stage efficiency

(ii) initial pressure and condition of steam

(iii) final pressure

It increases with the increase in number of stages. The value of reheat factor varies from 1.02 to 1.06.

5.8 (a)

Condenser size decreases due to regeneration not reheating.

5.10 (b)

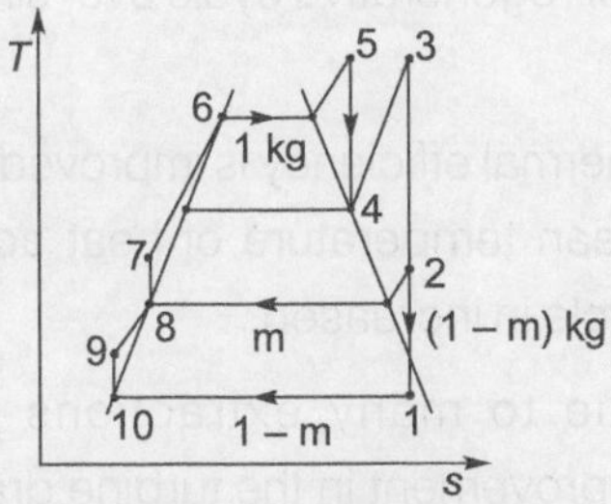

Work developed in the turbine

$= (h_5 - h_4) + (h_3 - h_2) + (1 - m)(h_2 - h_1)$

5.11 (b)

Diagram factor

$$= \frac{\text{Area of actual indicator diagram}}{\text{Area of hypothetical indicator diagram}}$$

In reaction turbine, pressure drop occur both in fixed and moving blade.

5.12 (c)

Blade speed ratio for

$$\text{50\% reaction stage} = \frac{2\cos^2\alpha}{1+\cos^2\alpha}$$

$$\text{Simple impulse stage} = \frac{\cos\alpha}{2}$$

5.13 (b)

For impulse turbine

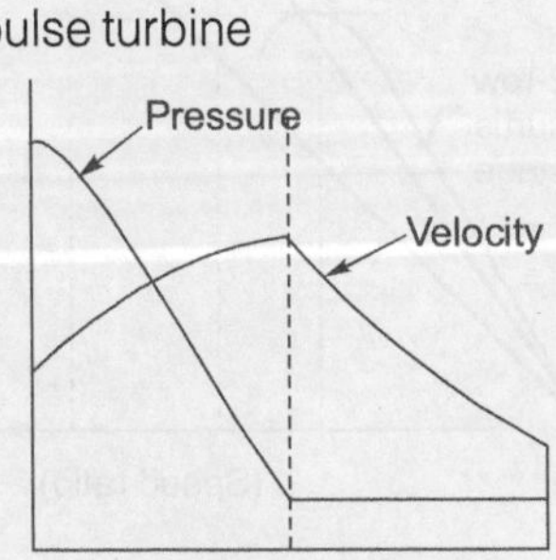

5.14 (b)

Reaction turbine is worked above the atmospheric pressure (greater than impulse turbine pressure) due to this reason reaction blading is commonly used at intermediate and low pressure parts of steam turbines.

5.15 (a)

$$\text{Work done on last stage} = \frac{1}{2^n} \times \text{total work where}$$

n = number of stages

Thus we see that as the number of moving blade rows increase, the work developed by last row decreases. Therefore number of rows are limited and now a days more than two rows are seldom used.

5.16 (a)

Throttle governor reduces the pressure and in turn enthalpy drop for useful work. It also increases initial superheat and delays the dew point. These cause reduction in wetness in pressure stage and reduction in efficiency

5.17 (d)

For maximum efficiency of reaction turbine

$$\frac{u}{V_1} = \cos\alpha_1$$

$$u = V_1 \cos\alpha_1$$

5.18 (b)

1. Anticipatory gear : Turbine governing system
2. Labyrinth : Sealing system
3. Inverted *T*-attachment : Blades
4. Deaerator : Steam power plants

5.19 (c)

$$\text{Degree of reaction} = \frac{(\Delta h)_{\text{moving}}}{(\Delta h)_{\text{stage}}}$$

$$= \frac{20 - 8}{20} \times 100 = \mathbf{60\%}$$

5.20 (a)

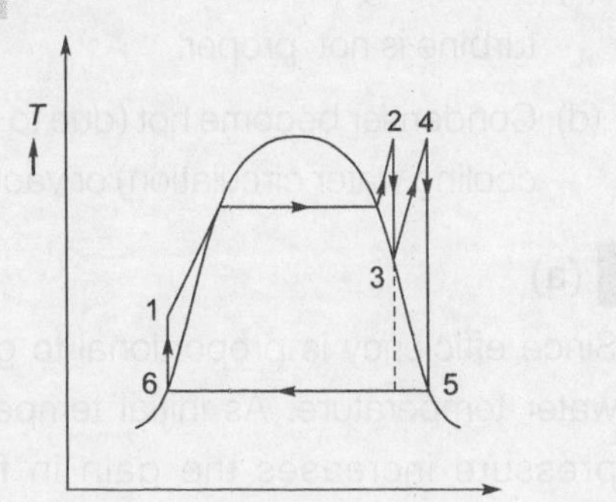

3-4 Reheating

5.21 (a)

$$\text{Work ratio} = \frac{\text{Net work}}{\text{Positive work}} = \frac{W_T - W_P}{W_T}$$

Work ratio unity means $W_P \approx 0$

5.22 (c)

With regeneration the entropy decreases at every extraction.

5.23 (b)

Superheated steam in turbines leads to decreases in erosion of blading, i.e. increase in the life of turbine blade.

5.24 (a)

$$\text{WD/kg of steam} = \frac{1}{2}(30^2 - 10^2) = \mathbf{400\ Nm}$$

5.25 (d)

$$\text{Work done} = u^2 = (75)^2 = \mathbf{5625\ m^2/s^2}$$

5.26 (c)

$$\text{Degree of reaction} = \frac{(\Delta h)_{rotor}}{(\Delta h)_{rotor} + (\Delta h)_{stator}}$$

$$= \frac{2.38}{2.38 + 4.62} = \mathbf{0.34}$$

5.27 (b)

Emergency governor: Every turbine is provided with some form of an emergency governor which trip the turbine (closes the stop valve and stops the steam supply) when,

(a) Shafts exceeds 110% of its rated values.

(b) Lubrication system fails.

(c) Balancing (static as well as dynamic) of turbine is not proper.

(d) Condenser become hot (due to inadequate cooling water circulation) or vacuum is less.

5.28 (a)

Since efficiency is proportional to gain in feed water temperature. As initial temperature and pressure increases the gain in feed water temperature decreases because gain follows the law of diminishing return with increase in the number of heater.

5.31 (a)

Effect of regenerative cycle over simple Rankine cycle

1. Thermal efficiency is improved because the mean temperature of heat addition to the cycle in increased.
2. Due to many extractions there is an improvement in the turbine drainage and it reduces erosion due to moisture.
3. A small size condenser is required.
4. For given power a large capacity boiler is required.

5.35 (d)

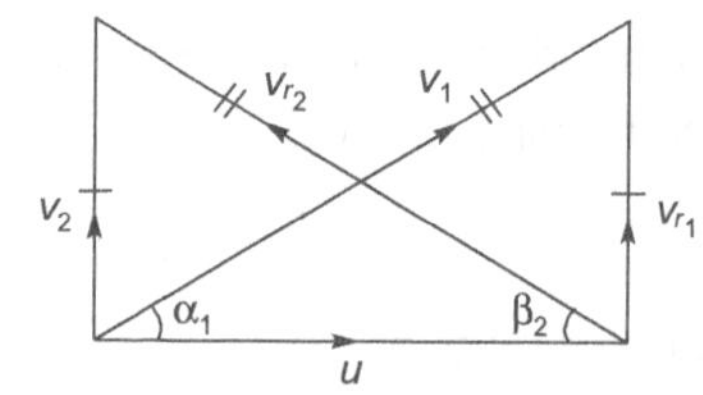

5.36 (b)

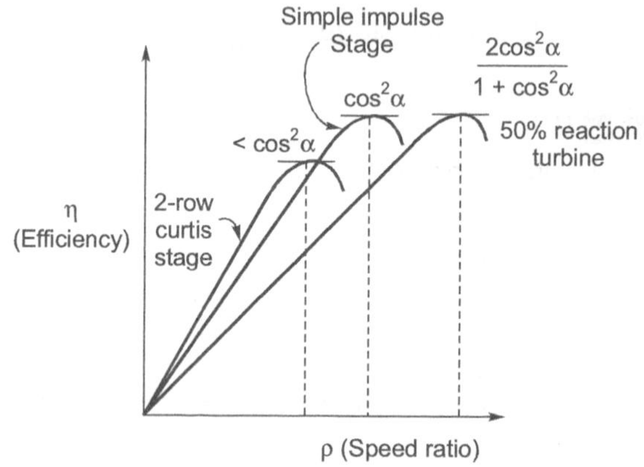

5.37 (a)

Reheat factor is always greater than one. Normally reheat factor is taken as 1.04

5.38 (c)

Regeneration process in steam power plant is accomplished by extracting or 'bleeding' steam from the turbine at various points.

5.39 (b)

Power developed by last stage

$$= \frac{\text{total power developed}}{2^n} \{n = \text{number of row}\}$$

$$\therefore P = \frac{6400}{(2)^4} = \mathbf{400\ kW}$$

5.40 (d)

A combined cycle power plant are those which have both gas and steam turbines supplying power. The idea of combined cycle has come out of the need to improve the simple Brayton cycle efficiency by utilizing the waste heat in the turbine exhaust gases.

5.41 (b)

Due to regeneration, work output decreases due to less amount of flow of steam. For a fixed output, high amount of steam is required, thus boiler capacity increases.

5.42 (d)

Throttle control governing is suitable for small, medium and large turbine. At the part load, there is a reduction in stage efficiency due to reduction in the wetness in the later stage.

5.43 (c)

Since no enthalpy drop takes place in moving blades in impulse turbine, its degree of reaction is zero.

5.44 (b)

Compounding of steam turbine is done to slow down the speed of steam turbine to practicle limit.

5.45 (b)

Velocity: $V = 44.72\,[h_i - h_o]^{1/2}$

$\therefore$ V will be very large for total enthalpy in a single stage.

$$\because \quad V = \frac{\pi DN}{60}$$

Thus rotational speed (N) will also be very large. Thus compounding is required.

5.46 (c)

Turbine speed does not depend upon reheating or regeneration.

5.47 (c)

Due to friction between the casing and the air carried around by the vanes and other losses, which have breaking effects such as disc friction or windage the actual work input is greater than this theoretical value. Work factor is introduced to take account of this.

$$\text{Work factor} = \frac{\text{Isentropic work}}{\text{Euler's work}}$$

The value of work factor is kept nearer to unity but its actual value is about 1.04.

5.48 (a)

Maximum efficiency of Parson's reaction turbine

$$\eta_{max} = \frac{2\cos^2\alpha}{1+\cos^2\alpha}$$

5.50 (a)

Degree of reaction,

$$R = \frac{(\Delta h)_{moving}}{(\Delta h)_{stage}} = \frac{32}{60} = \mathbf{0.533}$$

5.53 (b)

Near the exit of the turbine, steam is usually wet. The water particles in wet steam have no capacity to do work but they are to be carried along with the steam and some kinetic energy is lost in dragging them. Further the impact of water droplets reduce life of the turbine blade as they cause erosion of blades.

5.58 (d)

For maximum diagram efficiency

$$\text{Blade speed ratio} = \frac{\cos\alpha}{2} = \frac{\cos 30^\circ}{2}$$

$$= \frac{0.866}{2} = \mathbf{0.433}$$

5.60 (c)

$$\text{Degree of reaction} = \frac{10}{10+15} = \frac{10}{25} = 0.4 = \mathbf{40\%}$$

5.62 (c)

Single stage impulse steam turbine are not used in practice because as steam is expanded in single stage the rotor speed will be very-very high which is far away from the synchronus speed of generator.

5.63 (a)

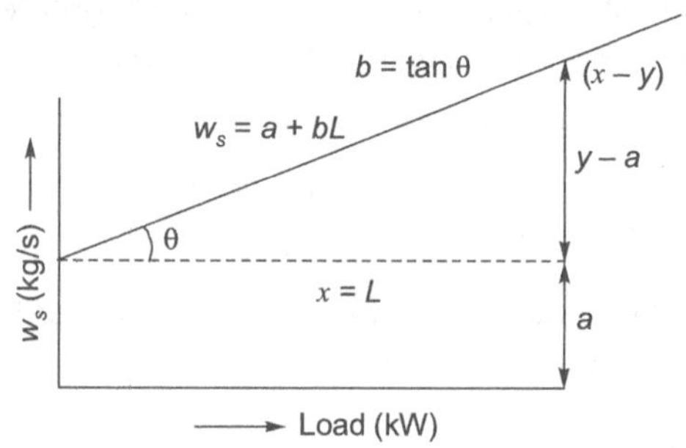

Willan's line shows the steam consumption plotted against the turbine load show a linear relationship with throttle governing

$$w_s = a + bL$$

where a = No load steam consumption(kg/s)
b = Steam rate (kg/kWs); L = Load (kW)

5.64 (b)

When super heated steam expands in nozzle, the condensation should begin when steam reaches the saturated vapour state but there was, delay in condensation because velocities are very high and time available for condensation is less and vapour does not undergo condensation, and it expands as if it is in superheater state. This is called super saturated flow.

5.65 (a)

Compounding of impulse turbine: This is done to reduce the rotational speed of the impulse turbine to practical limits. (A rotor speed of 30,000 rpm is not possible, which is pretty high for practical uses.)

5.67 (a)

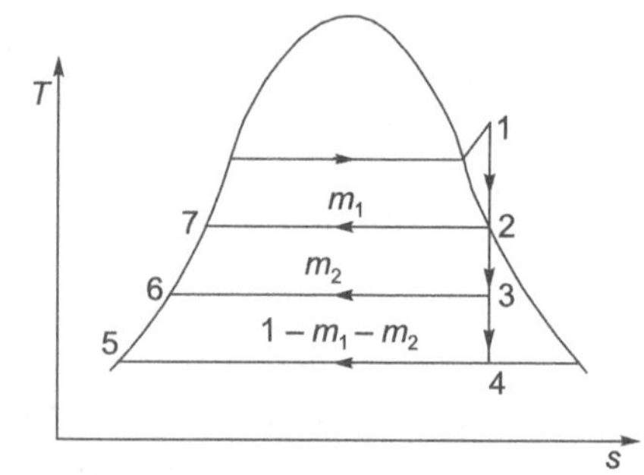

Following are the advantages of regenerative feed heating:

- The heating process in the boiler tends to become reversible.
- Reduced heat rate
- The thermal efficiency is improved because the average temperature of heat addition is increased.
- A small size condenser is required.

5.68 (d)

Compounding is done to reduce the turbine speed to practical limits.

5.69 (c)

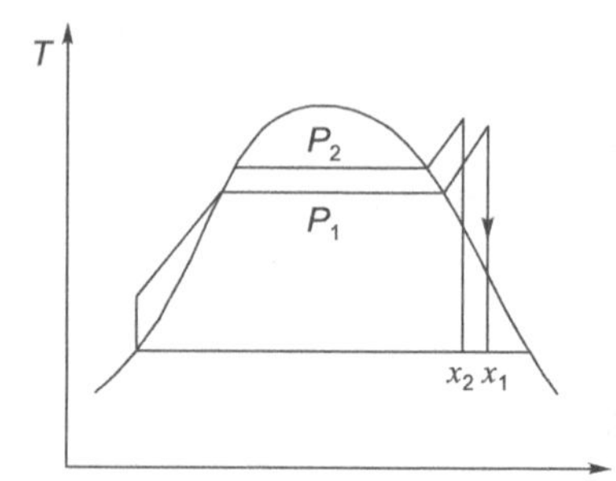

$P_2 > P_1$ and $x_2 < x_1$
So, statement (2) is not correct.

5.70 (c)

Degree of reaction,

$$R_d = \frac{\text{Heat drop in moving blades}}{\text{Heat drop in the stage}}$$

$$\therefore \quad \frac{1}{2} = \frac{\Delta h_m}{\Delta h_m + \Delta h_f}$$

$$\text{or} \quad \Delta h_m = \Delta h_f = \frac{\Delta h_{stage}}{2}$$

5.71 (d)

Advantages of reheating are:

(i) There is an increased output of the turbine

(ii) Erosion and corrosion problems in the steam turbine are avoided.

(iii) There is an increase in thermal efficiency.

5.72 (d)

Modern turbines have reaction stages instead of pressure compounding.

5.73 (b)

Stirling and Ericsson cycles are difficult to achieve in practice because they involve heat transfer through a differential temperature difference in all components including the regenerator. This would require providing infinitely large surface area for heat transfer or allowing infinitely long time for the process. Neither of two cycle is practical.

5.75 (b)

$$\text{Work done} = -\int vdP \text{ (for steady flow process)}$$

Work done in turbine is much greater than that of in pump because in turbine, working fluid is steam and in pump, working fluid is water. And specific volume of steam is much higher than that of water.

5.77 (d)

Statement 1 is incorrect and statements 2 and 3 are correct.

Stirling Cycle:

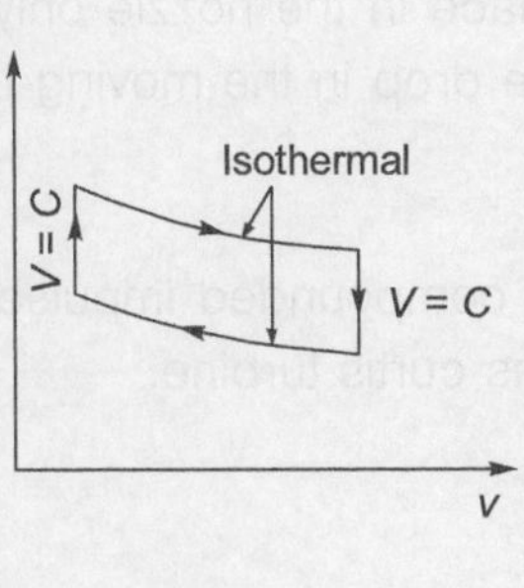

Vapour Compression Cycle:

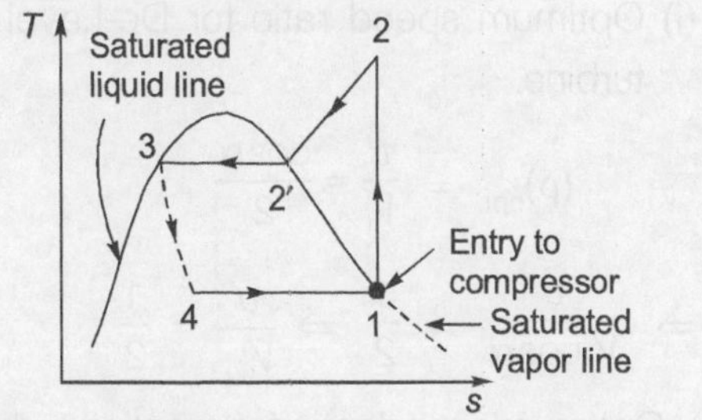

Diesel Cycle:

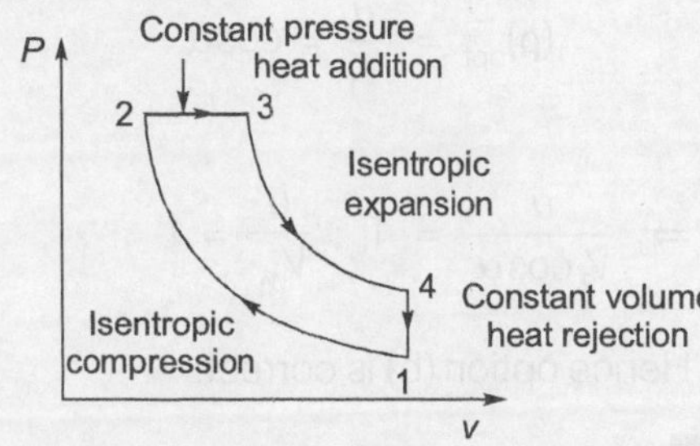

Option (d) is correct.

5.78 (d)

$$\eta_{\text{Rankine}} = 1 - \frac{T_L}{T_m}$$

- By increasing the mean temperature of heat addition, efficiency of Rankine cycle increase.
- By lowering the sink temperature also efficiency of Rankine cycle increase.
- By increasing the pressure of working fluid at which heat added, is practically not feasible due to erosion of turbine blade due to moisture content but mean temperature of heat addition increases and so the efficiency of Rankine cycle increases.

$$\eta_{\text{Rankine}} = f(T_m)$$

Option (d) is correct.

5.79 (b)

For a symmetric (50% R) reaction stage turbine:

$$\rho_{\text{optimum}} = \cos\alpha = \frac{V_b}{V_1}$$

$\therefore$ Blade velocity, $V_b = V_1 \times \cos\alpha$

So, statement 2 is wrong.

5.80 (b)

(i) Optimum speed ratio for De-Lavel impulse turbine.

$$(\rho)_{opt.} = \frac{u}{V_1} = \frac{\cos\alpha}{2}$$

$$\Rightarrow \frac{u}{V_1\cos\alpha} = \frac{1}{2} \Rightarrow \frac{u}{V_{w_1}} = \frac{1}{2}$$

(ii) Optimum speed ratio for reaction turbine:

$$(\rho)_{opt} = \frac{u}{V_1} = \cos\alpha$$

$$\Rightarrow \frac{u}{V_1\cos\alpha} = 1 \; ; \frac{u}{V_{w_1}} = 1$$

Hence option (b) is correct.

5.93 (d)

$$\rho_{opt} = \frac{v}{V_1} = \frac{\cos\alpha}{2} = \frac{\cos 18°}{2} = 0.48$$

$$W = 120 \text{ kJ/kg} = u\cdot\left(V_{w_1} + V_{w_2}\right)$$

$$= u\cdot\left(V_1\cos\alpha + V_{r_2}\cos\phi - u\right)$$

On solving $V_1 = 515.1$ m/s

5.96 (d)

Compoundings are of two types (1) velocity compounding and (2) Pressure compounding.

5.99 (a)

Steam turbines are governed by the following methods:

- throttle governing
- nozzle control governing
- by-pass governing

5.100 (a)

- Statement 1 is the definition of isentropic efficiency (not mechanical efficiency).
- Pressure compounded impulse turbine - Rateau type.
- Velocity compounded impulse turbine - Curtis type..

5.101 (c)

Given:

Mean blade velocity,

$$u = 144 \text{ m/s}$$

$$\text{Blade velocity ratio} = \frac{u}{V} = 0.48$$

$$\Rightarrow \quad V = \frac{144}{0.48} = 300 \text{ m/s}$$

$$\text{Actual enthalpy drop} = \frac{V^2}{2000} = \frac{300^2}{2000}$$

$$= 45 \text{ kJ/kg}$$

$$\text{Nozzle efficiency} = \frac{\text{Actual enthalpy drop}}{\text{Isentropic enthalpy drop}} \quad \ldots(i)$$

In this question nozzle efficiency is missing, Taking nozzle efficiency $\eta = 0.92$

From (i)

$$\text{Isentropic enthalpy drop} = \frac{45}{0.92} = 48.91 \text{ kJ/kg}$$

5.102 (d)

Given: $\alpha = 30°$; $K = 0.85$

$$\therefore \quad \eta_{max} = \frac{\cos^2\alpha(1+K)}{2}$$

$$= \frac{\cos^2 30(1+0.85)}{2} = 0.6937$$

$$\therefore \quad \eta_{max} = 69.37\%$$

5.103 (b)

In impulse turbine stage, the pressure drop takes place in the nozzle only so there is no pressure drop in the moving blades.

5.104 (b)

Velocity compounded impulse turbine is also known as curtis turbine.

5.105 (d)

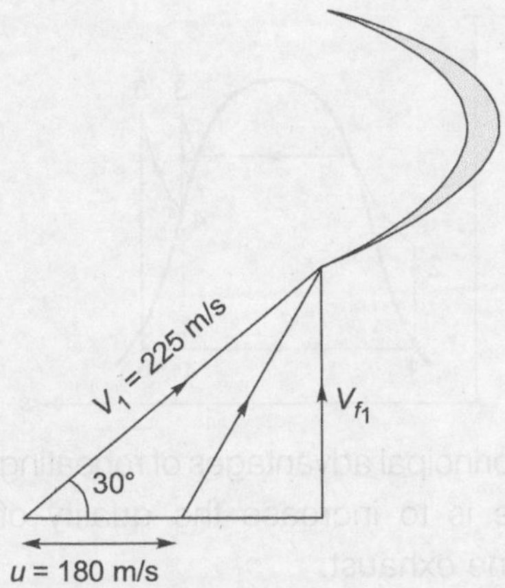

$$\frac{u}{V_1} = 0.8$$

$$\Rightarrow \quad \frac{180}{V_1} = 0.8$$

$$\Rightarrow \quad V_1 = \frac{180}{0.8} = 225 \text{ m/s}$$

Axial velocity component at the inlet;

$$V_{f_1} = V_1 \sin 30° = 112.5 \text{ m/s}$$

5.106 (a)

Multi-cylinder turbines which have their rotors mounted on one and the same shaft and coupled to a generator are known as single-shaft turbines. It is also known as tendem compounding turbines.

5.107 (b)

- In reaction turbine, the steam expands in the fixed blades and moving blades and attains high velocity.
- In steam turbine, the steam consumption increases with the increase in years of service.

5.108 (*)

- As an increase in the initial pressure of steam, results in increase in mean temperature of heat addition, thus increases the thermal efficiency.
- At same pressure, raising the temperature of steam results in increase in mean temperature of heat addition, thus increases the thermal efficiency.
- Intermediate reheating of steam can increase or decrease the thermal efficiency. As thermal efficiency in intermediate reheating is directly proportional to mean temperature of heat addition.
- An increase in the initial pressure of steam improves the thermal efficiency of steam power plant and increases the wetness fraction or decreases dryness fraction of steam at the end of expansion.

5.109 (c)

- Blade or diagram efficiency is the ratio of work done on the blade per second to energy entering the blade per second.
- $\eta_{internal} = \dfrac{\text{Heat converted into useful work}}{\text{Total adiabatic heat drop}}$
- Net efficiency or efficiency ratio

  $= \dfrac{\text{Brake thermal efficiency}}{\text{Thermal efficiency on the Rankine cycle}}$
- Actual thermal efficiency

  $= \dfrac{\text{Heat converted into useful work per kg of steam}}{\left[\begin{matrix}(\text{Total heat in steam at stop valve}) \\ -(\text{Waste heat in exhaust})\end{matrix}\right]}$
- Rankine efficiency

  $= \dfrac{\text{Adiabatic heat drop}}{\left[\begin{pmatrix}\text{Total heat in steam} \\ \text{at stop valve}\end{pmatrix} - \begin{pmatrix}\text{Waste heat in} \\ \text{exhaust}\end{pmatrix}\right]}$
- $\eta_{net} = \dfrac{\text{Heat converted into useful work}}{\text{Total adiabatic heat drop}}$

5.110 (a)

- Bleeding is the process of draining steam from the turbine, at certain points during its expansion and using this steam for heating the feed water supplied to the boiler.
- There is usual practice in bleeding installations to allow the bled steam to mix with the feed water. The mixture of steam and water then proceeds to the boiler.

5.111 (b)

We know that, thermal efficiency of the Rankine cycle is given by

$$\eta_{th} = 1 - \left(\frac{Q_R}{Q_s}\right)$$

Thus, Rankine cycle efficiency improves by increasing heat addition (Q_s) in boiler.

5.112 (d)

Isentropic expansion efficiency is the ratio of actual work in expansion to ideal work in expansion.

5.113 (a)

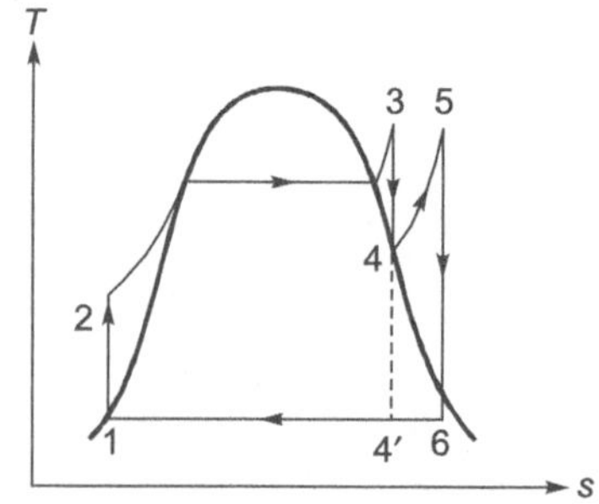

The principal advantages of reheating in Rankine cycle is to increase the quality of steam at turbine exhaust.

6 Gas Turbines

6.1 Assertion (A): The thermal efficiency of Brayton cycle would not necessarily increase with reheat.

Reason (R): Constant pressure lines on the T-s diagram slightly diverge with increase in entropy.

[ESE : 2000]

6.2 Assertion (A): The air-fuel ratio employed in a gas turbine is around 60 : 1.

Reason (R): A lean mixture of 60 : 1 in a gas turbine is mainly used for complete combustion.

[ESE : 2000]

6.3 The efficiency of a simple gas turbine can be improved by using a regenerator, because the

(a) work of compression is reduced
(b) heat required to be supplied is reduced
(c) work output of the turbine is increased
(d) heat rejected is increased **[ESE : 2000]**

6.4 The thermal efficiency of a gas turbine cycle with regeneration in terms of T_3 (maximum temperature), T_1 (minimum temperature), r_p (pressure ratio) and $k\left(=\frac{c_p}{c_v}\right)$ is given by

(a) $1-\frac{T_1}{T_3} r_p^{\left(\frac{k}{k-1}\right)}$ (b) $1-\frac{T_3}{T_1} r_p^{\left(\frac{k}{k-1}\right)}$

(c) $1-\frac{T_3}{T_1} r_p^{\left(\frac{k-1}{k}\right)}$ (d) $1-\frac{T_1}{T_3} r_p^{\left(\frac{k-1}{k}\right)}$

[ESE : 2001]

6.5 Brayton cycle with infinite intercooling and reheating stages would approximate a

(a) Stirling cycle (b) Ericsson cycle
(c) Otto cycle (d) Atkinson cycle

[ESE : 2002]

6.6 Consider the following statements about modification in a gas turbine power plant working on a simple Brayton cycle:

1. Incorporation of regeneration process increases specific work output as well thermal efficiency.
2. Incorporation of regeneration process increases thermal efficiency but specific work output remains unchanged.
3. Incorporation of intercooling process in a multi-stage compression increases specific work output but the heat input also increases.
4. Incorporation of intercooling process in a multi-stage compression system increases specific work output, the heat addition remains unchanged.

Which of these statements are correct?

(a) 1 and 3 (b) 1 and 4
(c) 2 and 3 (d) 2 and 4

[ESE : 2003]

6.7 Assertion (A): In constant pressure type gas turbines, large quantity of air is used, in excess of its combustion requirements.

Reason (R): Excess air is used to compensate for inevitable air-loss due to leakages in the system.

(a) Both **A** and **R** are individually true and **R** is the correct explanation of **A**
(b) Both **A** and **R** are individually true but **R** is not the correct explanation of **A**
(c) **A** is true but **R** is false
(d) **A** is false but **R** is true **[ESE : 2003]**

6.8 Consider the following statements:

1. The speed of rotation of the moving elements of gas turbines is much higher than those of steam turbines.
2. Gas turbine plants are heavier and larger in size than steam turbine plants.
3. Gas turbines require cooling water for its operations.
4. Almost any kind of fuel can be used with gas turbines.

Which of these statements are correct?

(a) 1 and 2 (b) 1 and 3
(c) 1 and 4 (d) 3 and 4

[ESE : 2004]

6.9 Consider the following statements:
Which of the following increase the work ratio in a simple gas turbine plant?
1. Heat exchanger
2. Inter cooling
3. Reheating

Select the correct answer using the code given below:
(a) 1 and 2 (b) 2 and 3
(c) 1 and 3 (d) 1, 2 and 3

[ESE : 2005]

6.10 In a Brayton cycle, the value of optimum pressure ratio for maximum net work done between temperatures T_1 and T_3, where T_3 is the maximum temperature and T_1 is the minimum temperature is

(a) $r_p = \left(\frac{T_3}{T_1}\right)^{\frac{\gamma}{\gamma-1}}$ (b) $r_p = \left(\frac{T_3}{T_1}\right)^{\frac{\gamma-1}{2\gamma}}$

(c) $r_p = \left(\frac{T_3}{T_1}\right)^{\frac{\gamma}{2(\gamma-1)}}$ (d) $r_p = \left(\frac{T_3}{T_1}\right)^{\frac{2(\gamma-1)}{\gamma}}$

[ESE : 2005]

6.11 Consider the following features for a gas turbine plant:
1. Intercooling
2. Regeneration
3. Reheat

Which of the above features in a simple gas turbine cycle increase the work ratio?
(a) 1, 2 and 3
(b) Only 1 and 2
(c) Only 2 and 3
(d) Only 1 and 3

[ESE : 2006]

6.12 Use of maximum pressure ratio, corresponding to maximum to minimum cycle temperature ratio in case of Joule cycle gives which one of the following?
(a) Maximum efficiency but very less specific work output.
(b) Maximum efficiency and very high specific work output
(c) Minimum efficiency and very less specific work output
(d) Minimum efficiency and very high specific work output

[ESE : 2007]

6.13 Consider the following statements in respect of gas turbines:
A gas turbine plant with reheater leads to a
1. considerable improvement in the work output.
2. considerable improvement in the thermal efficiency.

Which of these statements is/are correct?
(a) 1 only (b) 2 only
(c) Both 1 and 2 (d) Neither 1 nor 2

[ESE : 2007]

6.14 An open cycle constant pressure gas turbine uses a fuel of calorific value 40,000 kJ/kg, with air fuel ratio of 80 : 1 and develops a net output of 80 kJ/kg of air. The thermal efficiency of the cycle is
(a) 61 % (b) 16%
(c) 18% (d) None of these

[ESE : 2007]

6.15 For the same net power output
(a) the turbine used in gas turbine power plants is larger than that used in steam power plants
(b) the turbine used in gas turbine power plants is smaller than that used in steam power plants
(c) the same turbine can be used for both plants
(d) none of above

[ESE : 2007]

6.16 Consider the following statements in respect of gas turbine:
1. Supersonic flow leads to decreases in efficiency.
2. Supersonic flow leads to decrease in flow rate.

Which of these statements is/are correct?
(a) 1 only (b) 2 only
(c) Both 1 and 2 (d) Neither 1 nor 2

[ESE : 2007]

6.17 A gas turbine plant working on Joule cycle produces 4000 kW of power. If its work ratio is 40%. The power consumed by the compressor is
(a) 2000 kW (b) 4000 kW
(c) 6000 kW (d) 8000 kW

[ESE : 2008]

6.18 In a gas turbine cycle with regeneration
(a) pressure ratio increases
(b) work output decreases
(c) thermal efficiency increases
(d) heat input increases

[ESE : 2008]

6.19 Consider the following statements pertaining to gas turbines:

1. The degree of reaction of a reaction turbine is the ratio of energy transfer in fixed blade to the overall energy transfer across a stage.
2. The overall pressure drop in a turbine is the product of pressure drop per stage and number of stages.
3. Gas turbine cycle (Brayton cycle) is not as efficient as Rankine cycle for steam.

Which of these statements is/are correct?

(a) 1 only (b) 2 only
(c) 2 and 3 (d) 3 only

[ESE : 2009]

6.20 In a simple single stage gas turbine plant, if T_1 is the minimum temperature and T_3 is the maximum temperature then the work ratio in terms of r_p is

(a) $1-\frac{T_3}{T_1}(r_p)^{\frac{\gamma-1}{\gamma}}$ (b) $1-\frac{T_1}{T_3}(r_p)^{\frac{\gamma-1}{\gamma}}$

(c) $1-\frac{T_1}{T_3}(r_p)^{\frac{\gamma}{\gamma-1}}$ (d) $1-\frac{T_1}{T_3}(r_p)^{\frac{1}{\gamma}}$

[ESE : 2009]

6.21 Increasing the number of reheating stages in a gas turbine to infinity, makes the expansion tending:

(a) Reversible adiabatic
(b) Isothermal
(c) Isobaric
(d) Adiabatic

[ESE : 2009]

6.22 The *T-s* diagram for a gas turbine plant is drawn for the case where

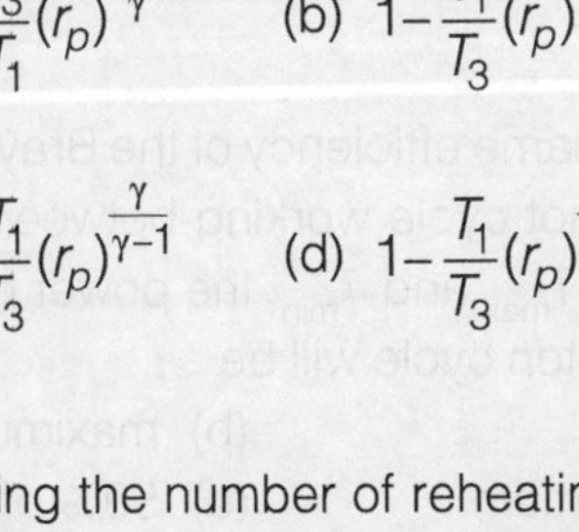

(a) Compression of air is done in two stages incorporating an intercooler between two.
(b) Expansion of gases is done in two stages followed by regeneration.
(c) Expansion of gases is done in two stages with a reheater between the two.
(d) Expansion of gases is done in two stages with a reheater between the two followed by regeneration.

[ESE : 2009]

6.23 In which modification of simple gas turbine cycle, is work ratio increased?

1. Regenerative gas turbine cycle.
2. Gas turbine cycle with reheating.

(a) Both 1 and 2 (b) 1 only
(c) 2 only (d) Neither 1 nor 2

[ESE : 2011]

6.24 Reheat between multistage expansions in Joule cycle increases

1. Overall work output.
2. The work ratio.
3. The thermal efficiency.

Which of the above are correct?

(a) 1, 2 and 3 (b) 1 and 2 only
(c) 2 and 3 only (d) 1 and 3 only

[ESE : 2011]

6.25 The gas in a cooling chamber of a closed-cycle gas turbine is cooled at

(a) constant volume (b) constant temperature
(c) constant pressure (d) None of the above

[ESE : 2012]

6.26 In a gas turbine power plant, reheating of gases between the high pressure and low pressure turbine stages will

(a) improve turbine output
(b) decrease turbine output
(c) increase compressor output
(d) decrease compressor work

[ESE : 2012]

6.27 In a gas turbine plant, regeneration is done to

(a) increase compression work
(b) decrease turbine work
(c) limit the maximum temperature
(d) improve plant efficiency

[ESE : 2012]

6.28 Consider a gas turbine supplied with gas at 1000 K and 5 bar to expand adiabatically to a 1 bar. The mean specific heat at constant pressure is 1.0425 kJ/kgK and constant volume is 0.7662 kJ/kgK. Calculate power developed in kW/kg of gas per second and exhaust gas temperature

(a) 462 kW/kg and 647 K
(b) 362 kW/kg and 653 K
(c) 462 kW/kg and 653 K
(d) 362 kW/kg and 647 K [ESE : 2013]

6.29 The thermal efficiency of a simple open gas turbine plant is improved by regeneration as this

(a) decreases the temperature of the gases at the turbine inlet
(b) decreases the quantity of heat supplied in combustion chamber
(c) increases the turbine output
(d) lowers the work input to compressor

[ESE : 2013]

6.30 In a two stage gas turbine plant, with intercooling and reheating

(a) Both work ratio and thermal efficiency increases
(b) Work ratio increases but thermal efficiency decreases
(c) Thermal efficiency increases but work ratio decreases
(d) Both work ratio and thermal efficiency decreases [ESE : 2014]

6.31 The maximum net specific work obtainable in an ideal Brayton cycle for T_{max} = 900 K and T_{min} = 400 K is given by

(a) 100 c_p (b) 500 c_p
(c) 700 c_p (d) 800 c_p

[ESE : 2015]

6.32 An open cycle pressure gas turbine uses a fuel of calorific value 40,000 kJ/kg with air-fuel ratio of 80:1 and develops a net output of 80 kJ/kg of air. The thermal efficiency of the cycle is

(a) 12% (b) 16%
(c) 20% (d) 18%

[ESE : 2015]

6.33 The maximum work developed by a closed cycle used in a gas turbine plant when it is working between 900 K and 289 K and using air as working substance is :

(a) 11 kJ/kg (b) 13 kJ/kg
(c) 17 kJ/kg (d) 21 kJ/kg

[ESE : 2016]

6.34 In a gas turbine cycle, the turbine output is 600 kJ/kg, the compressor work is 400 kJ/kg, and the heat supplied is 1000 kJ/kg. The thermal efficiency of the cycle is :

(a) 20% (b) 30%
(c) 40% (d) 50%

[ESE : 2016]

6.35 Which of the following units increase the work ratio in a gas turbine plant ?

1. Regeneration 2. Reheating
3. Intercooling

(a) 1 and 2 only (b) 2 and 3 only
(c) 1 and 3 only (d) 1, 2 and 3

[ESE : 2016]

6.36 The gas turbine blades are subjected to :

(a) High centrifugal stress and thermal stress
(b) Tensile stress and compressive stress
(c) High creep and compressive stress
(d) Compressive stress and thermal stress

[ESE : 2016]

6.37 For the same efficiency of the Brayton cycle and the Carnot cycle working between temperature limits of T_{max} and T_{min}, the power contribution of the Brayton cycle will be

(a) zero (b) maximum
(c) minimum (d) 50% of Carnot cycle

[ESE : 2017]

6.38 Consider the following statements in respect of ideal and practical gas turbine cycles:

1. In the ideal cycle case, the cycle efficiency depends on the pressure ratio only.
2. In the practical cycle case (with irreversibilities in the compression and expansion processes), the cycle efficiency depends on the maximum temperature as well as on the pressure ratio.
3. In the practical cycle case, at a given maximum temperature, the maximum efficiency and the maximum work done occur at a same pressure ratio.

Which of the above statements are correct?

(a) 1 and 2 only (b) 1 and 3 only
(c) 2 and 3 only (d) 1, 2 and 3

[ESE : 2017]

6.39 An ideal closed-cycle gas turbine plant is working between the temperatures 927°C and 27°C using

air as working fluid. The pressure ratio for maximum output is
(a) 11.3 (b) 13.3
(c) 15.3 (d) 17.3 [ESE : 2018]

6.40 Statement (I) : The air-standard efficiency of Brayton cycle depends only on the pressure ratio.
Statement (II) : For the same compression ratio, the air-standard efficiency of Brayton cycle is equal to that of Otto cycle.
[ESE : 2018]

6.41 Statement (I) : The air-fuel ratio employed in a gas turbine is around 60 : 1.
Statement (II) : A lean mixture of 60 : 1 in a gas turbine is mainly used for complete combustion.
[ESE : 2018]

6.42 Air enters the compressor of a gas turbine plant operating on Brayton cycle at 1 bar and 27°C. The pressure ratio in the cycle is 6. If the relation between the turbine work W_T and compressor work W_C is $W_T = 3W_C$ and $\gamma = 1.4$, the cycle efficiency will be nearly
(a) 40% (b) 50%
(c) 60% (d) 70%
[ESE : 2020]

6.43 The ideal thermodynamic cycle for the development of gas-turbine engine is
(a) Otto (b) Stirling
(c) Ericsson (d) Brayton
[ESE : 2020]

6.44 In a gas turbine plant, heat supplied is 667.2 kg/kg, and heat rejected is 391.43 kJ/kg. What is the thermal efficiency of the plant?
(a) 57.29% (b) 72.51%
(c) 41.33% (d) 32.83%
[ESE : 2021]

6.45 The constant pressure gas turbine works on
(a) Stirling Cycle (b) Atkinson Cycle
(c) Rankine Cycle (d) Brayton Cycle
[ESE : 2021]

6.46 Consider the following common failure mechanisms found in gas turbine blades:
1. Mechanical damage
2. High temperature damage
3. Creep failures

Which of the above failure mechanisms are correct?
(a) 1 and 2 only (b) 1 and 3 only
(c) 2 and 3 only (d) 1, 2 and 3
[ESE : 2023]

6.47 Consider the following features for a gas turbine plant:
1. Intercooling
2. Regeneration
3. Reheat

Which of the above features in a gas turbine cycle increase the specific output and thermal efficiency of the plant?
(a) 1, 2 and 3 (b) 1 and 2 only
(c) 2 and 3 only (d) 1 and 3 only
[ESE : 2024]

■■■■

Answers Gas Turbines

6.1 (b)	6.2 (c)	6.3 (b)	6.4 (d)	6.5 (b)	6.6 (c)	6.7 (c)	6.8 (c)	6.9 (b)
6.10 (c)	6.11 (d)	6.12 (a)	6.13 (a)	6.14 (b)	6.15 (a)	6.16 (a)	6.17 (c)	6.18 (c)
6.19 (c)	6.20 (b)	6.21 (b)	6.22 (c)	6.23 (c)	6.24 (b)	6.25 (c)	6.26 (a)	6.27 (d)
6.28 (b)	6.29 (b)	6.30 (b)	6.31 (a)	6.32 (b)	6.33 (*)	6.34 (a)	6.35 (b)	6.36 (a)
6.37 (a)	6.38 (a)	6.39 (a)	6.40 (b)	6.41 (c)	6.42 (a)	6.43 (d)	6.44 (c)	6.45 (d)
6.46 (d)	6.47 (a)							

Explanations Gas Turbines

6.2 (c)

A highly lean mixture is used to produce cooling effect in combustion chamber.

So temperature of flue gas at inlet of turbine will be within limit.

6.3 (b)

The efficiency of the turbine can be increased by utilizing part of the energy of the exhaust gas from the turbine in heating up the air leaving the compressor in a heat exchanger called regenerator, there by reducing the amount of heat supplied from an external source and also the amount of heat rejected.

6.4 (d)

Thermal efficiency of gas cycle with regeneration

$$\eta = \left[1 - \frac{T_{min}}{T_{max}}(r_p)^{\frac{k-1}{k}}\right]$$

6.5 (b)

In introducing multistage compression with intercooling and multistage expansion with reheating. It is seen from the *T-s* diagram that an approach is being made towards isothermal process and cycle obtained is known as Ericsson cycle.

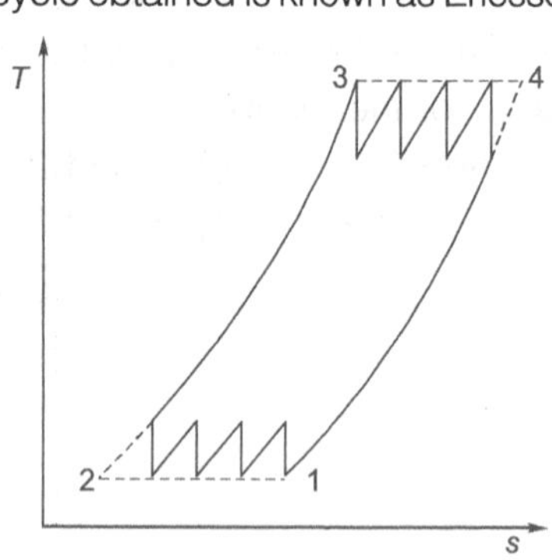

6.6 (c)

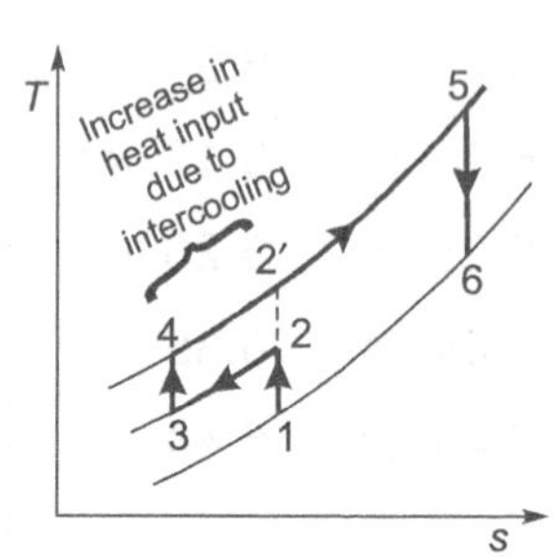

The effect of regeneration process is to reduce the amount of heat required from external source. The specific work and work ratio are unchanged but efficiency is increased. Intercooling increases heat input and decreases the compressor work, i.e. increases specific work out.

6.7 (c)

A highly lean mixture is used to produce cooling effect in combustion chamber.

So temperature of flue gas at inlet of turbine will be within limit.

6.8 (c)

Due to boiler and condenser, steam turbine plants are heavier than gas turbine plants. Gas turbine doesn't required cooling water for its operation.

6.9 (b)

$$\text{Work ratio : WR} = \frac{W_T - W_C}{W_T} = 1 - \frac{W_C}{W_T}$$

Reheating increases the turbine work hence work ratio increases. Similarly intercooling reduces the compression work hence work ratio increases. Heat exchanger increases the thermal efficiency but work ratio remains unchanged.

6.10 (c)

Optimum pressure ratio for maximum workout put

$$(r_p)_{opt} = \left(\frac{T_3}{T_1}\right)^{\gamma/2(\gamma-1)}$$

6.11 (d)

$$\text{Work ratio} = \frac{\text{Net work}}{\text{Positive work}} = \frac{W_T - W_C}{W_T}$$

6.12 (a)

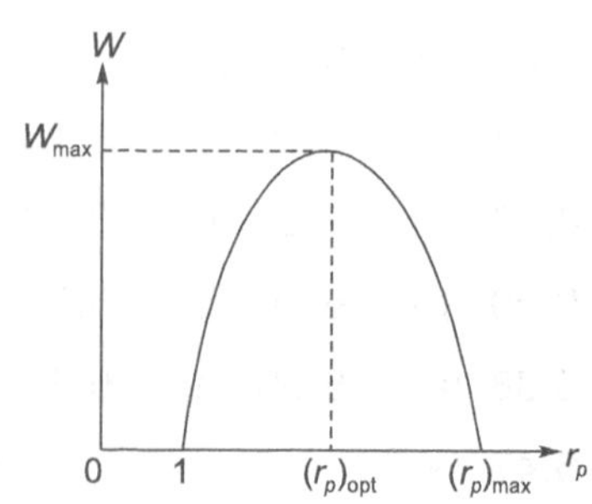

Hence for optimum pressure ratio work output is maximum and for maximum pressure ratio efficiency is maximum but work output is zero.

6.13 (a)

Reheating increases the work output but the thermal efficiency may be increased or decreased. It depends on mean temperature of heat addition.

6.14 (b)

$$\eta = \frac{W_{out}}{Q_{in}} = \frac{80 \times \dot{m}_a}{\dot{m}_f \times 40000}$$

Since $\dfrac{\dot{m}_a}{\dot{m}_f} = 80$

$$\eta = \frac{80 \times 80}{40000} \times 100 = 16\%$$

6.15 (a)

Expansion process in gas turbine is larger than steam turbine. Hence turbine used in gas turbine power plants is larger than that used in steam power plants, for same net power output.

6.17 (c)

$$\frac{W_T - W_C}{W_T} = 0.4$$

$$\Rightarrow \quad \frac{4000}{W_T} = 0.4$$

$$\Rightarrow \quad W_T = \frac{4000}{0.4} = 10{,}000 \text{ kW}$$

or $W_T - W_C = 4000$

$\Rightarrow 10{,}000 - W_C = 4000$

or $W_C =$ **6000 kW**

6.18 (c)

Regeneration increases the thermal efficiency of gas turbine plant.

6.21 (b)

Increasing the number of Reheating stages in a gas turbine to infinity makes the expansion tending to isothermal.

Increasing the intercooling in compressor to infinity also makes the compression tending to isothermal.

This is known as Ericssionisation of the Brayton cycle.

6.23 (c)

Work ratio is the ratio of the net work done to the turbine work.

6.24 (b)

The thermal efficiency depends on the heat supplied also. Because the efficiency is the ratio of network done to the heat supplied. The thermal efficiency may increase or decrease.

6.25 (c)

In closed-cycle gas turbine the cooling process occurs at constant pressure.

6.26 (a)

Reheating facilitates more work to be extracted from the same mass of flowing steam and hence the turbine output increases.

6.27 (d)

Regeneration increases the mean temperature of heat addition and hence it increases the plant efficiency.

6.28 (b)

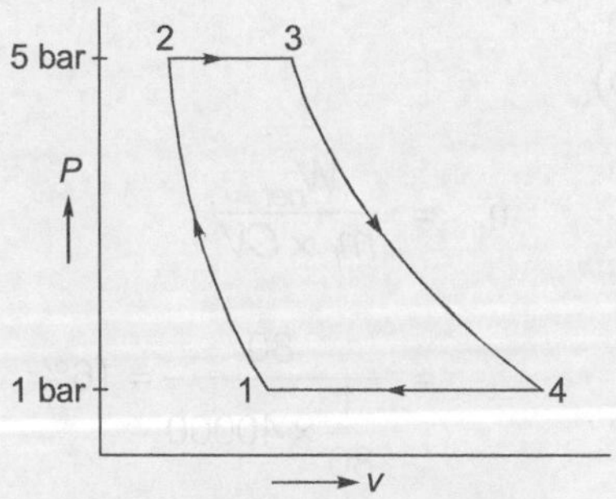

$c_p = 1.0425$ kJ/kgK

$c_v = 0.7662$ kJ/kgK

$$\gamma = \frac{c_p}{c_v} = \frac{1.0425}{0.7662} = 1.36$$

$$\frac{T_3}{T_4} = \left(\frac{P_3}{P_4}\right)^{\frac{\gamma-1}{\gamma}} = (5)^{\frac{0.36}{1.36}} = 1.531$$

$$\therefore \quad T_4 = \frac{1000}{1.531} = 653\text{K}$$

(T_4 is Exhaust gas temperature)

$$w_{net} = c_p(T_3 - T_4)$$

$$= 1.0425(1000 - 653)$$

$$= 362 \text{ kJ/kg}$$

$$\therefore \quad w_{net} = 362 \frac{\text{kJ/s}}{\text{kg/s}}$$

$$= \mathbf{362\ kW/kg\ of\ gas/s}$$

6.29 (b)

As the turbine output decreases, steam rate increases and so, quantity of heat supplied in the combustion chamber decreases.

6.30 (b)

Comparison of Various Cycles			
S.No.	**Addition to simple cycle**	η	$\frac{W}{c_P T_1}$
1	Heat exchange	+ve	NIL
2	Intercooling	–ve	+ve
3	Reheat	–ve	+ve
4	Reheat and intercooling	–ve	+ve

6.31 (a)

$$(w_{net})_{max} = c_p\left(\sqrt{T_{max}} - \sqrt{T_{min}}\right)^2$$

$$= c_p(30 - 20)^2 = \mathbf{100\, c_p}$$

Option (a) is correct.

6.32 (b)

$$\eta_{th} = \frac{W_{net}}{m_f \times CV}$$

$$= \frac{80}{\frac{1}{80} \times 40000} = 16\%$$

6.33 (*)

$$W_{max} = c_p\left(\sqrt{T_{max}} - \sqrt{T_{min}}\right)^2$$

$$= 1.005\left(\sqrt{900} - \sqrt{289}\right)^2$$

$$= \mathbf{169.87\ kJ/kg}$$

6.34 (a)

$$\eta_{Th} = \frac{W_T - W_C}{Q_s} = \frac{600 - 400}{1000} = \mathbf{20\%}$$

6.36 (a)

(Due to high rotational speeds of GT (No. of RPM 30000)), high centrifugal stresses induced.

6.39 (a)

$$(r_p)_{opt} = \left(\frac{T_{max}}{T_{min}}\right)^{\frac{\gamma}{2(\gamma-1)}} = \left(\frac{1200}{300}\right)^{\frac{1.4}{2\times(1.4-1)}}$$

$$= (4)^{\frac{1.4}{0.8}} = (4)^{\frac{14}{8}} = (4)^{1.75}$$

$$= (4)^1(4)^{\frac{3}{4}} = (4)^1(64)^{\frac{1}{4}}$$

$$= (4) \times 2\sqrt{2} = 4 \times 2 \times 1.414 = 11.31$$

6.43 (d)

Brayton cycle is the ideal gas turbine cycle.

6.44 (c)

$$\eta = 1 - \frac{Q_R}{Q_S} = 1 - \frac{391.43}{667.2} = 1 - 0.5866$$

$$= 0.4134 = 41.34\%$$

6.46 (d)

The following are the common failure mechanisms found in gas turbine blades:

- Mechanical damage
- High temperature damage
- Creep failures
- Fatigue failures
- Corrosion failures

6.47 (a)

- Intercooling and reheating in gas turbine results in decrease in work input to the compressor and increasing work output to the turbine. Thus, it will result in increase in net specific output.
- Whereas regeneration in gas turbine, increases the mean temperature of heat addition and decreases mean temperature of heat rejection. Thus regeneration increases the efficiency of gas turbine plant.
- The combined effect of intercooling, regeneration and reheating results in increase the specific output and thermal efficiency of the gas turbine cycle.

■■■■

7 Jet Propulsion and Compressible Flow

7.1 An aeroplane travels at 400 km/hr at sea level where the temperature is 15°C. The velocity of the aeroplane at the same Mach number at an altitude where a temperature of –25°C is prevailing, would be

(a) 126.78 km/hr (b) 130.6 km/hr
(c) 371.2 km/hr (d) 400.10 km/hr

[ESE : 2000]

7.2 If the upstream Mach number of a normal shock occurring in air (k = 1.4) is 1.68, then the Mach number after the shock is

(a) 0.84 (b) 0.646
(c) 0.336 (d) 0.564 [ESE : 2000]

7.3 In turbo prop, the expansion of gases takes place approximately

(a) 100% in the turbine
(b) 80% in the turbine and 20% in the nozzle
(c) 50% in the turbine and 50% in the nozzle
(d) 100% in the nozzle [ESE : 2000]

7.4 With reference to Turbo-jet and Rocket engines, consider the following statements:

1. Efficiency of Rocket engines is higher than that of Jet engines.
2. Exit velocities of exhaust gases in Rocket engines are much higher than those in Jet engines.
3. Stagnation conditions exists at the combustion chamber in Rocket engines.
4. Rocket engines are air-breathing engines.

Which of these statements are correct?

(a) 1 and 2 (b) 1, 3 and 4
(c) 2, 3 and 4 (d) 1, 2 and 3

[ESE : 2001]

7.5 In a normal shock in a gas:

(a) the stagnation pressure remains the same on both sides of the shock
(b) the stagnation density remains the same on both sides of the shock
(c) the stagnation temperature remains the same on both sides of the shock
(d) the Mach number remains the same on both sides of the shock [ESE : 2002]

7.6 For a jet propulsion unit, ideally the ratio of compressor work and turbine work is

(a) 2
(b) 1
(c) not related to each other
(d) unpredictable

[ESE : 2002]

7.7 A normal shock

(a) causes a disruption and reversal of flow pattern
(b) may occur only in a diverging passage
(c) is more serve than an oblique shock
(d) moves with a velocity equal to the sonic velocity

[ESE : 2002]

7.8 During subsonic, adiabatic flow of gases in pipes with friction, the flow properties go through particulars mode of changes.
Match **List-I** (Flow properties) with **List-II** (Mode of changes) and select the correct answer using the codes given below the lists:

List I

A. Pressure
B. Density
C. Temperature
D. Velocity

List II

1. Increase in flow direction
2. Decreases with flow direction

Codes:

	A	B	C	D
(a)	1	1	2	2

(b) 2 2 2 1
(c) 2 2 1 2
(d) 2 1 1 2

[ESE : 2002]

7.9 Acoustic velocity in an elastic gaseous medium is proportional to
(a) Absolute temperature
(b) Stagnation temperature
(c) Square root of absolute temperature
(d) Square root of stagnation temperature

[ESE : 2003]

7.10 The fluid property that remains unchanged across a normal shock wave is
(a) Stagnation enthalpy
(b) Stagnation pressure
(c) Static pressure
(d) Mass density

[ESE : 2003]

7.11 The absolute jet exit velocity from a jet engine is 2800 m/s and the forward flight velocity is 1400 m/s. The propulsive efficiency is
(a) 33.33% (b) 40%
(c) 66.67% (d) 90%

[ESE : 2003]

7.12 The efficiency of jet engine is
(a) higher at high speeds
(b) lower at low speeds
(c) higher at high altitudes
(d) same at all altitudes

[ESE : 2003]

7.13 The propulsive efficiency of a turbo-jet aircraft approaches 100% when the thrust approaches
(a) Maximum
(b) 50% of the maximum
(c) 25% of the maximum
(d) zero

[ESE : 2003]

7.14 Assertion (A): A normal shock always makes a supersonic flow of a compressible fluid subsonic, but an oblique shock may not ensure subsonic flow after the shock.

Reason (R): A normal shock reduces the stagnation pressure and stagnation enthalpy considerably whereas the loss at oblique shock is minimized.

(a) Both **A** and **R** are individually true and **R** is the correct explanation of **A**
(b) Both **A** and **R** are individually true but **R** is not the correct explanation of **A**
(c) **A** is true but **R** is false
(d) **A** is false but **R** is true

[ESE : 2003]

7.15 The relative jet exit velocity from a rocket is 2700 m/s, the forward flight velocity is 1350 m/s. The propulsive efficiency of the unit is
(a) 90 % (b) 66.66 %
(c) 50 % (d) 33.33 %

[ESE : 2004]

7.16 Consider the following statements regarding performance of turbo-jet engines:
1. The thrust decreases at higher altitudes due to reduced density of air and consequently lower mass flow of air.
2. At subsonic speeds, the effect of increased velocity is to increase the air flow and the thrust increases.
3. The relative velocity of jet with respect to the medium decreases at higher speeds which tends to reduce the thrust.
4. For turbo-jet engine the thrust of jet at subsonic speeds remains relatively constant.

Which of these statements are correct?
(a) 1, 2, 3 and 4 (b) 1 and 3
(c) 1, 2 and 4 (d) 2, 3 and 4

[ESE : 2004]

7.17 A compressible fluid flows through a passage as shown in the above diagram. The velocity of the fluid at the point A is 400 m/s.

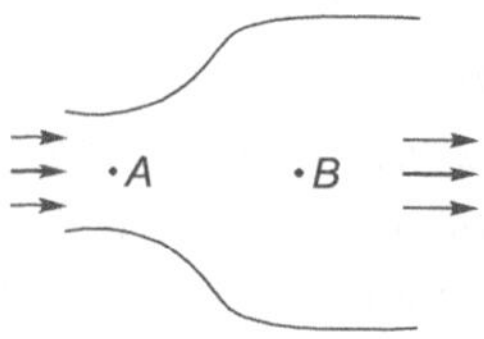

Which one of the following is correct?
At the point B, the fluid experiences
(a) an increase in velocity and decrease in pressure
(b) a decrease in velocity and increase in pressure
(c) a decrease in velocity and pressure
(d) an increase in velocity and pressure

[ESE : 2004]

7.18 In a perfect gas having ratio of specific heats as 1.4. The strength of a normal shock with upstream Mach number equal to 5.0 is

(a) 27 (b) 28
(c) 29 (d) 24 **[ESE : 2004]**

7.19 An aircraft is flying at a speed of 800 km/h at an altitude, where the atmospheric temperature is –20°C. What is the approximate value of the Mach number of the aircraft?

(a) 0.653 (b) 0.697
(c) 0.240 (d) 0.231

[ESE : 2004]

7.20 Which one of the following is the correct expression for the propulsion efficiency of a jet plane (neglecting the mass of fuel)?

(a) $\eta_p = \dfrac{2}{\left(\dfrac{V_a}{V_j}\right)+1}$ (b) $\eta_p = \dfrac{2}{\left(\dfrac{V_j}{V_a}\right)+1}$

(c) $\eta_p = \dfrac{2}{\left(\dfrac{V_a}{V_j}\right)-1}$ (d) $\eta_p = \dfrac{2}{\left(\dfrac{V_j}{V_a}\right)-1}$

[ESE : 2005]

7.21 Which of the following assumptions/conditions are true in the case of Rayleigh flow?

1. Perfect gas.
2. Constant area duct.
3. Steady one-dimensional real flow.
4. Heat transfer during the flow

Select the correct answer using the code given below:

(a) 1, 2 and 3 (b) 2, 3 and 4
(c) 1, 3 and 4 (d) 1, 2 and 4

[ESE : 2005]

7.22 Which of the following parameters decrease across a normal shock wave?

1. Mach number
2. Static pressure
3. Stagnation pressure
4. Static temperature

Select the correct answer using the code given below:

(a) Only 1 and 3 (b) Only 2 and 4
(c) 1, 2 and 3 (d) 2, 3 and 4

[ESE : 2006]

7.23 When are shock waves formed in air compressors?

(a) Mach number < 0.9
(b) Mach number > 0.9
(c) Mach number = 2
(d) Mach number changes suddenly from one value to another

[ESE : 2006]

7.24 Consider the following statements indicating a comparison between rocket and jet propulsion systems:

1. Both rocket and jet engines carry the fuel and oxidant.
2. Rockets do not employ compressor or propeller.
3. Rockets can operate in vacuum also.
4. Rockets can use solid fuels and oxidants.

Which of these statements are correct?

(a) 1, 2, 3 and 4 (b) Only 1 and 2
(c) Only 2, 3 and 4 (d) Only 1, 3 and 4

[ESE : 2006]

7.25 In a normal shock in a gas

(a) Both the downstream flow and the upstream flow are supersonic
(b) Only the upstream flow is supersonic
(c) The downstream flow is sonic
(d) The upstream flow is subsonic

[ESE : 2006]

7.26 Which one of the following statements is correct about the Fanno flow?

(a) For an initially subsonic flow, the effect of friction is to decrease the Mach number towards unity
(b) For an initially supersonic flow, the effect of friction is to increase the Mach number towards unity
(c) At the point of maximum entropy, the Mach number is unity
(d) Stagnation pressure always increases along the Fanno line

[ESE : 2007]

7.27 The turbine of the turbo-prop engine as compared to that of turbo-jet engine is

(a) similar (b) smaller
(c) bigger (d) unpredictable

[ESE : 2007]

7.28 The concentration of pressure pulses created by an object moving at Mach number of 0.5 is
(a) Larger ahead of the object
(b) Larger behind the object
(c) Uniform within Mach cone
(d) Uniform outside Mach cone [ESE : 2007]

7.29 In a normal shock wave in one dimensional flow
(a) the entropy remains constant
(b) the entropy increases across the shock
(c) the entropy decreases across the shock
(d) the velocity, pressure and density increase across the shock [ESE : 2008]

7.30 **Assertion (A):** A normal shock wave can occur at any section in a convergent-divergent nozzle.
Reason (R): A normal shock wave occurs only when the flow of the fluid is supersonic and the subsequent flow after the shock is subsonic.
(a) Both **A** and **R** are true and **R** is the correct explanation of **A**
(b) Both **A** and **R** are true but **R** is not a correct explanation of **A**
(c) **A** is true but **R** is false
(d) **A** is false but **R** is true [ESE : 2008]

7.31 Which one of the following is property of a normal shock wave?
(a) Reversible
(b) Irreversible
(c) Isentropic
(d) Occurs in a converging tube [ESE : 2009]

7.32 Which one of the following statements relates to expression 'ρvc'?
(a) Pressure rise in a duct due to normal closure of valve in the duct
(b) Pressure rise in a duct due to abrupt closure of valve in the duct
(c) Pressure rise in a duct due to slow opening of valve in the duct
(d) Pressure rise in a duct due to propagation of supersonic wave through the duct
[ESE : 2009]

7.33 Which one of the following is correct for tangential component of velocities before and after an oblique shock?
(a) Unity (b) Equal
(c) Unequal (d) None of these
[ESE : 2009]

7.34 Which of the following performance advantages does a rocket engine has as compared to a turbo-jet engine
1. No altitude limitation
2. Higher efficiency
3. Longer flight duration
4. No ram drag

Select the correct answer from the code given below:
(a) 1 and 2 only (b) 1 and 4 only
(c) 1, 2 and 3 (d) 2, 3 and 4
[ESE : 2009]

Direction: Each of the next questions consists of two statements, one labelled as the **'Assertion (A)'** and the other as **'Reason (R)'**. You are to examine these two statements carefully and select the answers to these items using the codes given below:

Codes:
(a) Both **A** and **R** are individually true and **R** is the correct explanation of **A**
(b) Both **A** and **R** are individually true but **R** is not the correct explanation of **A**
(c) **A** is true but **R** is false
(d) **A** is false but **R** is true

7.35 **Assertion (A) :** After burning increases the thrust of a jet engine.
Reason (R) : The air fuel ratio of jet engine is high. [ESE : 2009]

7.36 **Assertion (A):** Rocket engines are used for space research.
Reason (R): They have high specific impulse.
[ESE : 2010]

7.37 In rocket propulsion, the oxygen for combustion of its fuel is taken from
(a) Surrounding air
(b) The rocket itself
(c) Compressed atmospheric air
(d) Surrounding air and compressed atmospheric air [ESE : 2010]

7.38 Consider the following statements:
1. Static pressure increases.
2. Stagnation pressure decreases.
3. Static temperature increases.
4. Stagnation temperature decreases.

Which of these statements are correct for change in properties across normal shock wave?

(a) 1 and 3 only (b) 1, 2 and 4 only
(c) 1, 2 and 3 only (d) 1, 2, 3 and 4

[ESE : 2010]

7.39 Turbo-prop engine has the following additional feature over the turbojet
(a) Propeller
(b) Diffuser
(c) Starting engine
(d) Turbine and combustion chamber

[ESE : 2011]

7.40 Across a normal shock
(a) The pressure and temperature rise
(b) The density and temperature decrease
(c) The velocity and pressure increase
(d) The velocity and pressure decrease

[ESE : 2011]

7.41 What is the correct statement regarding normal shock wave in a gas?
(a) Normal shock is reversible
(b) The downstream flow as well as upstream flow is supersonic
(c) The entropy increases across the shock
(d) The entropy remains constant

[ESE : 2011]

7.42 In Rayleigh flow at subsonic Mach numbers, heat addition
(a) Decreases static pressure
(b) Decreases enthalpy
(c) Increases static pressure
(d) Decreases static temperature

[ESE : 2011]

7.43 **Assertion (A) :** In a Rayleigh flow of subsonic compressible flow heating causes the Mach number to reach a maximum of 1; but the static temperature at this point is less than that due to frictional effects alone.
Reason (R) : This is due to extra storage of energy as internal energy.

[ESE : 2011]

7.44 Normal shock wave occurs in a one-dimensional steady
(a) converging flow (b) diverging flow
(c) subsonic flow (d) supersonic flow

[ESE : 2012]

7.45 The Mach number of an aircraft when the dynamic (velocity) temperature of air at entry to the engine equals the static temperature is
(a) 1.236 (b) 1.736
(c) 2.236 (d) 2.736

[ESE : 2012]

7.46 Across the normal shock in a flow
(a) both *p* and *T* decrease
(b) *p* increases and *T* decreases
(c) *p* decreases and *T* increases
(d) both *p* and *T* increase

[ESE : 2012]

7.47 **Statement (I):** A normal shock wave cannot occur in convergent portion of a convergent-divergent nozzle.
Statement (II): Shock wave is generated only when the flow is supersonic.
(a) Both Statement (I) and Statement (II) are individually true and Statement (II) is the correct explanation of Statement (I).
(b) Both Statement (I) and Statement (II) are individually true but Statement (II) is **NOT** the correct explanation of Statement (I).
(c) Statement (I) is true but Statement (II) is false
(d) Statement (I) is false but Statement (II) is true.

[ESE : 2012]

7.48 In a compressible fluid flow field, the Mach number indicates the ratio of
(a) viscous force to elastic force
(b) inertia force to elastic force
(c) inertia force to viscous force
(d) viscous force to gravity force

[ESE : 2012]

7.49 Which of the following statements are correct for turbo-prop powered aircraft?
1. The propulsion efficiency of turbo-prop is higher than that of turbo-jet and rockets for low speeds up to about 800 km/hr.
2. For the same thrust the turbine in the turbo-prop aircraft is smaller than in the turbo-jet aircraft.
3. For the turbo-prop the flight velocity cannot exceed the jet velocity.

(a) 1, 2 and 3 (b) 1 and 2 only
(c) 2 and 3 only (d) 1 and 3 only

[ESE : 2013]

7.50 If absolute jet exit velocity from a jet engine is 2800 m/s and forward flight velocity is 1400 m/s, then propulsive efficiency is

(a) 33.33% (b) 40%
(c) 66.66% (d) 90%

[ESE : 2013]

7.51 An air-breathing aircraft is flying at an altitude where the air density is half the value at ground level. With reference to the ground level, the air-fuel ratio at this altitude will be

(a) $\frac{1}{2}$ (b) 1
(c) 2 (d) 4 [ESE : 2013]

7.52 Which of the following statement are correct for rockets?

1. Unlike the turbo-jet aircraft in rockets the flight velocity can exceed the jet velocity.
2. In rockets gases having lower molecular weight increase the specific thrust.
3. In rockets the gases are expanded in the nozzle up to the atmosphere pressure.

(a) 1, 2 and 3 (b) 1 and 2 only
(c) 2 and 3 only (d) 1 and 3 only

[ESE : 2013]

7.53 Fanno line flow is a flow in a constant area duct

(a) with friction and heat transfer but in the absence of work
(b) with friction and heat transfer accompanied by work
(c) with friction but in the absence of heat transfer or work
(d) without friction but accompanied by heat transfer and work

[ESE : 2013]

7.54 In an adiabatic flow with friction through a constant area duct, the following phenomenons are predicted:

1. The flow downstream tends to be sonic irrespective of the Mach-number upstream.
2. In the case of initially supersonic flow, any increase in duct length beyond chocked flow would make the exit flow subsonic.
3. In the case of initially subsonic flow, increase in duct length will not change sonic flow at the exit.
4. The optimum pipe length for sonic outlet in supersonic flow is considerably longer than that of subsonic flow for the same flow rate.

Which of these are correct?

(a) 1, 2, 3 and 4 (b) 1, 2 and 3 only
(c) 1, 3 and 4 only (d) 2, 3 and 4 only

[ESE : 2013]

7.55 In a nozzle designed for maximum discharge conditions, the flow velocity in the convergent section of the nozzle is

(a) subsonic
(b) sonic
(c) supersonic
(d) depends on initial pressure and condition of steam [ESE : 2013]

Directions: Each of the next consists of two statements, one labelled as the 'Statement (I)' and the other as 'Statement (II)'. You are to examine these two statements carefully and select the answers to these items using the codes given below:

Codes:

(a) Both Statement (I) and Statement (II) are individually true and Statement (II) is the correct explanation of Statement (I).
(b) Both Statement (I) and Statement (II) are individually true but Statement (II) is **NOT** the correct explanation of Statement (I).
(c) Statement (I) is true but Statement (II) is false
(d) Statement (I) is false but Statement (II) is true.

7.56 **Statement (I):** In an isentropic nozzle flow, discharge reaches a maximum value when the throat pressure reaches the critical value.

Statement (II): At the critical condition, throat Mach number remains at unity and the upstream side of throat is maintained low enough to give a continuous decrease of pressure and a continuous increase of velocity.

[ESE : 2013]

7.57 **Statement (I):** Mass flow through the convergent nozzle is maximum when the exit Mach number is 1.

Statement (II): The divergent section is added to convergent nozzle to increase the exit Mach number and not to increase the mass flow rate.

[ESE : 2013]

7.58 Statement (I): When a bluff body moves at supersonic speed in still air, an attached shock wave is produced at its nose causing the flow to be subsonic behind the shock plane.

Statement (II): A normal shock always transforms supersonic flow into subsonic flow causing increase in pressure, temperature and entropy.

[ESE : 2013]

7.59 Across the normal shockwave:

1. Stagnation pressure decreases whereas stagnation temperature remains constant
2. Mach number before the shockwave is always greater than one and after the shockwave, the Mach number need not to be less than one
3. Across the shockwave there is a rise in pressure and temperature
4. The product of Mach number downstream of normal shockwave and upstream of normal shockwave is always one

Which of the above statements are correct?

(a) 1 and 3 (b) 2 and 3
(c) 1 and 4 (d) 2 and 4

[ESE : 2014]

7.60 In supersonic flow of air, a diverging passage results in

(a) Increase in velocity and pressure
(b) Decrease in pressure and density
(c) Increase in velocity and density
(d) Decrease in velocity and pressure

[ESE : 2014]

Directions: Each of the next consists of two statements, one labelled as the 'Statement (I)' and the other as 'Statement (II)'. You are to examine these two statements carefully and select the answers to these items using the codes given below:

Codes:

(a) Both Statement (I) and Statement (II) are individually true and Statement (II) is the correct explanation of Statement (I).
(b) Both Statement (I) and Statement (II) are individually true but Statement (II) is **NOT** the correct explanation of Statement (I).
(c) Statement (I) is true but Statement (II) is false
(d) Statement (I) is false but Statement (II) is true.

7.61 Statement (I) : A rocket engine can operate even in vacuum and in any fluid medium.

Statement (II): Rocket engine is a pure reaction engine which produces propulsive thrust.

[ESE : 2014]

7.62 Statement (I) : Both pressure and temperature across the normal shock increase.

Statement (II): The stagnation pressure across the normal shock decreases. [ESE : 2014]

7.63 Across the normal shock, fluid properties change in such a manner that the

1. Velocity of flow is subsonic
2. Pressure increases
3. Specific volume decreases
4. Temperature decreases

Which of the above are correct?

(a) 1, 2, 3 and 4 (b) 2, 3 and 4
(c) 1 and 4 only (d) 1, 2 and 3 only

[ESE : 2014]

7.64 In a compressible flow with friction choking through a constant area duct with supersonic flow at inlet, if the pipe length is reduced with the same exit pressure

(a) Exit flow will still be sonic
(b) The velocity at exit is subsonic
(c) The flow will still be supersonic
(d) A shock will appear at the exit

[ESE : 2014]

7.65 The velocity of a gas flowing through a duct is 300 m/s; its temperature is 127°C, Gas constant R =0.25 kJ/kgK, the ratio of specific heat is $c_p/c_v = \gamma = 1.6$. What is the value of Mach number?

(a) 0.70 (b) 0.72
(c) 0.75 (d) 0.77

[ESE : 2015]

7.66 A shock wave which occurs in a supersonic flow represents a region in which :

(a) A zone of silence exists

(b) There is no change in pressure, temperature and density
(c) There is sudden change in pressure, temperature and density
(d) Analogy with a hydraulic jump is not possible

[ESE : 2016]

Directions: Each of the next consists of two statements, one labelled as the 'Statement (I)' and the other as 'Statement (II)'. You are to examine these two statements carefully and select the answers to these items using the codes given below:

Codes:

(a) Both Statement (I) and Statement (II) are individually true and Statement (II) is the correct explanation of Statement (I).
(b) Both Statement (I) and Statement (II) are individually true but Statement (II) is **NOT** the correct explanation of Statement (I).
(c) Statement (I) is true but Statement (II) is false
(d) Statement (I) is false but Statement (II) is true

7.67 **Statement (I) :** Commercial airplanes save fuel by flying at higher altitudes during long trips.
Statement (II) : At higher altitudes, the ambient temperature and the Carnot efficiency are low.

[ESE : 2016]

7.68 **Statement (I) :** In Fanno flow, heat transfer is neglected and friction is considered.
Statement (II) : In Rayleigh flow, heat transfer is considered and friction is neglected.

[ESE : 2016]

7.69 Consider the following statements regarding a Ram Jet:
1. The engine has neither a compressor nor a turbine.
2. It operates at much higher temperature than a gas turbine.
3. It cannot operate statically. It needs to be put in flight by some means at sufficiently high speed before it produces any thrust and propels itself.

Which of the above statements are correct?
(a) 1 and 2 only (b) 1 and 3 only
(c) 2 and 3 only (d) 1, 2 and 3

[ESE : 2018]

7.70 Air enters a turbojet engine at the rate of 40 kg/s with a velocity of 250 m/s relative to an aircraft which is moving at 300 km/hr. Exhaust of the engine has a velocity of 700 m/s relative to the moving aircraft. The thrust developed by the engine is
(a) 24 kN (b) 18 kN
(c) 12 kN (d) 9 kN

[ESE : 2018]

7.71 Which of the following are essential for a good combustion chamber of turbojet engine?
1. It should allow complete combustion of fuel.
2. It should maintain sufficiently high temperatures in the zone of combustion in addition to proper atomization of fuel thus leading to continuous combustion.
3. It should not have high rate of combustion.
4. The pressure drop should be as small as possible.

(a) 1, 2 and 4 only (b) 1, 2 and 3 only
(c) 1, 3 and 4 only (d) 2, 3 and 4 only

[ESE : 2020]

7.72 If m_f is the mass of fuel supplied per kg of air in one second, then the mass of gases leaving the nozzle of turbojet will be
(a) $(1 - m_f)$ kg/s (b) $\frac{1}{(1+m_f)}$ kg/s
(c) $(1 + m_f)$ kg/s (d) $\frac{1}{(1-m_f)}$ kg/s

[ESE : 2020]

7.73 Which one of the following may be considered as a single cylinder two-stroke reciprocating engine running at 2400 rpm to 2700 rpm for rapid chain of impulses?
(a) Turbo jet (b) Pulse jet
(c) Ram jet (d) Athodyd jet

[ESE : 2020]

7.74 In jet propulsion of ships, when the inlet orifices are at right angles to the direction of motion of the ships, the efficiency of propulsion η is

(a) $\frac{2u^2}{V+u}$ (b) $\frac{2Vu}{(V+u)^2}$

(c) $\frac{2u}{(V+u)^2}$ (d) $\frac{2Vu}{V+u}$

where; V = Absolute velocity of the issuing jet, u = Velocity of the moving ship

[ESE : 2020]

7.75 Lenoir cycle is used for

(a) Gas turbines (b) Pulse jet engines

(c) S.I. engines (d) C.I. engines

[ESE : 2021]

7.76 The water in a jet-propelled boat is drawn amidship and discharged at the back with an absolute velocity of 15 m/s. The boat speed is 30 km/hr. The cross-sectional area of the jet at the back is 0.03 m^2. What is the efficiency of the jet propulsion?

(a) 68.5% (b) 51.3%

(c) 45.9% (d) 39.4%

[ESE : 2022]

7.77 **Statement (I):** Fluid motion produced due to change in density resulting from the temperature gradients is called free convection.

Statement (II): The movement of fluid in free convection is due to the fact that the fluid particles in the immediate vicinity of the hot object become warmer than the surrounding fluid resulting in a local change of density.

(a) Both statement (I) and statement (II) are individually true statement (II) is the corect explanation of statement (I).

(b) Both statement (I) and statement (II) are individually true, but statement (II) is not the corect explanation of statement (I).

(c) Statement (I) is true, but Statement (II) is false.

(d) Statement (I) is false, but Statement (II) is true.

[ESE : 2023]

7.78 A jet propelled boat with an absolute velocity of 8.7 m/s is moving upstream in a river. The stream is flowing with a velocity of 2.3 m/s. A jet of water is ejected astern at a relative velocity of 18 m/s. If the flow in jet is 1.4 m^3/s, what is the efficiency of the propulsion device?

(a) 38.5% (b) 48.5%

(c) 58.5% (d) 68.5%

[ESE : 2023]

Directions for the following **two (02)** items:

Read the following information and answer the **two** items that follow:

A jet aeroplane flies at a speed of 900 km/h. The density of air at this altitude is 0.15 kg/m^3 and drag on plane is 6800 kW. Consider jet plane to have 2 jets and engine working on turbo-prop system with propulsive efficiency of 56%.

7.79 What is the absolute velocity of jet?

(a) 115.4 m/s (b) 250.4 m/s

(c) 392.86 m/s (d) 480.46 m/s

[ESE : 2023]

7.80 What is the volume flow rate?

(a) 86.6 m^3/s (b) 102.6 m^3/s

(c) 115.4 m^3/s (d) 131.8 m^3/s

[ESE : 2023]

■■■■

Answers Jet Propulsion and Compressible Flow

7.1 (c)	7.2 (b)	7.3 (b)	7.4 (d)	7.5 (c)	7.6 (b)	7.7 (a)	7.8 (b)	7.9 (d)
7.10 (a)	7.11 (c)	7.12 (a)	7.13 (d)	7.14 (c)	7.15 (c)	7.16 (b)	7.17 (a)	7.18 (b)
7.19 (b)	7.20 (b)	7.21 (d)	7.22 (a)	7.23 (c)	7.24 (c)	7.25 (b)	7.26 (c)	7.27 (c)
7.28 (b)	7.29 (b)	7.30 (d)	7.31 (b)	7.32 (d)	7.33 (b)	7.34 (b)	7.35 (b)	7.36 (b)
7.37 (b)	7.38 (c)	7.39 (a)	7.40 (a)	7.41 (c)	7.42 (c)	7.43 (d)	7.44 (d)	7.45 (c)
7.46 (d)	7.47 (a)	7.48 (b)	7.49 (d)	7.50 (c)	7.51 (a)	7.52 (b)	7.53 (c)	7.54 (c)
7.55 (a)	7.56 (b)	7.57 (b)	7.58 (a)	7.59 (a)	7.60 (b)	7.61 (a)	7.62 (b)	7.63 (d)
7.64 (c)	7.65 (c)	7.66 (c)	7.67 (c)	7.68 (b)	7.69 (d)	7.70 (b)	7.71 (a)	7.72 (c)
7.73 (b)	7.74 (b)	7.75 (b)	7.76 (c)	7.77 (c)	7.78 (c)	7.79 (c)	7.80 (c)	

Explanations Jet Propulsion and Compressible Flow

7.1 (c)

$$M = \frac{V}{\sqrt{\gamma RT}}$$

Hence $M \propto \dfrac{V}{\sqrt{T}}$

$$T_1 = 15°C = 288\text{ K}$$

$$T_2 = -25°C = 248\text{ K}$$

$$\frac{M_1}{M_2} = \frac{V_1/\sqrt{T_1}}{V_2/\sqrt{T_2}}$$

$$\frac{400}{\sqrt{288}} = \frac{V_2}{\sqrt{248}}$$

$$V_2 = 400\sqrt{\frac{248}{288}} = \mathbf{371.2\ km/hr}$$

7.2 (b)

$$M_y = \sqrt{\frac{M_x^2 + \frac{2}{\gamma-1}}{\frac{2\gamma}{\gamma-1}M_x^2 - 1}} = \sqrt{\frac{(1.68)^2 + \frac{2}{1.4-1}}{\left(\frac{2\times1.4}{1.4-1}\right)\times(1.68)^2 - 1}}$$

$= \mathbf{0.646}$

7.3 (b)

In turboprop, the expansion of gases takes place approximately 80% in turbine which drive compressor and propeller and 20% in nozzle which produce thrust from the jet.

7.4 (d)

Rocket properties efficiency

$$= \frac{2C_jC_a}{C_j^2 + C_a^2} = \frac{2\left(\frac{C_a}{C_j}\right)}{1+\left(\frac{C_a}{C_j}\right)^2}$$

where, C_a = Forward velocity of rocket

C_j = Velocity of jet

The jet propulsion efficiency of rocket approaches 100% as $\frac{C_j}{C_a}$ approaches unity and it lower on either side. But for the propeller and turbojet the flights velocity cannot exceed the jet velocity if the thrust is to be positive, for the rocket the jet velocity is independent of forward motion the C_j can be less than, equal to or greater than C_a.

7.5 (c)

Across the normal shock, the fluid properties change which are:

1. Stagnation temperature remains constant.
2. Stagnation pressure decreases.
3. Static pressure increases.
4. Static temperature increases.
5. Velocity decreases
6. Density of the fluid increases.

7.9 (d)

Acoustic velocity: $C = \sqrt{\gamma RT}$

Hence, $C \propto \sqrt{T}$

7.11 (c)

Propulsive efficiency

$$= \frac{2C_a}{C_J + C_a} = \frac{2 \times 1400}{2800 + 1400} \times 100 = \mathbf{66.67\%}$$

7.12 (a)

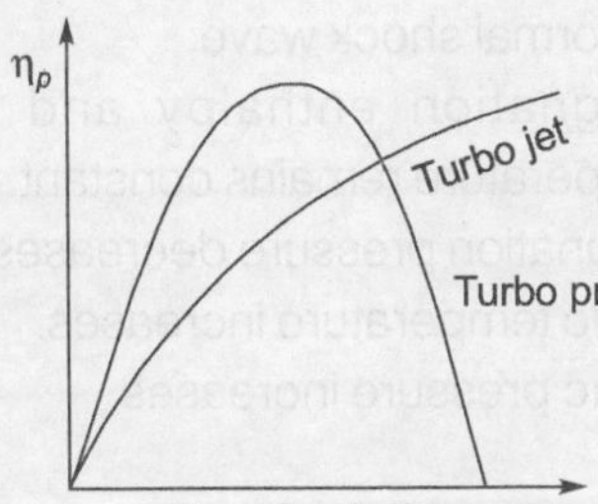

7.13 (d)

Propulsive efficiency is maximum when jet velocity (C_j) = forward velocity of aircraft (C_a)

Now thrust power = $(C_j - C_a)C_a = 0$

i.e. at maximum efficiency thrust power is zero.

7.14 (c)

Statement (II) is wrong as stagnation enthalpy remains unchanged in a normal shock wave.

7.15 (c)

$$C_j - C_a = 2700 \text{ m/s}$$
$$C_a = 1350 \text{ m/s}$$
$$C_j = 2700 + 1350 = 4050 \text{ m/s}$$

Propulsive efficiency

$$\eta_P = \frac{2C_a}{C_j + C_a} = \frac{2 \times 1350}{4050 + 1350}$$

$$= 0.50 = \mathbf{50\%}$$

7.16 (b)

Turbojet engines save fuel by flying at higher altitudes during long trips since the air at higher altitude is of less density and exerts a small drag forces on the Turbojet.

7.17 (a)

Velocity at A is 400 m/s $\therefore M > 1$

Now $\frac{dA}{A} = -\frac{dV}{V}(1 - M^2) = \frac{dP}{\rho v^2}(1 - M^2)$

$\therefore$ $dV > 0$ and $dP < 0$

7.18 (b)

Shock strength

$$p = \frac{p_y}{p_x} - 1$$

$$\frac{p_y}{p_x} = \left(\frac{\gamma - 1}{\gamma + 1}\right)\left[\frac{2\gamma}{\gamma - 1}M_x^2 - 1\right]$$

$$\gamma = 1.4$$
$$M_x = 5$$

$$\frac{p_y}{p_x} = \frac{0.4}{2.4}\left[\frac{2.8 \times 25}{0.4} - 1\right] = 29$$

$$p = 29 - 1 = \mathbf{28}$$

7.19 (b)

$$V = 800 \text{ km/h} = 222.22 \text{ m/s}$$

$$C = \sqrt{\gamma RT_0}$$

$$T_0 = -20°C = 253 \text{ K}$$

$$C = \sqrt{1.4 \times 287 \times 253} = 318.83 \text{ m/s}$$

Mach number,

$$M = \frac{V}{C} = \frac{222.22}{318.83} = 0.697$$

7.20 (b)

Propulsion efficiency of jet:

$$\eta_p = \frac{2V_a}{V_i + V_a} = \frac{2}{\left(\frac{V_j}{V_a}\right) + 1}$$

7.21 (d)

Real flow can not be friction less flow.

7.22 (a)

Across the normal shock, the fluid properties change which are

1. Stagnation temperature remains constant.
2. Stagnation pressure decreases.
3. Static pressure increases.
4. Static temperature increases.
5. Velocity decreases
6. Density of the fluid increases.

7.23 (c)

Shock occurs when the flow condition is supersonic and after shock it becomes subsonic.

7.24 (c)

A rocket propulsion device differs from other thermal jet engines in that its propellant carries both the fuels and the oxidizing agents. As a result, this type of the engine is independent of the atmospheric oxygen therefore it can operate at very high latitude and even in vacuum.

7.26 (c)

At the point of maximum entropy, the Mach number is unity. The state on the upper part of each curve is subsonic and lower part supersonic.

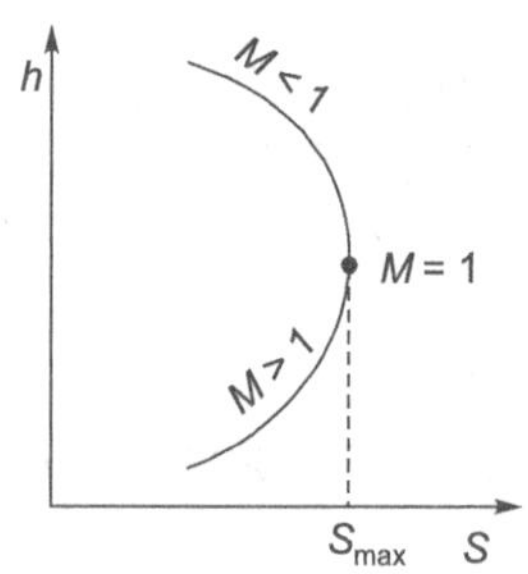

7.27 (c)

In turboprop engine turbine expansion process is greater than turbojet engine hence turbine of the turbo-prop engine is bigger than that of turbojet engine

7.29 (b)

The normal shock always involves a change from supersonic to subsonic speed with a consequent pressure rise and never the reverse. By second law, entropy always increases during irreversible adiabatic change. The entropy increases across the shock.

7.30 (d)

Normal shock occurs only in the divergent position. So Assertion is wrong.

A normal shock wave can occur only when the flow is supersonic and the subsequent flow after the shock becomes subsonic.

7.33 (b)

Oblique shock wave has two components of velocity. One is normal component and another is tangential component.

The normal component of velocity decreases during the oblique shocked and tangential component shock and tangential component of velocity remains same on both the sides of the shock wave because there is no pressure gradient along the shock wave in the tangent direction.

7.37 (b)

Rockets are equiped with both fuel cylinder and oxygen cylinder; they do not take oxygen from atmosphere.

7.38 (c)

During normal shock wave:

1. Stagnation enthalpy and stagnation temperature remains constant.
2. Stagnation pressure decreases.
3. Static temperature increases.
4. Static pressure increases.

7.39 (a)

Turbojet: Its basic design consists of an air intake, a gas generator unit and an exhaust nozzle. The gas generator in turn consists of the compressor, combustion chamber and turbine. In this design, the turbine drives only the compressor and various accessory systems; this leaves more energy to provide actual thrust than in other types.

Turboprop: The turboprop is similar to the turbojet, except that the turbine drives a propeller.

7.40 (a)

Across shock wave static pressure, temperature and density increases and velocity decreases.

7.41 (c)

Across the normal shock, the entropy increases across the shock. For the adiabatic irreversible process, entropy must increase.

7.42 (c)

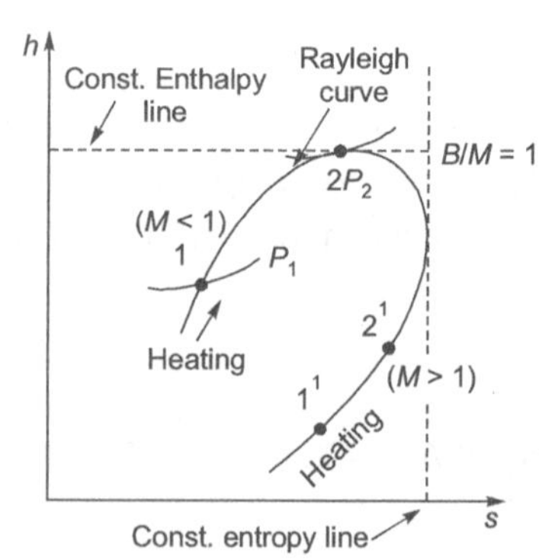

Rayleigh flow is a flow in constant area duct without friction with heat transfer.

7.43 (d)

Mach number is not unity, but $M = \frac{I}{\sqrt{e}}$

7.44 (d)

Normal shock wave occurs in supersonic flow.

7.46 (d)

Across a normal shock wave pressure, temperature increases and velocity decreases.

7.48 (b)

$$\text{Mach number, } M = \left(\frac{\text{Inertia force}}{\text{Elastic force}}\right)^{1/2}$$

7.49 (d)

(i) As compared to turbo prop engine, power to weight ratio of turbojet is about 4 times. Weight/power ratio of turboprop is higher.

(ii) The turboprop can operate economically over a wide range of speeds ranging from low speeds where as pure jet engine and rockets is uneconomical to high speeds of about 800 km/h.

(iii) For the same thrust, the turbine in the turboprop aircraft is larger than in the turbojet aircraft.

(iv) Thrust of turboprop engine is dependent on large mass flow rate and velocity difference between inlet and exhaust. Due to propeller, this velocity difference is incremental and flight speed can't exceed jet velocity.

7.50 (c)

Propulsive efficiency,

$$\eta_p = \frac{2C_i}{C_j + C_i}$$

We have, C_j = 2800 m/s; C_i = 1400 m/s

$$\therefore \quad \eta_p = \frac{2 \times 1400}{2800 + 1400} = 0.667 = \mathbf{66.66\%}$$

7.51 (a)

$$\text{Density} = \frac{\text{Mass}}{\text{Volume}}$$

If density is half, mass will also get halved. So, with reference to ground level, air-fuel ratio would be 1/2.

7.52 (b)

(i) Unlike, air breathing engine, flight velocity can exceed the jet velocity because expansion in nozzle can be done upto sub-atmospheric pressure.

(ii) Thrust produced in chemical rocket,

$F \propto \frac{1}{\sqrt{M}}$. (*M* is molecular weight)

So, to obtain high thrust, the molecular weight of the propellouts must be as low as possible.

(iii) Thrust developed by rocket,

$$F = \dot{m}C_j + (P_j - P_a)A_j$$

P_j is static pressure at nozzle exit and P_a is ambient pressure, for optimum thrust, $P_j = P_a$. So, expansion in nozzle can be done upto ambient pressure and ambient condition may be atmosphere or vacuum. Usually, rocket has no altitude limitation and even it can operate in vacuum.

7.53 (c)

<u>F</u> <u>a</u> *nno* line flow
<u>F</u> : frictional
<u>a</u> : adiabatic
It is frictional adiabatic flow.

7.54 (c)

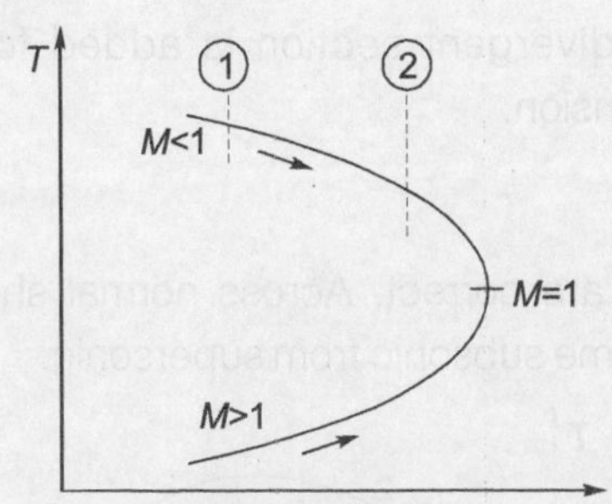

Fanno Line Flow
(Adiabatic flow with friction)

- Flow from section (1) to (2) tends to be sonic irrespective of Mach number upstream.
- If the inlet flow is subsonic, the additional length forces the sonic condition to move

down to the new exit. The flow rate also decreases.

- If the inlet flow is supersonic, the additional length causes a normal shock and flow become subsonic.

7.55 (a)

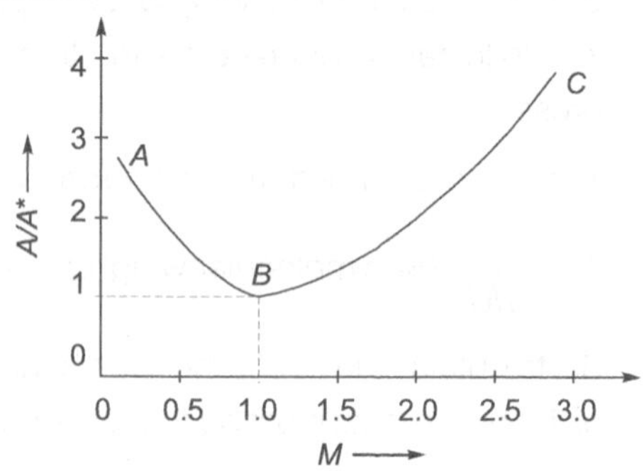

(Area ratio of isentropic nozzle vs. Mach number)

Convergent section AB has $M < 1$, so flow through convergent section is subsonic.

Divergent section BC has $M > 1$.

7.56 (b)

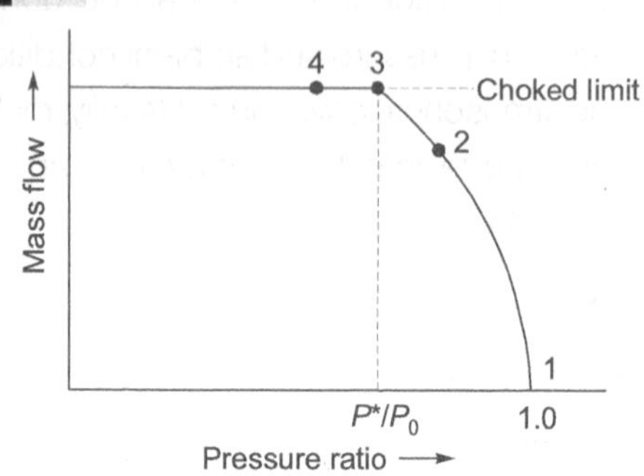

Point 3 is choke limit corresponding to particular throat pressure.

7.57 (b)

The divergent section is added for gradual expansion.

7.58 (a)

Both are correct. Across normal shock, flows become subsonic from supersonic.

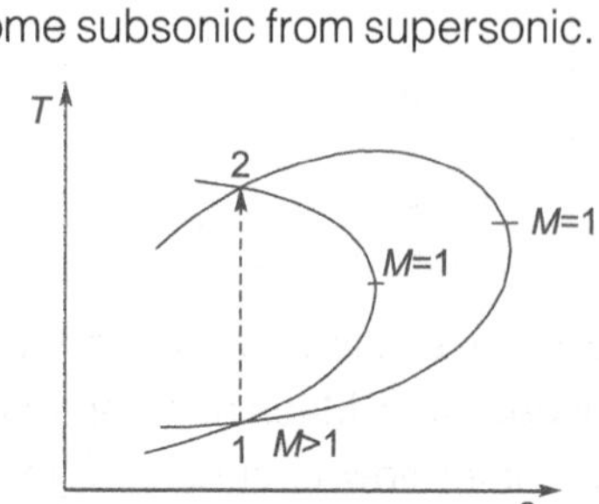

(1→ 2 is inclined line)

7.59 (a)

Mach number before the shock wave is always greater than one i.e. supersonic and it becomes essentially subsonic after the shock wave.

7.60 (b)

$$\frac{dA}{A} = \frac{dV}{V}(M^2 - 1)$$

For supersonic flow, $M > 1$,

For diverging passage, $\frac{dA}{A}$ is +ve

$\therefore \frac{dV}{V}$ = +ve, so pressure decreases.

As $\rho \propto P$,

∴ density also decreases.

7.61 (a)

A rocket engine can operate even in vacuum or in any fluid medium because it carries its own fuel and oxidiser. So reaction does not depend upon medium and this reaction only produces propulsive thrust.

7.62 (b)

Both pressure and temperature across the normal shock increase because of change of supersonic to subsonic flow. So statement (II) is not the correct explanation.

7.63 (d)

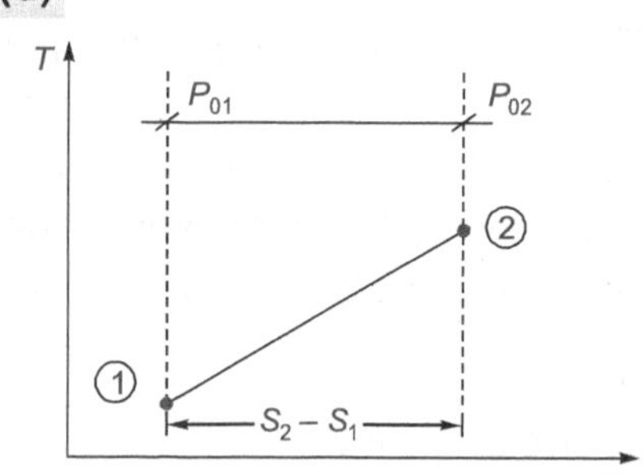

Across the normal shock:

- Velocity of flow decrease i.e it becomes subsonic from supersonic
- Pressure increases
- Stagnation pressure decreases
- Density increases so specific volume decreases
- Entropy increases
- Temperature increases

7.64 (c)

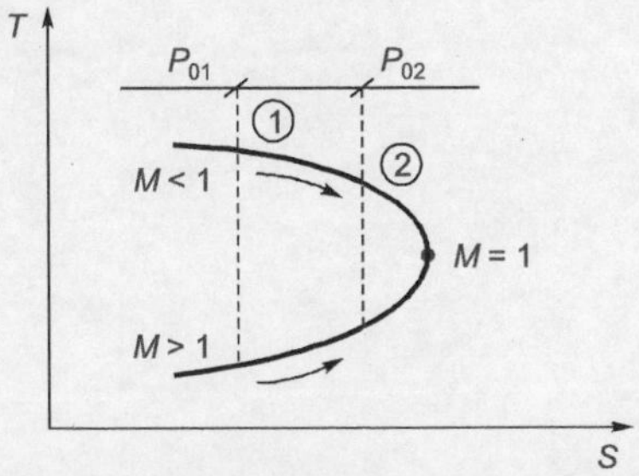

As given, flow is supersonic initially. If we reduce the pipe length, the friction will get reduced. So, maximum entropy point (i.e. $M = 1$) will not be reached and the flow will still be supersonic. This can be interpreted well from the fanno line curve as shown above.

7.65 (c)

$$M = \frac{V}{C} = \frac{V}{\sqrt{\gamma RT}}$$

$$= \frac{300}{\sqrt{1.6 \times 250 \times 400}} = 0.75$$

7.66 (c)

Across shockwave, there is abrupt deceleration of found from supersonic velocity to subsonic velocity with rise in P, T, ρ but entropy also greatly increases due to severe inversibilities.

7.70 (b)

$$\text{Thrust} = \dot{m}(C_j - C_a) = 40(700 - 250)$$

$$= 18000 \text{ N} = 18 \text{ kN}$$

7.71 (a)

Combustor should not have high rate of combustion, else the flame may die out.

7.72 (c)

For turbojet engine, total mass of gases coming out will be sum of fuel and air supplied. Hence,mass flow = $(1 + m_f)$ kg/s

7.73 (b)

Pulse jet engines have stable operation in range of 45 cycles per second. This can be correlated to a 2-stroke engine running at 2400 - 2700 rpm (40 - 45 cycles/sec).

7.76 (c)

Absolute velocity of jet, $V = 15$ m/s

Cross-sectional area of jet, $A = 0.03 \text{ m}^2$

Boat velocity, $U = 30$ km/hr $= 8.33$ m/s

$$\text{Efficiency, } \eta = \frac{2Vu}{(V+u)^2}$$

$$= \frac{2 \times 15 \times 8.33}{(15 + 8.33)^2} = 0.4591\% \text{ or } 45.9\%$$

7.77 (a)

Free convection: Circulation of bulk fluid motion is caused by changes in fluid density resulting from temperature gradients between the solid surface and the main mass of fluid.

7.78 (c)

Relative velocity of water = (8.7 + 2.3) = 11 m/s

Mass flow rate of water = $\rho \times V$ = 1000 × 1.4 = 1.4×10^3 kg/s.

Force, $F = 1.4 \times 10^3 \times (18 - 11)$ = 9.8 kN

Power, P = Force × Velocity = 9.8 × 8.7 = 85.26 kW

$$\frac{\text{KE of jet}}{\text{sec}} = \frac{1}{2} \times 1.4 \times 10^3 \times (18 - 8.7)^2$$

$$= 60.573 \text{ kW}$$

So, efficiency of the propulsive device,

$$\eta = \frac{85.26}{85.26 + 60.543} = 0.5847 = 58.5\%$$

7.79 (c)

Given : $V_a = 900$ km/h = 250 m/s

Propulsive efficiency;

$$\eta = 0.56$$

$$\eta = \frac{2V_a}{V_j + V_a}$$

$$\Rightarrow \quad 0.56 = \frac{2 \times 250}{V_j + 250}$$

$\therefore \quad V_j = 642.85$ m/s

Absolute velocity of jet;

$$V_j - V_a = 642.85 - 250$$
$$= 392.86 \text{ m/s}$$

7.80 (c)

As calculated in the above question

$$V_j - V_a = 392.86$$

$$\therefore \quad \text{Thrust} = \dot{m} \times (V_j - V_a)$$

$$6800 = \dot{m} \times (392.86)$$

(Here thrust is take as 6800 N)

$$\dot{m} = 17.31 \text{ kg}$$

$\therefore$Volume flow rate =

$$\frac{\dot{m}}{\rho} = \frac{17.31}{0.15} = 115.4\,\text{m}^3/\text{s}$$

■■■■

UNIT

VII

Power Plant Engineering

Syllabus

Rankine and Brayton cycles with regeneration and reheat, Fuels and their properties, Flue gas analysis, Boilers, steam turbines and other power plant components like condensers, air ejectors, electrostatic precipitators and cooling towers – their theory and design, types and applications, Compressors.

Contents

Sl.	Topic	Page No.
1.	Thermal Power Plant Components	502
2.	Nozzles	523
3.	Nuclear Power Plant	535
4.	Miscellaneous	540

OOOO

1 Thermal Power Plant Components

1.1 Consider the following statements:
1. Forced circulation is always used in high pressure power boilers.
2. Soot blowers are used for cleaning tube surfaces at regular intervals.
3. Electrostatic precipitator is used to remove fly ash from flue gases.

Which of these statements are correct?

(a) 1, 2 and 3 (b) 2 and 3
(c) 1 and 3 (d) 1 and 2

[ESE : 2000]

1.2 Once-through boilers operate at
(a) subcritical pressure
(b) supercritical pressure
(c) subcritical as well supercritical pressures
(d) critical pressure only [ESE : 2000]

1.3 Match **List-I** (Components) with **List-II** (Functions) and select the correct answer using the codes given below the lists:

List-I
- **A.** Steam trap
- **B.** Fusible plug
- **C.** Blow-off cock
- **D.** Feed-check valve

List-II
- **1.** Controls steam flow rate
- **2.** Controls rate of water flow to boiler
- **3.** Puts off furnace fire when water level reaches unsafe limit
- **4.** Removes mud and dirt collected at the bottom of boiler
- **5.** Drains off water collected by partial condensation of steam in pipes

Codes:

	A	B	C	D
(a)	5	1	4	2
(b)	1	3	5	4
(c)	5	3	4	2
(d)	1	2	5	4

[ESE : 2000]

1.4 **Assertion (A):** With the help of a Bomb calorimeter, the lower calorific value of a solid or liquid fuel can be determined, as the water vapour formed is carried away by the exhaust gases.

Reason (R): The lower calorific value of a fuel is the net value of heat available found by subtracting the latent heat of the water formed and carried away by exhaust gas from the higher calorific value.

(a) Both **A** and **R** are individually true and **R** is the correct explanation of **A**
(b) Both **A** and **R** are individually true but **R** is not the correct explanation of **A**
(c) **A** is true but **R** is false
(d) **A** is false but **R** is true

[ESE : 2000]

1.5 Consider the following statements
1. For the combustion of pulverized coal, 5 to 10% excess air is required.
2. Air contains 21% oxygen by weight.
3. The flue gases from a coal-fired furance contain around 70% nitrogen by volume.
4. In the combustion of liquid fuels, the number of moles of the products are invariably greater than the number of moles of the reactance.

Which of these statements are correct?

(a) 1, 2 and 4 (b) 1, 3 and 4
(c) 2, 3 and 4 (d) 1 and 3

[ESE : 2001]

1.6 Which one of the following sequences indicates the correct order for flue gas flow in the steam power plant layout?

(a) superheater, economiser, air preheater
(b) economiser, air preheater, superheater
(c) air preheater, economiser, superheater
(d) economiser, superheater, air preheater

[ESE : 2001]

1.7 Which one of the following statements is not correct?
In a fluidized-bed boiler
(a) the combustion temperatures are higher than those in the conventional boilers
(b) inferior grade of coal can be used without slagging problems
(c) the formation of NO_x is less than that in the conventional boilers
(d) the volumetric heat release rates are higher than those in the conventional boilers

[ESE : 2001]

1.8 Match **List-I** (Fuels) with **List-II** (Characteristics/ usages) and select the correct answer using the codes given below the lists:

List-I
A. Semi-bituminous coal
B. High-speed diesel oil
C. Biogas
D. LPG

List-II
1. Methane and carbon dioxide
2. Propane and butane
3. calorific value of 10,600 kCal/kg
4. Power plants

Codes:

	A	B	C	D
(a)	3	4	1	2
(b)	4	3	2	1
(c)	3	4	2	1
(d)	4	3	1	2

[ESE : 2001]

1.9 A device which is used to drain off water from steam pipes without escape of steam is called
(a) Steam separator
(b) Steam trap
(c) Pressure reducing valve
(d) Injector

[ESE : 2002]

1.10 Match **List-I** (Type of coal) with **List-II** (Coal properties) and select the correct answer using the codes given below the lists:

List-I
A. Lignite
B. Anthracite
C. Bituminous
D. Coke

List-II
1. Artificial fuel derived from coal
2. Contains inflammable gas (volatile matter) and burns with flame
3. Very hard and high heating value
4. High ash content and less volatile matter

Codes:

	A	B	C	D
(a)	2	3	4	1
(b)	4	1	2	3
(c)	2	1	4	3
(d)	4	3	2	1

[ESE : 2002]

1.11 Blowing down of boiler water is the process to
(a) reduce the boiler pressure
(b) increase the steam temperature
(c) control the solids concentration in the boiler water
(d) control the drum level

[ESE : 2002]

1.12 Match **List-I** (Boilers) with **List-II** (Type/ Description) and select the correct answer using the codes given below the lists:

List-I
A. Lancashire
B. Benson
C. Babcock and Wilcox
D. Stirling

List-II
1. Horizontal straight tube, fire-tube boiler
2. Horizontal straight tube, water-tube boiler
3. Bent tube, water tube boiler
4. High pressure boiler

Codes:

	A	B	C	D
(a)	4	2	1	3
(b)	1	4	2	3
(c)	4	2	3	1
(d)	1	4	3	2

[ESE : 2003]

1.13 Consider the following statements regarding the fluidized bed combustion boilers

1. The combustion temperatures are low, around 900°C.
2. The formation of oxides of nitrogen is low.
3. It removed sulphur from coal during combustion process.
4. It requires high quality of coal as fuel.

Which of these statements are correct?

(a) 1, 2, 3 and 4 (b) 1, 2 and 3
(c) 2, 3 and 4 (d) 1 and 4

[ESE : 2003]

1.14 The correct gas flow path in a typical large modern natural circulation boiler is

(a) Combustion chamber, Reheater, Superheater, Economiser, Air preheater, I.D. fan, Electrostatic precipitator, Stack
(b) Combustion chamber, Superheater, Reheater, Economiser, Air preheater, Electrostatic precipitator, I.D. fan, Stack
(c) Combustion chamber, Reheater, Superheater, Air preheater, Economiser, Electrostatic precipitator, I.D. fan, Stack
(d) Combustion chamber, Superheater, Reheater, Economiser, Air preheater, I.D. fan, Electrostatic precipitator, Stack

[ESE : 2003]

1.15 In a cooling tower the sum of range and approach is equal to twice the wet bulb depression. Then

(a) Dry bulb temperature is mean of water inlet temperature and wet bulb temperature
(b) Dry bulb temperature is mean of water outlet temperature and wet bulb temperature
(c) Water inlet temperature is mean of dry bulb temperature and wet bulb temperature
(d) Water inlet temperature is mean of water outlet temperature and wet bulb temperature

[ESE : 2003]

1.16 The draught in locomotive boilers is produced by

(a) Chimney (b) Centrifugal fan
(c) Steam jet (d) Locomotion

[ESE : 2003]

1.17 Consider the following statements regarding waste-heat boilers:

1. Waste-heat boilers are placed in the path of exhaust gases.
2. These are fire tube boilers.
3. The greater portion of the heat transfer in such boilers is due to convection.

Which of these statements are correct?

(a) 1, 2 and 3 (b) 1 and 2
(c) 2 and 3 (d) 1 and 3

[ESE : 2004]

1.18 Which one of the following statements is not true for a supercritical steam generator?

(a) It has a very small drum compared to a conventional boiler
(b) A supercritical pressure plant has higher efficiency than a subcritical pressure plant
(c) The feed pressure required is very high, almost about 1.2 to 1.4 times the boiler pressure
(d) As it requires absolutely pure feed water, preparation of feed water is more important than in a subcritical pressure boiler

[ESE : 2005]

1.19 Which one of the following is the fire-tube boiler?

(a) Babcock and Wilcox boiler
(b) Locomotive boiler
(c) Stirling boiler
(d) Benson boiler

[ESE : 2005]

1.20 Which of the following are boiler mountings?

1. Fusible plug 2. Blow-off cock
3. Steam trap 4. Feed check valve

Select the correct answer using the code given below:

(a) 1, 2 and 3 (b) 2, 3 and 4
(c) 1, 3 and 4 (d) 1, 2 and 4

[ESE : 2005]

1.21 Which one of the following is the correct path of water flow through various components of boiler of a modern thermal power plant?

(a) Economizer, boiler drum, water walls, boiler drum, superheater, turbine
(b) Economizer, boiler drum, water wall, superheater, turbine
(c) Economizer, water walls, boiler drum, superheater, turbine
(d) Economizer, water walls, superheater, turbine

[ESE : 2005]

1.22 Weight percentage of which one of the following is determined by Proximate analysis of coal?
(a) Fixed carbon, volatile matter, moisture and ash
(b) All solid and gaseous components
(c) All solid & gaseous components except volatile matter
(d) Fixed carbon and volatile matter
[ESE : 2005]

1.23 Consider the following:
1. Superheater 2. Economizer
3. Air pre-heater 4. Condenser
Which of the above improves overall steam power plant efficiency?
(a) Only 1, 2 and 3 (b) Only 2 and 3
(c) Only 1 and 4 (d) 1, 2, 3 and 4
[ESE : 2006]

1.24 Consider the following:
1. Injector 2. Economizer
3. Blow-off cock 4. Steam stop valve
Which of the above is/are not boiler mountings?
(a) Only 1 (b) Only 1 and 2
(c) 1, 2 and 3 (d) 2 and 4
[ESE : 2006]

1.25 Consider the following statements:
1. Forward swept blade impeller is used in draft fans.
2. Forward swept blade impeller is used in room air-conditioners.
3. Radial tipped blade impeller is used in draft fans.
4. Forward swept blade impeller is used in exhaust fans.
Which of these statements is/are correct?
(a) Only 1 (b) Only 2 and 3
(c) Only 4 (d) 1, 2, 3 and 4
[ESE : 2006]

1.26 Assertion (A): Fire tube boilers do not operate at high pressures while water tube boilers operate at high pressures.
Reason (R): Due to high temperature of flue gases, fire tubes may fail due to creep.
(a) Both **A** and **R** are individually true and **R** is the correct explanation of **A**
(b) Both **A** and **R** are individually true but **R** is not the correct explanation of **A**
(c) **A** is true but **R** is false
(d) **A** is false but **R** is true [ESE : 2006]

1.27 Which one of the following boiler accessories does not need 'flue-gas' for its operation?
(a) Economizer (b) Preheater
(c) Injector (d) Superheater
[ESE : 2007]

1.28 Which one of the following fittings is mounted on the boiler to put off the fire in the furnace, when water level falls to an unsafe limit?
(a) Feed check valve (b) Safety valve
(c) Fusible plug (d) Blow off cock
[ESE : 2007]

1.29 Which one of the following accessories is connected to the steam supply pipe line to maintain constant pressure?
(a) Pressure reducing valve
(b) Steam separator
(c) Steam trap
(d) Injector [ESE : 2007]

1.30 Performance of mechanical draft cooling tower is superior to natural draft with
(a) increase in air wet bulb temperature
(b) decrease in air wet bulb temperature
(c) increase in dry bulb temperature
(d) increase in recirculation of air
[ESE : 2007]

1.31 Water-tube boilers are preferred for
(a) high pressure and high output
(b) high pressure and low output
(c) low pressure and high output
(d) low pressure and low output
[ESE : 2008]

1.32 Which one of the following is not an accessory for the boiler?
(a) Condenser (b) Economizer
(c) Air preheater (d) Feed water pump
[ESE : 2009]

1.33 Match **List-I** (Type of boiler) with **List-II** (Features) and select the correct answer using the codes given below the lists:
List-I
A. Benson
B. Stirling

C. Babcock-Wilcox
D. Cochran

List-II

1. Vertical fire tube
2. Horizontal fire tube
3. Bent water tube
4. Once through flow
5. Inclined water tube

Codes :

	A	B	C	D
(a)	4	5	3	1
(b)	1	2	5	4
(c)	4	3	5	1
(d)	1	5	2	4

[ESE : 2009]

1.34 Water (c_p = 4 kJ/kgK) is fed to a boiler at 30°C, the enthalpy of vaporization at atmospheric pressure in the boiler is 2400 kJ/kg; the steam coming from the boiler is 0·9 dry. The net heat supplied in the boiler is
(a) 2160 kJ/kg (b) 2400 kJ/kg
(c) 2440 kJ/kg (d) 2280 kJ/kg
[ESE : 2009]

1.35 In a boiler, the air preheater is invariably located between:
(a) Forced draft fan and chimney
(b) Forced draft fan and furnace
(c) Economizer and feed pump
(d) Condenser and feed pump [ESE : 2009]

1.36 Consider the following statements:
For a fluidized bed boiler
1. There is less formation of NO_x.
2. It can burn inferior grades of coal.
Which of these statements is/are correct?
(a) 1 only (b) 2 only
(c) Both 1 and 2 (d) Neither 1 nor 2
[ESE : 2010]

1.37 Consider the following statements:
For a supercritical steam generator
1. It has no steam drum.
2. It has no circulation pump.
3. The phenomenon of boiling does not occur in it.
Which of these statements is/are correct?
(a) 1, 2 and 3 (b) 1 and 2 only
(c) 2 and 3 only (d) 3 only
[ESE : 2010]

1.38 **Assertion (A):** The modern electric power generating plants use only water tube boilers and not fire tube boilers.
Reason (R): The water tube boilers are comparatively cheaper in first cost than fire tube boilers.
(a) Both **A** and **R** are individually true and **R** is the correct explanation of **A**
(b) Both **A** and **R** are individually true but **R** is not the correct explanation of **A**
(c) **A** is true but **R** is false
(d) **A** is false but **R** is true [ESE : 2010]

1.39 Pertaining to a steam boiler, which of the following statements is correct?
(a) Primary boiler heat transfer surface includes evaporator section, economizer and air preheater
(b) Primary boiler heat transfer surface includes evaporator section, economizer and superheater section
(c) Secondary boiler heat transfer surface includes superheater, economizer and air preheater
(d) Primary boiler heat transfer surface includes evaporator section, superheater section and reheat section [ESE : 2011]

1.40 Out of the following impurities generally found in water, which one requires special consideration in case of very high pressure boilers?
(a) Hydrogen (b) Ammonia
(c) Silica (d) Dissolved salts
[ESE : 2011]

1.41 In a locomotive boiler, the draught is produced by
(a) Chimney (b) ID fan
(c) FD fan (d) Steam jet
[ESE : 2011]

1.42 In modern steam generator, the correct path of gases from boiler furnace to chimney is
(a) Boiler furnace, Economizer, Air preheater, Superheater and Chimney
(b) Boiler furnace, Superheater, Air preheater, Economizer and Chimney
(c) Boiler furnace, Air preheater, Superheater, Economizer and Chimney
(d) Boiler furnace, Superheater, Economizer, Air preheater and Chimney [ESE : 2011]

1.43 The function of economizer in a boiler is to
(a) Superheat the steam
(b) Reduce fuel consumption
(c) Increase steam pressure
(d) Maintain saturation temperature
[ESE : 2011]

Direction: Each of the next questions consists of two statements, one labelled as the **'Assertion (A)'** and the other as **'Reason (R)'**. You are to examine these two statements carefully and select the answers to these items using the codes given below:

Codes:
(a) Both **A** and **R** are individually true and **R** is the correct explanation of **A**
(b) Both **A** and **R** are individually true but **R** is not the correct explanation of **A**
(c) **A** is true but **R** is false
(d) **A** is false but **R** is true

1.44 **Assertion (A) :** In a modern coal burning steam generator, the temperature of exiting flue gases from the chimney should be below 100° C.
Reason (R) : The lower the temperature of exiting flue gases from the chimney, higher is the heat recovery and therefore higher the efficiency of the steam generator. [ESE : 2011]

1.45 **Assertion (A) :** A major shortcoming of a fire-tube boiler is that the maximum size of the unit and the maximum operating pressure are limited.
Reason (R) : Both large diameters and high pressures lead to prohibitively thick shells resulting in very high cost. [ESE : 2011]

1.46 The pump preferred to be used for pumping highly viscous fluids belongs to the category of
(a) screw pump (b) turbine pump
(c) plunger pump (d) centrifugal pump
[ESE : 2012]

1.47 When a hydrocarbon fuel burns in deficit air, Orsat analyzer will show the presence of following gases in the engine exhaust
(a) CO_2, N_2 and O_2
(b) CO, CO_2, N_2 and O_2
(c) CO, CO_2 and N_2
(d) CO_2, CO, N_2, O_2 and H_2O
[ESE : 2013]

1.48 The air pre-heater of a boiler is located between
(a) Forced draft fan and furnace
(b) Furnace and economizer
(c) Economizer and chimney
(d) Superheater and furnace
[ESE : 2014]

1.49 A super critical boiler requires
(a) Only preheater and superheater
(b) Preheater, evaporator and superheater
(c) Only preheater
(d) Only superheater
[ESE : 2014]

1.50 The correct sequence of location of equipment in the flue gas path from furnace exit up to chimney is
(a) Superheater, economizer, air heater, electrostatic precipitator and induced draft fans
(b) Superheater, economizer, electrostatic precipitator, induced draft fans and air heater
(c) Superheater, electrostatic precipitator, economizer, air heater and induced draft fans
(d) Superheater, electrostatic precipitator, induced draft fans, economizer and air heater
[ESE : 2014]

1.51 The main advantage of the water tube boiler over the fire-tube boiler is
(a) The water tube boiler can operate safely at higher pressure
(b) Soot deposition in the tubes is avoided
(c) Corrosion of the tubes is less
(d) Fouling of the tubes is reduced
[ESE : 2014]

1.52 A super critical boiler consists of only economizer and superheater and it does not have an evaporator because
(a) Water temperature can be raised to critical temperature in the economizer itself
(b) High evaporation rate is achieved through forced circulation of water through tubes
(c) Enthalpy of evaporation becomes zero at critical pressure or above that
(d) Flue gas used to run rotary compressor supply high pressure air to the furnace
[ESE : 2014]

1.53 The collection efficiency of cyclone separators increases with

1. Decreasing particle size
2. Increasing particle density
3. Decreasing gas velocity
4. Increasing number of gas revolutions
5. Increasing cyclone diameter

Which of the above statements are correct?

(a) 1, 3 and 4 only (b) 2 and 4 only
(c) 2, 4 and 5 only (d) 1, 2, 3, 4 and 5

[ESE : 2014]

Directions: The following items consist of two statements; one labelled as **'Statement (I)'** and the other as **'Statement (II)'**. You are to examine these two statements carefully and select the answers to these items using the codes given below:

Codes:

(a) Both Statement (I) and Statement (II) are individually true and Statement (II) is the correct explanation of Statement (I).
(b) Both Statement (I) and Statement (II) are individually true but Statement (II) is **NOT** the correct explanation of Statement (I).
(c) Statement (I) is true but Statement (II) is false
(d) Statement (I) is false but Statement (II) is true.

1.54 Statement (I) : The vertical boilers are used to save the floor space.
Statement (II): Horizontal boilers are more efficient than vertical boilers.

[ESE : 2014]

1.55 Statement (I) : The efficiency of a boiler is more if it is provided with mechanical draught rather than with natural draught.
Statement (II): Natural draught is very costly but highly efficient. [ESE : 2014]

1.56 Which of the following statements is correct for steam boiler?

(a) Boiler secondary heat transfer surface includes super-heater, economizer and air pre-heater
(b) Boiler primary heat transfer surface includes evaporator section, super-heater section and reheat section.
(c) Boiler primary heat transfer surface includes evaporator section, economizer and super-heater section
(d) Boiler secondary heat transfer surface includes evaporator section, economizer and air pre-heater

[ESE : 2015]

1.57 For maximum discharge of hot gases through a chimney, the height of hot column producing drought is

(a) Twice the height of chimney
(b) Equal to the height of chimney
(c) Half the height of chimney
(d) None of the above

[ESE : 2015]

1.58 The correct chronological order, in development of steam generators, is :

(a) Fire tube boiler, Monotube boiler and Water tube boiler
(b) Water tube boiler, Fire tube boiler and Monotube boiler
(c) Fire tube boiler, Water tube boiler and Monotube boiler
(d) Water tube boiler, Monotube boiler and Fire tube boiler

[ESE : 2016]

Directions: The following items consist of two statements; one labelled as **'Statement (I)'** and the other as **'Statement (II)'**. You are to examine these two statements carefully and select the answers to these items using the codes given below:

Codes:

(a) Both Statement (I) and Statement (II) are individually true and Statement (II) is the correct explanation of Statement (I).
(b) Both Statement (I) and Statement (II) are individually true but Statement (II) is **NOT** the correct explanation of Statement (I).
(c) Statement (I) is true but Statement (II) is false
(d) Statement (I) is false but Statement (II) is true.

1.59 Statement (I) : Heat carried away by hot gases in chimney draught is much greater than the work required for lifting the same gases through the height of the chimney. Yet artificial draught is not preferred.

Statement (II) : Artificial draught involves large initial cost as well as large maintenance cost. [ESE : 2017]

1.60 Statement (I) : The overall combustion efficiency of a fuel oil based plant is less as compared to that of a coal burning plant.

Statement (II) : Fuel oils contain comparatively larger percentage of hydrogen, which produce more moisture per kg of fuel burnt.

[ESE : 2017]

1.61 Statement (I) : Proximate analysis of coal is done to determine its calorific value.

Statement (II) : In proximate analysis of coal, the percentages of moisture, volatile matter, fixed carbon and ash are determined.

[ESE : 2017]

1.62 Statement (I) : Water entering into a condenser from the cooling tower has much dissolved impurities.

Statement (II) : In a closed cooling system, the water is continuously aerated, therefore, there is abundant dissolved oxygen in this water.

[ESE : 2017]

1.63 Consider the following statements in respect of natural-draft cooling towers:

1. Theoretically the water can be cooled to even below the dry-bulb temperature of the induced air flow.
2. Natural-draft cooling towers are 100 m or more in height.
3. The inner and outer surfaces are surfaces of revolution of a segment of a hyperbola about the vertical axis-affording improved strength rather than any thermodynamic augmentation.

Which of the above statements are correct?

(a) 1, 2 and 3 (b) 1 and 2 only
(c) 1 and 3 only (d) 2 and 3 only

[ESE : 2017]

1.64 Statement (I) : Direct condensers are more efficient than surface condensers.

Statement (II) : In condenser, the momentum pressure drop opposes the frictional pressure drop. [ESE : 2018]

1.65 The efficiency of superheat Rankine cycle is higher than that of simple Rankine cycle because

(a) the enthalpy of main steam is higher for superheat cycle
(b) the mean temperature of heat addition is higher for superheat cycle
(c) the temperature of steam in the condenser is high
(d) the quality of steam in the condenser is low

[ESE : 2019]

1.66 In steam power cycle, the process of removing non-condensable gases is called

(a) scavenging process
(b) deaeration process
(c) exhaust process
(d) condensation process

[ESE : 2019]

1.67 The internal irreversibility of Rankine cycle is caused by

1. fluid friction
2. throttling
3. mixing

Sele ct the correct answer using the code given below:

(a) 1 and 2 only (b) 1 and 3 only
(c) 2 and 3 only (d) 1, 2 and 3

[ESE : 2019]

1.68 A boiler is having a chimney of 35 m height. The draught produced in terms of water column is 20 mm. The temperature of flue gas inside the chimney is 365°C and that of air outside the chimney is 32°C. The mass of air used will be nearly

(a) 10.3 kg/kg of fuel
(b) 12.5 kg/kg of fuel
(c) 14.7 kg/kg of fuel
(d) 16.9 kg/kg of fuel

[ESE : 2019]

1.69 Which of the following statements is/are correct regarding superheater in boilers?

1. It is a heat exchanger in which heat is transformed to the saturated steam to increase its temperature.
2. It raises the overall efficiency.
3. It reduces turbine internal efficiency.

Select the correct answer using the code given below:

(a) 1 and 2 (b) 1 and 3
(c) 2 and 3 (d) 1 only

[ESE : 2019]

1.70 A steam turbine is supplied with steam at a pressure of 20 bar gauge. After expansion in the steam turbine, the steam passes to condenser which is maintained at a vacuum of 250 mm of mercury by means of pumps. The inlet and exhaust steam pressures will be nearly

(a) 2101 kPa and 68 kPa
(b) 2430 kPa and 78 kPa
(c) 2101 kPa and 78 kPa
(d) 2430 kPa and 68 kPa

[ESE : 2019]

1.71 In a power plant, the efficiencies of the electric generator, turbine, boiler, thermodynamic cycle and the overall plant are 0.97, 0.95, 0.92, 0.42 and 0.33 respectively. The total electricity generated for running the auxiliaries will be nearly

(a) 4.9% (b) 5.7%
(c) 6.5% (d) 7.3%

[ESE : 2019]

1.72 In a power plant, the efficiency of the electric generator, turbine, boiler, cycle and the overall plant are 0.97, 0.95, 0.92, 0.42 and 0.33 respectively. In the generated electricity, the auxiliaries will consumes nearly

(a) 7.3% (b) 6.5%
(c) 5.7% (d) 4.9%

[ESE : 2020]

1.73 The higher power requirements for compression in a steam power plant working on Carnot vapour cycle

(a) Increases the plant efficiency as well as work ratio
(b) Reduces the plant efficiency as well as work ratio
(c) Does not affect the plant efficiency as well as work ratio
(d) Increases the plant efficiency and reduces work ratio

[ESE : 2020]

1.74 An economizer in a steam generator performs the function of preheating the

(a) Combustion air
(b) Feed water
(c) Input fuel
(d) Combustion air as well as input fuel

[ESE : 2020]

1.75 The partial vacuum created by the fan in the furnace and flues, draws the products of the combustion from the main flue and allows them to pass up to the chimney. Such a draught is called

(a) Balanced draught
(b) Forced draught
(c) Induced draught
(d) Artificial draught

[ESE : 2020]

1.76 A steam ejector which removes air and other non-condensable gases from the condenser is known as

(a) Wet air pump
(b) Dry air pump
(c) Centrifugal pump
(d) Circulating pump

[ESE : 2020]

1.77 If the pressure at exhaust from the turbine is the saturation pressure corresponding to the temperature desired in the process heater, such a turbine is called

(a) Condensing turbine
(b) Extraction turbine
(c) Pass out turbine
(d) Back pressure turbine

[ESE : 2020]

1.78 The settling time for a unit step response of a second-order system is

(a) proportional to the natural frequency.
(b) inversely proportional to the natural frequency.
(c) equal to the damping ratio.
(d) proportional to the damping ratio.

[ESE : 2021]

1.79 In a power plant, the efficiencies of the electric generator, turbine (mechanical), boiler, cycle and the overall plant are 0.97, 0.95, 0.92, 0.42 and 0.33, respectively. What is the efficiency of auxiliaries?

(a) 98.14% (b) 92.68%
(c) 83.41% (d) 75.14%

[ESE : 2021]

1.80 Consider the following statements for analysis of steam cycles:

1. A steam power plant continuously converts the energy stored in fossil fuels or fissile fuels into shaft work.
2. Steam power plants work on Brayton cycle.
3. In supercritical steam cycle, steam is generated in a 'once-through' boiler at a pressure above the critical point of 27.5 bar.
4. Deaerator is used for the purpose of deaerating the feedwater.

Which of the above statements are correct?

(a) 2 and 3 only (b) 1 and 3 only
(c) 1 and 4 only (d) 2 and 4 only

[ESE : 2021]

1.81 Consider the following statements for solid fuels:

1. Peat is the first stage in the formation of coal from wood.
2. The average calorific value of bituminous coal is 1524 kJ/kg.
3. Anthracite is very hard coal and has a shining black lustre.
4. Wood charcoal is obtained by destructive distillation of wood.

Which of the above statements are correct?

(a) 1, 2 and 3 only (b) 1, 3 and 4 only
(c) 2 and 3 only (d) 1, 2 and 4 only

[ESE : 2021]

1.82 Consider the following statements for fluidized bed boilers:

1. Fluidized bed boilers produce steam from fossil and waste fuels by using a technique called fluidized bed combustion.
2. Cyclone separators are gas cleaning devices that utilize the centrifugal force created by a spinning gas stream to separate particles from a gas.
3. In a pressurized fluidized bed boiler, the combustion process takes place in a pressurized environment resulting in a compact furnace and improved combustion efficiency.

Which of the above statements are correct?

(a) 1 and 2 only (b) 2 and 3 only
(c) 1 and 3 only (d) 1, 2 and 3

[ESE : 2021]

1.83 Consider the following statements for cooling towers:

1. Cooling tower is an artificial device used to cool the hot cooling water coming out of condenser more effectively.
2. The amount of water usually lost with induced draft cooling towers ranges from 5% to 6% by evaporation.
3. The amount of water usually lost with induced draft cooling tower ranges from 7% to 8% by drift losses.
4. The rate of evaporation of water and its cooling effect on the remaining water depends upon the relative humidity of air passing through the tower.

Which of the above statements are correct?

(a) 1 and 4 only (b) 1 and 3 only
(c) 2 and 3 only (d) 1, 2, 3 and 4

[ESE : 2021]

1.84 All power plants use superheated steam due to which of the following advantages?

1. Superheating is mostly done from waste heat of boiler without additional cost of fuel.
2. The plant efficiency increases due to higher temperature of steam.
3. There is less corrosion and erosion of equipment due to absence of moisture in the steam.

Select the correct answer using the code given below:

(a) 1 and 2 only (b) 2 and 3 only
(c) 1 and 3 only (d) 1, 2 and 3

[ESE : 2021]

1.85 What are the effects of regenerative feedwater heating for the same turbine output?

1. It significantly increases the cycle efficiency and reduces the heat rate.
2. It increases the steam flow rate.
3. It increases the steam flow to the condenser.
4. If there is no change of boiler output, the turbine output drops.

Select the correct answer using the code given below:

(a) 1, 2 and 3 only (b) 1, 2 and 4 only
(c) 1, 2 and 4 only (d) 2, 3 and 4 only

[ESE : 2021]

1.86 Which of the following are the advantages of pulverized coal firing?

1. Higher boiler efficiency.
2. Fast response for no load changes.
3. Ability to use low preheated air reducing internal losses.
4. Ability to release large amounts of heat enabling it to generate about 2000 t/h of steam in one boiler.

Select the correct answer using the code given below

(a) 1 and 2 only (b) 1 and 3 only
(c) 1 and 4 (d) 2, 3 and 4 only

[ESE : 2021]

1.87 Which one of the following statements is NOT correct regarding Rankine cycle?

(a) For the steam boiler, the ideal process would be a reversible constant pressure heating process of water to form steam.
(b) For the turbine, the ideal process would be a reversible adiabatic expansion of steam.
(c) For the condenser, it would be a reversible constant pressure heat rejection as the steam condenses till it becomes saturated liquid.
(d) For the pump, the ideal process would be the reversible adiabatic expansion of the liquid ending at the final pressure.

[ESE : 2022]

1.88 Which one of the following statements is NOT correct?

(a) In surface condensers, the cooling water and exhaust steam do not come in direct contact with each other.
(b) The vacuum efficiency is the ratio of ideal vacuum to actual vacuum.
(c) The Hot well is a sump between the condenser and the boiler where the condensate coming from the condenser is collected.
(d) Condenser is a closed vessel heat exchanger in which the steam coming from turbine is condensed using a supply of cooling water at atmospheric temperature.

[ESE : 2022]

1.89 Which one of the following statements is NOT correct?

(a) Grate is the platform in the furnace of boiler upon which fuel is burnt and it is made of cast iron bars.
(b) The removal of the mud and other impurities of water from the lowest part of the boiler is termed as blowing off.
(c) Cochran boiler is one of the best types of vertical multi-tubular boiler.
(d) Babcock and Wilcox boiler is a fire tube boiler.

[ESE : 2022]

1.90 Which one of the following statements is correct?

(a) In water tube boiler, water surrounds the tubes and hot gases are inside the tubes.
(b) The boilers which produce steam at pressures of 10 bar and below are called high pressure boiler.
(c) Lancashire boilers are externally fired boiler.
(d) Stirling boilers are externally fired boiler.

[ESE : 2022]

1.91 In a power plant, the efficiencies of the electric generator, turbine (mechanical), boiler, cycle and overall plant are 0.97, 0.95, 0.92, 0.42 and 0.33 respectively. What percentage of the total electricity generated is consumed in running the auxiliaries?

(a) 9.29% (b) 8.50%
(c) 7.32% (d) 6.76%

[ESE : 2022]

1.92 An ideal cycle is impracticable because:

(a) there is an ideal fluid available which is considered as working fluid.
(b) it is not possible to transfer heat of expanding steam to the fluid in the turbine itself.
(c) it is always possible to transfer heat of expanding steam to the fluid in the turbine itself.
(d) there is no possibility of excessive humidity in low pressure stages of the turbine.

[ESE : 2022]

1.93 Match the following:

List-I (Boiler type)
A. La Mont boiler
B. Loeffler boiler
C. Benson boiler
D. Velox boiler

List-II (Pressure in kg/cm²)
1. 84
2. 170
3. 135
4. 230

Select the correct matching using the codes given below:

Codes:

	A	B	C	D
(a)	2	1	3	4
(b)	1	2	4	3
(c)	2	3	4	1
(d)	3	2	1	4

[ESE : 2022]

1.94 Which one of the following statements is correct?

(a) The Rankine cycle efficiency can be improved by increasing the average temperature at which heat is rejected.
(b) The Rankine cycle efficiency can be improved by decreasing/reducing the temperature at which heat is rejected.
(c) If the steam is superheated before allowing it to expand the Rankine cycle efficiency may be decreased.
(d) The thermal efficiency of the Rankine cycle can be amply improved by increasing the condenser pressure.

[ESE : 2022]

1.95 Match the following:

List-I
A. Boiler to turbine process
B. Turbine to condenser process
C. Condenser to pump process
D. Pump to boiler process

List-II
1. Reversible adiabatic expansion in the turbine
2. Constant pressure transfer of heat in the condenser
3. Reversible adiabatic pumping process in the feed pump
4. Constant pressure transfer of heat in the boiler

Select the correct matching using the code given below:

Codes:

	A	B	C	D
(a)	3	4	1	2
(b)	4	3	2	1
(c)	2	1	4	3
(d)	1	2	3	4

[ESE : 2022]

1.96 Consider the following statements regarding methods of compounding in steam engines:

1. Tandem type compounding of steam engines has the in-line cylinders having pistons mounted on the same piston rod which is further having crosshead and connecting rod providing power output at crankshaft.
2. Woolf compound engine is a cross-type compounding having two cylinders having pistons at 270° phase difference i.e., at some position one cylinder may have piston at inner dead centre and other cylinder has piston at outer dead centre.
3. Receiver compound engine is also a cross compound engine having two cylinders with out of phase pistons and receiver in between.

Which of the above statements are correct?

(a) 1 and 2 only (b) 2 and 3 only
(c) 1 and 3 only (d) 1, 2 and 3

[ESE : 2023]

1.97 The principal characteristic of an ash collector is the degree of collection (η) which is given in terms of quantity by

(a) $\frac{(G_1 - G_2)}{G_1}$ (b) $\frac{(G_1 + G_2)}{G_1}$

(c) $\frac{G_1}{(G_1 - G_2)}$ (d) $\frac{G_1}{(G_1 + G_2)}$

where G_1 = Quantity of ash entering an ash collector per unit time (kg/s)

G_2 = Quantity of uncollected ash passing through the collector per unit time (kg/s)

[ESE : 2023]

Directions for the following **four (04)** items:

Read the following information and answer the **four** items that follows:

In a condenser test, the following observations were made:

Vacuum = 720 mm of mercury;

Barometer = 765 mm of mercury;

Mean temperature of condensation = 34°C;

Hot well temperature = 29°C;

Inlet temperature of cooling water = 15°C;

Outlet temperature of cooling water = 25°C;

Absolute pressure of steam at 34°C is 0.0533 bar.

Take 760 mm of Hg = 1.013 bar.

1.98 What is the vacuum corrected to standard barometer of 760 mm?

(a) 348 mm of Hg (b) 424 mm of Hg
(c) 715 mm of Hg (d) 804 mm of Hg

[ESE : 2023]

1.99 What is the under cooling efficiency?

(a) 45.36 (b) 53.31
(c) 84.36 (d) 99.31

[ESE : 2023]

1.100 What is the under cooling of condenser?

(a) 2°C (b) 3°C
(c) 4°C (d) 5°C

[ESE : 2023]

1.101 What is the condenser pressure?

(a) 0.03 bar (b) 0.04 bar
(c) 0.05 bar (d) 0.06 bar

[ESE : 2023]

Directions for the following **two (02)** items:

Read the following information and answer the **two** items that follows:

In a surface condenser operating with steam turbine, the vacuum near inlet of air pump is 69 cm of Hg when barometer reading is 76 cm of Hg. Temperature at inlet of vacuum pump is 30°C. Air leakage occurs at the rate of 60 kg/hr. [Take 1 cm of Hg = 1333.22 Pa]

1.102 What is the absolute pressure at inlet to air pump?

(a) 4.33 kPa (b) 5.33 kPa
(c) 6.33 kPa (d) 9.33 kPa

[ESE : 2023]

1.103 What is the partial pressure of air if saturation pressure at 30°C is taken as 4.246 kPa?

(a) 3.08 kPa (b) 4.08 kPa
(c) 5.08 kPa (d) 6.08 kPa

[ESE : 2023]

Directions for the following **two (02)** items:

Read the following information and answer the **two** items that follows:

A simple Rankine cycle works between the boiler pressure of 3 MPa and condenser pressure of 4 kPa. The steam is dry saturated before the throttling in the turbine. From the steam tables at 3 MPa (30 bar) and saturated vapour condition, take enthalpy of steam entering into turbine as 2802.3 kJ/kg and enthalpy of steam leaving the turbine as 1862.04 kJ/kg. Consider pump work as 3 kJ/kg.

1.104 What is the work ratio?

(a) 0.677 (b) 0.799
(c) 0.845 (d) 0.997

[ESE : 2023]

1.105 What is the specific steam consumption?

(a) 2.83 kg/kWh (b) 3.83 kg/kWh
(c) 4.83 kg/kWh (d) 5.83 kg/kWh

[ESE : 2023]

1.106 In a power plant, the efficiencies of the electric generator, turbine (mechanical), boiler, cycle and the overall plant are 0.97, 0.95, 0.92, 0.42 and 0.33 respectively. What percentage of the total electricity generated is consumed in running the auxiliaries?

(a) 3.56% (b) 4.67%
(c) 5.67% (d) 7.32%

[ESE : 2023]

Directions for the following **three (03)** items:

Read the following information and answer the **three** items that follows:

A forced draught fan supplies air at 10 m/s against a draught of 20 mm of water across the fuel bed. 2500 kg/h of coal is consumed and 16 kg of air is supplied per kg of coal burned to run the fan. The temperature of the flue gas and the ambient air may be taken as 600 K and 300 K respectively. Take density of air as 1.176 kg/m^3.

1.107 What is the total pressure head to be produced by the fan?

(a) 176.50 N/m^2 (b) 255.00 N/m^2
(c) 310.05 N/m^2 (d) 412.00 N/m^2

[ESE : 2023]

1.108 What is the volume of air to be handled?

(a) 2.46 m^3/s (b) 4.79 m^3/s
(c) 8.29 m^3/s (d) 9.45 m^3/s

[ESE : 2023]

1.109 What is the power required to run the forced draught fan?

(a) 2.41 kW (b) 3.47 kW
(c) 4.36 kW (d) 5.33 kW

[ESE : 2023]

1.110 Consider the following statements regarding cooling towers in thermal power plants:

1. In a natural draught cooling tower, the cold water is allowed to fall over louvers.
2. In an atmospheric type cooling tower, the hot water from the condenser is pumped to the troughs and nozzles situated near the bottom.
3. In forced draught cooling towers, draught fans are installed at the bottom of towers.
4. Mechanical draught cooling towers may be forced draught cooling towers or induced draught cooling towers.

Which of the above statements are correct?

(a) 1 and 2 only (b) 2 and 3 only
(c) 3 and 4 only (d) 1 and 3 only

[ESE : 2024]

1.111 Consider the following statements regarding steam condensers:

1. Power plant cycle improves in efficiency as the turbine exhaust pressure drops with steam condenser
2. The use of condenser decreases the size of boiler installation.
3. The vacuum obtainable in a condenser is governed by the inlet water temperature which in turn varies with the amount of condensing water used per kg of steam and its initial temperature.

Which of the above statements are correct?

(a) 1 and 2 only (b) 2 and 3 only
(c) 1 and 3 only (d) 1, 2 and 3

[ESE : 2024]

■■■■

Answers Thermal Power Plant Components

1.1 (a)	1.2 (c)	1.3 (c)	1.4 (d)	1.5 (b)	1.6 (a)	1.7 (a)	1.8 (d)	1.9 (b)
1.10 (a)	1.11 (c)	1.12 (b)	1.13 (b)	1.14 (b)	1.15 (a)	1.16 (c)	1.17 (d)	1.18 (a)
1.19 (b)	1.20 (d)	1.21 (a)	1.22 (a)	1.23 (a)	1.24 (b)	1.25 (*)	1.26 (a)	1.27 (c)
1.28 (c)	1.29 (c)	1.30 (d)	1.31 (a)	1.32 (a)	1.33 (c)	1.34 (c)	1.35 (b)	1.36 (c)
1.37 (d)	1.38 (c)	1.39 (a)	1.40 (c)	1.41 (d)	1.42 (d)	1.43 (b)	1.44 (a)	1.45 (b)
1.46 (c)	1.47 (c)	1.48 (c)	1.49 (a)	1.50 (a)	1.51 (a)	1.52 (c)	1.53 (b)	1.54 (b)
1.55 (c)	1.56 (a)	1.57 (b)	1.58 (c)	1.59 (a)	1.60 (a)	1.61 (d)	1.62 (c)	1.63 (a)
1.64 (c)	1.65 (b)	1.66 (b)	1.67 (d)	1.68 (d)	1.69 (a)	1.70 (a)	1.71 (d)	1.72 (a)
1.73 (b)	1.74 (b)	1.75 (c)	1.76 (b)	1.77 (d)	1.78 (b)	1.79 (b)	1.80 (c)	1.81 (b)
1.82 (d)	1.83 (a)	1.84 (d)	1.85 (b)	1.86 (c)	1.87 (d)	1.88 (b)	1.89 (d)	1.90 (d)
1.91 (c)	1.92 (b)	1.93 (c)	1.94 (b)	1.95 (d)	1.96 (c)	1.97 (a)	1.98 (c)	1.99 (d)
1.100 (d)	1.101 (d)	1.102 (d)	1.103 (c)	1.104 (d)	1.105 (b)	1.106 (d)	1.107 (b)	1.108 (d)
1.109 (a)	1.110 (c)	1.111 (d)						

Explanations Thermal Power Plant Components

1.1 (a)

Forced circulation is required because at high pressure the density difference between steam and water is very less.

1.2 (c)

NTPC-Talcher (Orissa) plant has sub-critical once-through boiler.

1.3 (c)

Fusible plug is made of low melting point alloys. Which melt when temperature increases beyond a certain limit.

1.4 (d)

Bomb calorimeter measures the higher calorific value because the fuel sample is burnt at a constant volume in the bomb.

1.5 (b)

Air contains 23.20% of oxygen by weight and 21% by volume.

In the combustion the number of moles of the products are always greater than that of reactant.

1.6 (a)

The correct order are superheater, economiser, air preheater.

1.7 (a)

Fluidized bed boiler allow inferior grades of coal to be used without slagging problems. The carbon and ash carry over in the flue gases does not reach high temperature at which they become soft and foul heat transfer surface.

The low temperature combustion results in substantial reduction in the emission of oxides of nitrogen NO_2 a serious pollutant and cheaper alloy material are possible resulting in economy of combustion.

1.8 (d)

Composition of bio - gas	LPG Gas
CH_4 — 50-60%	Propane — 55-60%
CO_2 — 38-34%	Butane — 40-45%
N_2 — 4-5%	
O_2 — 1%	

1.9 (b)

Steam trap is used to discharge condensate and non-condensable gases with a negligible consumption or loss of line steam.

1.10 (a)

Coal may be of following types:

(i) Peat

(ii) Lignite

(iii) Subbituminous

(iv) Bituminous

(v) Anthracite, with increasing percentage of carbon. Anthracite is often difficult to burn as it contains more than 86% fixed carbon and less volatile matter.

Bituminous coal is the largest group containing 46 to 86% of fixed carbon and 20 to 40% of volatile matter which helps in burning. Lignite is the lowest grade of coal containing moisture as higher as 30% and high volatile matter.

1.11 (c)

It is used to periodically vent water from the boiler drum. This water contains the most concentrated precursor for sludge build-up.

1.13 (b)

Advantages of fluidised bed combustion:

1. It can use any cheap fuel without the support of any auxiliary fuel.
2. Due to superior mixing, the combustion efficiency is very high upto 99%.
3. Concurrent removal of SO_2 is possible.
4. Low NO_x emission.

1.14 (b)

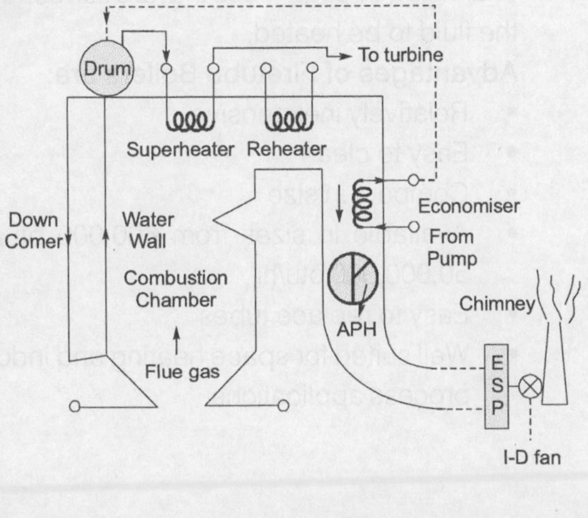

1.15 (a)

In case of cooling tower, approach (A) is defined as the difference between exit temperature of cooling water (t_{c_2}) and wet bulb temperature of the ambient air (t_{wb})

$$A = t_{c_2} - t_{wb}$$

And the range (R) is defined as the difference in the temperature of the incoming warm water (t_{c_1}) and the exiting cooled water (t_{c_2})

$$R = t_{c_1} - t_{c_2}$$

Given $$2(t_{db} - t_{wb}) = (t_{c_2} - t_{wb}) + (t_{c_1} - t_{c_2})$$

$$2t_{db} - 2t_{wb} = t_{c_1} - t_{wb}$$

$$2t_{db} = t_{c_1} + t_{wb}$$

$$t_{db} = \frac{t_{c_1} + t_{wb}}{2}$$

1.16 (c)

In induced draught system jet of steam is placed at the bottom of the chimney. It sucks in air and forces the same into the chimney. This type of draught is used in steam locomotives.

1.17 (d)

Modern waste heat recovery boilers are water tube boiler.

1.18 (a)

Super critical boilers does not have any drum.

1.19 (b)

Locomotive boiler: It is shell type, horizontal fire tube, internally fired, solid fuel boiler.

1.22 (a)

Proximate analysis indicate the behaviour of coal when it is heated. The proximate analysis of coal gives fixed carbon, volatile matter, moisture and ash and the summation of all these should be equal to the mass of coal taken for analysis.

1.23 (a)

APH: It increases the temperature of primary and secondary air, which is required for combustion.

Ecomonizer: It gives sensible heat to boiler feed water.

1.24 (b)

Following are some of the important accessories:

1. A feed water pump
2. An injector
3. A superheater
4. An air preheater

Economizer and injector are boiler accessories.

1.27 (c)

The function of an injector is same as that of a feed pump i.e. to force feed water into boiler that is operation under pressure. It is not operated by flue gases.

1.28 (c)

The function of a fusible plug is to extinguish the fire of the furnace when the water level inside the boiler falls to a dangerously low level. If this is not done the boiler tubes and shell may be damaged due to overheating or the boiler may explode.

1.31 (a)

Water tube boilers are high capacity boilers and preferred for high pressure and high output.

1.34 (c)

Heat added in boiler,

= 4 (100 – 30) + (0.9) (2400)

= 4(70) + (0.9) (2400) = **2440 kJ/kg**

1.36 (c)

Advantages of fluidized bed boiler

1. Any cheap fuel without the support of any auxiliary fuel can be used.
2. Due to superior mixing of solid fuels with air combustion efficiency is very high (up to 99%).
3. Due to low combustion temperature (around 900°C) and the staged combustion process, No_x emission is very low.
4. Compact furnace.

1.37 (d)

Supercritical steam generator has no drum and also no boiling occur in it. It has circulation pump.

1.39 (a)

The function of economizer is to reduce energy consumption or to perform function like preheating a fluid.

The function of airpreheater is to recover heat from the boiler flue gas which preheats the air used for combustion purpose.

1.40 (c)

Silica is the most critical impurity which require special consideration. Silica at higher operating pressure volatilizes and carries over to turbine blades, which will further effect the turbine blades.

1.41 (d)

In a locomotive boiler, draught is produced by steam jet.

1.42 (d)

Correct path of flue gases from boiler furnace to Chimney is

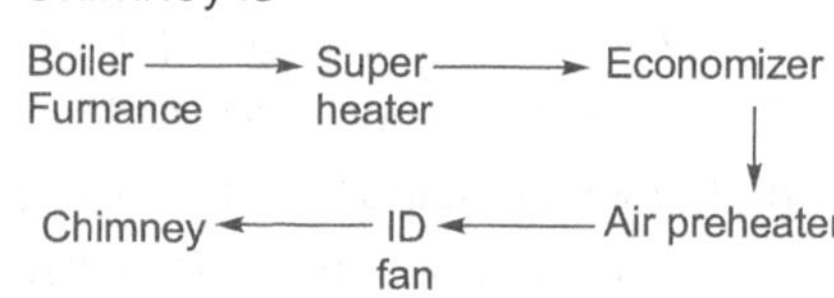

1.43 (b)

The economizer is a heat exchanger which raises the temperature of feed water leaving the highest pressure feed water heater to about the saturation temperature corresponding to boiler pressure. Thus reduce fuel consumption.

1.44 (a)

The modern coal power plant have exit temperature of flue gases from chimney should be below 100°C because in order to get the higher efficiency and also to reduce the exhaust emission.

1.45 (b)

The fire, or hot flue gases from the burner, is channeled through tubes that are surrounded by the fluid to be heated.

Advantages of Firetube Boilers are:

- Relatively inexpensive
- Easy to clean
- Compact in size
- Available in sizes from 600,000 btu/hr to 50,000,000 btu/hr
- Easy to replace tubes
- Well suited for space heating and industrial process applications

Disadvantages of Firetube Boilers include:

- Not suitable for high pressure applications 250 psig and above
- Limitation for high capacity steam generation

1.46 (c)

Plunger pump is used for pumping highly viscous fluids like slurries etc.

1.47 (c)

Orsat analyzer measures the volume or mole fractions of CO_2, CO and O_2 in the dry flue gas.

$CO_2 + CO + O_2 + N_2 = 100\%$ by volume

In deficit air, only CO_2, CO and N_2 will remain there.

1.48 (c)

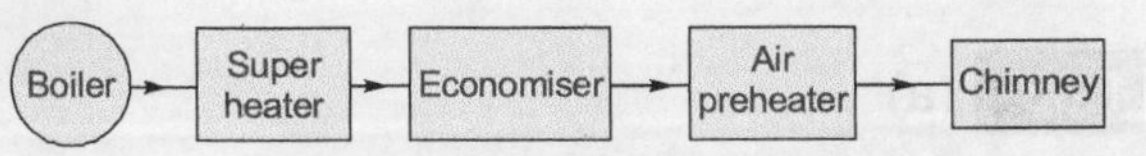

(Path of Flue Gas)

1.51 (a)

Operating pressure of fire tube boilers is limited to 16 bar while water tube boilers can work under as high pressure as 100 bar.

1.53 (b)

The cyclone separators collection efficiency increases with increasing

(a) particle size
(b) particle density
(c) inlet gas velocity
(d) cyclone body length
(e) number of gas revolutions
(f) smoothness of cyclone walls.
(g) $\dfrac{1}{\text{cyclone diameter}}$

1.55 (c)

Mechanical draught or forced draught fan is costly but highly effective as compared to natural draught or induced draught system.

1.56 (b)

The heat transfer sections of a large boiler include the primary heat transfer surfaces (the evaporator, superheater and reheater) and the secondary heat transfer surfaces (air preheater and economizer).

1.57 (b)

For maximum discharge of hot gases through a chimney, the height of hot column producing drought is equal to the height of chimney.

1.63 (a)

The lowest temperature to which water can be cooled in any cooling Tower is WBT of Ingoing air which is lessser than its DBT.

The hyperbolic shape is given to cooling tower for structural stability rather than increasing H.T. rate.

1.68 (d)

$$0.02 = \frac{353\times 35}{1000}\left\{\frac{1}{305} - \frac{m+1}{m}\left(\frac{1}{638}\right)\right\}$$

$$\frac{20}{353\times 35} = \left\{\frac{1}{305} - \frac{m+1}{m}\left(\frac{1}{638}\right)\right\}$$

$$\frac{1}{617.75} = \frac{1}{305} - \frac{m+1}{m}\left(\frac{1}{638}\right)$$

$$\frac{m+1}{m}\left(\frac{1}{638}\right) = \frac{1}{305} - \frac{1}{617.75}$$

$$\frac{m+1}{m} = \frac{312.75\times 638}{305\times 617.75} = 1.0590$$

$$\frac{1}{m} = 0.059$$

$$\Rightarrow \quad m = 16.9 \text{ kg/kg of fuel}$$

1.70 (a)

$$P_1 = 20 + 1.01325$$
$$= 21.01325 \text{ bar}$$
$$= 2101 \text{ kPa}$$

$$250 \text{ mm hg.} = \frac{250}{760}\times 101.325$$
$$= 33.33 \text{ kPa}$$
$$P_2 = 101.325 - 33.33$$
$$= 67.995 \approx 68 \text{ kPa}$$

1.71 (d)

$$\Delta_{Aux} = \frac{0.356 - 0.33}{0.356}$$

$$= 0.0730 = 7.3\%$$

1.72 (a)

The overall efficiency of plant is given by,

$$\eta_{overall} = \eta_{boiler} \times \eta_{cycle} \times \eta_{gen} \times \eta_{aux} \times \eta_{turbu}$$

$$0.33 = 0.92 \times 0.42 \times 0.95 \times 0.97 \times \eta_{aux}$$

$$\eta_{aux} = 0.926$$

Power consumed = 1 – 0.926 = 7.32%

1.78 (b)

Settling time, t_s : The settling time is the time required for the response curve to reach and stay within a range about the final value of size specified by absolute percentage of the final value (usually 2% or 5%). It is related to the largest time constant of the control system.

$$t_s = \frac{4}{\zeta\omega_n} \quad \text{(for 2\% criterion)}$$

$$t_s = \frac{3}{\zeta\omega_n} \quad \text{(for 5\% criterion)}$$

Therefore settling time $\propto \frac{1}{\omega_n}$ i.e., inversely proportional to the natural frequency.

1.79 (b)

$$0.33 = 0.97 \times 0.95 \times 0.92 \times 0.42 \times \eta$$

$$\eta = 92.69\%$$

1.80 (c)

1. Steam power plants work on rankine cycle.
2. In super-critical steam cycle, stream is generated in a 'once-through' boiler at a pressure above the critical point of 221.2 bar.

1.81 (b)

Coal Grade	Approximate lower Heating value (kJ/kg)
Anthracite	30080
Semi-anthracity	32084
Low volatile bituminous	33412
Medium volatile bituminous	32247
High volatile bituminous A	30499
High volatile bituminous B	28262
High volatile bituminous C	25047
Sub-bituminous B	21319
Sub-bituminous C	20830
Lignite	16077

So, statement 2 is wrong.

1.83 (a)

Evaporation loss ~ 1 to 1.5%

Drift loss ~ 0.03%

Therefore, statement (2) and (3) are not correct.

1.86 (c)

Advantages of Pulverized coal firing are:

1. Low excess air requirement.
2. Less fan power.
3. Ability to used highly preheated air reducing exhaust losses.
4. Higher boiler efficiency
5. Ability to burn wide variety of coals.
6. Fast response to load changing.
7. Ease of burning altemally with, or in combination with gas and oil.
8. Ability to release large amounts of heat enabling it to generate about 2000 t/h of steam or more in one boiler.
9. Ability to use flyash for making bricks etc.
10. Less pressure losses and draught need.

1.87 (d)

For pump ideal process would be reversible adiabatic compression. So option (d) is not correct.

1.88 (b)

$$\eta_{vacuum} = \frac{\text{Actual vacuum}}{\text{Ideal vacuum}}$$

1.89 (d)

Babcock and Wilcox boiler is a water tube boiler.

1.90 (d)

In water tube boiler water flows inside the tubes and hot gases surrounds the tubes. Low pressure boiler produces steam at pressure below 10 bar.
Lancashire boiler is a fire tube boiler which is internally fired.

1.91 (c)

$$\eta_{overall} = \eta_{gen} \times \eta_{turb} \times \eta_{boiler} \times \eta_{cycle} \times (1 - \eta_{auxi})$$

$$0.33 = 0.97 \times 0.95 \times 0.92 \times 0.42 \times (1 - \eta_{auxi})$$

$$\eta_{auxi} = 7.32\%$$

1.92 (b)

It is impossible to design heat exchanger inside turbine for transferring heat of expanding steam to feed water.

1.96 (c)

Methods of compounding in steam engines
(i) Tandem compound engines
(ii) Woolf compound engines
(iii) Receiver compound engines

Woolf compound engine is a cross type compounding having two cylinders have piston on 180° stage difference that is at several positions. One cylinder might have piston at inner dead centre with other cylinder have piston at outer dead centre. So, statement 2 is wrong.

1.97 (a)

$(G_1 - G_2)$ = Quantity of ash collected
G_1 = Quantity of ash entering

$$\therefore \text{ Degree of collection, } \eta = \frac{(G_1 - G_2)}{G_1}$$

1.98 (c)

Given :Vacuum pressure = 720 mm of Hg
Barometric pressure = 765 mm of Hg
∴ Vacuum corrected to standard barometer of 760 mm = 760 – (765 – 720)
= 715 mm of Hg

1.99 (d)

Under cooling efficiency,

$$\eta_V = \frac{\text{Actual vacuum}}{\text{Barometric pressure} - \text{Absolute pressure of steam}}$$

$$= \frac{720}{765 - 39.99}$$

(∵ 0.0533 bar = 39.99 mm of Hg)
= 0.9931 = 99.31%

1.100 (d)

Under cooling of condenser = Mean temperature of condensation – Hot well temperature
= (34 – 29)°C = 5°C

1.101 (d)

Absolute pressure, P_{abs} = (765 – 720)
= 45 mm of Hg

Absolute pressure inside the condenser = 45 mm of Hg
∴ 45 mm of Hg = 0.06 bar

1.102 (d)

Barometric pressure,
P_a = 76 cm of Hg

Vacuum pressure,
P_{vac} = 69 cm of Hg

Absolute pressure;
$P_{abs} = P_a - P_{vac}$
= (76 – 69) cm of Hg
= 7 cm of Hg

$$= \frac{7 \times 1333.22}{1000} = 9.33 \text{ kPa}$$

1.103 (c)

Given:
Saturation pressure of air at 30°C;
P_s = 4.246 kPa

Total pressure, $P = P_a + P_s$
= 9.33 – 4.246
= 5.08 kPa

1.104 (d)

Given:
Boiler pressure; P_b = 3 MPa
Condenser pressure;
P_c = 4 kPa

Enthalpy at turbine inlet;

$$h_1 = 2802.3 \text{ kJ/kg}$$

Enthalpy at turbine exit;

$$h_2 = 1862.04 \text{ kJ/kg}$$

Pump work, $W_P = 3$ kJ/kg

$$\therefore \quad W_{net} = W_T - W_P = (2802.3 - 1862.04) - 3 = 937.26$$

$$\text{Work ratio; } r_w = \frac{W_{net}}{W_T} = \frac{937.26}{940.26} = 0.997$$

1.105 (b)

Specific steam consumption;

$$\text{SSC} = \frac{3600}{W_{net}} = \frac{3600}{937.26} = 3.83 \text{ kg/kWh}$$

1.106 (d)

Overall efficiency,

$$\eta_0 = \eta_{boiler} \times \eta_{cycle} \times \eta_{mech} \times \eta_{aux}$$

$$\Rightarrow \quad 0.33 = 0.95 \times 0.92 \times 0.42 \times \eta_{aux}$$

$$\therefore \quad \eta_{aux} = 92.6\%$$

∴ Percentage of total electricity generated which is consumed in running the auxiliaries

$$= (100 - 92.6)\% = 7.32\%$$

1.107 (b)

Given:

Draught across fuel bed = 20 mm of H_2O = 20g N/m^2

$$\text{Draught velocity head} = \frac{V^2}{2g} = \frac{100}{2g} = \frac{100}{2 \times 9.81} = 5.1 \text{ m}$$

Pressure; $P = \rho g h = 1.176 \times g \times 5.1 = 5.59g$

∴ Total draught = $(20 + 5.59)g$ = 255 N/m^2

1.108 (d)

Volume of air to be handled,

$$V = \frac{m}{\rho} = \frac{(2500 \times 16)}{1.176 \times 3600} = 9.45 \text{ m}^3/\text{s}$$

1.109 (a)

Power required to run the forced draught fan;

$$P = \frac{255 \times 9.45}{1000 \times \eta}$$

Assuming efficiency, η = 100%

$$\therefore \quad P = \frac{255 \times 9.45}{1000} = 2.41 \text{ kW}$$

1.110 (c)

- In a natural draught cooling tower, the hot water is allowed to fall over louvers.
- In an atmospheric type cooling tower, the hot water from the condenser is pumped to the troughs and nozzles situated near the top.
- In forced draught cooling towers, draught fans are installed at the bottom of towers.
- Mechanical draught cooling towers may be forced draught cooling towers or induced draught cooling towers.

1.111 (d)

- As the turbine exhaust pressure drops, we can obtain more turbine output and thus efficiency of power plant improves.
- As we decrease condenser pressure, the net work output will increase, so the steam flow required will be less, thus it will result in smaller size of boiler.
- The vacuum obtainable in a condenser is governed by the inlet water temperature which in turn varies with the amount of condensing water used per kg of steam and its initial temperature.

■■■■

2 Nozzles

2.1 The plot for the pressure ratio along the length of the convergent divergent nozzle is shown in the given figure. The sequence of the flow conditions labelled (1), (2), (3) and (4) in the figure is respectively

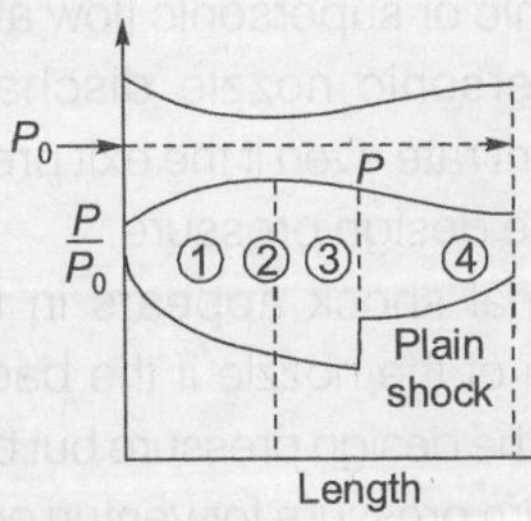

(a) supersonic, sonic, subsonic and supersonic
(b) sonic, supersonic, subsonic and supersonic
(c) subsonic, supersonic, sonic and subsonic
(d) subsonic, sonic, supersonic and subsonic

[ESE : 2000]

2.2 Consider the following statements:
For supersaturated flow through a steam nozzle, the
1. enthalpy drop reduces further.
2. exit temperature increases.
3. flow rate increases.
Which of these statements are correct?
(a) 1, 2 and 3 (b) 1 and 2
(c) 2 and 3 (d) 1 and 3

[ESE : 2000]

2.3 The critical pressure ratio for maximum discharge through a nozzle is given by

(a) $\left(\frac{n+1}{2}\right)^{\frac{n}{n-1}}$ (b) $\left(\frac{2}{n+1}\right)^{\frac{n}{n-1}}$

(c) $\left(\frac{n+1}{2}\right)^{\frac{n-1}{n}}$ (d) $\left(\frac{2}{n+1}\right)^{\frac{n-1}{n}}$

[ESE : 2000]

2.4 Match List-I (Name) with List-II (Figures) and select the correct answer using the codes given below the lists:

List-I / **List-II**

A. Subsonic nozzle — 1. 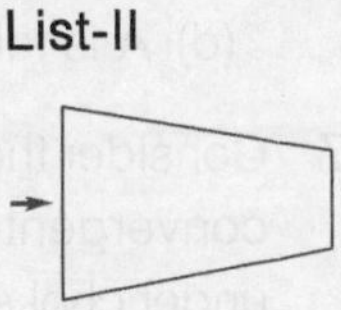

B. Supersonic nozzle — 2. 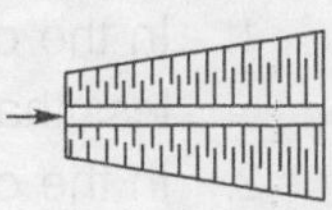

C. Subsonic diffuser — 3. 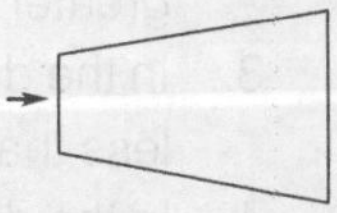

D. Centrifugal compressor — 4. 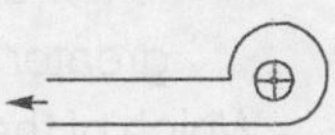

5. 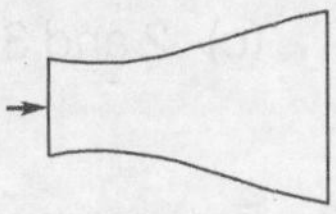

Codes:

	A	B	C	D
(a)	3	4	2	5
(b)	1	5	3	4
(c)	3	5	2	4
(d)	1	4	3	5

[ESE : 2001]

2.5 Which of the following statement(s) is/are relevent to critical flow through a steam nozzle?
1. Flow rate through the nozzle is minimum
2. Flow rate through the nozzle is maximum
3. Velocity at the throat is supersonic
4. Velocity at the throat is sonic
Select the correct answer using the codes given below:
Codes:
(a) 1 alone (b) 1 and 3
(c) 2 and 4 (d) 4 alone

[ESE : 2001]

2.6 **Assertion (A):** A convergent-divergent nozzle may give supersonic or subsonic flow at the exit even if the throat is choked.

Reason (R): Depending on the back pressure ratio p_b/p_o, the divergent part of the nozzle may act as a supersonic nozzle or a subsonic diffuser.

(a) Both **A** and **R** are individually true and **R** is the correct explanation of **A**
(b) Both **A** and **R** are individually true but **R** is not the correct explanation of **A**
(c) **A** is true but **R** is false
(d) **A** is false but **R** is true [ESE : 2001]

2.7 Consider the following statements in relation to a convergent-divergent steam nozzle operating under choked conditions:

1. In the convergent portion steam velocity is less than sonic velocity.
2. In the convergent portion steam velocity is greater than sonic velocity.
3. In the divergent portion the steam velocity is less than sonic velocity.
4. In the divergent portion the steam velocity is greater than sonic velocity.

Which of these statements are correct?

(a) 1 and 3 (b) 1 and 4
(c) 2 and 3 (d) 2 and 4

[ESE : 2002]

2.8 For maximum discharge through a convergent nozzle the pressure ratio p_2/p_1 should be (where n is the isentropic expansion index)

(a) $\left(\frac{n}{n+1}\right)^{\frac{2}{n+1}}$ (b) $\left(\frac{2}{n+1}\right)^{\frac{n}{n-1}}$

(c) $\left(\frac{n+1}{2}\right)^{\frac{n}{n-1}}$ (d) $\left(\frac{2}{n+1}\right)^{\frac{n}{n+1}}$

[ESE : 2002]

2.9 Which of the following statements is/are true in case of one-dimensional flow of perfect gas through a converging-diverging nozzle?

1. The exit velocity is always supersonic
2. The exit velocity can be subsonic or supersonic
3. If the flow is isentropic, the exit velocity must be supersonic
4. If the exit velocity is supersonic, the flow must be isentropic

Select the correct answer from the codes given below:

Codes:

(a) 2 and 4 (b) 2, 3 and 4
(c) 1, 3 and 4 (d) 2 alone

[ESE : 2002]

2.10 Consider the following statements pertaining to one-dimensional isentropic flow in a convergent-divergent passage:

1. A convergent-divergent passage may function as a supersonic nozzle or a venturi depending on the back pressure.
2. At the throat, sonic conditions exist for subsonic or supersonic flow at the outlet.
3. A supersonic nozzle discharges fluid at constant rate even if the exit pressure is lower than the design pressure.
4. A normal shock appears in the diverging section of the nozzle if the back pressure is above the design pressure but below a certain minimum pressure for venturi operation.

Which of these statements are correct?

(a) 1, 2, 3 and 4 (b) 1, 3 and 4
(c) 2, 3 and 4 (d) 1 and 2

[ESE : 2003]

2.11 Which one of the following is the correct expression for the critical pressure ratio of a nozzle?

(a) $\left(\frac{2}{n+1}\right)^{1/(n-1)}$ (b) $\left(\frac{1}{n+1}\right)^{n/(n-1)}$

(c) $\left(\frac{2}{n+1}\right)^{n/(n-1)}$ (d) $\left(\frac{1}{n+1}\right)^{1/(n-1)}$

[ESE : 2004]

2.12 What is the critical pressure ratio for isentropic nozzle flow with ratio of specific heats as 1.5?

(a) $(0.8)^3$ (b) $(0.8)^{0.6}$
(c) $(1.25)^{0.33}$ (d) $(1.25)^3$

[ESE : 2004]

2.13 **Assertion (A):** In a supersonic nozzle, with sonic condition at the throat, any reduction of downstream pressure will not be felt at the inlet of the nozzle.

Reason (R): The disturbance caused downstream of supersonic flow travels at sonic velocity which cannot propagate upstream by Mach cone.

(a) Both A and R are true and R is the correct explanation of A
(b) Both A and R are true but R is not a correct explanation of A
(c) A is true but R is false
(d) A is false but R is true [ESE : 2004]

2.14 Consider the following statements:
In the case of convergent nozzle for compressible flow
1. no shock wave can occur at any pressure ratio.
2. no expansion wave can occur below a certain pressure ratio.
3. expansion wave can occur below a certain pressure ratio.
4. shock wave can occur above a certain pressure ratio.

Which of these statements are correct?
(a) 1 and 2 (b) 3 and 4
(c) 1 and 3 (d) 2 and 4
[ESE : 2005]

2.15 To get supersonic velocity of steam at nozzle exit with a large pressure drop across it, the duct must
(a) converge from inlet to exit
(b) diverge from inlet to exit
(c) first converge to the throat and then diverge till exit
(d) remain constant in cross-section
[ESE : 2005]

2.16 If the cross-section of a nozzle is increasing in the direction of flow in supersonic flow, then in the downstream direction
(a) Both pressure and velocity will increase
(b) Both pressure and velocity will decrease
(c) Pressure will increase but velocity will decrease
(d) Pressure will decrease but velocity will increase [ESE : 2005]

Direction: Each of the next questions consists of two statements, one labelled as the **'Assertion (A)'** and the other as **'Reason (R)'**. You are to examine these two statements carefully and select the answers to these items using the codes given below:

Codes:
(a) Both **A** and **R** are individually true and **R** is the correct explanation of **A**
(b) Both **A** and **R** are individually true but **R** is not the correct explanation of **A**
(c) **A** is true but **R** is false
(d) **A** is false but **R** is true

2.17 Assertion (A): A correctly designed convergent divergent nozzle working at designed conditions is always choked.
Reason (R): In these conditions, the mass flow through the nozzle is minimum. [ESE : 2005]

2.18 Assertion (A): Gas and stream nozzles are shaped at inlet in such a way that the nozzle converges rapidly over the first portion of its length
Reason (R): This shape is provided so that velocity at inlet to the nozzle is negligibly small in comparison with the exit velocity
[ESE : 2005]

2.19 In a steam nozzle, to increase the velocity of steam above sonic velocity by expanding steam below critical pressure
(a) A vacuum pump is added
(b) Ring diffusers are used
(c) Divergent portion of the nozzle is necessary
(d) Abrupt change in cross-section is needed
[ESE : 2006]

2.20 Wilson line is associated with which one of the following?
(a) Total steam consumption with respect to power output
(b) Supersonic flow of steam through a nozzle
(c) Nozzle flow with friction
(d) Supersaturated flow of steam through a nozzle
[ESE : 2006]

2.21 A converging diverging nozzle is connected to a gas pipeline. At the inlet of the nozzle (converging section) the Mach number is 2. It is observed that there is a shock in the diverging section. The value of the Mach number at the throat is
(a) < 1 (b) Equal to 1
(c) > 1 (d) ≥ 1
[ESE : 2006]

2.22 In a subsonic diffuser
(a) Static pressure increases with Mach number
(b) Mach number decreases with increasing area ratio
(c) Static pressure decreases with Mach number
(d) Area ratio decreases in the flow direction
[ESE : 2007]

2.23 Across a normal shock wave in converging-diverging nozzle for adiabatic flow, which of the following relations are valid?

(a) Continuity and energy equations, equation of state, isentropic relation
(b) Energy and momentum equations, equation of state, isentropic relation
(c) Continuity, energy and momentum equations, equation of state
(d) Equation of state, isentropic relation, momentum equation, mass conservation principle

[ESE : 2007]

2.24 For incompressible flow a diverging section acts as a diffuser for upstream flow which is

(a) subsonic only
(b) supersonic only
(c) both subsonic and supersonic
(d) sonic [ESE : 2008]

2.25 Given,

η_s = Stage efficiency
η_n = Nozzle efficiency
η_b = Blade efficiency

Which one of the following is correct?

(a) $\eta_n = \eta_b \eta_s$ (b) $\eta_b = \eta_s \eta_n$
(c) $\eta_b \times \eta_n \times \eta_s = 1$ (d) $\eta_s = \eta_b \eta_n$

[ESE : 2008]

2.26 Consider the following statements for compressible flow through a varying area passage:

1. for a convergent nozzle, if the exit pressure is less than critical, external flow will not be isentropic
2. supersonic-subsonic diffuser would appear similar to nozzle and works without irreversibility.

Which of these statements is/are correct?

(a) 1 only (b) 2 only
(c) Both 1 and 2 (d) Neither 1 nor 2

[ESE : 2008]

2.27 Steam pressures at the inlet and exit of a nozzle are 16 bar and 5.2 bar respectively and discharge is 0.28 m^3/s. Critical pressure ratio is 0.5475. If the exit pressure is reduced to 3.2 bar then what will be the flow rate in m^3/s ?

(a) 0.280 (b) 0.328
(c) 0.356 (d) 0.455

[ESE : 2009]

2.28 Consider the following statements:

Choked flow through a nozzle means:

1. Discharge is maximum
2. Discharge is zero
3. Velocity at throat is supersonic
4. Nozzle exit pressure is less than or equal to critical pressure.

Which of these statements is/are correct?

(a) 1 only (b) 1 and 2
(c) 2 and 3 (d) 1 and 4

[ESE : 2009]

2.29 The index of expansion of dry saturated steam flowing through a nozzle is equal to 1.135, and then what is the critical pressure ratio for this flowing steam in the nozzle?

(a) 0.96 (b) 0.58
(c) 0.33 (d) 0.15 [ESE : 2009]

2.30 Match **List-I** with **List-II** and select the correct answer using the code given below the lists:

List-I

A. Steam turbine
B. Subsonic diffuser
C. Supersonic nozzle
D. Axial flow compressor

List-II

1. 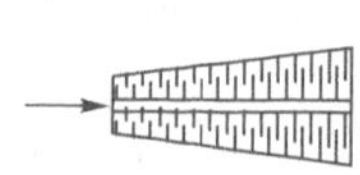

2. 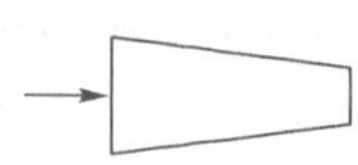

3. 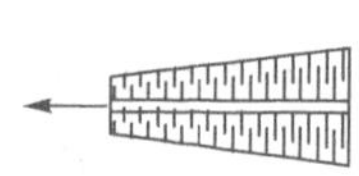

4. 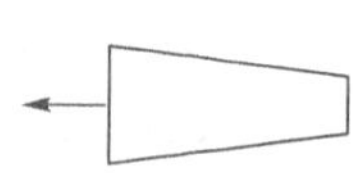

5. 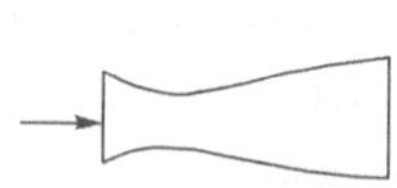

Codes:

	A	B	C	D
(a)	1	4	5	3
(b)	2	4	5	3

(c) 3 5 4 1
(d) 2 5 4 3

[ESE : 2010]

2.31 An isentropic nozzle is discharging steam through critical pressure ratio. If the back pressure is further decreased, the discharge will

(a) Decrease
(b) Increase
(c) Remain unaffected
(d) Come to a dead stop due to shock waves

[ESE : 2011]

2.32 Consider the following statements for an incompressible fluid flowing through a convergent-divergent duct

1. The convergent part acts as a nozzle.
2. The divergent part acts as a diffuser.
3. The maximum limiting speed of the fluid is the sonic velocity.

Which of these statements are correct?

(a) 1, 2 and 3 (b) 1 and 2 only
(c) 2 and 3 only (d) 1 and 3 only

[ESE : 2011]

2.33 In supersonic flow, if cross-section is increasing in the direction of flow

(a) both p and V increase
(b) p decreases and V increases
(c) both p and V decreases
(d) p increases and V decreases [ESE : 2012]

2.34 Supersaturated condition occurs in a steam expanding through a convergent-divergent nozzle because of the delay in

(a) change of dryness fraction
(b) change of superheat
(c) evaporation process
(d) condensation process [ESE : 2012]

2.35 In a nozzle designed to produce a supersonic jet, the flow in the convergent section will always be

(a) subsonic (b) sonic
(c) supersonic (d) transonic

[ESE : 2012]

Directions: Each of the next questions consists of two statements, one labelled as the 'Statement (I)' and the other as 'Statement (II)'. You are to examine these two statements carefully and select the answers to these items using the codes given below:

Codes:

(a) Both Statement (I) and Statement (II) are individually true and Statement (II) is the correct explanation of Statement (I).
(b) Both Statement (I) and Statement (II) are individually true but Statement (II) is **NOT** the correct explanation of Statement (I).
(c) Statement (I) is true but Statement (II) is false
(d) Statement (I) is false but Statement (II) is true.

2.36 Statement (I): To satisfy the momentum equation for flow with heat transfer, it is necessary to use a divergent duct or a duct with large cross-sectional area so that the changes in velocity are negligible.

Statement (II): Heat addition to a fluid in a constant area duct cannot be achieved without a drop in pressure. [ESE : 2012]

2.37 Statement (I): When a convergent-divergent nozzle operates at off design conditions, a shock wave will enter the exit and move towards the throat and vanish there.

Statement (II): The convergent section functioning as nozzle will have only subsonic flow where no shock can exist. [ESE : 2012]

2.38 When the fluid at sonic velocity enters a convergent-divergent passage, the velocity of the fluid

(a) increases continuously
(b) decreases continuously
(c) increases and then decreases
(d) decreases and then increases

[ESE : 2012]

2.39 A gas is flowing through an insulated nozzle. If the inlet velocity of gas is negligible and there is an enthalpy drop of 45 kJ/kg, the velocity of gas leaving of nozzle is

(a) 100 m/s (b) 200 m/s
(c) 300 m/s (d) 350 m/s

[ESE : 2013]

2.40 Consider the following statements:
Isentropic flow through a steam nozzle becomes 'choked' when

1. Discharge is maximum
2. Discharge is zero

3. Nozzle pressure ratio is ≤ critical pressure ratio
4. Throat velocity reaches sonic value

Which of these statements are correct?

(a) 1, 2, 3 and 4 (b) 1, 2 and 3 only
(c) 2, 3 and 4 only (d) 1, 3 and 4 only

[ESE : 2013]

2.41 Frictional losses in the nozzle
(a) reduces the enthalpy of the fluid
(b) increases the enthalpy of the fluid
(c) no effect on enthalpy of the fluid
(d) None of the above [ESE : 2013]

2.42 A converging-diverging nozzle is operated at a pressure difference which is not the design value for isentropic flow. As a consequence a normal shock is formed in the diverging portion. In this situation the Mach number at the throat is
(a) Less than 1
(b) More than 1
(c) Exactly 1
(d) Could be less or more than 1 depending on the pressure difference

[ESE : 2014]

2.43 When the pressure drop across a converging-diverging nozzle is different from the design value for isentropic flow, which of the following is possible?
(a) There is one normal shock in the converging part and one normal shock in the diverging part
(b) There is only one normal shock in the converging part and none in the diverging part
(c) There is only one normal shock in the diverging part and none in the converging part
(d) There are two or more normal shocks, depending on the pressure drop, in the diverging part and none in the converging part

[ESE : 2014]

2.44 When a converging-diverging nozzle is operated at off-design conditions, a normal shock forms in the diverging portion. The nozzle can be assumed to be perfectly insulated from the surroundings. Then across the shock.
(a) The velocity undergoes a jump but pressure and entropy remain unchanged
(b) The pressure undergoes a jump but velocity and entropy remain unchanged
(c) The velocity and pressure undergo a jump, but entropy remains unchanged because there is no heat transfer
(d) Velocity, pressure and entropy all undergo a jump

[ESE : 2014]

2.45 Choked flow through an isentropic nozzle implies
1. Discharge is maximum
2. Discharge is zero
3. Nozzle exit pressure ≤ critical pressure
4. Mach number at the throat is unity

Which of the above statements are correct?

(a) 1, 2, 3 and 4 (b) 1, 2 and 3 only
(c) 1, 3 and 4 only (d) 2, 3 and 4 only

[ESE : 2014]

2.46 The effect of considering friction in steam nozzle for the same pressure ratio leads to
(a) Increase in dryness fraction of exit steam
(b) Decrease in dryness fraction of exit steam
(c) No change in the quality of exit steam
(d) Decrease or increase of dryness fraction of exit steam depending upon inlet quality

[ESE : 2014]

2.47 In a convergent divergent nozzle, the velocity at throat of nozzle is given by

(a) $V=\left[\frac{2np_1v_1}{n-1}\right]^{1/2}$ (b) $V=\left[\frac{2np_1v_1}{n+1}\right]^{1/2}$

(c) $V=[2np_1v_1]^{1/2}$ (d) $V=[np_1v_1]^{1/2}$

[ESE : 2015]

2.48 Which of the following parameters of fluid will increase as a result of friction while flowing adiabatically through a convergent divergent nozzle?
(a) Available heat drop
(b) Entropy
(c) Stagnation pressure
(d) Stagnation temperature [ESE : 2015]

2.49 In a shock wave, the flow passes from a
(a) Subsonic to a sonic state
(b) Subsonic to a supersonic state
(c) Supersonic to a sonic state
(d) Supersonic to a subsonic state

[ESE : 2015]

2.50 Consider the flow of steam through a Convergent-Divergent nozzle under real conditions, where supersaturation occurs, the difference between the saturation temperature corresponding to the pressure and the supersaturated temperature is defined as degree of
(a) Under cooling (b) Superheat
(c) Reaction (d) Saturation [ESE : 2015]

2.51 For accelerating an ideal gas isentropically from rest to supersonic speed, we require a convergent-divergent nozzle. To decelerate an ideal gas isentropically from supersonic speed to subsonic speed the diffuser should be
(a) Diverging
(b) Converging
(c) Convergent-divergent
(d) Divergent-convergent [ESE : 2015]

2.52 **Assertion (A):** The supersaturated expansion in nozzle is a process in which the steam expands beyond the saturated vapour line in superheated condition.
Reason (R): Steam cannot exist in superheated state when the expansion process in nozzle reaches the saturated vapour line.
(a) Both **A** and **R** are true and **R** is the correct explanation of **A**
(b) Both **A** and **R** are true but **R** is not a correct explanation of **A**
(c) **A** is true but **R** is false
(d) **A** is false but **R** is true [ESE : 2015]

2.53 Consider the following statements pertaining to a convergent-divergent nozzle flow with Mach number 0.9 at the throat:
1. The flow is subsonic in both the converging and the diverging sections
2. The Mach number at the exit is less than one
3. In the diverging section, the flow is supersonic
4. There is a shock in the diverging section

Which of the above statements are correct ?
(a) 1 and 4 (b) 1 and 2
(c) 3 only (d) 3 and 4 [ESE : 2016]

2.54 Supersaturated flow occurs in a steam nozzle due to delay in :
(a) Throttling (b) Condensation
(c) Evaporation (d) Entropy drop
[ESE : 2016]

2.55 Under ideal conditions, the velocity of steam at the outlet of a nozzle for a heat drop of 450 kJ/kg from inlet reservoir condition upto the exit is:
(a) 649 m/s (b) 749 m/s
(c) 849 m/s (d) 949 m/s
[ESE : 2016]

2.56 A convergent-divergent nozzle is said to be choked when :
(a) Critical pressure is attained at the exit and Mach number at this section is sonic
(b) Velocity at the throat becomes supersonic
(c) Exit velocity becomes supersonic
(d) Mass flow rate through the nozzle reaches a maximum value
[ESE : 2016]

2.57 **Statement (I) :** In a choked flow in a convergent divergent nozzle, flow in the diverging section is supersonic.
Statement (II) : In a choked flow in a convergent divergent nozzle, the Mach number at the throat is larger than one.
(a) Both Statement (I) and Statement (II) are individually true and Statement (II) is the correct explanation of Statement (I).
(b) Both Statement (I) and Statement (II) are individually true but Statement (II) is **NOT** the correct explanation of Statement (I).
(c) Statement (I) is true but Statement (II) is false
(d) Statement (I) is false but Statement (II) is true.
[ESE : 2016]

2.58 Consider the following statements regarding steam nozzles:
1. Nozzle efficiency is defined as the ratio of the actual enthalpy drop to the isentropic enthalpy drop between the same pressures.
2. Velocity coefficient is defined as the ratio of the actual exit velocity to the exit velocity when the flow is isentropic between the same pressures.
3. Velocity coefficient is the square root of nozzle efficiency.

Which of the above statements are correct?
(a) 1 and 3 only (b) 2 and 3 only
(c) 1 and 2 only (d) 1, 2 and 3
[ESE : 2024]

■■■■

Answers Nozzles

2.1 (d)	2.5 (d)	2.3 (b)	2.4 (b)	2.5 (c)	2.6 (a)	2.7 (b)	2.8 (b)	2.9 (a)
2.10 (a)	2.11 (c)	2.12 (a)	2.13 (c)	2.14 (d)	2.15 (c)	2.16 (d)	2.17 (c)	2.18 (c)
2.19 (c)	2.20 (d)	2.21 (b)	2.22 (b)	2.23 (c)	2.24 (a)	2.25 (d)	2.26 (a)	2.27 (a)
2.28 (d)	2.29 (b)	2.30 (a)	2.31 (c)	2.32 (a)	2.33 (b)	2.34 (d)	2.35 (a)	2.36 (c)
2.37 (b)	2.38 (d)	2.39 (c)	2.40 (d)	2.41 (a)	2.42 (c)	2.43 (c)	2.44 (d)	2.45 (c)
2.46 (a)	2.47 (b)	2.48 (b)	2.49 (d)	2.50 (a)	2.51 (c)	2.52 (c)	2.53 (b)	2.54 (b)
2.55 (d)	2.56 (d)	2.57 (c)	2.58 (d)					

Explanations Nozzles

2.1 (d)

Between 3 and 4, flow is not isentropic in the diverging part, and it is accompanied by an irreversible phenomenon known as shock. Shock occur only when the flow is supersonic and after the shock the flow become subsonic and rest of duct act as diffuser.

2.2 (d)

Effect of supersaturation

(1) Increase in discharge by 2 to 5 percent due to increase in density.

(2) Decrease in heat drop and reduction in exit velocity.

(3) Improves dryness factor.

(4) Entropy increases.

2.3 (b)

$$\frac{T_0}{T} = 1+\frac{\gamma-1}{2}\cdot M^2$$

For critical condtion

$$M = 1$$

$$\frac{T_0}{T_*} = \frac{\gamma+1}{2}$$

$$\Rightarrow \left(\frac{T_*}{T_0}\right) = \frac{2}{\gamma+1}$$

$$\because \frac{P_2}{P_1} = \left(\frac{T_2}{T_1}\right)^{\gamma/(\gamma-1)}$$

$$\Rightarrow \frac{P^*}{P_0} = \left(\frac{T^*}{T_0}\right)^{\gamma/(\gamma-1)}$$

$$\therefore \frac{P^*}{P_0} = \left(\frac{2}{\gamma+1}\right)^{\gamma/(\gamma-1)}$$

2.5 (c)

For critical flow, flow rate through the nozzle is maximum. At this stage velocity at the throat is sonic i.e. Mach number is equal to 1.

2.7 (b)

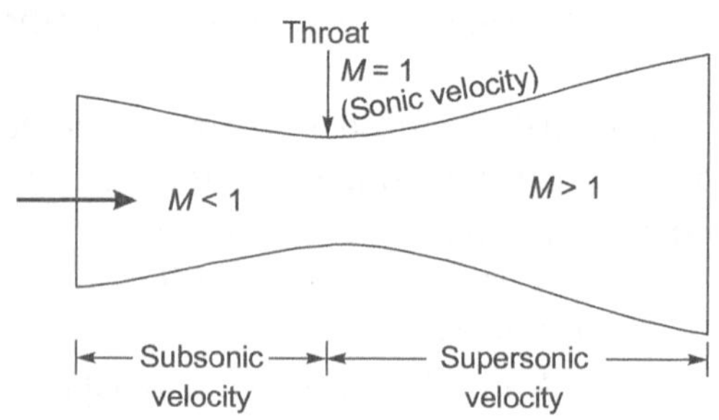

2.8 (b)

$$\frac{T_0}{T} = 1+\frac{\gamma-1}{2}\cdot M^2$$

For critical condtion

$$M = 1$$

$$\frac{T_0}{T^*} = \frac{\gamma+1}{2}$$

$$\Rightarrow \left(\frac{T^*}{T_0}\right) = \frac{2}{\gamma+1}$$

$$\because \frac{P_2}{P_1} = \left(\frac{T_2}{T_1}\right)^{\gamma/(\gamma-1)}$$

$$\Rightarrow \frac{P^*}{P_0} = \left(\frac{T^*}{T_0}\right)^{\gamma/(\gamma-1)}$$

$$\therefore \frac{P^*}{P_0} = \left(\frac{2}{\gamma+1}\right)^{\gamma/(\gamma-1)}$$

2.9 (a)

The exit velocity can be supersonic or subsonic it depends on back pressure.

2.11 (c)

$$\frac{P^*}{P_0} = \left(\frac{2}{n+1}\right)^{\frac{n}{n-1}}$$

2.12 (a)

$$\frac{P^*}{P_0} = \left(\frac{2}{n+1}\right)^{\frac{n}{n-1}} \text{ for } n = 1.5$$

$$\frac{P^*}{P_0} = (0.8)^3$$

2.13 (c)

When back pressure is lower than P_B then fluid achieves a sonic velocity at the throat continuous accelerating supersonic velocities in the diverging section as the pressure decreases. This acceleration comes to a sudden stop, however as a 'normal shock' develops at a section between the throat and the exit plane. Which causes a sudden drop in velocity to subsonic level and a sudden increase in pressure.

2.15 (c)

To get supersonic velocity the nozzle should be convergent divergent.

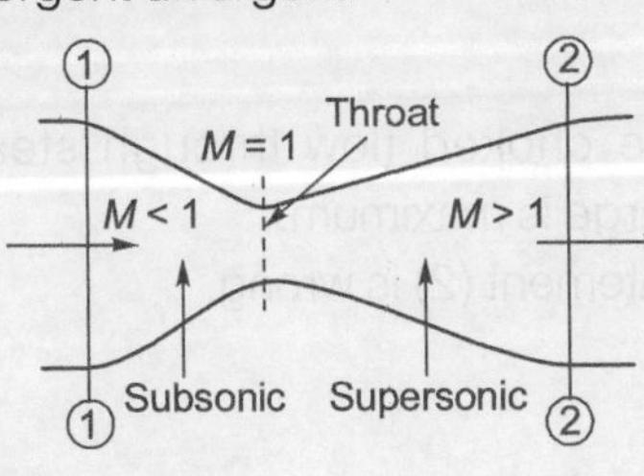

2.16 (d)

For supersonic flow

$$M > 1$$

We have $\frac{dA}{A} = \frac{dV}{V}(M^2 - 1)$

Since cross-section of nozzle is increasing i.e.

$$\frac{dA}{A} = +Ve$$

Hence $\frac{dV}{V} = +Ve$

Hence velocity will increase and pressure will decrease.

2.17 (c)

In chocking condition, the mass flow through the nozzle is maximum.

2.20 (d)

The limit of supersaturation is the Wilson line. The zone between the Wilson line and dry saturation line is called supersaturation field.

2.21 (b)

Mach number at throat is always unity.

2.22 (b)

$$\frac{dA}{A} = \frac{dV}{V}(M^2 - 1)$$

For diffuser, $\frac{dV}{V} = -\text{ve}$

For subsonic, $M < 1 \quad \therefore (M^2 - 1) = -\text{ve}$

$$\frac{dA}{A} = (-\text{ve}) \times (-\text{ve}) = +\text{ve}$$

For diffuser, velocity decreases, so Mach number decreases.

2.24 (a)

For incompressible flow a diverging section always acts as a diffuser for an upstream flow which is always subsonic.

2.25 (d)

Stage efficiency = Blade efficiency × Nozzle efficiency.

2.27 (a)

$$\frac{P_C}{P_1} = \left(\frac{2}{n+1}\right)^{\frac{n}{n-1}}$$

Assuming dry saturated steam

$$n = 1.135$$

$$\frac{P_C}{P_1} = \left(\frac{2}{1.135+1}\right)^{\frac{1.135}{1.135-1}}$$

$$= \left(\frac{2}{1.135+1}\right)^{8.407} = 0.5475$$

$$P_C = 0.5475 \times 16 = 8.76 \text{ bar}$$

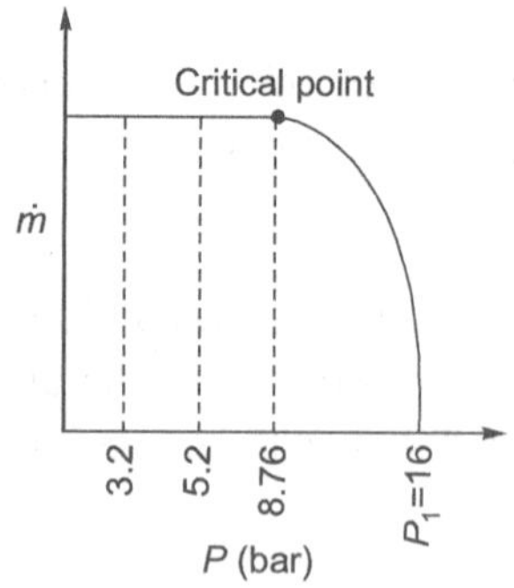

Since critical condition has already reached, hence mass flow rate will not be affected.

2.29 (b)

Critical pressure ratio

$$= \left(\frac{2}{n+1}\right)^{\frac{n}{n-1}} = \left(\frac{2}{1.135+1}\right)^{\left(\frac{1.135}{0.135}\right)} = \mathbf{0.577}$$

2.31 (c)

Isentropic nozzles is discharging steam through critical pressure ratio

$$\frac{P_2}{P_1} = \left(\frac{2}{k+1}\right)^{\frac{k}{k-1}}$$

If back pressure is further decreased, there will be no effect on discharge. At throat we get the maximum discharge.

For air : $\frac{P_2}{P_1} = 0.528$ $(k = 1.4)$

For superheated : $\frac{P_2}{P_1} = 0.545$ $(k = 1.3)$

For wet steam : $\frac{P_2}{P_1} = 0.577$ $(k = 1.135)$

2.32 (a)

Nozzle: A nozzle is a device designed to control the direction or characteristics of a fluid flow especially used to increase the velocity at its exit portion.

Diffuser: It is a device used to increase the pressure energy at the expense of kinetic energy.

2.33 (b)

For supersonic flow the divergent passage behaves as nozzle hence pressure decreases and velocity increases.

2.34 (d)

Since the transition is very fast thats why condensation cannot take place and hence the steam becomes supersaturated i.e., carries moisture more than the saturation limit.

2.35 (a)

In convergent section the flow is always less than sonic.

2.38 (d)

As the fluid is at sonic velocity the convergent passage will act as diffuser and hence the velocity will decrease and once the velocity is subsonic the convergent passage behaves as nozzle and increase the velocity upto sonic. This process is continued till the fluid remains in the convergent passage.

2.39 (c)

$$h_1 + \frac{V_1^2}{2} = h_2 + \frac{V_2^2}{2}$$

$$V_1 = 0 \text{ (Given)}$$

$$\Delta h = h_1 - h_2 = 45 \text{ kJ/kg}$$

$$V_2 = \sqrt{2(h_1 - h_2)}$$

$$V_2 = \sqrt{2 \times 45 \times 10^3} = \mathbf{300\ m/s}$$

2.40 (d)

For the choked flow through steam nozzle, discharge is maximum.

So, statement (2) is wrong.

2.42 (c)

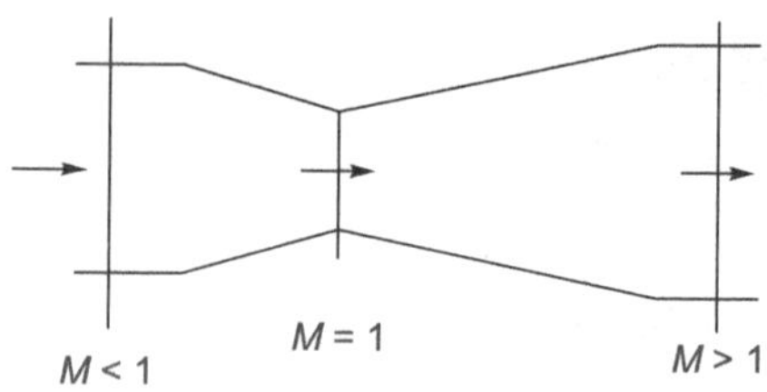

2.43 (c)

Flow in diverging part of *C-D* nozzle is always supersonic. So normal shock would occur in the diverging part not in the converging part.

2.44 (d)

Except stagnation temperature, every property undergoes a jump.

2.45 (c)

When $M = 1$ occurs at the throat, the discharge is maximum and the nozzle is said to be chocked. It is incapable of allowing more discharge even with further decrease in exit pressure. So, statement (2) is wrong.

2.46 (a)

The frictional losses entail the following effects:

(i) The expansion process is no more isentropic and enthalpy drop is reduced.

(ii) The final dryness fraction of steam is increased as the kinetic energy gets converted into heat due to friction and is absorbed by steam.

(iii) The specific volume of steam is increased as the steam becomes more dry due to this frictional reheating.

2.47 (b)

For isentropic flow through the nozzle, the enthalpy drop

$$(h_1 - h_2) = -\int v\,dP$$

$$\Rightarrow \quad \frac{V_2^2}{2} = \frac{n}{n-1}(P_1V_1 - P_2V_2)$$

$$= \frac{n}{n-1}P_1V_1\left(1 - \frac{P_2V_2}{P_1V_1}\right)$$

$$\Rightarrow \quad V_2 = \sqrt{\frac{2n}{n-1}P_1V_1\left[1 - \left(\frac{P_2}{P_1}\right)^{\frac{n-1}{n}}\right]}$$

$$\left[\frac{V_2}{V_1} = \left(\frac{P_2}{P_1}\right)^{\frac{-1}{n}}\right]$$

Velocity of steam at the throat corresponding to maximum mass flow rate can be obtained by using critical pressure ratio i.e.

$$\left(\frac{P_2}{P_1}\right) = \left(\frac{2}{n+1}\right)^{\frac{n}{n-1}}$$

$$\Rightarrow V_{max} = \sqrt{\frac{2n}{n-1}P_1V_1\left[1 - \left(\frac{2}{n+1}\right)^{\frac{n}{n-1}\times\frac{n-1}{n}}\right]}$$

$$= \sqrt{\frac{2nP_1V_1}{n+1}}$$

Option (b) is correct.

2.48 (b)

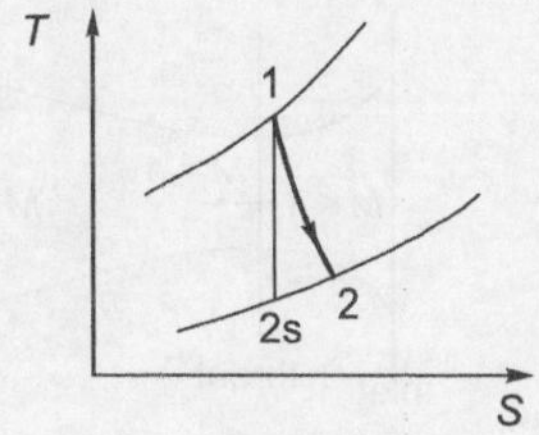

Effect of friction in a nozzle:

(i) Reduction in enthalpy drop (since $(h_1 - h_2) < (h_1 - h_{2s})$).

(ii) Reheating of fluid

(iii) Reduction in exit velocity.

(iv) Increase in specific volume.

(v) Decrease in mass flow rate.

(vi) Stagnation temperature remains same (since adiabatic).

(vii) Stagnation pressure reduces due to irreversibility.

2.49 (d)

In a shock wave, the flow passes from a supersonic to a subsonic state.

2.50 (a)

In a supersaturated flow the temperature of the supersaturated vapour at pressure when it begins to condense is less than the saturation temperature corresponding to aforesaid pressure. So the vapour said to be undercooled and the difference between the saturation temperature corresponding to the pressure and temperature is defined as degree of undercooling.

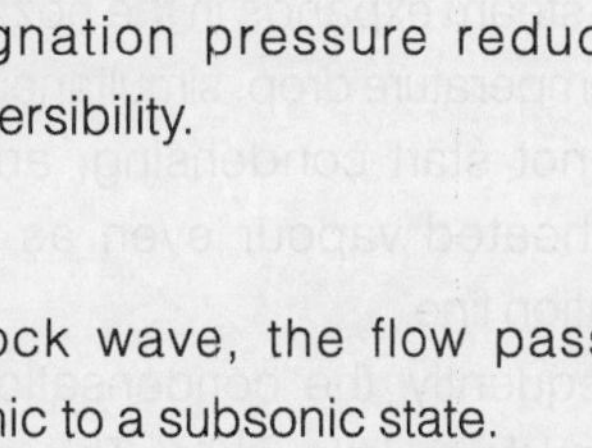

Degree of supercooling or undercooling

$$= T_2 - T_R$$

2.51 (c)

$$\frac{dA}{A} = \frac{dV}{V}(M^2 - 1)$$

Subsonic to supersonic $\rightarrow$ Converging diverging nozzle.

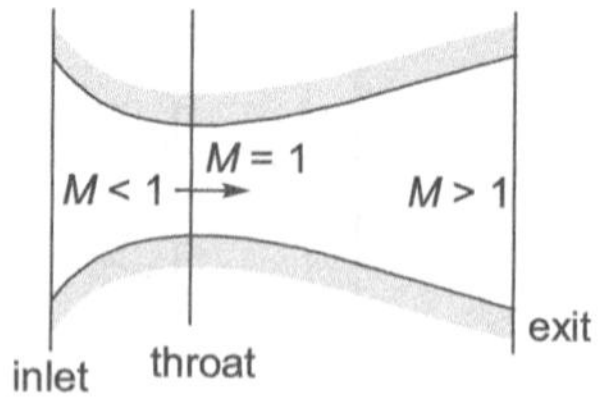

Supersonic to subsonic $\rightarrow$ converging diverging diffuser.

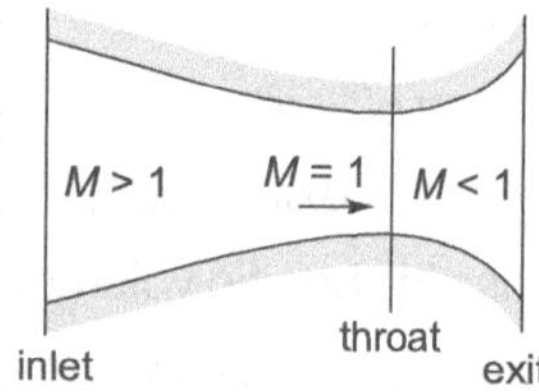

2.52 (c)

When steam expands in the nozzle, its pressure and temperature drop. simultaneously, but steam does not start condensing, and expands as superheated vapour even as it reaches the saturation line.

Consequently, the condensation of vapour is delayed for a little while. This phenomenon is known as supersaturation and steam, which exist in the wet region without containing any moisture is called supersaturated steam. Such expansion of steam is called a supersaturated expansion. The point at which condensation occurs may be within the nozzle or after the vapour leaves the nozzle.

2.53 (b)

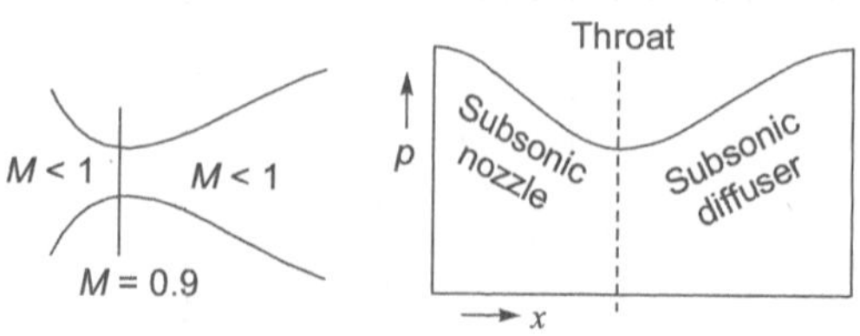

2.54 (b)

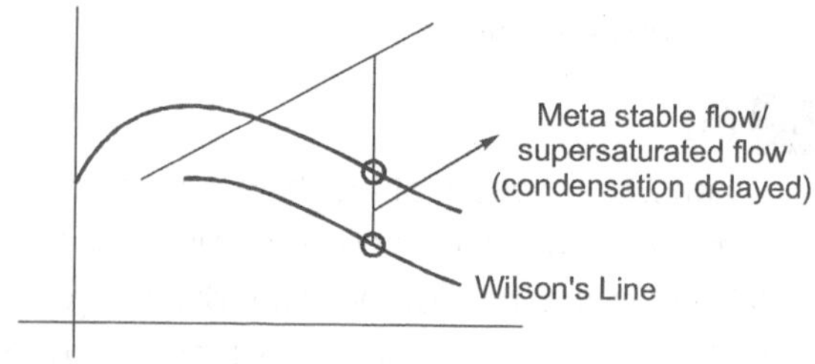

2.55 (d)

$$V_{exit} = \sqrt{2 \times (\Delta h) \times \eta_{isentropic} + V_1^2}$$

$$= \sqrt{2 \times 450 \times 10^3} = 949 \text{ m/s}$$

2.58 (d)

- Nozzle efficiency $= \dfrac{\text{Actual enthalpy drop}}{\text{Isentropic enthalpy drop}}$

$$= \frac{h_1 - h_2}{h_1 - h_{2s}} = \frac{(C_2^2 - C_1^2)/2}{(C_{2s}^2 - C_1^2)/2} = \frac{C_2^2 - C_1^2}{C_{2s}^2 - C_1^2}$$

- Velocity coefficient

$$= \frac{\text{Actual exit velocity at a given pressure}}{\text{Exit velocity when the flow is isentropic between the same given pressure}}$$

$$= \frac{C_2}{C_{2s}}$$

- The velocity coefficient is the square root of the nozzle efficiency, when the inlet velocity is assumed to be negligible.

■■■■

3 Nuclear Power Plant

3.1 The most commonly used moderator in nuclear power plants is
(a) heavy water (b) concrete and bricks
(c) steel (d) graphite
[ESE : 2000]

3.2 Shielding in a nuclear reactor is generally done to protect against
(a) excess electrons
(b) X-rays
(c) α- and β-rays
(d) neutron and gamma rays
[ESE : 2000]

3.3 Match **List-I** (Material) with **List-II** (Use) and select the correct answer using the codes given below the lists:

List-I	List-II
A. Graphite	1. Coolant
B. Thorium-233	2. Moderator
C. Molten Sodium	3. Fissionable material
D. Plutonium-239	4. Fissile material

Codes:

	A	B	C	D
(a)	1	4	2	3
(b)	2	4	1	3
(c)	2	3	1	4
(d)	1	3	2	4

[ESE : 2001]

3.4 **Assertion (A):** Pressurized water reactor (PWR) nuclear power plants use superheated steam
Reason (R): An increases in the superheat at constant pressure increases the cycle efficiency.
(a) Both **A** and **R** are individually true and **R** is the correct explanation of **A**
(b) Both **A** and **R** are individually true but **R** is not the correct explanation of **A**
(c) **A** is true but **R** is false
(d) **A** is false but **R** is true
[ESE : 2001]

3.5 In CANDU type nuclear reactor
(a) natural uranium is used as fuel and water as moderator
(b) natural uranium is used as fuel and heavy water as moderator
(c) enriched uranium is used as fuel and water as moderator
(d) enriched uranium is used as fuel and heavy water as moderator
[ESE : 2004]

3.6 Match **List-I** (Material) with **List-II** (Application) and select the correct answer using the codes given below the lists:

List-I	List-II
A. Plutonium-239	1. Fertile material
B. Thorium-232	2. Control rods
C. Cadmium	3. Moderator
D. Graphite	4. Fissile material

Codes:

	A	B	C	D
(a)	4	3	2	1
(b)	2	1	4	3
(c)	2	3	4	1
(d)	4	1	2	3

[ESE : 2004]

3.7 A nuclear unit becoming critical means
(a) It is generating power to rated capacity.
(b) It is capable of generating power much more than the related capacity.
(c) There is danger of nuclear spread.
(d) Chain reaction that causes automatic splitting of the fuel nuclei has been established.
[ESE : 2005]

3.8 What is the function of heavy water in a nuclear reactor?
(a) It serves as a coolant
(b) It serves as moderator

(c) It serves as a coolant as well as a moderator
(d) It serves as a neutron absorber

[ESE : 2006]

3.9 A nuclear fission is initiated when the critical energy as compared to neutron binding energy atoms is
(a) Less
(b) Same
(c) More
(d) Exactly two times

[ESE : 2006]

3.10 Uranium 238 is represented as $_{92}U^{238}$. What does it imply
(a) It has 92 protons and 146 neutrons
(b) It has 146 protons and 92 electrons
(c) It has 92 protons and 238 neutrons
(d) It has 92 neutrons and 238 protons

[ESE : 2007]

3.11 A nuclear reactor is said to be critical when the neutron population in the reactor core is
(a) rapidly increasing leading to the point of explosion
(b) decreasing from a specific value
(c) reduced to zero
(d) constant

[ESE : 2008]

3.12 Consider the following nuclear fuels:
1. Pu^{239} 2. U^{235}
3. U^{233} 4. Th^{232}

What is the correct sequences of the above nuclear fuels in order of increasing half life?
(a) 1-2-3-4 (b) 1-3-2-4
(c) 2-4-3-1 (d) 4-1-2-3

[ESE : 2008]

3.13 The mass defect is:
(a) a characteristic of certain elements
(b) a term used to prove the relationship between mass and energy
(c) a measure of fission energy
(d) the difference between mass of the nucleus and sum of the masses of the nucleons

[ESE : 2008]

3.14 Uranium 238 is represented by $_{92}U^{238}$. It represents
(a) 92 neutrons and 238 protons
(b) 92 protons and 238 neutrons
(c) 92 neutrons and 146 protons
(d) 92 protons and 146 neutrons [ESE : 2008]

3.15 With natural uranium, which of the following is used as moderator?
(a) Heavy water (b) Graphite
(c) Beryllium (d) All the above

[ESE : 2009]

3.16 The nuclear radiators produced in a reactor, which must be shielded, are:
(a) Electrons only
(b) Alpha, Beta and Gamma rays
(c) Neutrons and Gamma rays
(d) Electrons and Gamma rays [ESE : 2009]

3.17 Moderator in a nuclear plant is used to
(a) Protect against neutron and gamma rays
(b) Absorb excess neutrons
(c) Slow down the speed of fast moving neutrons
(d) Return the neutrons back into the core of the reactor [ESE : 2010]

3.18 The energy produced by 4.5 tons of high grade coal is equivalent to the energy produced by
(a) 1 kg of Uranium
(b) 1 gram of Uranium
(c) 100 grams of Uranium
(d) 10 grams of Uranium

[ESE : 2011]

3.19 Enriched Uranium is one in which
(a) Percentage of U^{235} has been artificially increased
(b) Percentage of U^{238} has been artificially increased
(c) Percentage of U^{234} has been artificially increased
(d) Extra energy is pumped from outside

[ESE : 2011]

3.20 Which one of the following statements is not correct?
(a) Fusion is the combustion of heavy nuclei of elements resulting in the release of heat energy

(b) Neutron bombardment is the most common and practiced method of initiation of reaction in fission materials
(c) When the multiplication factor is greater than 1, the chain reaction is uncontrollable
(d) The reactor produces α, β, γ and neutron radiations. Of all these γ radiation is the most dangerous

[ESE : 2011]

3.21 The reflectivity in a reactor depends upon
(a) Geometry of the reflector
(b) Energy of neutrons
(c) Properties of reflector
(d) All of these

[ESE : 2011]

3.22 Shielding in a nuclear power plant is done
(a) To protect against neutron and gamma rays
(b) To absorb excess neutrons
(c) To slow down the speed of fast moving neutrons
(d) To return the neutrons back into the core of the reactor

[ESE : 2011]

3.23 The risk of radioactive hazard is greatest in the turbine with following reactor
(a) pressurized water
(b) boiling water
(c) gas cooled
(d) liquid metal cooled

[ESE : 2013]

Directions: The following items consists of two statements; one labelled as **'Statement (I)'** and the other as **'Statement (II)'**. You are to examine these two statements carefully and select the answers to these items using the codes given below:
(a) Both Statement (I) and Statement (II) are individually true and Statement (II) is the correct explanation of Statement (I)
(b) Both Statement (I) and Statement (II) are individually true but Statement (II) is not the correct explanation of Statement (I)
(c) Statement (I) is true but Statement (II) is false
(d) Statement (I) is false but Statement (II) is true.

3.24 Statement (I) : A breeder reactor does not requires moderator.
Statement (II): The parasite absorption of neutrons is low.

[ESE : 2014]

3.25 Statement (I) : In Boiling Water Reactor (BWR) coolant serves the triple function of coolant, moderator and working fluid.
Statement (II): The steam flowing to the turbine is produced directly in the reactor core.

[ESE : 2014]

3.26 The source of energy which keeps the sun shining for billions of years is
(a) Combustion of Hydrogen
(b) Nuclear fusion of light elements
(c) Nuclear fission of heavy elements
(d) Intersection of currents in the interior of the sun with the galactic magnetic field.

[ESE : 2014]

■■■■

Answers Nuclear Power Plant

3.1 (a)	3.2 (d)	3.3 (c)	3.4 (a)	3.5 (b)	3.6 (d)	3.7 (d)	3.8 (c)	3.9 (c)
3.10 (a)	3.11 (d)	3.12 (b)	3.13 (d)	3.14 (d)	3.15 (d)	3.16 (c)	3.17 (c)	3.18 (b)
3.19 (a)	3.20 (a)	3.21 (d)	3.22 (a)	3.23 (b)	3.24 (b)	3.25 (b)	3.26 (b)	

Explanations Nuclear Power Plant

3.1 (a)

A moderator is a material which is used to slow down neutrons from high velocities. The moderator should have a high macroscopic neutrons scattering cross-section, a low macroscopic neutrons absorption cross-section and a low atomic mass to get maximum kinetic energy transfer from neutrons to the moderator atoms. Actually only four isotopes are used in material to moderate neutrons. They are ordinary hydrogen (H-1), heavy hydrogen (H-2), beryllium and carbon.

3.2 (d)

γ particles are electromagnetic radiation of extremely short wavelength and high frequency, resulting in high energy. Both γ and neutrons are highly penetrative.

3.3 (c)

Fissle Vs fissionable: Fissle is distinguished from fissionable. Fissionable are any material with atom that can under go nuclear fission. Fissile is defined to be material that are fissiable by neutrons with low kinetic energy. Fissile thus, is more restrictive than fissionable although all fissile material are fissionable, not all fissionable material are fissile. Notably, U-238 is fissionable but not fissile.

3.5 (b)

CANDU type nuclear reactor uses natural uranium which consist mostly U-238 and 0.72% (by weight) of fissile U-235.

3.6 (d)

1. Plutonium-239 : Fissible material
2. Thorium-232 : Fertile material
3. Cadmium : Control rods
4. Graphite : Moderator

3.7 (d)

The moment, nuclear power plant starts producing nuclear power, it is said that nuclear power plant goes critical.

3.8 (c)

The heavy water and graphite can be used as a moderator with natural uranium.

The water, heavy water, gas (He, CO_2) are used as coolant.

3.10 (a)

In ${}_{92}U^{238}$

Number of proton = Atomic number = 92

Number of neutron = 238 − 92 = 146

3.11 (d)

A nuclear reactor is said to be critical when the neutron population in the reactor core is constant.

3.12 (b)

Isotrope	Half - life $(\theta_{1/2})$
Plutonium 239	2.44×10^4 yr
Uranium 233	1.65×10^5 yr
Uranium 235	7.1×10^8 yr
Thorium 232	1.4×10^{10} yr

3.13 (d)

The mass defect is the difference between mass of the nucleus and sum of the mass of the nucleons.

3.14 (d)

${}_{92}U^{238}$ has 92 protons

So number of neutrons = 238 – 92 = 146

3.15 (d)

With Natural Uranium, Heavy Water, Graphite and Beryllium all there are used as moderator

3.17 (c)

Moderator is used to slow down the neutrons from high kinetic energy to low kinetic energy in a fraction of second.

3.18 (b)

The uranium is the most common element used to produce nuclear power today, has an energy content about 3 million times greater than that of fossil fuel. Consequently, 1 gram of uranium is equivalent to approximately 4.5 tonnes of high grade coal.

3.19 (a)

Enriched uranium is one in which percentage of uranium - 235 has been artificially increased. Ur-235 is basically a fissile material.

3.20 (a)

Since fusion is not combustion, it is combination of light elements to form heavy element & energy is released.

3.21 (d)

Reflectivity in a reactor depends upon geometry of reflector, energy of neutrons and properties of reflector also. The aim of reflector is to reflect back all the neutrons to the core so as to decrease the loss of neutrons.

3.22 (a)

Shielding in a nuclear power plant is done to protect the persons from radiation. The common radiation coming out are α, β, γ rays and fast neutrons. Generally concrete or steel linning is there.

3.23 (b)

Possibility of radioactive contamination in the turbine mechanism is more in BWR because there may be failure of fuel elements. More elaborate safety precaution is needed in BWR because same water is circulated through nuclear reactor, turbine, feed pump etc.

■■■■

4 Miscellaneous

4.1 A hydrocarbon fuel was burnt with air and the Orsat analysis of the dry products of combustion yielded the following data:

Initial volume of dry gas sample ...100 cc

Volume after absorption in pipette

1. containing potassium hydroxide solution ...89 cc

   Volume after absorption in pipette

2. containing solution of pyrogallic acid and potassium hydroxide ...84 cc

   Volume after absorption in pipette

3. containing cuprous chloride solution ...82 cc

The percentage (by volume) of CO_2 in the dry products was

(a) 2% (b) 5%

(c) 11% (d) 18%

[ESE : 2001]

4.2 Match **List-I** and (Cycles) with **List-II** (Processes) and select the correct answer using the codes given below the lists:

List-I

A. Joule cycle

B. Stirling cycle

C. Ericsson cycle

D. Diesel cycle

List-II

1. One constant pressure, one constant volume and two isentropic

2. Two constant pressure and two isentropic

3. Two constant pressure and two isothermal

4. Two constant volume and two isothermal

Codes:

	A	B	C	D
(a)	2	3	4	1
(b)	1	4	3	2
(c)	2	4	3	1
(d)	1	3	4	2

[ESE : 2003]

4.3 (1)

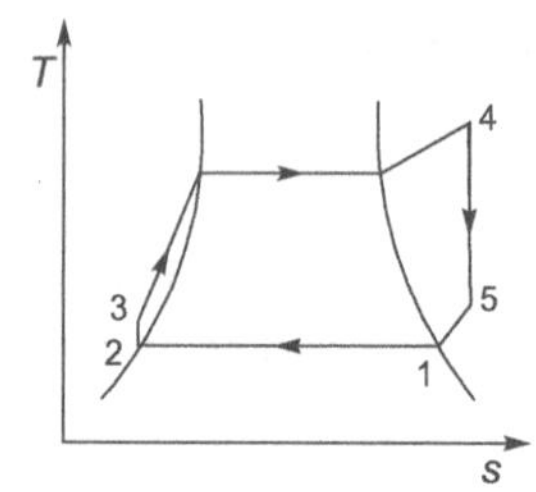

(2)

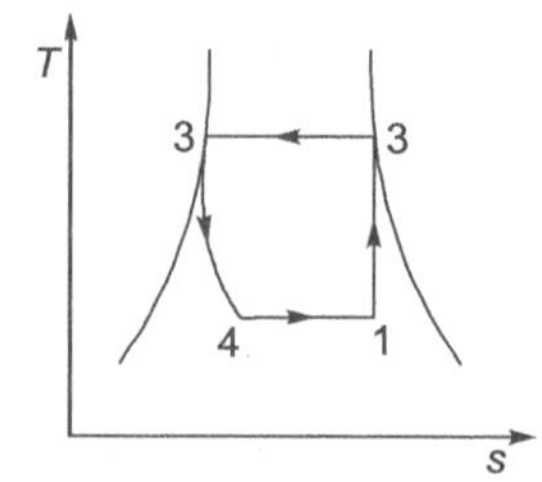

(3)

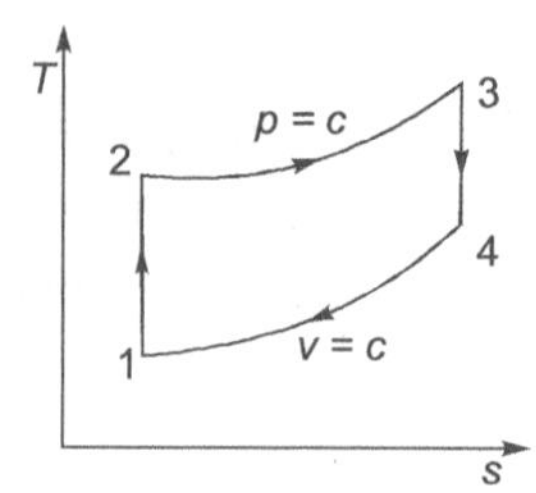

(4)

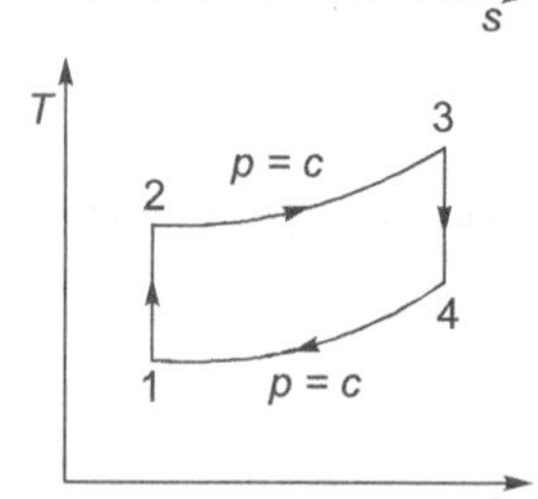

The correct sequence of the cycles given in the above *T-s* diagram is

(a) Vapour compression refrigeration, Rankine, Diesel, Otto

(b) Rankine, Vapour compression refrigeration, Diesel, Brayton

(c) Rankine, Carnot, Otto, Brayton

(d) Vapour compression refrigeration, Carnot, Diesel, Otto

[ESE : 2003]

4.4 Which one of the following cycles has the highest thermal efficiency for given maximum and minimum cycle temperatures?

(a) Brayton cycle (b) Otto cycle
(c) Diesel cycle (d) Stirling cycle

[ESE : 2005]

4.5 Match **List-I** (Prime mover) with **List-II** (Air standard cycles) and select the correct answer using the code given below the lists:

List-I

A. High speed diesel engine
B. IC engine having expansion ratio greater than compression ratio
C. Pulse jet engine
D. Gas turbine with multi-stage compression and multi-stage expansion

List-II

1. Atkinson cycle
2. Dual combustion (limited pressure) cycle
3. Ericsson cycle
4. Lenoir cycle

Codes:

	A	B	C	D
(a)	3	1	4	2
(b)	2	4	1	3
(c)	3	4	1	2
(d)	2	1	4	3

[ESE : 2007]

4.6 Thermodynamic cycle shown in figure below on the temperature-entropy diagram pertains to which one of the following?

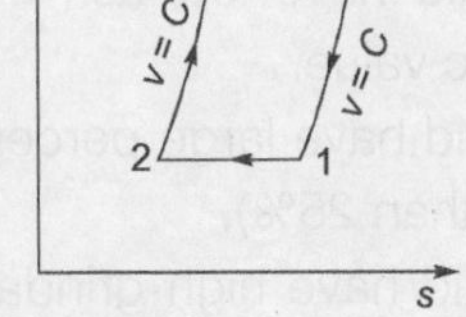

(a) Stirling cycle
(b) Ericsson cycle
(c) Vapour compression cycle
(d) Brayton cycle

[ESE : 2007]

4.7 Which gas power cycle consists of four processes during which work alone is transferred during two processes and heat alone is transferred during the other two processes?

(a) Atkinson cycle (b) Carnot cycle
(c) Diesel cycle (d) Otto cycle

[ESE : 2008]

4.8 Which cycle consists of two reversible isotherms and two reversible isobars?

(a) Carnot cycle (b) Stirling cycle
(c) Ericsson cycle (d) Brayton cycle

ESE : 2009]

4.9 Match **List-I** (Gas cycles) with **List-II** (Constants for different cycles) and select the correct answer using the code given below the lists:

List-I

A. Carnot cycle
B. Brayton cycle
C. Ericsson cycle
D. Stirling cycle

List-II

1. Pressure-Entropy
2. Pressure-Temperature
3. Temperature-Volume
4. Temperature-Entropy

Code :

	A	B	C	D
(a)	4	2	1	3
(b)	3	2	1	4
(c)	4	1	2	3
(d)	3	1	2	4

[ESE : 2009]

4.10 Isothermal compression of air in a Stirling engine is an example of

(a) open system
(b) steady flow adiabatic system
(c) closed system with a movable boundary
(d) closed system with fixed boundary

[ESE : 2010]

4.11 The cycle in which heat is supplied at constant volume and rejected at constant pressure is known as

(a) Dual combustion cycle
(b) Diesel cycle
(c) Atkinson cycle
(d) Rankine cycle

[ESE : 2011]

4.12 This reversible cycle consists of constant volume heat addition, reversible adiabatic expansion and constant pressure heat rejection

(a) Otto cycle (b) Lenoir cycle
(c) Atkinson cycle (d) Brayton cycle

[ESE : 2012]

4.13 In an engine working on air standard Stirling cycle the temperature at the beginning of isothermal compression is 127° C. The engine thermal efficiency is 50%. The specific heat of air at constant volume is C_P. The heat transferred to the regenerator is

(a) 200 C_V kJ/kg (b) 300 C_V kJ/kg
(c) 400 C_V kJ/kg (d) 500 C_V kJ/kg

[ESE : 2014]

4.14 Volumetric analysis of a certain flue gas gave CO_2 15%, O_2 5% and rest as N_2. The gas was at a temperature of 200°C and a pressure of 5 bar. The partial pressure of N_2 in the flue gas is

(a) 250 kN/m^2 (b) 300 kN/m^2
(c) 350 kN/m^2 (d) 400 kN/m^2

[ESE : 2017]

4.15 A fluidized bed combustion system having an output of 35 MW at 80% efficiency when using a coal of heating value 26 MJ/kg with a sulphur content of 3.6% requires a particular limestone to be fed to it at a calcium-sulphur molar ratio of 3.0 so as to limit emissions of SO_2 adequately. The limestone used contains 85% $CaCO_3$. The required flow rate of limestone will be

(a) 2405 kg/h (b) 2805 kg/h
(c) 3205 kg/h (d) 3605 kg/h

[ESE : 2020]

4.16 A fuel consists of 92% carbon, 7% hydrogen and remaining residual matter by mass. Working from first principles, the higher calorific value of the fuel is

(a) 40176 kJ/kg (b) 41176 kJ/kg
(c) 40876 kJ/kg (d) 41678 kJ/kg

[ESE : 2021]

4.17 Which one of the following statements is correct?

(a) Ultimate analysis is the determination of the percentages of fixed carbon, volatile matter, moisture and ash in fuel.
(b) The percentage of each constituent element in the fuel such as carbon, hydrogen, sulphur, oxygen, nitrogen, and ash, is determined by proximate analysis.
(c) Calorific value of solid or liquid fuel is defined as the heat evolved by the complete combustion of unit mass of fuel
(d) Proximate analysis value of gaseous fuel is expressed as heat developed by the complete combustion of one cubic metre of gas at standard temperature and pressure.

[ESE : 2022]

4.18 Consider the following statements regarding combustion of fuels:

1. Carbon combines with oxygen to form carbon monoxide.
2. Hydrogen burns with oxygen to give water as the product.
3. When methane burns in the presence of oxygen, the combustion products are carbon dioxide and water vapours.

Which of the above statements are correct?

(a) 1 and 3 only (b) 1 and 2 only
(c) 2 and 3 only (d) 1, 2 and 3

[ESE : 2024]

4.19 Consider the following statements :

1. Air contains 23% of oxygen and 77% of nitrogen by volume.
2. The main constituents of natural gas are methane and ethane.
3. Coal gas mainly consists of hydrogen, carbon monoxide and hydrocarbons.

Which of the above statements are correct?

(a) 1 and 2 only (b) 1 and 3 only
(c) 2 and 3 only (d) 1, 2 and 3

[ESE : 2024]

4.20 Consider the following statements regarding properties of good coal:

1. It should have low ash content and high calorific value.
2. It should have large percentage of sulphur (more than 25%).
3. It should have high grindability index.

Which of the above statements are correct?

(a) 1 and 2 only (b) 1 and 3 only
(c) 2 and 3 only (d) 1, 2 and 3

[ESE : 2024]

4.21 The yearly duration curve of a certain power plant can be considered as a straight line from 350 MW to 80 MW. Power is supplied with one generating unit of 250 MW capacity and two units of 150 MW capacity each. What is the utilization factor?

(a) 0.436 (b) 0.536
(c) 0.636 (d) 0.736

[ESE : 2024]

4.22 Which one of the following is Dulong's formula for the calculation of the calorific value of the solid or liquid fuels from their chemical composition? (Where H, O and S are hydrogen, oxygen and sulphur in percentage respectively in 100 kg of fuel)

(a) $HCV = \frac{1}{100}\left[13800 + 144000\left\{H - \frac{O}{8}\right\} + 7270S\right]$ kJ/kg

(b) $HCV = \frac{1}{100}\left[23800 + 144000\left\{H - \frac{O}{8}\right\} + 8270S\right]$ kJ/kg

(c) $HCV = \frac{1}{100}\left[33800 + 144000\left\{H - \frac{O}{8}\right\} + 9270S\right]$ kJ/kg

(d) One of these

[ESE : 2024]

4.23 The overall thermal efficiency of a 40 MW turbo alternator is 30%. If the load factor of the power station is 50% and the coal burnt has a calorific value of 6800 kcal, what is the energy produced per day?

(a) 44×10^4 kWh (b) 48×10^4 kWh
(c) 34×10^4 kWh (d) 38×10^4 kWh

[ESE : 2024]

■■■■

Answers Miscellaneous

4.1 (c)	4.2 (c)	4.3 (b)	4.4 (d)	4.5 (d)	4.6 (a)	4.7 (d)	4.8 (c)	4.9 (c)
4.10 (c)	4.11 (c)	4.12 (b)	4.13 (c)	4.14 (d)	4.15 (a)	4.16 (b)	4.17 (c)	4.18 (c)
4.19 (c)	4.20 (b)	4.21 (c)	4.22 (a)	4.23 (b)				

Explanations Miscellaneous

4.1 (c)

Amount of CO_2 absorbed in KOH = 100 – 89 = **11%**

4.6 (a)

Stirling cycle consists of two isothermal process and two isochoric process.

4.7 (d)

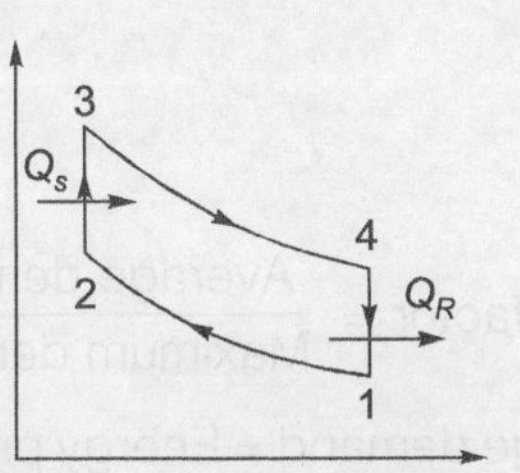

During constant volume process heat addition and heat rejection takes place and no work is transferred. During the adiabatic processes [compressions/expansion] only work transfer takes place but no heat transfer occurs.

4.8 (c)

Two reversible isotherms and two reversible isobars is represented by Ericsson cycle.

4.11 (c)

In case of **diesel cycle**, heat addition takes place at constant pressure and heat rejection takes place at constant volume.

In case of **Rankine cycle**, heat addition and heat rejection take place at constant pressure.

In case of **dual combustion** cycle, heat addition takes place at constant volume and constant pressure and heat rejection takes place at constant volume.

4.12 (b)

Lenoir cycle consist of only 3 process.

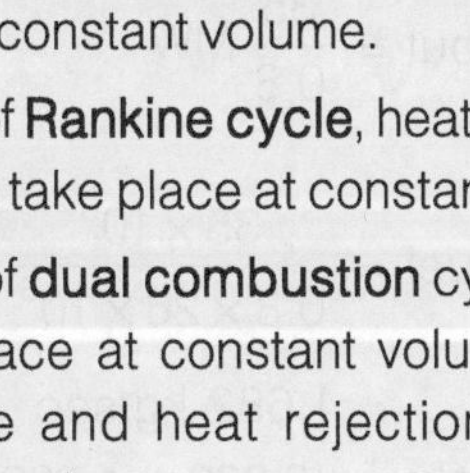

1-2 = Constant volume heat addition
2-3 = Reversible adiabatic expansion
3-1 = Constant pressure heat rejection

4.13 (c)

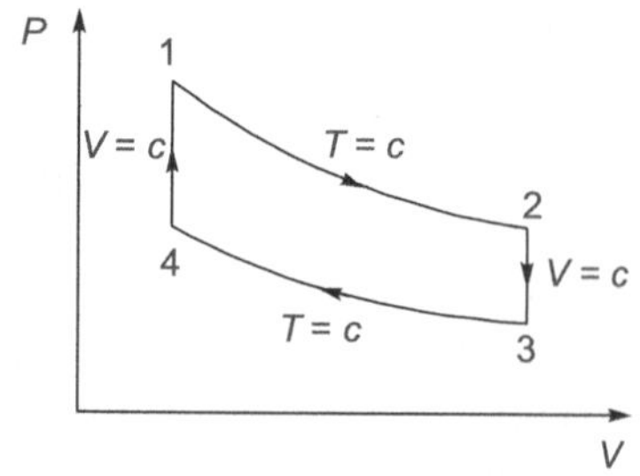

$$\eta = \frac{T_1 - T_3}{T_1}$$

$$T_3 = 127°C = 400 \text{ K}$$

$$1 - \frac{T_3}{T_1} = 0.5$$

or, $$\frac{T_3}{T_1} = 0.5$$

or $$T_1 = \frac{400}{0.5} = 800 \text{ K}$$

$$Q_s = C_V(T_1 - T_3) \qquad (\because T_4 = T_3)$$

$$= C_V(800 - 400) = \mathbf{400\ C_V}$$

4.15 (a)

Output = 35 MW

Efficiency = 0.8

$$\text{Heat input} = \frac{35}{0.8} \text{ MW}$$

$$\text{Mass of fuel} = \frac{35 \times 10^6}{0.8 \times 26 \times 10^6}$$

$$= 1.682 \text{ kg/sec}$$

Mass of sulphur = 0.036 × 1.682 = 0.06057 kg/s

$$\text{Moles of sulphur} = \frac{0.06057}{32} = 0.001893$$

Moles of $CaCO_3$ = 3 × 0.001893

= 0.005679 kmoles/s

= 20.44 kmoles/hr

Mass = 20.44 × 100 = 2044 kg/hr

$$\text{Mass of } CaCO_3 = \frac{2044}{0.85} = 2405 \text{ kg/hr}$$

4.18 (c)

- $C + O_2 \rightarrow CO_2$
- $C + \frac{1}{2}O_2 \rightarrow CO$ [It will occur only when there is insufficient supply of air]
- $H_2 + \frac{1}{2}O_2 \rightarrow H_2O$
- $CH_4 + 2O_2 \rightarrow CO_2 + 2H_2O$

4.19 (c)

Air contains 21% of oxygen and 79% of nitrogen by volume.

4.20 (b)

Desirable properties of coal:

- It should have high energy content or calorific value.
- It should contain low sulphur content, as burning coal produces SO_2 as an air pollutant which causes acid rain.
- It should have low ash content.
- It should have high grindability index.
- It should have good burning characteristics.
- It should have excellent weatherability.
- It should have high ash softening temperature which lower the risk of slagging.

4.21 (c)

Utilization factor

$$= \frac{\text{Maximum demand on the power station}}{\text{Rated capacity of the power station}}$$

$$= \frac{350}{250 \times 1 + 150 \times 2} = \frac{350}{550}$$

= 0.636

4.23 (b)

$$\text{Load factor} = \frac{\text{Average demand}}{\text{Maximum demand}}$$

∴ Average demand = Energy produced per day

= Load factor × Maximum demand

= $0.50 \times 40 \times 10^3 \times 24$

= 48×10^4 kWh

■■■■

UNIT

VIII

Renewable Sources of Energy

Syllabus

Solar Radiation, Solar Thermal Energy collection - Flat Plate and focusing collectors their materials and performance. Solar Thermal Energy Storage, Applications – heating, cooling and Power Generation; Solar Photovoltaic Conversion; Harnessing of Wind Energy, Bio-mass and Tidal Energy – Methods and Applications, Working principles of Fuel Cells.

Contents

Sl.	Topic	Page No.
1.	Renewable Sources of Energy	545-557

OOOO

1 Renewable Sources of Energy

1.1 In a solar collector, the function of the transparent cover is to :

(a) Transmit solar radiation only
(b) Protect the collector from dust
(c) Decrease the heat loss from collector beneath to atmosphere
(d) Absorb all types of radiation and protect the collector from dust

[ESE : 2016]

1.2 **Statement (I) :** Pyranometer is used to measure diffuse solar radiation by blocking the direct radiation with a shadow band.

Statement (II) : Pyrheliometer is used to measure diffuse radiation.

(a) Both Statement (I) and Statement (II) are individually true and Statement (II) is the correct explanation of Statement (I).
(b) Both Statement (I) and Statement (II) are individually true but Statement (II) is not the correct explanation of Statement (I).
(c) Statement (I) is true but Statement (II) is false.
(d) Statement (I) is false but Statement (II) is true.

[ESE : 2017]

1.3 Consider the following statements :

1. Wind velocity at about 20 m height above the ground is taken as the rated velocity for design of windmills.
2. The total power-of a wind stream is directly proportional to the cube of average velocity.
3. Wind turbine operates with variable load over a narrow range between cut-in and cut-out velocities.
4. Vertical wind machine operates in all wind directions, but it needs yaw adjustment.

Which of the above statements are correct?

(a) 1 and 2 (b) 1 and 4
(c) 3 and 4 (d) 2 and 3

[ESE : 2017]

1.4 Which fuel cell is suitable for spacecraft applications?

(a) Direct methanol fuel cell
(b) Proton exchange membrane fuel cell
(c) Alkaline fuel cell
(d) Phosphoric acid fuel cell

[ESE : 2017]

1.5 In solar flat-plate collectors, the absorber plate is painted with selective paints. The selectivity is the ratio of

(a) solar radiation-absorption to thermal infrared radiation-emission
(b) solar radiation-emission to thermal infrared radiation-absorption
(c) solar radiation-reflection to thermal infrared radiation-absorption
(d) solar radiation-absorption to thermal infrared radiation-reflection

[ESE : 2018]

1.6 The solar heat pipe works on the principle of

(a) heating and condensation cycle
(b) evaporation and condensation cycle
(c) cooling and condensation cycle
(d) heating and evaporation cycle

[ESE : 2019]

1.7 A good approximation of the measured solar spectrum is made by

(a) black-body energy distribution
(b) Planck's energy distribution
(c) inverse square law
(d) solar constant

[ESE : 2019]

1.8 Which one of the following types of tracker uses liquid contained in canisters that can turn easily into vapour?
(a) Active tracker (b) Passive tracker
(c) Single-axis tracker
(d) Altitude-azimuth tracker
[ESE : 2019]

1.9 Which type of flat-plate collector is used to heat the swimming pools with plastic panel, utilizing solar energy?
(a) Pipe and fin type
(b) Full water sandwich type
(c) Thermal traps type
(d) Corrugated plate with selective surface type
[ESE : 2019]

1.10 The edge loss U_e in a solar collector with respect to edge area A_e, collector area A_c and back loss coefficient U_b is
(a) $U_b\left(\frac{A_e}{A_c}\right)$ (b) $U_b\left(\frac{A_c}{A_e}\right)$
(c) $A_c\left(\frac{A_e}{U_b}\right)$ (d) $U_b\left(\frac{A_e}{2A_c}\right)$
[ESE : 2019]

1.11 In solar porous type air heater, the pressure drop is usually
(a) higher than non-porous type
(b) same an in non-porous type
(c) lower than non-porous type
(d) zero [ESE : 2019]

1.12 In a drainback solar water heating system
(a) the water in the heat exchange is recycled
(b) the water is heated in collectors only during times when there is available heat
(c) at the collector, the mixture of water and propylene-glycol is heated and returned to a solar storage tank
(d) there is an expansion tank with enclosed air chamber to assist water draining
[ESE : 2019]

1.13 A PV cell is illuminated with irradiance (E) of 1000 W/m^2. If the cell is 100 mm × 100 mm in size and produces 3 A at 0.5 V at the maximum power point, the conversion efficiency will be
(a) 15% (b) 19%
(c) 23% (d) 27%
[ESE : 2019]

1.14 In a barrage of 300000 m^2 area with a tide height of 3 m, barrage drain time of 10 hr, density of seawater as 1025 kg/m^3 and gravitational acceleration as 9.8 m/s^2, the average power will be
(a) 377 kW (b) 381 kW
(c) 388 kW (d) 396 kW
[ESE : 2019]

1.15 The platinum nano-coating is made on the anode of the fuel cell to
(a) create lighter and more efficient fuel cell membranes
(b) make the fuel effective
(c) create high thermal conductivity in the cell
(d) make the fuel cell non-corrosive
[ESE : 2019]

1.16 In a fuel cell, electric current is produced when
(a) hydrogen and oxygen react with each other and electrons are freed
(b) hydrogen reacts with water and electrons are freed
(c) oxygen reacts with water and electrons are freed
(d) electrons react with molecules of hydrogen and oxygen is freed
[ESE : 2019]

1.17 Which one of the following is suitable for fuel cell electric vehicle (FCEV)?
(a) Direct methanol fuel cell (DMFC)
(b) Alkaline fuel cell (AFC)
(c) Proton exchange membrane fuel cell (PEMFC)
(d) Solid oxide fuel cell (SOFC)
[ESE : 2019]

1.18 The angle through which the Earth must turn to bring the meridian of a point directly in line with the Sun's rays is called
(a) Altitude angle
(b) Hour angle
(c) Solar azimuth angle
(d) Zenith angle
[ESE : 2020]

1.19 In which type of collector is solar radiation focused into the absorber from the top, rather than from the bottom?
(a) Fresnel lens (b) Paraboloidal
(c) Concentrating (d) Compound parabolic
[ESE : 2020]

1.20 A flat plate collector is 150 cm wide and 180 cm high and is oriented such that it is perpendicular to the sun rays. Its active area is 90% of the panel size. If it is in a location that receives solar insolation of 1000 W/m^2 peak, the peak power delivered to the area of the collector will be
(a) 1.23 kW (b) 2.43 kW
(c) 4.46 kW (d) 6.26 kW
[ESE : 2020]

1.21 A surface having high absorptance for short wave radiation (less than 2.5 μm) and a low emittance of long-wave radiation (more than 2.5 μm), is called
(a) Absorber (b) Emitter
(c) Selective (d) Black [ESE : 2020]

1.22 In a solar tower power system, each mirror is mounted on a system called
(a) Regenerator (b) Linear Frensel
(c) Dish (d) Heliostat
[ESE : 2020]

1.23 The ratio of PV cell's actual maximum power output to its theoretical power output is called
(a) Quantum factor
(b) Fill factor
(c) Quantum efficiency
(d) PV factor [ESE.: 2020]

1.24 With respect to the wind turbine blades, TSR means
(a) Tip Swift Ratio
(b) Tip Sharp Ratio
(c) Tip Speed Ratio
(d) Tip Swing Ratio
[ESE : 2020]

1.25 For a wind turbine 10 m long running at 20 rpm in 12.9 kmph wind, the TSR will be nearly
(a) 3.6 (b) 5.8
(c) 7.6 (d) 9.8
[ESE : 2020]

1.26 Which one of the following is an enclosure or housing for the generator, gear box and any other parts of the wind turbine that are on the top of the tower?
(a) Turbine blade (b) Nacelle
(c) Turbine head (d) Gear box
[ESE : 2020]

1.27 The force required for producing tides in the ocean is
(a) 70% due to Moon and 30% due to Sun
(b) 30% due to Moon and 70% due to Sun
(c) 45% due to Moon and 55% due to Sun
(d) 55% due to Moon and 45% due to Sun
[ESE : 2020]

1.28 Which of the following are related to the Proton Exchange Membrane Fuel Cell (PEMFC)?
1. Polymer electrolyte
2. Hydrogen fuel and oxygen
3. Pure water and small amount of electricity
4. Nitrogen gas
(a) 1 and 3 only (b) 2 and 4 only
(c) 1 and 2 only (d) 3 and 4 only
[ESE : 2020]

1.29 Which of the following are the essential functions of fuel cells?'
1. The charging (or electrolyser) function in which the chemical AB is decomposed to A and B.
2. The storage function in which A and B are held apart.
3. The charge function in which A and B are charged with the simultaneous generation of electricity.
(a) 1 and 3 only (b) 2 and 3 only
(c) 1 and 2 only (d) 1, 2 and 3
[ESE : 2020]

1.30 In various solar energy storage systems, pumped hydro-electric storage system falls under which one of the following categories?
(a) Thermal energy storage
(b) Electrical energy storage
(c) Mechanical energy storage
(d) Electromagnetic energy storage
[ESE : 2021]

1.31 What is the standard value of solar constant adopted by World Radiation Centre?
(a) 1192 W/m² (b) 1084 W/m²
(c) 1927 W/m² (d) 1367 W/m²
[ESE : 2021]

1.32 What is the tip speed ratio of savonius wind turbine rotor?
(a) 1 (b) 3
(c) 5 (d) 7 [ESE : 2021]

1.33 What is the solidity of American multiblade wind turbine rotor?
(a) 0.4 (b) 0.7
(c) 0.9 (d) 1 [ESE : 2021]

1.34 The energy density of Bio-ethanol is
(a) 8.3 MJ/kg (b) 14.6 MJ/kg
(c) 26.9 MJ/kg (d) 34.7 MJ/kg
[ESE : 2021]

1.35 The percentage of hydrogen in producer gas is
(a) 34% (b) 27%
(c) 18% (d) 8%
[ESE : 2021]

1.36 In single basin, double effect scheme, power is generated
(a) during filling (b) during emptying
(c) on ebb only (d) on both flood and ebb
[ESE : 2021]

1.37 The operating temperature of alkaline fuel cell is
(a) 39°C (b) 90°C
(c) 127°C (d) 192°C [ESE : 2021]

1.38 The ideal emf produced by polymer electrolyte membrane fuel cell at 25°C is
(a) 3.57V (b) 2.94V
(c) 1.23V (d) 0.73V [ESE : 2021]

1.39 Which one of the following fuel cells has highest efficiency?
(a) PAFC (b) MCFC
(c) PEMFC (d) AFC
[ESE : 2021]

1.40 How many kilograms of steam per day is produced by 15 m diameter community solar cooker developed by Centre for Scientific Research, Auroville (Puducherry)?
(a) 100 kg (b) 300 kg
(c) 600 kg (d) 1000 kg
[ESE : 2021]

1.41 In a solar passive space heating system, the south-facing thick wall is called
(a) Vent wall (b) Trombe wall
(c) Damper wall (d) Ventilation wall
[ESE : 2021]

1.42 Solar thermal water pumps work on
(a) Rankine cycle (b) Otto cycle
(c) Carnot cycle (d) Diesel cycle
[ESE : 2022]

1.43 The concentration ratio of as high a value as 3000 can be obtained by
(a) modified flat plate collector
(b) compound parabolic concentrator
(c) cylindrical parabolic concentrator
(d) central tower receiver
[ESE : 2022]

1.44 The angle between the sun's ray and its projection on a horizontal surface is known as
(a) Inclination angle
(b) Zenith angle
(c) Solar azimuth angle
(d) Hour angle
[ESE : 2022]

1.45 The total solar radiation received at any point on the earth's surface is the sum of the direct and diffuse radiation and is referred as
(a) Total diffuse radiation
(b) Insolation
(c) Total radiation
(d) Total specular radiation
[ESE : 2022]

1.46 The rate at which solar energy arrives at the top of the atmosphere is called
(a) Total energy (b) Radiation
(c) Solar constant (d) Radiation constant
[ESE : 2022]

1.47 Consider the following statements regarding solar ponds:
1. It is possible to produce electricity from a solar pond by using a special 'High

temperature' heat engine coupled to an electric generator.

2. In a large solar pond, the thermal capacitance and resistance can be made large enough to retain the heat in the bottom layer from summer to winter and the pond can therefore be used for heating buildings in the winter.
3. A solar pond is an ingenious collector, which uses water as its top cover.

Which of the above statements are correct?

(a) 1 and 2 only (b) 2 and 3 only
(c) 1 and 3 only (d) 1, 2 and 3

[ESE : 2022]

1.48 Which one of the following is a vertical axis wind mill?

(a) Darrieus type wind mill
(b) Propeller type wind mill
(c) Sail type wind mill
(d) Multi blade type wind mill

[ESE : 2022]

1.49 Consider the following statements regarding solar radiation measurement:

1. Pyranometer collimates the radiation to determine the beam intensity as a function of incident angle.
2. Pyranometer measures the total hemi-spherical solar radiation.
3. Pyrheliometer collimates the radiation to determine the beam intensity as a function of incident angle.

Which of the above statements is/are correct?

(a) 1 only (b) 2 and 3 only
(c) 1 and 2 only (d) 3 only

[ESE : 2022]

1.50 Which one of the following materials is used for heat collection elements in parabolic trough plants of solar central receiver system?

(a) Copper (b) Mild steel
(c) Stainless steel (d) Aluminium

[ESE : 2022]

1.51 Which one of the following fuel cells has lowest operating temperature?

(a) PEMFC (b) MCFC
(c) PAFC (d) SOFC

[ESE : 2022]

1.52 Adjusting the nacelle about the vertical axis to bring the rotor facing wind is known as

(a) Pitch control (b) Hub control
(c) Rotor control (d) Yaw control

[ESE : 2022]

1.53 In a distributed collector solar thermal electric power plant, the heat collected in collectors is used to dissociate ammonia into nitrogen and hydrogen at approximately

(a) 300 atm pressure (b) 150 atm pressure
(c) 100 atm pressure (d) 40 atm pressure

[ESE : 2022]

1.54 Consider the following statements for vibration monitoring of the wind turbines:

1. Position sensors are used for low frequencies.
2. Velocity sensors are used for medium-range frequencies.
3. Accelerometers are used for high frequencies.

Which of the above statements are correct?

(a) 1 and 2 only (b) 1 and 3 only
(c) 2 and 3 only (d) 1, 2 and 3

[ESE : 2023]

1.55 Waves are created by the progressive transfer of energy from the wind as it blows over the surface of the water. Once created, waves can travel large distances without much reduction in energy. The energy in a wave is

(a) directly proportional to the height
(b) directly proportional to the height squared
(c) indirectly proportional to the height
(d) indirectly proportional to the height squared

[ESE : 2023]

1.56 A fuel cell converts chemical energy of a fuel into electricity

(a) indirectly, with conversion from fuel $\rightarrow$ heat $\rightarrow$ electricity
(b) directly, with no intermediate combustion cycle
(c) directly, with conversion from fuel $\rightarrow$ heat $\rightarrow$ work $\rightarrow$ electricity
(d) indirectly, with conversion from fuel $\rightarrow$ work $\rightarrow$ electricity

[ESE : 2023]

1.57 The efficiency of conversion from chemical energy to electricity by a fuel cell may

(a) theoretically be 100%
(b) practically be 50%
(c) theoretically be 50%
(d) practically be 75%

[ESE : 2023]

1.58 Consider the following statements related to Pyranometer:

1. Pyranometer is used to measure global (direct and diffuse) solar radiation on a surface.
2. This instrument cannot be used to measure the diffuse radiation by blocking out the direct radiation with a shadow band.

Which of the above statements is/are correct?

(a) 1 only (b) 2 only
(c) Both 1 and 2 (d) Neither 1 nor 2

[ESE : 2023]

1.59 Most of the reflection occurs from clouds, with a small proportion from the Earth's surface. This reflectance is called

(a) Scattering (b) Greenhouse effect
(c) Albedo (d) Climate change

[ESE : 2023]

1.60 What is the amount of solar heat that comes through a south-facing window (single pane) for January (70% sunshine)? (Consider vertical window, dimensions of 1.2 m × 2.5 m, single pane, transmission = 90%, area = 3 m^2 and insolation for January is 6 kWh/m^2 per clear day)

(a) 300 kWh (b) 351 kWh
(c) 400 kWh (d) 451 kWh

[ESE : 2023]

1.61 The temperature difference between absorber plate and outdoor air is –10°C, and then the flat-plate collector efficiency is higher for

(a) one pane (b) two panes
(c) three panes (d) more than three panes

[ESE : 2023]

1.62 The Hottel-Whillier-Bliss, equation, expresses the useful heat collected, Q, per unit area, in terms of two operating variables, the incident solar radiation normal to the collector plate, G_C, and the temperature difference between the mean temperature of the heat-removal fluid in the collector, T_m, transmittance-absorptance product ($\tau\alpha$) and the surrounding air temperature, T_a, as follows:

(a) $F[(\tau\alpha)G_C + U(T_m - T_a)]$
(b) $F[(\tau\alpha)G_C - U(T_m + T_a)]$
(c) $F[(\tau\alpha)G_C - U(T_m - T_a)]$
(d) $F[(\tau\alpha)G_C + U(T_a - T_m)]$

[ESE : 2023]

1.63 What is the approximate value of the energy output for a 0.5 kW PV system for Amarillo, Texas, for the month of January if the system is as follows : BP solar, crystalline silicon, 225-W module, 1.65 m × 1 m, area = 1.65 m^2, array of two modules tilted at latitude? [Consider the Amarillo data; January average day = 4.9 kWh/m^2/day, E_s = 70%, E_C = 17%]

(a) 60 kWh (b) 50 kWh
(c) 40 kWh (d) 30 kWh

[ESE : 2023]

1.64 What is the annual energy production (AEP) for a 3-MW wind turbine in a class 4 wind regime? (Take class 4 in a good wind regime, Capacity Factor = 40%)

(a) 8512 MWh/yr (b) 9512 MWh/yr
(c) 10512 MWh/yr (d) 11512 MWh/yr

[ESE : 2023]

1.65 Consider the following statements regarding tidal energy:

1. When the sun-earth and moon-earth directions are perpendicular, the solar and lunar tides are out of phase producing net tides of maximum range.
2. Neap tides occur twice per month at times of half-moon.
3. Spring tides occur twice per lunar month at times of both full and new moon.

Which of the above statements are correct?

(a) 1 and 2 only (b) 1 and 3 only
(c) 2 and 3 only (d) 1, 2 and 3

[ESE : 2024]

1.66 Consider the following statements regarding phosphoric acid fuel cell:

1. The basic phosphoric acid fuel cell consists of two electrodes of porous conducting material to collect charge.

2. At the negative electrode, hydrogen gas is converted to hydrogen ions.
3. Pure hydrogen or a hydrogen-rich gas is supplied at positive electrode and oxygen or air is supplied at negative electrode.

Which of the above statement are correct?
(a) 1 and 2 only74 (b) 1 and 3 only
(c) 2 and 3 only (d) 1, 2 and 3
[ESE : 2024]

1.67 Consider the following statements regarding solar energy:
1. The sun radiates energy uniformly in all directions in the form of electromagnetic waves.
2. The output of the sun is 2.8×10^{23} kW.
3. The energy reaching the earth is 1.1×10^{12} kWh/year.

Which of the above statements are correct?
(a) 1 and 2 only (b) 1 and 3 only
(c) 2 and 3 only (d) 1, 2 and 3
[ESE : 2024]

1.68 Consider the following statements regarding measurement of solar radiation:
1. Pyranometer cannot measure diffuse radiation.
2. Pyrheliometer is an instrument that measures beam radiation.
3. A sunshine recorder measures the sunshine hours in a day.

Which of the above statements are correct?
(a) 1 and 2 only (b) 1 and 3 only
(c) 2 and 3 only (d) 1, 2 and 3
[ESE : 2024]

1.69 Consider the following statements:
1. The sum of beam and diffuse radiation is referred to as global radiation.
2. Diffuse radiation does not have a unique direction.
3. Solar radiation propagating in a straight line and received at the earth surface without change of direction is called direct radiation.

Which of the above statements are correct?
(a) 1 and 2 only (b) 1 and 3 only
(c) 2 and 3 only (d) 1, 2 and 3
[ESE : 2024]

1.70 Consider the following statements regarding depletion of solar radiation:
1. Nitrogen gas absorbs the X-rays and extreme ultraviolet radiations.
2. Dust particles and air molecules absorb a part of solar radiant energy irrespective of wavelength
3. Ozone cannot absorb ultraviolet radiation.

Which of the above statements are correct?
(a) 1 and 2 only (b) 1 and 3 only
(c) 2 and 3 only (d) 1, 2 and 3
[ESE : 2024]

1.71 Consider the following statements :
1. Solar radiation incident on the outer atmosphere of the earth is known as extra radiation.
2. Solar radiation that reaches the earth surface after passing through the earth's atmosphere is known as terrestrial radiation.
3. The term 'solar insolation' is defined as solar radiation energy received on a given surface area in a given time.

Which of the above statements are correct?
(a) 1 and 2 only (b) 1 and 3 only
(c) 2 and 3 only (d) 1, 2 and 3
[ESE : 2024]

1.72 Consider the following statements regarding solar collectors:
1. Collector efficiency is defined as the ratio of the energy actually absorbed and transferred to heat- transporting fluid by the collector to the energy incident on the collector.
2. Concentration ratio is defined as the ratio of the area of the receiver to the area of aperture of the system.
3. Temperature range is the range of temperature to which the heat- transporting fluid is heated up by the collector.

Which of the above statements are correct?
(a) 1 and 2 only (b) 1 and 3 only
(c) 2 and 3 only (d) 1, 2 and 3
[ESE : 2024]

1.73 Consider the following statements:
1. The concentration ratio achieved from compound parabolic concentrator collector is in the range of 3-7.

2. The concentration ratio in the range of 70-80 may be achieved from cylindrical parabolic concentrator collector.
3. A concentration ratio of 10-30 may be achieved from linear Fresnel lens collector which yields temperatures between 150°C and 300 °C.

Which of the above statements are correct?
(a) 1 and 2 only (b) 1 and 3 only
(c) 2 and 3 only (d) 1, 2 and 3
[ESE : 2024]

1.74 Consider the following statements regarding solar greenhouse :
1. If natural means are adopted to collect, store and distribute the energy inside, it is known as active greenhouse.
2. In cold countries, winter 'greenhouses' provide supplementary heat to maintain adequate temperature during cold months when solar insolation is low.
3. Greenhouses for arid zone are designed to conserve water resources.

Which of the above statements are correct?
(a) 1 and 2 only (b) 1 and 3 only
(c) 2 and 3 only (d) 1, 2 and 3
[ESE : 2024]

1.75 Consider the following statements regarding wind energy :
1. The rate of change of wind speed with height is called wind shear.
2. The layer of air from ground to gradient height is known as planetary boundary layer.
3. In the surface layer, the mean wind speed with height can be represented by Prandtl logarithmic law model.

Which of the above statements are correct?
(a) 1 and 2 only (b) 1 and 3 only
(c) 2 and 3 only (d) 1, 2 and 3
[ESE : 2024]

1.76 Consider the following regarding biomass energy:
1. Charcoal is obtained by biomization process of woody biomass to achieve higher energy enthalpy per unit mass.
2. Biodiesel is simple to use, biodegradable and non-toxic.
3. The raw vegetable oil is upgraded as biodiesel through a chemical process called trans-esterification

Which of the above statements are correct?
(a) 1 and 2 only (b) 1 and 3 only
(c) 2 and 3 only (d) 1, 2 and 3
[ESE : 2024]

■■■■

Answers Renewable Sources of Energy

1.1 (c)	1.2 (c)	1.3 (d)	1.4 (c)	1.5 (a)	1.6 (b)	1.7 (b)	1.8 (b)	1.9 (b)
1.10 (d)	1.11 (c)	1.12 (b)	1.13 (a)	1.14 (a)	1.15 (a)	1.16 (a)	1.17 (c)	1.18 (b)
1.19 (a)	1.20 (b)	1.21 (c)	1.22 (d)	1.23 (b)	1.24 (c)	1.25 (b)	1.26 (b)	1.27 (a)
1.28 (c)	1.29 (c)	1.30 (c)	1.31 (d)	1.32 (a)	1.33 (b)	1.34 (c)	1.35 (c)	1.36 (d)
1.37 (b)	1.38 (c)	1.39 (d)	1.40 (c)	1.41 (b)	1.42 (a)	1.43 (d)	1.44 (a)	1.45 (b)
1.46 (c)	1.47 (b)	1.48 (a)	1.49 (b)	1.50 (c)	1.51 (a)	1.52 (d)	1.53 (a)	1.54 (d)
1.55 (b)	1.56 (b)	1.57 (a)	1.58 (a)	1.59 (c)	1.60 (b)	1.61 (a)	1.62 (c)	1.63 (a)
1.64 (c)	1.65 (c)	1.66 (a)	1.67 (a)	1.68 (c)	1.69 (d)	1.70 (a)	1.71 (d)	1.72 (b)
1.73 (b)	1.74 (c)	1.75 (d)	1.76 (c)					

Explanations Renewable Sources of Energy

1.2 (c)

Statement (I) correct.
Statement (II) wrong (as it measures only beam radiation)

1.3 (d)

1. Rated power,

$$P_T = \dot{m}(v_0 - v_2)v_1 = \frac{1}{2}\dot{m}(v_0^2 - v_2^2)$$

So, rated velocity is v_1, according to BETZ model $v_1 = \frac{v_0 + v_2}{2}$. So, rated velocity does not depend only upon wind velocity (v_0). It depends upon both wind velocity as well as down stream velocity (v_2).

2. $P_0 = 1/2\rho A v_0^3$ or $P_0 \propto v_0^3$

where, v_0 = Free wind velocity.

3. **Cut-in velocity:** The speed at which turbine first start to rotate.

Cut-out velocity: The speed beyond which it can be structural damage.

Wind turbine works between cut-in and cut-out velocity.

4. **Yaw Control Mechanism:** is used only in horizontal axis wind turbine.

1.4 (c)

Alkaline fuel cell:

- Alkaline fuel cell are primary source of electrical power on human space flight system for over 4 decades. However, AFC uses a costly technology.
- Proton exchange membrane fuel cell had water management problem, it is under development.

1.10 (d)

U_e - Edge loss coefficient
U_b - Back loss coefficient

$$U_e = \frac{(L_1 + L_2)L_3 k_i}{L_1 L_2 \delta_s} \quad \text{... (i)}$$

$$U_b = \frac{k_i}{\delta_b} \quad \text{... (ii)}$$

$$A_C = L_1 \times L_2 \quad \text{... (iii)}$$

$$A_e = 2(L_1 + L_2) \times L_3 \quad \text{... (iv)}$$

Assuming ($\delta_b = \delta_s$)

$$U_e = U_b \times \left(\frac{A_e}{2A_C}\right)$$

1.13 (a)

$$\eta_{conv} = \frac{P_{max}}{E \times A_c}$$

$E \rightarrow$ Total solar radiation incident flux

$$\eta_{conv} = \frac{P_{max}}{E \times A_c} = \frac{0.5 \times 3}{1000 \times 0.1 \times 0.1}$$

$$= 0.15 = 15\%$$

1.14 (a)

$$P_{avg} = \frac{1}{2}\frac{\rho A g R^2}{10 \times 3600}$$

$$= \frac{1}{2}\frac{1025 \times 300000 \times 9.8 \times 3^2}{10 \times 3600} = 376.68 \text{ kW}$$

1.20 (b)

Given: G_s = 100 W/m^2;
w = 150 mm = 1.5 m,
h = 180 cm = 1.8 mm

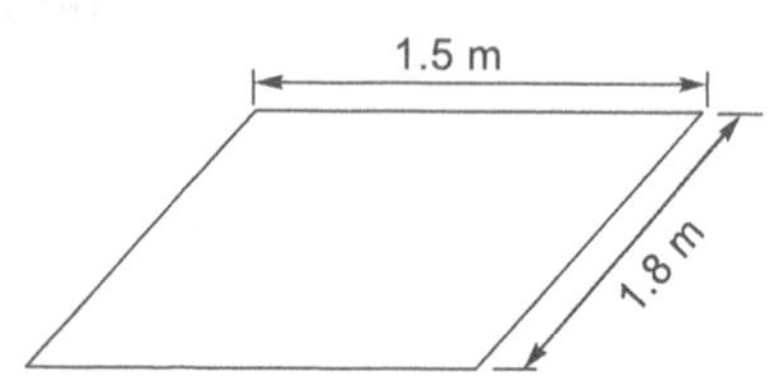

Effective area (A_e) = 0.9 × 1.5 × 1.8 m^2
= 0.9 × w × h

Incident solar radiation flux (G_s) = 1000 W/m^2

Peak power delivered to the area of the collector,

$$P_{max} = G_s \times A_e$$
$$= 1000 \text{ W/m}^2 \times 0.9 \times 1.5 \times 1.8 \text{ m}^2$$
$$P_{max} = 2430 \text{ Watt} = 2.43 \text{ kW}$$

1.23 (b)

Fill factor for PV cell is given by,

$$FF = \frac{P_{max}}{V_{oc} I_{sc}}$$

Theoretical power output = $V_{oc}\, I_{sc}$

1.24 (c)

TSR = Tip Speed Ratio

1.25 (b)

Given, R = 10 m,
N = 20 rpm, V_0 = 12.9 kmph
Tip speed ratio,

$$\text{TSR} = \frac{\text{Tip Speed}(R\omega)}{\text{Wind velocity}(V_o)}$$

$$R = 10 \text{ m}$$

$$\omega = \frac{2\times\pi\times 20}{60} \text{ rad/sec}$$

$$V_o = 12.9 \times \frac{5}{18} \text{ m/sec}$$

$$\text{TSR} = \frac{10\times\frac{2\times\pi\times 20}{60}}{12.9\times\frac{5}{18}} = 5.84$$

1.31 (d)

Solar constant is the total radiation energy received from SUN per unit time per unit area on a theoretical surface perpendicular to the SUN's rays and at Earth's mean distance from the Sun.

1.32 (a)

The tip speed ratio is given by dividing the speed of the tips of the turbine blades by the speed of the wind. The savonius or S-rotor consists of two half cylinders attached to a vertical axis and facing in opposite directions to form a two vaned rotor. It has high starting torque, low speed and low efficiency. TSR of savonious turbine is one. This turbine is used for low power applications.

1.33 (b)

Blade solidity is an important design parameter for the axial flow impeller and is defined as the ratio of blade chord length to spacing. High solidity rotors use drag force and turn slower. The solidity of the savonius rotor is unit and that of the American multiblade rotor is typically 0.7.

1.34 (c)

Ethanol (C_2H_5OH) is a colourless liquid biofuel. Its boiling point is 78°C and energy density is 26.9 MJ/kg.

1.35 (c)

Typical gas composition in producer gas is

CO ~ 19%
H_2 ~ 18%
CH_4 ~ 1%
CO_2 ~ 11%
N_2 ~ 51%

1.37 (b)

Fuel cell	Operating temperature
PEMFC	40 - 60°C
AFC	90°C
PAFC	150 - 200°C
MCFC	600 - 700°C
SOFC	600 - 1000°C

1.39 (d)

Fuel cell	Efficiency
PEMFC	48 - 58%
AFC	64%
PAFC	42%
MCFC	50%
SOFC	60 - 65%

1.40 (c)

A 15 m diameter community solar cooker has been developed at the centre for scientific research (CSR), Auroville (Puducherry). Around 600 kg of steam per day could be generated from this bowl, which is sufficient to cook two meals for about 1000 people. A very large community solar cooker has recently been installed at Tirumala Tripathi Devasthanan at Andhra Pradesh, under the demonstration scheme of Ministry of New and Renewable Energy, Government of India.

1.43 (d)

$$\frac{\text{Area of aperture}}{\text{Area of receiver}} = \text{CR}$$

Central tower receiver obtain very high concentration ratio by using reflecting mirrors and is approximately equal to 3000.

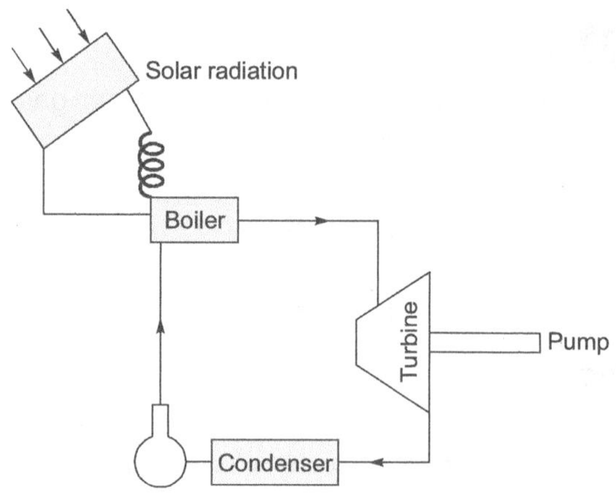

1.45 (b)

$$I_t = I_d + I_b$$

1.46 (c)

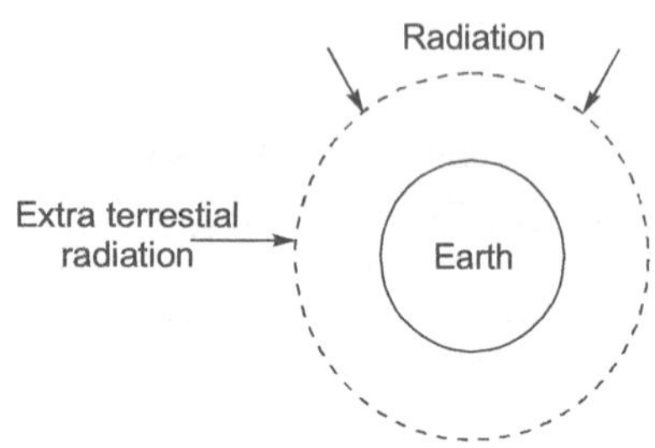

1.47 (b)

Solar ponds are used for low temperature applications, so statement 1 is wrong which implies option (b) is correct.

1.49 (b)

Pyranometer is used to measure the total hemispherical solar radiation whereas pyrheliometer measures beam radiations reaching earth surface.

1.51 (a)

Order of operating temperature for fuel cells.
SOFC > MCFC > PAFC > PEMFC
So, PEMFC has lowest operating temperature.

1.54 (d)

Vibration monitoring is used in wind turbines to determine the condition of rotating equipment, including the main bearing gearbox and generator. The vibration sensors may be placed in the axial and radial directions. Depending on the applicable frequency range, the position sensors (low range), velocity sensors (medium range) or accelerometers (high range) may be used.

1.55 (b)

Total energy per unit surface area in a wave

$$E = E_P + E_K$$

$$E = \frac{1}{4}\rho g a^2 + \frac{1}{4}\rho g a^2$$

$$= \frac{1}{2}\rho g a^2 \text{ J/m}^2$$

[where 'a' is the height of crest]
So, energy is directly proportional to the square of the height.

1.58 (a)

- A pyranometer is used to measure global (direct an diffuse) solar irradiance on a horizontal surface. A pyranometer with shading ring is used to measure diffuse solar irradiance.
- Direct radiation from sun is measured by a pyrheliometer.

1.60 (b)

Solar energy coming through window for a day

$$\text{in January} = \left(\frac{6kWhr}{m^2} \times 0.70\right) \times 3m^2 \times 0.9$$

$$= 11.34 \text{ kWhr}$$

For January month = 11.34 × 31 = 351.54 kWh

1.62 (c)

The Hottel-Whillier-Bliss equation is given by

$$Q = F[(\tau a)G_c - U(T_m - T_a)]$$

where, F is a factor related to the effectiveness of heat transfer from the collector plate to the heat removal fluid.

The transmittance-absorptance product ($\tau\alpha$) takes account of the complex interaction of optical properties in the solar radiation wavelegth.

1.63 (a)

$$\text{Energy output} = E_S E_C \text{(energy output per day)} \times \text{Area} \times \text{Number of days}$$

$$= 0.7 \times 0.17 \times 4.9 \times (2\times1.65) \times 31$$

$$= 59.64 \text{ kWh}$$

$$\simeq 60 \text{ kWh}$$

1.64 (c)

$$\text{AEP} = 3MJ/s \times \frac{3600s}{hour} \times \left(24 \times 365 \frac{hour}{year}\right) \times 0.40$$

$$\text{AEP} = 37843200 \frac{MJ}{year}$$

$$\text{AEP} = 37843200 \times \frac{MJ}{s} \times \frac{s}{year}$$

$$\text{AEP} = 37843200 MW \times \frac{3600^{-1} hr}{year}$$

$$= \frac{37843200}{3600} \frac{MW\text{-}hr}{\text{year}} = 10512 \frac{MW\text{-}hr}{\text{year}}$$

1.65 (c)

When the sun-earth and moon-earth directions are perpendicular, the solar and lunar tides are out of phase producing net tides of minimum range.

1.66 (a)

In the basic phosphoric acid fuel cell, pure hydrogen or a hydrogen rich gas is supplied at negative electrode (anode) and oxygen or air is supplied at positive electrode (cathode).

1.68 (c)

A pyranometer is designed to measure global radiation usually on a horizontal surface, but can also be used on an inclined surface, when shaded from beam radiation by using a shading ring, a pyranometer measures diffused radiation.

1.69 (d)

- The total radiation at any location on the surface of the earth is the sum of beam radiation and diffused radiation and is known as global radiation.
- Solar radiation scattered by aerosols, dust and molecules is known as diffused radiation. It does not have a unique direction.
- Solar radiation propagating in a straight line and received at the earth surface without change of direction i.e. in line with the sun is called beam or direct radiation.

1.70 (a)

Ozone absorbs ultraviolet radiation.

1.72 (b)

Concentration ratio is defined as the ratio of area of aperture of the system to the area of the reciever.

1.73 (b)

The concentration ratio in the range of 5-30 may be achieved from cylindrical parabolic concentrator collector.

1.74 (c)

When natural means are adopted to collect, store and distribute the energy inside, it is known as passive greenhouse.

1.76 (c)

Charcoal is obtained by carbonization process of woody biomass to achieve higher energy enthalpy per unit mass.

■■■■